This book covers the physics of magneto-optical recording, beginning with first principles and working through to contemporary state-of-the-art topics.

Both optics and magnetism are broad subjects, each with many applications in modern technology. The book prepares the reader for advanced work in either field. The first half of the book teaches the theory of diffraction using an original unified approach. It also covers the optics of multilayers (including dielectric, metal, birefringent, and magnetic layers), polarization optics, noise in photodetection, and thermal aspects. The second half of the book describes the basics of magnetism and magnetic materials, magneto-static field calculations, domains and domain walls, the mean-field theory, magnetization dynamics, the theory of coercivity, and the process of thermomagnetic recording. In each chapter, throughout the book, after coverage of the fundamentals the author explains the relevance of the topic to magneto-optical recording, and examples are given that emphasize this particular connection. Numerous examples based on real-world problems encountered in the engineering design of magneto-optical media and systems will give the reader valuable insights into the science and technology of optical recording. In addition, there are extensive problem sets at the ends of the chapters.

The book may be used as an advanced textbook on optical recording in both physics and engineering departments, and also as a reference work for scientists and engineers in industrial and academic laboratories.

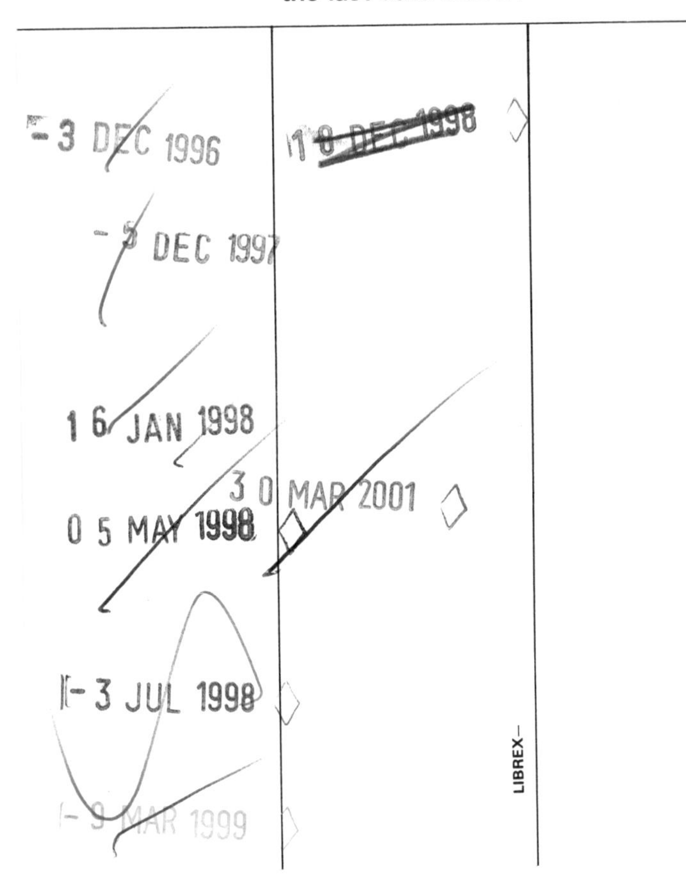

ANDERSONIAN LIBRARY
WITHDRAWN
FROM
LIBRARY
STOCK
UNIVERSITY OF STRATHCLYDE

THE PHYSICAL PRINCIPLES OF MAGNETO-OPTICAL RECORDING

The Physical Principles of Magneto-optical Recording

Masud Mansuripur

Optical Sciences Center
University of Arizona

Published by the Press Syndicate of the University of Cambridge
The Pitt Building, Trumpington Street, Cambridge CB2 1RP
40 West 20th Street, New York, ny 10011–4211, USA
10 Stamford Road, Oakleigh, Melbourne 3166, Australia

First published 1995

Printed in Great Britain at the University Press, Cambridge

A catalogue record for this book is available from the British Library

Library of Congress cataloging in publication data
Masud Mansuripur, 1955–
The physical principles of magneto-optical recording / Masud Mansuripur.
p. cm.
Includes bibliographical references.
ISBN 0 521 46124 3
1. Optical storage devices. 2. Magneto optical effects.
I. Title.
TA1635.M375 1995
621.39'767–dc20 93–48553 CIP

ISBN 0 521 46124 3 hardback

Cambridge University Press acknowledges the generous support of
the Hitachi Central Research Laboratory, the Hitachi Image and Media System Laboratory,
Sony Corporate Research Laboratories, and the University of Arizona
in the production of this book.

CR

Contents

Preface

This book has grown out of a course that I have taught at the Optical Sciences Center of the University of Arizona over the past five years. The idea has been to introduce graduate students from various backgrounds, either in physics or in one of the engineering disciplines, to the physical principles of magneto-optical (MO) recording. The topics selected for this course were of general interest, since both optics and magnetism are very broad and have many applications in modern science and technology. Each topic was treated in a self-contained and comprehensive manner. The students were first motivated by being told the relevance of a subject to optical data storage, then the subject was developed from basic principles, and examples were given along the way to show its application in quantitative detail. At the end of each chapter homework problems were assigned, so that students could learn certain details that were left out of the lectures or find out new directions in which their newly acquired knowledge could be applied.

The book essentially follows the format of the lectures, albeit with the addition of several chapters whose classroom coverage has usually been inhibited by the finite length of a semester. My primary goal in writing the book is to open the field to graduate students in physical sciences. Most college courses these days are based on single-issue topics that can be covered fairly comprehensively in the course of one or two semesters. Multidisciplinary subjects, such as magnetic or optical data storage, do not fit this description and, as a result, are left out of the curricula altogether. New graduates employed in research or product design, however, are often expected to educate themselves and master the relevant material in a short period of time. This state of affairs can be remedied if textbooks are available in selected multidisciplinary areas that will enable instructors (perhaps adjunct faculty with ties to the industry) to offer courses in those areas. I believe that this approach will be successful if the new courses satisfy the following criteria: first, the topics selected for coverage should be of general scientific interest for a broad range of specialties, so that students who end up working in some other field could also benefit from their acquired knowledge. Second, the treatment must be at a level compatible with the background and general knowledge of the students, so that they are given a physical understanding of the phenomena, rather than asked to memorize a collection of facts and formulas. Third, the multidisciplinary nature of the course must be emphasized at every point, and the students motivated to learn each new topic after understanding its relevance to the field as a whole. In this way, not only does the student appreciate each topic in its own right, but also he/she develops the proper perspective for dealing with complex systems and phenomena on a large scale.

The present book adheres to the above guidelines. It is rather fortunate that both optics and magnetism are broad and respected domains of scientific endeavor; their combination in the area of magneto-optical recording is fortuitous, but it is only a minor application area when compared with the wide range of applications commanded by each subject individually. Thus when the reader of this book learns about diffraction, polarization optics, antireflection multilayer structures, transition metal magnetism, magnetic domain walls, demagnetizing effects, magneto-resistance, Lorentz electron microscopy, etc., not only is he/she learning something that is directly related to MO recording, but also is gaining knowledge in areas that are independently important. The above topics (and several others) are not given a passing attention in this book; they are fully developed from the first principles and should be understandable by anyone with a sound undergraduate education in physical sciences. The glue that holds the book together, of course, is the application of each and every topic to the technology of magneto-optical recording. The relevance of each topic to the central theme is pointed out at appropriate points, and many problems are provided at the end of each chapter, so that the reader can further his/her understanding by exploring the applicability of the learned concepts.

Although intended as a textbook, the coverage of fundamentals as well as applied topics should make the present book appealing to industrial scientists and engineers as well. A number of other books in this field are presently available, and a few more are forthcoming. *Optical Recording* by Alan B. Marchant is introductory in nature and contains a wealth of information. Well-written and easy to understand, Marchant's book is a detailed exposition of the state of the art in optical recording technology, but treats the physics rather lightly. *Principles of Optical Disc Systems* by the staff of Philips Research Laboratories in Eindhoven (developers of the first optical disk) is also rich in technical detail; its focus is not on magneto-optical recording, but many related issues such as focus-error and track-error detection, the vector diffraction theory of grooves and preformat marks, mastering, signal processing, modulation and error correction coding, etc. are treated superbly. The present book complements those existing by placing the emphasis on the theoretical aspects of optical recording, and by delving more deeply into physical optics and magnetism-related issues.

Chapter 1 is introductory; here an overview of the entire field is given, and a broad range of topics is touched upon for the first time. After going through this chapter the reader will have an appreciation of the coverage of the rest of the book, and will recognize the relevance of various other chapters to the book's central theme.

Chapter 2 presents a simple treatment of Gaussian beam optics, and introduces the reader to the diffraction theory of light in a simple yet powerful context. Although important in its own right, the study of Gaussian beam optics clarifies the reasons for the divergence of diode laser beams, explains the focusing properties of the objective lens, and describes the concepts of spot size at focus and depth of focus, both of which are central issues in optical recording.

The general topic of diffraction is taken up in Chapter 3. Here we describe the near-field (Fresnel) regime of diffraction as well as the far-field (Fraunhofer) regime. The stationary-phase approximation is used not only to derive the far-field formulas but also to analyze the focusing of beams by lenses. Various primary (Seidel) aberrations are then considered and their effects on the intensity profiles at focus are described. Also covered in this chapter are the extension of scalar results to vector diffraction and the consideration of polarization-related effects.

An important problem in the diffraction theory of optical disks is diffraction from a sharp edge. Track-following, knife-edge focus error detection, preformat mark readout, and a number of other topics can be studied with the formalism of diffraction from sharp edges, developed in Chapter 4.

In Chapter 5 we turn our attention to the subject of multilayers, which play a vital role in many branches of optics. Chapter 5 is thus devoted to a comprehensive analysis of thin-film and multilayer optics. Starting from Maxwell's equations, we develop the mathematical tools for analyzing arbitrary multilayer structures that may contain metal, dielectric, birefringent, or magneto-optical layers. At the end of the chapter examples are given that pertain to optical disks with antireflection coatings and enhanced signal-to-noise properties.

Magneto-optical readout is the subject of Chapter 6. Polarization-related phenomena of interest in optical recording are described here in detail, and the operation of the differential detection scheme is discussed. In Chapter 7 we concentrate on the effects of high-numerical-aperture focusing on the read signal. When the objective lens, in its attempt to create a small focused spot, bends the incoming rays of the laser at sharp angles, the concomitant bending of the polarization vectors complicates the interaction between the focused beam and the storage medium; Chapter 7 illustrates these complications through several numerical studies.

Chapter 8 is a collection of examples aimed at clarifying optics-related phenomena in optical disk systems. Laser diode beam collimation, diffraction from periodic surface relief structures, the operation of the astigmatic focus-error detection scheme, and an analysis of cross-talk between focus and track servo channels are among the examples treated in this chapter.

Noise in readout is one of the main factors that limit the overall performance of any optical recording system. The various contributions to magneto-optical readout noise, namely, thermal and shot noise, laser fluctuations, surface roughness, depolarization at the groove-edges, domain boundary jaggedness, and domain-edge jitter are analyzed in Chapter 9. The concept of the carrier-to-noise ratio (CNR), which is widely used in practice for performance characterization of the media, is also described.

Chapter 10 is devoted to the subject of modulation coding and error-correction. In keeping with the spirit of the book, which is to expose ideas rather than to dwell on specific implementations, we describe a powerful algorithm for the modulation encoding and decoding of binary sequences. We then show that the existing methods of error correction can be applied after modulation in order to protect the data against random and burst errors. The technique we describe for modulation and our concept of post-modulation error correction are at variance with today's practice in the field of data storage. Readers interested in the actual practical issues should consult the vast and growing literature in this field. It is the author's belief, however, that with progress in the semiconductor technology and with the availability of powerful digital signal processing microchips, there will come a time when the methods described in this chapter will gain acceptance in practice. (The majority of the ideas in Chapter 10 have originated with Boris Fitingof whose work in this area remains largely unpublished.)

Optical recording in general and magneto-optical recording in particular rely for writing and erasure on the heating of thin-film layers by a focused laser beam. The absorption of light in multilayer disk structures, the generation and subsequent diffusion of heat in these media, and the evolution of temperature profiles in moving disk systems are subjects of great significance in optical data storage. The thermal analysis of magneto-optical recording media is substantially simplified by the fact that in these media melting and physical deformation do not occur. Chapter 11 describes methods of calculating the thermal profiles in MO disk systems. Through the examples in this chapter the reader can gain appreciation for the role of heat diffusion not only in the recording process but also in readout, where the laser power is reduced but the thermal effects are not negligible. Chapter 11 provides a bridge between the first half of the book, which is concerned mainly with readout, and the second half, which aims at elucidating the writing process.

Chapters 12 through 16 address the fundamental issues of thin-film magnetism as related to the process of thermomagnetic recording. In Chapter 12 preliminary notions and concepts of magnetism are described. This is followed by a comprehensive treatment of demagnetizing fields and field computation methods in Chapter 13. The mean-field theory of amorphous ferrimagnetic materials is the subject of Chapter 14; here the temperature dependence of magnetization in amorphous rare earth–transition metal (RE–TM) alloys is described from the perspective of the mean-field theory, and the relevant material parameters are extracted by comparing theory with experiment.

Chapter 15 is devoted to magnetization dynamics. After describing the Landau–Lifshitz–Gilbert equation and exploring its consequences for domain wall motion in uniform media, we turn to computer simulations and demonstrate several interesting phenomena that occur during the nucleation and growth of reverse-magnetized domains in inhomogeneous thin films. This line of investigation is continued in Chapter 16 which is concerned with the coercivity of thin-film media. Here we address both nucleation coercivity and wall motion coercivity, basing our arguments on certain assumptions about the nature of the inhomogeneity in amorphous films and the consequences of inhomogeneity as revealed by dynamic computer simulations.

The process of thermomagnetic recording is explored in Chapter 17, where thermal analysis, the mean-field theory of magnetization, and dynamic computer simulations are combined to yield a vivid picture of domain formation in the presence of local heating and external magnetic fields. (The computer simulations described in this and the preceding two chapters were done by Roscoe Giles on the Connection Machine at Boston University.)

The concern of the final chapter is the characterization of the media of magneto-optical recording. Chapter 18 describes bulk measurement techniques that provide insight into the magnetic behavior of thin-film media. The measurement of hysteresis loops and anisotropy constants by vibrating sample magnetometry, the magneto-optical Kerr effect, and the extraordinary Hall effect are among the topics discussed. Also described are magnetoresistance measurements on thin-film samples. (Most of the experimental data reported here were obtained by Roger Hajjar and Bruce Bernacki while working in my laboratory at the University of Arizona. The samples on which these measurements were made were fabricated by David Shieh, then at IBM's T.J. Watson Research Laboratory, and by Peter Carcia of Dupont Corporation.) Another topic covered in chapter 18 is Lorentz electron microscopy and its applications in MO media characterization. Finally, we describe the principle of magnetic force microscopy and its application as a non-intrusive probe for observation of domains and micromagnetic features in MO media.

There are many individuals and organizations to whom I am indebted for my education in magneto-optical recording, as well as for support and encouragement. My gratitude goes to Donald Ahonen, Pantelis Alexopoulos, Brian Bartholomeusz, Keith Bates, Alan Bell, Bernard Bell, Bruce Bernacki, Neal Bertram, Charles Brucker, James Burke, John Calkins, David Campbell, Peter Carcia, John Chapman, Scott Chase, Di Chen, Martin Chen, Tu Chen, Lu Cheng, David Cheng, Chi Chung, Neville Connell, Maarten deHaan, Anthony Dewey, Gary Eckhardt, Brad Engel, Edward Engler, Kevin Erwin, Charles Falco, Blair Finkelstein, Boris Fitingof, Hong Fu, Sergei Gadetsky, Richard Gambino, Ronald Gerber, Roscoe Giles, Joseph Goodman, Timothy Goodman, Franz Greidanus, Mool Gupta, Charles Haggans, Roger Hajjar, Peter Hansen, Dennis Howe, Yung-Chieh Hsieh, Sunny Hsu, Floyd Humphrey, Gordon Hughes, Jerry Hurst, Nobutake Imamura, Akiyoshi Itoh, Amit Jain, Victor Jipson, Masahiko Kaneko, David Kay, Soon Gwang Kim, Thomas Kincaid, Gordon Knight, Ray Kostuk, Dan Kowalski, Mark Kryder, Shigeo Kubota, Fumio Kugiya, James Kwiecien, Neville Lee, Seh Kwang Lee, Wilfried Lenth, Marc Levenson, Lev Levitin, C.J. Frank Lin, Robert Lynch, Terry McDaniel, Denis Mee, Pierre Meystre, Joseph Micelli, Thomas Milster, William Mitchell, Senri Miyaoka, Hiroshi Nishihara, Koichi Ogawa, Norio Ohta, Takeo Ohta, Masahiro Ojima, Hiroshi Ooki,

Louis Padulo, Nasser Peyghambarian, Richard Powell, Eric Rawson, Robert Rosenvold, Michael Ruane, Daniel Rugar, Yoshifumi Sakurai, Ingolf Sander, David Sellmyer, Robert Shannon, H.P. David Shieh, Hyun-Kuk Shin, John Simpson, Glenn Sincerbox, Douglas Stinson, the late William Streifer, Satoshi Sugaya, Yasuhiro Sugihara, Yutaka Sugita, Hirofumi and Hiroko Sukeda, Ryo Suzuki, Takao Suzuki, Masahiko Takahashi, Fujio Tanaka, Kunimaro Tanaka, David Treves, Stuart Trugman, Yoshito Tsunoda, Yuan-Sheng Tyan, Susumu Uchiyama, John Urbach, Randall Victora, Dieter Weller, Frank Whitehead, Peter Wolniansky, Te-ho Wu, Zheng Yan, Seiji Yonezawa, Takehiko Yorozu, Bas Zeper, and Fenglei Zhou. I am also grateful to the following institutions for the sponsorship of my projects: The Optical Data Storage Center at the University of Arizona, Advanced Research Projects Agency (ARPA), Boston University, Data General Corporation, Digital Equipment Corporation, Eastman Kodak Company, Hewlett–Packard Company, Hitachi Central Research Laboratory, Hitachi Image and Media System Laboratory, IBM Corporation, Komag Corporation, Korea Institute of Science and Technology (KIST), Laser Byte Corporation, Matsushita Electric Industrial Corporation, Nashua Corporation, the National Science Foundation, NEC Corporation, Samsung Advanced Institute of Technology, Sony Corporate Research Laboratories, Xerox Corporation, and 3M Corporation.

Dr. Simon Capelin, the editor at Cambridge University Press, and his professional staff have superbly transformed my original manuscript into its present form; I thank them for their support and advice during the course of production of the book. Much of the art work and drawings were done by James Abbott, to whom I am grateful not only for the fine work but also for putting up with all my revisions. Last but not least, I would like to thank my wife, Annegret, whose patience with my erratic working hours has been crucial in making this book a reality. It is to her and to our children, Kaveh and Tobias, that this book is dedicated.

Tucson, January 1995
Masud Mansuripur

1
Overview of Optical Data Storage

Introduction

Since the early 1940s, magnetic recording has been the mainstay of electronic information storage worldwide. Audio tapes provided the first major application for the storage of information on magnetic media. Magnetic tape has been used extensively in consumer products such as audio tapes and video cassette recorders (VCRs); it has also found application in the backup or archival storage of computer files, satellite images, medical records, etc. Large volumetric capacity and low cost are the hallmarks of tape data storage, although sequential access to the recorded information is perhaps the main drawback of this technology. Magnetic hard-disk drives have been used as mass-storage devices in the computer industry ever since their inception in 1957. With an areal density that has doubled roughly every other year, hard disks have been and remain the medium of choice for secondary storage in computers.† Another magnetic data storage device, the floppy disk, has been successful in areas where compactness, removability, and fairly rapid access to recorded information have been of prime concern. In addition to providing backup and safe storage, the inexpensive floppies with their moderate capacities (2 Mbyte on a 3.5"-diameter platter is typical) and reasonable transfer rates have provided the crucial function of file/data transfer between isolated machines. All in all, it has been a great half-century of progress and market dominance for magnetic recording devices which only now are beginning to face a potentially serious challenge from the technology of optical recording.

Like magnetic recording, a major application area for optical data-storage systems is the secondary storage of information for computers and computerized systems. Like the high-end magnetic media, optical disks can provide recording densities in the range of 10^7 bits/cm^2 and beyond. The added advantage of optical recording is that, like floppies, these disks can be removed from the drive and stored on the shelf. Thus the functions of the hard disk (i.e., high capacity, high data-transfer rate, rapid access) may be combined with those of the floppy (i.e., backup storage, removable

†At the time of writing, achievable densities on hard disks are in the range of 10^7 bits/cm^2. Random access to arbitrary blocks of data in these devices can take a time on the order of 10 ms, and individual read/write heads can transfer data at the rate of several megabits per second.

media) in a single optical disk drive. The applications of optical recording are not confined to computer data storage. The enormously successful compact audio disk (CD), which was introduced in 1983 and has since become the *de facto* standard of the music industry, is but one example of the tremendous potential of optical technology.

A strength of optical recording is that, unlike its magnetic counterpart, it can support read-only, write-once, and erasable/rewritable modes of data storage. Consider, for example, the technology of optical audio and video disks. Here the information is recorded on a master disk that is then used as a stamper to transfer the embossed patterns to a plastic substrate for rapid, accurate, and inexpensive reproduction. The same process is employed in the mass production of read-only files (CD-ROM, O-ROM), which are now being used to distribute software, catalogues, and other large data-bases. Or consider the write-once-read-many (WORM) technology, where one can permanently store massive amounts of information on a given medium, and have rapid, random access to it afterwards. The optical drive can be designed to handle read-only, WORM, and erasable media all in one unit, thus combining their useful features without sacrificing performance and ease of use, or occupying too much space. Moreover, the media can contain regions with prerecorded information as well as regions for read/write/erase operations, both on the same platter. These possibilities offer opportunities for applications that have heretofore been unthinkable; the hybrid optical disk and the interactive video-disk are perhaps good examples of such applications.

In this introductory chapter we present the conceptual basis for optical data-storage systems. Since the emphasis of the book will be on disk technology in general and magneto-optical disks in particular, section 1.1 will be devoted to a discussion of some elementary aspects of disk data storage including a description of the concept of track and the definition of a performance parameter known as the access time. The physical layout of data on tracks is the subject of section 1.2. The binary data of digital recording must be processed and converted into special waveforms prior to recording. This processing provides some measure of protection against random errors and also tunes the signals to special characteristics of the read/write channel; error-correction and modulation-coding schemes are covered in subsections 1.2.1 and 1.2.2, respectively. Section 1.3 describes the basic elements and function of the optical path; included in this discussion are the properties of the laser diode, the characteristics of beam-shaping optics, and various features of the focusing element. Because of the limited depth of focus of the objective lens and the eccentricity of tracks, optical disk systems must have a closed-loop feedback mechanism for maintaining the focused beam on the right track continuously. These mechanisms are described in sections 1.4 and 1.5 for automatic focusing and automatic track-following. The physical process of thermomagnetic recording in magneto-optical (MO) media is described in section 1.6, and this is followed by a discussion of the MO readout process in section 1.7. The properties of the media used in MO recording are described in section

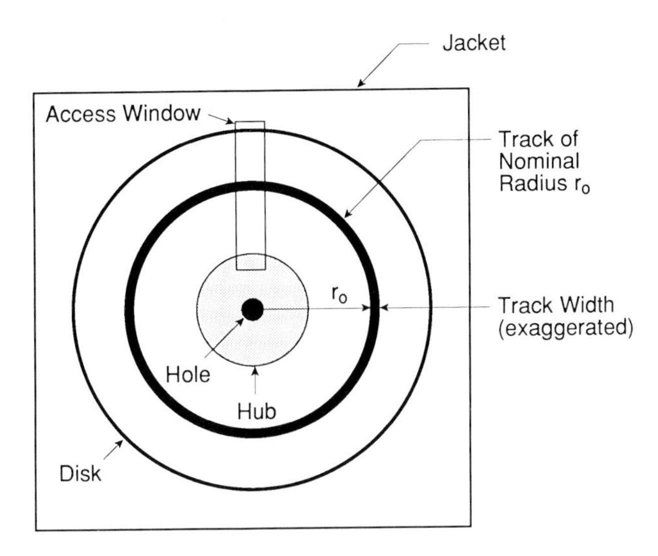

Figure 1.1. Schematic diagram showing the general features of an optical disk. The read/write head gains access to the disk through a window in the jacket; the jacket itself is for protection purposes only. The hub is the mechanical interface with the drive for mounting and centering the disk on the spindle. The track shown here is of the concentric-ring type, with radius r_0 and width W_t.

1.8. Concluding remarks and an examination of trends for the future of optical recording technology are the subjects of section 1.9.

1.1. Preliminaries and Basic Definitions

A disk, whether magnetic or optical, consists of a number of tracks along which information is recorded. These tracks, as shown in Fig. 1.1, may be concentric rings of a certain width, W_t. Neighboring tracks may be separated from each other by a guard-band whose width we shall denote by W_g. In the least sophisticated recording scheme, marks of length Δ_0 are recorded along these tracks. Now, if each mark can be in either of two states, say, present or absent, it may be associated with a binary digit, 0 or 1. When the entire disk surface of radius R is covered with such marks, its capacity C_0 (in bits per surface) is given by

$$C_0 = \frac{\pi R^2}{(W_t + W_g)\Delta_0} \tag{1.1}$$

Now, consider the parameter values typical of current optical-disk technology: R = 67 mm, corresponding to 5.25"-diameter platters, Δ_0 = 0.5 μm, which is roughly determined by the wavelength of the read/write laser diodes, and $W_t + W_g$ = 1 μm for the track-pitch. With these parameters the disk capacity will be around 28×10^9 bits, or 3.5 gigabytes. This is a reasonable estimate and one that is fairly close to reality, despite the many simplifying assumptions made in its derivation. In the following paragraphs we shall examine some of these assumptions in more detail.

(i) The disk is assumed to be fully covered with information-carrying marks. This is generally not the case in practice. Consider a disk rotating at $\mathcal{N}$ revolutions per second (rps). It is usually desirable to keep this rotation rate constant during the disk operation. Let the electronic circuitry have a fixed clock duration T_c. Then only pulses of length T_c (or an integer multiple thereof) may be used for writing. Now, a mark written along a track of radius r, with a pulse-width equal to T_c, will have length ℓ, where

$$\ell = 2\pi r \mathcal{N} T_c \tag{1.2}$$

Thus for a given rotational speed $\mathcal{N}$ and a fixed clock cycle T_c, the minimum mark length ℓ is a linear function of track radius r. ℓ decreases toward zero as r approaches zero. One must therefore pick a minimum usable track radius, r_{min}, where the spatial extent of individual marks is greater than or equal to the minimum allowed mark length, Δ_0. Equation (1.2) thus yields

$$r_{min} = \frac{\Delta_0}{2\pi \mathcal{N} T_c} \tag{1.3}$$

One may also define a maximum usable track radius r_{max}, although for present purposes $r_{max} = R$ is a perfectly good choice. The region of the disk used for data storage is thus confined to the area between r_{min} and r_{max}. The total number N of tracks in this region is given by

$$N = \frac{r_{max} - r_{min}}{W_t + W_g} \tag{1.4}$$

The number of marks on any given track in this scheme is independent of the track radius; in fact, the number is the same for all tracks, since the revolution period of the disk and the clock cycle uniquely determine the total number of marks on any individual track. Multiplying the number of usable tracks N with the capacity per track, we obtain for the usable disk capacity

$$C = \frac{N}{\mathcal{N} T_c} \tag{1.5}$$

Replacing N from Eq. (1.4) and $\mathcal{N} T_c$ from Eq. (1.3), we find that

$$C = \frac{2\pi r_{min}(r_{max} - r_{min})}{(W_t + W_g)\,\Delta_0} \tag{1.6}$$

If the capacity C in Eq. (1.6) is considered as a function of r_{min} with the remaining parameters held constant, it is not difficult to show that maximum capacity is achieved when

$$\boxed{r_{min} = \tfrac{1}{2}\, r_{max}} \tag{1.7}$$

With this optimum r_{min}, the value of C in Eq. (1.6) is only half that of C_0 in Eq. (1.1). In other words, the estimate of 3.5 Gbyte per side for 5.25" disks has been optimistic by a factor of two.

(ii) A scheme often proposed to enhance the capacity entails the use of multiple zones, where either the rotation speed $\mathcal{N}$ or the clock period T_c is allowed to vary from one zone to the next. Problem 1.1 describes one such zoning scheme. In general, zoning schemes can reduce the minimum usable track radius below that given by Eq. (1.7). More importantly, however, they allow tracks with larger radii to store more data than tracks with smaller radii. The capacity of the zoned disk is somewhere between C, Eq. (1.6), and C_0, Eq. (1.1), the exact value depending on the number of zones being implemented.

(iii) A fraction of the disk surface area is usually reserved for preformat information and cannot be used for data storage. Also, prior to recording, additional bits are generally added to the data for error correction as well as other house-keeping chores. These constitute a certain amount of overhead on the user data, and must be allowed for in determining the capacity. A good rule of thumb is that overhead consumes approximately 30% of the raw capacity of an optical disk, although the exact percentage may vary among the systems in use. Substrate defects and film contaminants during the deposition process can create bad sectors on the disk. These are typically identified by the disk manufacturer during the certification process, and marked for elimination from the sector directory. Needless to say, bad sectors must be discounted when evaluating the capacity.

(iv) Modulation codes may be used to enhance the capacity beyond what has been described so far. Modulation coding does not modify the minimum mark length of Δ_0, but frees the longer marks from the constraint of being integer multiples of Δ_0. The use of this type of code results in more efficient data storage and an effective number of bits per Δ_0 that is greater than unity. For example, the popular (2, 7) modulation code has an effective bit density of 1.5 bits per Δ_0. This and other modulation codes to be described in more detail in section 1.2, can increase the capacity of disks beyond the estimate of Eq. (1.6).

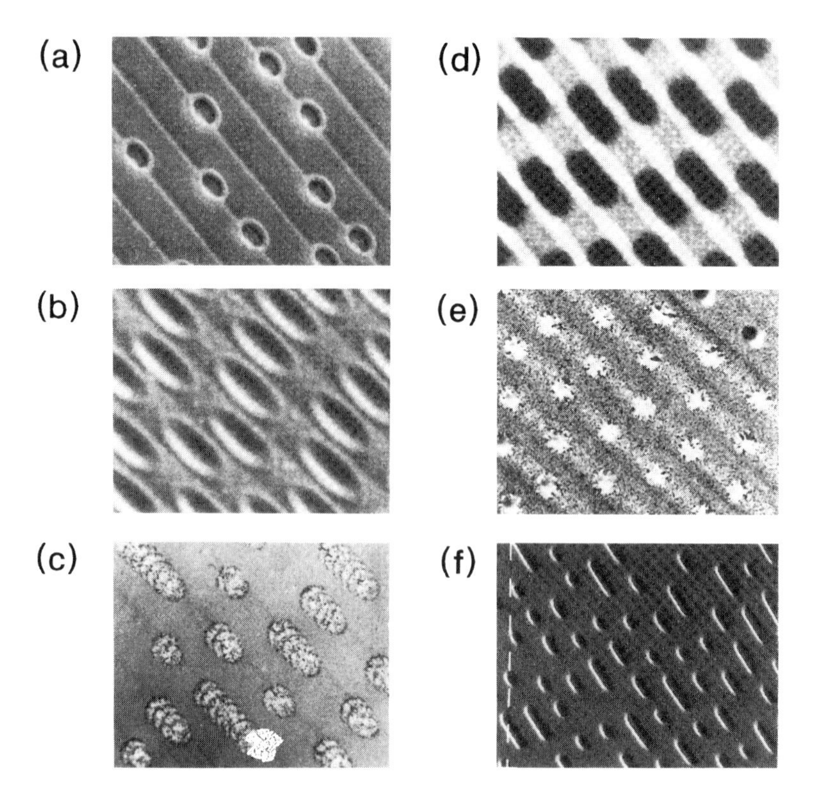

Figure 1.2. Micrographs of several types of optical storage media. The tracks are straight and narrow, with a pitch of 1.6 μm, running diagonally from the upper left corner to the lower right corner. (a) Ablative write-once tellurium alloy; (b) ablative write-once organic dye; (c) amorphous-to-crystalline write-once phase-change alloy GaSb; (d) erasable amorphous magneto-optic alloy GdTbFe; (e) erasable crystalline-to-amorphous phase-change tellurium alloy; (f) read-only CD audio, injection-molded from polycarbonate. Source: *Ullmann's Encyclopedia of Industrial Chemistry*, 5th edition, Vol. A14, copyright © 1994 by VCH publishers, reprinted by permission of the publisher.

1.1.1. The concept of a track

The information on magnetic and optical disks is recorded along tracks. Typically, a track is a narrow annulus at some distance r from the disk center. The width of the annulus is denoted by W_t, while the width of the guard-band, if any, between adjacent tracks is denoted by W_g. The track-pitch is the center-to-center distance between neighboring tracks and is therefore equal to $W_t + W_g$. A major difference between the magnetic floppy disk, the magnetic hard disk, and the optical disk is that their respective track-pitches are presently of the order of 100 μm, 10 μm, and 1 μm. Tracks may be imaginary entities, in the sense that no independent existence outside the pattern of recorded marks may be ascribed to them. This is the case, for example, with the compact audio disk format where prerecorded marks simply define their own tracks and help guide the laser beam during readout. At the other extreme are tracks that are physically engraved on the disk surface before any data is ever recorded. Examples of this type of track are provided by pregrooved WORM and magneto-optical disks. Figure 1.2 shows micrographs from several recorded optical disk surfaces. The tracks along which the data are written are clearly visible in these pictures.

It is generally desired to keep the read/write head stationary while the disk is spinning and reading from (or writing onto) a given track is taking place. Thus, in an ideal situation, not only should the track be perfectly circular but also the disk must be precisely centered on the spindle axis. In practical systems, however, tracks are neither precisely circular nor are they concentric with the spindle axis. These eccentricity problems are solved in low-performance floppy drives by making tracks wide enough to provide tolerance for misregistrations and misalignments. Thus the head moves blindly to a radius where the track center is nominally expected to be, and stays put until the reading or writing is over. By making the head narrower than the track-pitch, the track center is allowed to wobble around its nominal position without significantly degrading the performance during the read/write operation. This kind of wobble, however, is unacceptable in optical disk systems, which have a very narrow track, about the same size as the focused beam spot. In a typical situation arising in practice, the eccentricity of a given track may be as much as 50 μm while the track-pitch is only about 1 μm, thus requiring active track-following procedures.

A convenient method of defining tracks on an optical disk is provided by pregrooves, which are either etched, stamped, or molded onto the substrate (see Fig. 1.3(a)). The space between neighboring grooves is the so-called land. Data may be written in the grooves with the land acting as a guard band. Alternatively, the land regions may be used for recording while the grooves separate adjacent tracks. The groove depth is optimized for generating an optical signal sensitive to the radial position of the read/write laser beam. For the push–pull method of track-error detection (to be described in section 1.5) the groove depth is in the neighborhood of $\lambda/8$, where λ is the wavelength of the laser beam.

In digital data storage applications, each track is divided into small segments or sectors. A sector is intended for the storage of a single block of data which is typically either 512 or 1024 bytes. The physical length of a sector is thus a few millimeters. Each sector is preceded by header information such as the identity of the sector, identity of the corresponding track, synchronization marks, etc. The header information may be preformatted onto the substrate, or it may be written on the storage layer by the manufacturer. Tracks may be "carved" on the disk either as concentric rings or as a single continuous spiral. There are certain advantages to each format. A spiral track can contain a succession of sectors without interruption, whereas concentric rings may each end up with some empty space that is smaller than the required length for a sector. Also, large files may be written onto (and read from) spiral tracks without

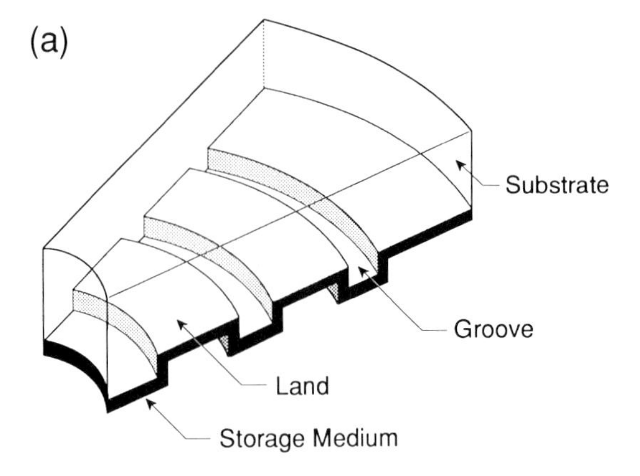

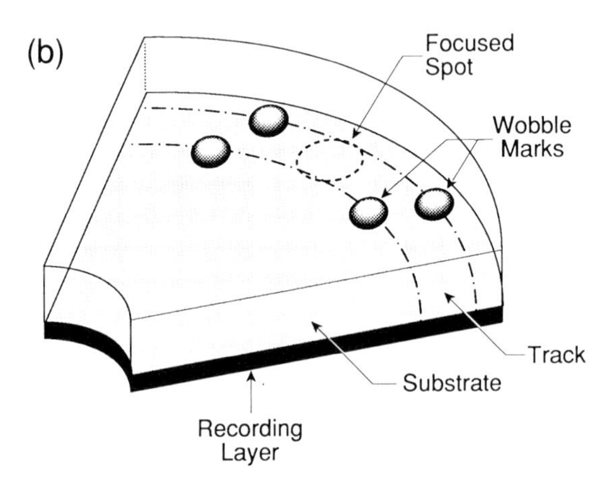

Figure 1.3. (a) Lands and grooves in an optical disk. The substrate is transparent, and the laser beam must pass through it to reach the storage medium. (b) Sampled-servo marks in an optical disk. These marks, which are offset from the track center, provide information regarding the position of the focused spot on the track during read/write operations.

jumping to the next track, which must occur when concentric tracks are used. On the other hand, multiple-path operations such as write-and-verify or erase-and-write, which require two paths each for a given sector, or still-frame video, are more conveniently handled on concentric-ring tracks.

Another track format used in practice is based on the sampled-servo concept. Here the tracks are identified by occasional marks that have been permanently placed on the substrate at regular intervals (see Fig. 1.3(b)). Details of track-following by the sampled-servo scheme will follow shortly (see subsection 1.5.2); suffice it to say at this point that servo marks help the system determine the position of the focused spot relative to the track within the plane of the disk. Once the relative position has been identified, it is rather straightforward to steer the beam and adjust the location of the focused spot on the track.

1.1.2. Disk rotation speed

When a disk rotates at a constant angular velocity ω, a track of radius r moves with the constant linear velocity $V = r\omega$. Ideally, one would like to have the same linear velocity for all the tracks, but this is impractical except in a limited number of situations. For instance, when the desired mode of access to the various tracks is sequential, such as in audio- and video-disk applications, it is possible to begin by placing the head at the inner radius, and move outward from the center thereafter while continuously decreasing the angular velocity. By keeping the product of r and ω constant, one can thus achieve constant linear velocity for all the tracks.† A sequential access mode, however, is the exception rather than the norm in data storage systems. In most applications, the tracks are accessed randomly with such rapidity that it becomes impossible to adjust the rotation speed for constant linear velocity. Under these circumstances the angular velocity is best kept constant during the normal operation of the disk. Typical rotation speeds are 1200 rpm and 1800 rpm for slower drives, and 3600 rpm for the high data-rate systems (in both 3.5" and 5.25" formats). Higher rotation rates (5000 rpm and beyond) are certainly feasible and likely to appear in the second- and third-generation optical storage devices.

1.1.3. Access time

The direct-access storage device or DASD, used in computer systems for the mass storage of digital information, is a disk drive capable of storing large quantities of data and accessing blocks of this data rapidly and in random order. In read/write operations it is often necessary to move the head to new locations on the disk in search of sectors containing specific data-items. Such random relocations are usually time consuming and can become the factor that limits performance in certain applications. The access time τ_a is defined as the average time spent in going from one randomly selected spot on the disk to another. τ_a can be considered as the sum of a seek time, τ_s, which is the average time needed to acquire the target track, and a latency, τ_l, which is the average time spent on the target track while waiting for the desired sector. Thus,

$$\tau_a = \tau_s + \tau_l \tag{1.8}$$

†In compact audio-disk players the linear velocity is kept constant at 1.2 m/s. The starting position of the head is at the inner radius $r_{\min} = 25$ mm, where the disk spins at 460 rpm. The spiral track ends at the outer radius $r_{\max} = 58$ mm, where the disk's angular velocity is 200 rpm.

The latency is half the revolution period of the disk, since a randomly selected sector is, on average, halfway along the track from the point where the head initially lands. Thus for a disk rotating at 1200 rpm τ_l = 25 ms, while at 3600 rpm we have $\tau_l \simeq 8.3$ ms. The seek time, on the other hand, is independent of the rotation speed of the disk, but is determined by the traveling distance of the head during an average seek, as well as by the mechanism of head actuation. It can be shown (see Problem 1.2) that the average length of travel in a random seek is one-third of the full stroke (full stroke = $r_{max} - r_{min}$). In magnetic disk drives where the head/actuator assembly is relatively light-weight (a typical Winchester head weighs about 5 grams) the acceleration and deceleration periods are short, and seek times are typically around 10 ms in small format drives (i.e., 5.25" and 3.5"). In optical disk systems, on the other hand, the head, being an assembly of discrete elements, is fairly large and heavy (typical weight $\simeq$ 50–100 grams), resulting in values of τ_s that are several times greater than those obtained in magnetic recording. The seek times reported for commercially available optical drives presently range from 20 ms in high-performance 3.5" drives to 80 ms in larger or slower drives. One must emphasize, however, that optical disk technology is still in its infancy; with the passage of time, the integration and miniaturization of elements within the optical head will surely produce light-weight devices capable of achieving seek times in the range of a few milliseconds.

1.2. Organization of Data on the Disk

For applications involving computer files and data, each track is divided into a number of sectors where each sector can store a fixed-length block of binary data. The size of the block varies among the various disk/drive manufacturers, but typically it is either 512 or 1024 bytes. As long as the disk is dedicated to a particular drive (such as in hard-disk Winchester drives) the sector size is of little importance to the outside world. However, if the disks are to be removable, then the sector size (among other things) needs to be standardized, since various drives must be able to read from and write onto the same disk.

A block of user-data cannot be directly recorded onto a sector. First, it must be coded for protection against errors (error-correction coding) and for the satisfaction of channel requirements (modulation coding). Also, it may be necessary to add synchronization bits or other kinds of information to the data before recording it. Thus a sector's capacity is somewhat larger than the block of raw data assigned to it. A sector also must have room for "header" information. The header is either recorded during the first use of the disk by the user, as in formatting a floppy disk, or is written by the manufacturer before shipping the disk. It typically contains the address of the sector plus synchronization and servo bits. In magnetic disks the header is recorded magnetically, which makes it erasable and provides for the option of reformatting at later times. On the negative side, formatting is time consuming and the information is subject to accidental erasure. In contrast, the optical disk's sector headers may be mass-produced from a master at the time of manufacture, eliminating the slow process of soft formatting. The additional space used by the codes and by the header within each sector constitutes the overhead. Depending on the quality of the disk, the degree of sophistication of the drive, and the particular needs of a given application, the overhead may take as little as 10% or as much as 30% of a disk's raw capacity.

As mentioned earlier, concentric-ring tracks are at a disadvantage when it comes to dividing them into fixed-length sectors: the track length may not be an integer multiple of the sector length, in which case the left-over piece in each track is wasted. If, on the other hand, a spiral track is used, the sectors can be placed contiguously and wasteful gaps avoided. The use of spiral tracks in optical disk systems is commonplace, but magnetic disks exclusively follow the concentric-ring format.

1.2.1. Error detection and correction

The binary sequences of 0's and 1's that constitute blocks of data in digital systems are prone to a multitude of possible errors. At one end of the spectrum are random errors that afflict individual bits randomly and independently. Such errors arise, for example, when additive white noise corrupts digital waveforms during processing and/or transmission. At the other extreme are "bursts" of error that also appear randomly, but affect a string of bits collectively and in a highly correlated fashion. An example of burst errors occurs when a scratch on the disk surface, a dust particle, or a fingerprint causes a substantial fraction of data to disappear from the read signal. In some cases an erroneous bit is readily identifiable, for example, when the read voltage on the line is close to neither V_{max} nor V_{min}; this type of error is called an "erasure" error. In other cases errors appear in disguise as, for example, when 0 is identified as 1, or vice versa. Naturally, erasure-type errors are easier to correct, because the system need only determine the correction for the erased bit and replace it. When errors are disguised, the system has the extra tasks of finding out whether errors have occurred and, if so, of identifying the location(s) of error within the bit-string. Depending on the type of error and its frequency of occurrence, different combat strategies may be devised with varying degrees of efficacy, as demonstrated by the following examples.

Example 1. The simplest error-detection technique is the so-called parity-check coding. Consider, for example, an eight-bit sequence of user data, protected by a ninth bit (the check-bit), which is chosen to make the overall number of 1's in the nine-bit block an even number. This scheme is capable of correcting a single erasure-type error, or detecting (but not correcting) a single disguised error† that might afflict any one of the nine

†A disguised error will make the overall number of 1's within the sequence an odd number, in which case the system is alerted to the existence of an error. Once the presence

bits. The price paid for this parity-check coding is an overhead of one extra bit for every byte of actual data, or approximately 11% of the overall capacity. The gain in performance may be quantified as follows. Let p_0 be the probability that a given bit is erased, and let erasures occur independently. Without protection, an eight-bit block has a probability $(1 - p_0)^8$ of being correct; thus the probability P_E of block error is

$$P_E = 1 - (1 - p_0)^8 \tag{1.9}$$

On the other hand, when the parity bit is appended to the byte of user-data, correct reception occurs when there is either no error or only one (erasure-type) error. The overall error probability in this case then becomes

$$P_E = 1 - (1 - p_0)^9 - 9p_0(1 - p_0)^8 \tag{1.10}$$

It is not difficult to verify that the P_E of Eq. (1.10) is always less than that of Eq. (1.9). For instance, when $p_0 = 10^{-3}$, we find $P_E = 7.97 \times 10^{-3}$ from Eq. (1.9) and $P_E = 3.58 \times 10^{-5}$ from Eq. (1.10), indicating that coding has improved the block-error rate by more than two orders of magnitude. □

Example 2. There is a generalization of the above parity-check coding scheme that can be used to correct bursts of erasure-type errors. We give a specific example of this code for a case in which the block of user-data consists of 12 bits, and the burst is expected to be at most three bits long; the extension of the idea to arbitrary-sized blocks with bursts of arbitrary length is trivial. Let the block of user-data be

$$b_1 b_2 b_3 b_4 b_5 b_6 b_7 b_8 b_9 b_{10} b_{11} b_{12}$$

where b_n is a binary digit (0 or 1). In order to correct a single burst of error with a maximum length of 3, we need to append three parity-check bits to the above block. The parity bits, which will be denoted by b_{13}, b_{14}, and b_{15} are chosen according to the following rules ($\oplus$ indicates modulo 2 summation):

$$\begin{aligned} b_{13} &= b_1 \oplus b_4 \oplus b_7 \oplus b_{10} \\ b_{14} &= b_2 \oplus b_5 \oplus b_8 \oplus b_{11} \\ b_{15} &= b_3 \oplus b_6 \oplus b_9 \oplus b_{12} \end{aligned}$$

The final 15-bit block thus consists of three interleaved sub-blocks, each containing an even number of 1's. Now, if a burst of erasure destroys three contiguous bits of this block then, since each of the above sub-blocks will

of an error is detected, the system may request a repeat of the process that yielded the erroneous data, for example, reread, rewrite, or retransmit. If the error persists, the system might produce an error message and halt, or might simply proceed to the next operation.

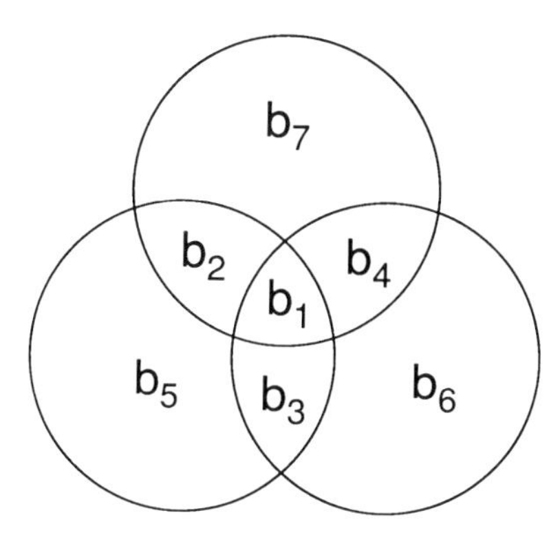

Figure 1.4. Circle diagram demonstrating Hamming's (7,4) error-correcting code. $b_1b_2b_3b_4$ are data bits, while $b_5b_6b_7$ are check bits.

contain only one of the erroneous bits, they can all be corrected independently and simultaneously.

This example shows a special instance of the use of interleaving techniques to correct long bursts of error. A more general case will be described in Example 4.

□

Example 3. In the case of disguised errors the job of correcting them is harder, since the error must first be pinpointed. Let us give an example of how random errors of this type may be corrected in practice. A simple example is provided by Hamming's (7,4) code. Consider the binary sequence of length 4 denoted by $b_1b_2b_3b_4$. Let the four bits of this sequence be placed, as shown in Fig. 1.4, within a circle diagram. The three additional (check) bits b_5, b_6 and b_7 in the diagram are chosen such that even parity is achieved for each circle; in other words, there is an even number of 1's within each circle. Thus b_5 is tied to b_1, b_2, b_3. Similarly b_6 is tied to b_1, b_3, b_4, and b_7 is tied to b_1, b_2, b_4. Now assume that upon transmission in a communication system (or upon readout after being recorded in some storage device) one of these seven bits is received in error, in the sense that if it were 1 to begin with it is changed to 0 and vice versa. If b_1 happens to be in error, then all three parity conditions are violated since b_1 belongs to all three circles. If b_2, b_3, or b_4 happens to be in error, then two of the parities will be violated. If b_5, b_6, or b_7 happens to be in error, then only one of the parities will be violated. In each case the specific violations uniquely identify the erroneous bit. Therefore, as long as there is either a single-bit error in the sequence $b_1b_2b_3b_4b_5b_6b_7$, or there are none at all, one can determine the existence of an error and also proceed to correct it. This is the essence of error-correcting codes.

The Hamming (7,4) code is not a powerful error-correcting code. It suffers from the facts that it applies only to short blocks of user-data (length = 4), it can correct single errors only, and it has a large overhead (43% of overall capacity is taken up by the parity bits). There exist many powerful codes for error correction from which system designers can choose nowadays, depending on the specific needs of their application. The class of Reed–Solomon codes is specially popular in the area of data storage, and many of these codes have been implemented in various magnetic and optical recording systems. Typically, these codes are designed to operate on bytes instead of bits, and, for correcting a specific number of erroneous bytes, they need twice as many parity-check bytes. For example, one of the two error-correction codes used in compact audio disk systems is the (28,24) Reed–Solomon code, which can correct up to two bytes of error in a 24-byte-long block of user-data, with the addition of four parity bytes. □

Example 4. Burst errors are usually too long to be handled efficiently by direct application of an error-correcting code. The technique of interleaving is designed to handle such errors by dispersing them among several blocks, thus reducing the required correcting power for each individual block. Interleaving can best be described by a simple example. Let $a_1a_2 \ldots a_{n-1}a_n$ be a block consisting of user-data and parity-check bits added for protection against errors. Similarly, let $b_1b_2 \ldots b_{n-1}b_n$ and $c_1c_2 \ldots c_{n-1}c_n$ be two other error-correction coded blocks of the same length. If these blocks are directly recorded on the disk, then any block affected by a burst longer than the correcting ability of the code will be unrecoverable. On the other hand, if the recorded sequence is arranged as

$$a_1b_1c_1a_2b_2c_2 \ldots a_{n-1}b_{n-1}c_{n-1}a_n b_n c_n$$

then the burst will affect all three original blocks, but the number of errors in each will be only one-third of the total length of the burst. The correcting power may now be sufficient to handle the reduced number of errors in each block. In principle there is no limit to the number of blocks that can be interleaved, although in practice the availability of memory and processing power as well as problems with the ensuing delays will constrain the extent of interleaving. □

1.2.2. Modulation coding

After the user-data is encoded for error correction, it goes through yet another stage, known as modulation, before it is suitable for recording. Traditionally, binary sequences have been converted into electrical waveforms using either the non-return-to-zero (NRZ) scheme, or a modified version of it known as NRZI. In NRZ each bit is alloted one unit of time (the clock cycle), during which the voltage is either high or low, depending on whether the bit is 1 or 0. In NRZI a 1 is represented as a transition in the middle of the clock cycle, while a 0 is represented by no transitions at all. Examples are given in Fig. 1.5 where the binary sequence 0011101 is shown converted to electrical waveforms using NRZ and NRZI

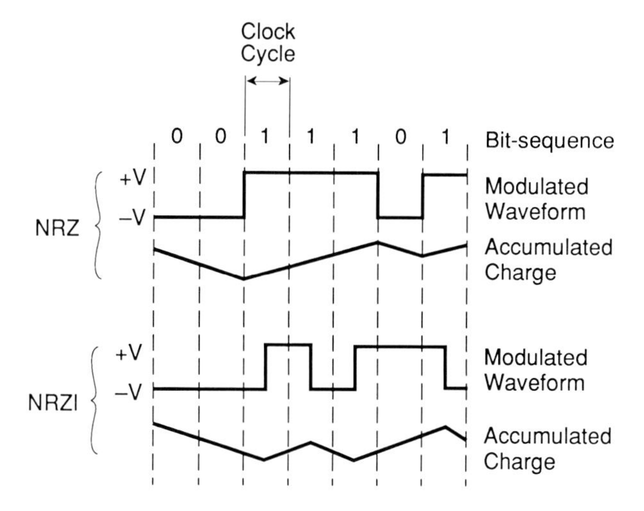

Figure 1.5. Conversion of binary sequence to electrical waveform. In NRZ during each clock cycle the signal stays high for 1's and low for 0's. In NRZI there is a transition in the middle of the clock cycle for each 1. Also shown are the integrated waveforms, which represent the accumulated electronic charge as a function of time.

methods. Note that in the NRZ scheme failure to detect a transition causes all subsequent bits to be detected erroneously (assuming that the electronic circuitry does not distinguish between "up" and "down" transitions). In contrast, in the NRZI waveform missing a transition amounts to missing a single bit only. In any event, both NRZ and NRZI have several shortcomings, some of which are listed below.

(i) In self-clocking systems, the clock, usually a voltage-controlled oscillator, is synchronized frequently by the stream of data itself. In such situations synchronization will be lost when long streams of 0's (in the case of NRZ, long streams of 1's also) create long intervals without transitions in the waveform.

(ii) The clock frequency is chosen in accordance with the physical limits imposed on the recorded marks by the properties of the media and by the read/write system. In particular, the length of the mark (along the track) has to be greater than a certain minimum value, or else the mark will not be resolved in readout. In NRZ and in NRZI, the spacing between transitions is always an integer multiple of the clock cycle. Thus, if Δ_0 is the minimum allowable separation between recorded transitions along the track, the only other separations that will occur will be $2\Delta_0$, $3\Delta_0$, $4\Delta_0$ and so on. Neither scheme will generate marks whose lengths are non-integer multiples of Δ_0, even though such marks may usually be accepted by the media and by the read/write system. Incorporation of such "non-integer" marks could potentially enhance the storage capacity of the system.

Table 1.1. The (2, 7) modulation code converts an unconstrained sequence of binary data to a sequence in which the minimum and maximum number of consecutive 0's are 2 and 7 respectively.

Bit-sequence	Modulation code-word
10	0100
11	1000
000	000100
010	100100
011	001000
0010	00100100
0011	00001000

(iii) It is often desired to shape the frequency spectrum of the electrical waveform in order to match it to certain features of the read/write channel. For instance, it is not uncommon to try to reduce the low-frequency content of the signal for the following reasons.
  (a) The electronic $1/f$ noise is strongest in this region.
  (b) The DC offset and drift of the amplifiers shift the signal levels.
  (c) Transformer-type coupling of the signal into and out of rotary heads (for magnetic tape recording) is impossible at very low frequencies.

  Neither NRZ nor NRZI offers a mechanism for shaping the spectrum of the signal; in particular, they impose no limits on the time average of the waveform. Figure 1.5 shows, in addition to the modulated waveforms, plots of the integrated signal for both cases. Note that, in the absence of constraints on the bit pattern, this so-called "charge" of the waveform can increase or decrease without limit.

The solution to the above problems is provided by modulation coding. Table 1.1 shows the code table for IBM's (2, 7) modulation code. This is an example of the so-called (d, k) codes where d is the minimum and k the maximum allowed number of 0's between successive 1's in the modulated sequence. A sequence of binary data consisting of random occurrences of 0's and 1's is converted into another binary sequence in the following way. Starting from the beginning of the sequence, the incoming bit pattern is parsed, i.e., it is broken up into sequences that appear in the left-hand column of Table 1.1. Each such block is then replaced by the corresponding block in the right-hand column of the table. For example, let the incoming stream of data be

$$0010101100110000110100011 10$$

After parsing, we have

0010 10 11 0011 000 011 010 0011 10

Encoding the above sequence using the (2,7) code table then yields

00100100010010000000100000010000100010010000000100000100

Observe that the resulting sequence has at least two and at most seven 0's between successive 1's. Careful examination of the code-words in Table 1.1 will reveal that this is a general property of any sequence formed in the prescribed manner. The recording waveform is now produced using the NRZI scheme, i.e., with a transition at every occurrence of 1 and with no transition at 0's.

In the (2,7)-coded sequence the number of consecutive 0's between 1's is limited to a maximum of seven, thus transitions in the waveform will occur frequently enough to allow proper clocking. At the same time, since the minimum distance between transitions is equal to three clock cycles, i.e., $(d + 1)T$, it is possible for the shortest possible mark that can be recorded on the storage medium to have a length of $\Delta_0 = 3VT$ where V is the velocity of the head relative to the disk. Other possible mark lengths are then $4\Delta_0/3$, $5\Delta_0/3$, $6\Delta_0/3$, $7\Delta_0/3$, and $8\Delta_0/3$. Note that in Table 1.1 the number of modulation bits is twice the number of data bits. But since each modulation bit now occupies only 1/3 of the space it would have occupied in the absence of modulation coding, the capacity must have increased by a factor of 3/2 = 1.5. In this way, by allowing fractions of the minimum mark-length into the process we have increased the capacity of the disk. The price one pays for this additional capacity is a reduced time window for observing the transitions: whereas in the absence of the modulation code the window was Δ_0/V, in its presence the window has shrunk by a factor of three.

There are several other modulation codes that have been developed for a variety of applications. We just mention the important class of $(d,k;c)$ codes, where, in addition to the minimum and maximum run-length constraints of d and k, the accumulated "charge" of the waveform is forced to stay within the range $(-c, +c)$. These codes have very small DC content, and are suitable in situations where the low-frequency content of the signal is to be filtered out. The interested reader is referred to the vast literature in this field for further information.

1.3. The Optical Path

The optical path begins at the light source which, in practically all laser disk systems in use today, is a semiconductor GaAs laser diode. Thanks to several unique features, the laser diode has become indispensable in optical recording, not only for readout of stored information but also for writing and erasure. The small size of this laser ($\simeq$ 300 μm $\times$ 100 μm $\times$ 50 μm) has made possible the construction of compact head assemblies, its coherence

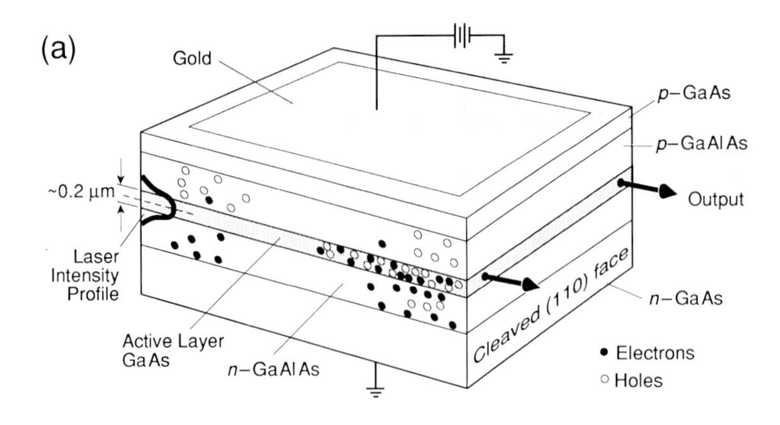

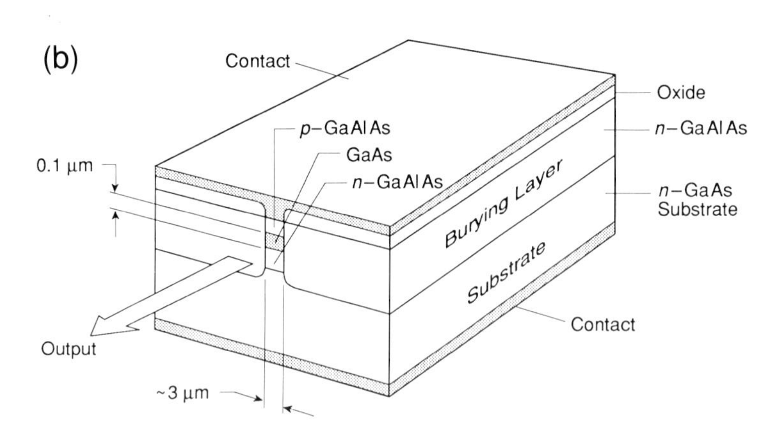

Figure 1.6. (a) Typical double heterostructure GaAs–GaAlAs laser. Electrons and holes are injected into the active GaAs layer from the n- and p-type GaAlAs layers. Frequencies near $\nu = E_g/h$ are amplified by stimulating electron–hole recombination. (b) The buried heterostructure laser provides an effective means for confining the beam in the lateral direction. After A. Yariv, *Optical Electronics*, 4th edition, copyright © 1991 by Saunders College Publishing, adapted with permission of the publisher.

properties have enabled diffraction-limited focusing to extremely small spots, and its direct modulation capability has eliminated the need for external light modulators. The basic structure of a GaAs/GaAlAs laser diode is shown in Fig. 1.6(a). The active GaAs layer is sandwiched between two doped layers of GaAlAs with proper composition. This double heterojunction structure provides the electronic band structure required for the confinement of charge carriers to the active layer. Also, by virtue of the fact that the refractive index of GaAs is somewhat greater than that of the surrounding GaAlAs layers, the structure produces the waveguiding action that confines the radiated photons within the active layer.

Pumping in the semiconductor laser diode is achieved by an external current source that injects the carriers across the heterojunctions into the active layer, thus establishing a population inversion. Electron–hole recombination by stimulated emission then provides the gain mechanism within the active medium. Cleaved facets in the front and back sides of the crystal provide mirrors for the resonant cavity; these facets may be coated to further increase their reflectivity and thereby improve the performance of the laser. The beam is confined in the vertical direction (normal to the plane of the junction) by optical trapping in the dense GaAs medium surrounded by the rare GaAlAs layers. In the lateral direction the beam may be confined by one of several techniques, for example, by confinement of the injection current (gain guiding), by controlling of the refractive index profile by ion implantation (index guiding), or by etching away the unnecessary material and growing a neutral crystal in its place (buried structure). The configuration of a buried heterostructure laser diode is depicted in Fig. 1.6(b).

The operating wavelength of the laser diode can be selected within a limited range by proper choice of material composition; presently available wavelengths from III–V semiconductor lasers for optical storage applications range from 670 nm to 880 nm. In the future, the wavelengths attainable from this class of materials might decrease further and perhaps even reach a lower limit of 600 nm.

Figure 1.7(a) shows a typical plot of laser power output versus input current for a GaAs-based laser diode. The lasing starts at the threshold current, and the output power rapidly increases beyond that point. Below threshold the diode operates in the spontaneous emission mode and the output light is incoherent. After threshold, stimulated emission takes place, yielding coherent radiation. The output power, of course, cannot increase indefinitely and, beyond a certain limit, the laser fails. Fortunately, the required optical power levels for the read/write/erase operations in present-day data storage systems are well below the failure levels of these lasers. Available lasers for data storage applications presently have threshold currents around 40 mA, maximum allowable currents of about 100 mA, and peak output powers (CW mode) around 50 mW. The relationship between the injection current and the output light power is very sensitive to the operating temperature of the laser, as evidenced by the various plots in Fig. 1.7(a). Also, because the semiconductor material's bandgap is a function of the ambient temperature, there is a small shift in the operating wavelength when the temperature fluctuates (see Fig. 1.7(b)). For best performance it is necessary to mount the laser on a good heat-sink or to try to steady its temperature by closed-loop feedback.

The output optical power of the laser can be modulated by means of controlling the injection current. One can apply pulses of variable duration to turn the laser on and off during the recording process. The pulse duration can be as short as a few nanoseconds, with rise and fall times that are typically less than one nanosecond. This direct-modulation capability of the laser diode is particularly useful in optical disk systems, considering that most other sources of coherent light (such as gas lasers) require bulky and expensive devices for external modulation. Although readout of optical

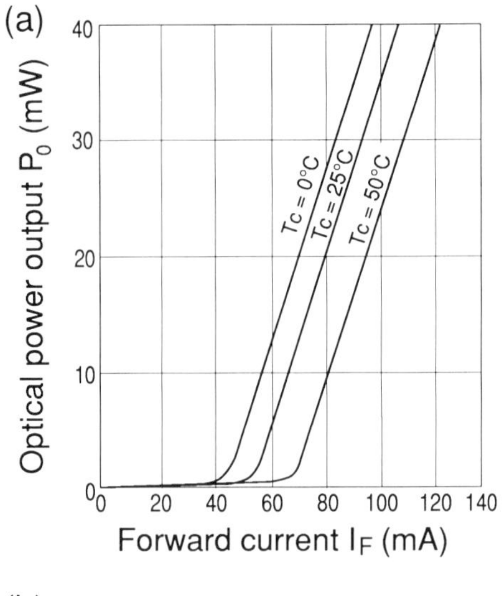

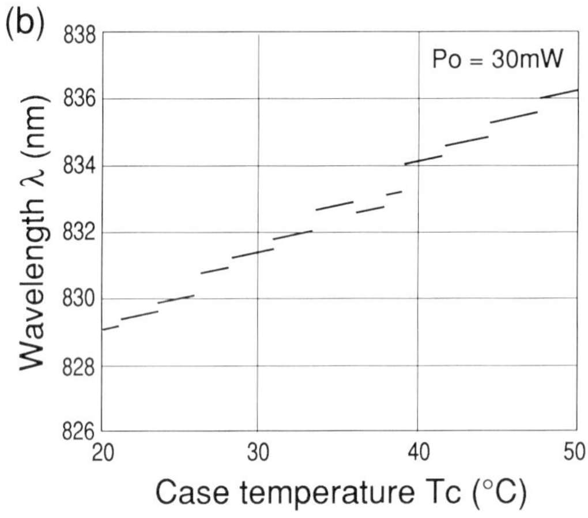

Figure 1.7. (a) Optical output power versus forward-bias current for a typical laser diode. Different curves correspond to different ambient temperatures. (b) Variation of wavelength as function of case temperature for a typical laser diode. The output power is fixed at P_0 = 30 mW. Source: Sharp Corporation's *Laser Diode User's Manual* (1986).

disks can be accomplished at constant power level in CW mode, it is customary (for noise-reduction purposes) to modulate the laser at a high frequency in the range of several hundred MHz.

1.3.1. Collimation and beam shaping

Since the cross-sectional area of the active region in a laser diode is only a fraction of a micrometer, diffraction effects cause the emerging beam to

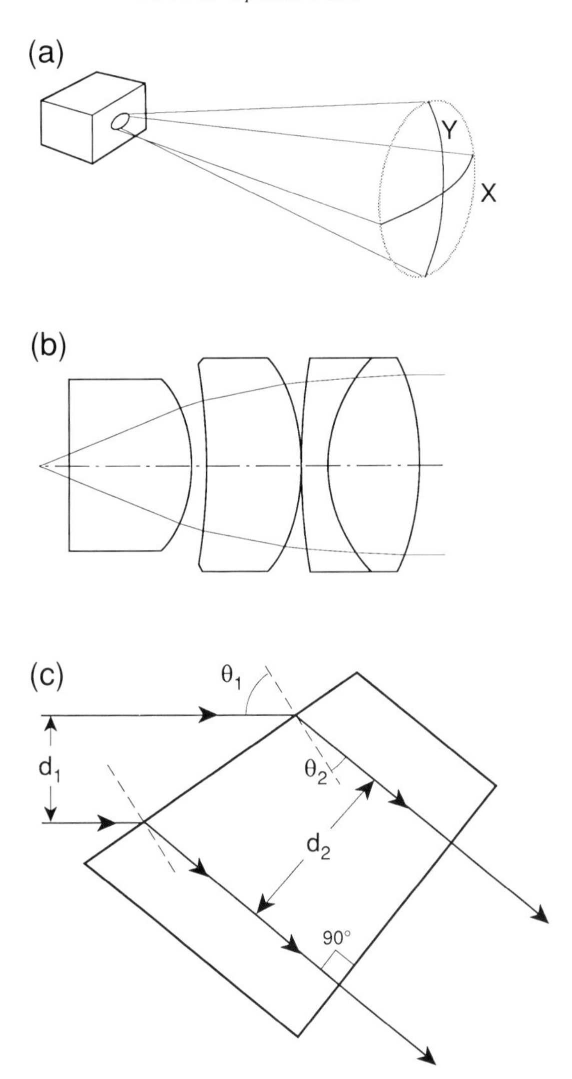

Figure 1.8. (a) Away from the facet, the output beam of a laser diode diverges rapidly. In general, the beam diameter along X is different from that along Y, which makes the cross-section of the beam elliptical. Also, the radii of curvature R_x and R_y are not the same, thus creating a certain amount of astigmatism in the beam. (b) Multi-element collimator lens for laser diode applications. Aside from collimating, this lens also corrects astigmatic aberrations of the beam. (c) Beam-shaping by deflection at a prism surface. θ_1 and θ_2 are related by Snell's law, and the ratio d_2/d_1 is the same as $\cos\theta_2/\cos\theta_1$. Passage through the prism circularizes the elliptical cross-section of the beam.

diverge rapidly. This phenomenon is depicted schematically in Fig. 1.8(a). In practical applications of the laser diode, the expansion of the emerging beam is arrested by a collimating lens, such as that shown in Fig. 1.8(b). If the beam happens to have aberrations (astigmatism is particularly severe in laser diodes), the collimating lens must be designed to correct this defect.

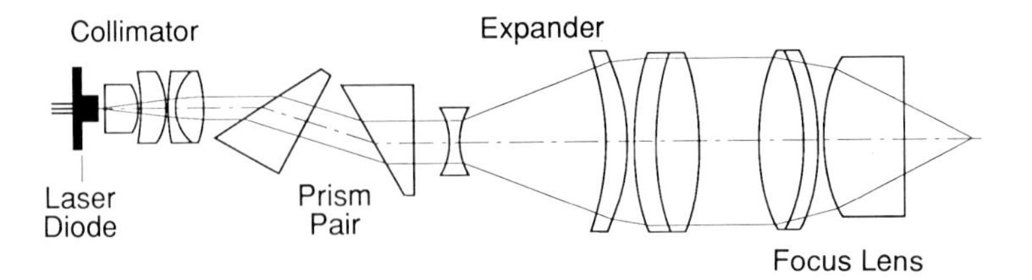

Figure 1.9. The special characteristics of laser diode radiation must be accommodated by the optics. This diagram shows the divergent laser output first collimated, then converted from elliptical to circular cross-section (in the anamorphic prism pair), expanded, and focused to a small spot. Source: D. Kuntz, *Specifying Laser Diode Optics*, Laser Focus (March 1984).

In optical recording it is most desirable to have a beam with a circular cross-section. The need for shaping the beam arises from the special geometry of the laser cavity with its rectangular cross-section. Since the emerging beam has different dimensions in the directions parallel and perpendicular to the junction, its cross-section at the collimator becomes elliptical, with the initially narrow dimension expanding more rapidly to become the major axis of the ellipse. The collimating lens thus produces a beam with elliptical cross-section. Circularization may be achieved by bending the beam at a prism, as shown in Fig. 1.8(c). The bending changes the beam's diameter in the plane of incidence, but leaves the diameter in the perpendicular direction intact. Figure 1.9 shows a possible arrangement of optical elements in the path from the laser diode to a focused spot. The path includes a collimating lens, beam-shaping anamorphic prism pair, beam expander, and a focusing (objective) lens. In optical recording practice, the collimated beam diameter is matched to that of the objective, thus obviating the need for the expander.

The output of the laser diode is linearly polarized in the plane of the junction. In some applications (such as the readout of compact disks or read/write on WORM media) the polarization state is immaterial as far as interaction with the storage medium is concerned. In such applications, one usually passes the beam through a polarizing beam splitter (PBS) and a quarter-wave plate, as shown in Fig. 1.10, in order to convert it into a circularly-polarized beam. Upon reflection from the disk, the beam passes through the quarter-wave plate once again, but this time emerges as linearly polarized in a direction perpendicular to its original direction of polarization. The returning beam is thus directed by the PBS away from the laser and towards the detection module, where its data content can be extracted and its phase/amplitude pattern can be used to generate error signals for automatic focusing and tracking. By thoroughly separating the returning beam from the incident beam, not only does one achieve efficiency in the use of optical power, but also one prevents the beam from going back to the laser, where it could cause instabilities in the cavity and increase the noise level. Unfortunately, there are situations where the

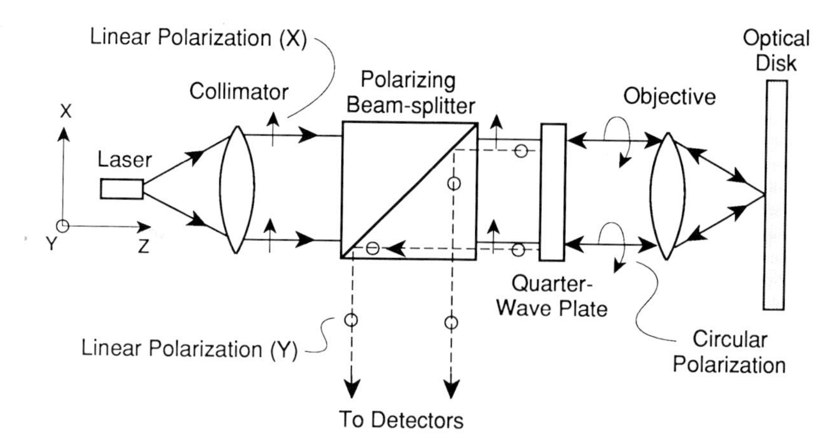

Figure 1.10. Separation of incident and reflected beams at the polarizing beam-splitter. The quarter-wave plate converts the linearly-polarized incident beam into a circularly polarized beam; in the return path, it converts the beam back to linear polarization, but now the direction of polarization is perpendicular to that of the incident beam. As a result of the 90° rotation of polarization, the reflected beam is diverted by the beam-splitter towards the detection channel.

polarization state is relevant to the interaction with the disk; magneto-optical readout that requires linear polarization is a case in point. In such instances the simple combination of PBS and quarter-wave plate becomes inadequate, and one resorts to other (less efficient) means of separating the beams. The issues attendant on beam separation in MO systems will be discussed in Chapter 9.

1.3.2. Focusing by the objective lens

The collimated and circularized beam of the laser diode is focused on the disk surface using an objective lens. Figure 1.11(a) shows the design of a typical objective made from spherical optics. The objective is designed to be aberration-free, so that its focused spot size is limited only by the effects of diffraction. According to the classical theory of diffraction, the diameter of the beam, d, at the objective's focal plane is given by

$$\boxed{d \simeq \frac{\lambda}{\mathrm{NA}}} \tag{1.11}$$

where λ is the wavelength of light, and NA is the numerical aperture of the objective.†

†The numerical aperture of a lens is defined as $\mathrm{NA} = n \sin \theta$, where n is the refractive index of the image space, and θ is the half-angle subtended by the exit pupil at the focal

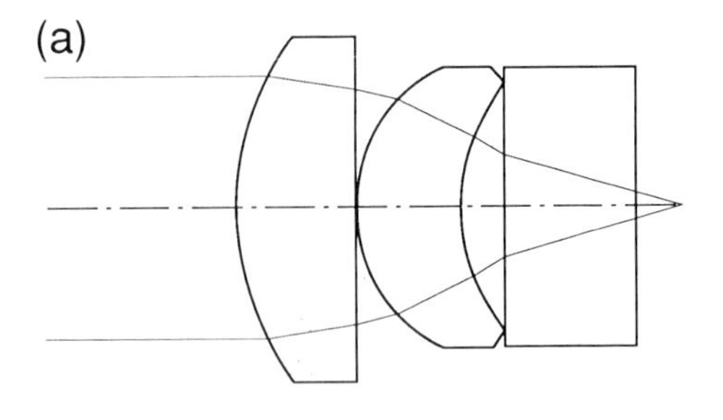

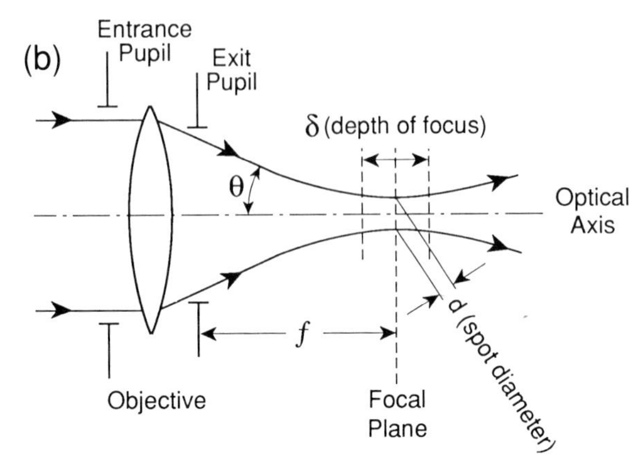

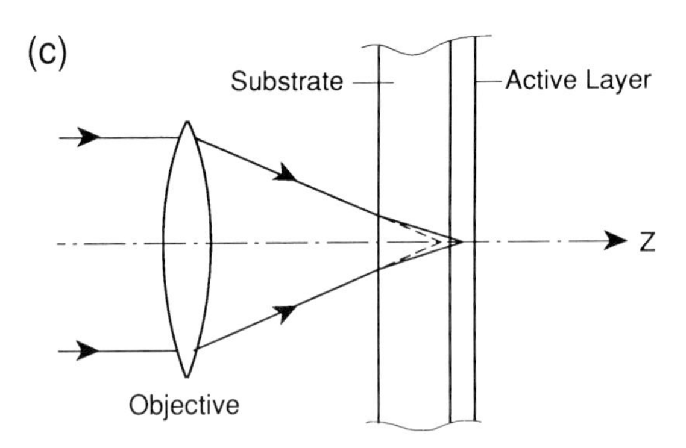

Figure 1.11. (a) Multi-element lens design for a high-NA video-disk objective. (b) Various parameters of the objective lens. The numerical aperture NA = $\sin\theta$. The spot diameter d and the depth of focus δ are given by Eqs. (1.11) and (1.12), respectively. (c) Focusing through the substrate can cause spherical aberration at the active layer.

point. In optical recording systems the image space is air, whose index is very nearly unity, and, therefore, NA = $\sin\theta$. The numerical aperture of a lens is closely related to its f-number which, by definition, is the ratio of the focal length to the clear-aperture diameter; namely, f-number = f/D. For small numerical apertures where $\sin\theta \simeq \tan\theta$, the relation f-number $\simeq 0.5/\text{NA}$ applies. For example, when NA = 0.1, the f-number is 5.

In optical recording it is desired to achieve the smallest possible spot, since the size of the spot is directly related to the size of marks recorded on the medium. Also, in readout the spot size determines the resolution of the system. According to Eq. (1.11) there are two ways to achieve a small spot: first, by reducing the wavelength of the light or, second, by increasing the numerical aperture of the objective. The wavelengths currently available from GaAs lasers are in the range of 670–840 nm. It is possible to use a nonlinear optical device to double the frequency of these laser diodes, thus achieving blue light. Good efficiencies have been demonstrated by frequency doubling. Also recent developments in II–VI materials have improved the prospects for obtaining green and blue light directly from semiconductor lasers. Consequently, there is hope that in the near future optical storage systems will operate in the wavelength range of 400–500 nm. As for the numerical aperture, current practice is to use a lens with NA $\simeq$ 0.5–0.6. Although this value might increase slightly in the coming years, much higher numerical apertures are unlikely, since they put strict constraints on the other characteristics of the system and limit the tolerances. For instance, the working distance at high NA is relatively short, making access to the recording layer through the substrate more difficult. The smaller depth of focus of a high-NA lens will make attaining and maintaining proper focus more difficult, while the limited field of view might restrict automatic track-following procedures. A small field of view also places constraints on the possibility of read/write/erase operations involving multiple beams.

The depth of focus of a lens, δ, is the distance away from the focal plane over which tight focus can be maintained (see Fig. 1.11(b)). According to classical diffraction theory,

$$\boxed{\delta \simeq \frac{\lambda}{(\mathrm{NA})^2}} \tag{1.12}$$

Thus for a wavelength of $\lambda = 700$ nm and for NA = 0.6, the depth of focus is about ± 1 μm. As the disk spins under the optical head at the rate of several thousand rpm, the objective lens must stay within a distance of $f \pm \frac{1}{2}\delta$ from the active layer if proper focus is to be maintained. Given the conditions under which drives usually operate, it is impossible to make the mechanical systems rigid enough to yield the required positioning tolerances. On the other hand, it is fairly simple to mount the objective lens in an actuator capable of adjusting its position with the aid of closed-loop feedback control. We shall discuss the techniques of automatic focusing later in section 1.4. For now, let us emphasize that by going to shorter wavelengths and/or larger numerical apertures (as is required for attaining higher data densities) one has to face a much stricter regime as far as automatic focusing is concerned. Increasing the numerical aperture is particularly worrisome, since δ drops with the square of NA.

A source of spherical aberration in optical disk systems is the substrate through which the light must pass in order to reach the active layer. Figure

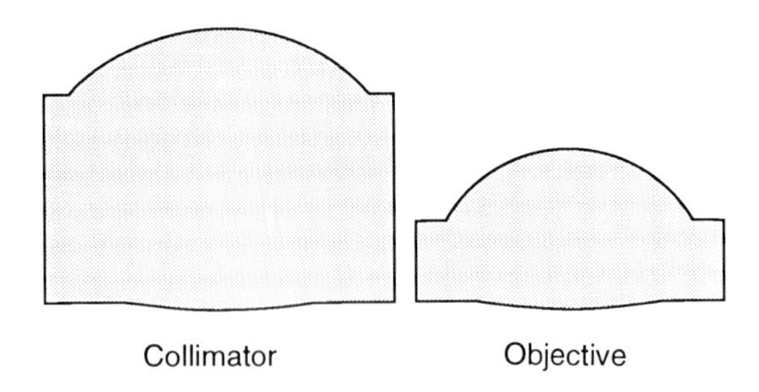

Figure 1.12. Molded glass aspheric lens pair. In recent years these singlets have replaced the multi-element spherical lenses shown in Figs. 1.8(b) and 1.11(a) in optical disk applications.

1.11(c) shows the bending of the rays at the surface of the disk, which causes the aberration. This problem can be solved by taking into account the effects of the substrate in the design of the objective, so that the lens is corrected for all aberrations including those arising at the substrate.†

Recent developments in the molding of aspheric glass lenses have gone a long way in simplifying the lens design problem. Figure 1.12 shows a pair of molded glass aspherics designed for optical disk system applications; both the collimator and the objective are single-element lenses, and have been corrected for all relevant aberrations.

1.3.3. Laser noise

Compared to other sources of coherent light such as gas lasers, laser diodes are noisy and unstable. Typically, within a semiconductor laser's cavity, several modes compete for dominance. Under these circumstances, small variations in the environment can cause mode-hopping and result in unpredictable power-level fluctuations and wavelength shifts. Unwanted optical feedback is specially troublesome, as even a small fraction of light returning to the cavity can cause a significant rise in the noise level. Fortunately, it has been found that high-frequency modulation of the injection current can be used to instigate power sharing among the modes and thereby reduce fluctuations of the output optical power. In general, a combination of efforts such as temperature stabilization of the laser,

†A possible approach to the problem of designing the objective is to ignore the substrate initially, and design the lens with a flat surface on the side facing the disk, as in Fig. 1.11(a). By subsequently removing from the lens a flat slab of material equivalent to the optical thickness of the substrate, one can compensate for the effects of the substrate. Whatever approach is taken, however, one is left with the problem that variations in the thickness or refractive index of the substrate (point-to-point variations on a given disk as well as disk-to-disk variations on a given drive) will introduce residual aberrations and cause deterioration in the quality of the focused spot. It is for this reason that the substrate and its optical properties are major concerns of the standardization committees.

antireflection coating of the various surfaces within the system, optical isolation of the laser, and high-frequency modulation of the injection current can yield acceptably low noise levels.

1.4. Automatic Focusing

In the preceding section we mentioned that since the objective lens has a large numerical aperture (NA ≥ 0.5), its depth of focus δ is very shallow ($\delta \simeq \pm 1\ \mu$m at $\lambda = 780$ nm). During all read/write/erase operations, therefore, the disk must remain within a fraction of a micrometer from the focal plane of the objective. In practice, however, disks are not flat and are not always mounted rigidly parallel to the focal plane, so that, during each revolution, movements away from focus occur a few times. The amount of this movement can be as much as a hundred micrometers. Without automatic adjustment of the position of the objective lens along the optical axis, this runout (or disk flutter) will be detrimental to the operation of the system. In practice, the objective is mounted on a small actuator (usually a voice-coil) and allowed to move back and forth in order to keep its distance within an acceptable range of the disk. The spindle turns at a few thousand rpm, which is a hundred or so revolutions per second. If the disk moves in and out of focus a few times during each revolution, then the voice coil must be fast enough to follow these movements in real time; in other words, its frequency response must have a bandwidth of several kilohertz.

The signal that controls the voice-coil is obtained from the light that has been reflected from the disk. There are several techniques for deriving the focus-error signal, one of which is depicted in Fig. 1.13(a). In this so-called obscuration method a secondary lens is placed in the path of the reflected light, one half of its aperture is covered, and a split detector is placed at its focal plane. When the disk is in focus, the returning beam is collimated and the secondary lens will focus it at the center of the split detector, giving a difference signal $\Delta S = 0$. If the disk now moves away from the objective, the returning beam will become converging, as in Fig. 1.13(b), sending all the light to detector 1. In this case ΔS will be positive and the voice-coil will push the lens towards the disk. On the other hand, when the disk moves closer to the objective, the returning beam becomes diverging and detector 2 receives the light (see Fig. 1.13(c)). This results in a negative ΔS which forces the voice-coil to pull back in order to return ΔS to zero.

A given focus-error detection scheme is generally characterized by the shape of its focus-error signal ΔS versus the amount of defocus Δz. One such curve is shown in Fig. 1.13(d). The slope of the FES curve near the origin is of particular importance since it determines the overall performance and the stability of the servo loop. In general, schemes with a large slope are preferred, although certain other aspects of system performance should also be taken into consideration. For instance, variations of the FES during seek operations (where multiple track-

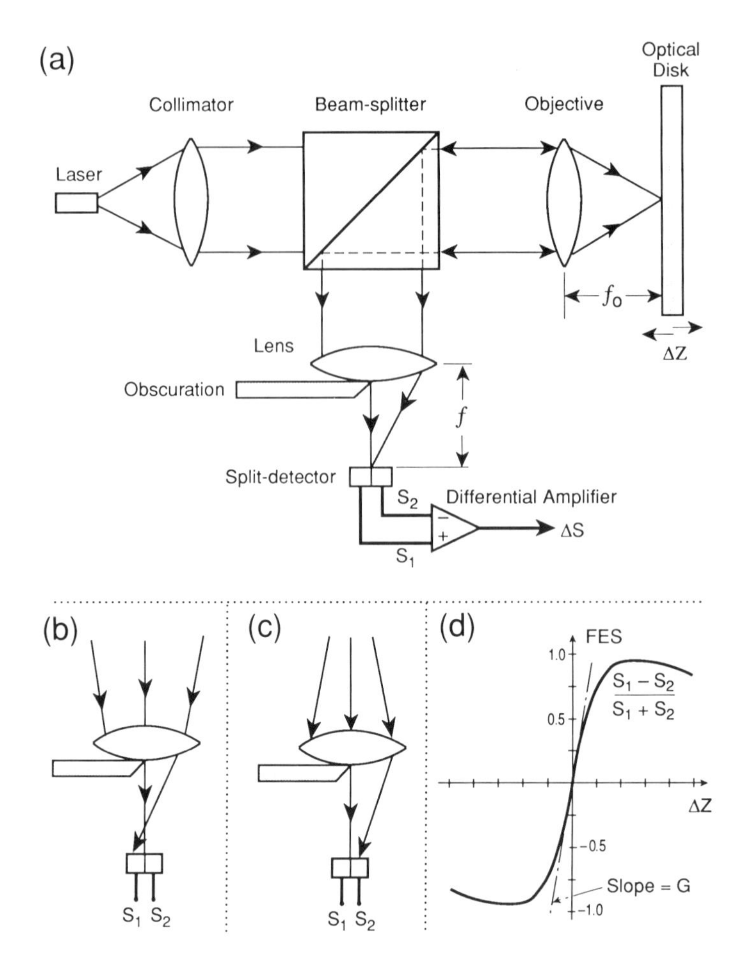

Figure 1.13. Focus-error detection by the obscuration method. In (a) the disk is in focus, and the two halves of the split detector receive equal amounts of light. When the disk is too far from the objective (b) or too close to it (c), the balance of detector signals shifts to one side or the other. A typical plot of the focus error signal versus defocus is shown in (d), and its slope near the origin is identified as the FES gain, G.

crossings occur) should be kept at a minimum, or else the resulting "feedthrough" signal might destabilize the focus servo. Also, it is important for a focus-error detection scheme to be insensitive to slight imperfections of the optical elements, as well as to positioning errors and misalignments, otherwise the manufacturing cost of the device may become prohibitively high. Finally, the focusing scheme must have a reasonable acquisition range, so that at startup (or on those occasions where focus is lost and needs to be re-acquired) the system can move in the correct direction in order to establish focus.

1.5. Automatic Tracking

Consider a circular track at a certain radius r_0, as in Fig. 1.1, and imagine viewing a portion of this track through the access window. (It is through this window that the read/write head gains access to the disk and, by moving in the radial direction, reaches arbitrarily selected tracks.) To a viewer looking through the window a perfectly circular track, centered on the spindle axis, will look stationary, irrespective of the rotational speed of the disk. However, any track eccentricity will cause an apparent motion towards or away from the center. The peak-to-peak radial distance travelled by a track (as seen through the window) might depend on a number of factors including centering accuracy of the hub, deformability of the disk substrate, mechanical vibrations, manufacturing tolerances, etc. For a 3.5" plastic disk, for example, this peak-to-peak motion can be as much as 100 μm during any period of rotation. Assuming a revolution rate of 3600 rpm, the apparent velocity of the track in the radial direction will be in the range of several mm/s. Now, if the focused spot (whose diameter is only about 1 μm) remains stationary while set to read or write this track (whose width is also about 1 μm), it is clear that the beam will miss the track for a sizeable fraction of every revolution cycle.

Practical solutions to the above problem are provided by automatic track-following techniques. Here the objective is placed in a fine actuator, typically a voice-coil, which is capable of moving the necessary radial distances and maintaining a lock on the desired track. The signal controlling the movement of this actuator is derived from the reflected light itself, which carries information about the position of the focused spot relative to the track. There exist several mechanisms for extracting the track-error signal (TES) from the reflected light; all these methods require some sort of structure on the disk in order to determine the relative position of the spot. In the case of read-only disks (CD, CD-ROM, and video disks) the embossed pattern of data provides ample information for tracking purposes. In the case of write-once and erasable disks, tracking guides are imprinted on the substrate during the manufacturing process. The two major formats for these tracking guides are pregrooves (for continuous tracking) and sampled-servo (for discrete tracking). A combination of the two schemes, known as the continuous–composite format, is often used in practice. This method is depicted schematically in Fig. 1.14, which shows a small section containing five tracks, each consisting of the tail-end of a groove, synchronization marks, a mirror area used for adjusting certain offsets, a pair of wobble marks for sampled tracking, and header information for sector identification.

1.5.1. Tracking on grooved regions

As shown in Fig. 1.3(a), grooves are continuous depressions that are either embossed or etched or molded onto the substrate prior to deposition of the

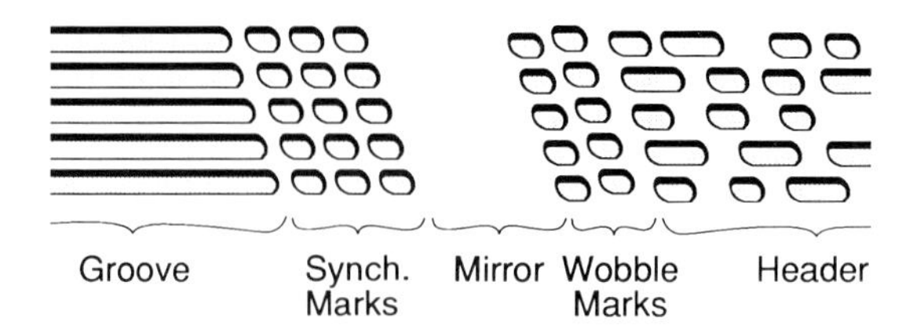

Figure 1.14. Servo offset fields in continuous–composite format contain a mirror area and offset marks for tracking. Source: A.B. Marchant, *Optical Recording*, © 1990 by Addison-Wesley Publishing Co., reprinted by permission of the publisher.

storage medium. If the data is recorded on the grooves, then the lands are not used except for providing a guard band between neighboring grooves. Conversely, the land regions may be used to record the information, in which case grooves provide the guard band. Typical track-widths are about one wavelength of the light. The guard bands are somewhat narrower than the tracks, their exact shape and dimensions depending on the beam size, the required track-servo accuracy, and the acceptable levels of cross-talk between adjacent tracks. The groove depth is usually around one-eighth of one wavelength ($\lambda/8$), since this depth can be shown to give the largest track-error signal (TES) in the push–pull method. The cross-sections of the land and the groove are, in general, trapezoidal; rectangular cross-sections are difficult (if not impossible) to fabricate. Triangular or V-shaped cross-sections are not uncommon for the guard band.

When the focused spot is centered on a given track, be it groove or land, it is diffracted symmetrically from the two edges of the track, resulting in a balanced far field pattern. As soon as the spot moves away from the track-center, the symmetry breaks down and the light distribution in the far-field tends to shift to one side or the other. A split photodetector placed in the path of the reflected light can therefore sense the relative position of the spot and provide the appropriate feedback signal (see Fig. 1.15(a)); this is the essence of the push–pull method. Figure 1.15 also shows several computed intensity plots at the detector plane after reflection from various locations on the grooved surface. Note how the intensity shifts to one side or the other, depending on which edge of the groove the spot is moving toward.

1.5.2. Sampled tracking

Since dynamic track runout is usually a slow and gradual process, there is actually no need for continuous tracking as done on grooved media. A pair of embedded marks, offset from the track-center as in Fig. 1.16(a), can provide the necessary information for correcting the relative position of the focused spot. The reflected intensity will indicate the positions of the two servo marks as two successive short pulses. If the beam happens to be on-

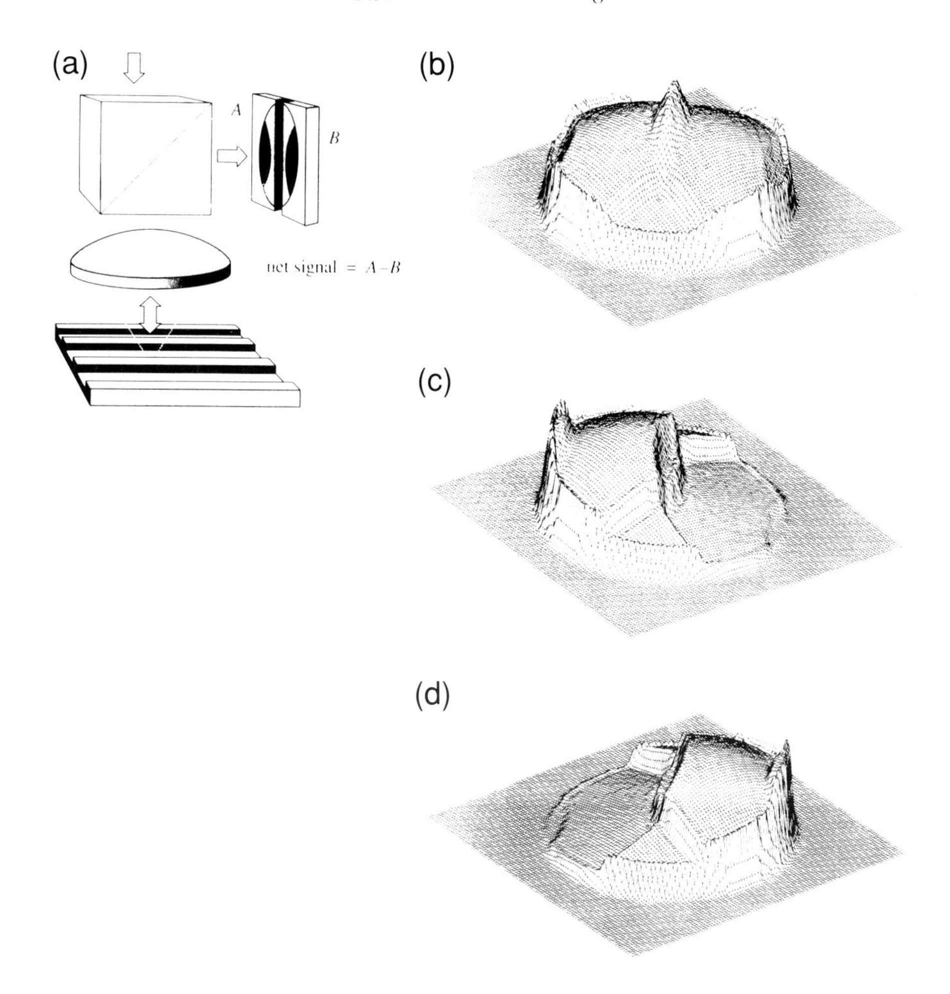

Figure 1.15. (a) Push–pull sensor for tracking on grooves. Source: A.B. Marchant, *Optical Recording*, ©1990 by Addison-Wesley Publishing Co., reprinted by permission of the publisher. (b) Computed intensity pattern at the detector with the disk in focus and the spot centered on track. (c) Same as (b) but with the spot centered on a groove edge. (d) Same as (c), with the spot centered on opposite edge.

track, the two pulses will have equal magnitudes and there is no need for correction. If, on the other hand, the beam is off-track, one of the pulses will be stronger than the other. Depending on which pulse is the stronger, the system will recognize the direction in which it has to move and will correct the error accordingly. Sampled-servo mark pairs must be provided frequently enough to ensure proper track-following. In a typical application, the track might be divided into groups of eighteen bytes, two bytes dedicated as servo offset areas and sixteen bytes filled with other format information or left blank for user data. Figure 1.16(b) shows a small section from a sampled-servo disk containing a number of tracks, three of which are recorded with user-data. The track servo marks in this case are preceded by synchronization marks (also prerecorded on the servo

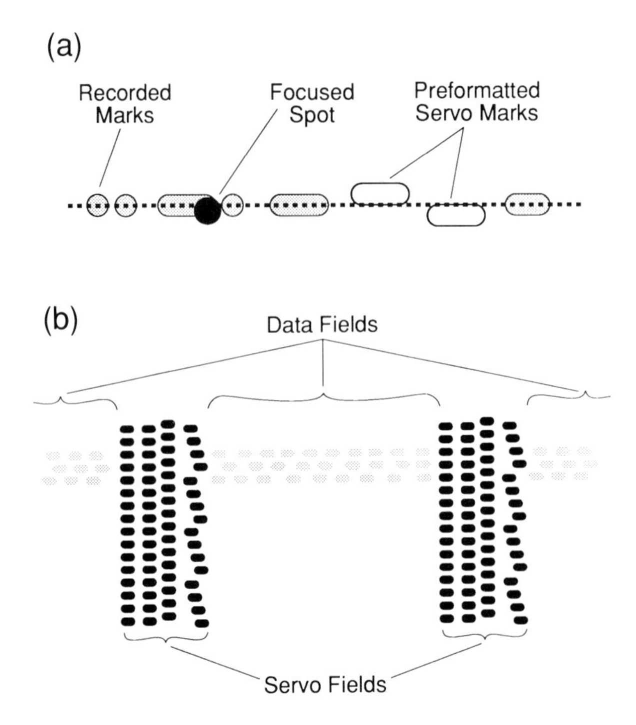

Figure 1.16. (a) In sampled-servo tracking a pair of preformatted servo marks help locate the focused spot relative to the track-center. (b) Servo fields occur frequently and at regular intervals in sampled-servo format. The data area shown here has data recorded on three tracks. Source: A.B. Marchant, *Optical Recording*, © 1990 by Addison-Wesley Publishing Co., reprinted by permission of the publisher.

offset area). Note in Fig. 1.16(b) that the format marks repeat a certain pattern every four tracks. This pattern is known as a "gray code", and allows the system to recognize and correct minor track-counting errors during the seek operation.

1.5.3. Track-counting during the seek operation

During the seek operation a coarse actuator moves the entire head assembly across the disk to a new location where the desired track is expected to be. In order to avoid landing on a nearby track and being forced to perform a second (fine) seek, most systems in use today count the tracks as they cross them. In this way the system can zero-in on the correct track while landing, and thereby minimize the overall seek time. The sampled-servo format is not suitable for this purpose, since the servo marks do not occur frequently enough to allow uninterrupted counting. In contrast, grooved media provide the necessary information for track-counting.

During seek operations the focus servo loop remains closed,

maintaining focus as the head crosses the tracks. The tracking loop, on the other hand, must be opened. The zero-crossings of the TES then provide the track count. Complications may arise in this process, however, due to the eccentricities of tracks. To an observer looking through the access window, an eccentric track moves in and out radially with a small (but not insignificant) velocity. As the head approaches the desired track and slows down to capture it, its velocity might fall just short of the apparent track velocity. Under these circumstances, a track that has already been counted may catch up with the head and be counted once again. Intelligence must be built into the system to recognize such problems and avoid them, if possible. Also, through the use of gray codes and similar schemes, the system can be made to correct its occasional miscounts before finally locking onto the destination track.

1.6. The Thermomagnetic Recording Process

The recording and erasure of information on a magneto-optical disk are both achieved by the thermomagnetic process. The essence of thermomagnetic recording is shown in Fig. 1.17. At the ambient temperature the film has a high magnetic coercivity† and therefore does not respond to the externally applied field. When a focused laser beam raises the local temperature on the film, the hot spot becomes magnetically soft (i.e., its coercivity decreases). As the temperature rises, the coercivity drops continuously until the field of the electromagnet finally overcomes the material's resistance to reversal and switches its magnetization. Turning the laser off brings the temperature back to normal, but the reverse-magnetized domain remains frozen in the film. In a typical situation, the film thickness may be around 300 Å, the laser power at the disk $\simeq$ 10 mW, the diameter of the focused spot $\simeq$ 1 μm, the laser pulse duration $\simeq$ 50 ns, the linear velocity of the track $\simeq$ 10 m/s, and the magnetic field strength $\simeq$ 200 gauss. The temperature may reach a peak of 500 K at the center of the spot, which is certainly sufficient for magnetization reversal but is not nearly high enough to melt or crystallize or in any other way modify the atomic structure of the material.

The materials used in MO recording have strong perpendicular magnetic anisotropy. This type of anisotropy favors the "up" and "down" directions of magnetization over all other orientations. The disk is initialized in one of these two directions, say up, and the recording takes place when small regions are selectively reverse-magnetized by the thermomagnetic process. The resulting magnetization distribution then represents the pattern of recorded information. For instance, binary

†Coercivity is a measure of the resistance of a magnetic medium to magnetization reversal. Consider a thin film with perpendicular magnetic moment saturated in the positive Z-direction, as in Fig. 1.17(a). A magnetic field applied along $-Z$ will succeed in reversing the magnetization only if the field is stronger than the coercivity of the film. (See subsection 1.8.3 for a more detailed description of coercivity.)

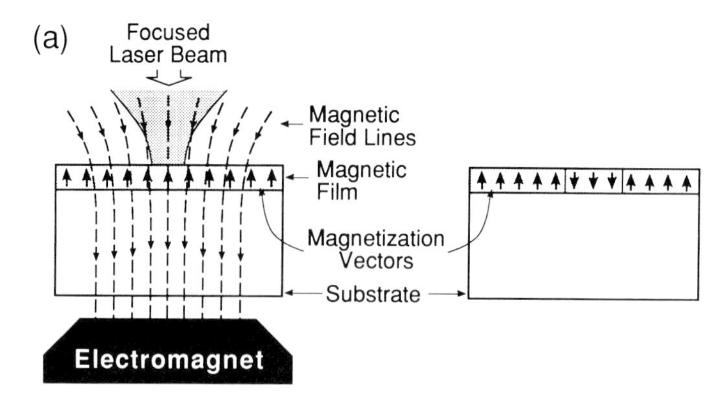

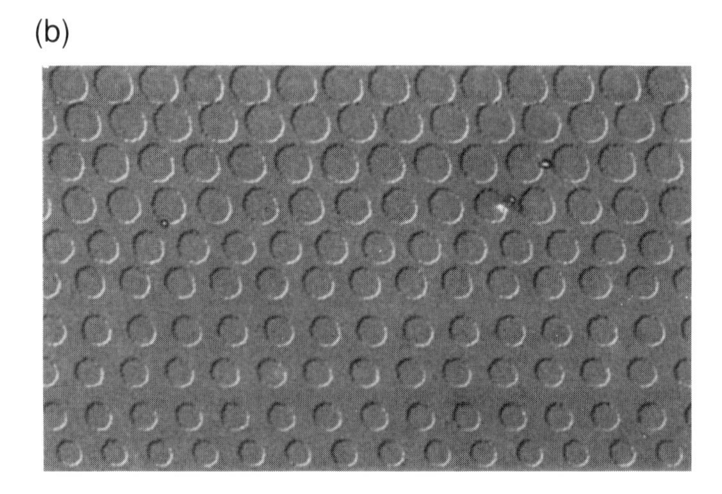

Figure 1.17. (a) The thermomagnetic recording process. The field of the electromagnet reverses the direction of magnetization in the area heated by the focused laser beam. (b) Lorentz micrograph of domains written thermomagnetically. The various tracks shown were written at different laser powers, with power level decreasing from top to bottom. Micrograph courtesy of F. Greidanus, Philips Research Laboratories.

sequences may be represented by a mapping of 0's to up-magnetized and 1's to down-magnetized regions (NRZ). Alternatively, the NRZI scheme might be used, whereby transitions (up-to-down and down-to-up) are used to represent the 1's in the bit-sequence.

1.6.1. Recording by laser power modulation (LPM)

In this traditional approach to thermomagnetic recording, the electromagnet produces a constant field, while the information signal is used to modulate the power of the laser beam. As the disk rotates under the focused spot, the on/off laser pulses create a sequence of up/down domains along the track. The Lorentz electron micrograph in Fig. 1.17(b) shows a number of domains recorded by LPM. The domains are highly stable and may be read over and over again without significant degradation. If, however, the user decides to discard a recorded block and to use the space for new data, the LPM scheme does not allow direct overwrite; the system must erase the old data during one revolution of the disk, and record the new data in a subsequent revolution cycle.

During erasure, the direction of the external field is reversed, so that the up-magnetized domains in Fig. 1.17(a) now become the favored ones. Whereas writing is achieved with a modulated laser beam, in erasure the laser is switched on at a constant power level until an entire sector is erased. Selective erasure of individual domains is not practical, nor is it desired, since mass data storage systems generally deal with data at the level of blocks, which are recorded onto and read from individual sectors. Note that at least one revolution period elapses between the erasure of an old block and its replacement by a new block. The electromagnet therefore need not be capable of rapid switchings. (When the disk rotates at 3600 rpm, for example, there will be a period of 16 ms or so between successive switchings.) This kind of slow reversal allows the magnet to be large enough to cover all the tracks simultaneously, thereby eliminating the need for a moving magnet and actuator. It also affords a relatively large gap between the disk and the magnet-tip, which enables the use of double-sided disks and relaxes the mechanical tolerances of the system without over-burdening the magnet's power supply.

The obvious disadvantage of LPM is its lack of direct overwrite capability. A more subtle concern is that it is perhaps unsuitable for the PWM (pulse-width modulation) scheme of representing binary waveforms. Due to fluctuations in the laser power, spatial variations of material properties, lack of perfect focusing and track-following, etc., the length of a recorded domain along the track may fluctuate in small but unpredictable ways. If the information is to be encoded in the distance between adjacent domain walls (i.e., PWM), then the LPM scheme of thermomagnetic writing may suffer from excessive domain-wall jitter. Laser power modulation works well, however, when the information is encoded in the position of domain centers (i.e., pulse-position modulation or PPM). In general, PWM is superior to PPM in terms of the recording density, and methods that allow PWM are preferred.

1.6.2. Recording by magnetic field modulation (MFM)

Another method of thermomagnetic recording is based on magnetic-field modulation, and is depicted schematically in Fig. 1.18(a). Here the laser power may be kept constant while the information signal is used to modulate the direction of the magnetic field. Photomicrographs of typical domain patterns recorded in the MFM scheme are shown in Fig. 1.18(b).

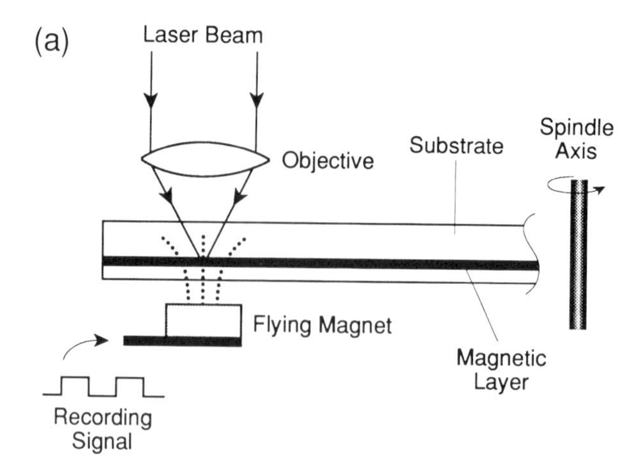
(a)
Laser Beam
Objective
Substrate
Spindle
Axis
Flying Magnet
Magnetic
Layer
Recording
Signal

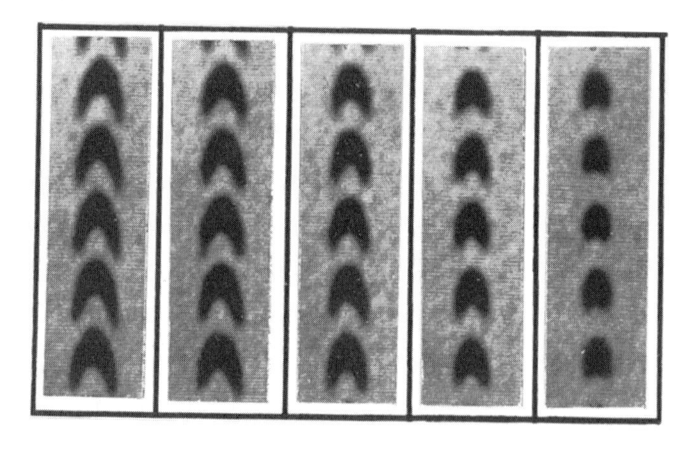
(b)

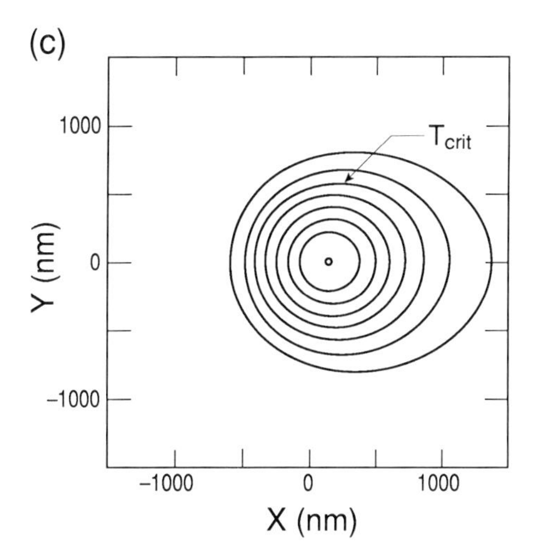

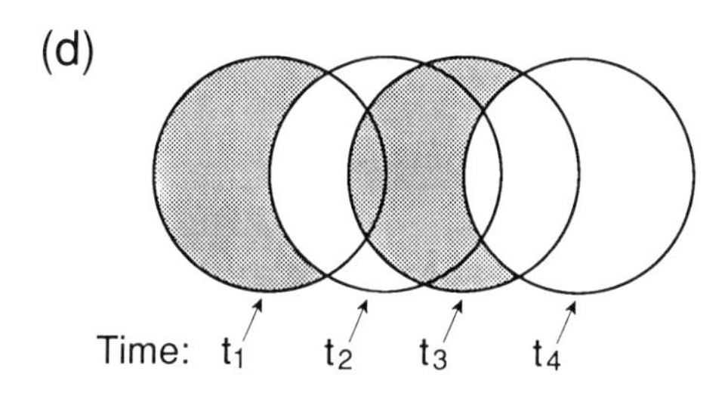

Figure 1.18. (a) Thermomagnetic recording by magnetic field modulation. The power of the beam is kept constant, while the magnetic field direction is switched by the data signal. (b) Polarized-light microphotograph of recorded domains. (c) Computed isotherms produced by a CW laser beam, focused on the magnetic layer of a disk. The disk moves with constant velocity under the beam. The region inside the isotherm marked as T_{crit} is above the critical temperature for writing, thus its magnetization aligns itself with the applied magnetic field. (d) Magnetization within the heated region (above T_{crit}) follows the direction of the applied field, whose switchings occur at times $_n$. The resulting domains (shaded) are crescent-shaped.

Crescent-shaped domains are the hallmark of the field modulation technique. If one assumes (using a much simplified model) that the magnetization aligns itself with the applied field within a region whose temperature has reached beyond a certain critical value, T_{crit}, then one can explain the crescent shape of these domains in the following way. With the laser operating in the CW-mode and the disk moving at constant velocity, the temperature distribution in the magnetic medium assumes a steady-state profile, such as that shown in Fig. 1.18(c). Of course, relative to the laser beam, the temperature profile is stationary, but in the frame of reference of the disk the profile moves along the track with the linear track velocity. The isotherm corresponding to T_{crit} is identified in the figure; within this isotherm the magnetization always aligns itself with the applied field. A succession of critical isotherms along the track, each obtained at the particular instant of time when the magnetic field switches direction, is shown in Fig. 1.18(d). From this diagram it is not difficult to see how the crescent-shaped domains arise, and also to understand the relation between the waveform that controls the magnet and the resulting domain pattern.

The advantages of magnetic-field modulation recording are twofold: (i) direct overwriting is possible, and (ii) domain wall positions along the track, being rather insensitive to defocus and to laser power fluctuations, are fairly accurately controlled by the timing of the magnetic field switchings. On the negative side, the magnet must now be small and must fly close to the disk surface, if it is to produce rapidly switched fields of the order of one hundred gauss. Systems that utilize magnetic field modulation often fly a small electromagnet on the opposite side of the disk from the optical stylus. Since mechanical tolerances are tight, this might compromise the removability of the disk in such systems. Moreover, the requirement of close proximity between the magnet and the storage medium dictates the use of single-sided disks in practice.†

1.6.3. Thermal optimization of the media; multilayer structures

The thermal behavior of an optical disk can be modified and improved if the active layer is incorporated into a properly designed multilayer structure, such as that shown in Fig. 1.19. In addition to thermal engineering, multilayers allow protective mechanisms to be built around the active layer; they also enable the enhancement of the signal-to-noise ratio in readout. (This latter feature will be explored in section 1.7.2.) Multilayers are generally designed to optimize the absorption of light by creating an antireflection structure, whereby a good fraction of the incident optical power is absorbed in the active layer. Whereas the reflectivity of bare metal films is typically over 50%, a quadrilayer structure can easily reduce that to 20% or even less, if so desired. Multilayers can also be designed to control the flow of heat generated by the absorbed light. The aluminum reflecting layer, for instance, may be used as a heat sink for the magnetic layer, thus minimizing the undesirable effects of lateral heat diffusion.

† In a slightly modified version of recording by magnetic-field modulation, the laser beam is turned off during the transition intervals when the magnetic field goes through zero. Since the magnetization follows the field when the latter is strong, but wavers when it is weak, de-emphasizing the field transition period is apt to improve the sharpness of the magnetization transitions. Indeed the modified method often results in better definition of domain walls and succeeds in producing playback signals with less jitter.

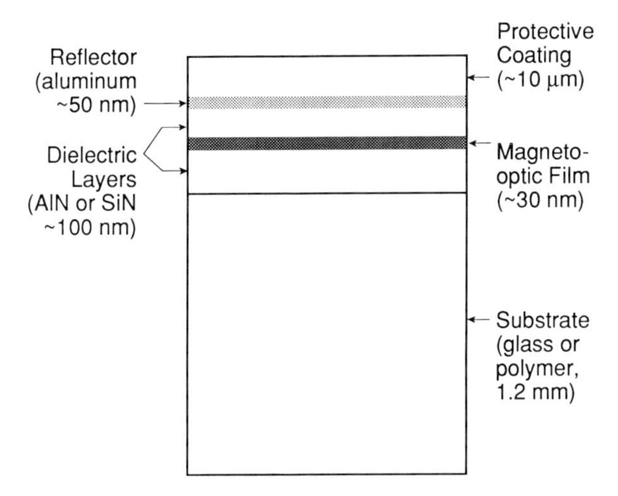

Figure 1.19. Quadrilayer magneto-optical disk structure. This particular design is for use in the substrate-incident mode, where the light goes through the substrate before reaching the MO layer. The thicknesses of the various layers can be optimized for enhancing the read signal, increasing the absorbed laser power, and controlling the thermal profile. Note in particular that the aluminum layer can play the dual role of light reflector and heat sink.

1.7. Magneto-optical Readout

The information recorded on a perpendicularly magnetized medium may be read with the aid of the polar magneto-optical Kerr effect. When linearly polarized light is normally incident on a perpendicular magnetic medium, its plane of polarization undergoes a slight rotation upon reflection. This rotation, whose sense depends on the direction of magnetization in the medium, is known as the polar Kerr effect. The schematic representation of this phenomenon in Fig. 1.20 shows that if the polarization vector suffers a counterclockwise rotation upon reflection from an up-magnetized region, then the same vector will rotate clockwise when the magnetization is down. A magneto-optical medium is characterized in terms of its reflectivity R and Kerr rotation angle θ_k.† R is a real number

†In reality, the reflected state of polarization is not linear, but has a certain degree of ellipticity. The reflected polarization consists of two linear components: $E_{\|}$, which is parallel to the incident polarization, and $E_{\perp}$, which is perpendicular to it. Now, if $E_{\|}$ is in phase with $E_{\perp}$, the net magneto-optic effect will be a pure rotation of the polarization vector. On the other hand, if $E_{\|}$ and $E_{\perp}$ are 90° out of phase, the reflected polarization will be elliptical, with no rotation whatsoever. In practice, the phase difference between $E_{\|}$ and $E_{\perp}$ is somewhere between 0 and 90°, resulting in a reflected beam that has some degree of ellipticity, ϵ_k, with the major axis of the polarization ellipse rotated by an angle θ_k (relative to the incident **E**-vector). By inserting a Soleil–Babinet compensator in the reflected beam's path, one can change the phase relationship between $E_{\|}$ and $E_{\perp}$ in such a way as to eliminate the beam's ellipticity; the emerging polarization then will become linear with an enhanced rotation angle. In this chapter, reference to the Kerr angle implies the effective angle that includes the above correction for ellipticity.

between zero and one that indicates the fraction of the incident light power reflected back from the surface of the medium (at normal incidence). θ_k is generally a positive number, although it is understood that the rotation angle might be positive or negative, depending on the direction of magnetization. In magneto-optical readout, it is the sign of the rotation angle that carries the information about the state of magnetization of the medium, i.e., the recorded bit pattern.

The laser used for readout is usually the same as that used for recording, but its output power level is substantially reduced in order to avoid erasing (or otherwise obliterating) the previously recorded information. For instance, if the power of the write/erase beam is 20 mW, then for the read operation the beam is attenuated to about 3 or 4 mW. The same objective lens that focuses the write-beam is now used to focus the read-beam, creating a diffraction-limited spot for resolving the recorded marks. Whereas in writing the laser was pulsed to selectively reverse-magnetize small regions along the track, in readout it operates with constant power, i.e., in CW-mode. Both up- and down-magnetized regions are read as the track passes under the focused light spot. The reflected beam, which is now polarization modulated, goes back through the objective and becomes collimated once again; its information content is subsequently decoded by polarization sensitive optics, and the scanned pattern of magnetization is reproduced as an electronic signal.

1.7.1. Differential detection

Figure 1.21(a) shows the differential detection system that is the basis of magneto-optical readout in practically all erasable optical storage systems in use today. A beam-splitter (BS) diverts half of the reflected beam away from the laser and into the detection module.† The polarizing beam-splitter (PBS) splits this latter beam into two components, each carrying the projection of the incident polarization along one axis of the PBS, as shown in Fig. 1.21(b). The component polarized along one of the axes goes straight through, while the component polarized along the other axis splits off to the side. The PBS is oriented such that in the absence of the Kerr effect its two branches will receive equal amounts of light. In other words, if the polarization did not undergo any rotations whatsoever upon reflection from the disk, then the beam entering the PBS would be polarized at 45° relative to the PBS axes, in which case it would split equally between the two branches. Under this condition, the two detectors would generate identical signals and the differential signal ΔS would be zero. Now, if the beam returns from the disk with its polarization rotated clockwise (rotation angle = θ_k), then detector 1 will receive more light than

†The use of an ordinary beam-splitter is an inefficient way of separating the incoming and outgoing beams, since half the light is lost in each pass through the beam-splitter. One can do much better by using a so-called "leaky" beam-splitter (see Chapter 9).

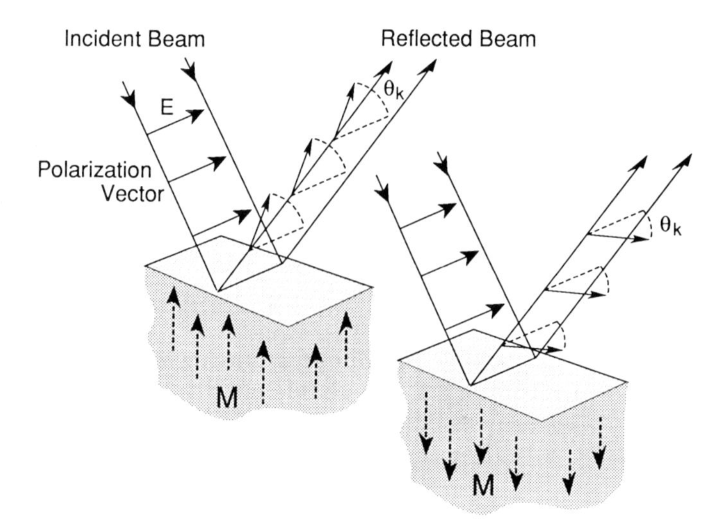

Figure 1.20. Schematic diagram describing the polar magneto-optical Kerr effect. Upon reflection from the surface of a perpendicularly magnetized medium, the polarization vector undergoes a rotation. The sense of rotation depends on the direction of the magnetization vector **M**, and switches sign when **M** is reversed.

detector 2, and the differential signal will be positive. Similarly, a counterclockwise rotated beam entering the PBS will generate a negative ΔS. Thus, as the disk rotates under the focused spot, the electric signal ΔS will reproduce the pattern of magnetization along the particular track being scanned at the time.

1.7.2. Enhancement of the signal-to-noise ratio by multilayering

At the present time, the materials suitable for optical recording have very small Kerr angles (typically $\theta_k \simeq 0.5°$), with the result that the signal ΔS is correspondingly small. Multilayering schemes designed for the enhancement of the MO signal increase the interaction between the light and the magnetic medium by encapsulating a thin film of the MO material in an antireflection-type structure. By providing a better index-match between the MO film and its surroundings, and also by circulating the light through the MO film, multilayered structures manage to trap a large fraction of the incident light within the magnetized medium, and thus increase the Kerr rotation angle. These efforts inevitably result in a reduced reflectivity R, but since the important parameter is the magneto-optically generated component of polarization, $E_\perp = \sqrt{R} \sin \theta_k$, it turns out that a net gain in the signal-to-noise ratio can be achieved by adopting multilayering schemes. Reported enhancements of $E_\perp$ have been as large as a factor of five. The popular quadrilayer structure depicted in Fig. 1.19 consists of a thin film of the MO material, sandwiched between two

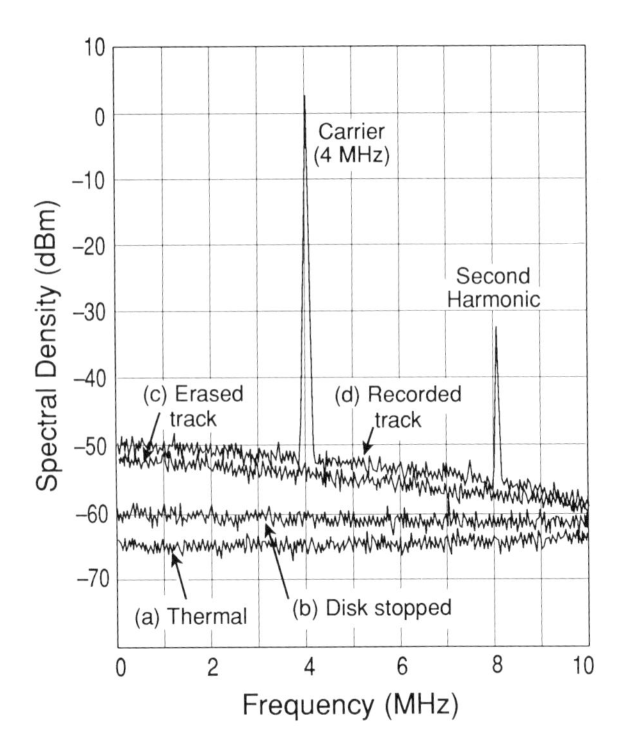

Figure 1.22. Spectra of various noise components in MO readout.

blocked and the trace on the analyzer screen is solely due to thermal noise. The trace in (b) where the beam reaches the detectors but the disk is stationary shows the combined effect of thermal, shot, and laser noise. Trace (c) corresponds to reading an erased track on a spinning disk; the noise here includes all the above plus media noise. When a single-frequency tone was recorded on the track and the readback signal was fed to the spectrum analyzer, trace (d) was obtained. The narrow pulse at frequency f_0 is the first harmonic (the fundamental) of the recorded signal; the corresponding second harmonic appears at $2f_0$. The noise level in (d) is somewhat greater than that in (c), which corresponds to the same track before any data was recorded. This difference is due to "data noise" and arises from jitter and nonuniformity of the recorded marks.

A commonly used measure of performance for optical recording media is the carrier-to-noise ratio (CNR). This is the ratio of the signal amplitude at the carrier frequency f_0 to the average level of noise. On a logarithmic scale the ratio is simply the difference between the two levels; in Fig. 1.22 the CNR is 53 decibels (dB). The noise level usually varies with frequency, and one must be careful in extracting an average value for the noise amplitude from the spectral measurements. The value of the CNR depends not only on the bandwidth of the scanning filter of the spectrum analyzer (typically 30 kHz), but also on the frequency of the recorded tone f_0, since at higher frequencies the signal amplitude drops. With currently achievable densities and disk rotation speeds, reasonable frequencies for comparing the performance of various media are in the neighborhood of 4 MHz. At the present time, good disks for application as erasable optical storage media exhibit CNR values in the range of 50–60 dB.

1.8. Materials used in Magneto-optical Recording

Amorphous rare earth–transition metal alloys are presently the media of choice for erasable optical data storage applications. The general formula for the composition of such an alloy is $(Tb_y Gd_{1-y})_x (Fe_z Co_{1-z})_{1-x}$ where terbium and gadolinium are the rare earth (RE) elements, while iron and cobalt are the transition metal (TM) elements. In practice, the transition metals constitute roughly 80 atomic percent of the alloy (i.e., $x \simeq 0.2$). In the transition-metal subnetwork the fraction of cobalt is usually small, typically around 10%, and iron is the dominant element ($z \simeq 0.9$). Similarly, in the rare-earth subnetwork terbium is the main element ($y \simeq 0.9$) while the gadolinium content is small or may even be absent in some cases. Since the rare-earth elements are highly reactive to oxygen, RE–TM films tend to have poor corrosion resistance and, therefore, to require protective coatings. In a disk structure such as that shown in Fig. 1.19, the dielectric layers that enable optimization of the medium for the best optical and thermal behavior also perform the crucial function of protecting the MO layer from the environment.

The amorphous nature of the material allows its composition to be varied continuously until a number of desirable properties are achieved. In other words, the fractions x, y, z of the various elements are not constrained by the rules of stoichiometry. Disks with very large areas can be coated uniformly with thin films of these media and, in contrast to polycrystalline films whose grains and grain-boundaries scatter the laser beam and cause noise, these amorphous films are continuous and smooth and substantially free from noise. The films are deposited either by sputtering from an alloy target, or by co-sputtering from multiple elemental targets. In the latter case, the substrate moves under the various targets and the fraction of a given element in the alloy film is determined by the time spent under the target as well as the power applied to that target. Substrates are usually kept at a low temperature (by water cooling, for instance) in order to reduce the mobility of deposited atoms and to inhibit crystal growth. The type of the sputtering gas (argon, krypton, xenon, etc.) and its pressure during sputtering, the bias voltage applied to the substrate, the deposition rate, the nature of the substrate and its pretreatment, and the temperature of the substrate can all have dramatic effects on the composition and short-range order of the deposited film. A comprehensive discussion of the factors that influence film properties is beyond our intended scope here; the interested reader, however, may consult the vast literature of this subject.

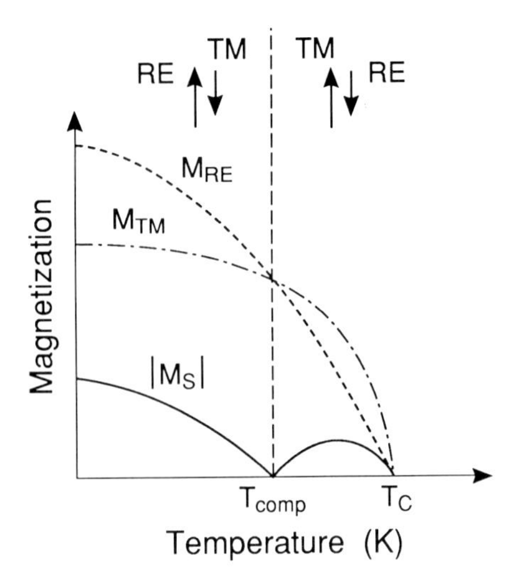

Figure 1.23. Temperature dependence of magnetization in amorphous RE–TM films. The moments of the RE and TM subnetworks decrease monotonically until they both vanish at the critical (Curie) temperature T_c. The net magnetization, being the difference between the two subnetwork moments, goes through zero at the compensation point T_{comp}.

1.8.1. Ferrimagnetism

The RE–TM alloys of interest in magneto-optical recording are ferrimagnetic, in the sense that the magnetization of the TM subnetwork is antiparallel to that of the RE subnetwork. The net magnetic moment exhibited by the material is the vector sum of the two subnetwork magnetizations. Figure 1.23 shows the typical temperature dependence of RE and TM magnetic moments, as well as the net saturation moment of the material. The exchange coupling between the two subnetworks is strong enough to give them the same critical temperature T_c. At $T = 0$ K the rare-earth moment is stronger than that of the transition metal, giving the material a net moment along the direction of the RE magnetization. As the temperature rises, thermal disorder competes with interatomic exchange forces that tend to align the individual atomic dipole moments. The decrease of M_{RE} with increasing temperature is faster than that of M_{TM}, and the net moment M_s begins to approach zero. At the compensation temperature, T_{comp}, the net moment vanishes. Between T_{comp} and T_c the net moment is dominated by the TM subnetwork and the material is said to exhibit TM-rich behavior (as opposed to when $T < T_{comp}$, where it exhibits RE-rich behavior). At the Curie temperature, thermal agitations finally break down the hold of the exchange forces on magnetic dipoles, and the magnetic order disappears. Beyond T_c the material is in the paramagnetic state.

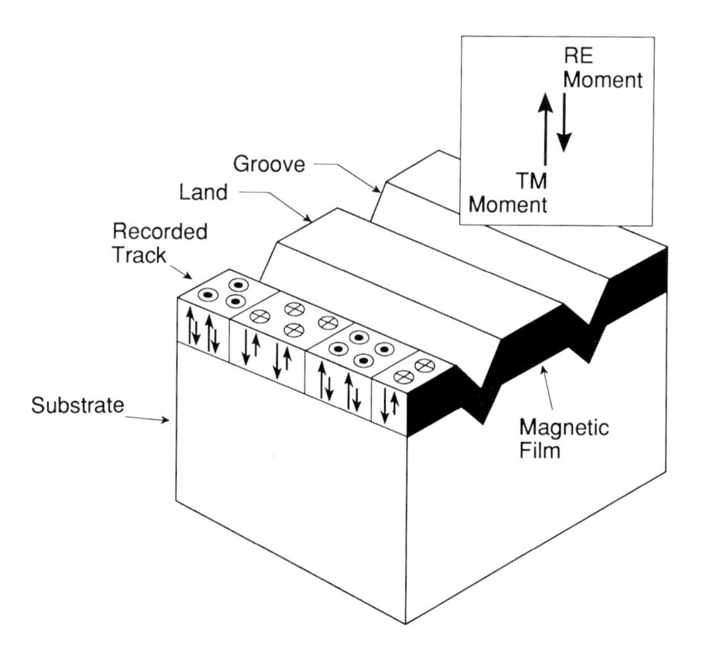

Figure 1.24. Schematic diagram of the pattern of magnetization along a recorded track. The rare earth and the transition metal couple antiferromagnetically, so that the net magnetization everywhere is small. However, since the read beam interacts mainly with the TM subnetwork, the readout signal is not necessarily small.

The composition of MO materials is chosen so that T_{comp} appears near the ambient temperature of $T_a \simeq 300$ K. Thus, under normal conditions, the net magnetization of the material is close to zero. Figure 1.24 shows a schematic diagram of the magnetization pattern in the cross-section of a recorded track. Note that although the net magnetization is nearly zero everywhere the moments of a given subnetwork are in opposite directions in adjacent domains. During readout, the light from the laser interacts mainly with the transition-metal subnetwork; thus the magneto-optic Kerr signal is strong, even though the net magnetization of the storage layer may be small. The magnetic electrons of Fe and Co are in the 3d electronic shell, which forms the outer layer of the ion once the 4s electrons have escaped into the sea of conduction electrons. The magnetic electrons of Tb and Gd, in contrast, are within the 4f shell, concealed by the 5s, 5p, and 5d shells, even after the 6s electrons have escaped to the conduction band. A red or near-infrared photon is not energetic enough to penetrate the outer shell and interact with the magnetic 4f electrons, but it readily interacts with the exposed 3d electrons that constitute the magnetic moment of the TM subnetwork. It is for this reason that the magneto-optical Kerr signal in the visible and the infrared is a probe of the state of magnetization of the transition-metal subnetwork.

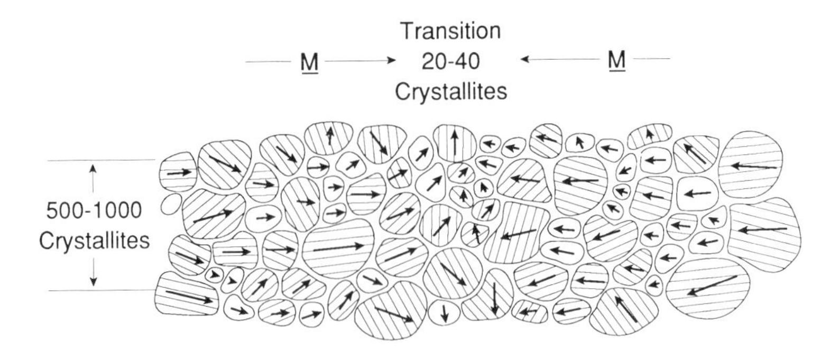

Figure 1.25. Schematic representation of a recorded transition on a polycrystalline film used in longitudinal magnetic recording. The head-to-head domain configuration is a state of high magnetostatic energy, which breaks down into a complex, jagged structure in order to lower this energy. Source: G.F. Hughes, Magnetization reversal in cobalt-phosphorus films, *J. Appl. Phys.* **54**, 5306 (1983).

1.8.2. Perpendicular magnetic anisotropy

An important property of amorphous RE–TM alloy films is that they possess perpendicular magnetic anisotropy. The magnetization in these films favors perpendicular orientation even though there is no discernible crystallinity or microstructure that might obviously be responsible for this sort of anisotropy. It is generally believed that atomic short-range order, established in the deposition process and aided by symmetry breaking at the surfaces of the film, gives preference to a perpendicular orientation. Experimental proof of this assertion, however, is sketchy due to lack of high-resolution observation instruments and material characterization techniques.

The perpendicular magnetization of the MO media is in sharp contrast to the in-plane orientation of the magnetization vector in ordinary magnetic recording. In magnetic recording, the neighboring domains are magnetized in head-to-head fashion, which is an energetically unfavorable situation since the domain walls are charged and highly unstable. The boundary between neighboring domains in fact breaks down into zigzags, vortices, and all manner of jagged, uneven structures in an attempt to reduce the magnetostatic energy (see Fig. 1.25). In contrast, the adjacent domains in MO media are highly stable, since the pattern of magnetization causes flux closure and reduces the magnetostatic energy.

1.8.3. Coercivity and the hysteresis loop

Typical hysteresis loops of an amorphous RE–TM thin film at various temperatures are shown in Fig. 1.26(a). These loops, obtained with a vibrating-sample magnetometer (VSM), show several characteristics of MO

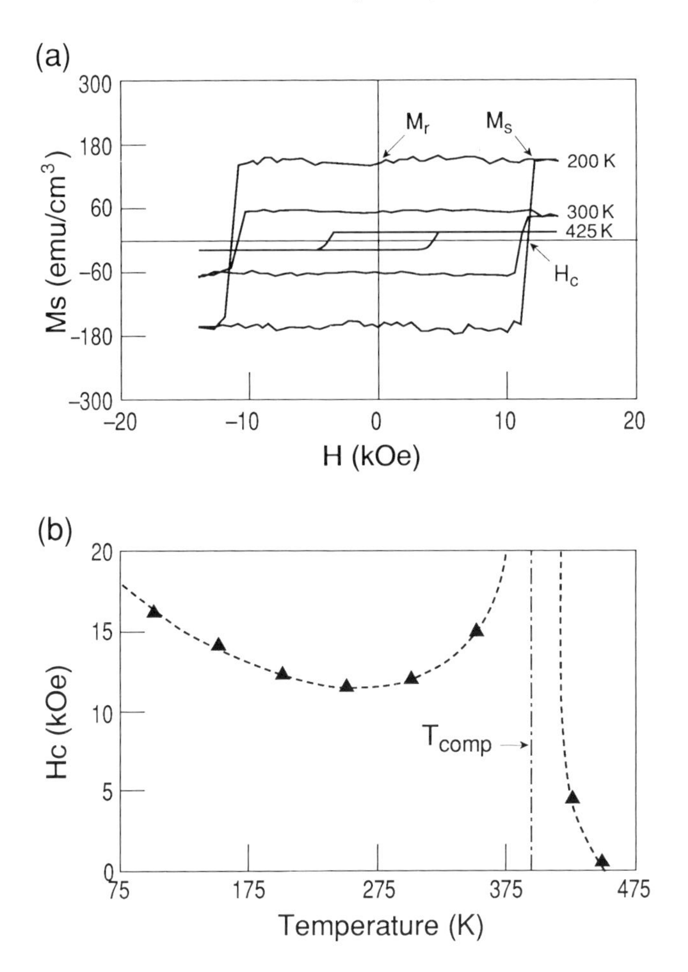

Figure 1.26. (a) Hysteresis loops of an amorphous $Tb_{27}(FeCo)_{73}$ film, measured by VSM at three different temperatures. The saturation moment M_s, the remanent moment M_r, and the coercive field H_c are identified for the loop measured at T = 200 K. (b) Coercivity as function of temperature for the above sample. At the compensation temperature, T_{comp} = 400 K, the coercivity is infinite; it drops to zero at the Curie point T_c = 450 K.

media. (The VSM applies a continuously varying magnetic field to the sample and measures its net magnetic moment as a function of the field.) The horizontal axis in Fig. 1.26(a) is the applied field, which varies from - 12 kOe to + 12 kOe, while the vertical axis is the measured magnetic moment per unit volume (in CGS units of emu/cm³). The high degree of squareness of the loops signifies the following:

(i) The remanent magnetization M_r and the saturation magnetization M_s are nearly equal. Thus, once the sample has been saturated in an applied magnetic field, removing the field does not cause a reduction of the magnetic moment.

(ii) Transitions of the magnetization from up to down (or from down to up) are very sharp. The reverse field does not affect the magnetization until the critical value of H_c, the coercive field, is reached. At the coercive field the magnetization suddenly reverses direction, and saturation in the opposite direction is almost immediate.

The fact that M_r is very nearly equal to M_s in MO media is significant, since it means that the recorded domains remain fully saturated and exhibit the maximum signal during readout. The coercivity H_c, in addition to being responsible for the stability of recorded domains, plays an important role in the processes of thermomagnetic recording and erasure. The coercivity at room temperature, being of the order of several kilo-oersteds, prevents fields weaker than H_c from disturbing the recorded data. With the increasing temperature, the coercivity decreases and drops to zero at the Curie point, T_c. Part (b) of Fig. 1.26 is the plot of H_c versus T for the same sample as in (a). Note that at the compensation point the coercivity goes to infinity simply because the magnetization vanishes and the external field does not see any magnetic moments to interact with. Above T_{comp} the coercive field decreases monotonically, which explains the reversal process during thermomagnetic recording: the magnetization vector **M** switches sign once the coercivity drops below the level of the applied field.

1.9. Recent Developments

In this chapter we have reviewed the basic characteristics of optical disk data storage systems, with emphasis on magneto-optical recording. The goal has been to convey the important concepts without getting distracted by secondary issues and less significant details. As a result, we have glossed over several interesting developments that have played a role in the technological evolution of optical data storage. In this section some of these developments are briefly described.

1.9.1. Multiple-track read/write with laser diode arrays

It is possible in an optical disk system to use an array of lasers instead of just one laser, focus all the lasers simultaneously through the same objective lens and perform parallel read/write/erase operations on multiple tracks. Since the individual lasers of an array can be modulated independently, the parallel channels thus obtained are totally independent of each other. In going from a single-channel drive to a multiple-channel one, the optics of the system (i.e., lenses, beam-splitters, polarization-sensitive elements, focus and track servos, etc.) remain essentially the same; only the lasers and detectors proliferate in number. Parallel-track operations boost the sustainable data rates in proportion to the number of channels used.

1.9.2. Diffractive optics

The use of holographic optical elements (HOEs) to replace individual refractive optics is a promising new development. Diffractive optical elements are relatively easy to manufacture, are lightweight and inexpensive, and can combine the functions of several elements on a single plate. These devices are therefore expected to help reduce the cost, size, and weight of optical heads, making optical drives more competitive in terms of price and performance.

An example of the application of HOEs in magneto-optical disk systems is given in Fig. 1.27, which shows a reflection-type element consisting of four separate holograms. The light, incident on the HOE at an angle of 60°, has a p-component that has the original polarization of the laser beam, and an s-component (parallel to the hologram's grooves) that is the magneto-optically generated polarization at the disk. Nearly 90% of the s- and 70% of the p-polarization in this case is reflected from the HOE without suffering any diffraction (i.e., in the zero-order beam); this light is captured in the differential detection module and yields the MO read signal. The four holograms deflect 20% of the incident p-polarized light in the form of first-order diffracted beams, and bring them to focus at four different spots on a multi-element detector. The two small holograms in the middle, H_3 and H_4, focus their first-order beams on detectors P_3 and P_4 to generate the push–pull tracking-error signal. The other two holograms, H_1 and H_2, send diffracted beams to a four-element detector in order to generate a focus-error signal based on the double knife-edge scheme. This HOE, therefore, combines the functions of beam splitting, masking, and focusing all in one compact unit.

Factors that must be taken into consideration in the design of HOEs include variations of the laser wavelength with time and temperature, misalignments during head assembly, and sensitivity to fluctuations in the ambient temperature.

1.9.3. Alternative storage media

The GaAlAs lasers of the present optical disk technology are likely to be replaced in the future by light sources that emit in the blue end of the spectrum. Shorter wavelengths allow smaller marks to be recorded, and also enable the resolving of those marks in readout. Aside from the imposition of tighter tolerances on focusing and tracking servos, operation in the blue will require storage materials that are sensitive to short wavelengths. The current favorites for erasable optical recording media, the amorphous RE–TM alloys, may not be suitable for readout in the blue, since their magneto-optical Kerr signal drops at short wavelengths. A new class of magnetic materials that holds promise for future-generation device applications is the class of TM/TM superlattice-type media. The best-known material in this class is the Co/Pt layered structure, which consists

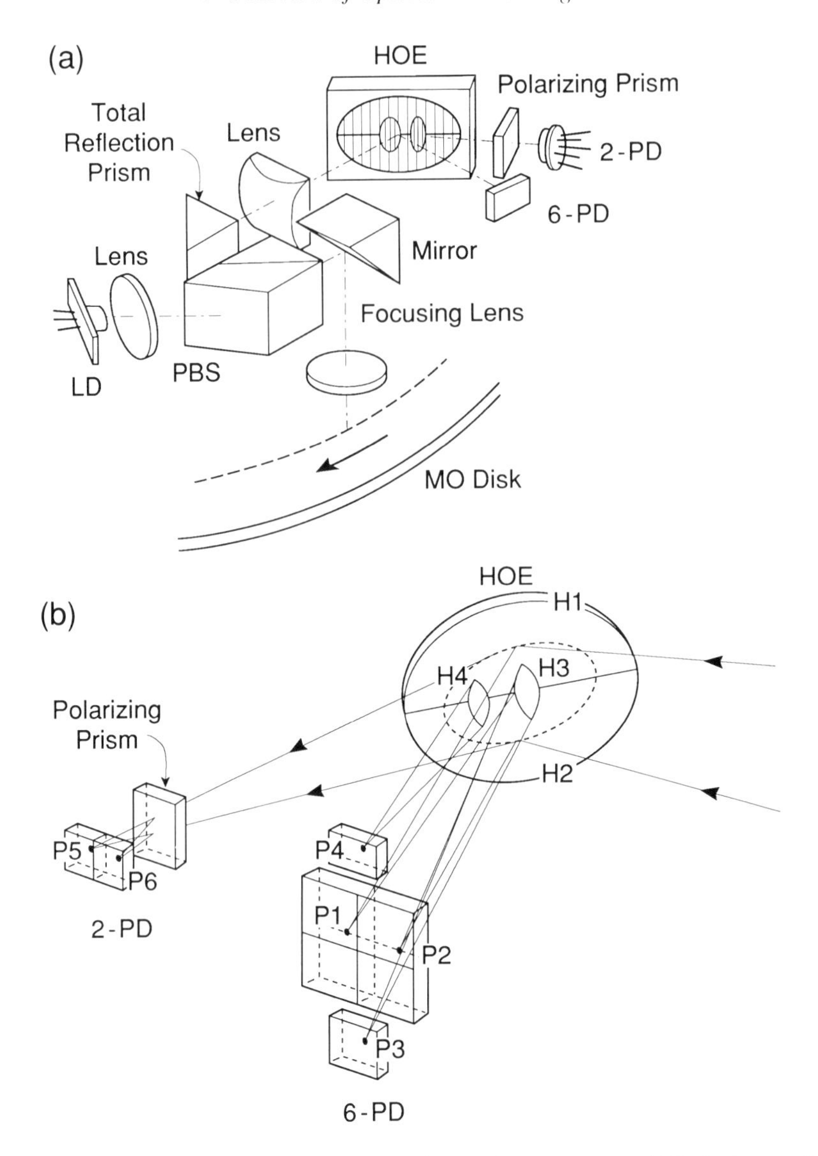

Figure 1.27. Application of a holographic optical element (HOE) in optical recording. (a) Configuration of the MO head using a polarization-sensitive reflection HOE. (b) Geometrical relation between holograms and detectors. 2-PD and 6-PD are two- and six-element photodetectors. Source: A. Ohba *et al.*, *SPIE Proceedings*, Vol. 1078 (1989).

of very thin layers of cobalt (typically one or two atomic layers) separated by several atomic layers of platinum. These polycrystalline films, which have very small crystallites (grain diameter $\simeq$ 200 Å), are prepared either by electron-beam evaporation or by sputtering. Co/Pt films have large perpendicular magnetic anisotropy, good signal-to-noise ratios in the blue, and sufficient sensitivity for write/erase operations; they are also more stable than RE–TM media in normal environments.

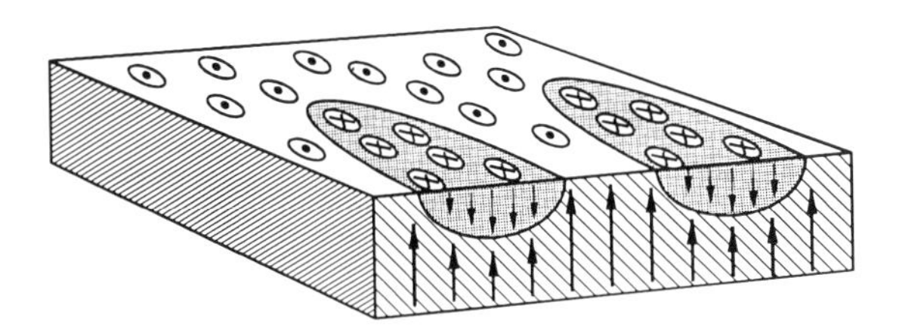

Figure 1.28. Direct overwrite in exchange-coupled magnetic multilayers involves the formation of domains that do not extend through the thickness of the medium.

1.9.4. Multifunctional drives

It is desirable to have drives that can handle all the various types of optical media, namely, drives that can read CD-ROMs, store archival files on WORM media, and use MO disks in applications that require rewritability. Such drives are not difficult to build, and the additional cost of multifunctionality is not expected to be out of bounds. The hope is that, in the near future, optical disk drives will be multifunctional, providing the means for storage, retrieval, transfer, and exchange of vast amounts of information in a truly universal fashion.

1.9.5. Direct overwrite

The problem of direct overwrite (DOW) in MO media has been the subject of extensive research in recent years. Some of the most promising solutions have been based on exchange-coupled magnetic multilayered structures. Later, in Chapter 17, we will discuss this subject in more detail; however, the basic idea of recording on exchange-coupled bilayers (or trilayers) is simple and involves the writing of reverse domains that do not extend through the entire film thickness, such as those shown schematically in Fig. 1.28. Such domains are under pressure from their excessive wall energies to collapse and can readily be erased with a moderate-power laser pulse. DOW on exchange-coupled media is thus achieved by writing (i.e., creating reverse domains) with a high-power pulse, and erasing (i.e., eliminating domains) with a moderate-power pulse. An external magnetic field is usually required for writing on such media, but neither the direction nor the magnitude of this field need change during erasure.

1.9.6. Small computer systems interface (SCSI)

SCSI (pronounced scuzzy) is a standard protocol for communication, handshaking, and data transfer between a computer (not necessarily small) and its peripheral devices. The I/O bus controller must be able to contact the storage device, establish the mode of communication, and transmit packets of data to (or receive them from) the device. Standard protocols for these transactions have been approved by the International Standards Organization (ISO) and by the American National Standards Institute (ANSI). These standards have been adopted by many computer manufacturers and peripheral-device vendors.

In conclusion, optical recording is an evolving technology that will undoubtedly see many innovations and improvements in the next few years. Some of the ideas and concepts described here will, it is hoped, remain useful for some time to come; others may have a shorter lifetime and limited usefulness. It is the author's hope, however, that they will all serve as stepping stones to more profound concepts and new ideas.

Problems

(1.1) Consider a disk of radius r_{max}, rotating at a constant angular velocity of $\mathcal{N}$ revolutions per second. The minimum allowed track-width is W_t, and the tracks must be separated by a guard band of width W_g. The minimum separation between transitions along the track is Δ_0.

(a) What is the total capacity of the disk?

(b) What is the capacity if the clock period T_c is fixed? (Use the optimum clock period for given $\mathcal{N}$ and Δ_0.)

(c) If the usable disk surface could be divided into three zones and the clock cycles were allowed to be different in each zone, what would be the optimum clock periods T_1, T_2, T_3 and the resulting disk capacity?

(1.2) Prove that the average number of tracks traversed in a random seek operation is one-third of the total number of tracks.

(1.3) Two methods of preparing binary data for storage on the disk are frequency modulation (FM) and modified frequency modulation (MFM). In FM there is a transition at the boundary of every bit cell and a 1 is represented by an additional transition in the cell center. MFM is similar except that transitions at the boundary occur only when a 0 follows another 0. For a sample bit-sequence 1001011100, these two modulation schemes are depicted in Fig. 1.29.

(a) Analyze the FM and MFM codes in terms of clock synchronization capability and storage capacity.

(b) Find the equivalent description of these codes in terms of the (d, k) modulation codes.

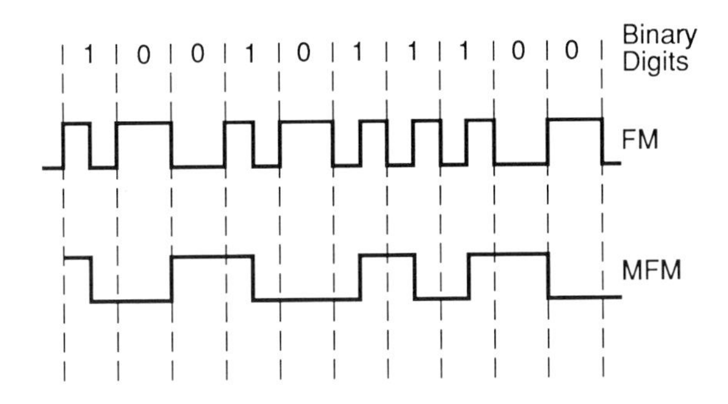

Figure 1.29.

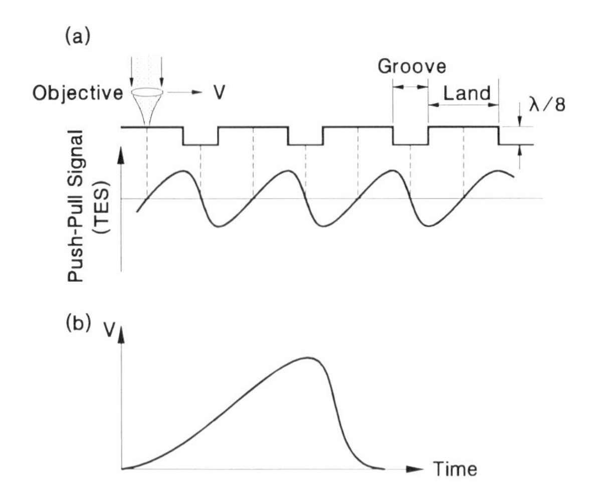

Figure 1.30.

(1.4) In the push–pull scheme of track-following, grooves of depth $\lambda/8$ produce a bipolar track-error signal (TES), as shown in Fig. 1.30(a). The TES is zero when the beam is centered either on the land or on the groove, but it develops a positive or negative value as soon as the beam moves away from the center toward one edge or the other. During read/write operations the tracking servo loop is closed, forcing the objective to follow a given track closely. On the other hand, during seek operations the head must move in the radial direction to reach a desired track; in this case the tracking servo loop is opened, and the system monitors the TES in order to count the number of crossed tracks.

(a) Assume that the tracks are concentric rings and that they are free from misalignment and eccentricity. Also assume that the velocity profile of the head during a seek operation is that shown in Fig. 1.30(b). Draw a qualitative picture of the TES as a function of time during the seek period, and describe briefly the various features of this signal.

(b) In practice the tracks have a certain amount of eccentricity that causes runout and tends to confuse the track-counting process. A good way to observe these eccentricities directly is to lock the head in place and open the tracking servo loop. From the head's point of view, each track moves in and out radially, performing a periodic motion that has the disk's revolution period. Draw a qualitative picture of the TES versus time in this case, and discuss its salient features.

(c) Explain how the eccentricity can cause errors in the track-count to occur. Pay particular attention to the initial and final stages of the seek operation, where the head velocity is comparable to the apparent radial velocity of the track.

(1.5) A linear array of nine laser diodes is being considered for incorporation into a 5.25 inch diameter laser disk system. The center-to-center spacing of the adjacent lasers is 60 μm and the cone of light from each laser requires a 0.3 NA collimator. The objective is a 0.6 NA lens, having a sufficiently wide field of view to bring the nine beams that emerge from the collimator into diffraction-limited focus on the

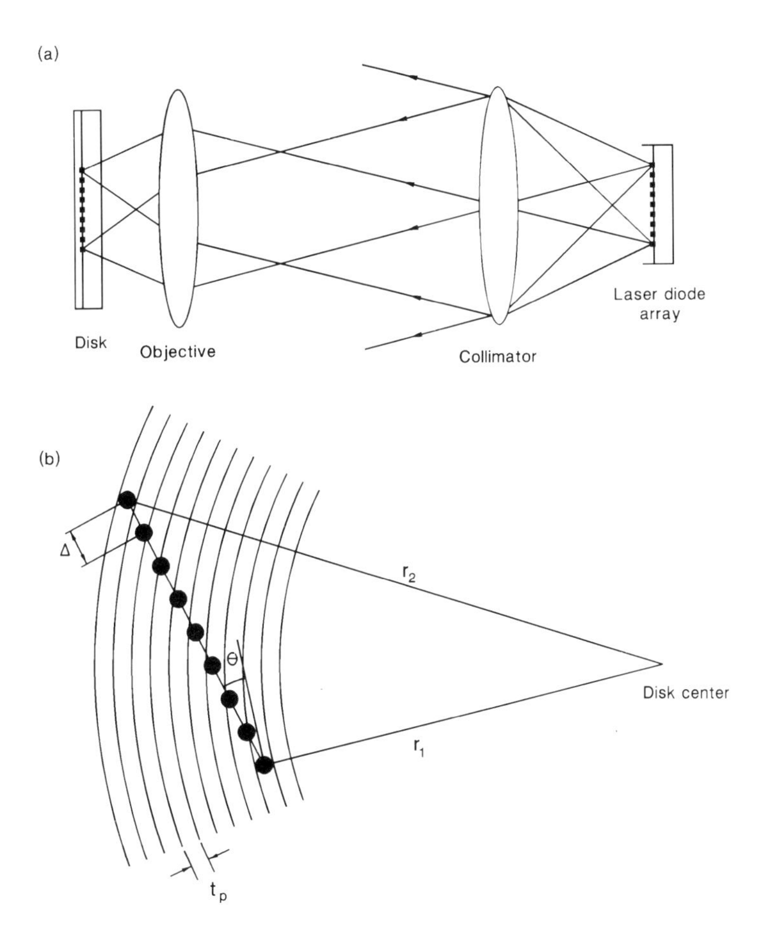

Figure 1.31.

disk surface. Figure 1.31(a) shows the arrangement of the laser diode array, the lenses, and the optical disk. For parallel read/write/erase operations it is desirable to rotate the image of the array until the focused spots fall on nine adjacent tracks, as shown in Fig. 1.31(b). (For numerical calculations you may assume the track-pitch $t_p = 1.6\ \mu$m.) The angle between the straight line that connects the focused spots and the track direction at $r = r_1$ is denoted by θ. Clearly, when the spots on both ends of the array are centered on their respective tracks, we will have $r_2 - r_1 = 8t_p$.

(a) What is the spacing Δ between adjacent focused spots on the disk?

(b) Derive a formula for θ as a function of r_1, and compute the numerical value of θ at the inner disk radius, $r_1 = 33$ mm, and at the outer disk radius, $r_1 = 66$ mm.

(c) Assuming that the angle θ is automatically adjusted by the head in its various positions along a radius of the disk, the first and the last focused spots will always be on-track. Which of the remaining spots is then going to be most off-track and by how much?

(d) In practical systems positioning errors cause a small separation between the actual disk center and the spindle axis. Assuming a 50 μm centering error, how does the angle between a track at radius r and the row of focused spots vary during one revolution of the disk?

2

Optics of Gaussian Beams

Introduction

A coherent, monochromatic beam whose complex amplitude distribution in a cross-sectional plane is described by a Gaussian function is known as a Gaussian beam. When propagating in free space (or in any other isotropic, homogeneous medium), a Gaussian beam's cross-sectional profile remains Gaussian at all times. Also, a Gaussian beam remains Gaussian upon passage through a lens. Simple algebraic expressions exist that describe these properties of Gaussian beams. These expressions are extremely useful when analyzing the divergence of collimated beams, or their diffraction-limited focusing by a lens; they also yield simple formulas for the beam divergence angle, spot size at focus, depth of focus, and so on. This chapter is devoted to the diffraction analysis of Gaussian beams, and derivation of formulas that describe their properties.

2.1. Definitions and Basic Properties

In general, a Gaussian beam propagating along the Z-axis has the following amplitude distribution in the XY-plane:

$$\boxed{A(x,y) = A_0 \exp(-\alpha x^2 - \beta y^2)} \tag{2.1}$$

Here α and β are complex numbers. The real part of α determines the width of the beam along the X-axis, while its imaginary part is related to the beam curvature in the X-direction. Similarly, β determines the width and the curvature of the beam along the Y-axis. Since the amplitude distribution in Eq. (2.1) is separable into the product of a function of x and a function of y, the following discussion will be confined to the X-direction only. Every statement concerning the X-direction can be modified to a statement about the Y-direction provided that α is replaced with β. The amplitude distribution along X may now be written

$$A(x) = A_0 \exp(-\alpha_1 x^2) \exp(-\mathrm{i}\alpha_2 x^2) \tag{2.2}$$

where α_1 and α_2 are the real and imaginary parts of α, respectively. Defining r_x as the 1/e radius of the Gaussian beam in the X-direction, we have

$$\alpha_1 = 1/r_x^{\ 2} \tag{2.3a}$$

In order to relate α_2 to the curvature of the beam, consider a spherical wavefront emanating from a distant point source on the Z-axis. Let R_0 be the distance between the source and the origin of the coordinate system ($R_0 > 0$ when the source is on the negative side of Z). Throughout this book, the time dependence of the electromagnetic waves will be denoted by $\exp(-\mathrm{i}\omega t)$; thus the spherical wavefront will have the following amplitude distribution in the XY-plane at $Z = 0$:

$$\begin{aligned}\frac{1}{r}\exp\left[\mathrm{i}\,\frac{2\pi}{\lambda}\,r\right] &\simeq \frac{1}{R_0}\exp\left[\mathrm{i}\,\frac{2\pi}{\lambda}\sqrt{x^2 + y^2 + R_0^{\ 2}}\,\right] \\ &\simeq \frac{1}{R_0}\exp\left[\mathrm{i}\,\frac{2\pi R_0}{\lambda}\left(1 + \frac{x^2 + y^2}{2R_0^{\ 2}}\right)\right] \\ &= \frac{1}{R_0}\exp\left[\mathrm{i}\,\frac{2\pi R_0}{\lambda}\right]\exp\left[\mathrm{i}\pi\,\frac{(x/\lambda)^2 + (y/\lambda)^2}{R_0/\lambda}\right]\end{aligned}$$

The final expression is written in a form that makes explicit the normalization of x, y, and R_0 by the wavelength λ. In what follows all lengths will be normalized by the wavelength and, therefore, λ will be dropped from the equations. Note in the above expression that, aside from a constant term, the curvature phase factor along X is given by $\exp(\mathrm{i}\pi x^2/R_0)$, provided that both x and R_0 are normalized by λ. Comparison with Eq. (2.2) now yields

$$\alpha_2 = -\pi/R_0 \tag{2.3b}$$

From Eqs. (2.3) one obtains the following expression for α in terms of the beam's 1/e radius r_x and its radius of curvature R_x:

$$\boxed{\alpha = \frac{1}{r_x^{\ 2}} - \mathrm{i}\,\frac{\pi}{R_x}} \tag{2.4}$$

In Eq. (2.4) both r_x and R_x are normalized by the wavelength and are, therefore, dimensionless. The sign convention is such that for diverging beams propagating in the positive Z-direction R_x is positive. (It is as if the beam originated at a point source on the negative Z-axis.) A beam converging towards a point on the positive Z-axis would have a negative radius of curvature.

2.1.1. Power of the beam

Let the beam have total power P_0. In order to express the amplitude A_0 of the beam in terms of P_0 we write

$$P_0 = \frac{1}{2}\,|A_0|^2 \iint_{-\infty}^{\infty} \exp\left[-2(\alpha_1 x^2 + \beta_1 y^2)\right] \mathrm{d}x\,\mathrm{d}y = \frac{\pi\,|A_0|^2}{4\sqrt{\alpha_1\beta_1}} \tag{2.5a}$$

Therefore

$$|A_0| = 2\sqrt{\frac{P_0}{\pi}}\left[\mathrm{Re}(\alpha)\ \mathrm{Re}(\beta)\right]^{1/4} \tag{2.5b}$$

In Eq. (2.5) only the real parts of α and β appear, which is consistent with the fact that their imaginary parts contribute to the phase of the beam only.

2.1.2. Fourier transform of the beam profile

Fourier transformation is an important tool in the theory of diffraction. In preparation for the diffraction analysis of the next section, we derive the Fourier transform of the complex amplitude distribution of the Gaussian beam in Eq. (2.1). Since the x- and y-dependences of $A(x,y)$ are separable, it is sufficient to find the transform of a one-dimensional function, namely,

$$\mathcal{F}\left\{\exp(-\alpha x^2)\right\} = \int_{-\infty}^{\infty} \exp(-\alpha x^2)\exp(-\mathrm{i}2\pi s x)\,\mathrm{d}x$$

$$= \int_{-\infty}^{\infty} \exp\left\{-\left(\sqrt{\alpha}\,x + \mathrm{i}\frac{\pi s}{\sqrt{\alpha}}\right)^2 - \frac{\pi^2 s^2}{\alpha}\right\}\mathrm{d}x$$

The variable of integration is now changed from x to $Z = \sqrt{\alpha}\,x + \mathrm{i}\pi s/\sqrt{\alpha}$. As shown in Fig. 2.1(a), when x varies from $-\infty$ to $+\infty$, the complex variable Z stays on the straight line L in the complex plane. The line L is completely identified by the two complex numbers $\sqrt{\alpha}$ and $\mathrm{i}\pi s/\sqrt{\alpha}$. The preceding equation is now written

$$\mathcal{F}\left\{\exp(-\alpha x^2)\right\} = \frac{1}{\sqrt{\alpha}}\exp\left(-\frac{\pi^2 s^2}{\alpha}\right)\int_L \exp(-Z^2)\,\mathrm{d}Z$$

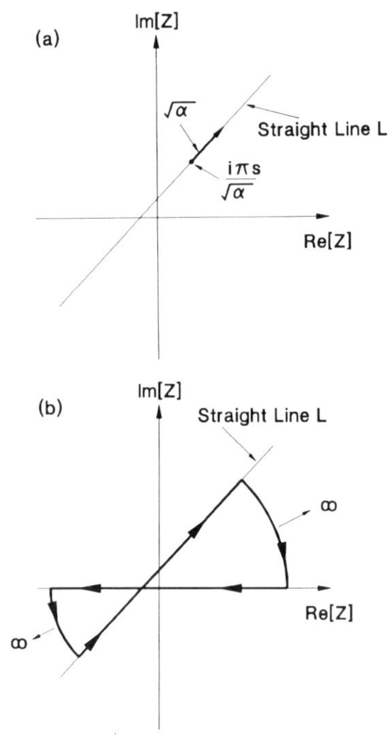

Figure 2.1. (a) Domain of integration in the complex Z-plane. The direction of the line L is determined by the complex number $\sqrt{\alpha}$. When $x = 0$ the corresponding point in the complex plane is $Z = i\pi s/\sqrt{\alpha}$. (b) Integration on a closed path. The integrand, $\exp(-Z^2)$, has no poles within the closed path. The sections of the path located at infinity do not contribute to the integral, since $\exp(-Z^2)$ is zero in these regions. Thus

$$\int_L \exp(-Z^2)\, dZ = \int_{-\infty}^{\infty} \exp(-x^2)\, dx = \sqrt{\pi}.$$

The integral on the straight line L in the complex plane is now converted to an integral on the real axis by considering the closed path in Fig. 2.1(b) and by invoking the fact that the integrand has no poles within this path. Using Cauchy's theorem of complex analysis, we find

$$\mathscr{F}\left\{\exp(-\alpha x^2)\right\} = \sqrt{\frac{\pi}{\alpha}} \exp\left(-\frac{\pi^2 s^2}{\alpha}\right) \tag{2.6}$$

Equation (2.6) states that the Fourier transform of a Gaussian function is another Gaussian function, albeit with different parameters. This equation will be used in the next section to derive formulas for the propagation of Gaussian beams in free space.

2.2. Gaussian-beam Propagation in Free Space

Consider the Gaussian beam described by Eq. (2.1) where x and y are normalized by the wavelength λ, and are therefore dimensionless. Using Eq. (2.6) we can write the Fourier transform of this amplitude distribution as follows:

$$\mathscr{F}\left\{A(x,y)\right\} = \iint_{-\infty}^{\infty} A(x,y)\exp\left[-i2\pi(x\sigma_x + y\sigma_y)\right] dx\,dy$$

$$= \frac{\pi A_0}{\sqrt{\alpha\beta}} \exp\left[-\pi^2\left(\frac{\sigma_x^2}{\alpha} + \frac{\sigma_y^2}{\beta}\right)\right] \qquad (2.7)$$

The corresponding inverse Fourier transform is

$$A(x,y) = \iint_{-\infty}^{\infty} \mathscr{F}\left\{A(x,y)\right\} \exp\left[i2\pi(x\sigma_x + y\sigma_y)\right] d\sigma_x\,d\sigma_y \qquad (2.8)$$

The right-hand side in Eq. (2.8) is the superposition of plane waves propagating along the unit vectors $\boldsymbol{\sigma}$, where

$$\boldsymbol{\sigma} = \sigma_x\,\hat{\mathbf{x}} + \sigma_y\,\hat{\mathbf{y}} + \sqrt{1 - \sigma_x^2 - \sigma_y^2}\,\hat{\mathbf{z}} \qquad (2.9)$$

The amplitude of each such plane wave is given by the Fourier transform of $A(x,y)$ at the frequency (σ_x, σ_y), which is the projection of the unit vector $\boldsymbol{\sigma}$ in the XY-plane. At a distance z from the origin along the Z-axis, the distribution is obtained by superimposing these plane waves, taking into account their different propagation paths. This results in

$$A(x,y,z) = \iint_{-\infty}^{\infty} \mathscr{F}\left\{A(x,y,z=0)\right\}$$

$$\times \exp\left\{i2\pi\left[x\sigma_x + y\sigma_y + z\sqrt{1 - \sigma_x^2 - \sigma_y^2}\right]\right\} d\sigma_x\,d\sigma_y \qquad (2.10)$$

Replacing in the above equation the Fourier transform of $A(x,y,z=0)$ given by Eq. (2.7), and using the first two terms in the Taylor series expansion of $\sqrt{1 - \sigma_x{}^2 - \sigma_y{}^2}$, one arrives at

$$A(x, y, z) = \frac{\pi A_0 \exp(\mathrm{i}2\pi z)}{\sqrt{\alpha\beta}} \iint_{-\infty}^{\infty} \exp\left\{-\pi^2 \left[\left(\frac{1}{\alpha} + \mathrm{i}\frac{z}{\pi}\right)\sigma_x{}^2 + \left(\frac{1}{\beta} + \mathrm{i}\frac{z}{\pi}\right)\sigma_y{}^2\right]\right\}$$

$$\times \exp\left[\mathrm{i}2\pi(x\sigma_x + y\sigma_y)\right] \mathrm{d}\sigma_x \, \mathrm{d}\sigma_y \tag{2.11}$$

We recognize the integral as the inverse Fourier transform of a Gaussian function. Defining the parameters $\alpha(z)$ and $\beta(z)$ for the beam at $Z = z$ as

$$\boxed{\frac{1}{\alpha(z)} = \frac{1}{\alpha} + \mathrm{i}\frac{z}{\pi}} \qquad \boxed{\frac{1}{\beta(z)} = \frac{1}{\beta} + \mathrm{i}\frac{z}{\pi}} \tag{2.12}$$

we notice that α and β are just $\alpha(0)$ and $\beta(0)$, respectively. Equation (2.11) is now written

$$\boxed{A(x, y, z) = \sqrt{\frac{\alpha(z)\beta(z)}{\alpha(0)\beta(0)}} \, A_0 \exp(\mathrm{i}2\pi z) \exp\left[-\alpha(z)x^2 - \beta(z)y^2\right]} \tag{2.13}$$

This is the general expression for Gaussian beam propagation in free space. Several features of Eq. (2.13) will now be elaborated.

(i) The phase factor $\exp(\mathrm{i}2\pi z)$ is usually insignificant. It represents the phase delay caused by propagation between the planes $Z = 0$ and $Z = z$. As long as one is interested in the distribution of light in a plane perpendicular to the Z-axis, this phase factor is constant and can be ignored. However, if one happens to be interested in a plane that is not perpendicular to Z, the effect of different propagation distances at different points of this oblique plane must be taken into consideration.

(ii) The beam at any distance z from the origin is still Gaussian, but its $1/e$ radii r_x and r_y, as well as its radii of curvature R_x and R_y, depend on the propagation distance z. These dependences can be obtained formally from Eqs. (2.12), but the latter expressions do not give much intuition into the nature of beam propagation. The circle diagram described in the next section contains much qualitative as well as quantitative information about the beam propagation process. The circle diagram is just a graphic way of representing the relationships embodied by Eqs. (2.12).

(iii) The power of the beam remains constant as the beam propagates through space. To see this, substitute the amplitude distribution function of

Eq. (2.13) into Eq. (2.5a). The conservation of power requires the following equality to be valid for all z:

$$\left|\frac{\alpha(z)}{\alpha(0)}\right| = \sqrt{\frac{\mathrm{Re}\left[\alpha(z)\right]}{\mathrm{Re}\left[\alpha(0)\right]}} \tag{2.14}$$

A similar relationship must also hold for β. Now, Eq. (2.14) will hold if the expression

$$\frac{\mathrm{Re}\left[\alpha(z)\right]}{\mathrm{Re}^2\left[\alpha(z)\right] + \mathrm{Im}^2\left[\alpha(z)\right]} \tag{2.15}$$

can be shown to be independent of z. To show this, rewrite the first of Eqs. (2.12) as follows:

$$\frac{1}{\mathrm{Re}\left[\alpha(z)\right] + \mathrm{i}\,\mathrm{Im}\left[\alpha(z)\right]} = \frac{1}{\mathrm{Re}\left[\alpha(0)\right] + \mathrm{i}\,\mathrm{Im}\left[\alpha(0)\right]} + \mathrm{i}\frac{z}{\pi}$$

or, equivalently,

$$\frac{\mathrm{Re}\left[\alpha(z)\right] - \mathrm{i}\,\mathrm{Im}\left[\alpha(z)\right]}{\mathrm{Re}^2\left[\alpha(z)\right] + \mathrm{Im}^2\left[\alpha(z)\right]} = \frac{\mathrm{Re}\left[\alpha(0)\right] - \mathrm{i}\,\mathrm{Im}\left[\alpha(0)\right]}{\mathrm{Re}^2\left[\alpha(0)\right] + \mathrm{Im}^2\left[\alpha(0)\right]} + \mathrm{i}\frac{z}{\pi}$$

The real part of the above equation establishes that expression (2.15) is independent of z. This completes the proof that the power of the beam does not vary as it propagates through space.

2.3. The Circle Diagram

In order to study Gaussian beam characteristics during propagation along Z, we establish a point of reference z_0 on the Z-axis. The most convenient point is one at which the curvature of the beam is zero. Generally speaking, since the curvature along X is different from that along Y, one must work with different points of reference for the X- and Y-directions. The following analysis is restricted to the properties of the beam along X as it propagates along Z. Although the same analysis applies to the beam parameters (width and curvature) along Y, the reader is warned that the point of reference z_0 may be different for the two cases.

At $Z = z_0$ the curvature along X is zero, i.e. $R_x(z_0) = \infty$; thus the first of Eqs. (2.12) yields

$$\frac{1}{\alpha(z_0)} = r_x^2(z_0) = \frac{1}{\alpha(0)} + i\,\frac{z_0}{\pi} = \frac{\dfrac{1}{r_x^2(0)} + i\,\dfrac{\pi}{R_x(0)}}{\dfrac{1}{r_x^4(0)} + \dfrac{\pi^2}{R_x^2(0)}} + i\,\frac{z_0}{\pi}$$

This equation can be solved to yield the following values for z_0 and $r_x(z_0)$:

$$z_0 = -\frac{R_x(0)}{1 + \left[\dfrac{R_x(0)}{\pi r_x^2(0)}\right]^2} \tag{2.16a}$$

$$r_x(z_0) = \frac{r_x(0)}{\sqrt{1 + \left[\dfrac{\pi r_x^2(0)}{R_x(0)}\right]^2}} \tag{2.16b}$$

According to Eq. (2.16a) z_0 is unique and its value is negative if at $Z = 0$ the beam happens to have a divergent phase front, i.e., if $R_x(0) > 0$. Conversely, z_0 is positive when the phase front is convergent, i.e., when $R_x(0) < 0$. By definition, the waist of the beam is at $Z = z_0$. At the waist the $1/e$ radius is given by Eq. (2.16b). This equation also indicates that $r_x(z_0)$ is smaller than $r_x(0)$, which is the statement of the fact that the beam is narrowest at its waist.

In the remainder of this section we shall assume that the beam's waist is located at $Z = 0$, so that all distances z are measured from the waist. Since at the waist $R_x = \infty$, one may simplify Eq. (2.12) by writing it as follows:

$$\frac{1}{\alpha(z)} = r_x^2(0) + i\,\frac{z}{\pi} \tag{2.17}$$

where $r_x(0)$ is now the $1/e$ radius at the beam's waist. Denoting the real and imaginary parts of $\alpha(z)$ by α_1 and α_2 respectively, one manipulates Eq. (2.17) to obtain two equations that relate α_1 and α_2 to $r_x(0)$ and z, namely,

$$\frac{\alpha_1 - i\alpha_2}{\alpha_1^2 + \alpha_2^2} = r_x^2(0) + i\,\frac{z}{\pi}$$

or, equivalently,

$$\left[\alpha_1 - \frac{1}{2r_x^2(0)}\right]^2 + \alpha_2^2 = \left[\frac{1}{2r_x^2(0)}\right]^2 \tag{2.18a}$$

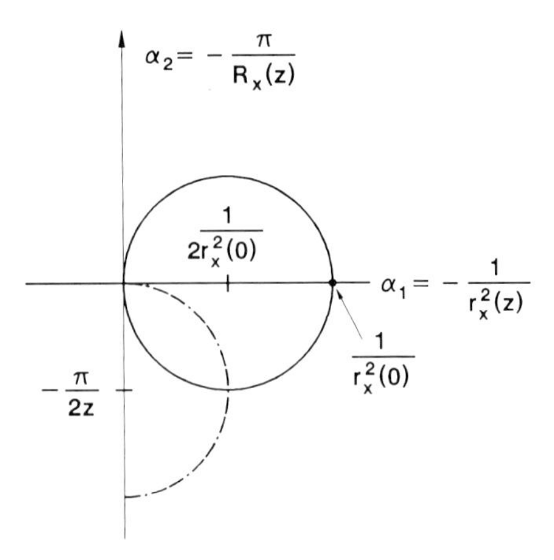

Figure 2.2. The circle diagram showing the locus of points in the complex plane corresponding to $\alpha(z)$ for a given waist radius $r_x(0)$ (solid circle) and the locus for a given propagation distance z (dashed semicircle). The coordinates of the point where the circles cross each other give the $1/e$ radius and the radius of curvature of the Gaussian beam at $Z = z$.

$$\alpha_1^{\,2} + \left(\alpha_2 + \frac{\pi}{2z}\right)^2 = \left(\frac{\pi}{2z}\right)^2 \tag{2.18b}$$

The above equations represent two circles in the $\alpha_1\alpha_2$-plane as shown in Fig. 2.2. The first circle is the locus of the pair (α_1, α_2) for given $r_x(0)$. The second circle is the locus of (α_1, α_2) for given z. The crossing point of the two circles gives the $1/e$ radius $r_x(z)$ and the radius of curvature $R_x(z)$ for the beam at $Z = z$. Several useful conclusions may be drawn from the circle diagram.

(i) As z goes from zero towards infinity the curvature of the beam first increases until it reaches a maximum and then it decreases again. The maximum curvature occurs when

$$z = \pi r_x^{\,2}(0) \tag{2.19}$$

at which point $r_x(z)$ and $R_x(z)$ are given by

$$r_x(z) = \sqrt{2}\, r_x(0) \tag{2.20a}$$

$$R_x(z) = 2\pi\, r_x^{\,2}(0) \tag{2.20b}$$

(ii) *A* sign change of z gives the same value for $r_x(z)$, but changes the sign of $R_x(z)$. Thus the intensity pattern is symmetric with respect to the waist, i.e., the intensity pattern in the plane $Z = z$ is the same as the pattern at

$Z = -z$. However, for positive z the curvature is positive, meaning that the beam diverges away from the waist, whereas for negative z the beam converges towards the waist.

(iii) For very large z one is only interested in the neighborhood of the origin in the $\alpha_1\alpha_2$-plane. Here the solid circle in Fig. 2.2 is almost tangent to the vertical axis. Therefore, at the crossing point of the two circles $\alpha_2 \simeq -\pi/z$, which results in

$$R_x(z) \simeq z \qquad (z >> 1) \tag{2.21}$$

In other words, far away from the waist the radius of curvature is equal to the distance from the waist. (It is as though the Gaussian beam originated from a point source located at the waist.) In the neighborhood of the origin in the $\alpha_1\alpha_2$-plane one can expand Eq. (2.18a), corresponding to the solid circle, to obtain

$$2r_x^2(0)\,\alpha_1 = 1 - \sqrt{1 - \left[2r_x^2(0)\,\alpha_2\right]^2} \simeq 2\left[r_x^2(0)\,\alpha_2\right]^2$$

or, equivalently,

$$\sqrt{\alpha_1} \simeq |r_x(0)\,\alpha_2|$$

Replacing the previously obtained value of $\alpha_2 \simeq -\pi/z$ in the above equation, one obtains

$$r_x(z) \simeq \frac{z}{\pi r_x(0)} \tag{2.22}$$

Thus, far away from the waist, the $1/e$ radius of the beam grows linearly with z. Based on Eq. (2.22), the divergence angle Θ of the beam may be defined as follows:

$$\Theta = 2\tan^{-1}\left[\frac{r_x(z)}{z}\right] = 2\tan^{-1}\left[\frac{1}{\pi r_x(0)}\right] \tag{2.23}$$

Among other things, Eq. (2.23) states that as the beam becomes narrower at its waist, its divergence angle increases.

2.4. Effect of Lens on Gaussian Beam

As shown in Fig. 2.3, a well-corrected lens of focal length f introduces the following phase factor on a normally incident beam:

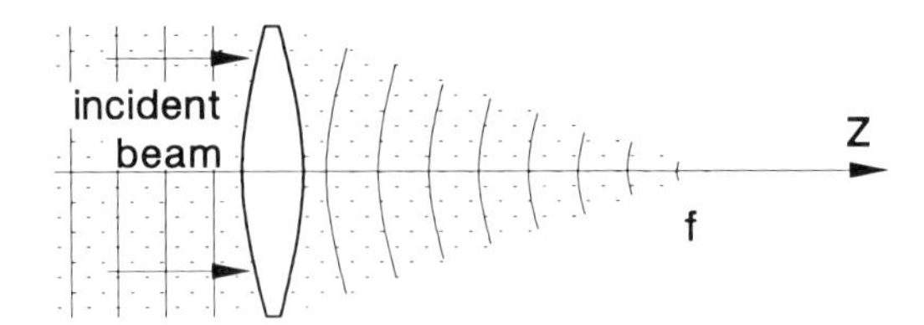

Figure 2.3. A flat wavefront incident on a lens acquires a curvature $C = 1/f$. For a positive spherical lens the acquired phase factor at the exit pupil is given by Eq. (2.24). When the incident beam's curvature is $1/R_x$, the modified curvature of the exiting beam will be given by Eq. (2.27).

$$\eta(x,y) = \exp\left(-\mathrm{i}\,\frac{2\pi}{\lambda}\,\sqrt{f^2 + x^2 + y^2}\right) \tag{2.24}$$

When the incident beam is not too wide, the above function can be approximated as follows:

$$\eta(x,y) \simeq \exp(-\mathrm{i}2\pi f)\,\exp\left(-\mathrm{i}\pi\,\frac{x^2 + y^2}{f}\right) \tag{2.25}$$

Notice that dimensionless quantities x, y, and f are used in Eq. (2.25). This equation, which is the paraxial phase factor of the lens, is separable in x and y. In fact the paraxial approximation can be generalized to include the effect of lens astigmatism by introducing different focal distances f_x and f_y for different lens curvatures along X and Y. Ignoring the constant prefactor $\exp(-\mathrm{i}2\pi f)$, and incorporating the possibility of astigmatism in the paraxial phase factor of Eq. (2.25), one arrives at the following generalized phase factor, for a lens with focal distances f_x and f_y, in the paraxial approximation:

$$\eta(x,y) \simeq \exp\left[-\mathrm{i}\pi\left(\frac{x^2}{f_x} + \frac{y^2}{f_y}\right)\right] \tag{2.26}$$

Next, assume that the Gaussian beam of Eq. (2.1) is incident on the above lens. The distribution at the exit pupil will then be the product of Eq. (2.1) and Eq. (2.26). Since in both equations the terms containing x and y are separable, in the following discussion only the x terms will be retained. The beam at the exit pupil is clearly Gaussian, and its $1/e$ radius has remained the same as that at the entrance pupil. The radius of curvature, however, has changed in accordance with the formula

$$\frac{1}{R'_x} = \frac{1}{R_x} - \frac{1}{f_x} \tag{2.27}$$

In this equation R_x and R'_x are the radii of curvature at the entrance and exit pupils respectively. The curvature caused by the lens is therefore added algebraically to the incident curvature.

2.4.1. The lens formula

As long as the waists of the beams on both sides of the lens are far from the lens, one may invoke the approximate relation between the curvature and the distance to the waist (see Eq. (2.21)) and write Eq. (2.27) in the following way:

$$\frac{1}{z} + \frac{1}{z'} = \frac{1}{f} \tag{2.28}$$

This is the well-known lens formula of geometrical optics. In order to understand the limits of its accuracy we use a numerical example.

Example 1. Let the incident beam have a flat wavefront at the entrance pupil. Then $R_x = \infty$ and $R'_x = -f_x$. Let $f_x = 10^6\lambda$, and choose the $1/e$ radius of the beam at the lens to be $r_x(0) = 10^3\lambda$. Equation (2.16a) can now be used to compute the position of the beam's waist behind the lens. The waist is at $z_0 = 9 \times 10^5\lambda$, which is about $10^5\lambda$ in front of the geometrical focal point. We conclude that, for a lens with large f-number, the spot with maximum brightness appears somewhat before the geometrical focal point. □

The formalism of Gaussian beam optics may be used to derive a more accurate lens formula, valid for all object and image distances from the lens. This formula is discussed in Problem 2.7 at the end of the chapter.

2.4.2. Numerical aperture, spot size at focus, and depth of focus

Thus far we have used the $1/e$ radius of the Gaussian beam as a measure of the beam width. Let us now define another radius, ρ_0, which will be useful in some other respects. For a beam with circular cross-section, ρ_0 is the radius of the circle that contains 99% of the beam's power. To determine ρ_0 in terms of r_x (for a circular beam $r_x = r_y$) we calculate the power within a circle of radius ρ as follows:

$$P(\rho) = \int_0^\rho \tfrac{1}{2} \left| A_0 \exp(-\alpha r^2) \right|^2 2\pi r \, \mathrm{d}r = |A_0|^2 \int_0^\rho \pi r \exp\left[-2(r/r_x)^2\right] \mathrm{d}r$$

$$= \frac{\pi}{4} r_x^2 \, |A_0|^2 \left[1 - \exp(-2\rho^2/r_x^2)\right] \tag{2.29}$$

Problems

(2.1) A laser diode operating at $\lambda = 830$ nm has a Gaussian beam with its waist at the center of the cavity. The $1/e$ radius of the beam along X is $r_{x0} = 4\ \mu$m and its curvature $1/R_{x0}$ is zero at the waist. Similarly, in the Y-direction $r_{y0} = 1\ \mu$m and $R_{y0} = \infty$. A collimating lens is placed a distance z from the cavity's center.

(a) Determine z such that the collimated beam's radius in the Y-direction, r_y, will be 2.5 mm.

(b) What is r_x, the radius of the collimated beam in the X-direction?

(c) Find the beam's radii of curvature R_x and R_y at the collimating lens.

(2.2) Explain how a pair of cylindrical lenses can produce a collimated Gaussian beam with $r_x = r_y$ from a collimated Gaussian beam with $r_x \neq r_y$.

(2.3) Analyze the effect of the beam-shaping prism in Fig. 2.4 on the elliptical cross-section of the Gaussian beam.

(2.4) A holographic optical element (transmission grating) is being proposed for simultaneously collimating, circularizing, and correcting the astigmatism of an arbitrary Gaussian beam, as shown in Fig. 2.5. What is the required transmission function of this element?

(2.5) In the example of subsection 2.4.1 try several other sets of parameters and derive the position of the beam waist behind the lens. Compare your answers with the predictions of geometrical optics (i.e., Eq. (2.28)).

(2.6) The astigmatic focus-error-detection scheme is shown in Fig. 2.6 (assume Gaussian beams throughout). The objective's focal length is f_0, the astigmat's focal lengths are f_x and f_y, and the separation between the two lenses is D. The reflecting disk surface is in the neighborhood of the objective's focal plane. Determine the beam's $1/e$ radii r_x and r_y in the region between the focal planes of the astigmat.

(2.7) Figure 2.7 shows a thin lens with focal length f, an incident Gaussian beam whose waist is a distance z in front of the lens, and an emerging beam whose waist is a distance z' behind the lens. The $1/e$ radii of the waists are r_{x0} and r'_x, respectively.

(a) Using the Gaussian beam formalism show that z' may be obtained from z, f, and r_{x0} as follows:

$$\frac{z'}{f} - 1 = \frac{\dfrac{z}{f} - 1}{\left(\dfrac{z}{f} - 1\right)^2 + \left(\dfrac{\pi r_{x0}^2}{f}\right)^2}$$

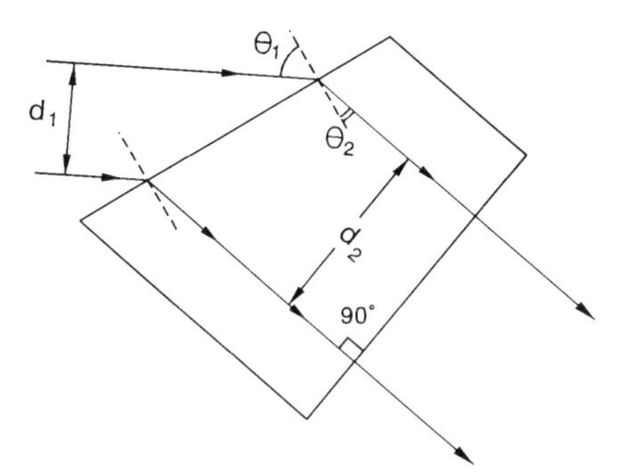

Figure 2.4. Beam-shaping prism. In the plane of the page, the diameter of the incident beam is d_1 while that of the exiting beam is d_2. The diameter perpendicular to the plane of the page remains unchanged upon passing through the prism.

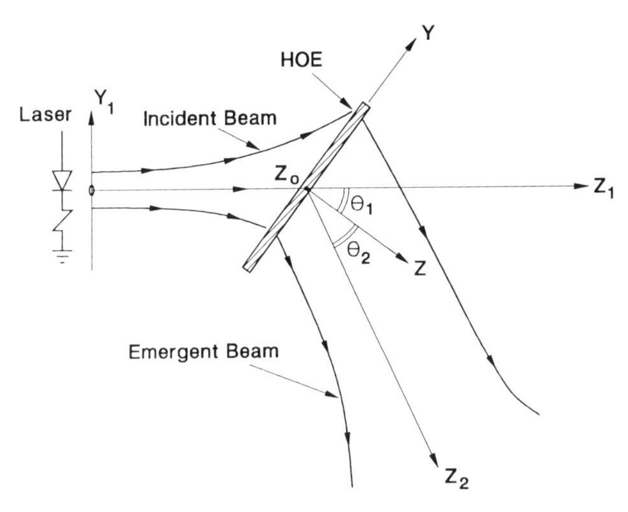

Figure 2.5. Schematic diagram of a holographic optical element (HOE) capable of collimating, circularizing, and correcting the astigmatism of a laser diode.

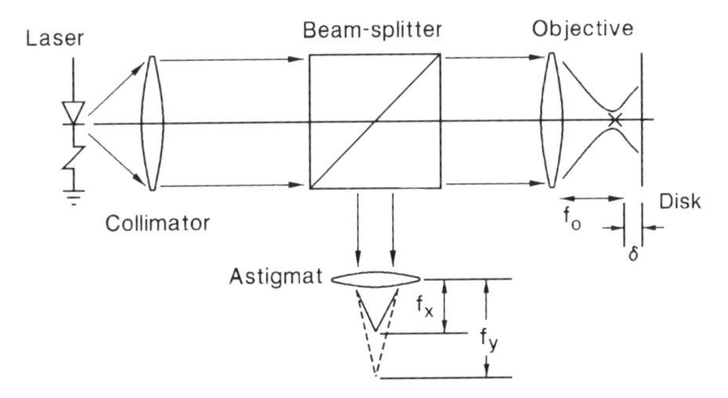

Figure 2.6. Schematic diagram of an astigmatic focus-error-detection system used in optical disk drives. A multi-element photodetector is placed halfway between the two foci of the astigmat, measuring, in effect, the dimensions r_x and r_y of the beam incident upon it. With the disk in focus (i.e., $\delta = 0$), $r_x = r_y$ at the detector. When $\delta \neq 0$, r_x will be either greater or less than r_y, depending on the sign of δ.

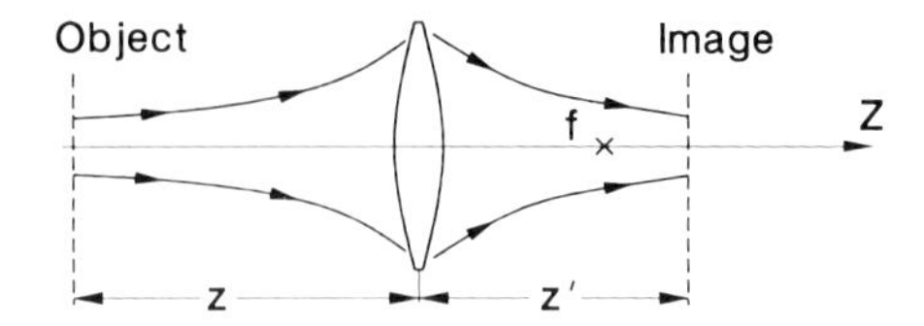

Figure 2.7. A Gaussian beam whose waist is in the object plane is refocused into a beam whose waist is in the image plane.

(b) Under what circumstances does this formula reduce to the well-known lens formula, $f/z + f/z' = 1$?

(c) Assuming that the waist of the incident beam is at the front focal plane, where do you expect to find the waist of the emerging beam?

(d) Show that the magnification m, defined as the ratio of the two waist diameters, r'_x / r_{x0}, is given by

$$m = \frac{1}{\sqrt{\left(\frac{z}{f} - 1\right)^2 + \left(\frac{\pi r_{x0}^2}{f}\right)^2}}$$

Explore the conditions under which the above formula conforms to the rules of geometrical optics.

(2.8) Consider a Gaussian beam with $r_x = r_y = 1.5$ mm, $R_x = R_y = \infty$, and wavelength $\lambda = 830$ nm, incident on a spherical lens. The lens is free from aberrations, has focal length $f = 4$ mm, and its aperture is wide enough to let the beam through without significant truncation.

(a) Find the minimum spot diameter (i.e. diameter at the waist in the neighborhood of focus) as well as the distance of the waist from the lens.

(b) A cover glass (refractive index $n = 1.5$) is placed a distance of 3 mm from the lens as shown in Fig. 2.8. Use Gaussian beam formulas to determine the new position and diameter of the minimum-size spot beneath the cover glass.

(c) According to the Gaussian treatment, the introduction of the cover glass does not produce aberrations. In practice, however, aberrations are known to occur. Why does the Gaussian treatment fail to predict these aberrations?

(2.9) A proposed method of reducing the focused spot size in an optical disk system involves the use of a solid immersion lens. In this problem we compare the characteristics of a spot focused through the substrate of an optical disk using first an ordinary lens and then a solid immersion lens.

(a) Figure 2.9(a) shows a Gaussian beam (vacuum wavelength $= \lambda_0$) focused through an optical disk substrate of refractive index n. The objective is corrected for all aberrations (including those induced by the substrate). The numerical aperture NA of the lens is defined as $\sin\theta$, where θ is the half-angle

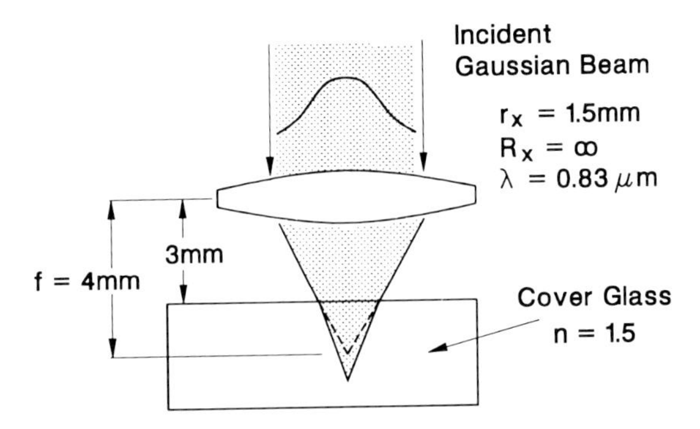

Figure 2.8. Focusing of a Gaussian beam through a cover glass.

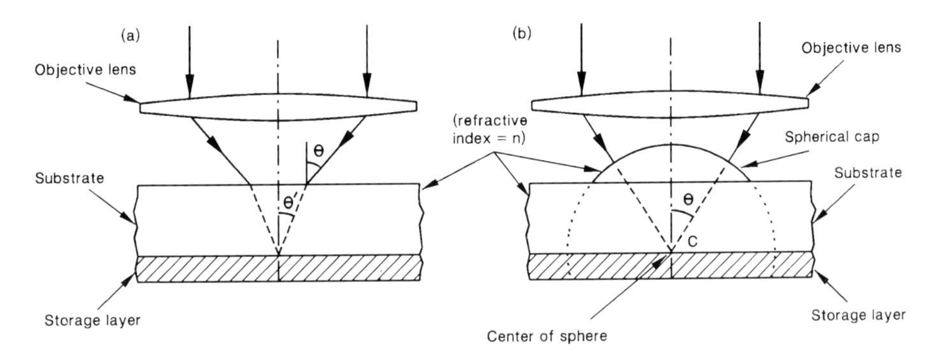

Figure 2.9.

of the focused cone *outside* the substrate. In terms of NA, λ_0, and n determine the diameter of the focused spot and the depth of focus of the beam.

(b) A spherical cap, cut from a solid glass sphere of the same index n as the substrate, is placed in contact with the disk, as shown in Fig. 2.9(b). The various dimensions are chosen to allow the center C of the glass sphere to coincide with the focal point of the objective at the storage layer. The objective, however, is not corrected for the substrate in this case; it is simply an aberration-free lens designed for operation in air (NA = sin θ). In terms of NA, λ_0, and n determine the diameter of the focused spot and the depth of focus of the beam.

(2.10) Consider a glass sphere of radius R and refractive index n, as shown in Fig. 2.10(a). A ray of light aimed at P, having angle θ with the Z-axis, is incident on the sphere at point A, where the angle of incidence is denoted by ϕ. The refracted ray crosses the Z-axis at P′, making an angle θ' with Z and an angle ϕ' with the surface normal. Let us now assume that the two triangles ACP and ACP′ are similar.

(a) What is the relationship between θ and ϕ?

(b) Determine the distances of P and P′ from the center of the glass sphere (namely, CP and CP′) in terms of R and n.

(c) Show that a cone of light converging towards P will be perfectly refocused by the glass sphere at P′, without any aberrations.

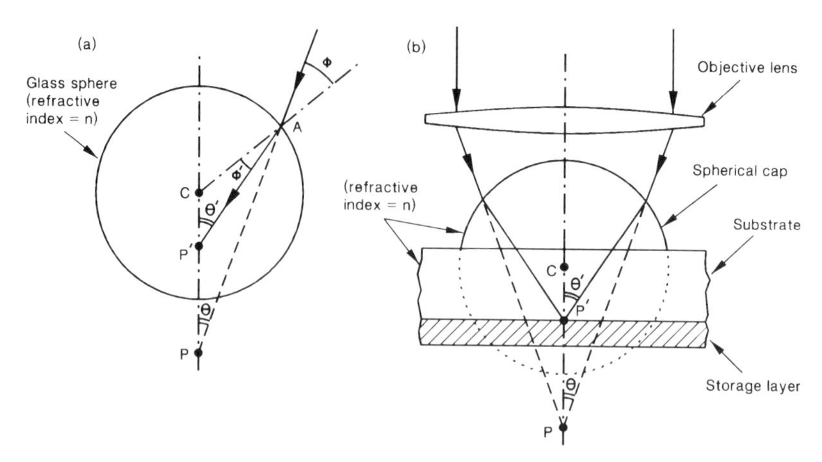

Figure 2.10.

(d) A spherical cap, cut from a solid glass sphere, is placed in contact with an optical disk substrate, as shown in Fig. 2.10(b). The glass sphere and the substrate material have the same refractive index n, the vacuum wavelength of the incident Gaussian beam is λ_0, and the objective's numerical aperture NA is defined as $\sin\theta$. The various dimensions are chosen to allow the focus of the objective in air, P, to fall below the disk, while its focus through the cap, P′, occurs at the storage layer. Determine the size of the focused spot at P′. What is the depth of focus at this point?

3

Theory of Diffraction

Introduction

The diffraction of light plays an important role in optical data storage systems. The rapid divergence of the beam as it emerges from the front facet of the diode laser is due to diffraction, as is the finite size of the focused spot at the focal plane of the objective lens. Diffraction of light from the edges of the grooves on an optical disk surface is used to generate the tracking-error signal. Some of the schemes for readout of the recorded data also utilize diffraction from the edges of pits and magnetic domains. A deep understanding of the theory of diffraction is therefore essential for the design and optimization of optical data storage systems.

Diffraction and related phenomena have been the subject of investigation for many years, and one can find many excellent books and papers on the subject. The purpose of the present chapter is to give an overview of the basic principles of diffraction, and to derive those equations that are of particular interest in optical data storage. We begin by introducing in section 3.1 the concept of the stationary-phase approximation as applied to two-dimensional integrals. Then in section 3.2 we approximate two integrals that appear frequently in diffraction problems using the stationary-phase method.

An arbitrary light amplitude distribution in a plane can be decomposed into a spectrum of plane waves. These plane waves propagate independently in space and, at the final destination, are superimposed to form a diffraction pattern. The mathematical statement of this decomposition and superposition procedure, which forms the starting point of section 3.3 was derived in the preceding chapter. Here we show that the application of the stationary-phase approximation to the *superposition* integral yields the Fraunhofer diffraction formula in the far field. A similar technique is used in section 3.4 to derive the diffraction pattern in the vicinity of the focal plane of a lens; here it is the *decomposition* integral that is being approximated. The primary aberrations of the lens and their effects on the focused spot are also covered in this section. Finally, in section 3.5 a simple concept is introduced that enables one to modify the scalar diffraction formulas to accommodate the vector nature of electromagnetic radiation. Examples are given throughout the chapter, and the accompanying intensity distribution plots should help the reader understand the physical meaning behind the mathematical formulas.

3.1. Stationary-phase Approximation

Consider the two-dimensional integral

$$I = \iint f(x,y) \exp\left[i\eta g(x,y) \right] \, dx dy \tag{3.1}$$

where, in general, $f(x,y)$ is a complex function, $g(x,y)$ is a real function, η is a large real number, and the domain of integration is a subset of the XY-plane. In the neighborhood of an arbitrary point (x_0, y_0) within the domain of integration, small variations of $g(x,y)$ will be amplified by η, which will subsequently result in rapid oscillations of the phase factor, $\exp[i\eta g(x,y)]$. Assuming that $f(x,y)$ in the same neighborhood of (x_0, y_0) is a slowly varying function, the oscillations result in a negligible contribution from this neighborhood to the integral. The main contributions to the integral then come from the regions in which $g(x,y)$ is nearly constant. These regions are in the neighborhood of stationary points, (x_0, y_0), which are defined by the following relation:

$$\frac{\partial}{\partial x} g(x_0, y_0) = \frac{\partial}{\partial y} g(x_0, y_0) = 0 \tag{3.2}$$

Around each stationary point one may expand $g(x,y)$ in a Taylor series up to the second-order term to obtain

$$g(x,y) \simeq g(x_0,y_0) + \tfrac{1}{2} g_{xx}(x_0,y_0)(x-x_0)^2 + g_{xy}(x_0,y_0)(x-x_0)(y-y_0) + \tfrac{1}{2} g_{yy}(x_0,y_0)(y-y_0)^2 \tag{3.3}$$

Replacing $g(x,y)$ in Eq. (3.1) with Eq. (3.3), and taking $f(x,y)$ outside the integral, one obtains

$$I \simeq \sum_{\substack{\text{all stationary} \\ \text{points } (x_0,y_0)}} f(x_0,y_0) \exp\left[i\eta g(x_0,y_0)\right] \iint_{-\infty}^{\infty} \exp\left\{ i\frac{\eta}{2} \left[g_{xx}(x-x_0)^2 + 2g_{xy}(x-x_0)(y-y_0) + g_{yy}(y-y_0)^2 \right] \right\} dx dy \tag{3.4}$$

Notice that the domain of integration is now extended to the entire plane, since the contribution to the integral from regions outside the immediate neighborhood of the stationary points is, in any event, negligible. The double integral in Eq. (3.4) can be readily carried out, yielding

$$I \simeq \frac{2\pi i}{\eta} \sum_{\substack{\text{all stationary} \\ \text{points } (x_0, y_0)}} \frac{\nu}{\sqrt{\left| g_{xx} g_{yy} - g_{xy}^2 \right|}} \exp\left[i\eta\, g(x_0, y_0) \right] f(x_0, y_0) \tag{3.5}$$

where the coefficient ν is given by

$$\nu = \begin{cases} -i & \text{if } g_{xx} g_{yy} < g_{xy}^2 \\ \pm 1 & \text{if } g_{xx} g_{yy} > g_{xy}^2 \text{ and } g_{xx} \gtrless 0 \end{cases}$$

Equation (3.5) is the final result of this section. If the numerical value of $(g_{xx} g_{yy} - g_{xy}^2)$ happens to be exactly zero at a particular stationary point, or if a stationary point occurs on the boundary of the domain of integration in Eq. (3.1), then Eq. (3.5) does not apply. In our analysis of diffraction problems, however, these special cases will not be encountered.

3.2. Application of Stationary-phase Method to Diffraction Problems

In many diffraction problems the function $g(x, y)$ appearing in Eq. (3.1) has the form

$$g(x, y) = \sqrt{1 + \alpha(x^2 + y^2)} + \beta x + \gamma y \tag{3.6}$$

β and γ are real and, in all cases of interest, α is either $+1$ or -1. The above function has only one stationary point, which is readily found to be

$$(x_0, y_0) = \left(\frac{-\beta/\alpha}{\sqrt{1 - (\beta^2 + \gamma^2)/\alpha}}, \frac{-\gamma/\alpha}{\sqrt{1 - (\beta^2 + \gamma^2)/\alpha}} \right) \tag{3.7}$$

At the stationary point (x_0, y_0) the value of the function is

$$g(x_0, y_0) = \sqrt{1 - (\beta^2 + \gamma^2)/\alpha} \tag{3.8}$$

We also find

$$g_{xx} g_{yy} - g_{xy}^2 = (\alpha - \beta^2 - \gamma^2)^2 \tag{3.9}$$

The coefficient ν is therefore either $+1$ or -1, depending on the sign of

g_{xx}. Since the value of ν results only in an insignificant phase factor, we shall omit it in subsequent analysis. The resulting stationary-phase approximation to the integral is thus written

$$I = \iint f(x,y) \exp\left\{ \mathrm{i}\eta\left[\sqrt{1+\alpha(x^2+y^2)} + \beta x + \gamma y\right]\right\} \mathrm{d}x\mathrm{d}y$$

$$\simeq \frac{2\pi\mathrm{i}}{\eta} \exp\left[\mathrm{i}\eta\sqrt{1-(\beta^2+\gamma^2)/\alpha}\right]$$

$$\times \frac{f\left(\dfrac{-\beta/\alpha}{\sqrt{1-(\beta^2+\gamma^2)/\alpha}}, \dfrac{-\gamma/\alpha}{\sqrt{1-(\beta^2+\gamma^2)/\alpha}}\right)}{\alpha\left[1-(\beta^2+\gamma^2)/\alpha\right]} \tag{3.10}$$

We shall be interested in situations where α and η are constants and the integral is considered as a function of β and γ. Since the only values of α which are of interest are +1 and -1, we consider the corresponding two cases separately.

Case 1. Here $\alpha = +1$. Equation (3.10) simplifies as follows:

$$\iint f(x,y) \exp\left[\mathrm{i}\eta\left(\sqrt{1+x^2+y^2} + \beta x + \gamma y\right)\right] \mathrm{d}x\mathrm{d}y$$
$$\simeq \frac{2\pi\mathrm{i}}{\eta}\left[\exp\left(\mathrm{i}\eta\sqrt{1-\beta^2-\gamma^2}\right)\right] f_c(-\beta,-\gamma) \tag{3.11}$$

where $f_c(\beta,\gamma)$, the *compressed* version of the function $f(x,y)$, is given by

$$f_c(\beta,\gamma) = \frac{f\left(\dfrac{\beta}{\sqrt{1-\beta^2-\gamma^2}}, \dfrac{\gamma}{\sqrt{1-\beta^2-\gamma^2}}\right)}{1-\beta^2-\gamma^2} \tag{3.12}$$

In order to understand the nature of the transformation that carries $f(x,y)$ to $f_c(\beta,\gamma)$ (the transformation that we have named *compression*) consider the diagram in Fig. 3.1. This diagram shows the XY-plane on which the functions are defined, and the angles θ and ϕ corresponding to a given point (x,y) on this plane. If the distance OO′ is chosen to be unity, the relationship between (x,y) and (θ,ϕ) may be written

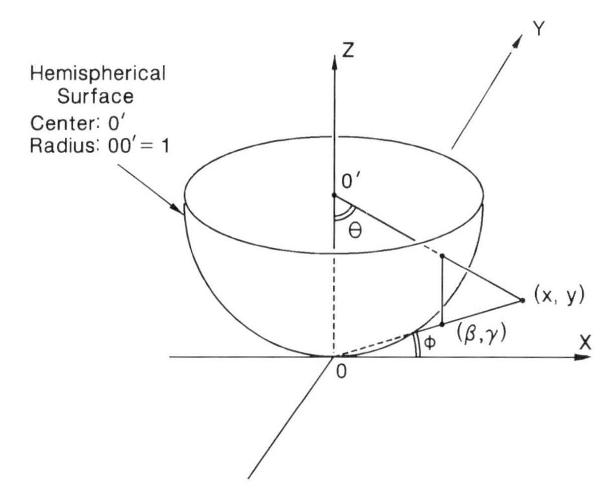

Figure 3.1. Geometry of the compression transformation. The hemispherical surface is centered at O′ and its radius is unity. Compression maps the point (x, y) to the point (β,γ).

$$(x, y) = (\tan\theta\,\cos\phi,\ \tan\theta\,\sin\phi) \tag{3.13}$$

On the same XY-plane, the point (β,γ) that corresponds to the above point (x, y) (and to which a value proportional to $f(x, y)$ is assigned according to Eq. (3.12)) is given by

$$(\beta,\gamma) = (\sin\theta\,\cos\phi,\ \sin\theta\,\sin\phi) \tag{3.14}$$

It is thus seen that the domain of the function $f(x, y)$ must be radially compressed toward the origin in order to produce the domain of $f_c(\beta,\gamma)$. The compression is rather weak for points near the origin O, but as (x, y) moves further away from the center it becomes stronger, so much so that the points at infinity are brought onto the unit circle. The domain of $f_c(\beta,\gamma)$ is therefore the unit disk, containing all the points within the unit circle.

The factor $1 - \beta^2 - \gamma^2$ by which the function $f(x, y)$ is divided in Eq. (3.12) is a normalization factor that preserves the integrated *intensity* of the function in the following sense:

$$\iint_{-\infty}^{\infty} |f(x, y)|^2\, \mathrm{d}x\,\mathrm{d}y = \iint_{\text{unit disk}} |f_c(\beta,\gamma)|^2\, \mathrm{d}\beta\,\mathrm{d}\gamma \tag{3.15}$$

Equation (3.11) represents an important instance of the stationary-phase approximation, and is widely used in the theory of diffraction.

Case 2. Here $\alpha = -1$. Equation (3.10) simplifies as follows:

$$\iint_{\text{unit disk}} f(x,y) \exp\left[i\eta \left(\sqrt{1 - x^2 - y^2} + \beta x + \gamma y\right)\right] dx\,dy \simeq -\frac{2\pi i}{\eta}\left[\exp\left(i\eta\sqrt{1 + \beta^2 + \gamma^2}\right)\right] f_s(\beta,\gamma) \tag{3.16}$$

Here $f_s(\beta,\gamma)$ is a *stretched* version of the function $f(x,y)$ defined by

$$f_s(\beta,\gamma) = \frac{f\left(\dfrac{\beta}{\sqrt{1 + \beta^2 + \gamma^2}}, \dfrac{\gamma}{\sqrt{1 + \beta^2 + \gamma^2}}\right)}{1 + \beta^2 + \gamma^2} \tag{3.17}$$

Stretching is the inverse of compression, which was described earlier; it is applied radially to a function defined on the unit disk and yields a function spread out on the entire plane. Like compression, the stretching transformation preserves the integrated intensity of the function.

3.3. Near-field and Far-field Diffraction

Consider the complex-amplitude distribution $A(x,y)$ in the XY-plane. For reasons that were explained in Chapter 2, x and y will be assumed to be normalized by the wavelength λ of the light. According to Eq. (2.10) the amplitude distribution in a plane parallel to XY and at a distance $Z = z$ from it can be written as the following superposition integral:

$$A(x,y,z) = \iint_{-\infty}^{\infty} \mathscr{F}\left\{A(x,y,z=0)\right\} \exp\left\{i2\pi\left[x\sigma_x + y\sigma_y + z\sqrt{1 - \sigma_x^{\,2} - \sigma_y^{\,2}}\right]\right\} d\sigma_x\, d\sigma_y \tag{3.18}$$

(Like x and y, z is normalized by λ.) In the above equation $\mathscr{F}$ is the Fourier transform operator and z could be either positive or negative, depending on whether the desired distribution is to the right or the left of the initial XY-plane. The domain of integration in Eq. (3.18) consists of two regions separated by the unit circle. When (σ_x, σ_y) happens to be within the unit circle the corresponding plane waves propagate without attenuation. On the other hand, when $\sigma_x^2 + \sigma_y^2 > 1$ the plane waves become evanescent and their amplitudes decay exponentially with propagation distance. Thus, for most practical purposes, one can safely ignore the evanescent waves and confine attention to the integral over the unit disk.

Let us first consider the case of far-field (Fraunhofer) diffraction where the distance z is sufficiently large to justify the application of the stationary-phase technique. Comparison of Eq. (3.18) with Eq. (3.16) shows that the two integrals are identical provided that one makes the following associations:

$$f(\sigma_x, \sigma_y) \rightarrow \mathscr{F}\left\{A(x, y, z = 0)\right\}; \qquad \eta \rightarrow 2\pi z\,; \qquad \beta \rightarrow x/z\,; \qquad \gamma \rightarrow y/z$$

The superposition integral in Eq. (3.18) is therefore approximated as follows:

$$A(x, y, z) \simeq -\frac{i}{z} \exp\left[i2\pi \sqrt{x^2 + y^2 + z^2} \right] \mathscr{F}_s\left\{A(x, y, z = 0)\right\}_{\substack{\beta = x/z \\ \gamma = y/z}} \tag{3.19}$$

This is the basic equation of scalar diffraction theory in the far-field regime. The exponential term in Eq. (3.19) is a simple curvature phase factor, corresponding to a radius of curvature z. The decay of the amplitude with inverse distance, i.e., the $1/z$ term, is also consistent with our expectations since, far from the origin, the initial distribution should behave as a point source. As for the remaining term containing the Fourier transform of the initial distribution $A(x, y, z = 0)$, note that the stretching operation (as defined by Eq. (3.17)) must be applied before the function is evaluated at $(x/z,\ y/z)$. Once again, all spatial coordinates appearing in Eq. (3.19) are normalized by the wavelength λ and are therefore dimensionless.

In the near-field (Fresnel) regime where the stationary-phase approximation does not apply, one must calculate the integral in Eq. (3.18) directly. We rewrite this equation using the notation for the Fourier transform $\mathscr{F}$ and the inverse Fourier transform $\mathscr{F}^{-1}$:

$$A(x, y, z) = \mathscr{F}^{-1}\left\{\mathscr{F}\left\{A(x, y, z = 0)\right\} \exp\left[i2\pi z \sqrt{1 - \sigma_x^2 - \sigma_y^2}\right]\right\} \tag{3.20}$$

Unless z is excessively large, the complex exponential term in Eq. (3.20) may be accurately represented in the $\sigma_x \sigma_y$-plane using a discrete mesh with a reasonable number of samples, and the two transforms are subsequently evaluated by a computer. Numerical computations of this type are usually fast, reliable, and applicable to a wide variety of distributions.

Example 1. Let the initial distribution of light amplitude be uniform over a rectangular aperture of area $\lambda\Delta_x \times \lambda\Delta_y$. Assuming that the integrated intensity over the aperture is unity and that the coordinates (x, y) are normalized by λ, we have

$$A(x, y, z = 0) = \begin{cases} \dfrac{1}{\lambda\sqrt{\Delta_x \Delta_y}}, & |x| < \frac{1}{2}\Delta_x \text{ and } |y| < \frac{1}{2}\Delta_y \\ 0 & \text{otherwise} \end{cases} \tag{3.21}$$

The Fourier transform of the above distribution is

$$\mathcal{A}(\sigma_x, \sigma_y) = \mathcal{F}\left\{A(x, y, z = 0)\right\} = \frac{1}{\lambda}\,\frac{\sin(\pi\Delta_x\sigma_x)}{\pi\sqrt{\Delta_x}\,\sigma_x} \times \frac{\sin(\pi\Delta_y\sigma_y)}{\pi\sqrt{\Delta_y}\,\sigma_y} \tag{3.22}$$

After stretching, this function becomes

$$\mathcal{A}_s(x/z,\, y/z) = \frac{\mathcal{A}\left(\dfrac{x}{\sqrt{x^2 + y^2 + z^2}}, \dfrac{y}{\sqrt{x^2 + y^2 + z^2}}\right)}{1 + (x/z)^2 + (y/z)^2} \tag{3.23}$$

Finally, the far-field distribution (at $Z = z$) is obtained by substituting the above in Eq. (3.19), which yields

$$A(x, y, z) \simeq -\frac{i}{\lambda z}\exp\left[i2\pi\sqrt{x^2 + y^2 + z^2}\right] \times \frac{\sin\left(\dfrac{\pi\Delta_x x}{\sqrt{x^2 + y^2 + z^2}}\right)}{\pi\sqrt{\Delta_x}\,(x/z)} \times \frac{\sin\left(\dfrac{\pi\Delta_y y}{\sqrt{x^2 + y^2 + z^2}}\right)}{\pi\sqrt{\Delta_y}\,(y/z)} \tag{3.24}$$

Patterns of the far-field intensity distribution at $z = 10^4$ (in units of λ) for two different rectangular apertures are shown in Fig. 3.2. In (a) the aperture dimensions are $\Delta_x = 100$, $\Delta_y = 75$; in (b) $\Delta_x = 50$, $\Delta_y = 100$. In both cases Δ_x and Δ_y are kept at or below 100 in order to ensure that the distance of 10^4 is sufficiently far from the aperture to justify the stationary-phase approximation. (Here is a good rule of thumb for

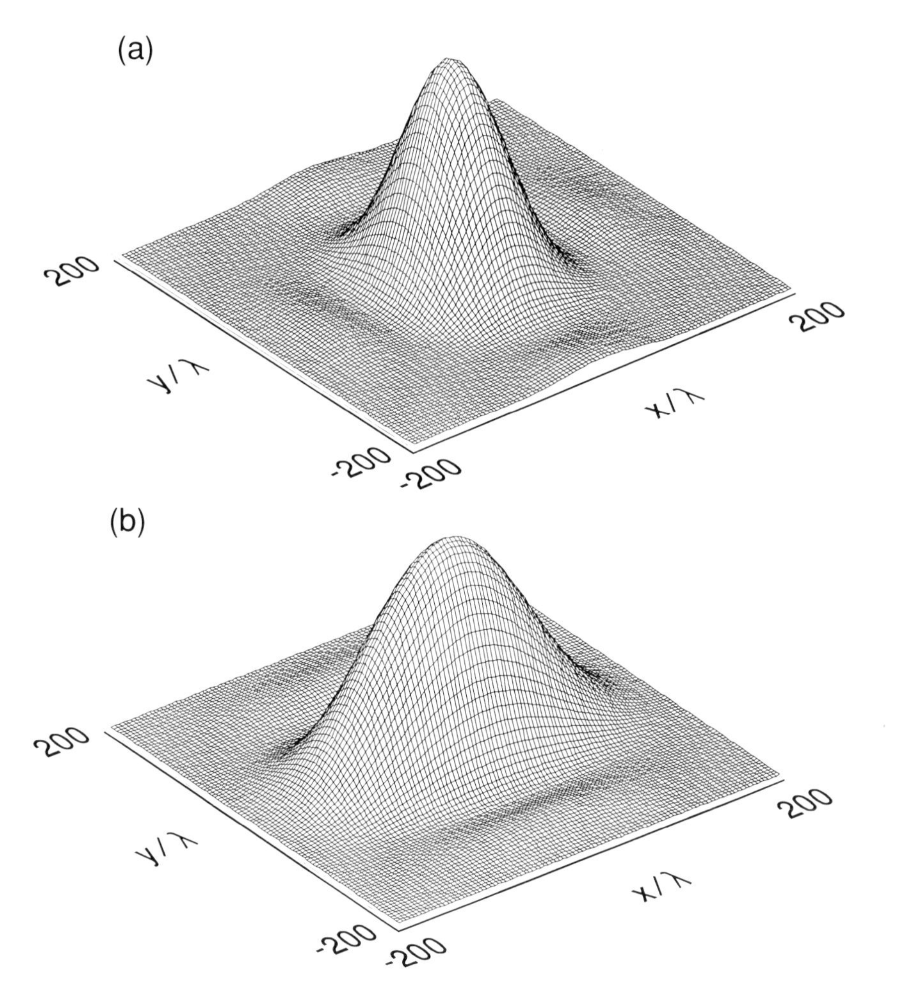

Figure 3.2. Far-field (Fraunhofer) patterns of intensity distribution for two rectangular apertures. In (a) the aperture dimensions (Δ_x, Δ_y) are (100, 75); in (b) they are (50, 100). The propagation distance in both cases is $z = 10000$.

estimating the distance to the far field: if the initial distribution is relatively smooth and is confined to an aperture with diameter D, then the far field occurs at $z \gtrsim D^2$, where both D and z are in units of λ.) □

Example 2. Figure 3.3 shows a sequence of intensity distributions at various distances, computed for a uniformly illuminated square aperture of dimension $\Delta_x = \Delta_y = 100$. In (a), (b), and (c) where $z = 1000$, 2500, and 5000 respectively, we are in the near-field regime and Fresnel's formula, Eq. (3.20), applies. (A discrete 512 × 512 mesh was sufficient for these computations.) In (d) the propagation distance is $z = 10^4$ and the Fraunhofer formula (Eq. (3.19) or Eq. (3.24)) becomes applicable. Of

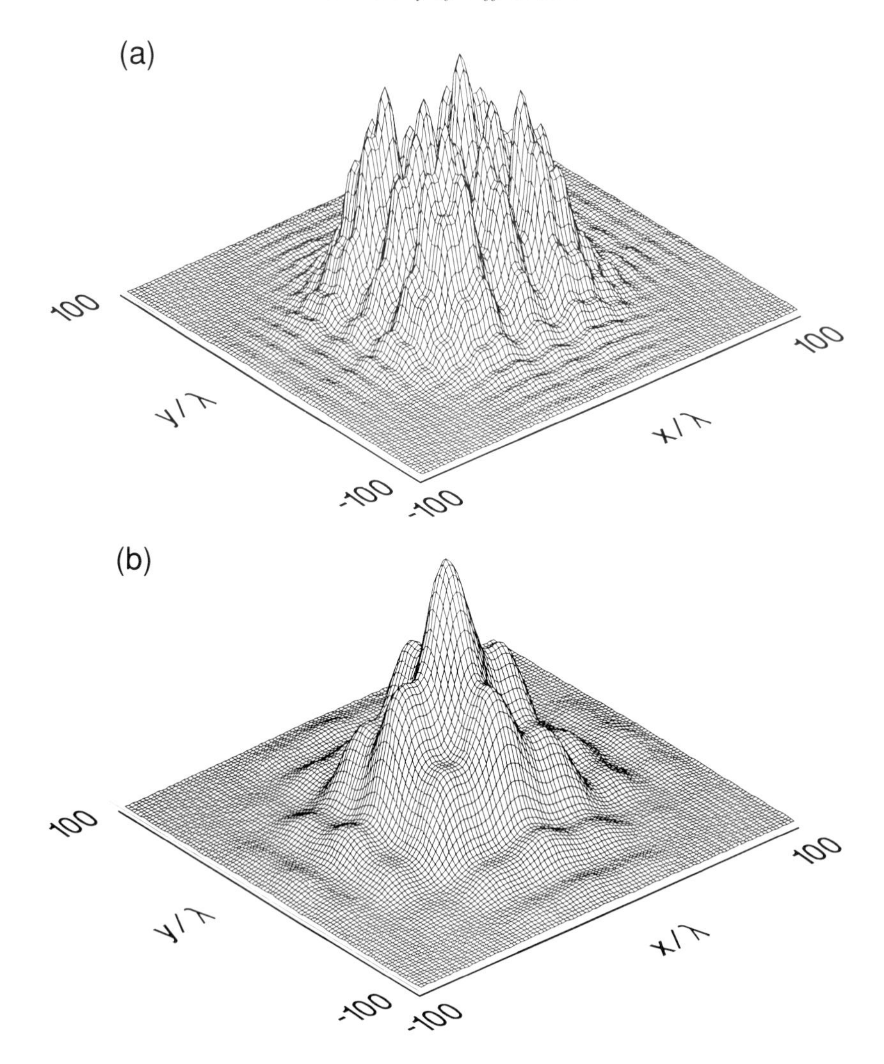
(a)
100
y/λ
-100
-100
x/λ
100
(b)
100
y/λ
-100
-100
x/λ
100

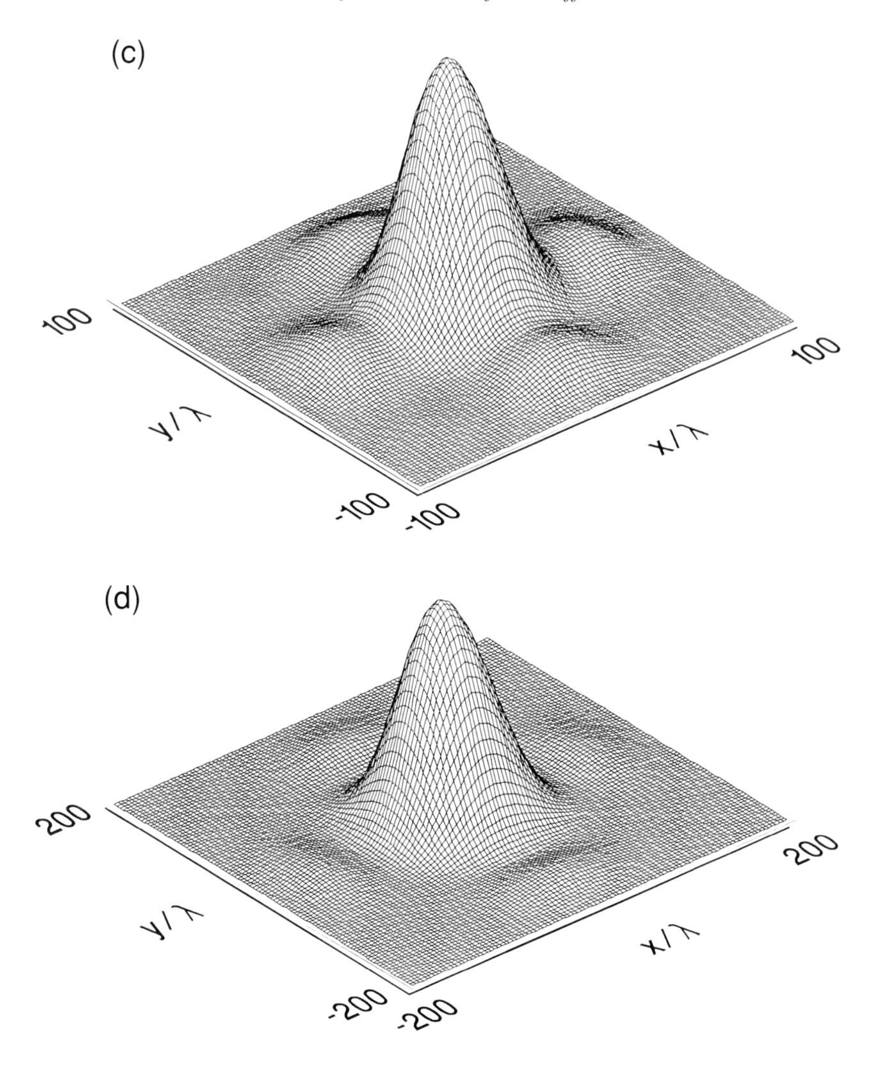

Figure 3.3. Intensity distribution at various distances from a square aperture. The aperture dimensions (Δ_x, Δ_y) are (100, 100) and the incident distribution is uniform. The vertical scale varies among the plots and the actual peak intensities in panels (a)–(d) are in the ratio of 0.87 : 1 : 1 : 0.39 respectively. The volume under each curve is a measure of the total power of the beam, which remains constant during propagation.

course, one could continue to use Eq. (3.20) even in the far-field regime, but the required mesh size becomes prohibitively large. □

Example 3. A sequence of intensity patterns, computed for a uniformly illuminated circular aperture of diameter $D = 10^3\lambda$, is shown in Fig. 3.4. The propagation distance z is $8.5 \times 10^4\lambda$ in (a), $12.5 \times 10^4\lambda$ in (b),

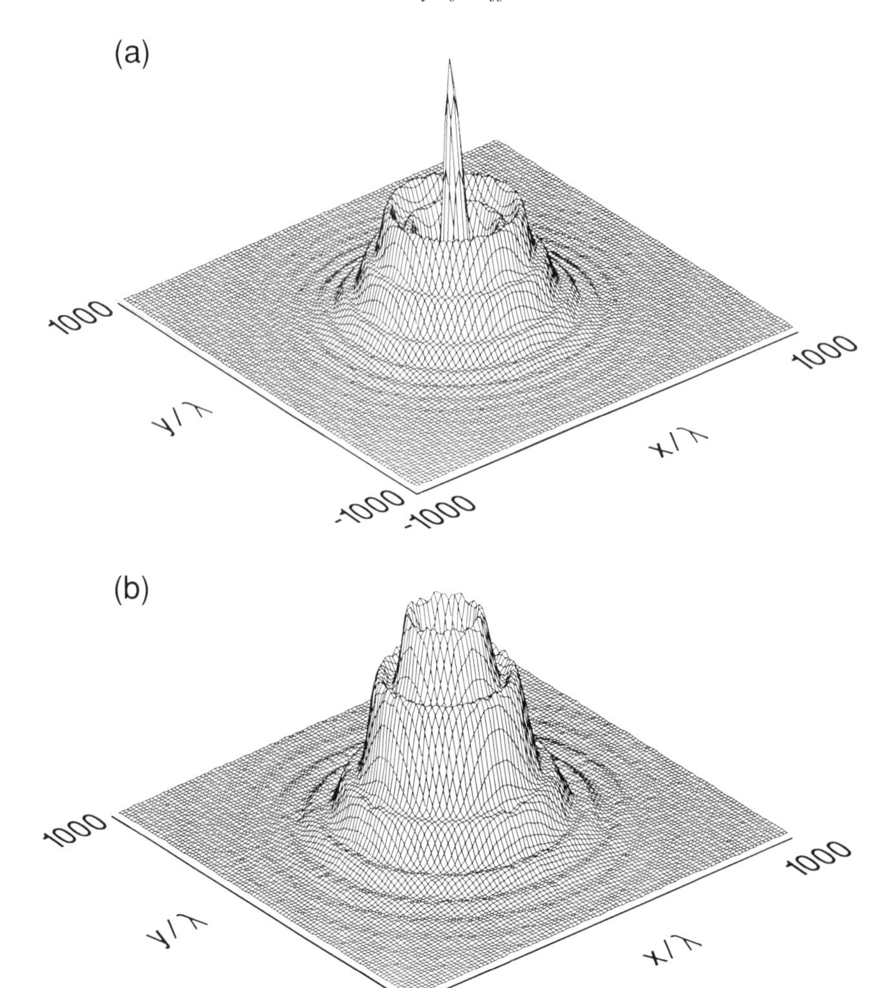
(a)
1000
y/λ
-1000
-1000
x/λ
1000
(b)
1000
y/λ
-1000
-1000
x/λ
1000

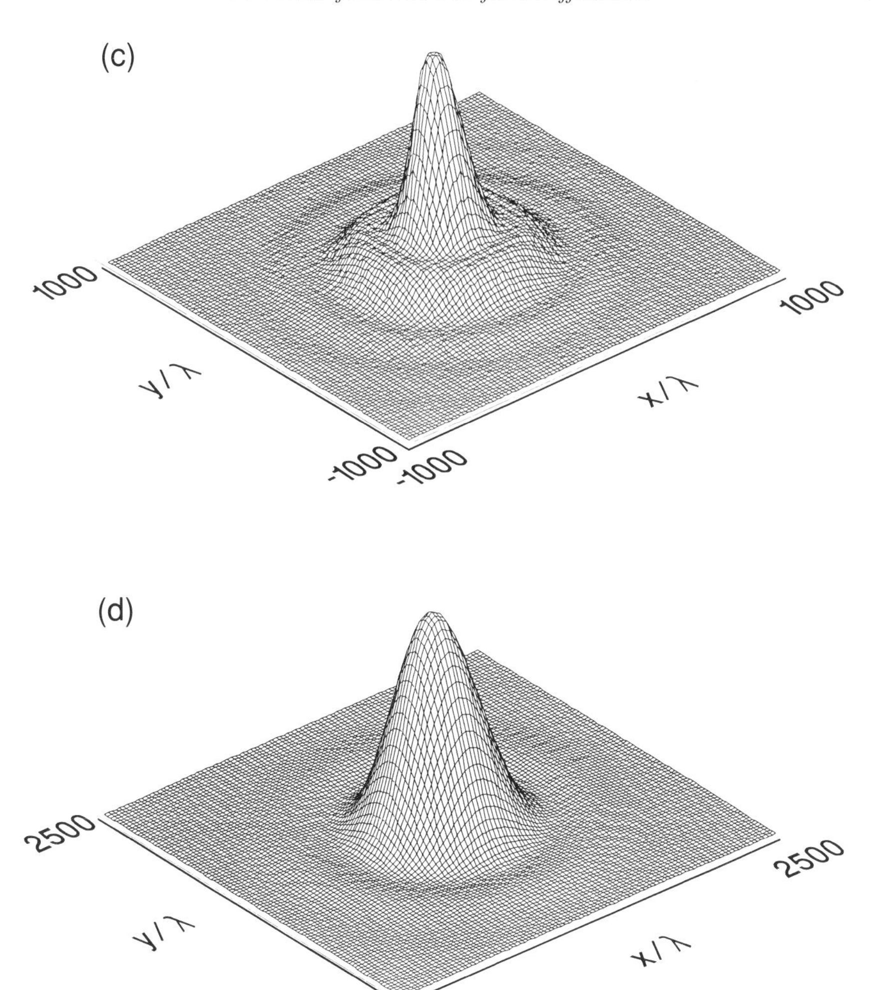

Figure 3.4. Intensity distribution at various distances from a uniformly illuminated circular aperture of diameter D = 1000. The vertical scale varies among the plots and the actual peak intensities in panels (a)–(d) are in the ratio of 1 : 0.49 : 0.99 : 0.16 respectively.

$25 \times 10^4\lambda$ in (c), and $10^6\lambda$ in (d). The vertical scale is adjusted in each case for the best display. The volume under each curve, which corresponds to the total power of the beam, is a constant of propagation. Note that as z increases the near-field patterns, computed numerically in accordance with Eq. (3.20), approach the far-field pattern, obtained from Eq. (3.19). The mesh size in all these calculations was 512 × 512. □

3.4. Diffraction in the Presence of a Lens

With reference to Fig. 3.5, let the complex amplitude distribution $A_0(x, y)$ represent the pattern of light at the entrance pupil of a lens, and assume that the lens has focal length f_0.† (Like all spatial dimensions, f_0 is normalized by the vacuum wavelength and is therefore dimensionless.) As discussed in section 2.4, the main effect of a lens on the incident distribution is the introduction of a curvature phase factor. All other effects (including aberrations) may be incorporated into the exit pupil distribution by modifying $A_0(x, y)$ and calling it $\hat{A}_0(x, y)$. Thus the distribution at the exit pupil (assumed to be at $z = 0$) is

$$A(x, y, z = 0) = \hat{A}_0(x, y) \exp\left[-\mathrm{i}2\pi \sqrt{f_0^2 + x^2 + y^2}\right] \tag{3.25}$$

This initial distribution will propagate beyond the lens according to the basic propagation formula of Eq. (3.18). The Fourier transform of the initial distribution that appears in Eq. (3.18) is written

$$\mathscr{F}\left\{A(x, y, z = 0)\right\}$$

$$= \iint_{-\infty}^{\infty} \hat{A}_0(x, y) \exp\left\{-\mathrm{i}2\pi\left(\sqrt{f_0^2 + x^2 + y^2} + \sigma_x x + \sigma_y y\right)\right\} \mathrm{d}x\,\mathrm{d}y$$

$$= f_0^2 \iint_{-\infty}^{\infty} \hat{A}_0(f_0 x', f_0 y') \exp\left\{-\mathrm{i}2\pi f_0\left(\sqrt{1 + x'^2 + y'^2} + \sigma_x x' + \sigma_y y'\right)\right\} \mathrm{d}x'\mathrm{d}y' \tag{3.26}$$

The last integral in Eq. (3.26) may be evaluated in the stationary-phase approximation, provided that f_0 is sufficiently large. Comparing Eq. (3.26) with Eq. (3.11), we find the following correspondences:

†For simplicity we have assumed that the entrance and exit pupils are at the principal planes. The positions of the principal planes along the optical axis are determined in the paraxial ray approximation. Although for non-paraxial rays these are no longer conjugate planes with unit magnification, they can still be useful as reference planes. The complex amplitude distribution $\hat{A}_0(x, y)$ at the exit pupil may be obtained by ray tracing from the distribution $A_0(x, y)$ at the entrance pupil. In general, A_0 and $\hat{A}_0$ will differ from each other, even in the absence of aberrations, except in the vicinity of the optical axis where the paraxial approximation is valid.

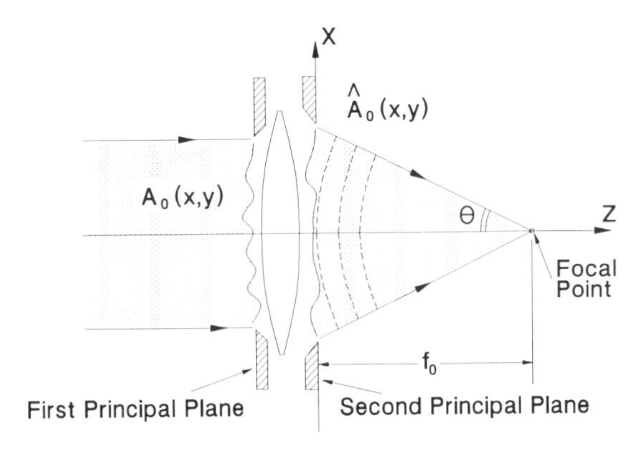

Figure 3.5. Diffraction from a lens with numerical aperture NA = sin Θ and focal length f_0. The incident distribution at the entrance pupil is $A_0(x, y)$. Aside from a curvature phase factor corresponding to a perfectly converging spherical wave-front, the distribution at the exit pupil has the (complex) amplitude $\hat{A}_0(x, y)$.

$$f(x', y') \rightarrow \hat{A}_0(f_0 x', f_0 y'); \qquad \eta \rightarrow -2\pi f_0 ; \qquad \beta \rightarrow \sigma_x ; \qquad \gamma \rightarrow \sigma_y$$

The approximation thus yields

$$\mathscr{F}\left\{A(x, y, z = 0)\right\} \simeq \mathrm{i} f_0 \exp\left[-\mathrm{i}2\pi f_0 \sqrt{1 - \sigma_x^2 - \sigma_y^2}\right] \hat{A}_{0c}(-\sigma_x, -\sigma_y) \tag{3.27}$$

where $\hat{A}_{0c}(\sigma_x, \sigma_y)$ is the compressed version of $\hat{A}_0(f_0 x, f_0 y)$, namely,

$$\hat{A}_{0c}(\sigma_x, \sigma_y) = \frac{\hat{A}_0\left[\dfrac{f_0\,\sigma_x}{\sqrt{1 - \sigma_x^2 - \sigma_y^2}}, \dfrac{f_0 \sigma_y}{\sqrt{1 - \sigma_x^2 - \sigma_y^2}}\right]}{1 - \sigma_x^2 - \sigma_y^2} \tag{3.28}$$

Note that the compression is not applied to $\hat{A}_0(x, y)$ itself, but to its scaled version, $\hat{A}_0(f_0 x, f_0 y)$. The exit pupil in Fig. 3.5 has radius $r = f_0 \tan \Theta$, where Θ is the half-angle subtended by the cone of light at focus. The scaled function $\hat{A}_0(f_0 x, f_0 y)$ is therefore confined to a disk of radius $r = \tan \Theta$, while the domain of the compressed function $\hat{A}_{0c}(\sigma_x, \sigma_y)$ is a disk of radius $r = \sin \Theta$ in the $\sigma_x \sigma_y$-plane. Since in the present case the numerical aperture of the lens, NA, equals sin Θ, one might say that the domain of $\hat{A}_{0c}(\sigma_x, \sigma_y)$ is a disk of radius NA.

Digression. The compression of $\hat{A}_0(f_0x, f_0y)$ described by Eq. (3.28) is simplified if the lens happens to be aplanatic. An aplanatic lens satisfies Abbe's sine condition which, in the absence of aberrations, may be stated as follows. The scaled distribution at the exit pupil, $\hat{A}_0(f_0x, f_0y)$, is the stretched version of the scaled distribution at the entrance pupil, $A_0(f_0x, f_0y)$. Thus for an aplanatic lens one may avoid the compression operation at the exit pupil by using the distribution at the entrance pupil instead. For an aberration-free aplanatic lens, Eq. (3.28) becomes

$$\hat{A}_{0c}(\sigma_x, \sigma_y) = A_0(f_0\sigma_x, f_0\sigma_y) \tag{3.28'}$$

The entrance pupil of an aplanatic lens has aperture radius $r = f_0 \sin \Theta$, causing the domain of the scaled function $A_0(f_0\sigma_x, f_0\sigma_y)$ in the $\sigma_x \sigma_y$-plane to be a disk of radius NA. □

The scaling of the domain of $\hat{A}_0(x, y)$ by f_0, when combined with the multiplication of the function by f_0 (as in Eq. (3.27)), leaves the integrated intensity of the function unchanged. Compression does not affect the integrated intensity either; thus the power content of $f_0\hat{A}_{0c}(\sigma_x, \sigma_y)$ is identical to that of the initial function $\hat{A}_0(x, y)$.

Continuing now with the problem of propagation beyond the lens, we substitute in Eq. (3.18) the expression for the Fourier transform of the initial distribution from Eq. (3.27) and obtain

$$A(x, y, z) \simeq \mathrm{i} f_0 \mathscr{F}\left\{\exp\left\{\mathrm{i}2\pi(z - f_0)\sqrt{1 - \sigma_x^2 - \sigma_y^2}\right\}\hat{A}_{0c}(\sigma_x, \sigma_y)\right\} \tag{3.29}$$

Equation (3.29) gives the distribution in the vicinity of the focal plane in terms of the distribution at the exit pupil, when the latter is properly scaled and compressed (see Eq. (3.28)). Note that x, y, z, and f_0 in these equations have all been normalized by λ. At $z = f_0$ the exponential phase factor in Eq. (3.29) disappears, making the focal plane distribution simply the Fourier transform of the scaled and compressed distribution at the exit pupil.

If $\hat{A}_0(x, y)$ happens to be a real function, Eq. (3.29) predicts symmetry with respect to the focal point in the following sense:

$$A(x, y, z = f_0 + \Delta z) = A^*(-x, -y, z = f_0 - \Delta z) \tag{3.30}$$

This symmetry with respect to the focal point is indeed observed in the majority of practical situations. There are, however, circumstances in which it breaks down. For example, for lenses with very small numerical aperture, it is known that the peak intensity occurs somewhat closer to the lens than the point of geometrical focus. The presence of asymmetry with respect to the focal point in a given situation is an indication of the inadequacy of the stationary-phase approximation for that situation.

Example 4. As an example of computation using Eq. (3.29), consider Fig. 3.6, which shows plots of the intensity distribution in the vicinity of the focal plane of an aberration-free, aplanatic lens with NA = 0.5 and $f_0 = 4000\lambda$. The assumed incident beam is a uniform plane wave propagating along Z, and the numerical computations are performed with a discrete version of Eq. (3.29) on a 512 × 512 mesh. Panels (a) to (d) of this figure correspond, respectively, to defocus distances of $z - f_0 = 0$, 3λ, 5λ, and 20λ. The symmetry property expressed by Eq. (3.30) ensures that the results apply to either side of the focal plane. □

3.4.1. Primary aberrations

Because of the various compromises and imperfections inherent in the design and manufacturing processes of a lens, and also because of mounting and alignment errors, wavelength shifts, etc., a lens is seldom free from aberrations. The amplitude distribution function $\hat{A}_0(x,y)$ at the exit pupil must therefore represent not only the input distribution $A_0(x,y)$ but also deviations from an ideally converging wave-front. Such deviations usually occur in the phase of the distribution function, and may be described as follows:

$$\hat{A}_{0c}(\sigma_x,\sigma_y) = \hat{A}_{0c}^{(\text{ideal})}(\sigma_x,\sigma_y)\exp\left[\,\mathrm{i}2\pi W(\sigma_x,\sigma_y)\,\right] \tag{3.31}$$

$\hat{A}_{0c}^{(\text{ideal})}$ is the scaled and compressed exit pupil distribution in the absence of aberrations. (For an aplanatic lens this is simply the scaled distribution at the entrance pupil.) $W(\sigma_x,\sigma_y)$ is the aberration function, representing deviations from the perfect wave-front in units of the wavelength λ. Note that the domain of $W(\sigma_x,\sigma_y)$ is the disk of radius NA in the $\sigma_x\sigma_y$-plane. The aberration function may be expanded in the complete set of orthogonal functions known as Zernike's circle polynomials. The individual terms of this series expansion correspond to different types of aberration, and the first few terms usually suffice to express all aberrations of practical interest. Zernike's polynomials are generally written in polar coordinates $(\rho,\phi) = (\sqrt{\sigma_x{}^2+\sigma_y{}^2},\ \tan^{-1}(\sigma_y/\sigma_x))$, where $0 \le \rho \le \mathrm{NA}$ and $0 \le \phi \le 2\pi$. Certain combinations of these polynomials correspond to the so-called primary or Seidel aberrations. Since in this book our attention will be confined to primary aberrations, we shall bypass Zernike's polynomials altogether and give the expressions for primary aberrations directly. These are written as follows:

$$\text{Spherical:}\qquad W(\rho,\phi) = C_1(\rho/\mathrm{NA})^4 \tag{3.32a}$$

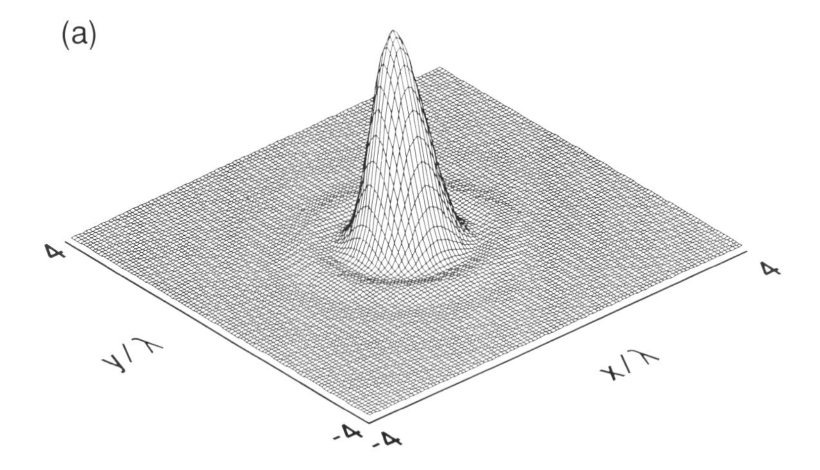
(a)
4
4
y/λ
x/λ
-4
-4

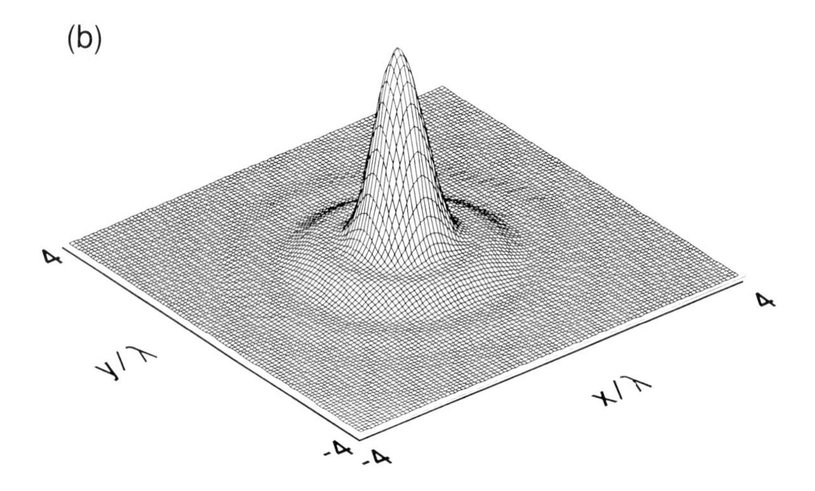
(b)
4
4
y/λ
x/λ
-4
-4

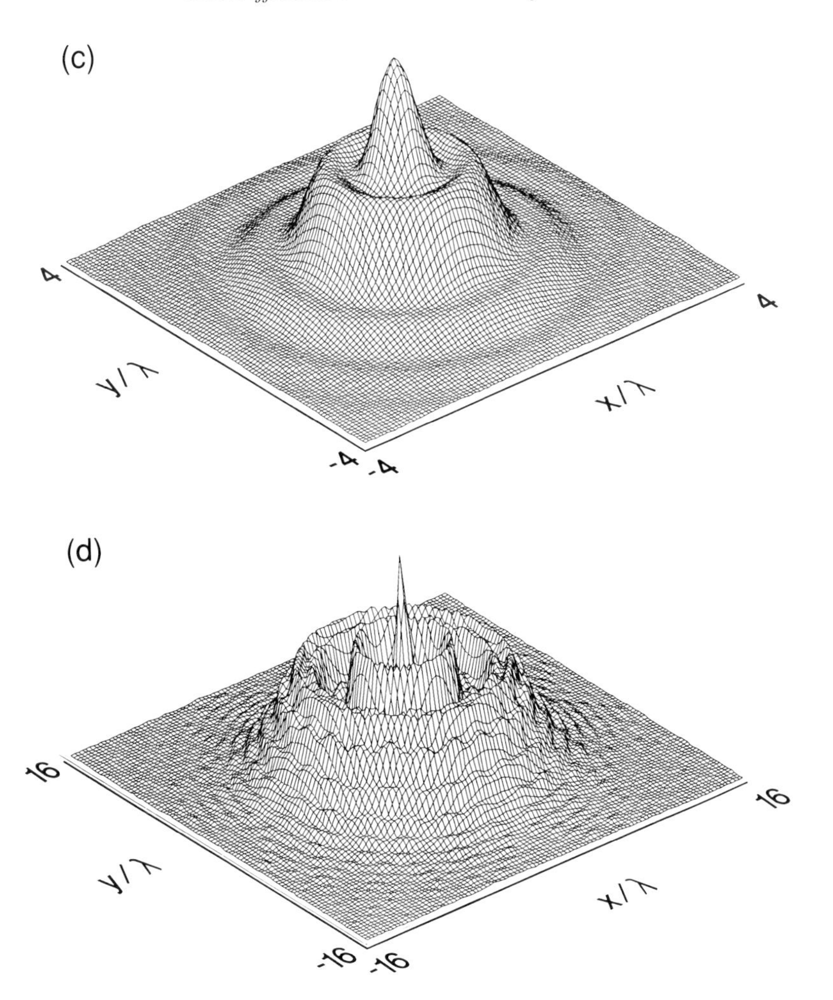

Figure 3.6. Patterns of intensity distribution in the vicinity of the focal plane of a lens with NA = 0.5 and $f_0 = 4000\lambda$. (a) shows the distribution in the focal plane, while (b), (c), and (d) correspond to defocus distances of 3, 5, and 20 wavelengths. The vertical scale varies among the plots and the actual peak intensities in (a)–(d) are in the ratio of 1 : 0.57 : 0.17 : 0.01 respectively.

Coma : $$W(\rho, \phi) = C_2 (\rho/\text{NA})^3 \cos\phi \tag{3.32b}$$

Astigmatism : $$W(\rho, \phi) = C_3 (\rho/\text{NA})^2 \cos 2\phi \tag{3.32c}$$

Curvature : $$W(\rho, \phi) = C_4 (\rho/\text{NA})^2 \tag{3.32d}$$

Distortion : $$W(\rho, \phi) = C_5 (\rho/\text{NA}) \cos\phi \tag{3.32e}$$

The coefficient C_i in each of the above expressions determines the severity

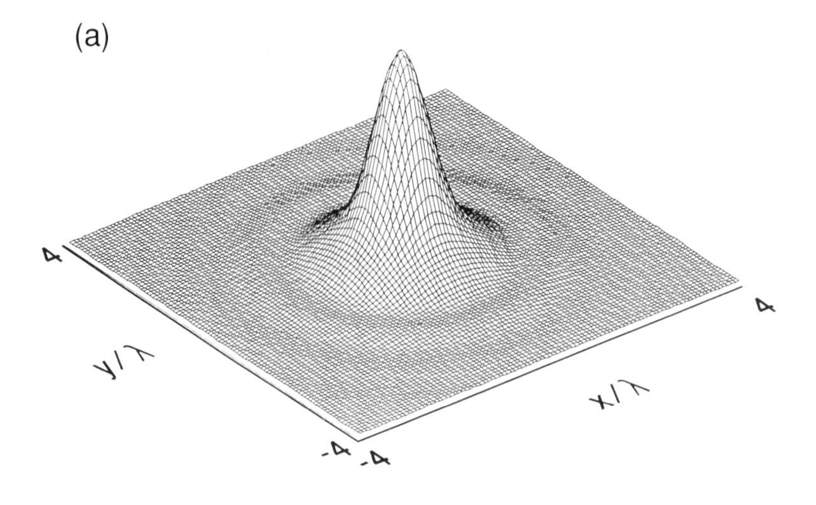
(a)
4
y/λ
x/λ
4
-4
-4

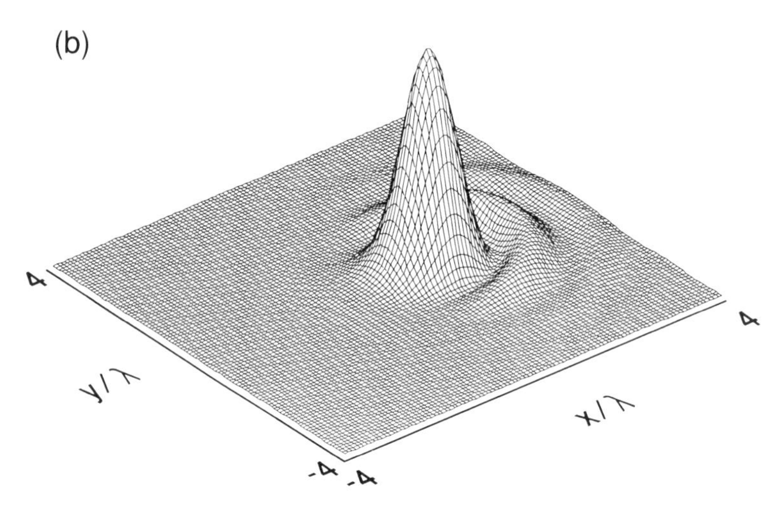
(b)
4
y/λ
x/λ
4
-4
-4

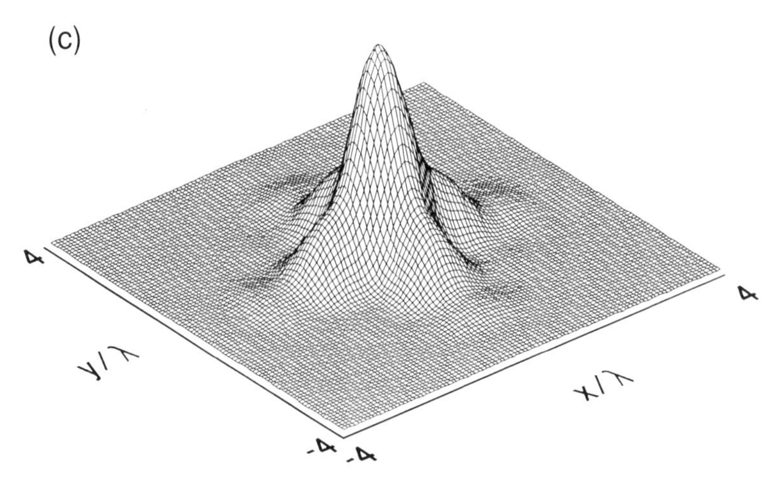
(c)
4
y/λ
x/λ
4
-4
-4

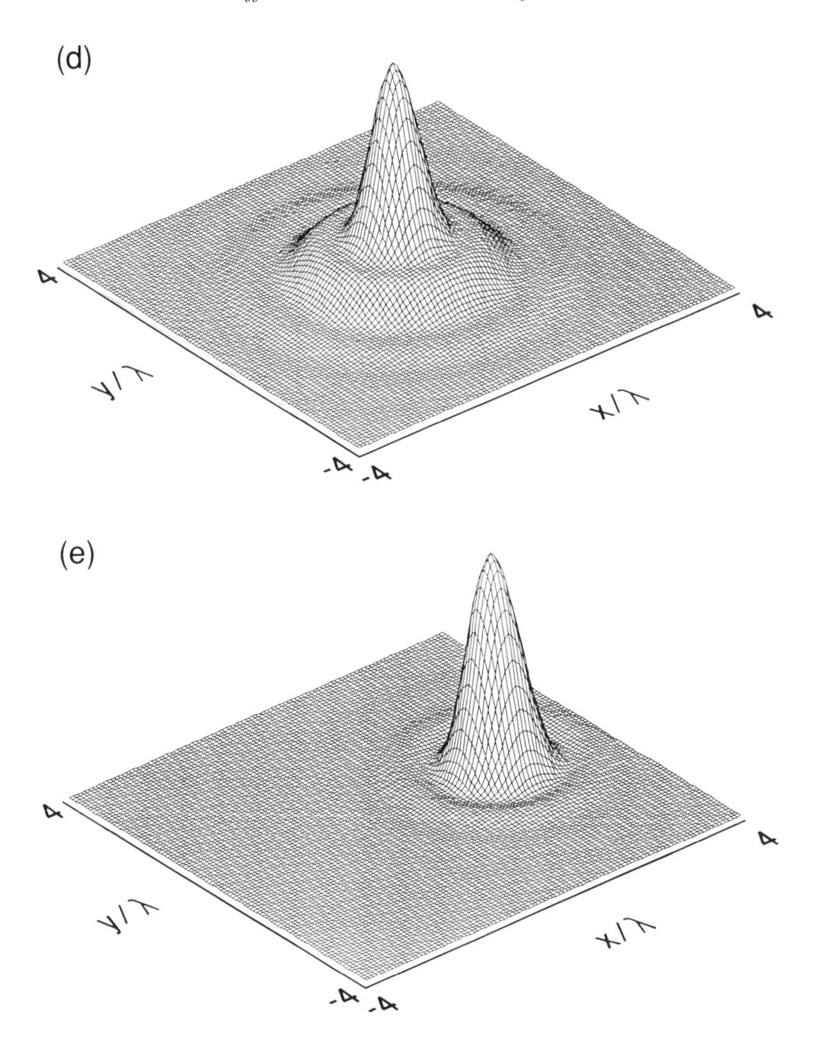

Figure 3.7. Effects of the various primary (Seidel) aberrations on the intensity distribution in the focal plane of a lens with NA = 0.5 and f_0 = 4000. (a) Spherical aberration; (b) coma; (c) astigmatism; (d) curvature; (e) distortion.

of the particular aberration; its numerical value (in units of λ) is the maximum deviation of the wave-front at the exit pupil from the ideally converging spherical shape.

Example 5. The effects of various primary aberrations on the focused spot are shown in Fig. 3.7. The lens for which these patterns are computed was assumed to be aplanatic with NA = 0.5 and $f_0 = 4000\lambda$. The incident beam was uniform with propagation direction along Z. As in previous examples, the discrete mesh was 512 × 512. (a) shows the intensity pattern in the focal plane of the lens in the presence of spherical aberration with

$C_1 = 0.5$. The ratio of the peak intensity to that in the absence of aberrations is known as the *Strehl ratio*, and is found in this case to be 0.39. (b) shows the effect of primary coma with $C_2 = 0.75$; the Strehl ratio in this case is 0.72. The effect of astigmatism on the focal plane intensity pattern is shown in (c); here $C_3 = 0.35$ and the Strehl ratio is found to be 0.44. (d) corresponds to a value of C_4 equal to half a wavelength of primary curvature, and its computed Strehl ratio is 0.4. To a good approximation, primary curvature is the same as defocus. Finally, (e) shows the effect of a value of C_5 equal to one wavelength of primary distortion on the focal plane distribution. Clearly this kind of aberration results in a simple shift of the spot in the focal plane, as should be expected from the linear phase shift introduced in the wave-front by the corresponding aberration function in Eq. (3.32e).

□

3.5. Vector Effects in Diffraction

Up to this point we have ignored the vector nature of electromagnetic radiation in neglecting to include any polarization effects. To be sure, such effects are usually small and, in many cases of practical interest, calculations based on the scalar theory give accurate predictions. However, when one works with an optical system which forces severe bending of the light rays (such as high-NA lenses), one expects the polarization effects to be significant. A simple treatment of polarization-related phenomena within the theory of diffraction will be given in this section. This will not be a rigorous treatment based on Maxwell's equations, rather it will be rooted in reasonable physical arguments based on the bending of plane waves by prisms. Our approach to vector diffraction is in keeping with the spirit of diffraction theory, which is not exact as far as Maxwell's equations are concerned, but incorporates intuitive ideas about the propagation of electromagnetic waves.

Consider a plane wave propagating along the unit vector $\boldsymbol{\sigma}_0 = (0, 0, 1)$ (i.e., the Z-axis), and let the beam be linearly polarized along X, as shown in Fig. 3.8. Let a prism with proper orientation be placed in the path of this beam, such that the emerging beam propagates in the direction specified by the unit vector $\boldsymbol{\sigma}_1 = (\sigma_x, \sigma_y, \sigma_z)$. Now, the incident polarization vector $\mathbf{E}_0 = (1, 0, 0)$ may be decomposed into two components; one, the so-called p-polarization, is in the plane of $\boldsymbol{\sigma}_0$ and $\boldsymbol{\sigma}_1$; the other, known as the s-polarization, is perpendicular to this plane. As the latter component (perpendicular to the $\sigma_0\sigma_1$-plane) emerges from the prism, it will have suffered no deviations in its direction. The p-component, on the other hand, will have reoriented itself in order to remain perpendicular to the emergent direction. If we further assume that there are no losses (due to surface reflections or otherwise) involved in this refraction process, we can use simple geometry to determine the emerging polarization. A similar calculation can be performed for an incident plane wave linearly polarized along the Y-axis. Details of these calculations are left to the reader (see

Table 3.1. Polarization $\mathbf{E}_1$ of a refracted beam. The original polarization is $\mathbf{E}_0$ and the refraction (from $\boldsymbol{\sigma}_0$ to $\boldsymbol{\sigma}_1$) is lossless

Incident polarization $\boldsymbol{\sigma}_0 = (0,0,1)$	Emergent polarization $\boldsymbol{\sigma}_1 = (\sigma_x, \sigma_y, \sigma_z)$
$\mathbf{E}_0 = (1,0,0)$	$\mathbf{E}_1 = \left(1 - \frac{\sigma_x^2}{1+\sigma_z}, \frac{-\sigma_x\sigma_y}{1+\sigma_z}, -\sigma_x\right)$
$\mathbf{E}_0 = (0,1,0)$	$\mathbf{E}_1 = \left(\frac{-\sigma_x\sigma_y}{1+\sigma_z}, 1 - \frac{\sigma_y^2}{1+\sigma_z}, -\sigma_y\right)$

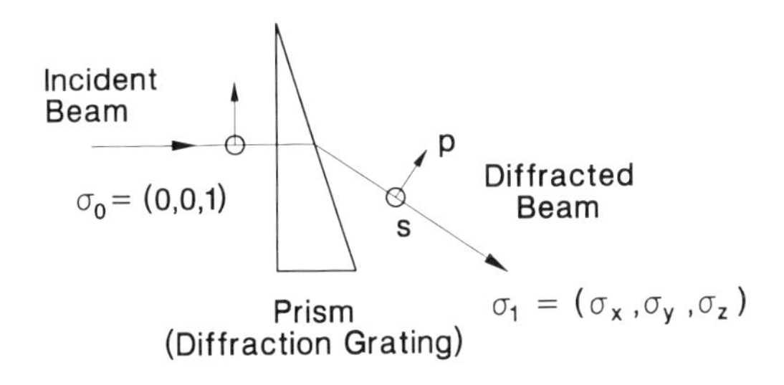

Figure 3.8. Lossless refraction of polarized plane wave by a prism. The original direction of propagation is $\boldsymbol{\sigma}_0 = (0,0,1)$ and the corresponding polarization vector is $\mathbf{E}_0$. After refraction the beam assumes a new direction $\boldsymbol{\sigma}_1 = (\sigma_x, \sigma_y, \sigma_z)$, and its new polarization state becomes $\mathbf{E}_1$.

Problem 3.9), but the final results are listed in Table 3.1. Notice that the reorientation of the polarization vector described in Table 3.1, while a consequence of the refraction of the propagation direction, is independent of the particular mechanism responsible for the refraction. Given an initial direction $\boldsymbol{\sigma}_0$ and a direction for the emerging beam $\boldsymbol{\sigma}_1$, one can use Table 3.1 to identify the emerging components of polarization for an arbitrary state of incident polarization.

Another consequence of plane-wave refraction is a change in beam amplitude by virtue of the fact that the beam cross-section changes upon refraction. In Fig. 3.8 the cross-sectional area of the beam propagating along $\boldsymbol{\sigma}_1$ is reduced by a factor σ_z and, as a result, its amplitude must have increased by $1/\sqrt{\sigma_z}$ in order to preserve its power. Thus the correct amplitude for the emergent beam is not $\mathbf{E}_1$ as indicated in Table 3.1, but rather $\mathbf{E}_1/\sqrt{\sigma_z}$.

Going back to Eq. (3.18) we observe that the amplitude distribution at $z = 0$ is treated as a superposition of plane waves traveling in various directions. If the incident beam happens to propagate along Z with certain (known) polarization components along X and Y, then the polarization state of a plane wave diffracted along $(\sigma_x, \sigma_y, \sigma_z)$ will be uniquely specified with the aid of Table 3.1. One can incorporate these results into Eq. (3.18) and obtain the following formula for vector diffraction:

$$\begin{pmatrix} A_x(x,y,z) \\ A_y(x,y,z) \\ A_z(x,y,z) \end{pmatrix} = \iint_{\text{unit disk}} \frac{1}{\sqrt{\sigma_z}} \begin{pmatrix} 1 - \dfrac{\sigma_x^2}{1+\sigma_z} & -\dfrac{\sigma_x \sigma_y}{1+\sigma_z} \\ -\dfrac{\sigma_x \sigma_y}{1+\sigma_z} & 1 - \dfrac{\sigma_y^2}{1+\sigma_z} \\ -\sigma_x & -\sigma_y \end{pmatrix}$$

$$\times \begin{pmatrix} \mathscr{F}\left\{A_x(x,y,z=0)\right\} \\ \mathscr{F}\left\{A_y(x,y,z=0)\right\} \end{pmatrix} \exp\left[\mathrm{i}2\pi(x\sigma_x + y\sigma_y + z\sigma_z)\right] \mathrm{d}\sigma_x \, \mathrm{d}\sigma_y \qquad (3.33)$$

In the above equation, A_x, A_y, and A_z are the components of polarization along the X-, Y-, and Z-axes respectively. The incident beam, assumed to be well collimated and propagating along Z, contains only the X- and Y-components of polarization; that is why $A_x(x,y,z=0)$ and $A_y(x,y,z=0)$ are the only components that appear on the right-hand side of the equation. σ_z, of course, is related to σ_x and σ_y through the relation $\sigma_z = \sqrt{1 - \sigma_x^2 - \sigma_y^2}$. The coordinates x, y, and z are normalized by λ.

Equation (3.33) is the vector version of the scalar Eq. (3.18). The latter equation was the basis of the discussions in the preceding sections that produced the far-field diffraction equation, Eq. (3.19), the near-field diffraction equation, Eq. (3.20), and the equation of diffraction in the presence of a lens, Eq. (3.29). Similar results can be derived from Eq. (3.33) by following precisely the same lines of reasoning as before. These derivations are straightforward and will not be given here.

Example 6. Numerical computation results incorporating the above formulation of vector diffraction are shown in Fig. 3.9. Again we have considered an aplanatic lens with NA = 0.5 and $f_0 = 4000\lambda$, and have assumed an incident plane wave propagating along Z and linearly polarized in the X-direction. The incident power (integrated intensity) in the aperture of the lens was set to unity. The intensity plots for the X-, Y-, and Z-components of polarization at the focal plane of the lens are shown in

Figs. 3.9(a), (b), and (c) respectively. ((a′) is the same as (a) but on a logarithmic scale.) Note that the vertical scale, which has been adjusted for the best display, varies significantly from one plot to another. The integrated intensity (area under the curve) for the X-component of polarization in (a) is 0.9361. The same quantity for the Y-component in (b) is 0.0007, and for the Z-component in (c) is 0.0627. Note that the Y-component is concentrated equally in the four quadrants of the XY-plane, whereas the Z-component is mainly in two lobes, split along the Y-axis. These observations are also consistent with geometric-optical arguments based on ray bending. □

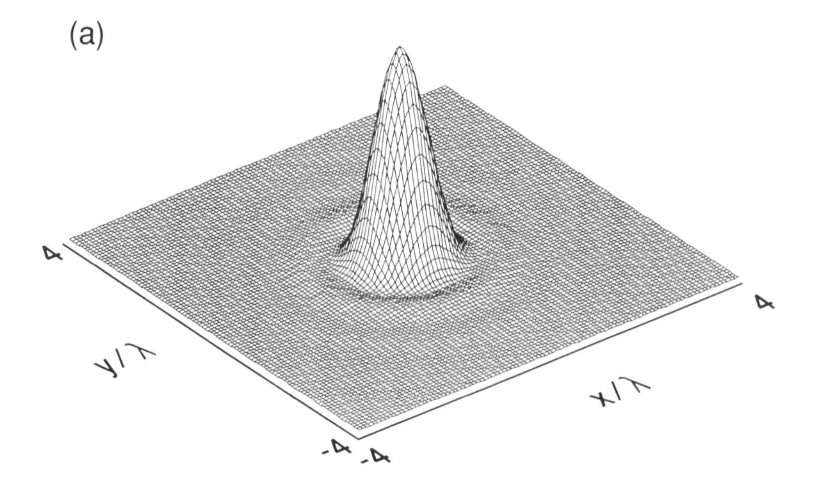
(a)
4
y/λ
-4
-4
x/λ
4

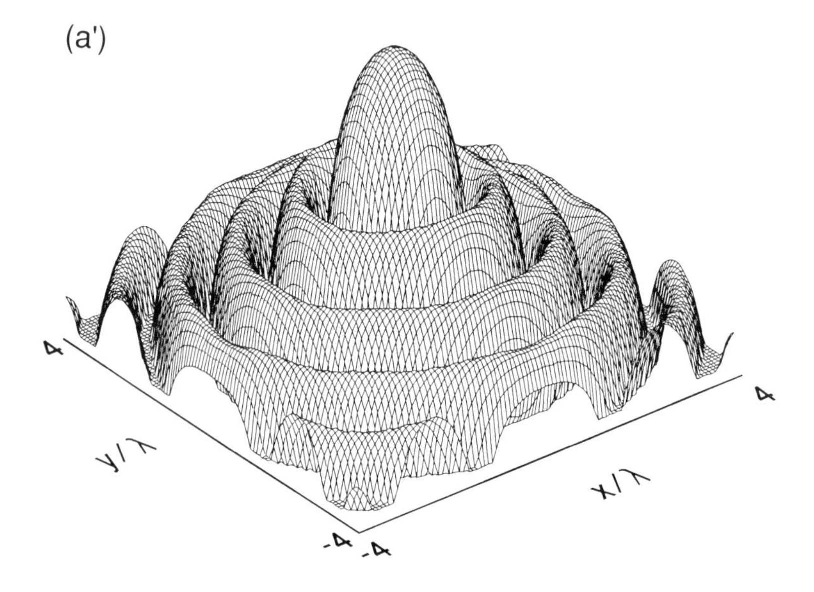
(a')
4
y/λ
-4
-4
x/λ
4

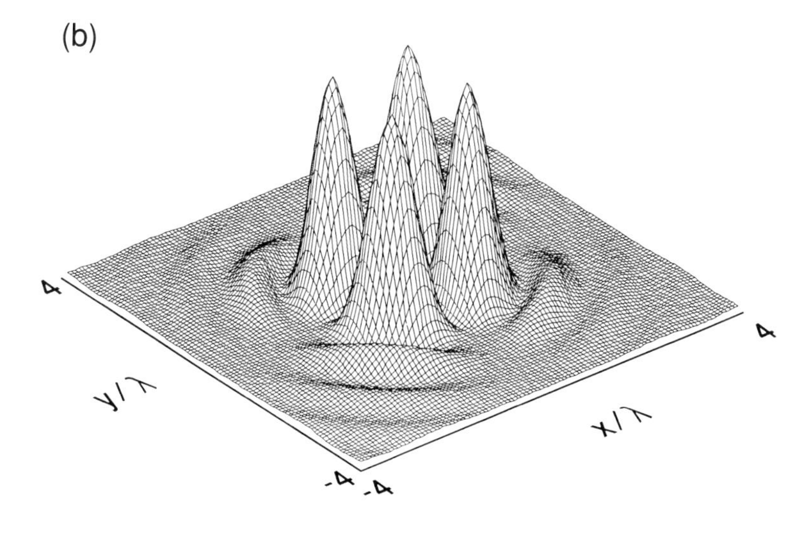

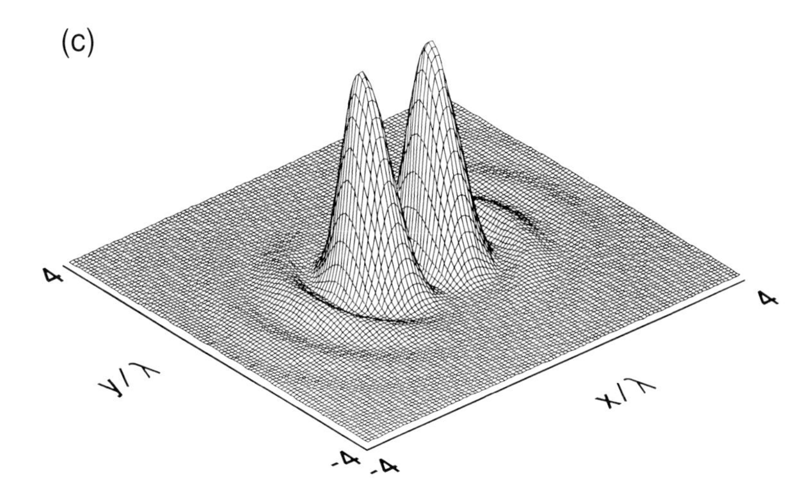

Figure 3.9. Intensity distribution for the three components of polarization in the focal plane of a lens with NA = 0.5 and $f_0 = 4000$. The incident beam has been assumed uniform and linearly polarized along the X-axis. The vertical scale varies among the plots and the actual peak intensities in (a)–(c) are in the ratio of 10000 : 2.3 : 341 respectively. (a′) is the same as (a) but plotted on a logarithmic scale.

Problems

(3.1) Apply the stationary-phase technique to approximate the following one-dimensional integral in the limit of large η:

$$I = \int_{x_1}^{x_2} f(x) \exp\left[\mathrm{i}\eta\, g(x)\right] \mathrm{d}x$$

Discuss the special case when a stationary point coincides with either x_1 or x_2.

(3.2) Verify Eq. (3.5) by carrying out the two-dimensional integral in Eq. (3.4).

(3.3) Show that the *compression* operation, defined by Eq. (3.12) for an arbitrary function $f(x, y)$, preserves the integrated intensity of the function.

(3.4) Let an initial amplitude distribution $A(x, y)$, centered at the origin of the XY-plane, be translated in the same plane to a new center (x_0, y_0) such that the new distribution is described by $A(x - x_0,\ y - y_0)$. On physical grounds one expects the far field to be translated by an equivalent shift to the new center. However, Eq. (3.19) indicates that the far field is essentially the Fourier transform of $A(x, y)$, in which case, rather than being shifted, the far field will be multiplied by a phase factor according to the shift theorem of the theory of Fourier transforms, namely,

$$\mathscr{F}\left\{A(x - x_0,\ y - y_0)\right\} = \mathscr{F}\left\{A(x, y)\right\} \exp\left[-\mathrm{i}2\pi\,(x_0\sigma_x + y_0\sigma_y)\right]$$

How does one resolve this discrepancy?

(3.5) Compare the far-field pattern of a Gaussian beam, as obtained from Eq. (3.19), with the pattern predicted by the laws of Gaussian beam propagation described in Chapter 2.

(3.6) A plane monochromatic beam (vacuum wavelength = λ_0) is normally incident on a transmission diffraction grating, as shown in Fig. 3.10. The grating is made from glass ($n = 1.5$) and its orientation is such that its grooves are parallel to the Y-axis. The depth of the grooves is d, their period is Δ (split equally between groove and land), and the position of the grating along the X-axis is identified by the location of the edge of one of the grooves at $X = x_0$. The incident beam is confined by a slit aperture of diameter D, centered at $X = 0$, as shown in the figure. (d, Δ, x_0, and D have all been normalized by λ_0.)

(a) Assuming an infinite aperture (i.e. $D = \infty$) find the diffraction pattern in the far field. In particular, specify the positions and the complex amplitudes of the first few orders of diffraction.

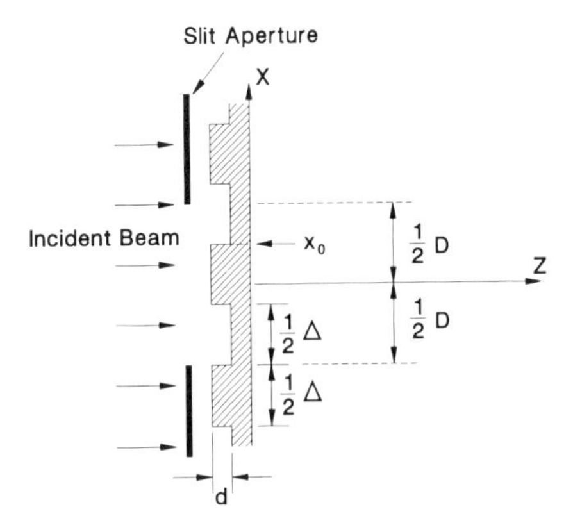

Figure 3.10. Transmission grating with rectangular grooves. The incident beam is confined to a slit aperture with diameter D.

(b) What is the effect of the aperture's finite diameter D on the far-field pattern? Give a graphical sketch of this pattern, taking into account the first few orders of diffraction.

(c) Let $x_0 = 0$ and $D = \Delta$. Analyze the symmetry properties of the far-field pattern in the two cases $d = 1$ and $d = 0.5$.

(3.7) When the exit pupil distribution $\hat{A}_0(x,y)$ is a real function of x and y, Eq. (3.29) predicts one kind of symmetry for the light amplitude distribution around the focal point (see Eq. (3.30)). It is possible, however, to have other kinds of symmetry (with respect to the focal point) depending on the properties of the distribution function at the exit pupil. Identify some of these symmetries.

(3.8) A cylindrical lens has a circular aperture of radius R_a, and its focal length is f_0. With reference to the geometry of Fig. 3.5, let the cylinder axis be parallel to the Y-axis. The exit pupil distribution function for this lens may be written as follows:

$$A(x,y,z=0) = \hat{A}_0(x,y) \exp\left(-i2\pi\sqrt{f_0^2 + x^2}\right)$$

Using the method of stationary-phase approximation, determine the distribution of light in the neighborhood of the focus.

(3.9) As shown in Fig. 3.8, the polarization vector $\mathbf{E}_0$ of a linearly polarized plane wave propagating along $\boldsymbol{\sigma}_0 = (0,0,1)$ will become $\mathbf{E}_1$ when the beam is refracted in a new direction $\boldsymbol{\sigma}_1 = (\sigma_x, \sigma_y, \sigma_z)$. Assuming that the refraction process is lossless, determine $\mathbf{E}_1$ and thereby verify the expressions given in Table 3.1.

(3.10) A comparison of Eq. (3.29) with Eq. (3.31) indicates that defocus may be expressed as an aberration. By expanding the function $\sqrt{1-\sigma_x^2-\sigma_y^2}$ in a Taylor series, determine the various amounts of primary (Seidel) aberration associated with a given amount $z - f_0$ of defocus.

(3.11) Consider a circularly symmetric Gaussian beam having $1/e$ radius r_0 and zero curvature (i.e., $r_x = r_y = r_0$ and $R_x = R_y = \infty$). The beam goes through a lens of numerical aperture NA and focal length f. Assume that the clear aperture radius of the lens is $1.5 r_0$, so that approximately 99% of the beam gets through. Also assume that the parameters are chosen to yield a focused spot with its waist at the geometrical focus of the lens.

(a) Let the lens have C_4 waves of curvature aberration (see Eq. (3.32d)). What is the shift of the focused spot from the nominal focal plane?

(b) Let the lens have C_3 waves of astigmatism (see Eq. (3.32c)). What is the separation between the two astigmatic line foci?

4

Diffraction of Gaussian Beams from Sharp Edges

Introduction

In this chapter, as a first application of the mathematical results derived in Chapter 3, we describe the far-field pattern obtained when a Gaussian beam is reflected from (or transmitted through) a surface with a sharp discontinuity in its reflection (or transmission) function. A good example of situations in which such phenomena occur is the knife-edge method of focus-error detection in optical disk systems; here a knife-edge partially blocks the column of light, allowing a split detector in the far field to sense the sign of the beam's curvature. Another example is the diffraction of the focused spot from the sharp edge of a groove. The far-field pattern in this case is used to derive the track-error signal, which drives the actuator responsible for track-following. Readback of the embossed pattern of information on the disk surface (e.g., data marks on CD and CD-ROM, preformat marks on WORM and magneto-optical media) also involves diffraction from the edges of small bumps and/or pits.

This chapter begins with a general description of the problem and a derivation of all relevant formulas in section 4.1. In subsequent sections several specific cases of the general problem are treated; the emphasis will be on those instances where diffraction from a sharp edge finds application in optical-disk data storage systems.

4.1. Formulation of the Problem

Let a Gaussian beam be incident on a plate with transmission (or reflection) function

$$\tau(x, y) = f(x)\, g(y) \tag{4.1}$$

Since the Gaussian beam itself is also separable into a function of x and a function of y, the diffraction pattern can be analyzed for the X- and Y-directions separately. Therefore we confine this discussion to the one-dimensional problem, keeping in mind that the full two-dimensional pattern is simply the product of two independent patterns along X and Y.

With reference to Fig. 4.1, let the transmission (reflection) function along X, i.e., $f(x)$, be zero when $x < x_0$, and allow the function to have a constant complex value ζ_+ for $x \geq x_0$. Using the notation $U(x)$ for the unit

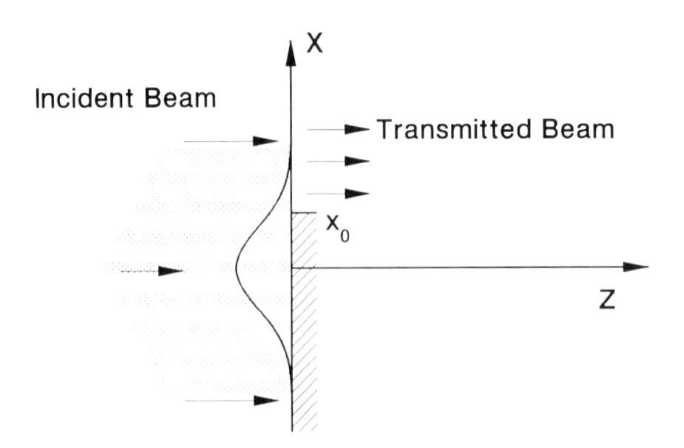

Figure 4.1. Gaussian beam incident on a plate with a step-like transmission function. The section of the plate corresponding to $x < x_0$ is opaque, while the region $x \geq x_0$ has a constant (complex) amplitude transmission ζ_+. The incident beam is centered at the origin ($x = 0$) and has 1/e radius r_x, and radius of curvature R_x.

step function, one can write the transmitted (reflected) amplitude just behind the plate as follows:

$$A_+(x) = \zeta_+ U(x - x_0) \exp(-\alpha x^2) \tag{4.2}$$

α is the complex parameter that identifies the Gaussian beam profile along X. All lengths are assumed to be normalized by the wavelength λ.

The far-field pattern is obtained from the Fourier transform of $A_+(x)$ in Eq. (4.2), namely

$$\mathscr{F}\left\{A_+(x)\right\} = \int_{-\infty}^{\infty} \zeta_+ U(x - x_0) \exp(-\alpha x^2 - i2\pi\sigma_x x)\, dx$$

$$= \zeta_+ \exp\left(-\frac{\pi^2\sigma_x^{\,2}}{\alpha}\right) \int_{x_0}^{\infty} \exp\left\{-\left(\sqrt{\alpha}\, x + i\frac{\pi\sigma_x}{\sqrt{\alpha}}\right)^2\right\} dx \tag{4.3}$$

Changing the variable from x to Z, where $Z = \sqrt{\alpha}\, x + i\pi\sigma_x/\sqrt{\alpha}$, converts the integral along X to one along the straight line L_0 in the complex plane of Z, as shown in Fig. 4.2. Thus

$$\mathscr{F}\left\{A_+(x)\right\} = \frac{\zeta_+}{\sqrt{\alpha}} \exp\left(-\frac{\pi^2\sigma_x^{\,2}}{\alpha}\right) \int_{L_0} \exp(-Z^2)\, dZ \tag{4.4}$$

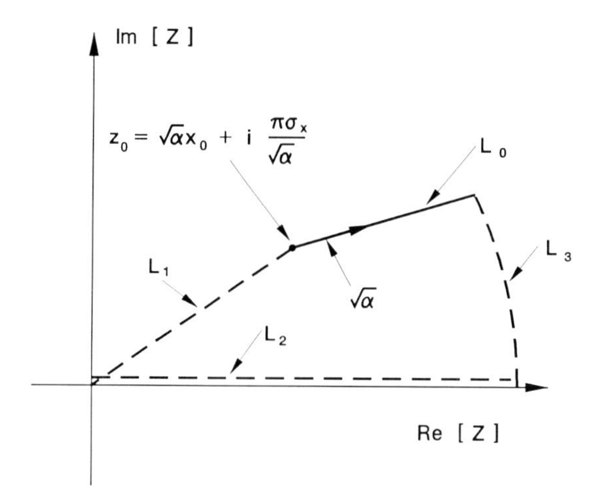

Figure 4.2. Integration path in the complex plane of Z corresponding to Eq. (4.4). At $x = x_0$ the starting point of the path L_0 is at Z_0. The straight line L_0 is along the direction of the complex number $\sqrt{\alpha}$ and extends from Z_0 to ∞. L_1 is a straight line between the origin and Z_0, while L_2 is the positive real axis. L_3 is at infinity, where the integrand is zero over its entire length.

The integral over L_0 can be written in terms of the integrals over L_1, L_2 and L_3, using the fact that the integrand has no poles within the closed contour. Now the contribution of the last leg, L_3, is zero because the integrand is zero over this path. L_2 contributes a finite amount given by

$$\int_{L_2} \exp(-Z^2)\, dZ = \int_0^\infty \exp(-x^2)\, dx = \frac{\sqrt{\pi}}{2} \tag{4.5}$$

As for the contribution of the path L_1, we define the function $F(Z)$ as

$$\boxed{F(Z) = \frac{2}{\sqrt{\pi}} \int_0^Z \exp(-Z'^2)\, dZ'} \tag{4.6}$$

The path of integration in Eq. (4.6) is the straight line from the origin to the point Z. The preceding results are now combined to yield

$$\mathscr{F}\left\{A_+(x)\right\} = \frac{\sqrt{\pi}\zeta_+}{2\sqrt{\alpha}} \exp\left(-\frac{\pi^2\sigma_x^{\ 2}}{\alpha}\right)\left\{1 - F\left(\sqrt{\alpha}\,x_0 + i\frac{\pi\sigma_x}{\sqrt{\alpha}}\right)\right\} \tag{4.7}$$

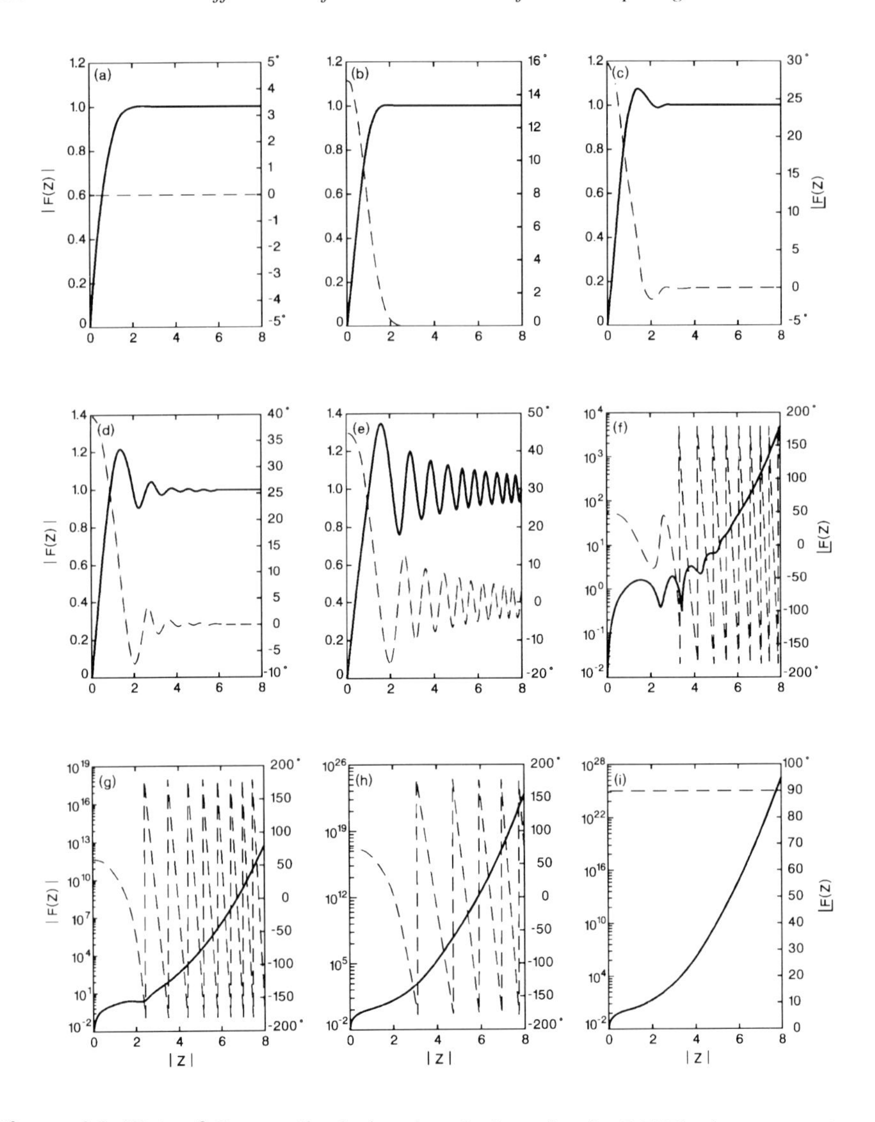

Figure 4.3. Plots of the amplitude (——) and phase (---) of $F(Z)$ along several rays in the complex plane of Z. Each ray is the locus of points $Z = |Z| \exp(i\phi)$ with ϕ = constant. (The ray with $\phi = 0$ is the positive real axis, while $\phi = 90°$ corresponds to the positive imaginary axis.) Frame (a) shows the magnitude and phase of $F(Z)$ versus $|Z|$ on a ray with $\phi = 0$. Frames (b)–(i) show the same function along rays with ϕ = 15°, 30°, 40°, 45°, 50°, 60°, 75°, and 90° respectively.

applied with the following parameters: $\zeta_+ = 1$ for complete transmission above the edge, $\zeta_- = 0$ for no transmission below the edge, and $\alpha = (1/r_x^{\,2} - i\pi/R_x)$ in the plane $Z = 0$. With x_0 representing the coordinate of the edge, the far-field pattern will be given by

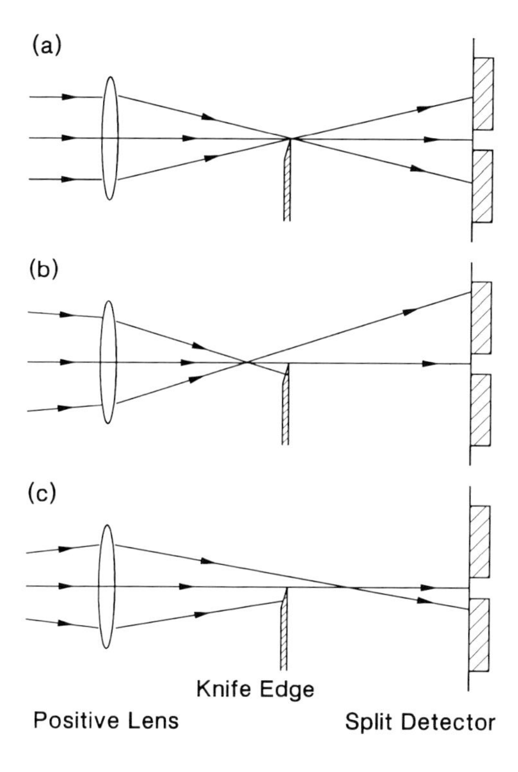

Figure 4.4. Geometric-optical description of the knife-edge method for detecting the sign of a beam's curvature. In (a) the incoming beam is collimated and both detectors receive equal amounts of light. In (b) the beam is converging (i.e., it has negative curvature) and the upper detector receives more light. In (c) the beam is diverging (i.e., it has positive curvature) and the lower detector receives more light.

$$\mathscr{F}\left\{A(x)\right\} = \frac{1}{2}\sqrt{\frac{\pi}{\alpha}}\,\exp\left(-\frac{\pi^2\sigma_x^{\,2}}{\alpha}\right)\left\{1 - F\left[\sqrt{\alpha}\,x_0 + i\frac{\pi\sigma_x}{\sqrt{\alpha}}\right]\right\} \qquad (4.15)$$

The general features of this pattern are readily understood if one sets $x_0 = 0$, i.e., one centers the beam on the edge, and uses the fact that $F(Z)$ is an odd function of Z. For instance, when the incident beam has zero curvature, $\alpha = 1/r_x^{\,2}$, which is real, and therefore $F(i\pi r_x\,\sigma_x)$ in Eq. (4.15) is purely imaginary. In this case the amplitude of the far-field pattern at σ_x is the conjugate of the amplitude at $-\sigma_x$ and thus the intensity pattern is an even function of σ_x. The two halves of the split detector in Fig. 4.4 then receive equal amounts of light. On the other hand, when the incident curvature is nonzero, $\sqrt{\alpha}$ becomes a complex number with nonzero real and imaginary parts, and $F(i\pi\sigma_x/\sqrt{\alpha})$ in Eq. (4.15) acquires a nonzero real part as well. This nonzero real part, whose sign changes with the sign of σ_x, is

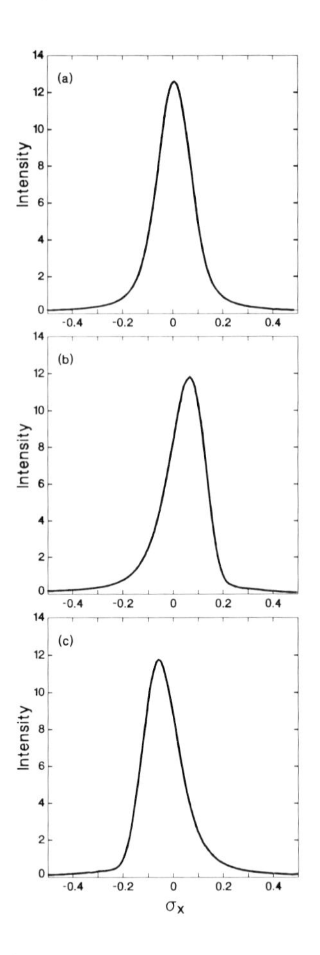

Figure 4.5. Far-field intensity pattern of a Gaussian beam diffracted from a knife-edge. The knife-edge is at the center of the beam, i.e., $x_0 = 0$, and the beam's $1/e$ radius is $r_x = 4$. The three curves correspond to different curvatures of the incident beam: (a) $R_x = \infty$, (b) $R_x = +50$, (c) $R_x = -50$.

ultimately responsible for the asymmetry of the far-field intensity distribution, which results in an imbalance of the detector signals.

More quantitative results can be obtained by numerical evaluation of Eq. (4.15). Figures 4.5 and 4.6 show several far-field intensity patterns, computed for different combinations of α and x_0. In Fig. 4.5 the edge of the knife is at $x_0 = 0$ and the incident beam has $1/e$ radius $r_x = 4$; the three curves in (a), (b), and (c) differ in the curvature of the incident beam, $C_x = 1/R_x$, which has been set to 0, +0.02, and -0.02, respectively. Note how the peak intensity shifts to the right or left, depending on the sign of the incident curvature. Figure 4.6 corresponds to a similar case, except for the position of the knife-edge, which is now at $x_0 = 1$. Since in this case more of the light has been blocked, the intensity levels are comparatively lower, but the general features of the patterns are similar.

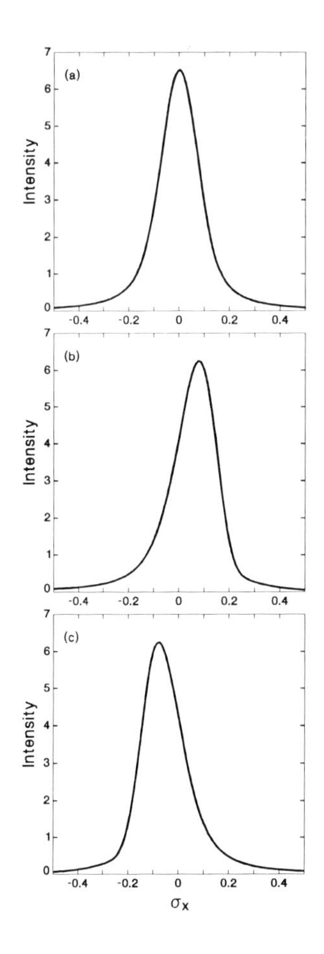

Figure 4.6. Far-field intensity pattern of a Gaussian beam diffracted from a knife-edge. The beam's 1/e radius is $r_x = 4$ and the knife-edge is at $x_0 = +1$. The radius of curvature of the incident beam, R_x, is ∞ in (a), +50 in (b), and −50 in (c).

4.3. Diffraction from 180° Phase-step

Assume that $\zeta_+ = 1$ and $\zeta_- = -1$. Then, according to Eq. (4.10)

$$\mathscr{F}\left\{A(x)\right\} = -\sqrt{\frac{\pi}{\alpha}}\exp\left(-\frac{\pi^2\sigma_x{}^2}{\alpha}\right)F\left(\sqrt{\alpha}\,x_0 + \mathrm{i}\,\frac{\pi\sigma_x}{\sqrt{\alpha}}\right) \tag{4.16}$$

This situation may arise in optical disk readout, when a bump or pit with sharp edges is encountered on the disk surface. Let us assume that the feature, i.e., the bump (pit) has a flat top (bottom) and a height (depth) which is a quarter of one wavelength. Then the reflected beam, which is twice delayed (advanced) by the feature, will experience a 180° phase-step as it crosses the edge.

To obtain a qualitative picture of the situation, let us specialize for the moment to the case where the incident beam has zero curvature. In this case $\alpha = 1/r_x^{\,2}$ and we have

$$\mathscr{F}\left\{A(x)\right\} = -\sqrt{\pi}\, r_x \exp\left(-\pi^2 r_x^{\,2} \sigma_x^{\,2}\right) F\left(\frac{x_0}{r_x} + i\pi r_x \sigma_x\right) \tag{4.17}$$

Equation (4.17) shows that a change of sign of σ_x will change only the phase of the function, keeping the intensity intact; property (ii) of $F(Z)$, given by Eq. (4.12), should help clarify this point. Thus the far-field intensity pattern remains symmetric (i.e., an even function of σ_x) for all values of x_0.

Two examples of diffraction from 180° phase-steps are computed from Eq. (4.16) and shown in Fig. 4.7. In both cases the incident beam has $r_x = 3$ and $C_x = 0$. In (a) the discontinuity is at $x_0 = 0$, while in (b) it has been shifted to $x_0 = -1$. Note that the far-field pattern is an even function of σ_x, irrespective of the position of the beam relative to the edge. Also notice that, as the beam moves across the phase step, its far-field pattern develops a dip near the center. Thus, a small detector in the central region of the far field can detect the passage of a 180° phase-step.

4.4. Diffraction from 90° Phase-step

A commonly used technique for obtaining the track-error signal from grooved optical disks is the push–pull method. In this approach the focused beam, reflected from the pregrooved surface of the disk, produces an asymmetric distribution in the far field that contains information about the position of the focused spot relative to the edge of the track. Typically, the groove's optical depth is one-eighth of the wavelength ($\lambda/8$), which causes a 90° phase difference between the two parts of the beam reflected from the land and the groove. The asymmetry of the far-field pattern enables a split detector to sense the position of the beam relative to the groove edge. (This signal is subsequently fed back to the actuator that controls the radial position of the objective lens.) Let us, therefore, study Eq. (4.10) for the case of $\zeta_+ = \exp(i\pi/4)$ and $\zeta_- = \exp(-i\pi/4)$, corresponding to a 90° phase-step at $x = x_0$. We find

$$\mathscr{F}\left\{A(x)\right\} = \sqrt{\frac{\pi}{2\alpha}} \exp\left(-\frac{\pi^2 \sigma_x^{\,2}}{\alpha}\right) \left\{1 - iF\left(\sqrt{\alpha}\, x_0 + i\frac{\pi\sigma_x}{\sqrt{\alpha}}\right)\right\} \tag{4.18}$$

For a qualitative discussion of the light distribution pattern described by

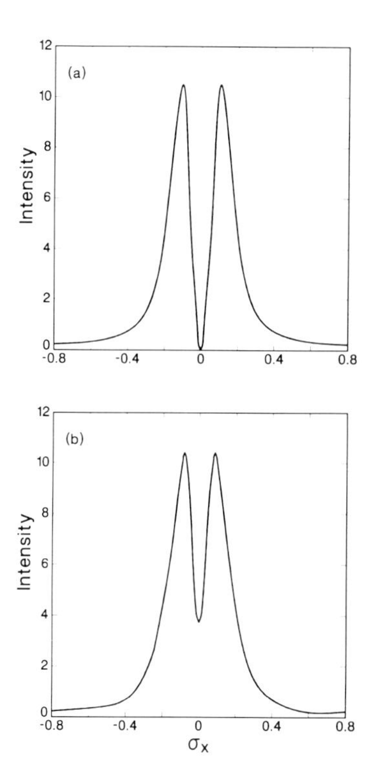

Figure 4.7. Far-field intensity pattern of a Gaussian beam diffracted from a 180° phase-step. The incident beam's parameters are $r_x = 3$ and $R_x = \infty$. In (a) the center of the beam coincides with the discontinuity, i.e., $x_0 = 0$, whereas in (b) the edge is at $x_0 = -1$.

Eq. (4.18) we confine our attention to the case where the incident beam has zero curvature and is centered on the groove edge, i.e., $\alpha = 1/r_x^{\,2}$ and $x_0 = 0$. Then

$$\mathscr{F}\left\{A(x)\right\} = \sqrt{\frac{\pi}{2}}\, r_x \exp\left[-\pi^2 r_x^{\,2} \sigma_x^{\,2}\right] \left[1 - iF(i\pi r_x \sigma_x)\right] \qquad (4.19)$$

Since $F(\cdot)$ is now purely imaginary (see Eq. (4.14)), a change of sign of σ_x produces a change in the far-field intensity pattern. Consequently, when a

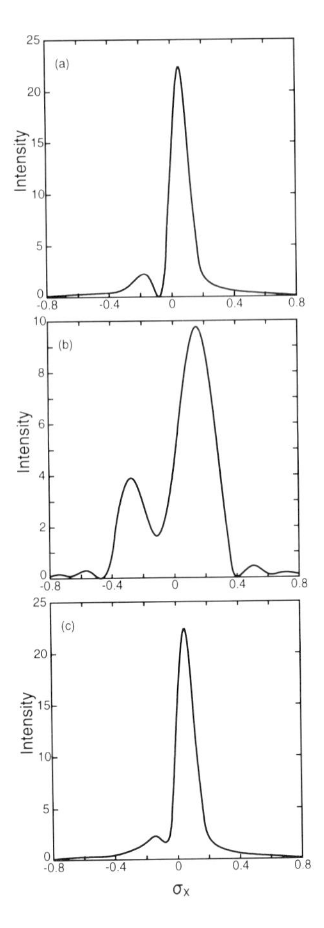

Figure 4.8. Far-field intensity pattern of a Gaussian beam diffracted from a 90° phase-step. The incident beam's 1/e radius is $r_x = 3$, and the parameters describing the phase object are $\zeta_+ = i$ and $\zeta_- = 1$. In (a) $x_0 = 0$ and $R_x = \infty$, in (b) $x_0 = 0$ and $R_x = +10$, whereas in (c) $x_0 = -1$ and $R_x = \infty$.

beam crosses a 90° phase-step, its far-field pattern becomes asymmetric. This asymmetry is employed in the push–pull method to determine the position of the beam relative to the groove edge. Figure 4.8 shows several intensity plots in the far field of a Gaussian beam transmitted through a 90° phase-step; these plots have been obtained numerically from Eq. (4.18).

4.5. Detecting Phase-steps by Spatial Matched Filtering

Consider a one-dimensional sequence of marks recorded on a flat surface, as shown in Fig. 4.9. The complex amplitude reflectivity of the surface in the absence of marks is unity. Where there is a mark, the reflection

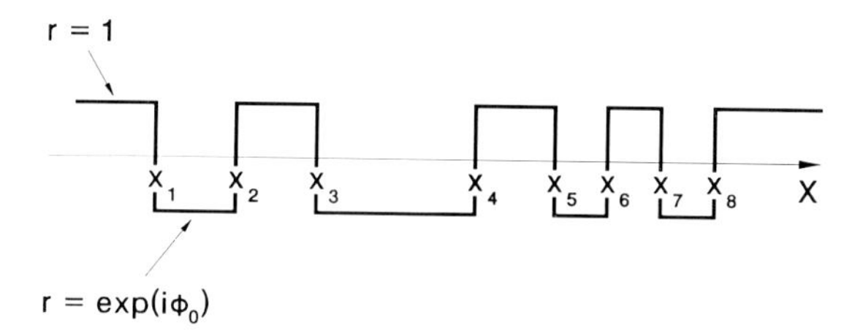

Figure 4.9. Amplitude reflectivity $r(x)$ from a one-dimensional sequence of pits on a flat surface. The transition points x_n are arbitrarily distributed along the X-axis.

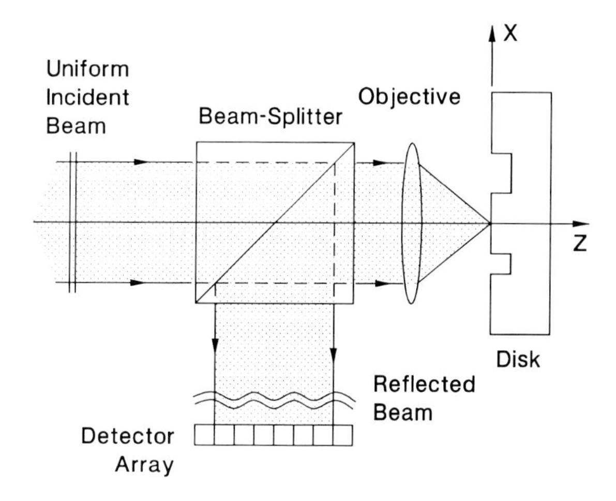

Figure 4.10. Schematic diagram of the readout system used in conjunction with the one-dimensional pattern of recorded marks in Fig. 4.9. The objective lens is cylindrical (cylinder axis = Y), with numerical aperture NA and focal length f. The detector array placed in the path of the reflected beam is a hypothetical detector of complex-amplitude; that is, it detects both amplitude and phase.

coefficient changes and becomes $r = \exp(i\phi_0)$, with ϕ_0 being the phase depth of the mark. As a function of the position x along the track, the reflection coefficient r may be written as

$$r(x) = 1 + \left[\exp(i\phi_0) - 1\right]\left[U(x - x_1) - U(x - x_2) + U(x - x_3) - U(x - x_4) + \cdots\right] \quad (4.20)$$

In this equation $U(x)$ is the unit step function. We assume that x is normalized by the wavelength λ of the light, and shall proceed to treat it as a dimensionless quantity. λ, therefore, does not appear in any of the following equations.

Let a lens with numerical aperture NA and focal length f be used to focus a uniform beam on the disk surface, as shown in Fig. 4.10. In this one-dimensional analysis the lens is cylindrical, with the cylinder axis parallel to the Y-axis. The incident amplitude $A_1(x)$ at the entrance pupil may be written as

$$A_1(x) = \frac{1}{\sqrt{2f\mathrm{NA}}} \mathrm{Rect}\left(\frac{x}{2f\mathrm{NA}}\right) \tag{4.21}$$

where $\mathrm{Rect}(x)$ is a rectangular function whose magnitude is unity when $|x| < 0.5$, and zero elsewhere. The function in Eq. (4.21) is normalized to have unit power in the aperture. With proper scaling, the Fourier transform of $A_1(x)$ yields the distribution incident on the disk. This distribution is then multiplied by the reflectivity $r(x)$ of the disk and propagated back to the lens. The reflected distribution at the exit pupil is thus the convolution between $A_1(x)$ in Eq. (4.21) and the Fourier transform of $r(x)$ in Eq. (4.20). The transform of $r(x)$ is readily found to be

$$\mathscr{F}\left\{r(x)\right\} = \delta(\sigma) - \frac{\mathrm{i}}{2\pi\sigma}\left[\exp(\mathrm{i}\phi_0) - 1\right]\left[\exp(-\mathrm{i}2\pi x_1\sigma) - \exp(-\mathrm{i}2\pi x_2\sigma) + \exp(-\mathrm{i}2\pi x_3\sigma) - \cdots\right] \tag{4.22}$$

Here σ is the variable of the Fourier transform, and $\delta(\sigma)$ is Dirac's delta function. The right-hand side of Eq. (4.22) may be thought of as the angular spectrum of the reflected light, with σ playing the role of $\sin\theta$. At the aperture of the lens, therefore, σ must be replaced in Eq. (4.22) with x/f in order to yield a spatial distribution. Alternatively, one can scale the function $A_1(x)$ in Eq. (4.21) by replacing x/f with σ, in order to convert to an angular spectrum. In either case, once the two functions have the same scale, they can be convoluted to yield the reflected distribution $A_2(x)$ at the exit pupil. We find

$$A_2(\sigma) \propto \mathrm{Rect}\left(\frac{\sigma}{2\mathrm{NA}}\right) * \delta(\sigma) + \frac{\mathrm{i}}{2\pi}\left[\exp(\mathrm{i}\phi_0) - 1\right] \mathrm{Rect}\left(\frac{\sigma}{2\mathrm{NA}}\right) * \left(-\frac{\exp(-\mathrm{i}2\pi x_1\sigma)}{\sigma} + \frac{\exp(-\mathrm{i}2\pi x_2\sigma)}{\sigma} - \cdots\right) \tag{4.23}$$

The symbol $*$ in the above equation stands for the convolution operation. Because the rays with $|\sigma| \geq \mathrm{NA}$ are blocked by the lens, we are interested only in the result of convolution for $|\sigma| < \mathrm{NA}$. Let us take advantage of the symmetry of the functions involved, and define $F_x(\sigma)$ as follows:

$$F_x(\sigma) = \frac{\mathrm{i}}{2\pi} \mathrm{Rect}\left(\frac{\sigma}{2\mathrm{NA}}\right) * \frac{\exp(-\mathrm{i}2\pi x\sigma)}{\sigma}; \qquad |\sigma| < \mathrm{NA}$$

Subsequently,

$$F_x(\sigma) = \left\{ \frac{1}{2\pi} \int_0^{NA+\sigma} \frac{\sin(2\pi x \sigma')}{\sigma'} \, d\sigma' + \frac{1}{2\pi} \int_0^{NA-\sigma} \frac{\sin(2\pi x \sigma')}{\sigma'} \, d\sigma' \right\} + i \left\{ \frac{1}{2\pi} \int_{NA-\sigma}^{NA+\sigma} \frac{\cos(2\pi x \sigma')}{\sigma'} \, d\sigma' \right\} \tag{4.24}$$

Note that, as a function of σ, $F_x(\sigma)$ is Hermitian, that is, $F_x(-\sigma) = F_x^*(\sigma)$. On the other hand, when considered as a function of x, $F_x(\sigma)$ is anti-Hermitian, namely, $F_{-x}(\sigma) = -F_x^*(\sigma)$. Plots of $F_x(\sigma)$ for several values of x and for NA = 0.5 are shown in Fig. 4.11. With the aid of $F_x(\sigma)$, Eq. (4.23) may be rewritten as follows:

$$A_2(\sigma) = 1 + \left[\exp(i\phi_0) - 1\right] \sum_n (-1)^n F_{x_n}(\sigma) \,; \quad |\sigma| < \text{NA} \tag{4.25}$$

Let us assume that this complex amplitude of the reflected beam at the exit pupil can somehow be measured, and that $A_2(\sigma)$ is made fully available to the read system for further processing.† One can thus form the following function based on a priori knowledge of the phase depth ϕ_0, and on the measurements performed at the exit pupil:

$$\hat{A}_2(\sigma) = \frac{A_2(\sigma) - 1}{\exp(i\phi_0) - 1} = \sum_n (-1)^n F_{x_n}(\sigma) \tag{4.26}$$

$\hat{A}_2(\sigma)$ is the superposition of the responses of the system to individual mark edges. Since the disk moves at a constant velocity, each edge produces the same response at the detectors, albeit with a characteristic delay that depends on the position of the edge along the track. A filter, matched to the function $F_x(\sigma)$, can therefore pick out different edges as they move under the focused beam.

† Determination of the reflected complex amplitude at the exit pupil requires highly accurate interferometric measurements. Although this may seem prohibitively difficult in practice, there should be no objection to its feasibility in principle.

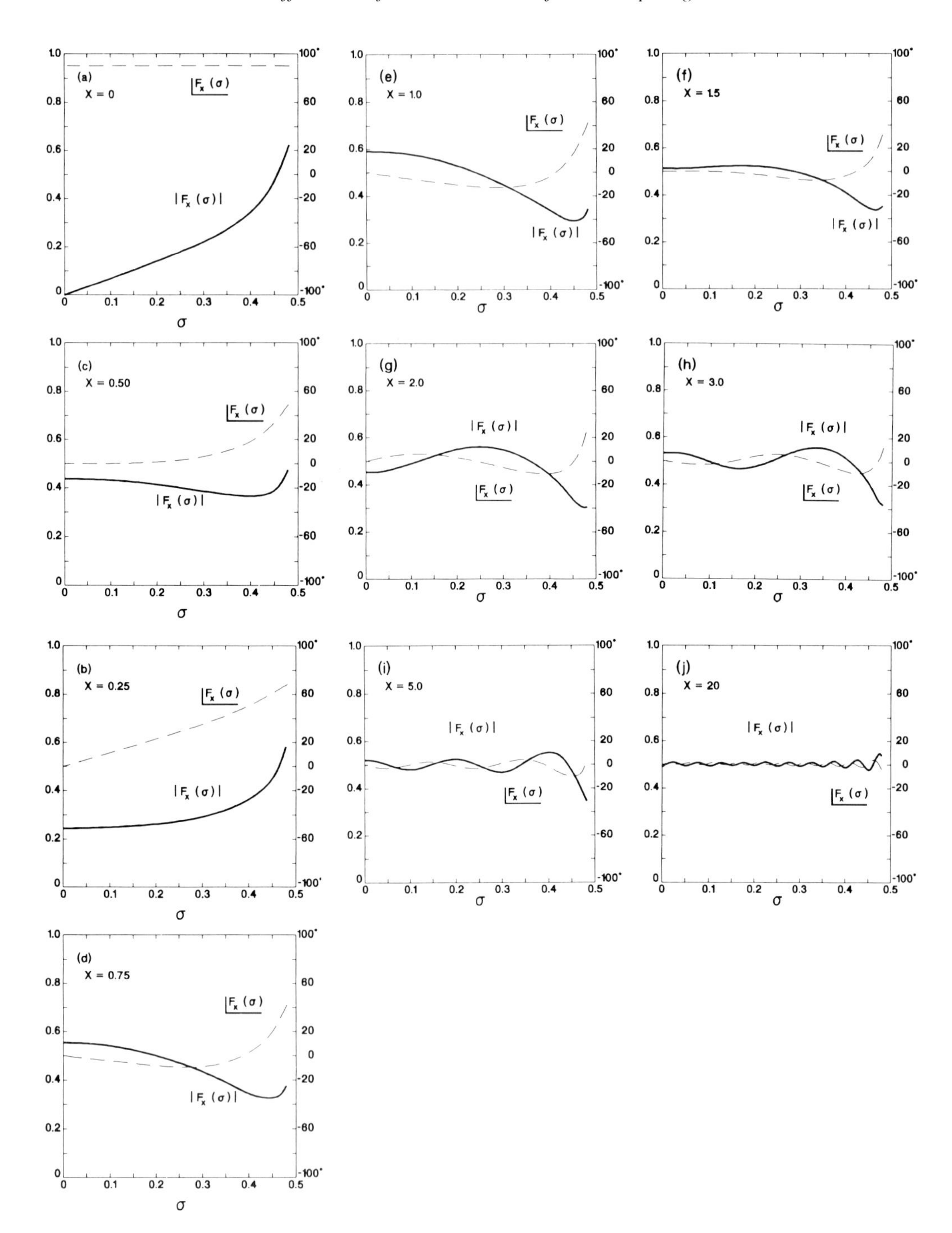

Figure 4.11. Plots of amplitude and phase for the complex function $F_x(\sigma)$ defined in Eq. (4.24). σ is the normalized coordinate of the aperture, while x is the distance between the pit edge and the center of the focused spot. For these plots NA is fixed at 0.5. (a) $x = 0$; (b) $x = 0.25$; (c) $x = 0.50$; (d) $x = 0.75$; (e) $x = 1.0$; (f) $x = 1.5$; (g) $x = 2.0$; (h) $x = 3.0$; (i) $x = 5.0$; (j) $x = 20$.

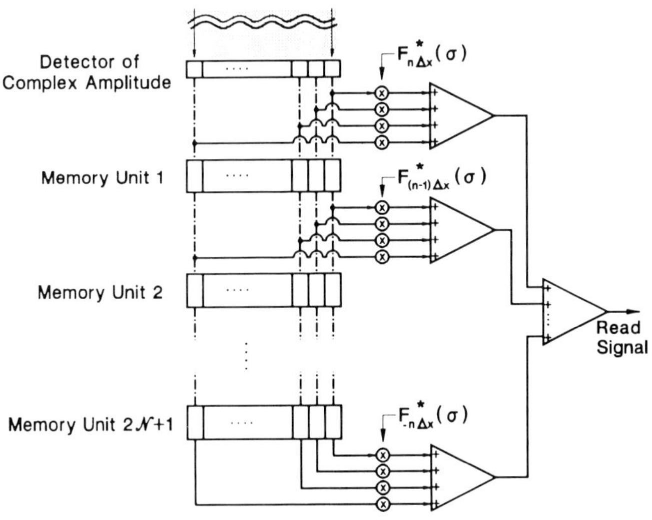

Figure 4.12. A possible implementation of the matched filter.

A possible implementation of this matched-filtering scheme is shown in Fig. 4.12. Here the complex function $\hat{A}_2(\sigma)$ is detected during a time interval Δt, then the function is shifted to memory unit 1, while detection continues. Another Δt seconds later, the contents of the unit are shifted to memory unit 2, and a newly detected function is transferred to memory unit 1. The process thus continues and snapshots of $\hat{A}_2(\sigma)$ at intervals Δt are successively stored in the memory units. Assuming that the track velocity is V, each one of these snapshots corresponds to a distance of the edge from the origin equal to an integer multiple of $\Delta x = V\Delta t$. If there are $2\mathcal{N}+1$ memory units, their contents at a given instant of time correspond to the following positions of the edge:

$$x = \mathcal{N}\Delta x, (\mathcal{N}-1)\Delta x, \ldots, \Delta x, 0, -\Delta x, \ldots, -(\mathcal{N}-1)\Delta x, -\mathcal{N}\Delta x$$

The matched-filter coefficients $F_x^*(\sigma)$ are then multiplied by $\hat{A}_2(\sigma)$ corresponding to edge position x. The final read signal is simply the total summed signal, as indicated in Fig. 4.12. If an edge actually passes under the beam, it will create a peak in the read signal; the sign of the peak will depend on whether the edge is up-going or down-going.

4.5.1. Numerical results

The strength of the read signal in the system of Fig. 4.12 may now be calculated for the following set of parameters.

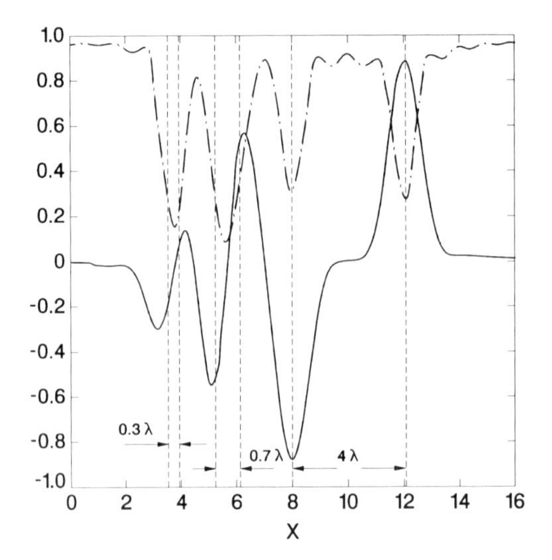

Figure 4.13. The computed read signal as a function of the coordinate along the track for three successive marks. The marks are 0.3λ, 0.7λ, and 4λ wide, having a phase depth of 180°. The solid curve is obtained by matched filtering, while the dashed curve is the total integrated intensity at the detector.

Numerical aperture:	NA = 0.5
Phase depth:	$\phi_0 = 180°$
Elements of detector array:	51
Memory units:	$2\mathcal{N} + 1 = 21$
Window of observation:	$x \in [-0.8\lambda, 0.8\lambda]$ in steps of $\Delta x = 0.08\lambda$

Note that with the above NA, the diameter of the spot (distance between the first zeros on each side of the focused beam center) is 2λ; therefore, the chosen window of observation is somewhat smaller than the Airy disk.†

We assumed the presence of three marks along a track, their widths being 0.3λ, 0.7λ, and 4λ, as shown at the bottom of Fig. 4.13. The calculated read signal for this pattern of marks is shown as the solid curve in Fig. 4.13. All the mark edges are resolved, but for smaller marks some peak-shift is observed. The dashed curve is the integrated intensity in the objective's aperture, which apparently also can resolve the two edges of the long mark, but fails to identify the edges of the short marks. By observing only the total reflected intensity (dashed curve), it is hard to decide whether the first two peaks are individual marks whose edges are not being resolved or whether they belong to opposite edges of a longer mark.

Figure 4.14 is similar to Fig. 4.13, except for the assumed depth of marks, ϕ_0, being 90°. The read signal obtained by matched-filtering is identical to that in Fig. 4.13, since ϕ_0 disappears at the outset when $\hat{A}_2(\sigma)$ is

† In this one-dimensional analysis a better term for the Airy disk may be the Airy strip.

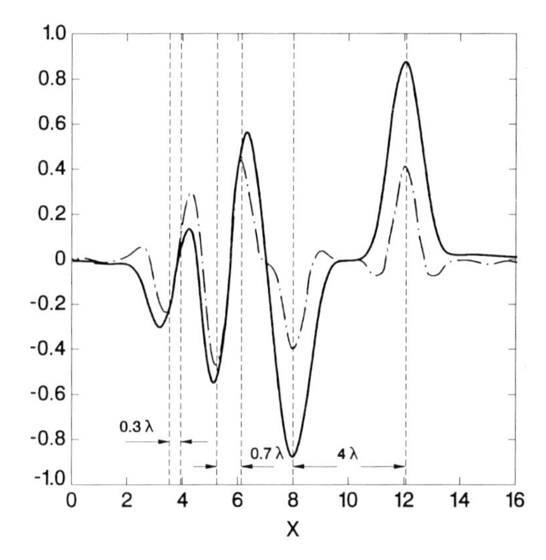

Figure 4.14. Computed read signal as function of coordinate along the track for three successive marks. The marks are 0.3λ, 0.7λ, and 4λ wide, having a phase depth of 90°. The solid curve is the result of matched filtering, while the dashed curve is the differential signal obtained from a simple split detector.

determined. The dashed curve in Fig. 4.14 is the differential signal at the exit pupil obtained with a split detector. It is seen that the differential signal is as good as that obtained by matched filtering; in fact, differential detection may be considered superior, since it has less peak-shift.

Based on these and numerous other calculations, one may conclude that spatial matched-filtering is superior to direct detection of the reflected intensity, which is the conventional method for marks with a phase depth of 180°. However, when $\phi_0 = 90°$, differential detection is as good as (or sometimes even better than) the filtering scheme.

Problems

(4.1) Numerical evaluation of $F(Z)$ as depicted in Fig. 4.3 seems to indicate that in the limit when $|Z| \to \infty$ the value of $F(Z)$ approaches unity, provided that $-45° \leq \phi \leq +45°$ (ϕ is the complex phase of Z). Prove this fact analytically.

(4.2) Analyze the diffraction of a Gaussian beam from a Fresnel biprism, as shown in Fig. 4.15. Investigate the effect of the curvature of the incident beam on the far-field pattern.

(4.3) A planar object with the following amplitude transmission function is placed in the XY-plane at $Z = 0$:

$$\tau(x, y) = \begin{cases} \zeta_1 & x < x_1 \\ \zeta_2 & x_1 \leq x \leq x_2 \\ \zeta_1 & x > x_2 \end{cases}$$

Let the Gaussian beam $A(x, y) = \exp(-\alpha x^2 - \beta y^2)$, propagating along the Z-axis, be incident on the above transparency.

(a) Obtain an expression for the far-field pattern.

(b) Analyze a few simple cases, using symmetry properties of $F(Z)$.

(c) Write a computer program to evaluate the expression obtained in (a). Investigate the effects of varying x_1, x_2, ζ_1, ζ_2, and α on the far-field pattern.

(4.4) The detection of magnetic domain walls is an important function of any magneto-optical readout system. Figure 4.16 shows a simplified model of one such detection scheme based on diffraction. It depicts a sharp transition in the magnetic state of a thin, perpendicularly magnetized film. A circularly polarized Gaussian beam is incident on the film, and the reflection (or transmission) coefficients are written as $\zeta_+ = \zeta \exp(i\phi_1)$ and $\zeta_- = \zeta \exp(i\phi_2)$; that is, the two coefficients have the same magnitude but different phases. Analyze the far-field diffraction pattern and suggest a method for detecting the presence of such domain walls.
Hint: You may assume that the incident beam has zero curvature (i.e., $R_{x0} = \infty$) and $1/e$ radius r_{x0}, and that it is centered on the X-axis (i.e., $x_0 = 0$).

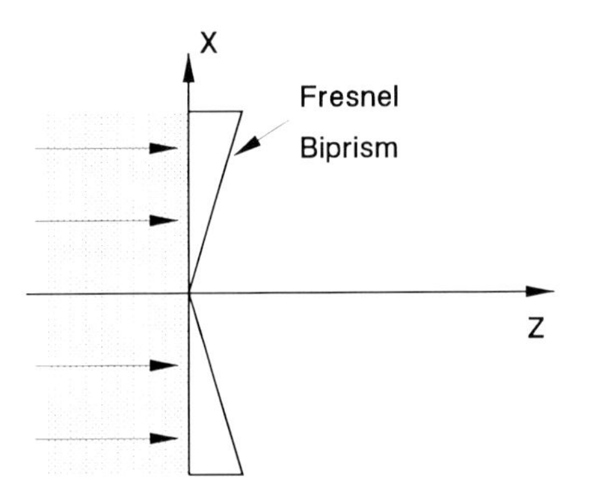

Figure 4.15.

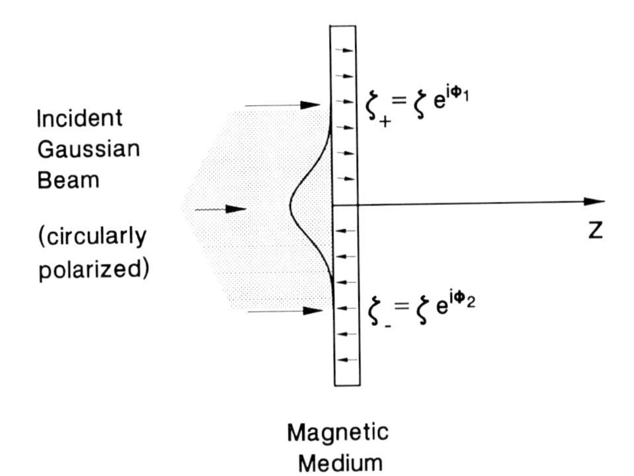

Figure 4.16.

5.1. Notation and Formalism

Let $\mathbf{u} = u_x \hat{\mathbf{x}} + u_y \hat{\mathbf{y}} + u_z \hat{\mathbf{z}}$ be a unit vector with real components in three-dimensional Cartesian space. We define the following row vector and matrix based on the components of $\mathbf{u}$:

$$\underline{u} = (u_x, \; u_y, \; u_z) \tag{5.1a}$$

$$\underline{\underline{u}} = \begin{bmatrix} 0 & -u_z & u_y \\ u_z & 0 & -u_x \\ -u_y & u_x & 0 \end{bmatrix} \tag{5.1b}$$

The dot product of two vectors $\mathbf{u} \cdot \mathbf{v}$ can be written as $\underline{u}\,\underline{v}^{\mathrm{T}}$ or $\underline{v}\,\underline{u}^{\mathrm{T}}$. Similarly, the cross product of these vectors, $\mathbf{u} \times \mathbf{v}$, is written $\underline{\underline{u}}\,\underline{v}^{\mathrm{T}}$ or $\underline{u}\,\underline{\underline{v}}$. The concept is not restricted to unit vectors or to vectors with real components: it may be extended to arbitrary vectors with complex components. Thus, in general, one can define the matrices $\underline{u}$ and $\underline{\underline{u}}$ even when u_x, u_y, and u_z are arbitrary complex numbers. As an example, consider the propagation vector $\mathbf{k}$ for a generalized plane wave, which may be written

$$\mathbf{k} = (k_x + \mathrm{i}k'_x)\,\hat{\mathbf{x}} + (k_y + \mathrm{i}k'_y)\,\hat{\mathbf{y}} + (k_z + \mathrm{i}k'_z)\,\hat{\mathbf{z}} \tag{5.2a}$$

The real and imaginary parts of the components of $\mathbf{k}$ are explicitly identified in this expression. Defining $\mathscr{K}$ as $\sqrt{k_x{}^2 + k_y{}^2 + k_z{}^2}$ and $\mathscr{K}'$ as $\sqrt{k'^2_x + k'^2_y + k'^2_z}$, one can write $\mathbf{k}$ equivalently as

$$\mathbf{k} = \mathscr{K}\mathbf{u} + \mathrm{i}\mathscr{K}'\mathbf{u}' \tag{5.2b}$$

where $\mathbf{u}$ and $\mathbf{u}'$ are two unit vectors having, in general, different orientations. With this notation we have

$$\underline{k} = \mathscr{K}\underline{u} + \mathrm{i}\mathscr{K}'\underline{u}' \tag{5.3a}$$

$$\underline{\underline{k}} = \mathscr{K}\underline{\underline{u}} + \mathrm{i}\mathscr{K}'\underline{\underline{u}}' \tag{5.3b}$$

When defining the dot and cross products of two complex vectors, one must be careful in generalizing concepts from the domain of real vectors to the domain of complex vectors. For instance, the complex electric field vector $\mathbf{E}_0$ and the complex propagation vector $\mathbf{k}$ specify the generalized plane wave

$$\boxed{\mathbf{E}(\mathbf{r},t) = \mathbf{E}_0 \exp\left[\mathrm{i}(\mathbf{k}\cdot\mathbf{r} - \omega t)\right]} \tag{5.4}$$

It will be shown later that the magnetic field vector $\mathbf{H}_0$ is proportional to the cross product of $\mathbf{k}$ and $\mathbf{E}_0$, and can be written as

$$\underline{H}_0^{\mathrm{T}} \propto \underline{\underline{k}}\,\underline{E}_0^{\mathrm{T}} \tag{5.5a}$$

The concept of orthogonality, however, cannot be applied to these complex vectors. This becomes evident as one expresses Eq. (5.5a) in terms of the real and imaginary components of the vectors:

$$(\mathcal{H}_0\,\underline{w} + \mathrm{i}\mathcal{H}_0'\,\underline{w}')^{\mathrm{T}} \propto (\mathcal{K}\underline{\underline{u}} + \mathrm{i}\mathcal{K}'\underline{\underline{u}}')\,(\mathcal{E}_0\underline{v} + \mathrm{i}\mathcal{E}_0'\underline{v}')^{\mathrm{T}} \tag{5.5b}$$

Clearly $\mathbf{w}$ and $\mathbf{w}'$ are linear combinations of $\mathbf{u}\times\mathbf{v}$, $\mathbf{u}\times\mathbf{v}'$, $\mathbf{u}'\times\mathbf{v}$ and $\mathbf{u}'\times\mathbf{v}'$ and, as such, no simple relationship (such as orthogonality) can be identified among pairs of these vectors.

Both $\mathbf{k}$ and $\mathbf{E}_0$ are complex vectors in three-dimensional space, but we will find it useful to interpret them differently. $\mathbf{k}$ is the propagation vector and, when written as $\mathcal{K}\mathbf{u} + \mathrm{i}\mathcal{K}'\mathbf{u}'$, specifies two directions in space: $\mathbf{u}$ is the direction of phase propagation, whereas $\mathbf{u}'$ is the direction along which the wave amplitude undergoes an exponential decay. In other words, the planes perpendicular to $\mathbf{u}$ have constant phase while those perpendicular to $\mathbf{u}'$ have constant amplitude. The interpretation of $\mathbf{E}_0 = \mathcal{E}_0\mathbf{v} + \mathrm{i}\mathcal{E}_0'\mathbf{v}'$ is as follows. At any given point in space the electric field vector is $\mathbf{E}_0 \exp(-\mathrm{i}\omega t)$. One may consider either the real part or the imaginary part of this vector and show that, as time progresses, the tip of the vector travels around an ellipse confined to the plane of $\mathbf{v}$ and $\mathbf{v}'$ (see Problem 5.1). The unit vectors $\mathbf{v}$ and $\mathbf{v}'$ therefore define the plane of polarization of the wave described by Eq. (5.4).

We shall find it convenient to define the divergence ($\nabla\cdot$) and curl ($\nabla\times$) operators in our matrix notation as follows:

divergence:
$$\underline{\nabla} = \left(\frac{\partial}{\partial x}, \frac{\partial}{\partial y}, \frac{\partial}{\partial z}\right) \tag{5.6}$$

curl:
$$\underline{\underline{\nabla}} = \begin{bmatrix} 0 & -\partial/\partial z & \partial/\partial y \\ \partial/\partial z & 0 & -\partial/\partial x \\ -\partial/\partial y & \partial/\partial x & 0 \end{bmatrix} \tag{5.7}$$

This allows for certain simplifications in later discussions. In particular, when Maxwell's equations are written in matrix notation, their solutions may be found by simple linear-algebraic techniques.

5.2. Maxwell's Equations and Plane-wave Propagation

In a medium with neither free charges nor currents Maxwell's equations are written as follows:

$$\nabla \cdot \mathbf{D} = 0 \tag{5.8a}$$

$$\nabla \times \mathbf{H} = \frac{\partial \mathbf{D}}{\partial t} \tag{5.8b}$$

$$\nabla \times \mathbf{E} = -\frac{\partial \mathbf{B}}{\partial t} \tag{5.8c}$$

$$\nabla \cdot \mathbf{B} = 0 \tag{5.8d}$$

Let the time-dependence be expressed as exp $(-\mathrm{i}\omega t)$ with ω in the optical frequency range. It is generally believed that in this regime $\mathbf{B} = \mu_0 \mathbf{H}$ where μ_0 is the permeability of free space. Under these circumstances the constitutive relation that describes the interaction of light with the medium of propagation will be

$$\mathbf{D} = \epsilon_0 \underset{\sim}{\epsilon}\, \mathbf{E} \tag{5.8e}$$

Here ϵ_0 is the permittivity of free space and $\underset{\sim}{\epsilon}$, a 3×3 matrix, is the dielectric tensor of the propagation medium. In the matrix notation of section 5.1, Maxwell's equations may be written as follows:

$$\underline{\nabla}\, \underset{\sim}{\epsilon}\, \underline{E}^{\mathrm{T}} = 0 \tag{5.9a}$$

$$\underline{\underline{\nabla}}\, \underline{H}^{\mathrm{T}} = -\mathrm{i}\omega \epsilon_0 \underset{\sim}{\epsilon}\, \underline{E}^{\mathrm{T}} \tag{5.9b}$$

$$\underline{\underline{\nabla}}\, \underline{E}^{\mathrm{T}} = \mathrm{i}\omega \mu_0 \underline{H}^{\mathrm{T}} \tag{5.9c}$$

$$\underline{\nabla}\, \underline{H}^{\mathrm{T}} = 0 \tag{5.9d}$$

In the MKSA system of units the electric field is in volts/meter, the magnetic field is in amps/meter, $\mu_0 = 4\pi \times 10^{-7}$ henrys/meter and $\epsilon_0 = 8.82 \times 10^{-12}$ farads/meter. We shall normalize the electric field by the impedance of free space, $Z_0 = \sqrt{(\mu_0/\epsilon_0)} \simeq 377$ ohms, and use $\hat{\mathbf{E}}$ instead of $\mathbf{E}$ to indicate the normalized electric field. Also introducing the propagation constant in free space, k_0, defined as

$$k_0 = \frac{2\pi}{\lambda_0} = \sqrt{\mu_0 \epsilon_0}\, \omega \tag{5.10}$$

one can rewrite Eqs. (5.9) as follows:

$$\underline{\nabla}\underset{\sim}{\epsilon}\,\underline{\hat{E}}^{\mathrm{T}} = 0 \tag{5.11a}$$

$$\underline{\underline{\nabla}}\,\underline{H}^{\mathrm{T}} = -ik_0\underset{\sim}{\epsilon}\,\underline{\hat{E}}^{\mathrm{T}} \tag{5.11b}$$

$$\underline{\underline{\nabla}}\,\underline{\hat{E}}^{\mathrm{T}} = ik_0\underline{H}^{\mathrm{T}} \tag{5.11c}$$

$$\underline{\nabla}\,\underline{H}^{\mathrm{T}} = 0 \tag{5.11d}$$

In this streamlined version of Maxwell's equations both $\hat{\mathbf{E}}$ and $\mathbf{H}$ are in units of amps/meter, k_0 is in m^{-1}, and $\underset{\sim}{\epsilon}$ is dimensionless. The generalized plane wave described by Eq. (5.4), together with a similar expression for the magnetic field $\mathbf{H}(\mathbf{r},t)$, will be a solution of Maxwell's equations provided that the following equations are satisfied:

$$\underline{k}\underset{\sim}{\epsilon}\,\underline{\hat{E}}_0^{\mathrm{T}} = 0 \tag{5.12a}$$

$$\underline{\underline{k}}\,\underline{H}_0^{\mathrm{T}} = -k_0\underset{\sim}{\epsilon}\,\underline{\hat{E}}_0^{\mathrm{T}} \tag{5.12b}$$

$$\underline{\underline{k}}\,\underline{\hat{E}}_0^{\mathrm{T}} = k_0\underline{H}_0^{\mathrm{T}} \tag{5.12c}$$

$$\underline{k}\,\underline{H}_0^{\mathrm{T}} = 0 \tag{5.12d}$$

Only two of the above four equations are independent; to see this, note that for any complex vector $\mathbf{k}$,

$$\underline{k}\,\underline{\underline{k}} = 0 \tag{5.13}$$

Therefore, if Eq. (5.12b) is multiplied on both sides by $\underline{k}$, Eq. (5.12a) is obtained. Similarly, multiplication of Eq. (5.12c) by $\underline{k}$ leads to Eq. (5.12d). Thus the first and the last of equations (5.12) need no longer be considered. Now if Eq. (5.12c) is multiplied by $\underline{\underline{k}}$ and if the right-hand side of the resulting equation is replaced using Eq. (5.12b), we obtain

$$\boxed{\left[(\underline{\underline{k}}/k_0)^2 + \underset{\sim}{\epsilon}\right]\underline{\hat{E}}_0^{\mathrm{T}} = 0} \tag{5.14a}$$

$$\boxed{\underline{H}_0^{\mathrm{T}} = (\underline{\underline{k}}/k_0)\,\underline{\hat{E}}_0^{\mathrm{T}}} \tag{5.14b}$$

Equations (5.14) are the fundamental equations of plane-wave propagation in a homogeneous medium characterized by the dielectric tensor $\underset{\sim}{\epsilon}$. The first one, commonly referred to as the Helmholtz equation, will have a nontrivial solution when the determinant of the coefficient matrix vanishes, yielding the characteristic equation

$$\left| (\underline{\underline{k}}/k_0)^2 + \underset{\sim}{\epsilon} \right| = 0 \tag{5.15}$$

Equation (5.15) imposes certain restrictions on the propagation vector **k**, depending on the dielectric tensor of the medium. To give a simple example, consider propagation in an isotropic medium where $\underset{\sim}{\epsilon} = \epsilon I$, ϵ being a complex constant and I the identity matrix. Equation (5.15) in this case reduces to

$$k_x^2 + k_y^2 + k_z^2 = k_0^2 \epsilon \tag{5.16}$$

In other words, any wave-vector **k** satisfying Eq. (5.16) is an acceptable solution to Maxwell's equations for propagation in the corresponding isotropic medium. In practice, the restrictions imposed by the characteristic equation are augmented by other restrictions (such as those imposed by Snell's law), limiting the acceptable values of **k** to a handful of vectors. For each acceptable wave vector, the Helmholtz equation (5.14a) must be solved for the components of $\hat{\underline{E}}_0$. Since the three linear equations implied by Eq. (5.14a) are not independent, they do not yield a unique solution for $\hat{\underline{E}}_0$; rather, they place certain constraints on the electric field vector. Complete identification of $\hat{\underline{E}}_0$ is made possible only after additional constraints (such as the continuity of fields at the interfaces) are taken into account. Finally, Eq. (5.14b) is used to determine the magnetic field vector $\mathbf{H}_0$ for acceptable wave vectors **k** once the corresponding electric field vector $\hat{\mathbf{E}}_0$ has been identified.

5.3. Plane Wave in an Isotropic Medium

Consider a homogeneous isotropic medium with dielectric constant ϵ where, in general, ϵ is complex. Without loss of generality, a plane wave propagating in this medium can be taken to have its wave vector **k** confined to the YZ-plane, i.e., $k_x = 0$. Moreover, we assume that $k_y = k_0 \sin\theta$, where θ is a real angle between 0 and 180°. The reason for this choice of k_x and k_y will become clear later. The characteristic equation (5.15) for the chosen plane wave in this isotropic medium now becomes

$$\left| (\underline{\underline{k}}/k_0)^2 + \epsilon I \right| = 0 \tag{5.17}$$

where I is a 3 × 3 identity matrix. Equation (5.17) yields a fourth-order polynomial equation for k_z as follows:

$$(k_z/k_0)^4 - 2(\epsilon - \sin^2\theta)(k_z/k_0)^2 + (\epsilon - \sin^2\theta)^2 = 0 \tag{5.18}$$

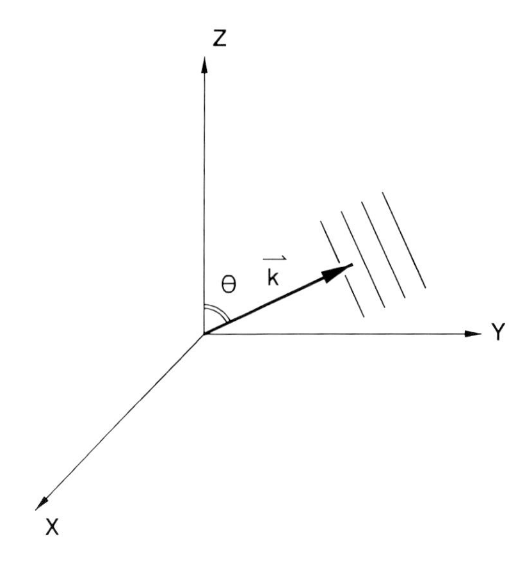

Figure 5.1. Plane-wave propagation in free space. The wave vector **k** is confined to the YZ-plane and makes an angle θ with the Z-axis.

The general solutions to this equation are

$$k_z = \pm k_0 \sqrt{\epsilon - \sin^2\theta} \tag{5.19}$$

By convention, the square root in Eq. (5.19) is in the first quadrant of the complex plane. When $0 \leq \theta \leq 90°$ the plane wave propagates in the positive Z-direction and the value of k_z with plus sign is accepted. When $90° < \theta \leq 180°$, the propagation direction is $-Z$ and the only acceptable k_z is the one with minus sign.

In the special case of propagation in free space, where $\epsilon = 1$, Eq. (5.19) reduces to $k_z = k_0 \cos\theta$. For this special case the diagram in Fig. 5.1 shows the propagation vector relative to the Cartesian coordinate system, identifying θ as the angle between **k** and the Z-axis.

Once the characteristic equation has been solved, one must go back to the Helmholtz equation (5.14a) and solve it for the components of the electric field $\hat{\underline{E}}$. Thus we have

$$\begin{bmatrix} 0 & 0 & 0 \\ 0 & \sin^2\theta & \pm\sin\theta\sqrt{\epsilon - \sin^2\theta} \\ 0 & \pm\sin\theta\sqrt{\epsilon - \sin^2\theta} & \epsilon - \sin^2\theta \end{bmatrix} \begin{bmatrix} \hat{E}_x \\ \hat{E}_y \\ \hat{E}_z \end{bmatrix} = \begin{bmatrix} 0 \\ 0 \\ 0 \end{bmatrix} \tag{5.20}$$

Equation (5.20) gives us three equations for the three unknowns $\hat{E}_x$, $\hat{E}_y$,

$\hat{E}_z$. However, these equations are not independent and yield no unique solution. Since the first column of the coefficient matrix is zero, $\hat{E}_x$ does not even appear in these equations. Thus, as far as the Helmholtz equation is concerned, $\hat{E}_x$ can have an arbitrary value. The second and third equations are identical and, as such, do not yield a unique solution for $\hat{E}_y$ and $\hat{E}_z$ either; all they do is specify the following relation:

$$\hat{E}_z = \mp \frac{\sin\theta}{\sqrt{\epsilon - \sin^2\theta}} \hat{E}_y \tag{5.21}$$

The minus sign in Eq. (5.21) applies when $0 \le \theta \le 90°$, while the plus sign applies when $90° < \theta \le 180°$. In general, ϵ is complex and there will be a phase difference between $\hat{E}_y$ and $\hat{E}_z$.

It is instructive to study the simple case of propagation in free space ($\epsilon = 1$). In this case

$$\hat{E}_z = -\tan\theta\, \hat{E}_y \tag{5.22}$$

(The ± sign has been removed because $\tan\theta$ carries its own sign.) Equation (5.22) simply states that in free space the component of the electric field in the YZ-plane, $\hat{E}_y \hat{\mathbf{y}} + \hat{E}_z \hat{\mathbf{z}}$, must be orthogonal to the propagation vector $\mathbf{k}$. Of course $\hat{E}_x \hat{\mathbf{x}}$ is already orthogonal to $\mathbf{k}$; thus the total electric field in free space is in a plane perpendicular to $\mathbf{k}$.

Another useful relationship concerning plane waves in isotropic media is the relation between the magnetic field H and the electric field $\hat{E}$. Substituting in Eq. (5.14b) the values of k_x, k_y, k_z described earlier in this section as well as $\hat{E}_z$ from Eq. (5.21), we find

$$H_x = \mp \frac{\epsilon}{\sqrt{\epsilon - \sin^2\theta}} \hat{E}_y \tag{5.23a}$$

$$H_y = \pm \sqrt{\epsilon - \sin^2\theta}\, \hat{E}_x \tag{5.23b}$$

$$H_z = -\sin\theta\, \hat{E}_x \tag{5.23c}$$

Usually H_x and H_y are the components of interest, while H_z can always be derived from H_y:

$$H_z = \mp \frac{\sin\theta}{\sqrt{\epsilon - \sin^2\theta}} H_y \tag{5.24}$$

The relation between $(\hat{E}_x, \hat{E}_y)$ and (H_x, H_y) can thus be written in the following compact form:

$$\begin{pmatrix} H_x \\ \\ H_y \end{pmatrix} = \pm \begin{pmatrix} 0 & -\dfrac{\epsilon}{\sqrt{\epsilon - \sin^2\theta}} \\ \\ \sqrt{\epsilon - \sin^2\theta} & 0 \end{pmatrix} \begin{pmatrix} \hat{E}_x \\ \\ \hat{E}_y \end{pmatrix} \tag{5.25}$$

For the special case of propagation in free space, the above equation reduces to

$$\begin{pmatrix} H_x \\ \\ H_y \end{pmatrix} = \begin{pmatrix} 0 & -1/\cos\theta \\ \\ \cos\theta & 0 \end{pmatrix} \begin{pmatrix} \hat{E}_x \\ \\ \hat{E}_y \end{pmatrix} \tag{5.26}$$

The $\pm$ sign is carried by $\cos\theta$.

The power of the beam may be computed using Poynting's theorem. For the plane wave in isotropic medium the Poynting vector $\mathbf{S}$ is given by

$$\mathbf{S} = \tfrac{1}{2} \operatorname{Re}\left(\mathbf{E} \times \mathbf{H}^*\right) \tag{5.27}$$

Thus,

$$\underline{S}^{\mathrm{T}} = \tfrac{1}{2} \operatorname{Re}\left\{ \underline{\underline{E}}_0 \, \underline{H}_0^{*\,\mathrm{T}} \exp\left[\mathrm{i}\,(k_z - k_z^*)\, z \right] \right\}$$

$$= \tfrac{1}{2} \operatorname{Re}\left\{ \begin{pmatrix} 0 & -E_z & E_y \\ E_z & 0 & -E_x \\ -E_y & E_x & 0 \end{pmatrix} \begin{pmatrix} H_x^* \\ H_y^* \\ H_z^* \end{pmatrix} \right\} \exp\left[\mp 2k_0 \operatorname{Im}\left(\sqrt{\epsilon - \sin^2\theta}\right) z \right] \tag{5.28}$$

Equation (5.28) is quite general and can be used to compute the power of the beam in arbitrary situations. For concreteness, we study the cases of s- and p-polarization separately.

(i) The case of s-polarization. Here $E_y = E_z = H_x = 0$ and Eq. (5.28) yields

$$\underline{S}^{\mathrm{T}} = \tfrac{1}{2} \frac{|E_x|^2}{Z_0} \begin{pmatrix} 0 \\ \\ \sin\theta \\ \\ \pm \operatorname{Re} \sqrt{\epsilon - \sin^2\theta} \end{pmatrix} \exp\left[\mp 2k_0 \operatorname{Im}\left(\sqrt{\epsilon - \sin^2\theta}\right) z \right] \tag{5.29}$$

where $Z_0 \simeq 377\ \Omega$ is the impedance of free space. Clearly the Poynting vector $\mathbf{S}$ and $\mathrm{Re}(\mathbf{k})$ are parallel, implying that phase and energy propagate in the same direction. The s-polarized plane wave in isotropic medium may thus be considered as the *ordinary* ray.

(ii) The case of p-polarization. Here $E_x = 0$ and, consequently, $H_y = H_z = 0$. Equation (5.28) reduces to

$$\underset{\sim}{S}^{\mathrm{T}} = \frac{1}{2}\,\frac{|E_y|^2}{Z_0\,|\epsilon - \sin^2\theta|}\begin{pmatrix} 0 \\ \mathrm{Re}(\epsilon)\sin\theta \\ \pm\,\mathrm{Re}\left(\epsilon^*\sqrt{\epsilon - \sin^2\theta}\right)\end{pmatrix}\exp\left[\mp 2k_0\,\mathrm{Im}\left(\sqrt{\epsilon - \sin^2\theta}\right)z\right] \tag{5.30}$$

If ϵ happens to be real then $\mathbf{S}$ and $\mathbf{k}$ will be parallel, i.e., the wave vector will be aligned with the direction of energy propagation. When ϵ is complex, however, $\mathbf{S}$ will have a different direction than $\mathrm{Re}(\mathbf{k})$, which is the direction of phase propagation. The p-polarized plane wave in an isotropic medium may thus be considered as the *extraordinary* ray.

5.4. Reflection at the Interface of Free Space and an Isotropic Medium

The reflection of a plane wave from the interface of free space and a homogeneous medium is depicted schematically in Fig. 5.2. Throughout the section we shall assume that this medium is isotropic with dielectric tensor $\underset{\sim}{\epsilon} = \epsilon I$. The incident beam propagates along $\mathbf{k}^{(i)}$, whose angle to Z is $\pi - \theta$ and which is entirely contained within the YZ-plane. Thus $k_x^{(i)} = 0$ and $k_y^{(i)} = k_0 \sin\theta$. (The superscript (i) identifies the incident beam; below we use superscripts (r) and (t) to refer to the reflected and transmitted beams.)

To satisfy Maxwell's equations, the tangential components of $\hat{\mathbf{E}}$ and $\mathbf{H}$ must be continuous across the interface which, in the present case, is the XY-plane at $Z = 0$. Since the incident, reflected, and transmitted fields all have the same form as in Eq. (5.4), continuity at the interface is achieved when the exponential factors cancel out. At $Z = 0$ the term in the exponent containing k_z disappears, but the terms containing k_x and k_y remain. The continuity requirement then imposes the following conditions (Snell's Law):

$$k_x^{(r)} = k_x^{(t)} = k_x^{(i)} = 0 \tag{5.31}$$

$$k_y^{(r)} = k_y^{(t)} = k_y^{(i)} = k_0 \sin\theta \tag{5.32}$$

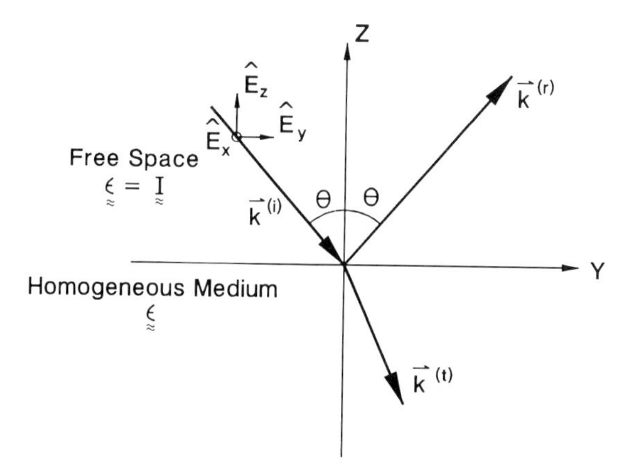

Figure 5.2. Geometry of the reflection of a plane wave from the interface between free space and a homogeneous medium with dielectric tensor $\tilde{\epsilon}$. The **k**-vectors of the incident, reflected, and transmitted beams are identified by the superscripts (i), (r), and (t) respectively. The angle of incidence is θ, the plane of incidence is YZ, and the **E**-field is specified by its Cartesian components E_x, E_y, and E_z.

The three beams (incident, reflected, transmitted) thus have the same k_x and k_y, and the relations derived in section 5.3 (with the appropriate value of ϵ for each medium) apply to every one of them. Next we define a reflection matrix R as follows:

$$\begin{pmatrix} \hat{E}_x^{(r)} \\ \hat{E}_y^{(r)} \end{pmatrix} = \begin{pmatrix} r_{xx} & r_{xy} \\ r_{yx} & r_{yy} \end{pmatrix} \begin{pmatrix} \hat{E}_x^{(i)} \\ \hat{E}_y^{(i)} \end{pmatrix} \tag{5.33}$$

The continuity of the $\hat{\mathbf{E}}$-field then gives

$$\begin{pmatrix} \hat{E}_x^{(i)} \\ \hat{E}_y^{(i)} \end{pmatrix} + \begin{pmatrix} r_{xx} & r_{xy} \\ r_{yx} & r_{yy} \end{pmatrix} \begin{pmatrix} \hat{E}_x^{(i)} \\ \hat{E}_y^{(i)} \end{pmatrix} = \begin{pmatrix} \hat{E}_x^{(t)} \\ \hat{E}_y^{(t)} \end{pmatrix} \tag{5.34}$$

As for the continuity of the **H**-field, we use Eq. (5.25) to write

$$\begin{pmatrix} 0 & 1/\cos\theta \\ -\cos\theta & 0 \end{pmatrix} \begin{pmatrix} \hat{E}_x^{(i)} \\ \hat{E}_y^{(i)} \end{pmatrix} + \begin{pmatrix} 0 & -1/\cos\theta \\ \cos\theta & 0 \end{pmatrix} \begin{pmatrix} \hat{E}_x^{(r)} \\ \hat{E}_y^{(r)} \end{pmatrix}$$

$$= \begin{pmatrix} 0 & \dfrac{\epsilon}{\sqrt{\epsilon - \sin^2\theta}} \\ -\sqrt{\epsilon - \sin^2\theta} & 0 \end{pmatrix} \begin{pmatrix} \hat{E}_x^{(t)} \\ \hat{E}_y^{(t)} \end{pmatrix} \tag{5.35}$$

Using Eqs. (5.33) and (5.34) to substitute for the reflected and transmitted $\hat{\mathbf{E}}$-components in Eq. (5.35), we obtain

$$\begin{pmatrix} 0 & 1/\cos\theta \\ -\cos\theta & 0 \end{pmatrix} (I - R) \begin{pmatrix} \hat{E}_x^{(i)} \\ \hat{E}_y^{(i)} \end{pmatrix}$$

$$= \begin{pmatrix} 0 & \dfrac{\epsilon}{\sqrt{\epsilon - \sin^2\theta}} \\ -\sqrt{\epsilon - \sin^2\theta} & 0 \end{pmatrix} (I + R) \begin{pmatrix} \hat{E}_x^{(i)} \\ \hat{E}_y^{(i)} \end{pmatrix} \tag{5.36}$$

In this equation I is a 3×3 identity matrix. Equation (5.36) is valid for all values of $\hat{E}_x^{(i)}$ and $\hat{E}_y^{(i)}$; thus the coefficient matrices on the two sides must be identical. Solving Eq. (5.36) for the components of the reflectivity matrix R, we find

$$r_{xx} = \frac{\cos\theta - \sqrt{\epsilon - \sin^2\theta}}{\cos\theta + \sqrt{\epsilon - \sin^2\theta}} \tag{5.37a}$$

$$r_{xy} = 0 \tag{5.37b}$$

$$r_{yx} = 0 \tag{5.37c}$$

$$r_{yy} = \frac{\sqrt{\epsilon - \sin^2\theta} - \epsilon\cos\theta}{\sqrt{\epsilon - \sin^2\theta} + \epsilon\cos\theta} \tag{5.37d}$$

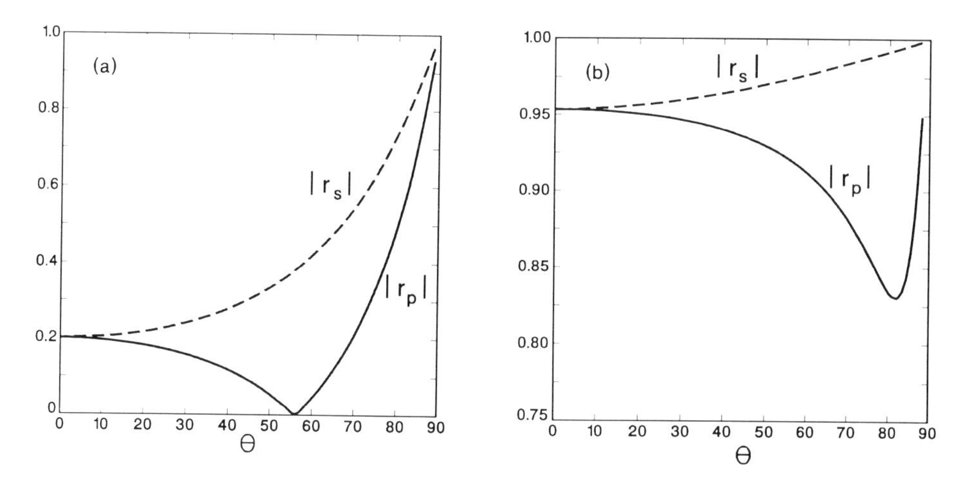

Figure 5.3. Reflectivity versus the angle of incidence for p-polarized and s-polarized incident beams. $r_s = r_{xx}$ and $r_p = r_{yy}$. In (a) the reflecting medium is glass with $n = 1.5$. In (b) the medium is aluminum with $n = 1.4 + 7.6i$ (at $\lambda = 633$ nm).

Some interesting features of these reflection coefficients are noted below.

(i) $r_{xy} = r_{yx} = 0$ means that there is no cross-polarization. If the incident beam is s-polarized then the reflected beam will also be s-polarized. Similarly, if the incident beam is p-polarized, the reflected beam will also be p-polarized.

(ii) When $\theta = 0$ (i.e., at normal incidence), we have

$$r_{xx} = r_{yy} = \frac{1 - \sqrt{\epsilon}}{1 + \sqrt{\epsilon}} \tag{5.38}$$

which is the well-known formula for reflectivity at normal incidence. It is customary to define the complex refractive index n for the material with dielectric contant ϵ as $n = \sqrt{\epsilon}$. The real part of n is the usual dielectric refractive index, whereas its imaginary part is the absorption coefficient.

Figure 5.3 shows plots of reflectivity for the p and s components of polarization for two different materials. In (a) the material is glass with $\epsilon = 2.25$; the p-polarized light shows a reflectivity null at the Brewster angle $\theta_B = 56.31°$. In (b) the material is aluminum with $\epsilon = -55.8 + 21.3i$; this is the measured value of the dielectric constant at $\lambda = 633$ nm.

5.5. Reflection at the Interface of Free Space and a Birefringent Medium

With reference to Fig. 5.2 let us assume that the homogeneous medium is birefringent, with dielectric tensor

$$\underset{\sim}{\epsilon} = \begin{bmatrix} \epsilon_1 & 0 & 0 \\ 0 & \epsilon_2 & 0 \\ 0 & 0 & \epsilon_3 \end{bmatrix} \tag{5.39}$$

In general, ϵ_1, ϵ_2, ϵ_3 are complex constants. The characteristic equation for this medium is

$$\left(\frac{k_z}{k_0}\right)^4 + \left\{\left(1 + \frac{\epsilon_2}{\epsilon_3}\right)\sin^2\theta - \epsilon_1 - \epsilon_2\right\}\left(\frac{k_z}{k_0}\right)^2$$

$$+ \left\{\frac{\epsilon_2}{\epsilon_3}\sin^4\theta - \epsilon_2\left(1 + \frac{\epsilon_1}{\epsilon_3}\right)\sin^2\theta + \epsilon_1\epsilon_2\right\} = 0 \tag{5.40}$$

The solutions of Eq. (5.40) are readily obtained as follows:

$$k_{z1} = -k_0\sqrt{\epsilon_1 - \sin^2\theta} \tag{5.41a}$$

$$k_{z2} = -k_0\sqrt{\epsilon_2 - (\epsilon_2/\epsilon_3)\sin^2\theta} \tag{5.41b}$$

The minus sign in the above equations is chosen so that the wave vectors correspond to beams traveling in the negative Z-direction. The medium clearly supports two separate beams for a single value of θ, that is, each incident plane wave at the interface is broken into two beams in the birefringent medium. We now solve the Helmholtz equation for the two beams separately.

(i) Beam 1 with $\mathbf{k_z} = \mathbf{k_{z1}}$. The Helmholtz equation for this beam is

$$\begin{pmatrix} 0 & 0 & 0 \\ 0 & \epsilon_2 - \epsilon_1 + \sin^2\theta & -\sin\theta\sqrt{\epsilon_1 - \sin^2\theta} \\ 0 & -\sin\theta\sqrt{\epsilon_1 - \sin^2\theta} & \epsilon_3 - \sin^2\theta \end{pmatrix} \begin{pmatrix} \hat{E}_x \\ \hat{E}_y \\ \hat{E}_z \end{pmatrix} = 0 \qquad (5.42)$$

This equation does not impose any restrictions on $\hat{E}_x$ and, therefore, $\hat{E}_x$ can be chosen arbitrarily. However, the remaining two equations in $\hat{E}_y$ and $\hat{E}_z$ are independent of each other and their only solution is $\hat{E}_y = \hat{E}_z = 0$. Consequently, beam 1 (with $k_z = k_{z1}$) must be linearly polarized along X; we shall denote its **E**-field amplitude by $\hat{E}_x^{(t_1)}$ where (t_1) stands for transmitted beam 1. The magnetic field for this beam is readily found from Eq. (5.14b) as follows:

$$\underline{H}^{\mathrm{T}} = \begin{pmatrix} 0 & \sqrt{\epsilon_1 - \sin^2\theta} & \sin\theta \\ -\sqrt{\epsilon_1 - \sin^2\theta} & 0 & 0 \\ -\sin\theta & 0 & 0 \end{pmatrix} \begin{pmatrix} \hat{E}_x^{(t_1)} \\ 0 \\ 0 \end{pmatrix}$$

$$= \begin{pmatrix} 0 \\ -\sqrt{\epsilon_1 - \sin^2\theta}\ \hat{E}_x^{(t_1)} \\ -\sin\theta\ \hat{E}_x^{(t_1)} \end{pmatrix} \qquad (5.43)$$

As for the power of the beam, the situation is the same as for the s-polarized ray in the isotropic medium; therefore, Eq. (5.29) (with ϵ replaced by ϵ_1) gives the Poynting vector for the *ordinary* ray in this birefringent medium.

(ii) Beam 2 with $\mathbf{k_z = k_{z2}}$. The Helmholtz equation for this beam is

$$\begin{pmatrix} \epsilon_1 - \epsilon_2 + \left(\frac{\epsilon_2}{\epsilon_3} - 1\right)\sin^2\theta & 0 & 0 \\ 0 & (\epsilon_2/\epsilon_3)\sin^2\theta & -\sin\theta\sqrt{\epsilon_2 - \frac{\epsilon_2}{\epsilon_3}\sin^2\theta} \\ 0 & -\sin\theta\sqrt{\epsilon_2 - \frac{\epsilon_2}{\epsilon_3}\sin^2\theta} & \epsilon_3 - \sin^2\theta \end{pmatrix} \begin{pmatrix} \hat{E}_x \\ \hat{E}_y \\ \hat{E}_z \end{pmatrix} = 0 \tag{5.44}$$

The first row of Eq. (5.44) yields $\hat{E}_x = 0$. The second and third rows yield identical equations and impose a relationship between $\hat{E}_y$ and $\hat{E}_z$. If $\hat{E}_y$ is considered arbitrary, then $\hat{E}_z$ must be given by

$$\hat{E}_z = \frac{(\epsilon_2/\epsilon_3)\sin\theta}{\sqrt{\epsilon_2 - (\epsilon_2/\epsilon_3)\sin^2\theta}}\hat{E}_y \tag{5.45}$$

We conclude that beam 2 is p-polarized (i.e., $\hat{E}_x = 0$) and that only $\hat{E}_y$ can be arbitrarily specified, since the Helmholtz equation then requires the remaining component, $\hat{E}_z$, to have the value assigned to it by Eq. (5.45). The $\hat{\mathbf{E}}$-field amplitude for beam 2 will hereinafter be denoted by $\hat{E}_y^{(t_2)}$. The magnetic field for this beam is obtained from Eq. (5.14b) as follows:

$$\underline{H}^{\mathrm{T}} = \begin{pmatrix} 0 & \sqrt{\epsilon_2 - \frac{\epsilon_2}{\epsilon_3}\sin^2\theta} & \sin\theta \\ -\sqrt{\epsilon_2 - \frac{\epsilon_2}{\epsilon_3}\sin^2\theta} & 0 & 0 \\ -\sin\theta & 0 & 0 \end{pmatrix} \begin{pmatrix} 0 \\ \hat{E}_y^{(t_2)} \\ \frac{(\epsilon_2/\epsilon_3)\sin\theta\,\hat{E}_y^{(t_2)}}{\sqrt{\epsilon_2 - (\epsilon_2/\epsilon_3)\sin^2\theta}} \end{pmatrix}$$

$$= \begin{pmatrix} \frac{\epsilon_2\,\hat{E}_y^{(t_2)}}{\sqrt{\epsilon_2 - (\epsilon_2/\epsilon_3)\sin^2\theta}} \\ 0 \\ 0 \end{pmatrix} \tag{5.46}$$

As for the power of the beam, we use Eq. (5.27) and find the following expression for the Poynting vector:

$$\underline{S}^{\mathrm{T}} = \frac{\frac{1}{2}\,|E_y|^2}{Z_0\,|\epsilon_2 - (\epsilon_2/\epsilon_3)\sin^2\theta|} \begin{pmatrix} 0 \\ \mathrm{Re}(\epsilon_3)\,|\epsilon_2/\epsilon_3|^2 \sin\theta \\ -\mathrm{Re}\left(\epsilon_2^* \sqrt{\epsilon_2 - (\epsilon_2/\epsilon_3)\sin^2\theta}\right) \end{pmatrix}$$

$$\times \exp\left\{2k_0\,\mathrm{Im}\left(\sqrt{\epsilon_2 - (\epsilon_2/\epsilon_3)\sin^2\theta}\right)z\right\} \tag{5.47}$$

In contrast to the case of p-polarized light in isotropic media, in a birefringent medium the directions of **S** and **k** will be different, even when ϵ_2 and ϵ_3 are real.

Having found the components of the two beams in the birefringent medium, we now match the tangential components of the electric and magnetic fields at the interface with free space, and proceed to find the reflectivity matrix. The continuity of the $\hat{\mathbf{E}}$-field gives

$$\begin{pmatrix} \hat{E}_x^{(i)} \\ \hat{E}_y^{(i)} \end{pmatrix} + \begin{pmatrix} r_{xx} & r_{xy} \\ r_{yx} & r_{yy} \end{pmatrix} \begin{pmatrix} \hat{E}_x^{(i)} \\ \hat{E}_y^{(i)} \end{pmatrix} = \begin{pmatrix} \hat{E}_x^{(t_1)} \\ \hat{E}_y^{(t_2)} \end{pmatrix} \tag{5.48}$$

From the requirement of continuity for the **H**-field we find

$$\begin{pmatrix} 0 & 1/\cos\theta \\ -\cos\theta & 0 \end{pmatrix} \begin{pmatrix} \hat{E}_x^{(i)} \\ \hat{E}_y^{(i)} \end{pmatrix} + \begin{pmatrix} 0 & -1/\cos\theta \\ \cos\theta & 0 \end{pmatrix} \begin{pmatrix} r_{xx} & r_{xy} \\ r_{yx} & r_{yy} \end{pmatrix} \begin{pmatrix} \hat{E}_x^{(i)} \\ \hat{E}_y^{(i)} \end{pmatrix}$$

$$= \begin{pmatrix} 0 & \dfrac{\epsilon_2}{\sqrt{\epsilon_2 - (\epsilon_2/\epsilon_3)\sin^2\theta}} \\ -\sqrt{\epsilon_1 - \sin^2\theta} & 0 \end{pmatrix} \begin{pmatrix} \hat{E}_x^{(t_1)} \\ \hat{E}_y^{(t_2)} \end{pmatrix} \tag{5.49}$$

Substituting in Eq. (5.49) the transmitted field amplitude from Eq. (5.48), eliminating the incident field from both sides of the resulting equation, and solving for R we find

$$r_{xx} = \frac{\cos\theta - \sqrt{\epsilon_1 - \sin^2\theta}}{\cos\theta + \sqrt{\epsilon_1 - \sin^2\theta}} \tag{5.50a}$$

$$r_{xy} = 0 \tag{5.50b}$$

$$r_{yx} = 0 \tag{5.50c}$$

$$r_{yy} = \frac{\sqrt{\epsilon_2 - (\epsilon_2/\epsilon_3)\sin^2\theta} - \epsilon_2\cos\theta}{\sqrt{\epsilon_2 - (\epsilon_2/\epsilon_3)\sin^2\theta} + \epsilon_2\cos\theta} \tag{5.50d}$$

The main features of these results are summarized below.

(i) Since $r_{xy} = r_{yx} = 0$, there can be no cross-polarization. Thus a s-polarized incident beam, upon reflection at the interface of a birefringent medium whose dielectric tensor is diagonal, emerges as s-polarized. Similarly, p-polarized incident beams result in p-polarized reflected beams.

(ii) r_{xx} is independent of ϵ_2 and ϵ_3, because the s-polarized component of the incident beam is refracted into the *ordinary* ray (t_1), which carries only the X-component of the **E**-field. This beam does not interact with the material along the Y- and Z-axes and, therefore, does not "feel" the corresponding dielectric constants ϵ_2 and ϵ_3. Similarly, r_{yy} is independent of ϵ_1, since the p-polarized component is refracted into the *extraordinary* ray (t_2), which does not "feel" the dielectric constant of the material along X.

(iii) If ϵ_2 becomes equal to ϵ_3, then r_{yy} in Eq. (5.50d) reduces to a similar expression obtained for isotropic media in the previous section (see Eq. (5.37d)).

(iv) At normal incidence $\theta = 0$; the reflectivities in Eq. (5.50) are then given by

$$r_{xx} = \frac{1 - \sqrt{\epsilon_1}}{1 + \sqrt{\epsilon_1}} \tag{5.51a}$$

$$r_{yy} = \frac{1 - \sqrt{\epsilon_2}}{1 + \sqrt{\epsilon_2}} \tag{5.51b}$$

The reason that ϵ_3 does not appear in Eq. (5.51) is that at normal incidence both ordinary and extraordinary rays propagate along Z and, as such, their **E**-fields interact with the material along the X- and Y-directions only.

5.6. Reflection at the Interface of Free Space and a Magnetic Medium; the Magneto-optical Kerr Effect

This section is devoted to an analysis of the polar MO Kerr effect. The polar effect is observed when the magnetization vector **M** of the medium is perpendicular to the plane of the interface, as shown in Fig. 5.4(a). The polar effect is of particular importance in optical data storage, since it is the basis for readout of information from MO media. Other interesting geometries for the Kerr effect are the longitudinal and transverse geometries, shown in Figs. 5.4(b) and 5.4(c) respectively. In the longitudinal (transverse) case **M** lies in the plane of the interface and parallel (perpendicular) to the plane of incidence (i.e., the plane containing the incident **k**-vector and the normal **N** at the interface). The analysis of the longitudinal and transverse effects is very similar to that of the polar effect, and is therefore left as an exercise for the reader (see Problem 5.10). Also, the general case in which the magnetization vector is arbitrarily oriented relative to the interface plane may be studied in similar fashion, but the algebra is tedious (see Appendix A).

When a homogeneous medium is uniformly magnetized along the Z-axis, as in Fig. 5.4(a), its dielectric tensor may be written

$$\underset{\sim}{\epsilon} = \begin{pmatrix} \epsilon & \epsilon' & 0 \\ -\epsilon' & \epsilon & 0 \\ 0 & 0 & \epsilon \end{pmatrix} \tag{5.52}$$

The diagonal element ϵ of the tensor represents the ordinary interactions between light and the medium, while the off-diagonal element ϵ' is representative of the MO activity. Considering the definition of the dielectric tensor, $\mathbf{D} = \underset{\sim}{\epsilon}\,\mathbf{E}$, one observes that E_x contributes a component $-\epsilon' E_x$ to D_y, while E_y contributes a component $\epsilon' E_y$ to D_x. The differing signs of ϵ' in Eq. (5.52) are thus a manifestation of the symmetry of Cartesian space. In general, both ϵ and ϵ' may be complex numbers.

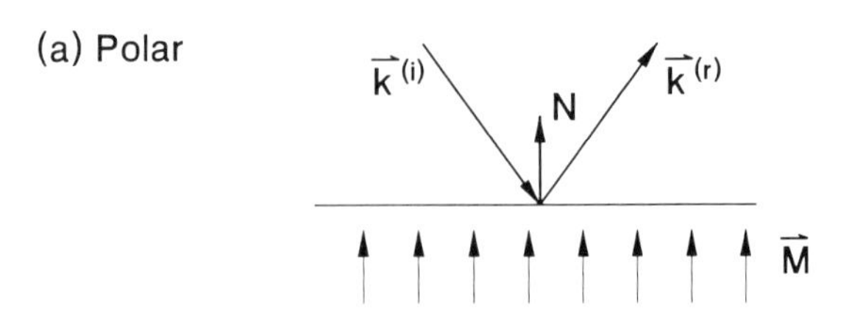

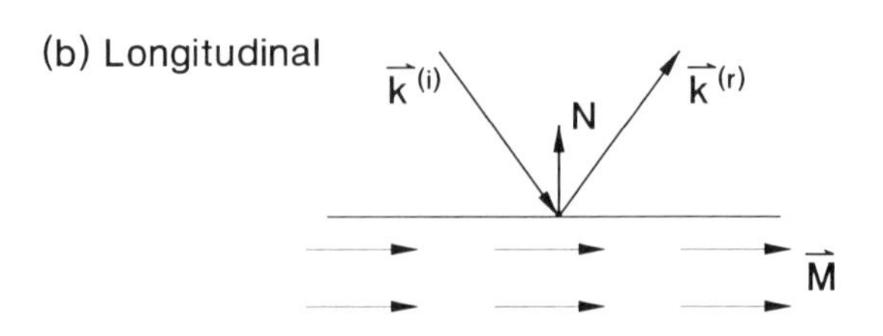

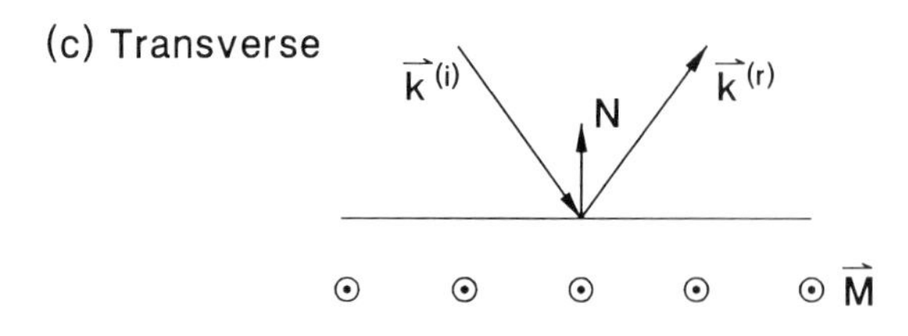

Figure 5.4. Various geometries of the magneto-optical Kerr effect. In (a) the magnetization **M** is perpendicular to the interface plane and the effect is referred to as polar. For the longitudinal Kerr effect observed in (b) the vector **M** is parallel to the interface while completely contained within the plane of incidence. In the transverse geometry of (c) the magnetization is parallel to the interface but perpendicular to the plane of incidence.

The characteristic equation (5.15) for the dielectric tensor in Eq. (5.52) is as follows:

$$\left(\frac{k_z}{k_0}\right)^4 - 2(\epsilon - \sin^2\theta)\left(\frac{k_z}{k_0}\right)^2 + \left\{(\epsilon - \sin^2\theta)^2 - \epsilon'^2\left(\frac{\sin^2\theta}{\epsilon} - 1\right)\right\} = 0 \tag{5.53}$$

The solutions to Eq. (5.53) are:

$$k_{z1} = \pm\, k_0 \sqrt{\epsilon - \sin^2\theta + \epsilon' \sqrt{\frac{\sin^2\theta}{\epsilon} - 1}} \tag{5.54a}$$

$$k_{z2} = \pm k_0 \sqrt{\epsilon - \sin^2\theta - \epsilon' \sqrt{\frac{\sin^2\theta}{\epsilon} - 1}} \tag{5.54b}$$

(We are at present interested in solutions with the minus sign in front since they correspond to plane waves propagating in the negative Z-direction.) Next we solve the Helmholtz equation (5.14a) for each beam separately. For beam 1 (with $k_z = k_{z1}$) we have

$$\begin{pmatrix} -\epsilon'\sqrt{\frac{\sin^2\theta}{\epsilon} - 1} & \epsilon' & 0 \\ -\epsilon' & \sin^2\theta - \epsilon'\sqrt{\frac{\sin^2\theta}{\epsilon} - 1} & -\sin\theta\sqrt{\epsilon - \sin^2\theta + \epsilon'\sqrt{\frac{\sin^2\theta}{\epsilon} - 1}} \\ 0 & -\sin\theta\sqrt{\epsilon - \sin^2\theta + \epsilon'\sqrt{\frac{\sin^2\theta}{\epsilon} - 1}} & \epsilon - \sin^2\theta \end{pmatrix} \begin{pmatrix} \hat{E}_x \\ \hat{E}_y \\ \hat{E}_z \end{pmatrix} = 0 \tag{5.55}$$

Equation (5.55) yields the other **E**-field components in terms of $\hat{E}_x$:

$$\hat{E}_y^{(t_1)} = \sqrt{\frac{\sin^2\theta}{\epsilon} - 1}\ \hat{E}_x^{(t_1)} \tag{5.56a}$$

$$\hat{E}_z^{(t_1)} = -\frac{\sin\theta\sqrt{\epsilon - \sin^2\theta + \epsilon'\sqrt{\frac{\sin^2\theta}{\epsilon} - 1}}}{\epsilon\sqrt{\frac{\sin^2\theta}{\epsilon} - 1}}\ \hat{E}_x^{(t_1)} \tag{5.56b}$$

The **H**-field components are then obtained from Eq. (5.14b) as follows:

$$H_x^{(t_1)} = -\frac{\sqrt{\epsilon - \sin^2\theta + \epsilon'\sqrt{\frac{\sin^2\theta}{\epsilon} - 1}}}{\sqrt{\frac{\sin^2\theta}{\epsilon} - 1}}\ \hat{E}_x^{(t_1)} \tag{5.57a}$$

$$H_y^{(t_1)} = -\sqrt{\epsilon - \sin^2\theta + \epsilon'\sqrt{\frac{\sin^2\theta}{\epsilon} - 1}}\ \hat{E}_x^{(t_1)} \tag{5.57b}$$

$$H_z^{(t_1)} = -\sin\theta \; \hat{E}_x^{(t_1)} \tag{5.57c}$$

The same procedure is repeated for beam 2 (with $k_z = k_{z2}$). Here, however, we choose to relate all field components to $\hat{E}_y$. We find

$$\hat{E}_x^{(t_2)} = -\frac{1}{\sqrt{\dfrac{\sin^2\theta}{\epsilon} - 1}} \; \hat{E}_y^{(t_2)} \tag{5.58a}$$

$$\hat{E}_z^{(t_2)} = \frac{\sin\theta}{\epsilon - \sin^2\theta} \sqrt{\epsilon - \sin^2\theta - \epsilon' \sqrt{\dfrac{\sin^2\theta}{\epsilon} - 1}} \; \hat{E}_y^{(t_2)} \tag{5.58b}$$

$$H_x^{(t_2)} = \frac{\epsilon}{\epsilon - \sin^2\theta} \sqrt{\epsilon - \sin^2\theta - \epsilon' \sqrt{\dfrac{\sin^2\theta}{\epsilon} - 1}} \; \hat{E}_y^{(t_2)} \tag{5.59a}$$

$$H_y^{(t_2)} = \frac{\sqrt{\epsilon - \sin^2\theta - \epsilon' \sqrt{\dfrac{\sin^2\theta}{\epsilon} - 1}}}{\sqrt{\dfrac{\sin^2\theta}{\epsilon} - 1}} \; \hat{E}_y^{(t_2)} \tag{5.59b}$$

$$H_z^{(t_2)} = \frac{\sin\theta}{\sqrt{\dfrac{\sin^2\theta}{\epsilon} - 1}} \; \hat{E}_y^{(t_2)} \tag{5.59c}$$

Having found the various components of the transmitted beams, we are now prepared to match the tangential components of **E** and of **H** at the interface. (The geometry again is that of Fig. 5.2 with the homogeneous medium being a magnetic medium whose direction of magnetization is perpendicular to the plane of the interface.) One obtains the following equations from the continuity requirements:

$$\begin{pmatrix} \hat{E}_x^{(i)} \\ \hat{E}_y^{(i)} \end{pmatrix} + \begin{pmatrix} r_{xx} & r_{xy} \\ r_{yx} & r_{yy} \end{pmatrix} \begin{pmatrix} \hat{E}_x^{(i)} \\ \hat{E}_y^{(i)} \end{pmatrix} = \begin{pmatrix} 1 & -\dfrac{1}{\sqrt{\dfrac{\sin^2\theta}{\epsilon} - 1}} \\ \sqrt{\dfrac{\sin^2\theta}{\epsilon} - 1} & 1 \end{pmatrix} \begin{pmatrix} \hat{E}_x^{(t_1)} \\ \hat{E}_y^{(t_2)} \end{pmatrix} \tag{5.60}$$

$$\begin{pmatrix} 0 & 1/\cos\theta \\ -\cos\theta & 0 \end{pmatrix} (I - R) \begin{pmatrix} \hat{E}_x^{(i)} \\ \hat{E}_y^{(i)} \end{pmatrix} =$$

$$\begin{pmatrix} -\dfrac{\sqrt{\epsilon - \sin^2\theta + \epsilon' \sqrt{\dfrac{\sin^2\theta}{\epsilon} - 1}}}{\sqrt{\dfrac{\sin^2\theta}{\epsilon} - 1}} & -\dfrac{\sqrt{\epsilon - \sin^2\theta - \epsilon' \sqrt{\dfrac{\sin^2\theta}{\epsilon} - 1}}}{\dfrac{\sin^2\theta}{\epsilon} - 1} \\ -\sqrt{\epsilon - \sin^2\theta + \epsilon' \sqrt{\dfrac{\sin^2\theta}{\epsilon} - 1}} & \dfrac{\sqrt{\epsilon - \sin^2\theta - \epsilon' \sqrt{\dfrac{\sin^2\theta}{\epsilon} - 1}}}{\sqrt{\dfrac{\sin^2\theta}{\epsilon} - 1}} \end{pmatrix} \begin{pmatrix} \hat{E}_x^{(t_1)} \\ \hat{E}_y^{(t_2)} \end{pmatrix} \tag{5.61}$$

Equations (5.60) and (5.61) can now be combined to yield a solution for the reflectivity matrix R. The solution is straightforward but the algebra is somewhat tedious.

An important special case of this problem occurs at normal incidence. Setting $\theta = 0$ results in the following characteristics for the two beams:

$$k_{z1} = -k_0 \sqrt{\epsilon + i\epsilon'} \quad (5.62a)$$

$$\hat{E}_y^{(t_1)} = i\hat{E}_x^{(t_1)} \quad (5.62b)$$

$$\hat{E}_z^{(t_1)} = 0 \quad (5.62c)$$

$$\hat{H}_x^{(t_1)} = i\sqrt{\epsilon + i\epsilon'}\, \hat{E}_x^{(t_1)} \quad (5.62d)$$

$$\hat{H}_y^{(t_1)} = -\sqrt{\epsilon + i\epsilon'}\, \hat{E}_x^{(t_1)} \quad (5.62e)$$

$$H_z^{(t_1)} = 0 \quad (5.62f)$$

$$k_{z2} = -k_0 \sqrt{\epsilon - i\epsilon'} \quad (5.63a)$$

$$\hat{E}_x^{(t_2)} = i\hat{E}_y^{(t_2)} \quad (5.63b)$$

$$\hat{E}_z^{(t_2)} = 0 \quad (5.63c)$$

$$\hat{H}_x^{(t_2)} = \sqrt{\epsilon - i\epsilon'}\, \hat{E}_y^{(t_2)} \quad (5.63d)$$

$$\hat{H}_y^{(t_2)} = -i\sqrt{\epsilon - i\epsilon'}\, \hat{E}_y^{(t_2)} \quad (5.63e)$$

$$H_z^{(t_2)} = 0 \quad (5.63f)$$

These results indicate that, at normal incidence, the natural modes for the transmitted beam in the magnetic medium are right circularly polarized (RCP) and left circularly polarized (LCP). The effective refractive indices of the medium for these modes are then given by

$$n_1 = \sqrt{\epsilon + i\epsilon'} \quad (5.64a)$$

$$n_2 = \sqrt{\epsilon - i\epsilon'} \quad (5.64b)$$

Thus in order to calculate the reflectivity of polar Kerr MO media (at normal incidence) one should decompose the state of incident polarization into its constituent RCP and LCP components. For each component the reflectivity is calculated using the effective refractive indices n_1 and n_2 given by Eqs. (5.64). The state of polarization of the reflected beam is then determined by superimposing the resulting circularly polarized components.

5.7. Plane Wave in a Medium with Arbitrary Dielectric Tensor

In this section we develop a general solution for plane-wave propagation in homogeneous media with arbitrary dielectric tensors. We have already seen instances of this solution for isotropic, birefringent, and magneto-optical media. The goal of the present section is to generalize the notions of ordinary and extraordinary beams, and to develop a systematic algorithm for identifying the various beams that coexist in a given medium. The ultimate goal is to obtain solutions for the multilayer stack shown in Fig. 5.5. This multilayer consists of an arbitrary number of layers, each of which can have an arbitrary dielectric tensor, that is, each layer could be dielectric, metal, birefringent, magneto-optical, etc. In general, because of reflections at the interfaces, there will be four propagating plane waves in

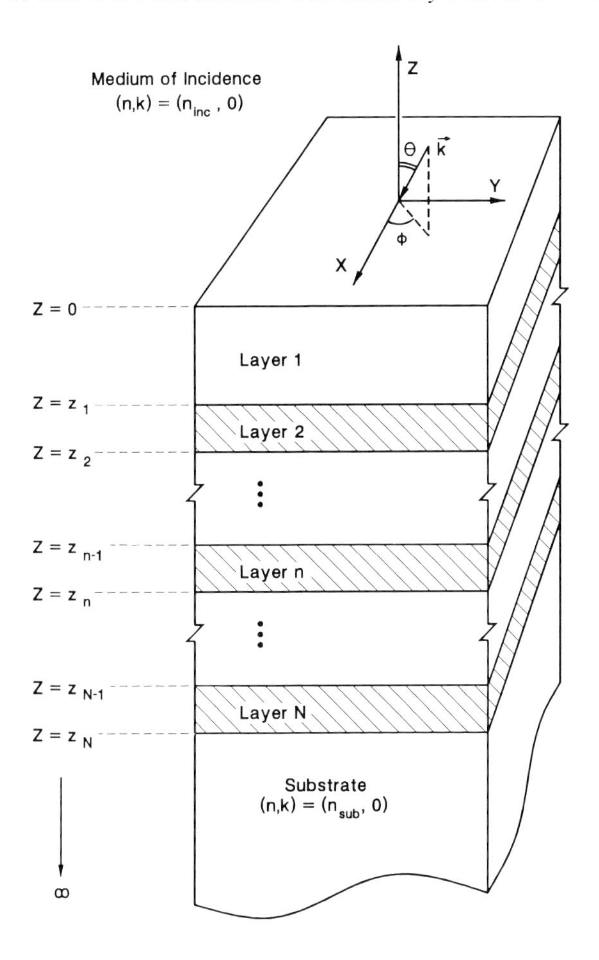

Figure 5.5. Schematic diagram of a multilayer structure. Both the substrate and the medium of incidence are semi-infinite, with real refractive indices n_{sub} and n_{inc} (i.e., their absorption coefficients are zero). The number of layers N, the individual layer thicknesses d_n, and the dielectric tensors $\underset{\sim}{\epsilon}_n$ of the various layers may be chosen arbitrarily. The incident beam's propagation vector is $\mathbf{k}$.

each layer. Of these, two will be traveling up (along $+Z$) and two will be traveling down (along $-Z$). Certain parameters of these beams can be computed directly for each layer (with no attention paid to the other layers in the stack) from the corresponding characteristic and Helmholtz equations; this will be done in the present section. The remaining parameters of the beams are obtained by matching the tangential components of the fields at the interfaces; this will be the subject of the next section.

With reference to Fig. 5.5, a plane wave at oblique incidence on the surface of the multilayer stack, with propagation direction characterized by angles θ and ϕ, has the following wave vector:

$$\mathbf{k}^{(i)} = k_0 \mathbf{u}^{(i)} = -k_0 (\sin\theta \cos\phi \hat{\mathbf{x}} + \sin\theta \sin\phi \hat{\mathbf{y}} + \cos\theta \hat{\mathbf{z}}) \tag{5.65}$$

(The superscript (i) stands for incident.) The continuity of **E** and **H** at the interface between adjacent layers requires that the components k_x and k_y of the propagation vector along $\hat{\mathbf{x}}$ and $\hat{\mathbf{y}}$ be the same on both sides of the interface. Therefore, for all plane waves in all layers,

$$k_x = -k_0 \sin\theta \cos\phi \tag{5.66a}$$

$$k_y = -k_0 \sin\theta \sin\phi \tag{5.66b}$$

Let us now examine the characteristic equation (5.15) for a medium having the general dielectric tensor

$$\underset{\sim}{\epsilon} = \begin{pmatrix} \epsilon_{xx} & \epsilon_{xy} & \epsilon_{xz} \\ \epsilon_{yx} & \epsilon_{yy} & \epsilon_{yz} \\ \epsilon_{zx} & \epsilon_{zy} & \epsilon_{zz} \end{pmatrix} \tag{5.67}$$

Substituting in Eq. (5.15) the tensor $\underset{\sim}{\epsilon}$ from Eq. (5.67) and the values of k_x, k_y from Eq. (5.66), one obtains after some algebraic manipulations the following fourth-order equation for k_z:

$$(k_z/k_0)^4 + A(k_z/k_0)^3 + B(k_z/k_0)^2 + C(k_z/k_0) + D = 0 \tag{5.68}$$

The coefficients A, B, C, D of Eq. (5.68) are listed in Table 5.1. In general, this equation has four complex solutions for k_z (see Appendix B). Of these, two solutions are in the upper half and two in the lower half of the complex plane. The solutions with positive imaginary part propagate in the positive Z-direction, while those with negative imaginary part propagate in the negative Z-direction. A good strategy (and one that we shall adopt) is to arrange the four solutions so that k_{z1} and k_{z2} are in the lower half of the complex plane (propagating downward) while k_{z3} and k_{z4} are in the upper half (propagating upward).

For each value of k_z the corresponding electric field must satisfy the Helmholtz equation, Eq. (5.14a). Since the determinant of the coefficient matrix has been set to zero, the three equations in the three unknowns $\hat{E}_x$, $\hat{E}_y$, $\hat{E}_z$ can be solved for two of the **E**-field components in terms of the third one. We adopt the following strategy regarding these field-component assignments. For beam 1 (with $k_z = k_{z1}$) and beam 3 (with $k_z = k_{z3}$) we express $\hat{E}_y$ and $\hat{E}_z$ in terms of the corresponding $\hat{E}_x$. Similarly, for beams 2 and 4 (with $k_z = k_{z2}$ and k_{z4}) we express $\hat{E}_x$ and $\hat{E}_z$ in terms of the corresponding $\hat{E}_y$. Table 5.2 summarizes this strategy. The coefficients a_m and b_m in Table 5.2 must be obtained from the Helmholtz equation using

Table 5.1. Coefficients of the characteristic equation (5.68) corresponding to the general dielectric tensor defined in Eq. (5.67). k_x and k_y are given by Eq. (5.66)

$$A = \frac{k_x}{k_0}\left(\frac{\epsilon_{xz}+\epsilon_{zx}}{\epsilon_{zz}}\right) + \frac{k_y}{k_0}\left(\frac{\epsilon_{yz}+\epsilon_{zy}}{\epsilon_{zz}}\right)$$

$$B = \frac{k_x^2}{k_0^2}\left(1+\frac{\epsilon_{xx}}{\epsilon_{zz}}\right) + \frac{k_y^2}{k_0^2}\left(1+\frac{\epsilon_{yy}}{\epsilon_{zz}}\right) + \frac{k_x k_y}{k_0^2}\left(\frac{\epsilon_{xy}+\epsilon_{yx}}{\epsilon_{zz}}\right) + \left(\frac{\epsilon_{xz}\epsilon_{zx}+\epsilon_{yz}\epsilon_{zy}}{\epsilon_{zz}} - \epsilon_{xx} - \epsilon_{yy}\right)$$

$$C = \frac{k_x^2+k_y^2}{k_0^2}\left\{\frac{k_x}{k_0}\left(\frac{\epsilon_{xz}+\epsilon_{zx}}{\epsilon_{zz}}\right) + \frac{k_y}{k_0}\left(\frac{\epsilon_{yz}+\epsilon_{zy}}{\epsilon_{zz}}\right)\right\}$$

$$+ \frac{k_x}{k_0}\left(\frac{\epsilon_{xy}\epsilon_{yz}+\epsilon_{yx}\epsilon_{zy}}{\epsilon_{zz}} - \frac{\epsilon_{yy}}{\epsilon_{zz}}(\epsilon_{xz}+\epsilon_{zx})\right) + \frac{k_y}{k_0}\left(\frac{\epsilon_{xy}\epsilon_{zx}+\epsilon_{yx}\epsilon_{xz}}{\epsilon_{zz}} - \frac{\epsilon_{xx}}{\epsilon_{zz}}(\epsilon_{yz}+\epsilon_{zy})\right)$$

$$D = \frac{k_x^2+k_y^2}{k_0^2}\left\{\frac{k_x^2}{k_0^2}\left(\frac{\epsilon_{xx}}{\epsilon_{zz}}\right) + \frac{k_y^2}{k_0^2}\left(\frac{\epsilon_{yy}}{\epsilon_{zz}}\right) + \frac{k_x k_y}{k_0^2}\left(\frac{\epsilon_{xy}+\epsilon_{yx}}{\epsilon_{zz}}\right) - \frac{\epsilon_{xx}\epsilon_{yy}}{\epsilon_{zz}}\right\}$$

$$+ \frac{k_x^2}{k_0^2}\left(\frac{\epsilon_{xy}\epsilon_{yx}+\epsilon_{xz}\epsilon_{zx}}{\epsilon_{zz}} - \epsilon_{xx}\right) + \frac{k_y^2}{k_0^2}\left(\frac{\epsilon_{xy}\epsilon_{yx}+\epsilon_{yz}\epsilon_{zy}}{\epsilon_{zz}} - \epsilon_{yy}\right)$$

$$+ \frac{k_x k_y}{k_0^2}\left(\frac{\epsilon_{xz}\epsilon_{zy}+\epsilon_{zx}\epsilon_{yz}}{\epsilon_{zz}} - \epsilon_{xy} - \epsilon_{yx}\right)$$

$$+ \left\{\epsilon_{xx}\epsilon_{yy} + \frac{\epsilon_{xy}\epsilon_{yz}\epsilon_{zx}+\epsilon_{yx}\epsilon_{zy}\epsilon_{xz}}{\epsilon_{zz}} - \epsilon_{xy}\epsilon_{yx} - \left(\frac{\epsilon_{xx}}{\epsilon_{zz}}\right)\epsilon_{yz}\epsilon_{zy} - \left(\frac{\epsilon_{yy}}{\epsilon_{zz}}\right)\epsilon_{xz}\epsilon_{zx}\right\}$$

$k_z = k_{zm}$. For each beam, the magnetic field components, obtained from Eq. (5.14b), are also given in Table 5.2.

The number of parameters for each layer that now remain to be identified has been reduced to four: E_{x1}, E_{y2}, E_{x3}, and E_{y4}. We will show in the next section how to determine these parameters by matching the tangential field components at the various interfaces.

Table 5.2. Relationships between the various components of **E** and **H** for plane waves in homogeneous media. The four beams described have the same k_x and k_y

	Beam 1	Beam 2	Beam 3	Beam 4
k_z	k_{z1}, lower half	k_{z2}, lower half	k_{z3}, upper half	k_{z4}, upper half
E_x	adjustable	$a_2 E_{y2}$	adjustable	$a_4 E_{y4}$
E_y	$a_1 E_{x1}$	adjustable	$a_3 E_{x3}$	adjustable
E_z	$b_1 E_{x1}$	$b_2 E_{y2}$	$b_3 E_{x3}$	$b_4 E_{y4}$
$k_0 H_x$	$(-k_{z1} a_1 + k_y b_1) E_{x1}$	$(-k_{z2} + k_y b_2) E_{y2}$	$(-k_{z3} a_3 + k_y b_3) E_{x3}$	$(-k_{z4} + k_y b_4) E_{y4}$
$k_0 H_y$	$(k_{z1} - k_x b_1) E_{x1}$	$(k_{z2} a_2 - k_x b_2) E_{y2}$	$(k_{z3} - k_x b_3) E_{x3}$	$(k_{z4} a_4 - k_x b_4) E_{y4}$
$k_0 H_z$	$(-k_y + k_x a_1) E_{x1}$	$(-k_y a_2 + k_x) E_{y2}$	$(-k_y + k_x a_3) E_{x3}$	$(-k_y a_4 + k_x) E_{y4}$

5.8. Boundary Conditions at the Interface Between Adjacent Layers; Iterative Formula for Computing the Reflectivity of Multilayers

Figure 5.5 shows a multilayer structure with interface planes at $Z = z_1, z_2, \ldots$. The surface is at $Z = 0$ and the bottom layer (substrate) extends to infinity along the negative Z-axis. We develop the boundary conditions for layer n whose upper and lower surfaces are at z_{n-1} and z_n respectively. We shall use the plane $Z = z_n$ as the reference plane for the four beams in layer n, so that each beam is described by the equation

$$\mathbf{E} = \mathbf{E}_0 \exp\left\{\mathrm{i}\left[k_x x + k_y y + k_z (z - z_n)\right]\right\} \tag{5.69}$$

The reflectivity matrix R_n at the lower interface is defined as follows:

$$\begin{pmatrix} E_{x3} \\ E_{y4} \end{pmatrix} = R_n \begin{pmatrix} E_{x1} \\ E_{y2} \end{pmatrix} = \begin{pmatrix} r_{11} & r_{12} \\ r_{21} & r_{22} \end{pmatrix}_n \begin{pmatrix} E_{x1} \\ E_{y2} \end{pmatrix} \tag{5.70}$$

The components of **E** parallel to the plane at z_n are thus given by

$$\begin{pmatrix} E_x \\ E_y \end{pmatrix}_{z_n^+} = \sum_{m=1}^{4} \begin{pmatrix} E_{xm} \\ E_{ym} \end{pmatrix}$$

$$= \begin{pmatrix} 1 & a_2 \\ a_1 & 1 \end{pmatrix} \begin{pmatrix} E_{x1} \\ E_{y2} \end{pmatrix} + \begin{pmatrix} 1 & a_4 \\ a_3 & 1 \end{pmatrix} \begin{pmatrix} E_{x3} \\ E_{y4} \end{pmatrix} \tag{5.71}$$

where z_n^+ stands for points just above z_n. Defining the 2×2 matrices in the preceding equation as A_{12} and A_{34} and using Eq. (5.70), the preceding equation is written as

$$\begin{pmatrix} E_x \\ E_y \end{pmatrix}_{z_n^+} = \left[A_{12} + A_{34} R \right]_n \begin{pmatrix} E_{x1} \\ E_{y2} \end{pmatrix}_n \tag{5.72}$$

The tangential components of the magnetic field at z_n^+ are given by

$$k_0 \begin{pmatrix} H_x \\ H_y \end{pmatrix}_{z_n^+} = \left[B_{12} + B_{34} R \right]_n \begin{pmatrix} E_{x1} \\ E_{y2} \end{pmatrix}_n \tag{5.73}$$

where

$$B_{12} = \begin{pmatrix} -k_{z1} a_1 + k_y b_1 & -k_{z2} + k_y b_2 \\ k_{z1} - k_x b_1 & k_{z2} a_2 - k_x b_2 \end{pmatrix} \tag{5.74}$$

with a similar expression for B_{34}. The tangential components of **E** and **H** at the upper boundary of the layer n, namely at z_{n-1}, are

$$\begin{pmatrix} E_x \\ E_y \end{pmatrix}_{z_{n-1}^-} = \left[A_{12} C_{12} + A_{34} C_{34} R \right]_n \begin{pmatrix} E_{x1} \\ E_{y2} \end{pmatrix}_n \tag{5.75}$$

$$\begin{pmatrix} H_x \\ H_y \end{pmatrix}_{z_{n-1}^-} = \left[B_{12} C_{12} + B_{34} C_{34} R \right]_n \begin{pmatrix} E_{x1} \\ E_{y2} \end{pmatrix}_n \tag{5.76}$$

The matrix C_{12} in Eqs. (5.75)–(5.76) is

$$C_{12} = \begin{bmatrix} \exp(\mathrm{i}k_{z1}d_n) & 0 \\ 0 & \exp(\mathrm{i}k_{z2}d_n) \end{bmatrix} \tag{5.77}$$

where d_n is the thickness of layer n. A similar expression is obtained for C_{34}. The continuity of the tangential components of **E** and of **H** at the interface $Z = z_n$ implies that

$$\left[A_{12} + A_{34}R\right]_n \begin{bmatrix} E_{x1} \\ E_{y2} \end{bmatrix}_n = \left[A_{12}C_{12} + A_{34}C_{34}R\right]_{n+1} \begin{bmatrix} E_{x1} \\ E_{y2} \end{bmatrix}_{n+1} \tag{5.78}$$

$$\left[B_{12} + B_{34}R\right]_n \begin{bmatrix} E_{x1} \\ E_{y2} \end{bmatrix}_n = \left[B_{12}C_{12} + B_{34}C_{34}R\right]_{n+1} \begin{bmatrix} E_{x1} \\ E_{y2} \end{bmatrix}_{n+1} \tag{5.79}$$

By eliminating from these equations the vector $(E_{x1}\ E_{y2})_{n+1}$, and equating the coefficients of $(E_{x1}\ E_{y2})_n$ that appear on both sides of the resulting equation, we obtain

$$\left[A_{12}C_{12} + A_{34}C_{34}R\right]_{n+1}^{-1} \left[A_{12} + A_{34}R\right]_n$$
$$= \left[B_{12}C_{12} + B_{34}C_{34}R\right]_{n+1}^{-1} \left[B_{12} + B_{34}R\right]_n \tag{5.80}$$

Let us define the matrix F_{n+1} as follows:

$$F_{n+1} = \left[B_{12}C_{12} + B_{34}C_{34}R\right]_{n+1} \left[A_{12}C_{12} + A_{34}C_{34}R\right]_{n+1}^{-1} \tag{5.81}$$

Then, Eq. (5.80) is written as

$$F_{n+1}\left[A_{12} + A_{34}R\right]_n = \left[B_{12} + B_{34}R\right]_n \tag{5.82}$$

Or, equivalently,

$$R_n = \left[F_{n+1}A_{34}^{(n)} - B_{34}^{(n)}\right]^{-1} \left[B_{12}^{(n)} - F_{n+1}A_{12}^{(n)}\right] \tag{5.83}$$

Equation (5.83) gives R_n in terms of R_{n+1}. Realizing that R for the substrate is zero, one can calculate R_n iteratively, starting at the substrate and moving up the multilayer until the surface reflectivity R_0 is obtained. This is a very general algorithm for computing the reflectivity of multilayers and one that is easily programmed on a personal computer. Note that all the matrices involved in this calculation are 2 × 2.

In practice it is useful to express the state of polarization of the incident and reflected beams at the surface in terms of the p- and s-components of polarization. We now express the relationship between the p- and s-components of the reflected and incident beams in terms of the matrix R_0. Assuming that the incident **k**-vector has spherical coordinates θ and ϕ, one may write

$$\begin{pmatrix} E_x \\ E_y \end{pmatrix} = \begin{pmatrix} \cos\theta\cos\phi & -\sin\phi \\ \cos\theta\sin\phi & \cos\phi \end{pmatrix} \begin{pmatrix} E_p \\ E_s \end{pmatrix} \tag{5.84}$$

We denote the coefficient matrix in Eq. (5.84) by P_{inc}, and define the reflectivity matrix R for the multilayer by the relation

$$\begin{pmatrix} E_p \\ E_s \end{pmatrix}_{\text{ref}} = R \begin{pmatrix} E_p \\ E_s \end{pmatrix}_{\text{inc}} \tag{5.85}$$

Then

$$R = P_{\text{inc}}^{-1} R_0 P_{\text{inc}} \tag{5.86}$$

Given R and the polarization state of the incident beam, one can calculate the polarization state of the reflected beam from Eq. (5.85).

5.9. Plane-wave Transmission through Multilayers

Having found the matrices R_n from the recursive relation in Eq. (5.83) we go back to Eq. (5.78) and rewrite it as follows:

$$\begin{pmatrix} E_{x1} \\ E_{y2} \end{pmatrix}_{n+1} = T_n \begin{pmatrix} E_{x1} \\ E_{y2} \end{pmatrix}_n \tag{5.87}$$

where

$$T_n = \left[A_{12}C_{12} + A_{34}C_{34}R \right]_{n+1}^{-1} \left[A_{12} + A_{34}R \right]_n \tag{5.88}$$

All the matrices on the right-hand side of Eq. (5.88) are known; thus it is possible to calculate T_n for all layers. The total transmission matrix is then given by

$$\hat{T} = T_N \cdots T_2 T_1 T_0 \tag{5.89}$$

$\hat{T}$ relates the incident (E_x, E_y) to the transmitted (E_x, E_y). To find the relation between the p- and s-components of the incident wave and those of the transmitted wave we must first find the direction of transmission, namely the angles θ_s and ϕ_s that describe the propagation direction in the substrate. Assuming that the substrate is transparent with refractive index n_{sub}, and that the medium of incidence (also transparent) has refractive index n_{inc}, one obtains the following relations from Snell's law:

$$\sin\theta_s = \frac{n_{\text{inc}}}{n_{\text{sub}}} \sin\theta \tag{5.90a}$$

$$\phi_s = \phi \tag{5.90b}$$

If the right-hand side of Eq. (5.90a) happens to be greater than unity, the wave in the substrate will be evanescent and no power transmission to the substrate will occur. On the other hand, when Eq. (5.90a) has a real solution for θ_s we define the matrix P_{sub} as

$$P_{\text{sub}} = \begin{pmatrix} \cos\theta_s \cos\phi & -\sin\phi \\ \cos\theta_s \sin\phi & \cos\phi \end{pmatrix} \tag{5.91}$$

The total transmission matrix T for the p- and s-components of polarization is defined as

$$\begin{pmatrix} E_p \\ E_s \end{pmatrix}_{\text{trans}} = T \begin{pmatrix} E_p \\ E_s \end{pmatrix}_{\text{inc}} \tag{5.92}$$

This matrix is related to $\hat{T}$, Eq. (5.89), as follows:

$$T = P_{\text{sub}}^{-1}\, \hat{T}\, P_{\text{inc}} \tag{5.93}$$

Given T and the polarization state of the incident beam, one can calculate the polarization state of the transmitted beam from Eq. (5.93).

5.10. Power Computation using Poynting's Theorem

The total electric field at a point (x, y, z) within the layer n is the sum of the four plane wave amplitudes at that point, namely,

$$\underline{E}(z) = \sum_{m=1}^{4} \underline{E}_0^{(m)} \exp\left\{i\left[k_x\, x + k_y\, y + k_{zm}(z - z_n)\right]\right\} \tag{5.94}$$

The magnetic field is also the sum of four magnetic amplitudes as follows:

$$\left[\underline{H}(z)\right]^{\mathrm{T}} = \frac{1}{k_0} \sum_{m=1}^{4} \underline{\underline{k}}^{(m)} \left[\underline{E}_0^{(m)}\right]^{\mathrm{T}} \exp\left\{i\left[k_x\, x + k_y\, y + k_{zm}(z - z_n)\right]\right\} \tag{5.95}$$

$\underline{\underline{k}}$ and $\underline{E}_0$ for the various beams in each layer can be calculated with the methods of the preceding sections. Therefore, the total electromagnetic field can be computed from Eqs. (5.94) and (5.95). The Poynting theorem (Eq. (5.27)) now yields

$$\underline{S}^{\mathrm{T}} = \frac{1}{2k_0} \mathrm{Re}\left\{\sum_{m=1}^{4} \sum_{m'=1}^{4} \underline{\underline{E}}_0^{(m)} \underline{\underline{k}}^{*(m')} \left[\underline{E}_0^{*(m')}\right]^{\mathrm{T}} \times \exp\left[i(k_{zm} - k^*_{zm'})(z - z_n)\right]\right\} \tag{5.96}$$

As expected, the Poynting vector **S** is independent of x and y. The component S_z of the Poynting vector along Z is the time-averaged rate of flow of energy per unit area in the Z-direction. S_z is the quantity of interest in most applications.

5.11. Numerical Results

We now present numerical results based on the algorithm developed in the preceding sections. We study a quadrilayer MO device whose schematic diagram is shown in Fig. 5.6. This device has a glass substrate of refractive index $n_{sub} = 1.5$, coated with a 500 nm-thick layer of aluminum with complex index $(n, k) = (2.75, 8.31)$. A quarter-wave-thick layer of SiO_x (thickness = 143.2 nm) with $(n, k) = (1.449, 0)$ separates the aluminum

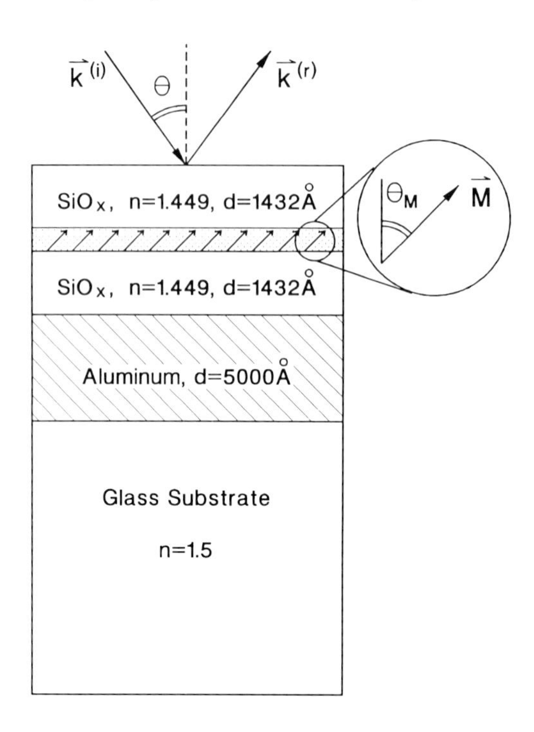

Figure 5.6. Schematic diagram of a quadrilayer magneto-optical device used in numerical computations. The beam is incident from the surface side. The magnetic film is uniformly magnetized, but the direction of its magnetization **M** is chosen differently in different examples. The computation results in Figs. 5.7 and 5.8 are obtained with a perpendicular magnetization (i.e., $\Theta_M = 0$), while for Fig. 5.10 the magnetization is assumed to be parallel to the *Y*-axis (i.e., $\Theta_M = 90°$).

layer from a 20 nm-thick MO film whose tensor elements are $\epsilon_{xx} = \epsilon_{yy} = \epsilon_{zz} = -4.8984 + 19.415i$ and $\epsilon_{xy} = -\epsilon_{yx} = 0.4322 + 0.0058i$. The magnetic film is coated with another quarter-wave-thick layer of SiO_x and the medium of incidence is air ($n_{inc} = 1$). The incident beam is a monochromatic plane wave with $\lambda = 830$ nm.

Figure 5.7(a) shows the various reflectivities as functions of the angle of incidence θ. (Only the magnitudes of the complex reflectivities are shown). r_{pp} and r_{ps} are, respectively, the p- and s-components of the reflected beam when the incident polarization is p. Similarly, r_{sp} and r_{ss} correspond to s-polarized incident light. Notice that r_{ps} and r_{sp} are identical. Note also that, unlike reflection from a single interface where r_{pp} shows a dip around the Brewster angle, the dip for this multilayer appears in the r_{ss} curve. Figure 5.7(b) shows the Kerr rotation angle and the ellipticity as functions of θ for p-polarized incident light. (See Chapter 6 for definitions of the Kerr rotation angle and the ellipticity.) The corresponding curves for s-polarization are shown in Fig. 5.7(c). The Kerr rotation angle and the ellipticity in (b) are very different from those in (c), even though the effective MO signals in the two cases are the same, i.e., $r_{ps} = r_{sp}$.

Figure 5.8 shows the magnitude of the Poynting vector, S_z, when a linearly polarized beam with unit power density illuminates the quadrilayer at normal incidence. This is a plot of the average power density that crosses planes parallel to the XY-plane at various positions along the Z-axis. The power density is constant in the transparent layers but drops rather sharply in the absorbing layers, as expected.

Next we consider the MO signal at normal incidence in the case where the magnetization vector **M** is being pulled into the plane of the film, a situation that arises, for example, in optical measurements of magnetic anisotropy.[9] For a linearly polarized, normally incident beam, Fig. 5.9 shows the calculated values of the Kerr rotation angle versus the angle Θ_M between **M** and the Z-axis. It is found that the polar Kerr signal is proportional to $\cos\Theta_M$ and that this result is independent of the relative orientation of the magnetization vector **M** and the polarization vector.

Finally, in Fig. 5.10 we show certain manifestations of the longitudinal Kerr effect when the magnetic film in our quadrilayer becomes in-plane magnetized. (The longitudinal Kerr effect occurs when the magnetization vector **M** lies in the plane of incidence parallel to the film surface.) Figure 5.10(a) shows the magnitudes of r_{sp} and r_{ps}, which are again identical, as was the case with the polar Kerr effect in Fig. 5.7(a). The largest MO signal in the longitudinal case, however, occurs at $\theta \simeq 65°$. Figure 5.10(b) shows plots of the Kerr rotation angle and the ellipticity versus θ for p-polarized incident light. Similar results for s-polarization are shown in Fig. 5.10(c). Notice that these values are smaller than the corresponding polar Kerr values in Fig. 5.7 by more than an order of magnitude.

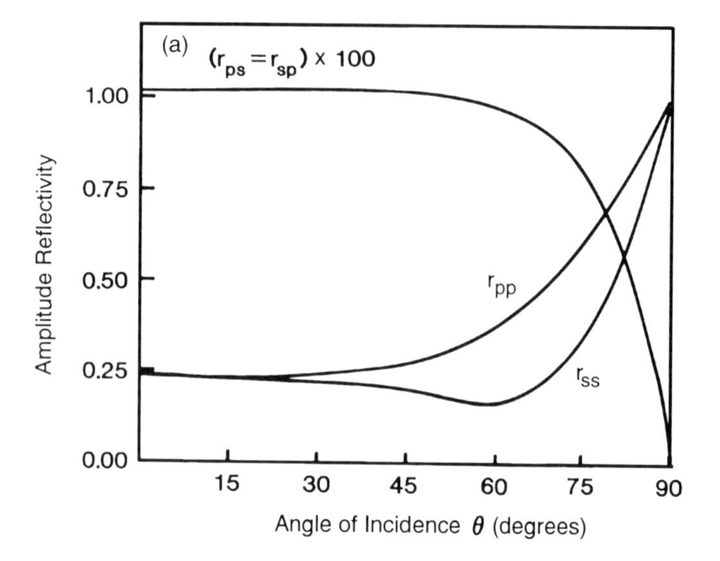
(a)
$(r_{ps} = r_{sp}) \times 100$
1.00
0.75
0.50
0.25
0.00
Amplitude Reflectivity
r_{pp}
r_{ss}
15
30
45
60
75
90
Angle of Incidence θ (degrees)

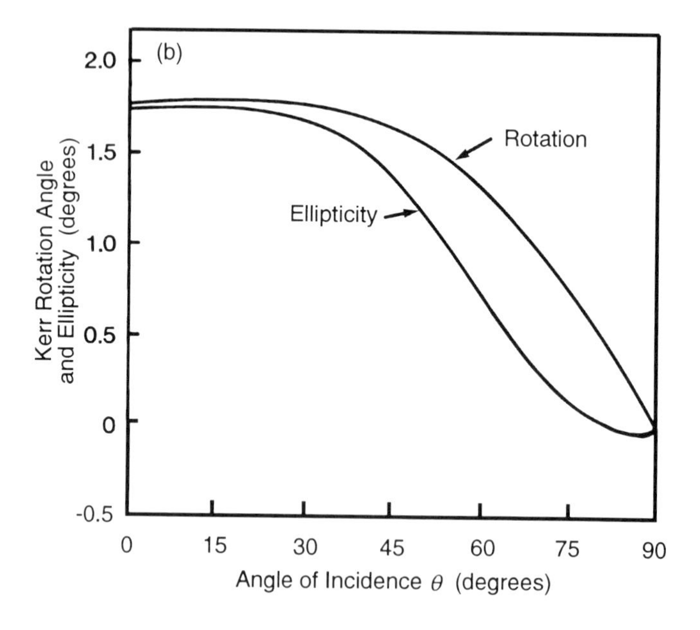
(b)
2.0
1.5
1.0
0.5
0
-0.5
Kerr Rotation Angle
and Ellipticity (degrees)
Rotation
Ellipticity
0
15
30
45
60
75
90
Angle of Incidence θ (degrees)

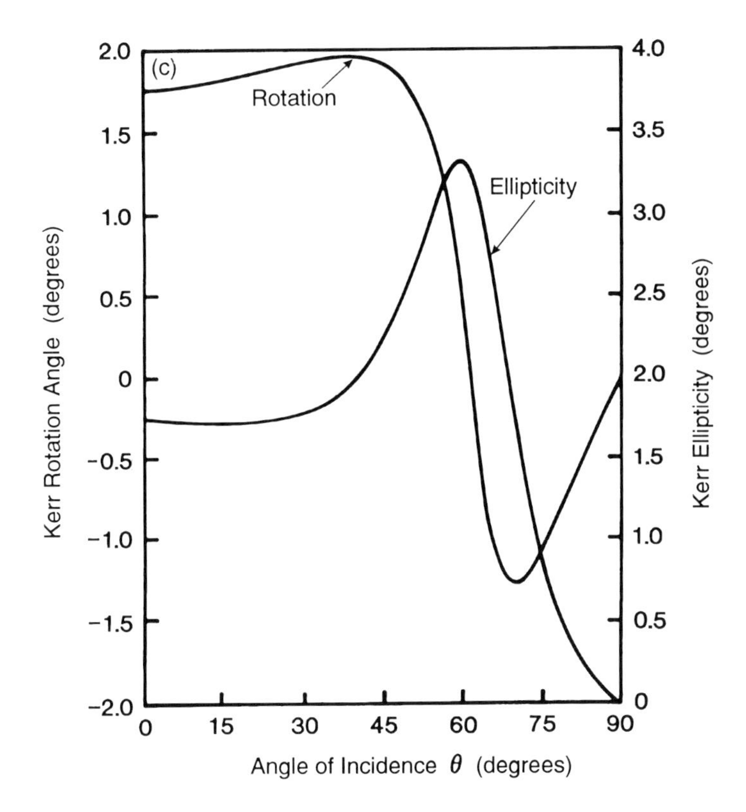

Figure 5.7. Magneto-optical Kerr effect signals for the quadrilayer of Fig. 5.6 with perpendicular magnetization. (a) Magnitudes of r_{pp}, r_{ss}, r_{ps}, and r_{sp} versus angle of incidence θ. Note that $r_{ps} = r_{sp}$ and that the corresponding curve is magnified by a factor of 100. (b) Kerr rotation angle and ellipticity versus θ for p-polarized incident light. (c) Kerr rotation angle and ellipticity for s-polarized incident light. Note that the two curves have different scales.

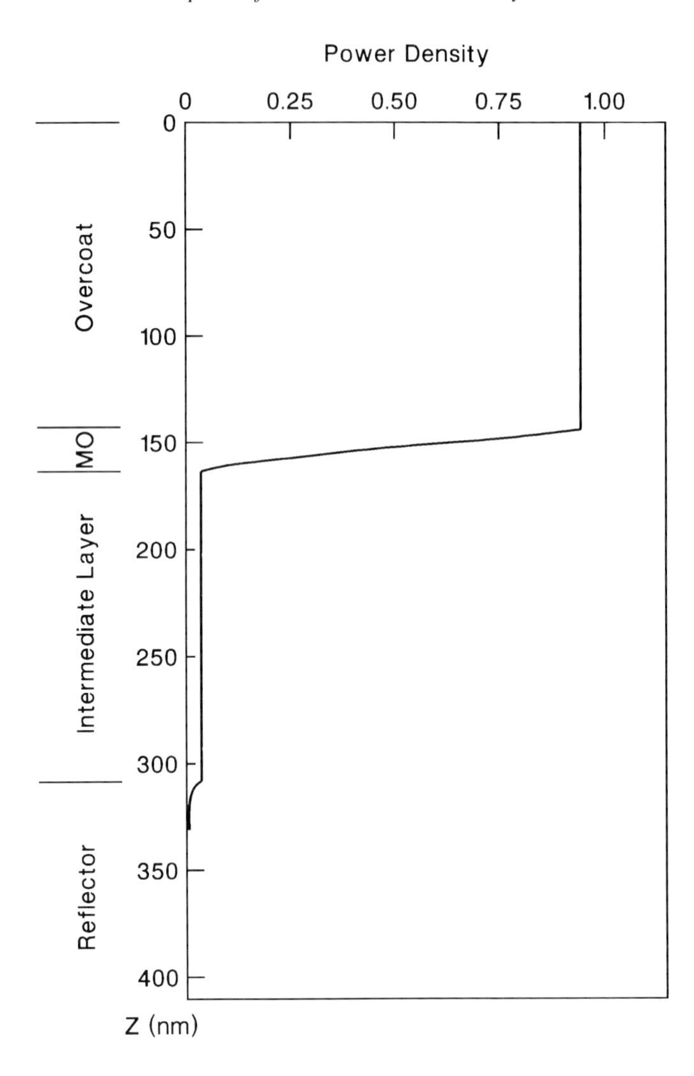

Figure 5.8. Poynting vector throughout the quadrilayer of Fig. 5.6 (with perpendicular magnetization). The normally incident beam is assumed to have unit power density and linear polarization. Note that the bulk of the absorption (about 91% of the total incident power) takes place within the magneto-optical layer, while the reflector absorbs only 3%. The remaining 6% of the incident power is reflected back into the medium of incidence.

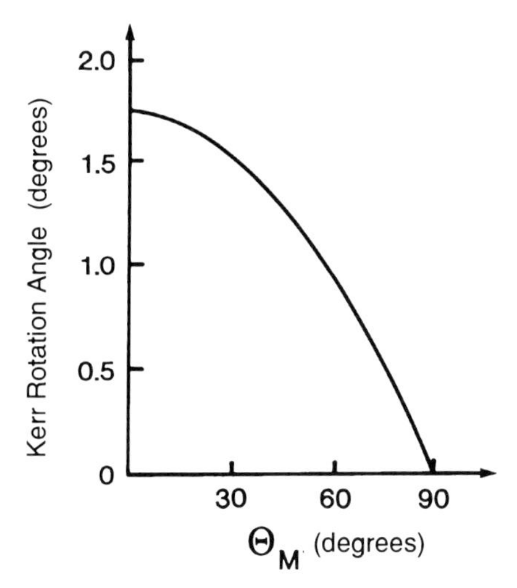

Figure 5.9. The magneto-optical Kerr rotation angle versus Θ_M (the angle between the magnetization vector **M** and the Z-axis) for linearly polarized light normally incident on the quadrilayer of Fig. 5.6. The curve can be precisely matched to a plot of $\cos\Theta_M$.

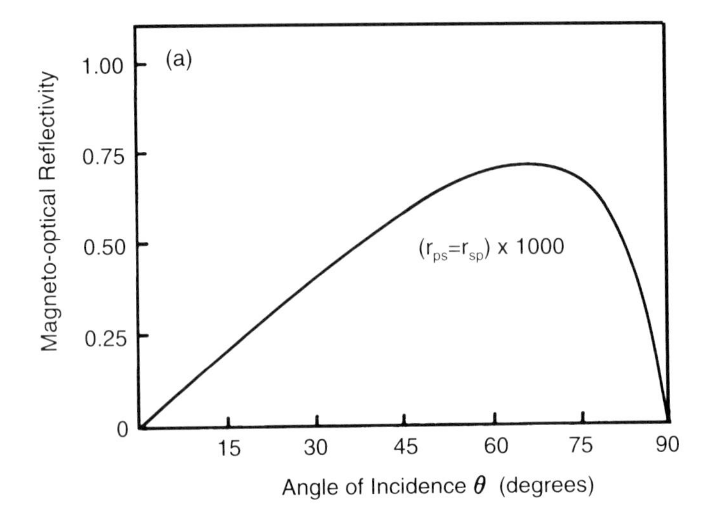
(a)
Magneto-optical Reflectivity
1.00
0.75
0.50
0.25
0
$(r_{ps}=r_{sp})$ x 1000
15
30
45
60
75
90
Angle of Incidence θ (degrees)

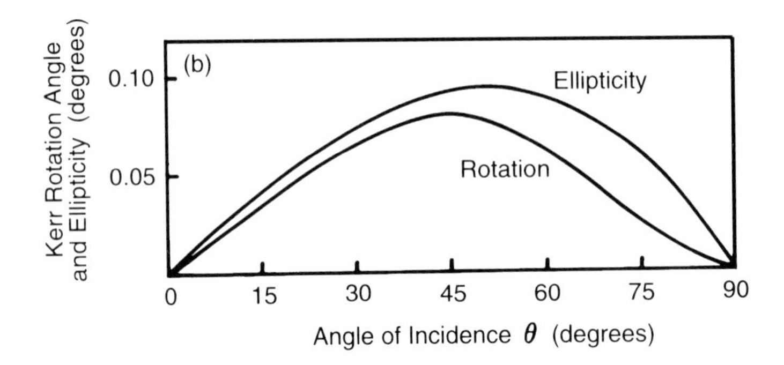
(b)
Kerr Rotation Angle and Ellipticity (degrees)
0.10
0.05
Ellipticity
Rotation
0
15
30
45
60
75
90
Angle of Incidence θ (degrees)

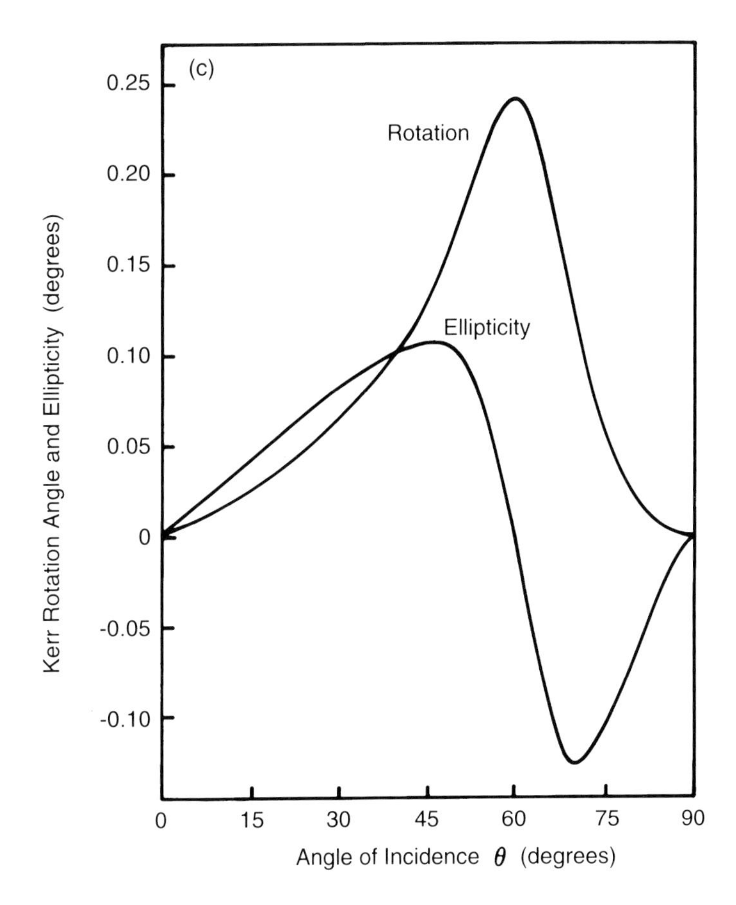

Figure 5.10. Longitudinal magneto-optical Kerr effect signals for the quadrilayer of Fig. 5.6 with in-plane magnetization. (a) Magnitudes of r_{ps} and r_{sp} versus angle of incidence θ (note that $r_{ps} = r_{sp}$). The curve is magnified by a factor of 1000. (b) Polarization rotation angle and ellipticity versus θ for p-polarized incident light. (c) Polarization rotation angle and ellipticity versus θ for s-polarized incident light.

Appendix 5.A. Transformation of Dielectric Tensor under Rotation of Coordinates

In this appendix we describe the transformation of dielectric tensors caused by a rotation of the coordinate system. Consider two Cartesian coordinate systems XYZ and $X'Y'Z'$, which share the origin O. In general three angles Θ, Φ, and Ψ specify the relative orientation of these systems. Θ and Φ are the spherical coordinates of Z' in the XYZ system. Imagine a rotation of $X'Y'Z'$ where Φ remains constant while Θ goes to zero. After this rotation Z and Z' will be aligned but the angle between X and X' will not necessarily be zero. This angle is denoted by Ψ and is measured counter-clockwise from the positive X-axis. Keeping Z and Z' aligned while rotating $X'Y'Z'$ through the angle Ψ will bring the two systems into complete alignment. (Ψ, Θ, and $\Phi - \Psi$ are the Euler angles.) It is thus possible to express the unit vectors $\hat{\mathbf{x}}'$, $\hat{\mathbf{y}}'$, $\hat{\mathbf{z}}'$ in terms of the unit vectors $\hat{\mathbf{x}}$, $\hat{\mathbf{y}}$, $\hat{\mathbf{z}}$ as follows:

$$\hat{\mathbf{x}}' = a_{11}\hat{\mathbf{x}} + a_{12}\hat{\mathbf{y}} + a_{13}\hat{\mathbf{z}} \quad (5.A1)$$

$$\hat{\mathbf{y}}' = a_{21}\hat{\mathbf{x}} + a_{22}\hat{\mathbf{y}} + a_{23}\hat{\mathbf{z}} \quad (5.A2)$$

$$\hat{\mathbf{z}}' = a_{31}\hat{\mathbf{x}} + a_{32}\hat{\mathbf{y}} + a_{33}\hat{\mathbf{z}} \quad (5.A3)$$

where

$$a_{11} = \cos\Theta \cos\Phi \cos(\Phi - \Psi) + \sin\Phi \sin(\Phi - \Psi) \quad (5.A4)$$

$$a_{12} = \cos\Theta \sin\Phi \cos(\Phi - \Psi) - \cos\Phi \sin(\Phi - \Psi) \quad (5.A5)$$

$$a_{13} = -\sin\Theta \cos(\Phi - \Psi) \quad (5.A6)$$

$$a_{21} = \cos\Theta \cos\Phi \sin(\Phi - \Psi) - \sin\Phi \cos(\Phi - \Psi) \quad (5.A7)$$

$$a_{22} = \cos\Theta \sin\Phi \sin(\Phi - \Psi) + \cos\Phi \cos(\Phi - \Psi) \quad (5.A8)$$

$$a_{23} = -\sin\Theta \sin(\Phi - \Psi) \quad (5.A9)$$

$$a_{31} = \sin\Theta \cos\Phi \quad (5.A10)$$

$$a_{32} = \sin\Theta \sin\Phi \quad (5.A11)$$

$$a_{33} = \cos\Theta \quad (5.A12)$$

Now, let $\mathbf{E}$ and $\mathbf{D}$ be the electric field vector and the displacement vector in the XYZ coordinate system. The dielectric tensor $\underset{\sim}{\epsilon}$ in this system relates $\mathbf{E}$ and $\mathbf{D}$ as follows:

$$\mathbf{D} = \underset{\sim}{\epsilon}\, \mathbf{E} \quad (5.A13)$$

Similarly, in the $X'Y'Z'$ system we have

$$\mathbf{D}' = \underset{\sim}{\epsilon}' \, \mathbf{E}' \tag{5.A14}$$

But $\underline{E} = \underline{E}'P$ and $\underline{D} = \underline{D}'P$ where

$$P = \begin{pmatrix} a_{11} & a_{12} & a_{13} \\ a_{21} & a_{22} & a_{23} \\ a_{31} & a_{32} & a_{33} \end{pmatrix} \tag{5.A15}$$

is the rotation matrix between XYZ and $X'Y'Z'$. Therefore, Eqs. (5.A13) and (5.A14) yield

$$\underset{\sim}{\epsilon} = P^{\mathrm{T}} \underset{\sim}{\epsilon}' P \tag{5.A16}$$

Here $P^{\mathrm{T}} = P^{-1}$ because P is unitary. Equation (5.A16) is the transformation of the dielectric tensor from the $X'Y'Z'$ system to the XYZ system.

As an example consider a MO material with direction of magnetization along Z'. Then in $X'Y'Z'$ the dielectric tensor is that of a polar Kerr material, i.e.,

$$\underset{\sim}{\epsilon}' = \begin{pmatrix} \epsilon_{xx} & \epsilon_{xy} & 0 \\ -\epsilon_{xy} & \epsilon_{xx} & 0 \\ 0 & 0 & \epsilon_{xx} \end{pmatrix} \tag{5.A17}$$

The rotation angles Θ and Φ specify the magnetization vector in the XYZ system but Ψ is arbitrary. Using Eq. (5.A16) the dielectric tensor in the new XYZ coordinates is found to be

$$\underset{\sim}{\epsilon} = \begin{pmatrix} \epsilon_{xx} & \cos\Theta\, \epsilon_{xy} & -\sin\Theta \sin\Phi\, \epsilon_{xy} \\ -\cos\Theta\, \epsilon_{xy} & \epsilon_{xx} & \sin\Theta \cos\Phi\, \epsilon_{xy} \\ \sin\Theta \sin\Phi\, \epsilon_{xy} & -\sin\Theta \cos\Phi\, \epsilon_{xy} & \epsilon_{xx} \end{pmatrix} \tag{5.A18}$$

Note that the dielectric tensor in XYZ is independent of Ψ and that it reduces to the familiar forms for the longitudinal and transverse Kerr effects when $\Theta = 90°$, $\Phi = 0$ and when $\Theta = 90°$, $\Phi = 90°$ respectively.

Appendix 5.B. Method of Computing the Roots of Fourth-order Polynomials

In this appendix we describe a method for computing the roots of the fourth-order polynomial equation

$$S^4 + AS^3 + BS^2 + CS + D = 0 \tag{5.B1}$$

The first step is to calculate the roots of the following third-order polynomial equation:

$$8S^3 - 4BS^2 + 2(\mathrm{AC} - 4D)S - [C^2 + D(A^2 - 4B)] = 0 \tag{5.B2}$$

To this end we define y_1, y_2, x_1, and x_2 as follows:

$$y_1 = \frac{AC}{12} - \frac{D}{3} - \frac{B^2}{36} \tag{5.B3}$$

$$y_2 = \left(\frac{B}{6}\right)^3 + \frac{A^2D + C^2}{16} - \frac{ABC}{48} - \frac{BD}{6} \tag{5.B4}$$

$$x_1 = \left\{y_2 + \sqrt{y_1{}^3 + y_2{}^2}\right\}^{1/3} \tag{5.B5}$$

$$x_2 = \left\{y_2 - \sqrt{y_1{}^3 + y_2{}^2}\right\}^{1/3} \tag{5.B6}$$

Since a complex number has three possible cube roots, x_1 and x_2 can each have three different values. Of these, the only acceptable ones are those that satisfy the equality

$$x_1x_2 + y_1 = 0 \tag{5.B7}$$

It is not difficult to verify that only three pairs (x_1, x_2) satisfy Eq. (5.B7). The solutions of Eq. (5.B2) in terms of the acceptable pairs (x_1, x_2) are

$$S_0, S_1, S_2 = x_1 + x_2 + \tfrac{1}{6}B \tag{5.B8}$$

The next step is to choose one of the solutions of the third-order equation (say S_0) and form two quadratic equations as follows:

$$S^2 + \left(\tfrac{1}{2}A + \sqrt{\tfrac{1}{4}A^2 - B + 2S_0}\right)S + \left(S_0 + \sqrt{S_0^2 - D}\right) = 0 \qquad (5.B9)$$

$$S^2 + \left(\tfrac{1}{2}A - \sqrt{\tfrac{1}{4}A^2 - B + 2S_0}\right)S + \left(S_0 - \sqrt{S_0^2 - D}\right) = 0 \qquad (5.B10)$$

The four roots of the original equation (5.B1) will be the roots of the quadratic equations (5.B9) and (5.B10). Because square roots of complex numbers appear in these equations, the question arises as to which root belongs in which equation. The answer is found by setting the product of Eqs. (5.B9) and (5.B10) equal to Eq. (5.B1). It turns out that the complex square roots must satisfy the relation

$$\sqrt{S_0^2 - D}\,\sqrt{\tfrac{1}{4}A^2 - B + 2S_0} = \tfrac{1}{2}(AS_0 - C) \qquad (5.B11)$$

Although the procedure described here is exact, in practice numerical errors rapidly accumulate and give inaccurate results. To overcome this difficulty, one often uses an iterative technique that pushes the inexact solutions towards the correct roots. Since the analytic solutions obtained are fairly close to the desired roots, the algorithm does not need to be highly sophisticated. For instance, let $\hat{S}$ be an estimated root of Eq. (5.B1). By moving along the slope of the polynomial at $\hat{S}$ a better estimate $\hat{\hat{S}}$ may be obtained as follows:

$$\hat{\hat{S}} = \hat{S} - \frac{\hat{S}^4 + A\hat{S}^3 + B\hat{S}^2 + C\hat{S} + D}{4\hat{S}^3 + 3A\hat{S}^2 + 2B\hat{S} + C} \qquad (5.B12)$$

One can get arbitrarily close to the exact roots by repeating the procedure described in Eq. (5.B12).

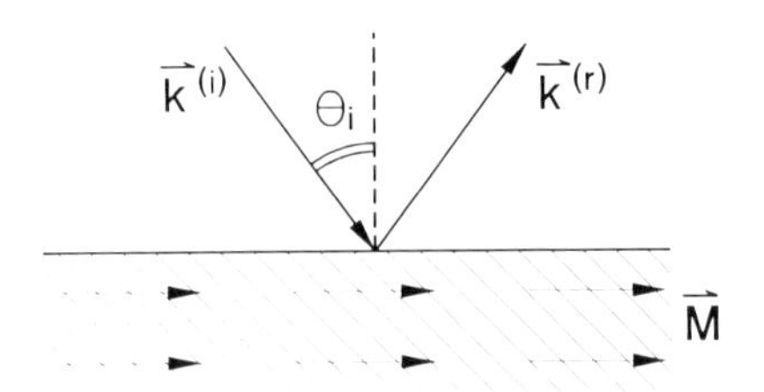

Figure 5.11.

(5.8) A beam is normally incident on a homogeneous, semi-infinite medium with the following dielectric tensor:

$$\underset{\sim}{\epsilon} = \begin{pmatrix} -2.3 + 27.8i & -0.2 + 0.8i & 0 \\ 0.2 - 0.8i & -2.3 + 27.8i & 0 \\ 0 & 0 & -3 + 28i \end{pmatrix}$$

(a) Does this medium exhibit any MO effects? If so, which one?
(b) The incident beam is linearly polarized. Determine the state of polarization of the reflected beam by identifying the Kerr rotation angle and the ellipticity.
(c) Assuming an incident beam with unit intensity, what is the intensity of the reflected beam?

(5.9) Write the dielectric tensor of a non-absorbing polar Kerr MO medium and determine the corresponding Kerr rotation angle.

(5.10) In Fig. 5.11 a plane wave is incident on a magnetic medium whose magnetization vector **M** is along the Y-axis; the plane of incidence is YZ. (This is the geometry for the longitudinal Kerr effect.) Determine the reflection coefficients as functions of the angle of incidence θ_i.

(5.11) The plane wave in Fig. 5.12 is normally incident on a magnetic medium with a dielectric coating layer. The magnetization vector of the medium, **M**, is in the Z-direction and the coating layer has thickness d and refractive index n where $n = \sqrt{\epsilon}$ is real. Determine the reflection coefficients at the surface for a linearly polarized incident beam.

(5.12) The plane monochromatic beam in Fig. 5.13 has wavelength λ_0 and is normally incident on a coated substrate. The medium of incidence is air ($\epsilon = 1$), the film and the substrate are isotropic with dielectric constants ϵ_1 and ϵ_2, the film has thickness d, and the substrate is semi-infinite.
(a) Find the reflectivity matrix R_0 at the surface.
(b) Verify the result of part (a) for some trivial cases.
(c) If the coating layer is a non-absorbing dielectric with refractive index $n_0 = \sqrt{\epsilon_1}$ and thickness $d = \lambda_0/4n_0$, what will be the reflectivity of the structure?

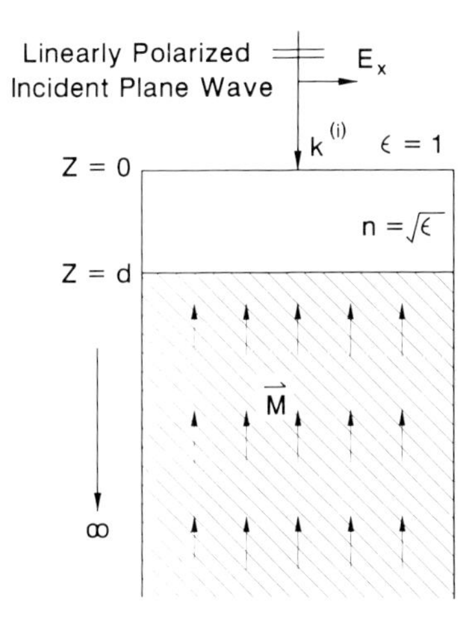

Figure 5.12.

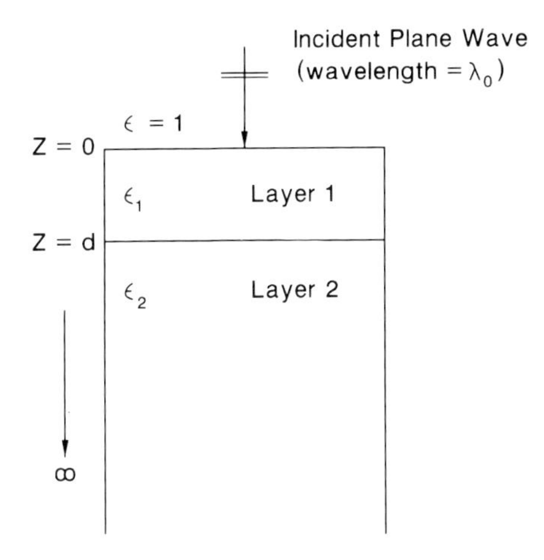

Figure 5.13.

(d) Generalize the result of part (c) to the case where the substrate exhibits the MO polar Kerr effect. Let the dielectric constants for right and left circular polarizations be ϵ_2^- and ϵ_2^+, respectively, and denote the corresponding reflectivities by r^- and r^+. Explain how the MO signal $|r^+ - r^-|$ is enhanced as a result of coating with the quarter-wave-thick dielectric layer.

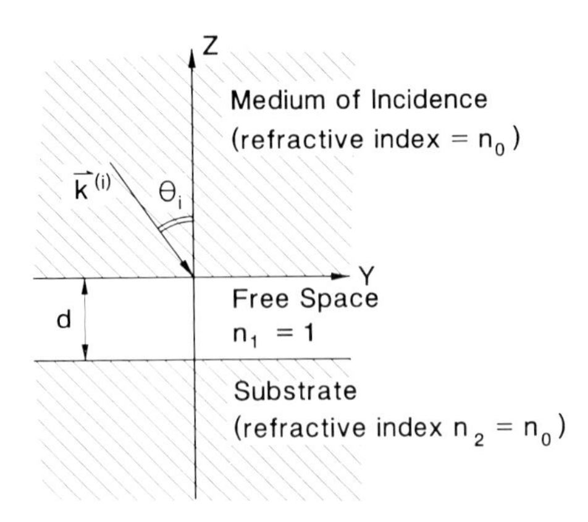

Figure 5.14.

(5.13) Two semi-infinite dielectric media, both having the same real refractive index n_0, are separated by a distance d as shown in Fig. 5.14. The angle of incidence θ_i is greater than the critical angle for total internal reflection. Determine the coupling of light into the second medium as a function of d.

(5.14) The schematic diagram in Fig. 5.15(a) shows a plane monochromatic beam at oblique incidence on a transparent, uniaxially birefringent medium. The interface is the XY-plane, the medium thickness is d, and the refractive indices n_x and n_y are identical, being somewhat different from n_z.

(a) What is the optical path-length difference between the p- and s-polarized states of the beam after one passage through the medium?

(b) A linearly polarized beam is focused on an optical storage medium through a birefringent substrate, as shown in Fig. 5.15(b). Assume that the objective is corrected for focusing through a cover plate of thickness d and isotropic refractive index n_x; you may thus ignore all spherical aberrations caused by the substrate. What type of aberration does the birefringence produce on the focused spot?

(5.15) Consider a non-absorbing slab of birefringent material with thickness d and principal refractive indices n_x, n_y, n_z along the X-, Y-, Z-axes of a Cartesian coordinate system. The facets of the slab are parallel to the XY-plane and the surrounding environment is free space. A column of coherent, circularly polarized, monochromatic light with wavelength λ_0 and diameter D, propagating along the unit vector $\boldsymbol{\sigma} = (0, \sin\theta, \cos\theta)$, is incident on one facet of the birefringent slab and emerges from the opposite facet. Assuming that $D >> \lambda_0$, decompose the incident beam into its spatial Fourier spectrum, allow each component to propagate through the slab, and superimpose the resulting plane waves to reconstruct the two beams that exit the slab. Determine the location and diameter of the emerging beams.

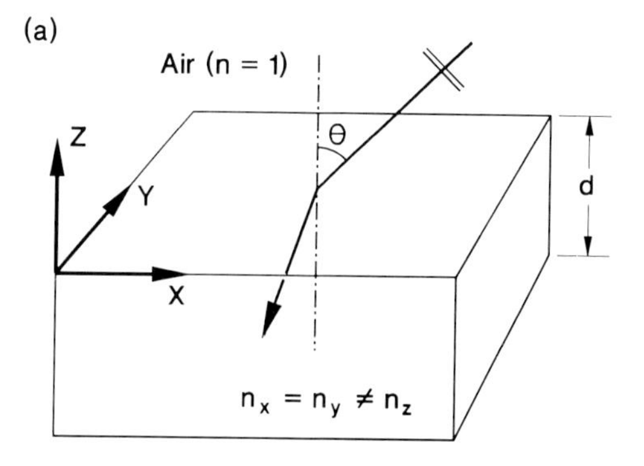
(a)
Air (n = 1)
Z
Y
X
θ
d
n x = n y ≠ n z

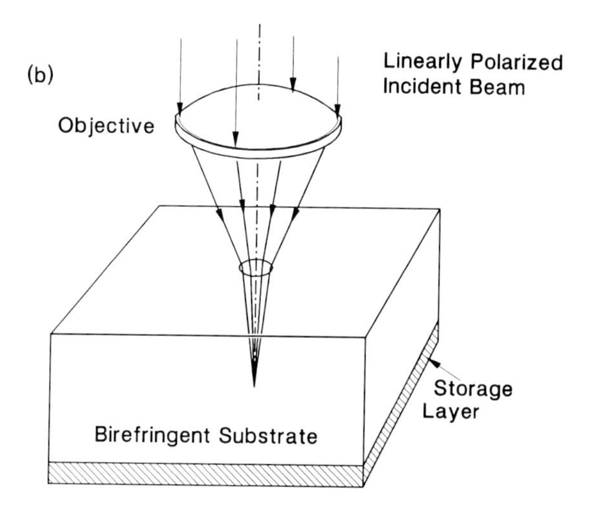
(b)
Linearly Polarized
Incident Beam
Objective
Storage
Layer
Birefringent Substrate

Figure 5.15.

6

Magneto-optical Readout

Introduction

In magneto-optical (MO) data storage systems the readout of recorded information is achieved by means of a focused beam of polarized light. Conventional systems today utilize a linearly polarized beam, whereas some suggested alternative methods rely instead on circular or elliptical polarization for readout.

Although the Jones calculus is the standard vehicle for analyzing the polarization properties of optical systems, we believe the alternative approach used in this chapter provides a better, more intuitive explanation for the operation of the readout system. The chapter begins by introducing the two basic states of circular polarization: right (RCP) and left (LCP). It then proceeds to show that the other two polarization states, linear (LP) and elliptical (EP), may be constructed by superposition of the two circular states. The sections that follow describe the actions of a quarter-wave plate (QWP) and a polarizing beam-splitter (PBS) on the various states of polarization. These two optical elements are of primary importance in detection schemes aimed at probing the state of polarization of a beam.

The classical scheme of differential detection for MO readout will be analyzed in section 6.4, first in its simple form for detecting the polarization rotation angle and then, with the addition of phase-compensating elements, for detecting ellipticity as well. In section 6.5 we describe an extension of the differential detection scheme. This extended scheme is used for spectral characterization (i.e., measurement of the wavelength-dependence of the Kerr angle and the ellipticity) of MO media. Section 6.6 describes an alternative method of MO readout based on diffraction from the magnetic domain boundaries. In the final section, section 6.7, we derive a figure of merit (FOM) for the magneto-optical media based on their dielectric tensor, and argue that this FOM places an upper bound on the performance of the readout system.

6.1. States of Circular, Linear, and Elliptical Polarization

When a light beam's electric field vector **E** rotates in the plane of polarization, its tip tracing a circle with angular velocity ω of the light, the beam is said to be circularly polarized. If the observer is in a position to see the beam head-on, then clockwise (counterclockwise) rotation of the

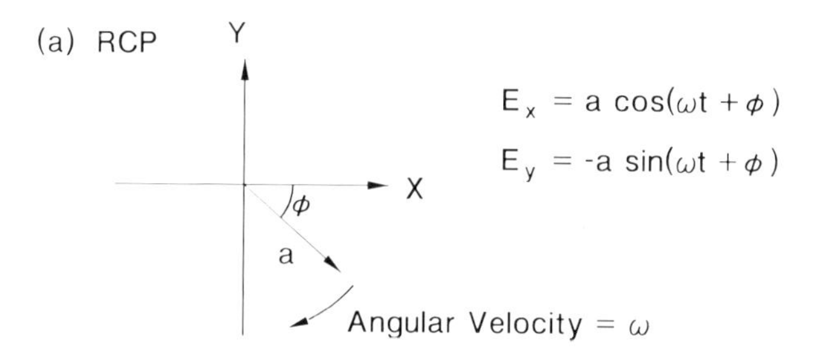

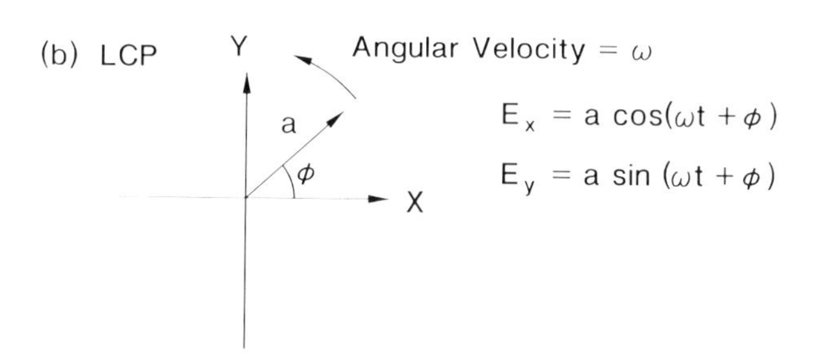

Figure 6.1. Electric field vector in the plane of polarization for (a) right circularly polarized and (b) left circularly polarized light. In both cases ϕ is assumed to be positive. The position of the vector shown in each case corresponds to time $t = 0$.

E-field is defined as right (left) circular polarization or RCP (LCP). An RCP beam (usually identified by a superscript minus sign) with amplitude a and phase ϕ, has the following projections on the X- and Y-axes in the plane of polarization:

$$\text{RCP:} \qquad a^{(-)} \underline{/\phi} = a \cos(\omega t + \phi)\, \hat{\mathbf{x}} - a \sin(\omega t + \phi)\, \hat{\mathbf{y}} \tag{6.1}$$

An LCP beam, on the other hand, is identified by a superscript plus sign:

$$\text{LCP:} \qquad a^{(+)} \underline{/\phi} = a \cos(\omega t + \phi)\, \hat{\mathbf{x}} + a \sin(\omega t + \phi)\, \hat{\mathbf{y}} \tag{6.2}$$

Figure 6.1 shows vector diagrams representing RCP and LCP in the XY-plane of polarization. Notice that, aside from the direction of rotation, the two modes of circular polarization also differ in the initial positions of their **E** vectors relative to the X-axis.

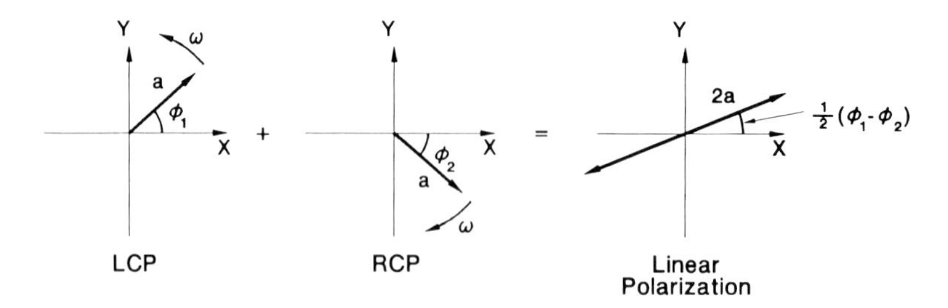

Figure 6.2. When an LCP beam with amplitude a and phase ϕ_1 is superimposed on an RCP beam with amplitude a and phase ϕ_2, the result is a linearly polarized beam of amplitude $2a$ and phase $\frac{1}{2}(\phi_1 + \phi_2)$. The angle of the resulting **E**-field to the X-axis is $\frac{1}{2}(\phi_1 - \phi_2)$.

In an alternative representation using complex vectors, **E** may be written as follows:

$$\text{RCP:} \qquad \mathbf{E}^{(-)} = a\left[\exp(-\mathrm{i}\phi)\right](\hat{\mathbf{x}} - \mathrm{i}\hat{\mathbf{y}}) \tag{6.3}$$

$$\text{LCP:} \qquad \mathbf{E}^{(+)} = a\left[\exp(-\mathrm{i}\phi)\right](\hat{\mathbf{x}} + \mathrm{i}\hat{\mathbf{y}}) \tag{6.4}$$

If Eq. (6.3) is multiplied by the time-dependence factor $\exp(-\mathrm{i}\omega t)$, the real part of the resulting complex function will be the **E**-field in Eq. (6.1). Similarly, Eq. (6.4) will lead to Eq. (6.2).

In general, two circularly polarized beams with opposite senses of rotation, equal amplitudes a, and different phases ϕ_1 and ϕ_2, may be superimposed to yield a linearly polarized beam, that is,

$$a^{(+)}\underline{/\phi_1} + a^{(-)}\underline{/\phi_2} = 2a\cos\left[\tfrac{1}{2}(\phi_1 - \phi_2)\right]\cos\left[\omega t + \tfrac{1}{2}(\phi_1 + \phi_2)\right]\hat{\mathbf{x}}$$
$$+ 2a\sin\left[\tfrac{1}{2}(\phi_1 - \phi_2)\right]\cos\left[\omega t + \tfrac{1}{2}(\phi_1 + \phi_2)\right]\hat{\mathbf{y}} \tag{6.5}$$

Clearly, the X- and Y-components of the resultant polarization are in phase, which is characteristic of linear polarization. The new **E**-field vector has a length of $2a$ and is oriented at angle $\frac{1}{2}(\phi_1 - \phi_2)$ relative to the X-axis. Figure 6.2 gives a geometrical interpretation of Eq. (6.5). The angle of the resulting LP vector to X is dependent only on the phase difference between its LCP and RCP components. When the circular beams are in phase, that is, when $\phi_1 = \phi_2$, the LP vector will be parallel to the X-axis.

By a reverse procedure, it is always possible to decompose a linear beam into circularly polarized components. Consider, for example, an LP beam with amplitude E_0, phase $\phi = 0$, and angle Δ relative to the X-axis. From the preceding discussion it follows that

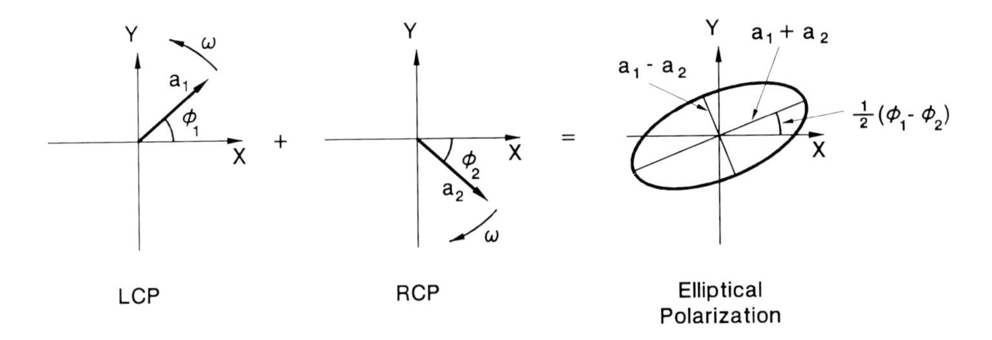

Figure 6.3. Superposition of LCP and RCP beams with different amplitudes results in elliptical polarization. The length of the major axis of the ellipse is $2(a_1 + a_2)$, while the length of the minor axis is $2(a_1 - a_2)$. The major axis makes an angle $\frac{1}{2}(\phi_1 - \phi_2)$ with X. The sense of rotation of the **E**-vector around the ellipse is counterclockwise if $a_1 > a_2$, and clockwise if $a_1 < a_2$.

$$2a = E_0 \; ; \qquad \tfrac{1}{2}(\phi_1 + \phi_2) = 0 \; ; \qquad \tfrac{1}{2}(\phi_1 - \phi_2) = \Delta \tag{6.6}$$

Consequently, $a = \frac{1}{2} E_0$, $\phi_1 = \Delta$, and $\phi_2 = -\Delta$. In general, superposition of RCP and LCP beams results in elliptical polarization. Conversely, any beam with elliptical polarization may be resolved into RCP and LCP components; the geometrical construction shown in Fig. 6.3 should help explain this point. The angle of the major axis of the ellipse with X is $\frac{1}{2}(\phi_1 - \phi_2)$, and the ellipticity ϵ of the beam is given by

$$\epsilon = \tan^{-1}\left(\frac{a_1 - a_2}{a_1 + a_2}\right) \tag{6.7}$$

When the amplitudes of the RCP and LCP beams are identical, the resulting beam has zero ellipticity and is therefore linearly polarized; any difference between a_1 and a_2 results in nonzero ellipticity. The maximum ellipticity, $\epsilon = 45°$, occurs when either $a_1 = 0$ or $a_2 = 0$. It is thus seen that the state of elliptical polarization is the most general state, yielding linear polarization in one extreme and circular polarization in the other.

6.2. Quarter-wave Plate (QWP)

A quarter-wave (or $\lambda/4$) plate is an optical element with two orthogonal axes η and ζ, as shown in Fig. 6.4. If the incident polarization is resolved into linear components along η and ζ, then the emerging components along these axes will suffer a relative phase shift of 90°. We have for the **E**-fields:

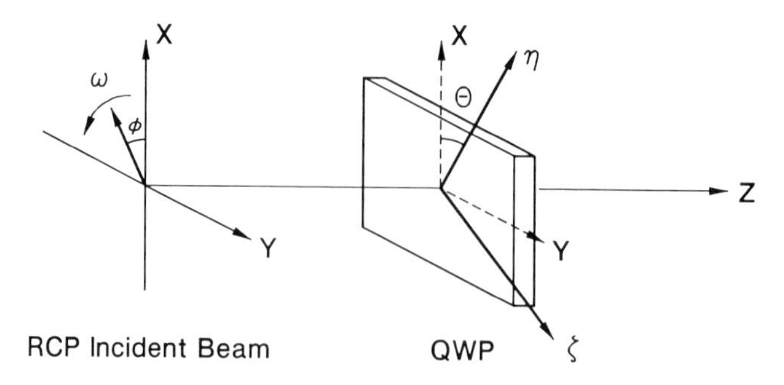

Figure 6.4. Quarter-wave plate in the XY-plane. η is the slow axis and ς the fast axis of the QWP. The angle between η and X is denoted by Θ. The incident beam shown in this figure is RCP with amplitude a and phase ϕ. The transmitted beam (not shown) is linearly polarized at $-45°$ to η.

$$\text{incident,} \quad a_1 \cos(\omega t + \phi_1)\, \hat{\boldsymbol{\eta}} + a_2 \cos(\omega t + \phi_2)\, \hat{\boldsymbol{\varsigma}}\,; \tag{6.8a}$$

$$\text{emergent,} \quad a_1 \cos(\omega t + \phi_1 + \phi_0)\, \hat{\boldsymbol{\eta}} + a_2 \cos(\omega t + \phi_2 + \phi_0 + 90°)\, \hat{\boldsymbol{\varsigma}} \tag{6.8b}$$

In Eq. (6.8) ϕ_0 is the common phase shift experienced by both components of polarization; this phase shift is usually of little significance and may be ignored in practice. The axis ς along which the phase has been advanced is generally referred to as the fast axis.

The QWP converts a circularly polarized incident beam into a beam with linear polarization. To observe this consider again the situation depicted in Fig. 6.4, where an RCP beam along the Z-axis with amplitude a and phase ϕ (in the reference XY-plane) goes through a QWP whose axis η is fixed at angle Θ relative to X. We have for the **E**-fields:

$$\text{incident, RCP in the } XY\text{-plane,} \quad a \cos(\omega t + \phi)\, \hat{\mathbf{x}} - a \sin(\omega t + \phi)\, \hat{\mathbf{y}}\,; \tag{6.9a}$$

$$\text{incident, RCP in the } \eta\varsigma\text{-plane,} \quad a \cos(\omega t + \phi + \Theta)\, \hat{\boldsymbol{\eta}} - a \sin(\omega t + \phi + \Theta)\, \hat{\boldsymbol{\varsigma}}\,; \tag{6.9b}$$

emergent, LP in the $\eta\varsigma$-plane,

$$a \cos(\omega t + \phi + \Theta + \phi_0)\, \hat{\boldsymbol{\eta}} - a \sin(\omega t + \phi + \Theta + \phi_0 + 90°)\, \hat{\boldsymbol{\varsigma}}$$
$$= a \cos(\omega t + \phi + \Theta + \phi_0)\, (\hat{\boldsymbol{\eta}} - \hat{\boldsymbol{\varsigma}})\,; \tag{6.9c}$$

emergent, LP in the XY-plane,

$$\sqrt{2}\, a \cos(\omega t + \phi + \Theta + \phi_0) \left[\cos(45° - \Theta)\, \hat{\mathbf{x}} - \sin(45° - \Theta)\, \hat{\mathbf{y}} \right] \tag{6.9d}$$

The QWP circularizes the polarization of an incident LP beam provided the angle between the incident polarization and either axis of the plate is 45°. As an example, let $\Theta = 45°$ in Fig. 6.4, and assume that the incident beam is LP with polarization along X. We have for the **E**-fields:

incident, LP in the XY-plane, $\quad E_0 \cos(\omega t + \phi)\, \hat{\mathbf{x}}\,;$ (6.10a)

incident, LP in the $\eta\varsigma$-plane, $\quad \frac{E_0}{\sqrt{2}} \cos(\omega t + \phi)\, \hat{\boldsymbol{\eta}} - \frac{E_0}{\sqrt{2}} \cos(\omega t + \phi)\, \hat{\boldsymbol{\varsigma}}\,;$

(6.10b)

emergent, LCP in the $\eta\varsigma$-plane,

$$\frac{E_0}{\sqrt{2}} \cos(\omega t + \phi + \phi_0)\, \hat{\boldsymbol{\eta}} - \frac{E_0}{\sqrt{2}} \cos(\omega t + \phi + \phi_0 + 90°)\, \hat{\boldsymbol{\varsigma}}$$

$$= \frac{E_0}{\sqrt{2}} \cos(\omega t + \phi + \phi_0)\, \hat{\boldsymbol{\eta}} + \frac{E_0}{\sqrt{2}} \sin(\omega t + \phi + \phi_0)\, \hat{\boldsymbol{\varsigma}}$$

$$= \frac{1}{\sqrt{2}} E_0^{(+)} \underline{/\phi + \phi_0}\,; \tag{6.10c}$$

emergent, LCP in the XY-plane, $\quad \frac{1}{\sqrt{2}} E_0^{(+)} \underline{/\phi + \phi_0 + 45°}$ (6.10d)

In a similar fashion one may obtain an RCP beam by passing a linearly polarized beam through a QWP. Once again consider the diagram of Fig. 6.4 and assume that $\Theta = 45°$. This time, however, the incident beam is linearly polarized along Y. We have for the **E**-fields:

incident, LP in the XY-plane, $\quad E_0 \cos(\omega t + \phi)\, \hat{\mathbf{y}}\,;$ (6.11a)

incident, LP in the $\eta\varsigma$-plane, $\quad \frac{E_0}{\sqrt{2}} \cos(\omega t + \phi)\, \hat{\boldsymbol{\eta}} + \frac{E_0}{\sqrt{2}} \cos(\omega t + \phi)\, \hat{\boldsymbol{\varsigma}}\,;$

(6.11b)

emergent, RCP in the $\eta\varsigma$-plane,

$$\frac{E_0}{\sqrt{2}} \cos(\omega t + \phi + \phi_0)\, \hat{\boldsymbol{\eta}} - \frac{E_0}{\sqrt{2}} \sin(\omega t + \phi + \phi_0)\, \hat{\boldsymbol{\varsigma}} = \frac{1}{\sqrt{2}} E_0^{(-)} \underline{/\phi + \phi_0}\,;$$

(6.11c)

emergent, RCP in the XY-plane, $\quad \frac{1}{\sqrt{2}} E_0^{(-)} \underline{/\phi + \phi_0 - 45°}$ (6.11d)

Other states of incident polarization and/or values of the QWP orientation angle Θ may occur in practice, but the representative cases studied here are sufficiently general to provide the basis for analysis in the most general case.

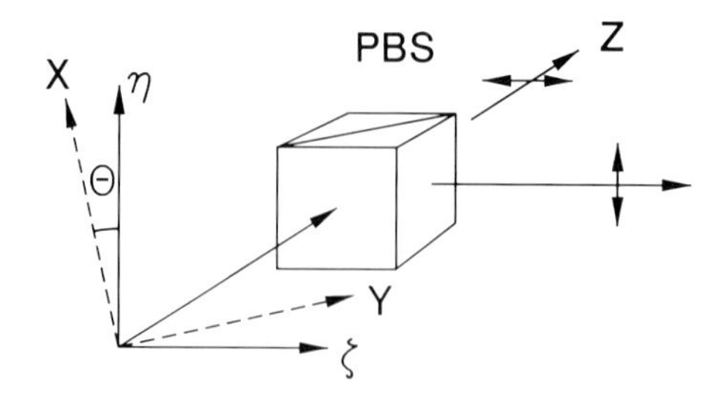

Figure 6.5. Polarizing beam-splitter (PBS) cube with axes along η and ζ. The $\eta\zeta$-plane is parallel to XY, and the angle between η and X is Θ. The incident polarization is resolved along the PBS axes, and the two emergent beams are linearly polarized.

6.3. Polarizing Beam-splitter (PBS)

As shown in Fig. 6.5, the PBS resolves the components of incident polarization along two mutually orthogonal axes η and ζ; the two beams transmitted through the PBS are always linearly polarized. It is not difficult to find the characteristics of the outgoing beams for arbitrarily polarized incident beams. An interesting special case is that of an incident beam with circular polarization, which we shall now treat in some detail. First consider an incident RCP beam in the XY-plane of Fig. 6.5, and assume that the axis η of the PBS makes an angle Θ with X. We have for the **E**-fields:

incident, RCP in the XY-plane, $\quad a^{(-)} \underline{/\phi} \;;$ (6.12a)

incident, RCP in the $\eta\zeta$-plane,

$$a^{(-)} \underline{/\phi + \Theta} = a\cos(\omega t + \phi + \Theta)\,\hat{\boldsymbol{\eta}} - a\sin(\omega t + \phi + \Theta)\,\hat{\boldsymbol{\zeta}} \;; \qquad (6.12b)$$

transmitted, LP along η, $\quad +a\cos(\omega t + \phi + \Theta)\;;$ (6.12c)

transmitted, LP along ζ, $\quad -a\sin(\omega t + \phi + \Theta)$ (6.12d)

Similarly, one may obtain expressions for transmitted beams corresponding to an LCP incident beam. We have for the **E**-fields:

incident, LCP in the XY-plane, $\quad a^{(+)} \underline{/\phi} \;;$ (6.13a)

incident, LCP in the $\eta\zeta$-plane,

$$a^{(+)} \underline{/\phi - \Theta} = a\cos(\omega t + \phi - \Theta)\,\hat{\boldsymbol{\eta}} + a\sin(\omega t + \phi - \Theta)\,\hat{\boldsymbol{\zeta}} \;; \qquad (6.13b)$$

transmitted, LP along η, $\quad a \cos(\omega t + \phi - \Theta)$; (6.13c)

transmitted, LP along ζ, $\quad a \sin(\omega t + \phi - \Theta)$ (6.13d)

Note that the orientation angle Θ of the PBS appears as $+\Theta$ in the outputs for the RCP beam, but as $-\Theta$ in the outputs for the LCP beam. This difference in the response of the PBS to RCP and LCP beams will be exploited in the next section in order to explain how, in differential detection of the MO Kerr effect, the beam's ellipticity may be converted into a useful signal.

6.4. Differential Detection Scheme and Magneto-optical Readout

Before embarking on a description of the differential detection scheme, let us analyze a mixture of RCP and LCP light going through a polarizing beam-splitter, such as the one in Fig. 6.5. We have for the **E**-fields:

incident, a mixture of RCP and LCP in the XY-plane,

$$a_1^{(-)} \angle \phi_1 + a_2^{(+)} \angle \phi_2 \; ; \tag{6.14a}$$

transmitted, LP along η, $\quad a_1 \cos(\omega t + \phi_1 + \Theta) + a_2 \cos(\omega t + \phi_2 - \Theta)$; (6.14b)

transmitted, LP along ζ, $\quad -a_1 \sin(\omega t + \phi_1 + \Theta) + a_2 \sin(\omega t + \phi_2 - \Theta)$ (6.14c)

We assume that each transmitted beam is captured by a separate detector and that the resulting signals S_1 and S_2 are fed to a differential amplifier. The output is then proportional to the difference between the intensities of the transmitted beams whose amplitudes are given by Eqs. (6.14b) and (6.14c). Using brackets $\langle \cdot \rangle$ to indicate time averaging, we have

$$\begin{aligned} S_1 &= a_1^2 \langle \cos^2(\omega t + \phi_1 + \Theta) \rangle + a_2^2 \langle \cos^2(\omega t + \phi_2 - \Theta) \rangle \\ &\quad + a_1 a_2 \langle [\cos(2\omega t + \phi_1 + \phi_2) + \cos(2\Theta + \phi_1 - \phi_2)] \rangle \\ &= \tfrac{1}{2} (a_1^2 + a_2^2) + a_1 a_2 \cos(2\Theta + \phi_1 - \phi_2) \end{aligned} \tag{6.15a}$$

$$\begin{aligned} S_2 &= a_1^2 \langle \sin^2(\omega t + \phi_1 + \Theta) \rangle + a_2^2 \langle \sin^2(\omega t + \phi_2 - \Theta) \rangle \\ &\quad - a_1 a_2 \langle [\cos(2\Theta + \phi_1 - \phi_2) - \cos(2\omega t + \phi_1 + \phi_2)] \rangle \\ &= \tfrac{1}{2} (a_1^2 + a_2^2) - a_1 a_2 \cos(2\Theta + \phi_1 - \phi_2) \end{aligned} \tag{6.15b}$$

the detection process is best understood if one expresses the reflected polarization from the MO medium as the sum of two orthogonal LP components:

$$a_1^{(-)} \angle \phi_1 + a_2^{(+)} \angle \phi_2$$

$$= [a_1 \cos(\omega t + \phi_1) + a_2 \cos(\omega t + \phi_2)]\, \hat{\mathbf{x}} + [-a_1 \sin(\omega t + \phi_1) + a_2 \sin(\omega t + \phi_2)]\, \hat{\mathbf{y}}$$

$$= \sqrt{(a_1 \cos\phi_1 + a_2 \cos\phi_2)^2 + (a_1 \sin\phi_1 + a_2 \sin\phi_2)^2}$$

$$\times \cos\left\{\omega t + \tan^{-1}\left[(a_1 \sin\phi_1 + a_2 \sin\phi_2)/(a_1 \cos\phi_1 + a_2 \cos\phi_2)\right]\right\} \hat{\mathbf{x}}$$

$$\pm \sqrt{(a_1 \cos\phi_1 - a_2 \cos\phi_2)^2 + (a_1 \sin\phi_1 - a_2 \sin\phi_2)^2}$$

$$\times \sin\left\{\omega t + \tan^{-1}\left[(a_1 \sin\phi_1 - a_2 \sin\phi_2)/(a_1 \cos\phi_1 - a_2 \cos\phi_2)\right]\right\} \hat{\mathbf{y}} \tag{6.22}$$

The ± symbol in the above equation indicates that the Y-component of the reflected polarization switches sign when the magnetization of the MO medium is reversed. Remembering that the incident beam on the medium is linearly polarized along X, one may write

$$\text{reflectivity of MO medium for } X\text{-polarized light} = r_x \hat{\mathbf{x}} \pm r_y \hat{\mathbf{y}} \tag{6.23}$$

In Eq. (6.23) r_x and r_y are complex quantities, each with its own amplitude and phase, that is, $r_x = |r_x| \exp(i\phi_x)$ and $r_y = |r_y| \exp(i\phi_y)$. Using a similar complex notation for the RCP and LCP reflectivities, one may deduce from Eqs. (6.18), (6.22), and (6.23) that

$$\boxed{r_x = \tfrac{1}{2}\left[r_1 \exp(i\phi_1) + r_2 \exp(i\phi_2)\right]} \tag{6.24a}$$

$$\boxed{r_y = -\tfrac{1}{2} i \left[r_1 \exp(i\phi_1) - r_2 \exp(i\phi_2)\right]} \tag{6.24b}$$

The differential signal, which was written in Eqs. (6.19) and (6.21) in terms of the RCP and LCP reflectivities, may now be derived in terms of

r_x and r_y as follows:

$$\text{PBS-transmitted LP amplitude along } \eta = (r_x \cos\Theta \pm r_y \sin\Theta)\, E_0 \qquad (6.25a)$$

$$\text{PBS-transmitted LP amplitude along } \zeta = (r_x \sin\Theta \mp r_y \cos\Theta)\, E_0 \qquad (6.25b)$$

$$\Delta S = S_1 - S_2 = \tfrac{1}{2}\, |r_x \cos\Theta \pm r_y \sin\Theta|^2\, E_0{}^2 - \tfrac{1}{2}\, |r_x \sin\Theta \mp r_y \cos\Theta|^2\, E_0{}^2$$

$$= P_0\,(|r_x|^2 - |r_y|^2)\cos 2\Theta \;\pm\; 2P_0\,|r_x\, r_y|\,\cos(\phi_x - \phi_y)\sin 2\Theta \qquad (6.26)$$

The $\pm$ symbol in Eq. (6.26) indicates that switching the direction of magnetization will alter the sign of the second term. The best choice of orientation for the PBS is obviously $\Theta = 45°$. The readout signal is seen to be proportional to $\cos(\phi_x - \phi_y)$, whose magnitude is always less than or equal to unity. Thus any phase difference between r_x and r_y (other than 0 and π, of course) is harmful and must be eliminated. One way to achieve this goal is to insert a retardation plate (an adjustable Soleil–Babinet compensator, for instance) in the path of the beam, as shown in Fig. 6.6. The compensator is set with its axes parallel to X and Y, giving a phase shift of $\Delta\phi = \phi_x - \phi_y$ to the reflected Y-component. This phase shift is the same for both states of magnetization of the MO medium. The net swing of the differential signal will now be

$$\Delta S(\uparrow) - \Delta S(\downarrow) = 4P_0\,|r_x\, r_y| \qquad (6.27)$$

where the arrows refer to the direction of magnetization. The elimination of $\cos(\phi_x - \phi_y)$ by the retardation plate and the consequent increase of the readout signal is essentially due to the conversion of ellipticity to rotation in the reflected state of polarization.

The same effect may be obtained with a $\lambda/4$-plate in place of the compensator, but the mechanism is somewhat more subtle. Let the QWP axes be at 45° to the X-axis. As was pointed out in section 6.2, both the X- and Y-components of polarization will become circular after the QWP and, in accordance with Eqs. (6.10d) and (6.11d), we shall have for the **E**-fields:

$$\text{emergent, LCP in the } XY\text{-plane,} \quad \frac{1}{\sqrt{2}}\,|r_x|\,E_0{}^{(+)}\ \underline{/\ \phi_x + 45°}\ ; \qquad (6.28a)$$

$$\text{emergent, RCP in the } XY\text{-plane,} \quad \frac{1}{\sqrt{2}}\,|r_y|\,E_0{}^{(-)}\ \underline{/\ \phi_y - 45°} \qquad (6.28b)$$

After the QWP, these circularly polarized beams will pass through the PBS, whose orientation in the XY-plane is given by the angle Θ. The PBS splits the amplitudes of each of the two circular beams equally between the two detectors, but the phase angles will be different. We have for the **E**-fields:

beam 1, LP along η-axis of PBS,

$$\frac{1}{\sqrt{2}}|r_x|E_0\cos(\omega t+\phi_x+45^\circ-\Theta)+\frac{1}{\sqrt{2}}|r_y|E_0\cos(\omega t+\phi_y-45^\circ+\Theta)\;; \tag{6.29a}$$

beam 2, LP along ς-axis of PBS,

$$\frac{1}{\sqrt{2}}|r_x|E_0\sin(\omega t+\phi_x+45^\circ-\Theta)-\frac{1}{\sqrt{2}}|r_y|E_0\sin(\omega t+\phi_y-45^\circ+\Theta) \tag{6.29b}$$

Now, the angle Θ may be chosen to eliminate the phase difference between the two components impinging on each detector. This is achieved when

$$\Theta = 45^\circ + \tfrac{1}{2}(\phi_x - \phi_y) \tag{6.30}$$

Then

$$S_1 = \tfrac{1}{2}(|r_x|+|r_y|)^2 E_0^{\,2}\,\langle\cos^2[\omega t+\tfrac{1}{2}(\phi_x+\phi_y)]\rangle \tag{6.31a}$$

$$S_2 = \tfrac{1}{2}(|r_x|-|r_y|)^2 E_0^{\,2}\,\langle\sin^2[\omega t+\tfrac{1}{2}(\phi_x+\phi_y)]\rangle \tag{6.31b}$$

Consequently,

$$\Delta S = 2P_0|r_x r_y| \tag{6.32}$$

When the magnetization is reversed ΔS becomes $-\Delta S$ and, as seen earlier in Eq. (6.27), the net excursion of the output signal becomes $4P_0|r_x r_y|$. The action of the QWP in conjunction with differential detection has thus been to convert the ellipticity of the reflected beam into a useful signal.

Digression. It is not difficult to show that in terms of r_x and r_y the Kerr rotation angle θ_k and ellipticity ϵ_k of the reflected beam are given by

$$\tan 2\theta_k = \frac{2|r_y/r_x|}{1-|r_y/r_x|^2}\cos(\phi_y-\phi_x)$$

$$\sin 2\epsilon_k = -\frac{2|r_y/r_x|}{1+|r_y/r_x|^2}\sin(\phi_y-\phi_x)$$

Alternatively r_y/r_x may be derived from θ_k and ϵ_k as follows:

$$\left| \frac{r_y}{r_x} \right|^2 = \frac{\tan^2 \theta_k + \tan^2 \epsilon_k}{1 + \tan^2 \theta_k \tan^2 \epsilon_k}$$

$$\tan(\phi_y - \phi_x) = -\frac{\tan 2\epsilon_k}{\sin 2\theta_k}$$

It is sometimes useful to define the complex Kerr angle as $(\theta_k - i\epsilon_k)$. In the light of the above relations, this complex quantity may be approximated by r_y / r_x, provided that $|r_y / r_x| << 1$. □

6.5. Wavelength-dependence of Polar Magneto-optical Kerr Effect

The emergence of magneto-optical data storage technology has enhanced the need for accurate measurements of MO effects. Because data storage devices utilize diode lasers for read/write/erase operations, and since the trend in semiconductor diode laser development is towards shorter wavelengths, it is important to characterize the storage media over a broad range of optical frequencies. The range of interest may in fact span the visible wavelengths in addition to the near infrared and near ultraviolet. Progress in understanding the physics of MO interactions is also dependent upon the availability of experimental data on the wavelength-dependence of Kerr effect signals.

In this section a simple and accurate method for the spectroscopy of MO media is described.[1] The measurement system, shown in Fig. 6.7, consists of the following elements: a monochromatic light source, a polarizer, a Fresnel rhomb (acting as a quarter-wave plate), a polarizing beam-splitter, and two photodetectors. Also required are an electromagnet capable of switching the magnetization of the sample, and a stepper-motor to rotate the detection module in small steps between -90° and +90°. With regard to the photodiodes, there is a finite range of wavelengths over which they operate efficiently. Silicon photodiodes, for instance, operate between 200 nm and 1100 nm, with a responsivity ranging from 0.1 to 0.6 amps/watt. The signals thus generated in this range of optical frequencies will be of sufficient strength for our purposes. Since the difference signal from detectors 1 and 2 will eventually be normalized by their sum signal, there is no need for calibration with respect to the responsivity of the photodetectors.

6.5.1. The Fresnel rhomb

This is a quarter-wave plate with a relatively wide bandwidth. Commercial products (made from BK7 glass) typically have a phase retardation of 90° ± 4° in the wavelength range of 350–1100 nm. For the purposes of

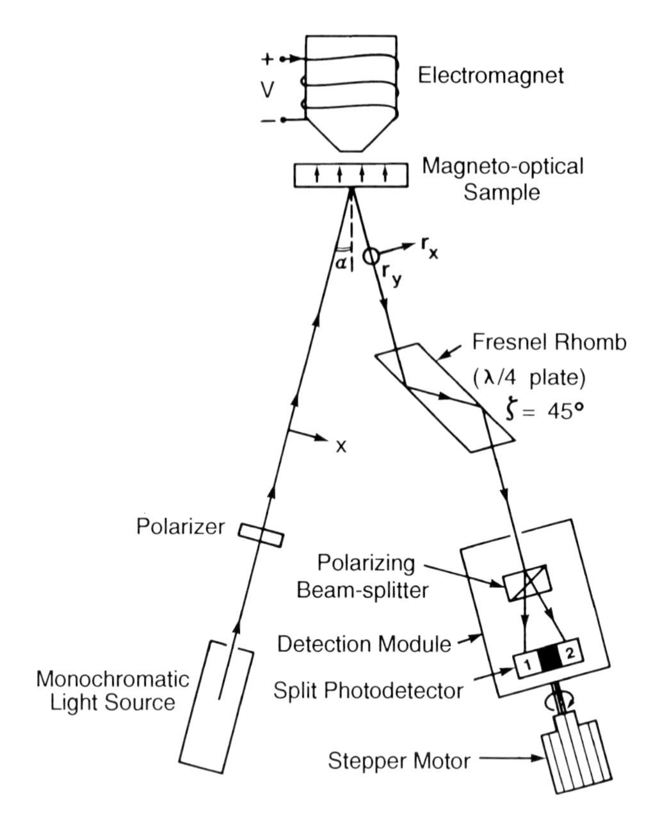

Figure 6.7. Schematic diagram of the magneto-optical Kerr spectrometer (MOKS) system, used for wavelength-dependence measurements of the Kerr rotation angle and ellipticity.

this work, the QWP need not be perfect; in fact a phase error of several degrees may be tolerated. This phase error, however, should be measured in advance for the wavelengths of interest, so that the final results may be corrected. We will show how the system of Fig. 6.7 may be used to measure the phase error of the QWP. The operation of the rhomb is based on total internal reflection from the glass–air interface which, when the angle of incidence is around 54°, introduces nearly 45° of phase shift between the p- and s-components of polarization. Two reflections of this type are thus required if the device is to operate as a QWP. For a medium of refractive index n at angle of incidence Θ, the phase difference Δ between p and s is given by:[2]

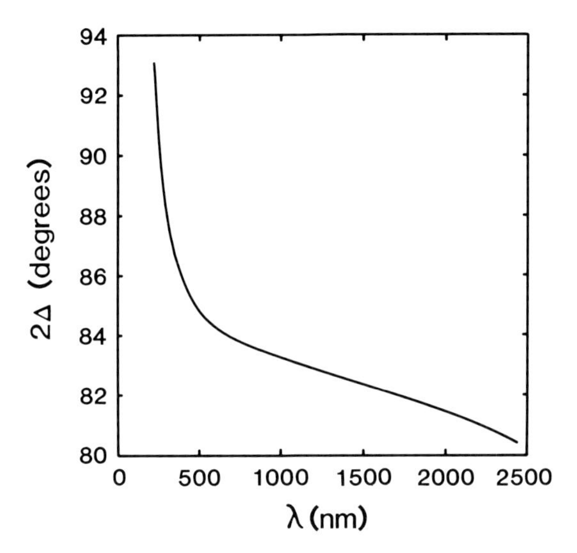

Figure 6.8. Calculated values of phase shift versus wavelength for a Fresnel rhomb made from fused silica. The assumed angle of incidence at the glass–air interface is $\Theta = 54°$.

$$\tan\left(\tfrac{1}{2}\,\Delta\right) = \frac{\cos\Theta\,\sqrt{n^2 \sin^2\Theta - 1}}{n \sin^2\Theta} \tag{6.33}$$

Figure 6.8 shows the total phase shift (2Δ) versus wavelength, calculated for a fused silica rhomb using Eq. (6.33) with $\Theta = 54°$. (The wavelength-dependence of the refractive index n is taken from the *Handbook of Optics*.[3]) Clearly, a QWP made in this way will be accurate only around $\lambda_0 = 250$ nm but, as will be shown, it may be used for the proposed measurements over a wide range of wavelengths in the neighborhood of λ_0. Denoting the phase deviation from 90° by δ, where δ is a function of the wavelength, the Jones matrix of an imperfect QWP is written as follows:

$$\begin{pmatrix} 1 & 0 \\ 0 & \mathrm{i}\exp(\mathrm{i}\delta) \end{pmatrix} \tag{6.34}$$

6.5.2. Mathematical analysis

With the aid of the Jones calculus, one can readily determine the photodetector signals in the system of Fig. 6.7. Let the power of the incident beam be P_0, and assume that the direction of incident linear polarization is X. As before the MO reflectivities will be denoted by r_x and

r_y. The phase difference between r_y and r_x will be encountered frequently in the following discussion; we find it useful to refer to the "up" state (↑) of magnetization as the state with $\phi_y - \phi_x$ in the interval $[-\frac{1}{2}\pi, \frac{1}{2}\pi]$; the corresponding phase difference for the "down" state (↓) will then be in the interval $[\frac{1}{2}\pi, \frac{3}{2}\pi]$. The imperfect QWP has its fast axis rotated from X by an angle ξ, while the axis of the PBS is rotated from X by ψ. The sum and difference signals from the photodetectors are found to be

$$S_1 + S_2 = P_0(|r_x|^2 + |r_y|^2) \tag{6.35a}$$

$$\begin{aligned} S_1 - S_2 = {} & P_0(|r_x|^2 - |r_y|^2)\left[\cos 2\xi \cos(2\psi - 2\xi) + \sin 2\xi \sin(2\psi - 2\xi)\sin\delta\right] \\ & \pm 2P_0|r_x r_y|\Big\{\sin 2\xi \cos(2\psi - 2\xi)\cos(\phi_y - \phi_x) - \sin(2\psi - 2\xi) \\ & \times\left[\sin^2\xi \sin(\phi_y - \phi_x - \delta) + \cos^2\xi \sin(\phi_y - \phi_x + \delta)\right]\Big\} \end{aligned} \tag{6.35b}$$

The ± signs in Eq. (6.35b) correspond respectively to ↑ and ↓ states of magnetization of the sample. Let us fix the orientation of the QWP at $\xi = 45°$, and normalize the difference signal by the sum signal. We obtain

$$\begin{aligned} \frac{S_1 - S_2}{S_1 + S_2} = {} & -\frac{|r_x|^2 - |r_y|^2}{|r_x|^2 + |r_y|^2}\cos 2\psi \sin\delta \pm \frac{2|r_x r_y|}{|r_x|^2 + |r_y|^2} \\ & \times\left[\sin 2\psi \cos(\phi_y - \phi_x) + \cos 2\psi \sin(\phi_y - \phi_x)\cos\delta\right] \end{aligned} \tag{6.36}$$

An immediate consequence of Eq. (6.36) is that if the sample happens to be non-magnetic (e.g., an aluminum mirror), then the system automatically measures the phase error δ of the rhomb. To see this, we set $r_y = 0$ in Eq. (6.36) and obtain

$$\boxed{\frac{S_1 - S_2}{S_1 + S_2} = -\cos 2\psi \sin\delta} \tag{6.37}$$

Measuring $(S_1 - S_2)/(S_1 + S_2)$ as a function of ψ (the orientation angle of the detection module) thus produces a cosine function whose amplitude will be equal to $\sin\delta$. In this way, one determines the phase error δ of the QWP over the wavelength range of interest, and uses the results later on to correct the Kerr effect data.

We proceed by defining two additional parameters, the angle γ and the coefficient C, as follows:

$$\tan\gamma = \tan(\phi_y - \phi_x)\cos\delta\,; \qquad -\frac{\pi}{2} < \gamma \leq \frac{\pi}{2} \tag{6.38a}$$

$$C = \frac{1}{\sqrt{1 + \tan^2\delta\,\sin^2\gamma}} \tag{6.38b}$$

Note that for $\delta = 0$ the angles γ and $\phi_y - \phi_x$ are the same, but when δ deviates from zero there will be a small difference between the two. Equation (6.36) may now be written as follows:

$$\frac{S_1 - S_2}{S_1 + S_2} = -\frac{1 - |r_y/r_x|^2}{1 + |r_y/r_x|^2}\cos 2\psi\,\sin\delta \pm \frac{2C\,|r_y/r_x|}{1 + |r_y/r_x|^2}\sin(2\psi + \gamma) \tag{6.39}$$

Thus, upon measuring $(S_1 - S_2)/(S_1 + S_2)$ as a function of ψ for both directions of magnetization and subtracting the resulting curves we obtain

$$\left[\frac{S_1 - S_2}{S_1 + S_2}\right]_{\text{up}} - \left[\frac{S_1 - S_2}{S_1 + S_2}\right]_{\text{down}} = \frac{4C\,|r_y/r_x|}{1 + |r_y/r_x|^2}\sin(2\psi + \gamma) \tag{6.40}$$

When the sinusoidal function in Eq. (6.40) is plotted against 2ψ, its zero-crossing will immediately give the value of γ. Since δ is already known for a particular wavelength, we obtain the phase difference $\phi_y - \phi_x$ and the coefficient C from Eqs. (6.38). The remaining unknown $|r_y/r_x|$ can now be determined from the amplitude of the sinusoidal signal in Eq. (6.40).

6.5.3. Results and discussion

Calculated plots of $(S_1 - S_2)/(S_1 + S_2)$ versus 2ψ for several values of δ are shown in Fig. 6.9. In each set the dashed curve corresponds to "up" magnetization, the dotted curve to "down" magnetization, and the solid curve is the difference between the two. The assumed values of the MO signal are $r_x = 0.7$, $r_y = 0.005$, and $\phi_y - \phi_x = 23°$. Figure 6.9(a) represents the situation when the QWP is perfect, i.e., $\delta = 0$. The zero-crossing occurs at $2\psi = -23°$ and the amplitude of the difference signal is 0.028, yielding $|r_y/r_x| = 0.007$. Figure 6.9(b) corresponds to a QWP with $\delta = 4°$; as before, the MO parameters may be determined from the difference curve. Note, however, that the magnitude of the difference signal is only half the magnitude of its parent signals; the electronic circuitry in this case

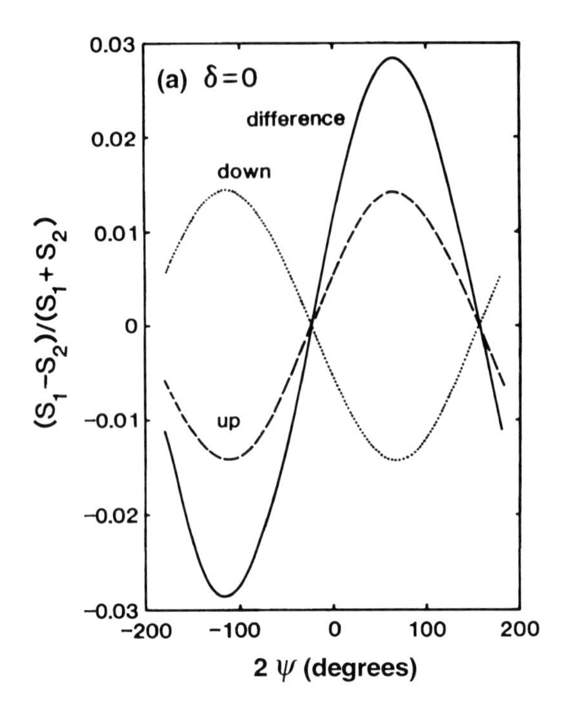
(a) δ=0
difference
down
up
0.03
0.02
0.01
0
−0.01
−0.02
−0.03
−200
−100
0
100
200
2 ψ (degrees)
(S1 −S2)/(S1 + S2)

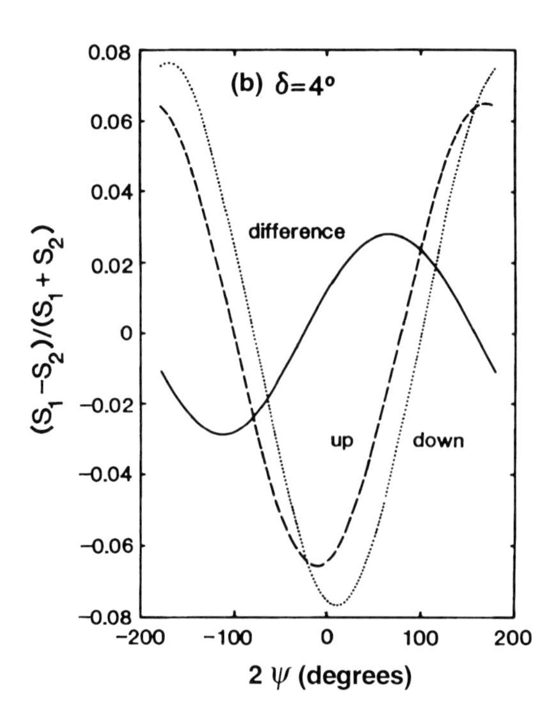
(b) δ=4°
difference
up
down
0.08
0.06
0.04
0.02
0
−0.02
−0.04
−0.06
−0.08
−200
−100
0
100
200
2 ψ (degrees)
(S1 −S2)/(S1 + S2)

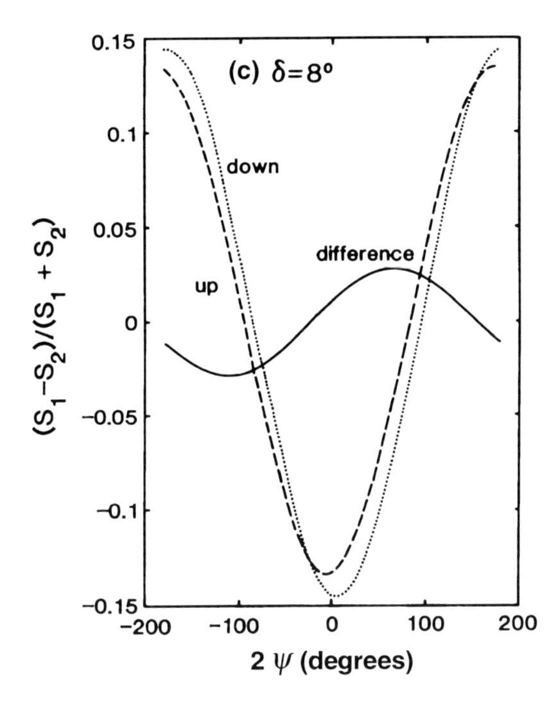

Figure 6.9. Calculated signals $(S_1 - S_2)/(S_1 + S_2)$ for "up" and "down" states of magnetization of the sample and the difference between them. The assumed magneto-optical parameters of the sample are $|r_x| = 0.7$, $|r_y| = 0.005$ and $\phi_y - \phi_x = 23°$. In (a) the quarter-wave plate is assumed to be perfect, i.e., $\delta = 0$. In (b) and (c) the values of the QWP phase error are $\delta = 4°$ and $\delta = 8°$ respectively.

therefore needs to have a wider dynamic range. Similar considerations apply to the case of $\delta = 8°$, shown in Fig. 6.9(c).

Figure 6.10 shows the results of an experiment on a Co/Pt sample at $\lambda = 1000$ nm. If we ignore all corrections (and they are minor in this case), the difference curve yields $|r_y/r_x| \simeq 0.004$ and $\phi_y - \phi_x \simeq 0$. These correspond to a rotation angle $\theta_k \simeq 0.22°$ and an ellipticity $\epsilon_k \simeq 0$.

The measured spectra of the MO Kerr effect for two different samples are shown in Fig. 6.11. In (a) the sample is a layer (1350 Å thick) of amorphous $Tb_{28}Fe_{72}$ alloy, sputtered on glass substrate and coated with 1250 Å of SiO_x for protection. The sample in (b) is a superlattice-type film of Co/Pt, whose bilayer of 4 Å cobalt and 10 Å platinum is repeated 21 times for a total thickness of 294 Å. It is observed that the Kerr angle of TbFe decreases at shorter wavelengths, whereas the Co/Pt sample shows

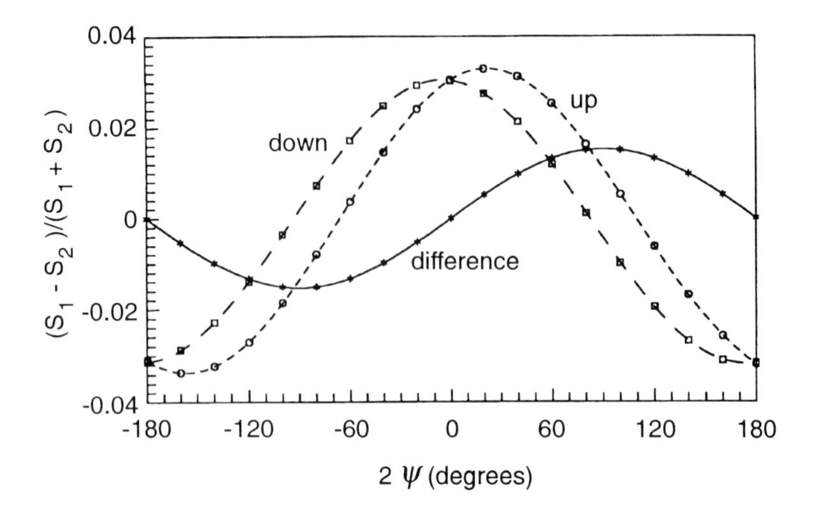

Figure 6.10. Measured signals $(S_1 - S_2)/(S_1 + S_2)$ for "up" and "down" states of magnetization of a Co/Pt sample, and the difference between them. The continuous curves are obtained by matching the experimental data to the appropriate sinusoids.

enhancement in the blue and ultraviolet regions. This property of the latter is one reason why it is considered a promising candidate for future generations of MO media.

6.6. Edge Detection using Diffraction from Domain Walls

Readout by differential detection for MO data storage systems was described in section 6.4. To put the matters in perspective, in the following paragraph we summarize the salient features of differential detection.

Differential detection requires that a linearly-polarized beam of light be reflected from a perpendicularly magnetized medium. Upon reflection, the state of polarization becomes elliptical, with the major axis of the ellipse rotated away from the direction of incident polarization. The chiralities of both rotation and ellipticity depend upon the state of magnetization of the medium (in the region immediately under the focused light spot), and, therefore, constitute the information signal. The differential detector converts these chiralities into an electric waveform that tends to be a good reproduction of the pattern of magnetization on the storage medium.

In this section we describe a different kind of scheme for MO readout, with potential advantages over differential detection. The new scheme,[4,5] depicted in Fig. 6.12, utilizes circularly polarized light and relies on diffraction from domain boundaries for the detection of transitions of the

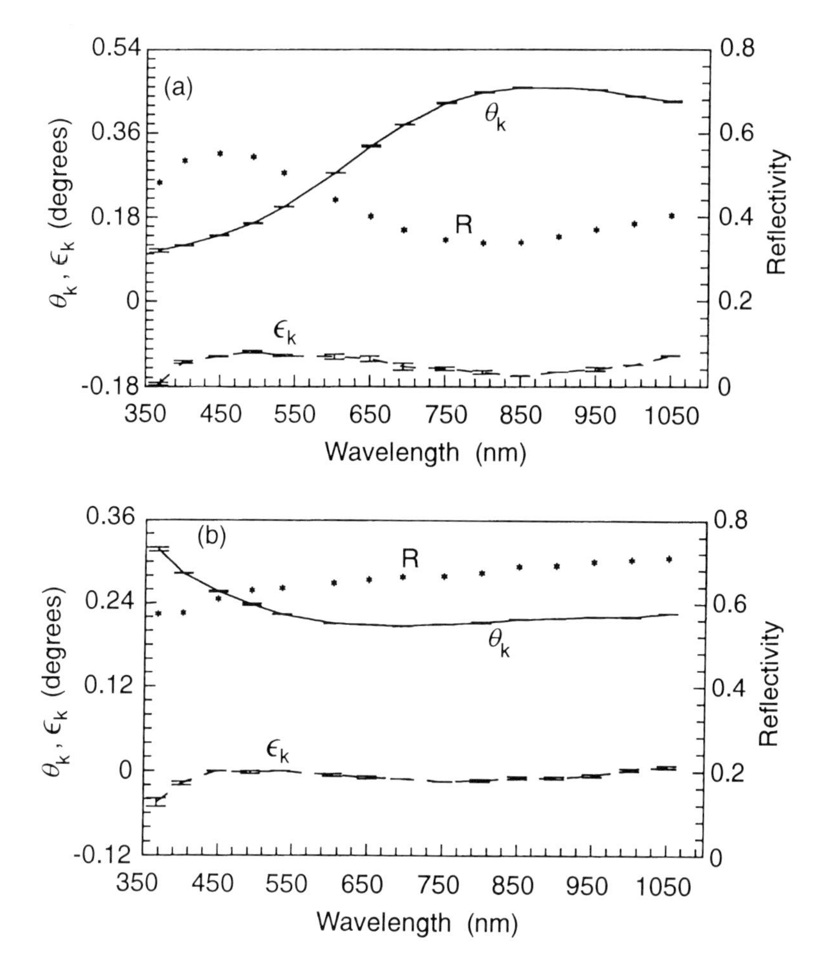

Figure 6.11. Measured plots of the Kerr rotation angle θ_k, ellipticity ϵ_k, and reflectivity R in the wavelength range 350 nm–1050 nm. (a) $Tb_{28}Fe_{72}$ sample; (b) Co/Pt sample.

magnetization state (from ↑ to ↓ and vice versa). One advantage of this new method is that, as in read-only and write-once systems, the beam reflected from the disk can be fully diverted from the laser; in this way the laser feedback noise is substantially reduced. Other advantages include compatibility with read-only and write-once pickup systems, total utilization of the laser power during recording and erasure, simpler optics, and direct detection of transitions (i.e., no need for differentiation of the read signal). A disadvantage of the new scheme over differential detection is its smaller signal by a factor of about two. However, the advantages outlined above may outweigh this particular disadvantage in practice. What is more, the factor of two difference in signal amplitudes is for isolated transitions only. As the recording density increases and the transitions

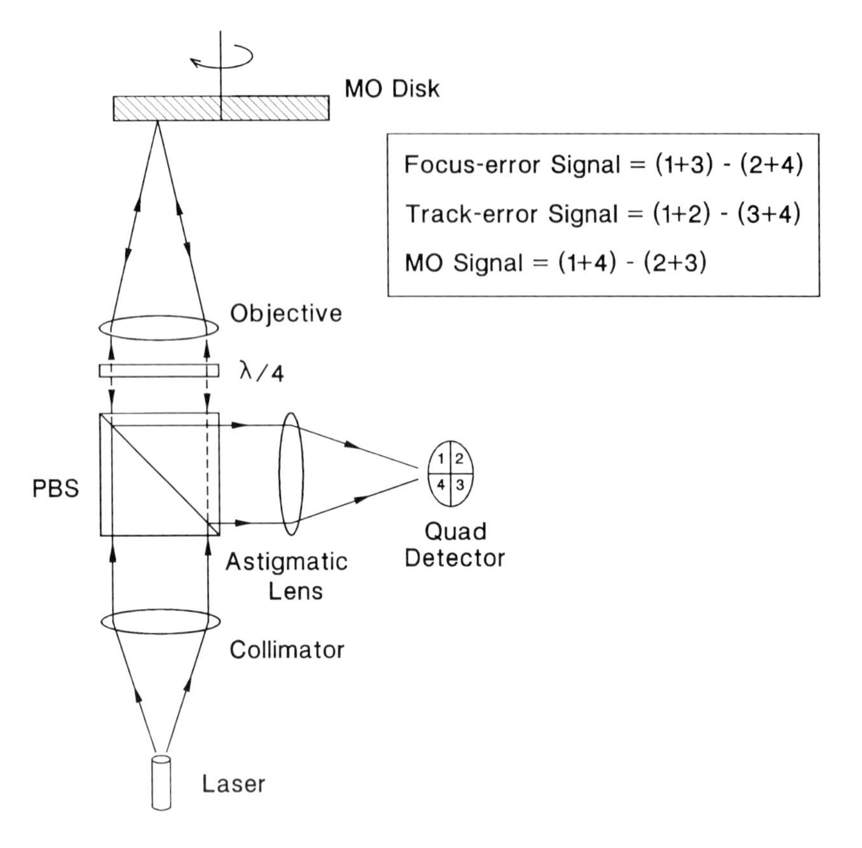

Figure 6.12. Schematic diagram of the readout system utilizing circularly polarized light. The same quad detector as is used for generating the focus-error and track-error signals detects the MO signal.

crowd together, the new scheme begins to look even better than differential detection.

For the sake of simplicity, we restrict the analysis of diffraction from a domain wall to one dimension, concentrating on the geometry of Fig. 6.13. Here the lens is cylindrical (with cylinder axis parallel to the domain wall), and the storage medium reflects right and left circularly polarized light differently. For the up-magnetized state of the medium, let $r_1 \exp(i\phi_1)$ and $r_2 \exp(i\phi_2)$ be the reflection coefficients corresponding to RCP and to LCP beams; the same coefficients in reverse order apply to the down-magnetized state. In Fig. 6.13 the direction of magnetization of the medium is up when $x > 0$, and down when $x < 0$; the transition, therefore, is sharp and parallel to the Y-axis. Using the reflection coefficients r_x and r_y defined in Eq. (6.24), and with the aid of $\operatorname{sgn}(x) = x/|x|$, the amplitude reflectivity $r(x)$ for the entire medium may be written as follows:

$$r(x) = r_x \pm i \operatorname{sgn}(x)\, r_y \tag{6.41}$$

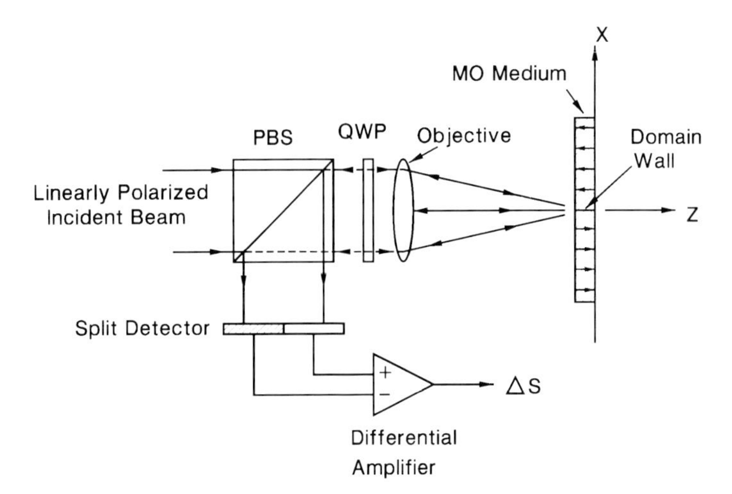

Figure 6.13. Sharp magnetic transition in a thin-film magnetic medium under a focused beam of light. The geometry is one-dimensional, so that the lens is a cylindrical one with the cylinder axis running perpendicular to the plane of the page (i.e., along the *Y*-axis). The sharp domain wall, which also runs parallel to *Y*, is centered under the focused spot at $x = 0$. The incident beam is circularly polarized (either RCP or LCP). Because of diffraction, some of the reflected light spills over the aperture of the lens and is therefore missed by the detector. The MO read signal is the difference between the signals from the two halves of the split detector.

The upper and lower signs in Eq. (6.41) correspond, respectively, to RCP and LCP incident beams.

The focusing lens has focal length f and numerical aperture NA. Let the amplitude distribution at the entrance pupil of the lens be uniform, and assume a total optical power (i.e., integrated intensity) P_0. The amplitude distribution $A_1(x)$ at the entrance pupil is thus written

$$A_1(x) = \sqrt{\frac{P_0}{2f\text{NA}}}\ \text{Rect}\left(\frac{x}{2f\text{NA}}\right) \tag{6.42}$$

$\text{Rect}(x)$ is unity when $|x| < 0.5$, and zero otherwise. The distribution $A_2(x)$ at the focal plane is obtained by scaling and Fourier-transforming $A_1(x)$:

$$A_2(x) = \sqrt{\frac{\lambda P_0}{2\,\text{NA}}}\ \frac{\sin(2\pi\text{NA}\,x/\lambda)}{\pi x} \tag{6.43}$$

The reflected distribution at the surface of the medium is obtained by multiplying $r(x)$, Eq. (6.41), by $A_2(x)$, Eq. (6.43). This product must be Fourier transformed again to yield the pattern of reflected light back at the exit pupil of the lens. The Fourier transform of the product of two functions, however, is the convolution of the individual transforms, which, in this case, are given by

$$\mathscr{F}\left\{r(\lambda x)\right\} = r_x\,\delta(\sigma) \pm \frac{1}{\pi\sigma}\,r_y \tag{6.44}$$

$$\mathscr{F}\left\{A_2(\lambda x)\right\} = \sqrt{\frac{P_0}{2\mathrm{NA}\lambda}}\;\mathrm{Rect}\left(\frac{\sigma}{2\mathrm{NA}}\right) \tag{6.45}$$

Here σ is the independent variable of the Fourier transform, and $\delta(\sigma)$ is Dirac's delta function. Convolution of the above functions yields

$$\mathscr{F}\left\{r(\lambda x)\,A_2(\lambda x)\right\} = \sqrt{\frac{P_0}{2\mathrm{NA}\lambda}}\left\{r_x\;\mathrm{Rect}\left(\frac{\sigma}{2\mathrm{NA}}\right) \pm \frac{1}{\pi}\,r_y\,\ln\left|\frac{\mathrm{NA}+\sigma}{\mathrm{NA}-\sigma}\right|\right\} \tag{6.46}$$

With proper scaling and normalization, the above expression represents the far-field pattern of the reflected beam. Noting that the diffracted rays beyond the lens aperture are blocked, we obtain the distribution $A_3(x)$ of the beam immediately after the lens, as follows:

$$\boxed{A_3(x) = \sqrt{\frac{P_0}{2f\mathrm{NA}}}\left\{r_x \pm \frac{1}{\pi}\,r_y\,\ln\left|\frac{f\mathrm{NA}+x}{f\mathrm{NA}-x}\right|\right\}\mathrm{Rect}\left(\frac{x}{2f\mathrm{NA}}\right)} \tag{6.47}$$

This is the reflected amplitude distribution at the exit pupil of the lens. (Remember that the upper and lower signs correspond to RCP and LCP incident light, respectively.) The total reflected power P can now be obtained from Eq. (6.47) by integration:

$$P = \int_{-\infty}^{\infty} |A_3(x)|^2\,\mathrm{d}x = P_0\left\{|r_x|^2 + \frac{1}{\pi^2}|r_y|^2\int_0^1 \ln^2\left(\frac{1+x}{1-x}\right)\mathrm{d}x\right\}$$

$$= P_0\left\{|r_x|^2 + \tfrac{1}{3}|r_y|^2\right\} \tag{6.48}$$

Apparently, a fraction of the reflected power equal to $\frac{2}{3}|r_y|^2 P_0$ is diffracted out of the aperture of the lens and is never collected. The MO read signal ΔS is the difference between the incident powers on the two halves of the split detector (see Fig. 6.13); thus

$$\Delta S = \int_{-\infty}^{0} |A_3(x)|^2 \, dx - \int_{0}^{\infty} |A_3(x)|^2 \, dx$$

$$= \pm \frac{2}{\pi} P_0 |r_x r_y| \cos(\phi_x - \phi_y) \int_0^1 \ln\left(\frac{1+x}{1-x}\right) dx$$

When the above integral is evaluated, we find

$$\boxed{\Delta S = \pm \frac{4P_0 \ln 2}{\pi} |r_x r_y| \cos(\phi_x - \phi_y)} \qquad (6.49)$$

We must emphasize that the above ΔS is only the peak value of the read signal, obtained when the wall is exactly in the middle of the focused spot; as the beam moves away from the wall, the magnitude of the signal declines. The $\pm$ symbol in front of ΔS in Eq. (6.49) indicates that the polarity of the read signal depends on whether the incident beam is RCP or LCP, or, what amounts to the same thing, whether the magnetization pattern of the wall is ↑↓ or ↓↑.

ΔS as given by Eq. (6.49) is maximized when $\phi_x - \phi_y = 0$, which means that, for best results, the storage medium must be designed to exhibit as little ellipticity as possible. The magnitude of the above ΔS should be compared with the peak signal of differential detection, $2P_0|r_x r_y|$, which is given in Eq. (6.32); the two peak signals are seen to differ by a factor of $\pi/(2\ln 2) \simeq 2.27$. Note, however, that the quoted value of the read signal for differential detection does not take account of the losses incurred at the regular beam-splitter (see Fig. 6.6), nor of the fact that the recorded domains are typically smaller than the focused spot. For these reasons, at high recording densities under realistic conditions the signal produced by the system of Fig. 6.12 is comparable to or better than that obtained from differential detection.

One drawback of the diffraction-based method of detecting domain walls is its ultra-sensitivity to surface roughness. Even with polished glass substrates, the readout system depicted in Fig. 6.12 produces an inordinate amount of noise. This type of noise, however, has the same magnitude whether the incident beam is RCP or LCP. The MO signal, on the other hand, switches sign when the chirality of the incident polarization is reversed. By mixing equal amounts of RCP and LCP in the readout beam, and separating them after reflection from the medium, it is thus possible to eliminate roughness-induced noise.[5]

6.7. Figure of Merit for Magneto-optical Media

The analyses of the read process in sections 6.4 and 6.6 have identified $|r_x r_y|$ as the media-related parameter that determines the strength of the read signal. On the other hand, shot noise, which is the fundamental noise in photodetection, has an rms amplitude proportional to $\sqrt{(|r_x|^2 + |r_y|^2)}$ (see Chapter 9). Thus, provided that $|r_y| << |r_x|$ (a condition almost always satisfied in practice) the signal-to-shot-noise ratio will be proportional to $|r_y|$. Media that exhibit large values of $|r_y|$ are therefore desirable and, for this reason, the figure of merit (FOM) is defined as $|r_y|$. When $|r_y| << |r_x|$ the reflectivity R of the medium may be approximated as follows:

$$R = |r_x|^2 + |r_y|^2 \simeq |r_x|^2$$

Also, the complex Kerr angle may be written as

$$\theta_k - i\epsilon_k \simeq r_y / r_x$$

Consequently, the FOM is related to R, θ_k, and ϵ_k as follows:

$$\boxed{\mathrm{FOM} = |r_y| \simeq \sqrt{R(\theta_k^2 + \epsilon_k^2)}} \tag{6.50}$$

The above FOM is often used to characterize the performance of MO media. This is a reasonable practice, provided that one's interest is in, say, comparing various disks fabricated by different manufacturers. On the other hand, if one is given a material and asked to evaluate it for use as an MO storage medium, then R, θ_k, and ϵ_k can no longer be uniquely identified; under these circumstances the FOM of Eq. (6.50) ceases to be useful. The values of R, θ_k, and ϵ_k depend on the structure in which the magnetic material is embedded, and are ultimately determined by the interference-type phenomena taking place within the layered structure. A question thus arises concerning the maximum possible $|r_y|$ a given material can exhibit, provided of course that all interference-type enhancement schemes are allowed.

In this section we argue for a figure of merit for MO materials based on their dielectric tensor. It appears that this FOM is the least upper bound of the set of all $|r_y|$ obtained by incorporating the given material into multilayer structures of arbitrary design.[6] We also show, by means of examples, that it is possible to design practical structures whose $|r_y|$ approaches the proposed FOM arbitrarily closely.

Let us begin by considering the simple structure shown in Fig. 6.14, which consists of a thin MO film (thickness Δ) on a semi-infinite substrate. A light beam with wavelength λ_0 is normally incident on this

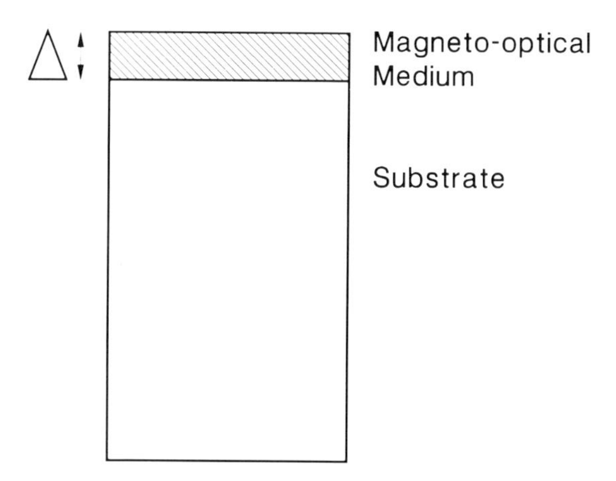

Figure 6.14. Thin magneto-optical film on a semi-infinite substrate of refractive index n_s. The medium of incidence is air and the normally incident beam is linearly polarized along X.

structure from the film side, the film is very thin (i.e., $\Delta \ll \lambda_0$), the complex refractive index of the substrate is n_s, and the magnetic material's dielectric tensor is

$$\underset{\sim}{\epsilon} = \begin{pmatrix} \epsilon & \epsilon' & 0 \\ -\epsilon' & \epsilon & 0 \\ 0 & 0 & \epsilon \end{pmatrix} \tag{6.51}$$

Using the methods of Chapter 5, the reflection coefficient $|r_y|$ for the structure under consideration is found to be

$$|r_y| \simeq \frac{4\pi\Delta}{\lambda_0} \frac{|\epsilon'|}{|1 + n_s|^2}; \qquad \Delta \ll \lambda_0 \tag{6.52}$$

In order to calculate the absorbed power P_{abs} in the magnetic layer, we ignore the off-diagonal elements of its dielectric tensor, and treat it as a metal film with (complex) refractive index $\sqrt{\epsilon}$. Again using the methods of Chapter 5, calculation of the Poynting vector yields

$$\frac{P_{abs}}{P_{inc}} \simeq \frac{8\pi\Delta}{\lambda_0} \frac{\text{Im}(\epsilon)}{|1 + n_s|^2}; \qquad \epsilon' \ll \epsilon \tag{6.53}$$

Here P_{inc} is the power of the incident beam, and the assumption $\Delta \ll \lambda_0$ is

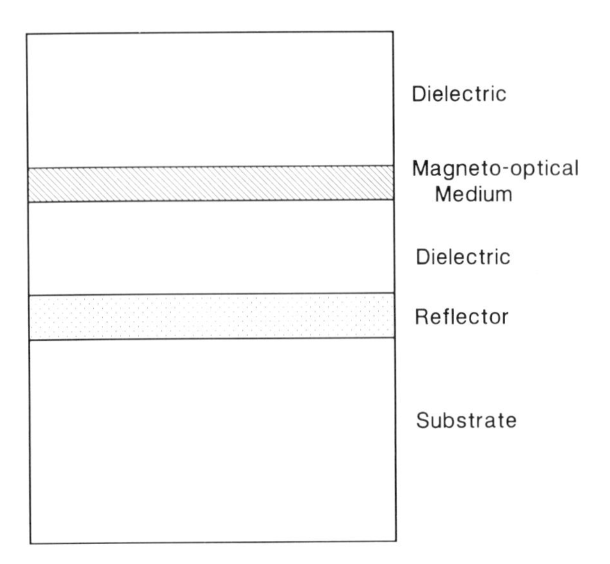

Figure 6.15. Multilayer structure with one magneto-optical layer. The other layers are metal or dielectric.

implicit. Comparing Eqs. (6.52) and (6.53), one arrives at the following conclusion:

$$|r_y| \simeq \frac{|\epsilon'|}{2\,\mathrm{Im}(\epsilon)} \left(\frac{P_{abs}}{P_{inc}} \right) \tag{6.54}$$

Although the above arguments have been confined to the simple structure of Fig. 6.14, the final result in Eq. (6.54) is believed to be general, applicable to thin MO films within multilayers of arbitrary structure. There is an intuitive justification for this belief that should become clear in the course of the following discussion.

Consider a multilayer structure, such as that in Fig. 6.15, where an MO film of thickness t_f is sandwiched among several layers of various non-magnetic metals and dielectrics. To be specific, we choose the magnetic material to be an amorphous alloy of TbFeCo, with $\epsilon = -1.4 + 28.3\mathrm{i}$ and $\epsilon' = 0.57 - 0.12\mathrm{i}$ at $\lambda_0 = 820$ nm. The structures that we shall consider are listed in Table 6.1.

Figure 6.16 shows plots of $|r_y|P_{inc}/P_{abs}$ versus t_f for the listed structures. Structure *a* is TbFeCo on a "matched" substrate, that is, the refractive index of the substrate is given by $n_s = \sqrt{\epsilon}$, ϵ being the diagonal tensor element for the magnetic film. Curve *a* in Fig. 6.16, which corresponds to structure *a* in Table 6.1, approaches the value of $|\epsilon'|/2\,\mathrm{Im}(\epsilon) = 1.03 \times 10^{-2}$ as t_f tends to zero, in agreement with Eq. (6.54). As t_f increases, however, the function decreases, indicating that for thicker films

Table 6.1. Various multilayer structures whose performance curves are shown in Fig. 6.16. n_s, n_i are the refractive indices of the substrate and of the medium from which the beam is incident. t_2, n_2 are the thickness and refractive index of layer 2. t_4, n_4 are the corresponding parameters for layer 4

Structure		Layer 1	Layer 2		Layer 3	Layer 4		
	n_s		t_2(nm)	n_2		t_4(nm)	n_4	n_i
a	3.67 + 3.86 i	TbFeCo	---	---	---	---	---	1.0
b_1	1.5	TbFeCo	---	---	---	---	---	1.0
b_2	1.5	TbFeCo	30	3.67 + 3.86 i	---	---	---	1.0
b_3	1.5	TbFeCo	100	1.5	---	---	---	1.0
b_4	1.5	TbFeCo	100	1.5	Al, 30 nm	---	---	1.0
b_5	1.5	TbFeCo	---	---	---	---	---	1.5
c_1	1.5	Al, 100 nm	100	1.5	TbFeCo	---	---	1.0
c_2	1.5	Al, 100 nm	75	2.0	TbFeCo	100	2.0	1.0

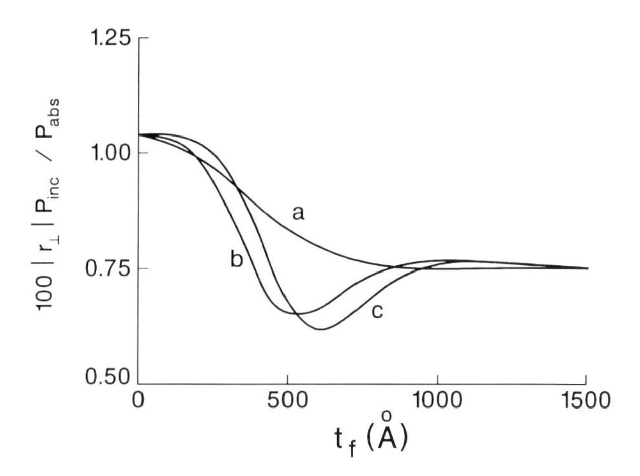

Figure 6.16. Performance curves for the structures of types a, b, c listed in Table 6.1. The abscissa is the thickness t_f of the magnetic layer.

the magneto-optically generated polarization vectors at different depths in the MO layer are not in phase.

Structures b_1 through b_5 in Table 6.1 all share the same characteristic curve b in Fig. 6.16. The common feature of these structures is a TbFeCo layer that is directly deposited on a glass substrate; the differences are in

Table 6.2. Single and multilayer structures and their performance characteristic $|r_y|$, as compared with the figure of merit $|r_y|_{max}$ of the MO material. The last column gives the reflectivity R of the structure as a percentage of the incident power. t_1, t_2, t_3 are the thicknesses of layers 1, 2, 3

Substrate	Layer 1		Layer 2		Layer 3		$\lvert r_y\rvert / \lvert r_y\rvert_{max}$	R(%)
	t_1(nm)	type	t_1(nm)	type	t_1(nm)	type		
glass	500	TbFeCo	---	---	---	---	0.29	60
glass	10	TbFeCo	---	---	---	---	0.39	33
glass	75	TbFeCo	85	$n = 2$	---	---	0.60	13
glass	22	TbFeCo	90	$n = 2$	---	---	0.69	5
aluminum	190	$n = 2$	50	TbFeCo	80	$n = 2$	0.76	12
aluminum	75	$n = 2$	20	TbFeCo	125	$n = 2$	0.80	16
aluminum	75	$n = 2$	20	TbFeCo	100	$n = 2$	0.93	2

the coating layers, and in the refractive indices of the medium of incidence. At small t_f the curve obeys Eq. (6.54), and thus coincides with curve a, but as t_f increases the effect of the substrate becomes apparent. Initially the reflection at the film–substrate interface creates an MO signal in phase with that from the surface, thus enhancing r_y; at larger t_f, however, destructive interference occurs and the signal drops, bringing curve b well below curve a. Eventually at very large t_f the light no longer sees the substrate and thus the two curves coincide once again.

Structures c_1 and c_2 in Table 6.1 share the curve c in Fig. 6.16. These structures consist of a glass substrate overcoated with 100 nm of aluminum; a layer of transparent dielectric follows and this lies under the TbFeCo layer. Structure c_2 is overcoated with another dielectric, and the medium of incidence in both cases is air. Note that the optical thickness of the intermediate dielectric layer is the same for c_1 and c_2. It is observed that when $t_f \to 0$ the curve approaches the limit of Eq. (6.54). Also, as in curve b, interference among the signals generated throughout the MO layer is responsible for the enhancement at small t_f and the depression at intermediate t_f. The large reflectivity at the aluminum layer in the case of curve c accounts for its differences from curve b.

Based on the above considerations, we conjecture that the value of $|r_y|$ given by Eq. (6.54) is the maximum possible value that can be obtained from a given material. The fact that P_{abs} is generally less than P_{inc} leads to the following conclusion:

$$\boxed{\text{FOM} = |r_y|_{max} = \frac{|\epsilon'|}{2\,\text{Im}(\epsilon)}} \tag{6.55}$$

The upper bound on $|r_y|$ as given by Eq. (6.55) is achieved if two conditions are satisfied: (i) the incident beam is fully absorbed by the MO medium; (ii) the magneto-optically induced components of polarization at various depths within the magnetic layer(s) are added in phase.

Table 6.2 lists several structures of varying complexity, and compares their performance against the upper bound given by Eq. (6.55). Note, in particular, that the structure in the last row can achieve as much as 93% of the maximum possible signal. In conclusion, the figure of merit defined in Eq. (6.55) places an upper bound on the MO signal $|r_y|$ achievable from a given material. It also appears that this upper limit can be closely approached in properly designed multilayer structures.

Problems

(6.1) Let a linearly polarized beam of light be reflected from a uniformly magnetized medium. The reflected components of polarization along X and Y are denoted by $|r_x|\exp(i\phi_x)$ and $|r_y|\exp(i\phi_y)$ respectively. Express the Kerr rotation angle and the ellipticity of the reflected beam in terms of $|r_x|$, ϕ_x, $|r_y|$, and ϕ_y.

(6.2) Use the Jones matrix formalism to investigate the passage of light from the disk to the detectors in the differential detection system of Fig. 6.6 with the $\lambda/4$ plate in place. Express the difference signal as a function of $|r_x|$, $|r_y|$, ϕ_x, ϕ_y and the orientation angles of the $\lambda/4$ plate and the polarizing beam-splitter.

(6.3) Describe the effect of a half-wave ($\lambda/2$) plate on the polarization states of normally incident RCP and LCP light. Use vector diagrams and be specific about the phase angles.

(6.4) Apply the Jones matrix calculus to the system shown in Fig. 6.7, and derive the expressions for the sum and difference signals given in Eq. (6.35).

(6.5) Verify Eq. (6.46) by performing the convolution operation on the two functions in Eqs. (6.44) and (6.45).

(6.6) Consider the MO film of thickness Δ supported by the substrate of (complex) refractive index n_s shown in Fig. 6.14. Find the reflection coefficient r_y for this structure by finding first the reflectivities for RCP and LCP beams, namely, $r_1\exp(i\phi_1)$ and $r_2\exp(i\phi_2)$, and then combining them according to Eqs. (6.24). Verify Eq. (6.52) in the limit $\Delta \ll \lambda_0$, where λ_0 is the incident wavelength.

(6.7) Consider the thin-film structure shown in Fig. 6.14, and assume that the off-diagonal element ϵ' of the magnetic layer's tensor is negligible; in other words, treat the magnetic film as a simple metal layer with (complex) refractive index $n = \sqrt{\epsilon}$. For a normally incident beam of light, calculate the ratio of the power absorbed by the film to the incident power. Verify Eq. (6.53) in the limit $\Delta \ll \lambda_0$.

(6.8) A plane monochromatic wave is obliquely incident at the interface between an isotropic dielectric medium of refractive index n and air, as shown in Fig. 6.17. $n = \sqrt{\epsilon}$ is real (i.e., there is no absorption), the refractive index of air is unity, and the angle of incidence is Θ. When Θ is large enough, total internal reflection takes place. Determine the reflection coefficients r_p and r_s at the interface and verify Eq. (6.33) for the phase difference between r_p and r_s under the condition of total internal reflection.

(6.9) A simplified diagram of a polarized-light microscope is shown in Fig. 6.18. A beam from the light source is linearly polarized and brought to focus on the sample. The reflected light goes through the half-silvered mirror and an analyzer before

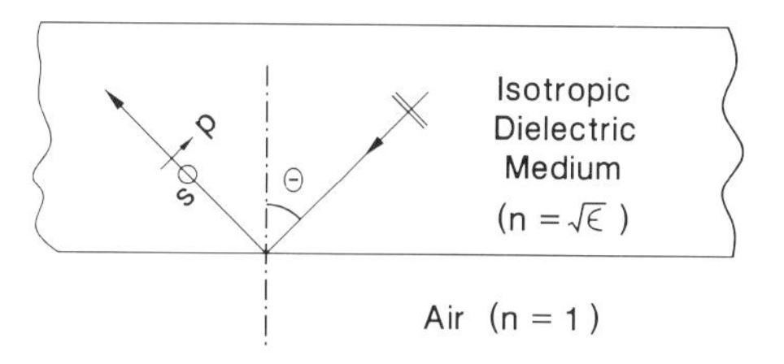

Figure 6.17.

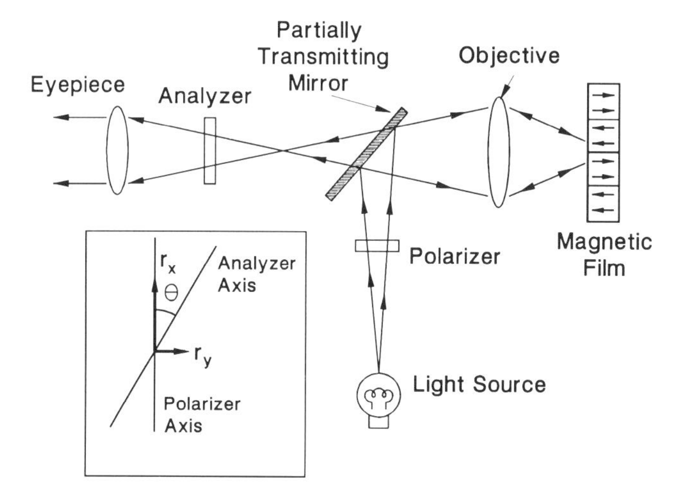

Figure 6.18.

reaching the viewer. Let the angle between the transmission axes of the analyzer and the polarizer be Θ, and assume that a magnetic sample consisting of up- and down-magnetized regions is placed under the microscope. The sample's complex reflection coefficients r_x and r_y correspond respectively to light polarized along the direction of incident polarization (X) and perpendicular to it (Y).

(a) In terms of $|r_x|$, $|r_y|$ and the phase difference between them $\phi_x - \phi_y$, derive expressions for the light intensities $I(\uparrow)$ and $I(\downarrow)$ originating in the up- and down-magnetized regions of the sample and arriving at the observation point.

(b) The goal in microscopy is usually not to maximize the signal or even the signal-to-noise ratio, but to enhance the contrast ratio C defined as

$$C = \frac{I(\uparrow) - I(\downarrow)}{I(\uparrow) + I(\downarrow)}$$

Using the results obtained in part (a) of this problem, find the optimum setting of the analyzer angle Θ for achieving the best contrast ratio.

7

Effects of High-numerical-aperture Focusing on the State of Polarization

Introduction

It has been suggested that a focused beam in the neighborhood of the focal point behaves just like a plane wave. This misconception is based on the fact that the focused beam at its waist has a flat wave-front. The finite size of the beam, however, causes the plane-wave analogy to fail: Fourier analysis shows that a focused light spot is the superposition of a multitude of plane waves with varying directions of propagation. At the high values of numerical aperture typically used in optical recording, the fraction of light having a sizeable obliquity factor is simply too large to be ignored. The reflection coefficients and magneto-optical conversion factors of optical disks are rather strong functions of the angle of incidence; thus the presence of oblique rays within the angular spectrum of the focused light must influence the outcome of the readout process. The goal of the present chapter is to investigate the consequences of sharp focus in optical disk systems, and to clarify the extent of departure of the readout signals from those predicted in the preceding chapter.

The vector diffraction theory of Chapter 3 and the methods of Chapter 5 for computing multilayer reflection coefficients are used here to analyze the effects of high-NA focusing. The focused incident beam is decomposed into its spectrum of plane waves, and the reflected beam is obtained by the superposition of these plane waves after they are independently reflected from the multilayer surface. In section 7.1 qualitative arguments are put forward that help classify the relevant phenomena, that suggest methods of analysis, and that justify the assumptions made. The results of numerical analysis for a number of multilayer structures are then presented in section 7.2. Section 7.3 summarizes the results and contains a few concluding remarks.

7.1. Focused Beams, Oblique Rays, and Polarization Conversion

When a collimated beam of light is brought to focus with a high-NA lens, its spectrum of spatial frequencies broadens to include plane waves propagating at angles up to arcsin(NA) relative to the original propagation direction. From the standpoint of geometrical optics these plane waves correspond to rays that are bent by the lens to various degrees in order to

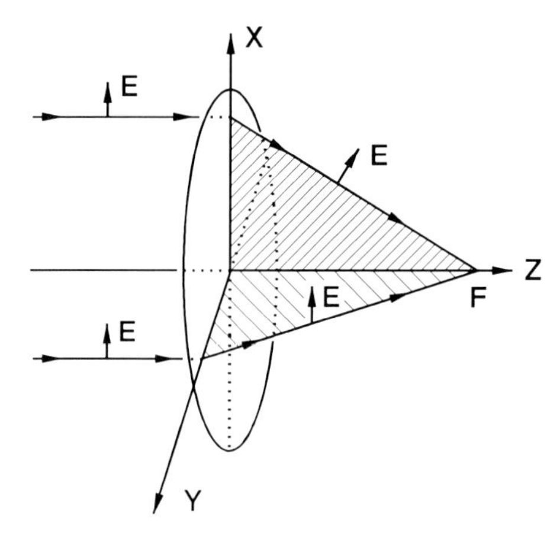

Figure 7.1. A collimated, linearly polarized beam of light is brought to focus by a lens. **E** is the electric field vector, showing the direction of linear polarization for the various rays.

form the focused spot. There are at least three factors that must be taken into consideration when dealing with these oblique rays:

(i) Because the p- and s-components of polarization are reflected differently from the various regions on the surfaces of the lens, one must include the appropriate Fresnel reflection coefficients for these regions in order to account for the variations in the transmission function of the lens over its clear aperture. In principle, the p and s transmission functions may be computed provided that structural details of the lens are available.

(ii) The polarization of a ray after passing through the lens depends on the incident state of polarization and the emerging direction of propagation. For instance, in Fig. 7.1 where the incident beam is linearly polarized along X, rays in the YZ-plane retain their polarization direction after the lens, whereas those in the XZ-plane have to bend their polarization vector in order to keep it orthogonal to the new direction of propagation. In general, the incident polarization must be resolved locally into s- and p-components; while the direction of the s-component remains unchanged after the lens, the p-component must be rotated to satisfy the requirement of orthogonality. It is important to recognize that, in the absence of reflection and transmission losses at the lens, the p- and s-components of each ray may be identified independently of the specific structure of the lens; all that is needed is knowledge of the state of polarization before the lens and the direction of propagation of the ray after the lens.

(iii) Rays that propagate along different directions and have different states of polarization will be reflected differently from a given flat surface. Thus in analyzing the reflection of a focused beam, one must decompose the beam into its spectrum of plane waves (i.e., rays), compute the reflection coefficients for individual rays, and superimpose the resulting plane waves to obtain an accurate distribution of the reflected beam.

Usually the reflected beam goes back through the same objective and encounters phenomena (i) and (ii) again, this time in the reverse direction, before it reappears at the exit pupil of the lens. At this point the beam is collimated and a beam-splitter may be used to isolate it from the incident beam for further processing.

Several investigators have analyzed the effects of phenomena (i) and (ii) above on the state of polarization in the focal region.[1-4] There are also published reports concerning the effect of the first phenomenon on the image quality in polarized-light microscopes.[5] Phenomenon (iii), however, has not received much attention in the past. Since in practice the three phenomena occur simultaneously, it might seem desirable at first glance to combine them all in one model and study their collective effects. However, the first phenomenon is dependent on the structure of the lens, and models that incorporate a particular lens design will, of necessity, have limited usefulness. It is also conceivable that, with special antireflection coatings applied to the surfaces of the lens, the variations in the p and s transmission factors across the aperture could be eliminated. The second and third phenomena, on the other hand, are universal, in the sense that their effects are independent of the particular structure of the lens used; moreover, these phenomena are unavoidable. It is for these reasons that our attention in this chapter will be confined to the second and third phenomena only. Ignoring the first phenomenon is tantamount to assuming that the lens has uniform and equal transmissivities for both p- and s-polarizations over its entire aperture.

The purpose of the present chapter is to evaluate the combined effects of phenomena (ii) and (iii) above for several optical and magneto-optical disk structures. Schematic diagrams of the basic systems studied here are shown in Fig. 7.2. The collimated incident beam is uniform and monochromatic, with vacuum wavelength λ_0, unit power across the lens aperture, and linear polarization along the X-axis. The objective is corrected for all aberrations and has numerical aperture NA and focal length f. The optical disk surface is located in the focal plane of the lens. In the substrate-incident configuration of Fig. 7.2(a) the front facet of the substrate is out of focus, thus allowing one to ignore it in subsequent analysis.† The vector diffraction theory of Chapter 3 has been used to

† It is assumed here that the lens is designed to correct automatically the spherical aberrations introduced by the substrate. In this sense, the front facet of the substrate belongs to the lens and not to the disk. Also, multiple reflections and interference-type phenomena at this facet are negligible provided that the substrate thickness is greater than the beam's coherence length, which is generally the case in practice. One must be aware of the fact that reflection (and transmission) coefficients at the substrate's front facet are both angle- and polarization-dependent, just as are those at the various surfaces of the lens. In the present chapter these dependences are ignored (100% transmission is assumed for all the rays at the substrate's front facet). The refractive index of the substrate, of course, has been taken into account when computing the disk reflectivity, since the substrate is now the medium of incidence for the rest of the disk.

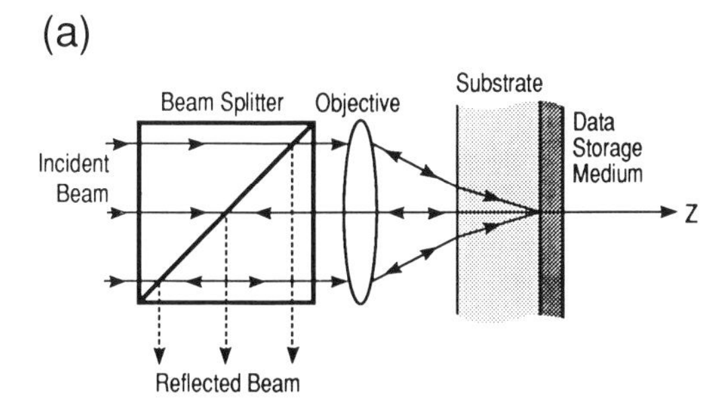

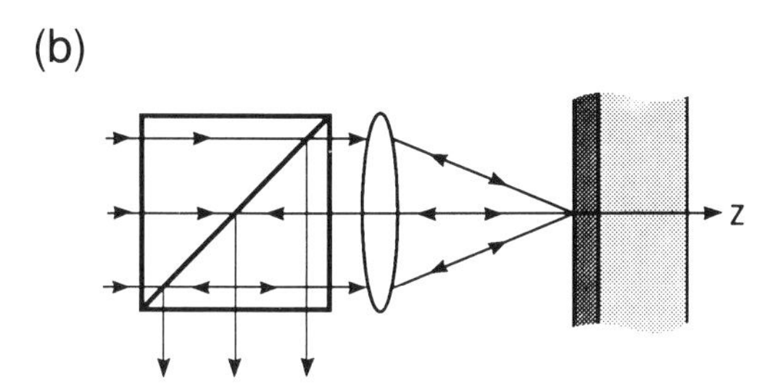

Figure 7.2. Schematic diagram of the objective lens and the storage medium in an optical disk system. The incident beam at the objective is collimated and linearly polarized along the X-axis. The substrate in configuration (a) is the medium of incidence, while in (b) the beam is incident from the air side. The lens is corrected for all aberrations, including those caused by the substrate in configuration (a).

calculate the propagation of the beam from the lens to the disk.[1,2,6,7] For each plane wave contained in the spectrum of the focused beam, the reflectivity has been calculated using the algorithm described in Chapter 5.[8] Immediately after reflection, the emerging plane waves are combined to create the distribution of the reflected beam at the focal plane. This distribution is propagated back to the lens using the far-field (Fraunhofer) diffraction formula. Finally, the polarization vectors are reoriented locally to account for the reverse propagation through the lens and the subsequent recollimation.

7.2. Numerical Analysis

The various procedures described in section 7.1 have been incorporated into a computer program. A typical run of this program consists of the following steps:

(i) computing the distribution of light at the focal plane of the objective;
(ii) Fourier decomposition of the focused beam into its plane-wave constituents;

(iii) calculation of the reflectivity matrix for individual plane waves at the multilayer surface;
(iv) superposition of the reflected plane waves;
(v) propagation of the reconstructed beam back to the objective lens;
(vi) final passage of the beam through the lens and recollimation.

Results of numerical analysis for several cases of practical interest are presented in the following subsections. In each case a particular disk structure is chosen and the patterns of intensity and polarization upon reflection from the disk are computed. In all cases, the beam entering the lens is linearly polarized along X, as shown in Fig. 7.1; the angle Θ of polarization rotation is thus measured relative to the X-axis.

7.2.1. The case of a perfect reflector

In order to confirm the validity and the accuracy of the computer program, in this first example we analyze the simple case of reflection from a perfect reflector. Let a uniform plane wave propagating along Z and linearly polarized along X go through a lens of numerical aperture NA = 0.5 and focal length $f = 6000\lambda_0$. At the focal plane the beam will be focused to a small spot (the Airy pattern) and the bending of the polarization vectors by the lens will create, in addition to the original polarization along X, a small component of polarization along Y and a not-so-small component along Z. Fig. 7.3 shows the intensity plots I_x, I_y, I_z for these three components of polarization at the focal plane. The X-component shown in Fig. 7.3(a) contains a total power (integrated intensity) of 0.9367, while the power content of the Y and Z components are 0.0007 and 0.0626 respectively. (Remember that the total incident power is unity, and that thus far in the system there are no losses.) Figure 7.3 also shows the contour plots for these intensity distributions.

Let us now assume that a perfect reflector is placed in the focal plane, that is, the various plane waves contained in the spectrum of the focused beam return with a reflection coefficient of unity and with no relative phase shifts. The distribution of the reflected beam after recollimation is shown in Fig. 7.4. The total power content of the X-, the Y-, and the Z-component of this beam are now 0.98, 0.45×10^{-8}, and 0.74×10^{-7} respectively. In this example, the Y- and the Z-component are much smaller than the X-component and, for all practical purposes, may be ignored. The total power content of the beam is reduced by about 2%; this results from the spread of the beam during its round trip and the subsequent cutoff by the finite aperture of the lens. Notice the uniform distribution of the X-polarization within the aperture which, as expected, is identical with the distribution of the incident beam, thus confirming the accuracy of the computational method.

In general, whether the disk is perfectly reflecting or not, the Z-component of the recollimated beam resembles that in Fig. 7.4(c); its magnitude being negligibly small, this component will be omitted in the following examples. In contrast, the *Y*-component becomes important in other cases and so will henceforth be retained. Note that the distributions of the *Y*- and the *Z*-component in the present example are concentrated around the rim of the lens; the assumed sharp cutoff of the fields at the rim is responsible for this behavior.

7.2.2. The case of a front-surface aluminum mirror

We consider the case of an aluminum mirror with complex refractive index $(n, k) = (2.75, 8.31)$ at $\lambda_0 = 830$ nm. The surface reflectivity at normal incidence is

$$R = \left| \frac{1 - n - ik}{1 + n + ik} \right|^2 = 0.868$$

When this mirror is placed in the focal plane of an objective lens (NA = 0.5, $f = 6000\lambda_0$) and the reflected distribution in the exit pupil is computed, the patterns of Fig. 7.5 are obtained. As in the preceding case, I_x is fairly uniform over the aperture, but the power content of the *X*-component is now 0.85. The power of the *Y*-component is 0.14×10^{-3}, which is substantially greater than the corresponding quantity in the case of the perfect reflector. The plot of I_y in Fig. 7.5(b) shows four peaks in the four quadrants of the *XY*-plane. There is perfect symmetry among these quadrants, as can be observed from the corresponding contour plot in Fig. 7.5(c). As expected, the power content of the *Z*-component (distribution not shown) is negligible ($\simeq 10^{-7}$). Also shown in Fig. 7.5 are plots of the polarization rotation angle Θ and ellipticity ϵ for the reflected beam at the exit pupil.† The maximum and minimum values of Θ and ϵ are $\pm 0.58°$ and $\pm 1.76°$ respectively.

7.2.3. Thin dielectric layer on a glass substrate

Figure 7.6(a) shows a simple, partially reflecting structure consisting of a glass substrate ($n_0 = 1.53$) coated with a dielectric thin film of thickness d_1 and refractive index $n_1 = 2.00$. This structure is used in the substrate-incident configuration of Fig. 7.2(a), where the substrate faces the objective and its interface with the film is at the focal plane. Figures 7.6(b) and (c) show plots of reflected intensity, I_x and I_y, at the objective's exit

†The symbol ϵ is used in the present chapter to denote not only the ellipticity of the beam but also the various elements of the dielectric tensors of the media. The possibility of confusion is remote, however, since the intended meaning of the symbol in each case should be clear from the context.

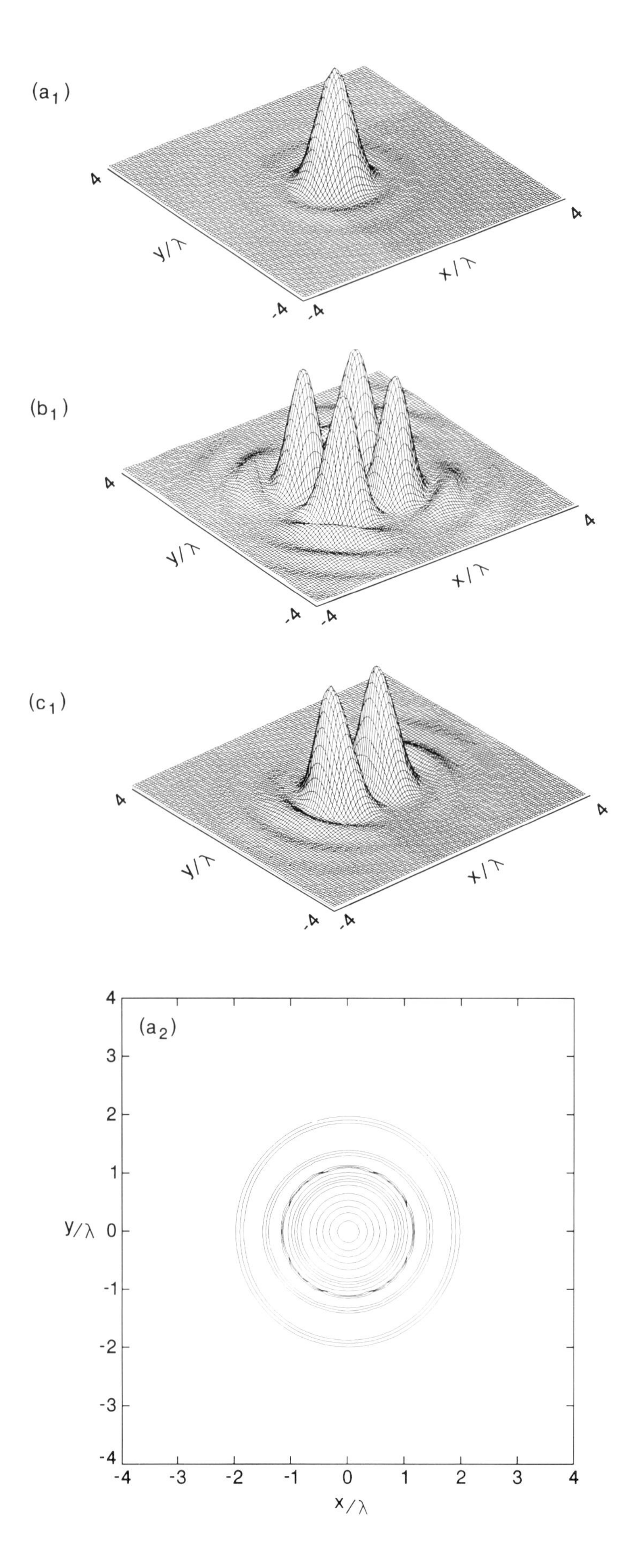

(a1)
y/λ
x/λ
4
-4
(b1)
(c1)
(a2)

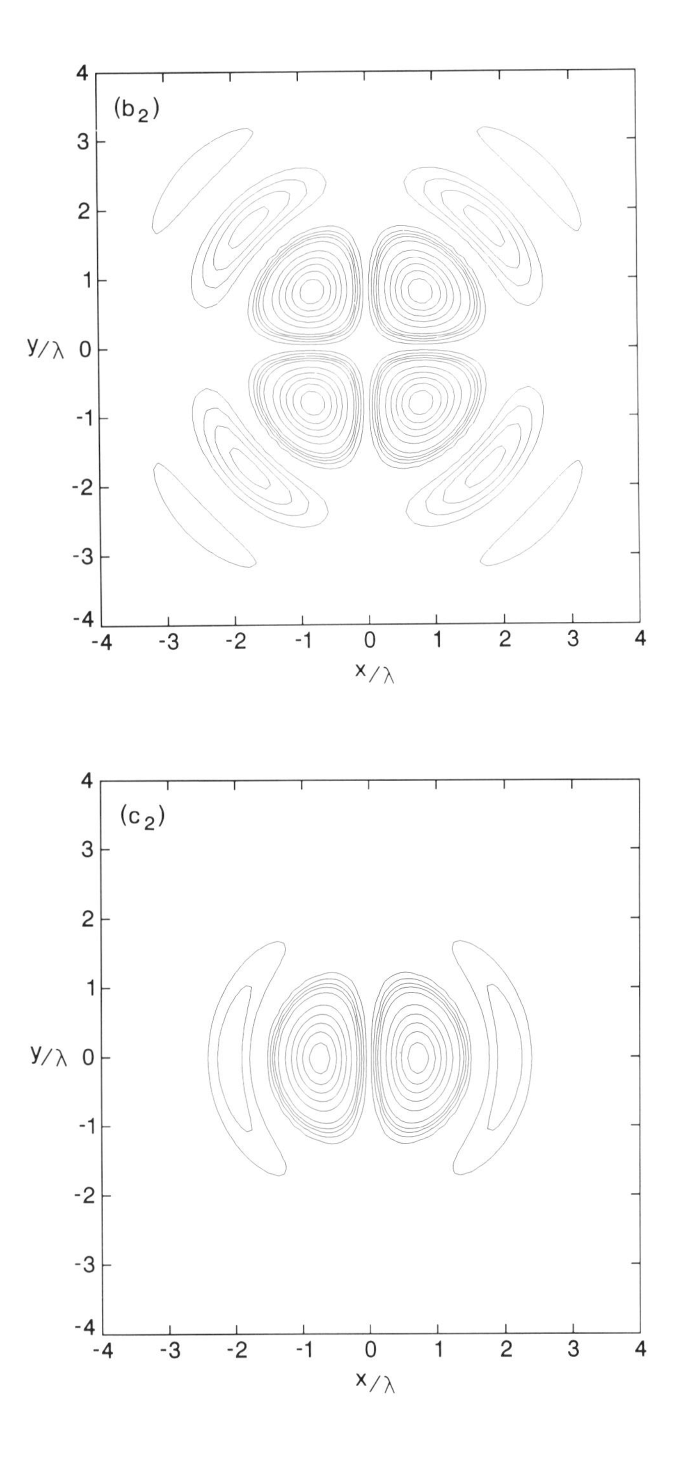

Figure 7.3. Plots of intensity in the focal plane of the lens for the components of polarization along the X-, the Y-, and the Z-axis. Also shown are the contour plots of intensity in each case. The maximum values of I_x, I_y, and I_z are 0.74, 0.00016, and 0.025 respectively.

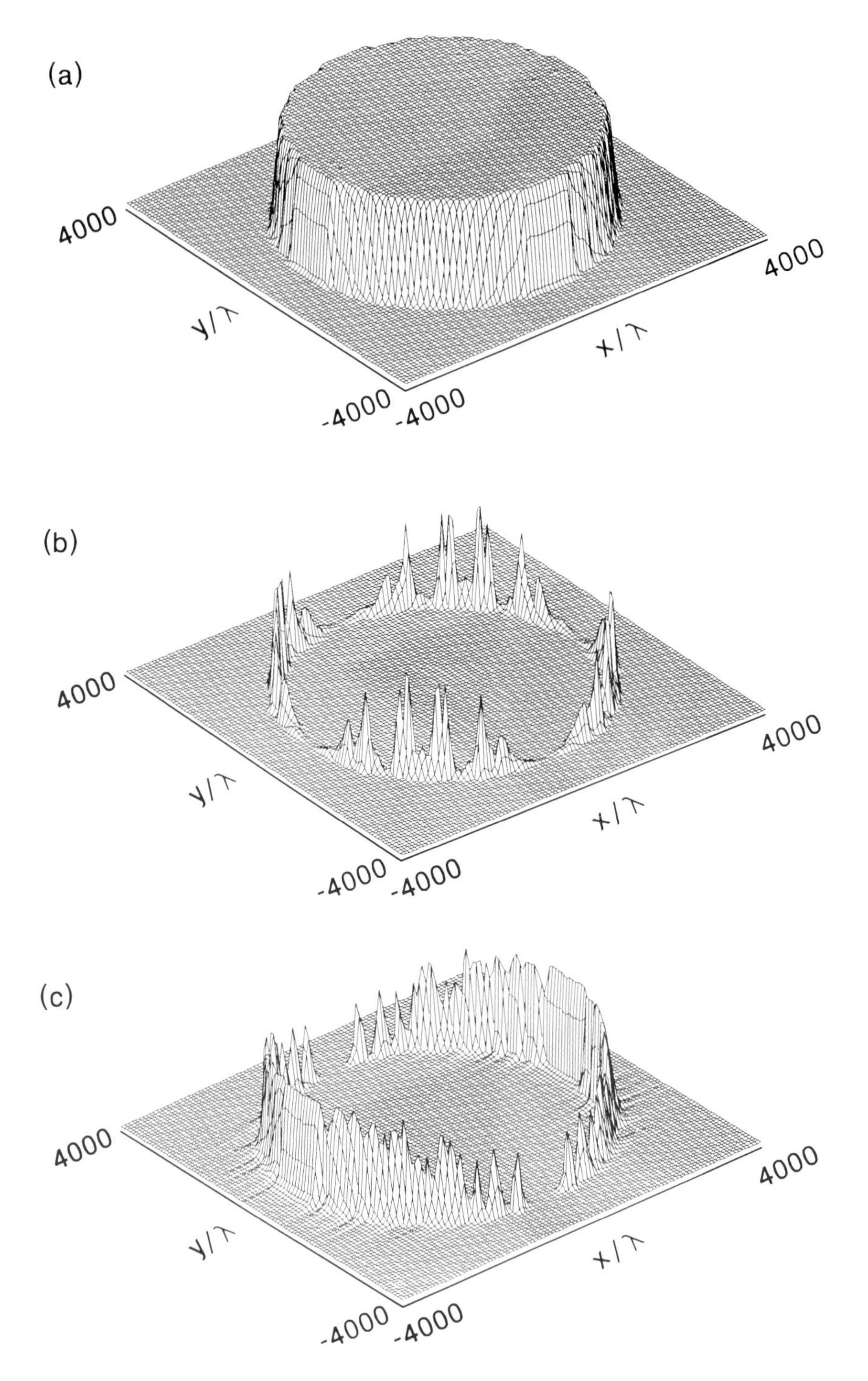

Figure 7.4. Plots of reflected intensity in the exit pupil of the objective for the three components of polarization; the disk is assumed to be a perfect reflector. (a) I_x, (b) I_y, (c) I_z. The maximum values of I_x, I_y, and I_z are 0.36×10^{-7}, 0.58×10^{-14}, and 0.29×10^{-13} respectively.

Table 7.1. Calculated values of reflectivity for the structure of Fig. 7.6(a). R_1 is the reflectivity for a normally incident plane wave. R_2 corresponds to a beam focused by a lens having NA = 0.5, $f = 6000\lambda_0$. R_3 corresponds to a beam focused by a lens having NA = 0.9, $f = 3333\lambda_0$

n_1d_1/λ_0	R_1	R_2	R_3
0.000	0.0439	0.0430	0.0571
0.100	0.1041	0.0100	0.1129
0.200	0.1869	0.1804	0.1917
0.225	0.1963	0.1901	0.2023
0.250	0.1995	0.1940	0.2076
0.275	0.1963	0.1920	0.2077
0.300	0.1869	0.1840	0.2025
0.400	0.1041	0.1074	0.1373
0.500	0.0439	0.0436	0.0640

pupil (NA = 0.5, $f = 6000\lambda_0$, $\lambda_0 = 830$ nm, $d_1 = 207.5$ nm). The saddle-like structure of I_x can be explained as follows. The rays in the XZ-plane are p-polarized and, at higher angles of incidence, have a reduced reflectivity. The rays in the YZ-plane, on the other hand, are s-polarized and their reflectivity increases with increasing angle of incidence. As in the previous example, the plot of I_y shows perfect symmetry among its four quadrants.

Calculated values of the reflectivity versus d_1, the thickness of the coating layer, are listed in Table 7.1. The first column of the table gives values of reflectivity for a plane wave normally incident through the substrate (ignoring the front facet). The maximum reflectivity ($\simeq 20\%$) obtains at $d_1 = \lambda_0/4n_1$, and the reflectivity values are symmetric around this peak. The second column lists reflectivity as calculated at the exit pupil of a lens (NA = 0.5, $f = 6000\lambda_0$). There is a slight asymmetry in these values but, by and large, the maximum reflectivity still occurs for the same quarter-wave-thick dielectric layer. The asymmetry, of course, is attributable to the optical path differences amongst the various oblique rays within the layer. The third column gives reflectivity as calculated at the exit pupil of another lens (NA = 0.9, $f = 3333\lambda_0$). The peak is now shifted to slightly larger values of d_1 and the asymmetry is more pronounced.

7.2.4. Magneto-optical film with a dielectric coating

We study the case of the magneto-optical bilayer shown in Fig. 7.7(a). The structure consists of a substrate of refractive index $n_0 = 1.5$, 83 nm of a magnetic material, and 103.75 nm of a dielectric coating having $n = 2.0$. The operating wavelength is $\lambda_0 = 830$ nm and the beam is incident on the

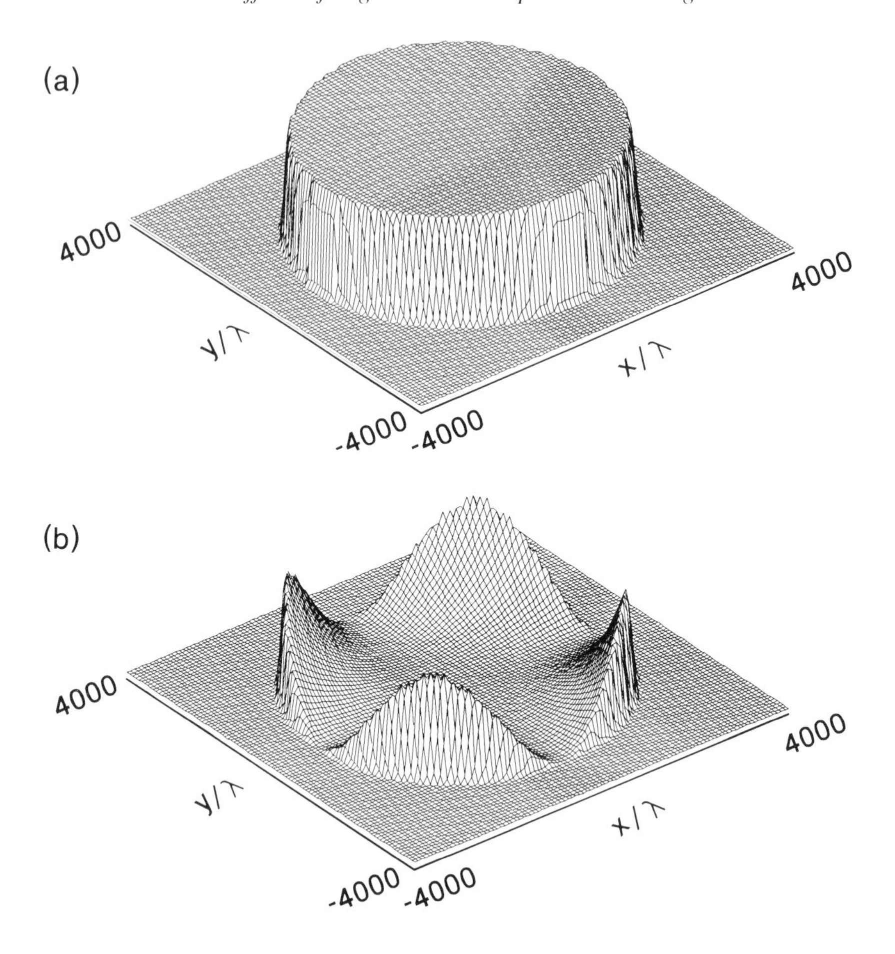

Figure 7.5. Reflected intensity and polarization distributions in the exit pupil. The disk is an aluminum reflector with $(n, k) = (2.75, 8.31)$. (a) Intensity of the X-component of polarization; $(I_x)_{max} = 0.31 \times 10^{-7}$. (b) Intensity pattern of the Y-component; $(I_y)_{max} = 0.28 \times 10^{-10}$. (c) Contour plot of I_y. (d) Polarization rotation angle Θ (relative to X); $\Theta_{min} = -0.58°$, $\Theta_{max} = +0.58°$. (e) Polarization ellipticity ϵ; $\epsilon_{min} = -1.76°$, $\epsilon_{max} = +1.76°$.

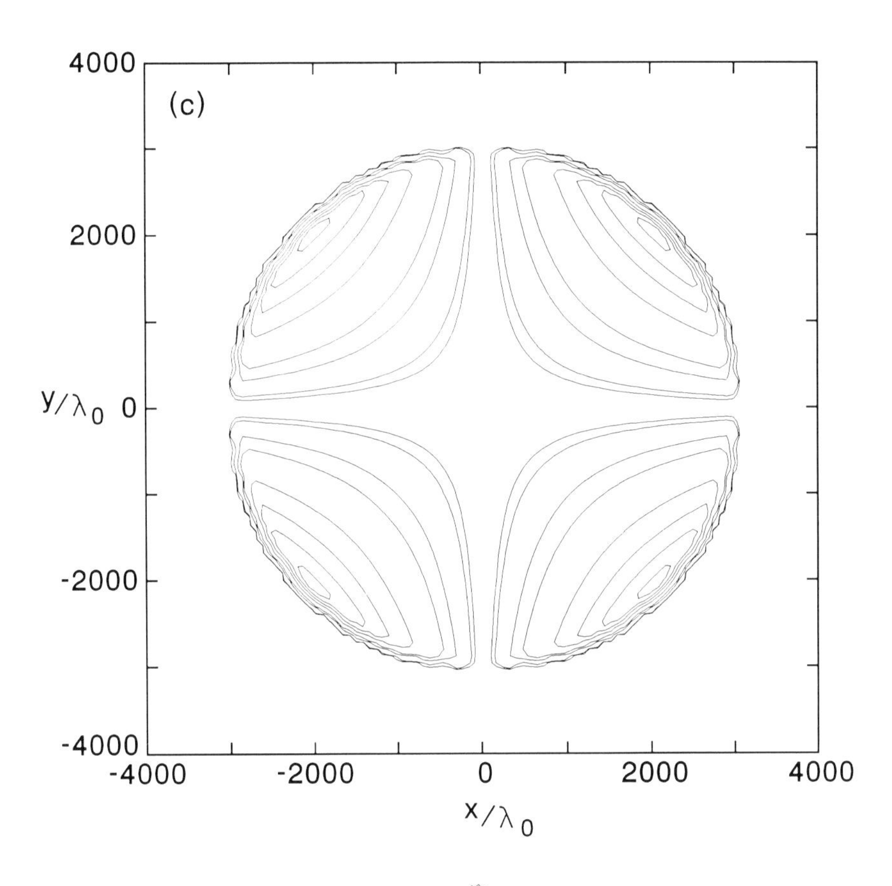
(c)
4000
2000
0
-2000
-4000
y/λ_0
-4000
-2000
0
2000
4000
x/λ_0

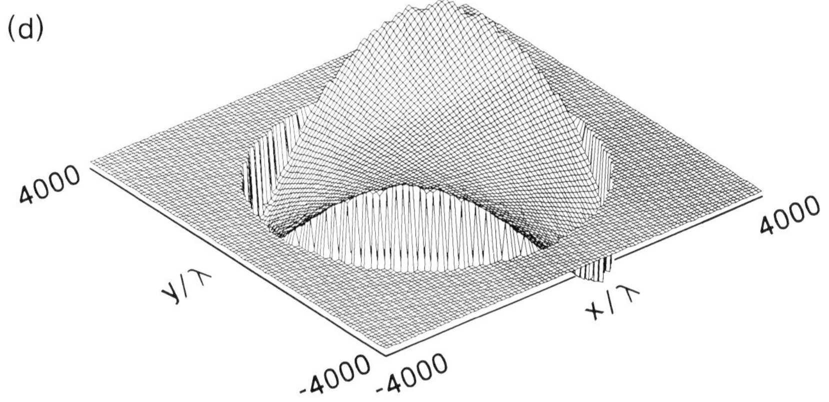
(d)
4000
4000
y/λ
x/λ
-4000
-4000

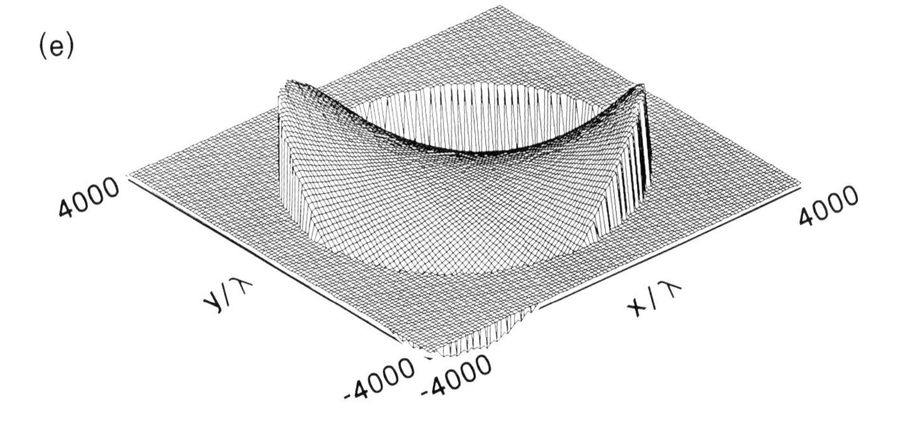
(e)
4000
4000
y/λ
x/λ
-4000
-4000

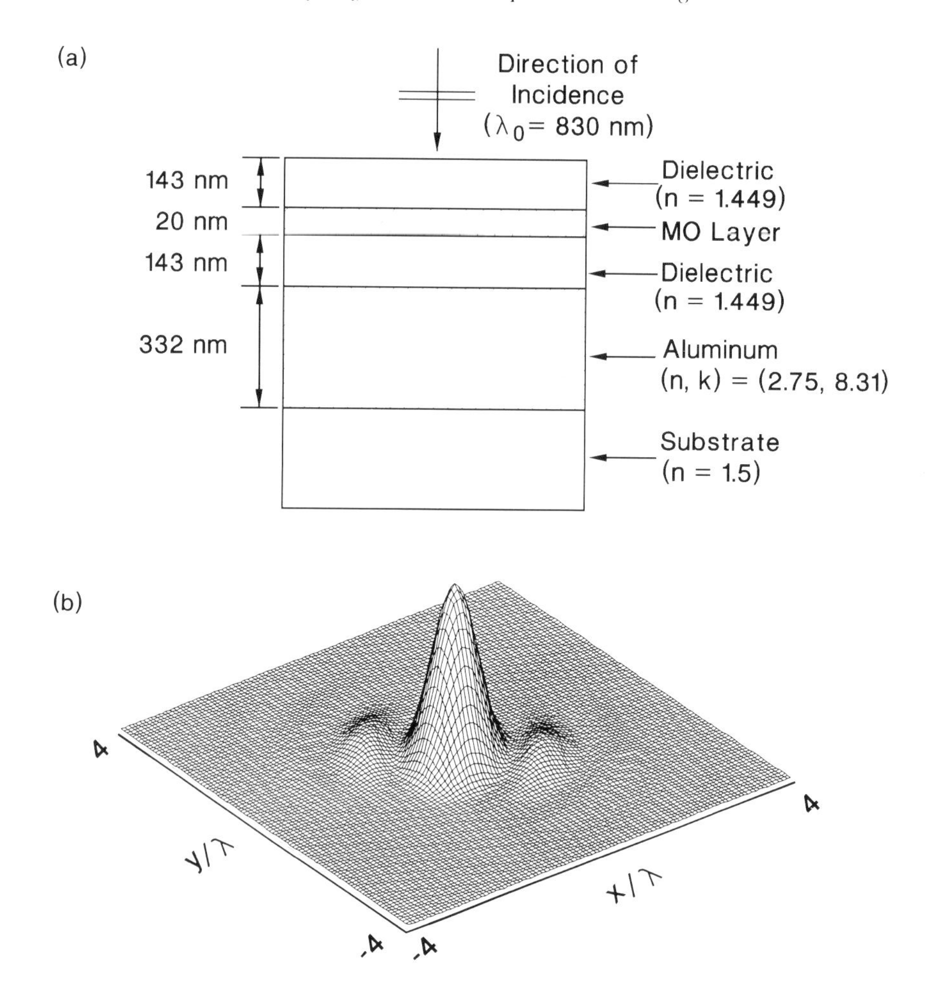

Figure 7.9. (a) Quadrilayer magneto-optical disk structure. The magnetic layer's dielectric tensor is the same as that given in the caption to Fig. 7.7(a). (b) Reflected intensity in the plane of the disk (i.e., the focal plane of the lens) for the Y-component of polarization. The maximum value of I_y is 0.76×10^{-4} and the power content of the Y-component is 0.12×10^{-3}.

7.2.6. Plastic substrate and the effects of birefringence

We shall study the effects of substrate birefringence on the focused spot as well as on the quality of the read signal. Plastic disk substrates used in practice typically have three different refractive indices n_r, n_t, n_z along the radial, tangential, and vertical directions. These differences are caused in the injection molding process where preferential alignment of polymer molecules occurs in the plane of the substrate. Observed values of birefringence are on the order of $n_r - n_t \simeq 10^{-5}$, and $n_r - n_z \simeq 10^{-4}$. Substrate birefringence introduces astigmatism on the focused spot and affects the polarization state of the beam, both of which cause performance degradations in the optical disk system.[9-11]

For technical reasons having to do with computational procedures, we treat the substrate as an isotropic, semi-infinite medium in contact with a birefringent layer of finite thickness, as indicated in Fig. 7.14(a). In the figure a thin metallic film is shown supported by a substrate whose birefringence is concentrated within a 20 μm-thick layer having $n_x = 1.5$, $n_y = 1.50$, and $n_z = 1.47$. The direction of beam propagation, Z, is perpendicular to the plane of the substrate, and the state of incident polarization is linear along the X-axis. The beam is focused on the metallic film through the substrate by a lens of NA = 0.55 and $f = 4700\lambda_0$. It was found (by trial and error) that in order to compensate the spherical aberration caused by the 20 μm-thick birefringent layer, it is necessary to place its interface with the substrate at 13 μm before the focus; Fig. 7.14(b) shows the computed intensity profile at this interface. The beam reaching the metallic layer is now very close to the ideal Airy pattern, as can be seen from its intensity profile in Fig. 7.14(c).

In the first series of calculations we assume that the metallic layer is a perfect reflector. Figure 7.15(a) shows the computed phase distribution at the exit pupil of the objective; its cross-sections along X and Y are shown in Fig. 7.15(b). The peaks and valleys of the curves in Fig. 7.15(b) are produced by the residual amounts of spherical aberration and defocus, but the astigmatism, amounting to nearly $0.2\lambda_0$, is a direct consequence of birefringence. In essence, s-polarized rays within the focused cone, which are not affected by the n_z of the birefringent layer, suffer a greater phase shift than the p-polarized rays, which do have a component of **E** in the Z-direction. Note that the astigmatism of $0.2\lambda_0$ is twice the amount produced in each passage through the substrate. Figure 7.16 shows contour plots of intensity at the metallic layer for three values of disk defocus: $-2\lambda_0$, 0 and $+2\lambda_0$. At best focus the spot is circularly symmetric, but at either side of focus it is elongated along X or Y. This elongation is another manifestation of the astigmatism produced by vertical birefringence. In a disk drive, if the polarization happens to be parallel to the tracks (tangential) or perpendicular to the tracks (radial), then birefringence of the type described here will enlarge the focused spot either along the track or perpendicular to it, depending on the sign of defocus. Tangential elongation will degrade the resolution of MO readout, while radial elongation will reduce the magnitude of the push–pull track-crossing signal.

Next we assume that the metallic layer is magneto-optical, with diagonal and off-diagonal elements $\epsilon = -5 + 20\,\mathrm{i}$ and $\epsilon' = 0.5 + 0.005\,\mathrm{i}$ respectively. Figure 7.17 shows plots of polarization rotation angle Θ and ellipticity ϵ at the exit pupil of the objective. Despite the nonuniform distributions across the aperture, the differential signal will not suffer excessively from birefringence, provided that a quarter-wave plate is used to convert ellipticity into rotation. In this case, the maximum signal

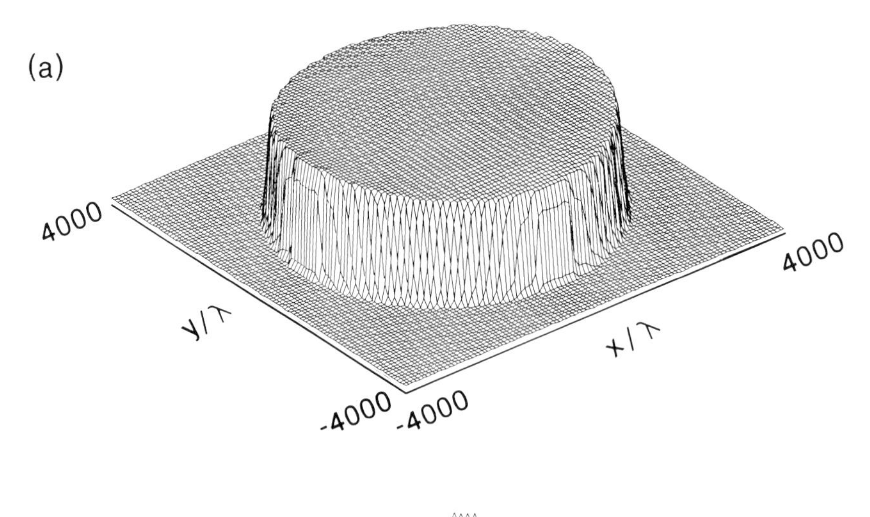

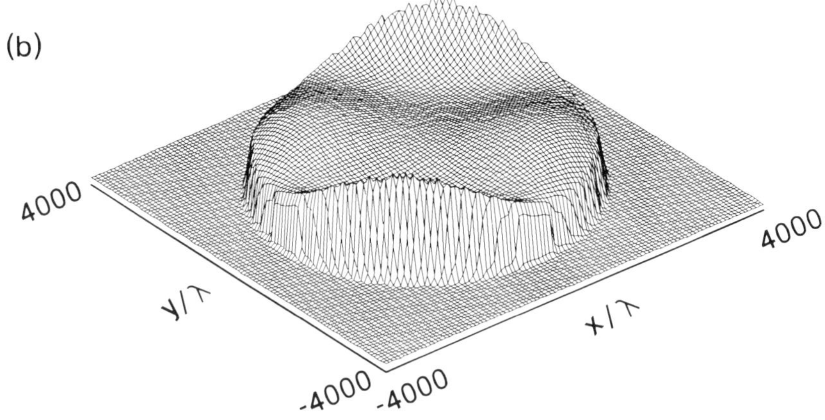

Figure 7.10. Distributions of the reflected intensity and polarization in the exit pupil for the disk structure shown in Fig. 7.9(a). (a) Intensity pattern for the X-component of polarization; $(I_x)_{max} = 0.2 \times 10^{-8}$. (b) Intensity pattern for the Y-component of polarization; $(I_y)_{max} = 0.7 \times 10^{-11}$. (c) Contour plot of I_y. (d) Polarization rotation angle Θ (relative to X); $\Theta_{max} = +3.52°$. (e) Polarization ellipticity ϵ; $\epsilon_{max} = +2.49°$.

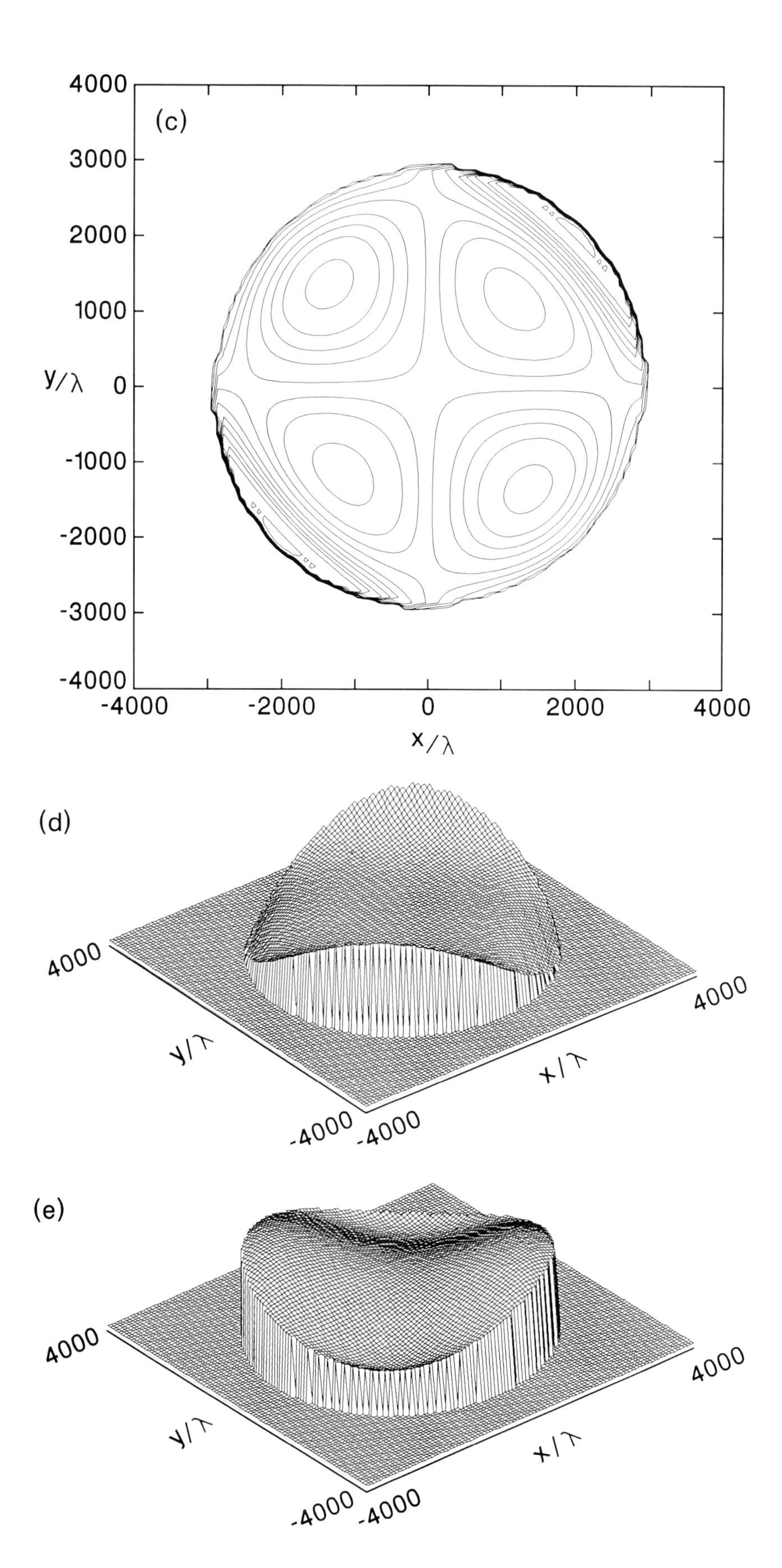

(c)
4000
3000
2000
1000
0
-1000
-2000
-3000
-4000
y/λ
-4000
-2000
0
2000
4000
x/λ
(d)
4000
y/λ
-4000
-4000
x/λ
4000
(e)
4000
y/λ
-4000
-4000
x/λ
4000

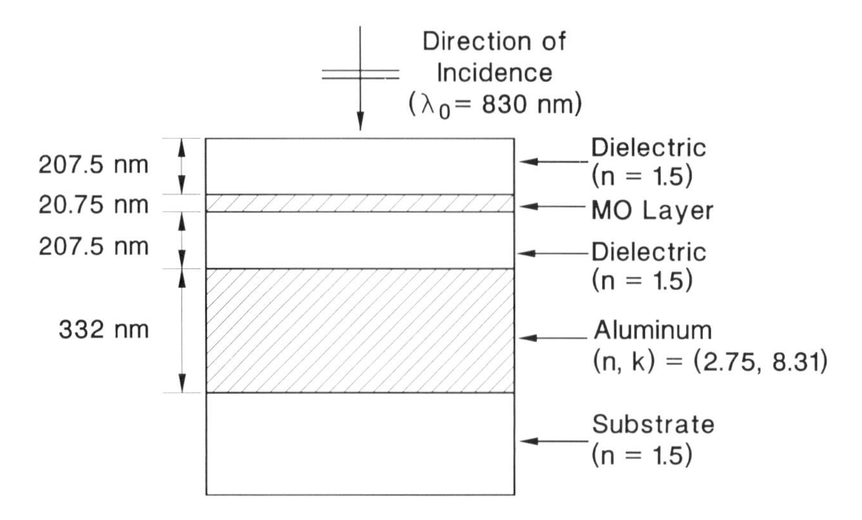

Figure 7.11. Quadrilayer magneto-optical disk structure. The magnetic layer's dielectric tensor is the same as that given in the caption to Fig. 7.7(a).

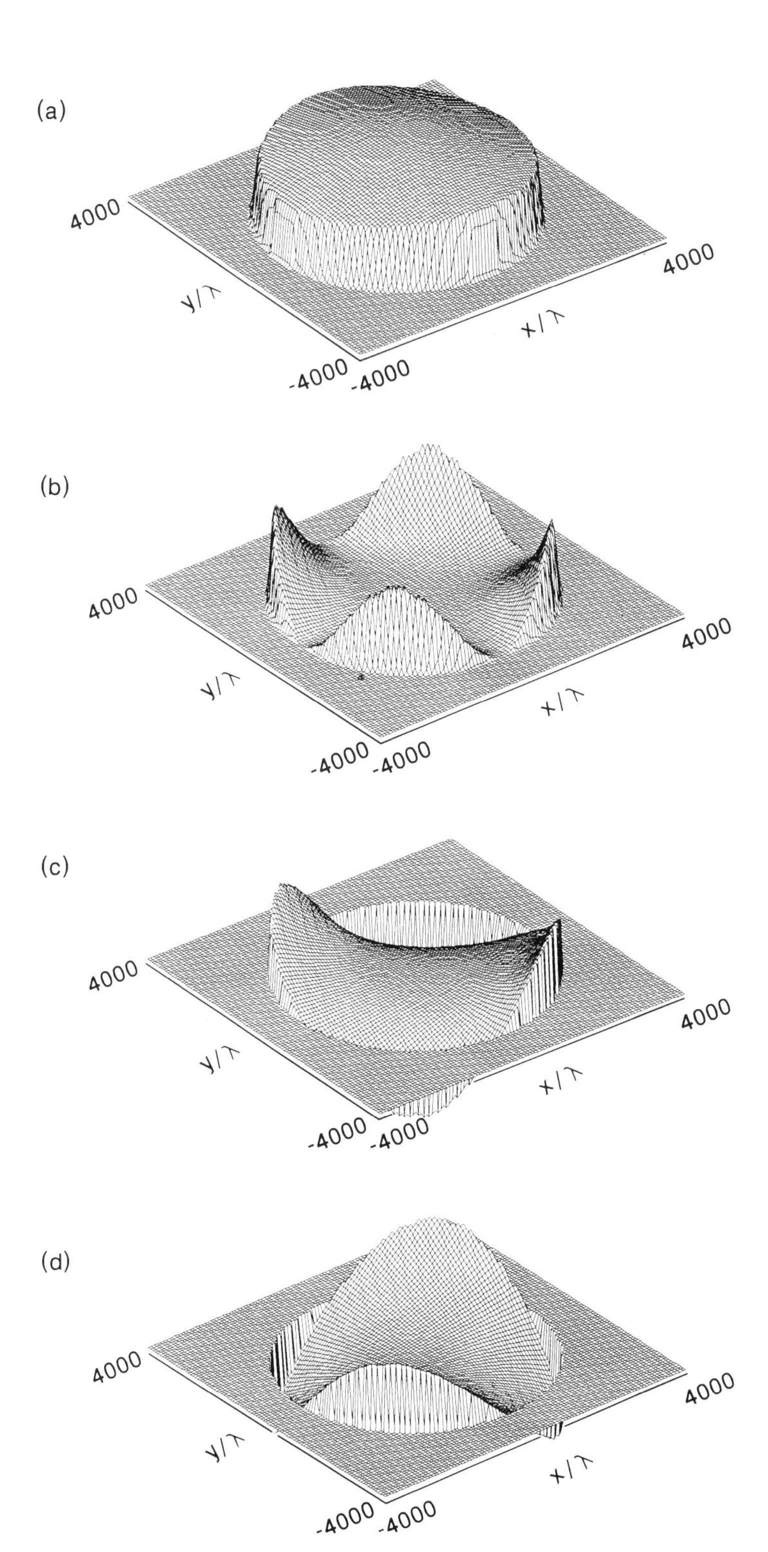

Figure 7.12. Reflected intensity and polarization patterns in the exit pupil for the quadrilayer disk shown in Fig. 7.11. These plots are obtained with the off-diagonal element ϵ_{xy} of the dielectric tensor set to zero. (a) Intensity for the X-component of polarization; $(I_x)_{max} = 0.95 \times 10^{-8}$. (b) Intensity for the Y-component of polarization; $(I_y)_{max} = 0.12 \times 10^{-9}$. (c) Polarization rotation angle Θ; $\Theta_{min} = -2.8°$, $\Theta_{max} = +2.8°$. (d) Polarization ellipticity ϵ; $\epsilon_{min} = -7.72°$, $\epsilon_{max} = +7.72°$.

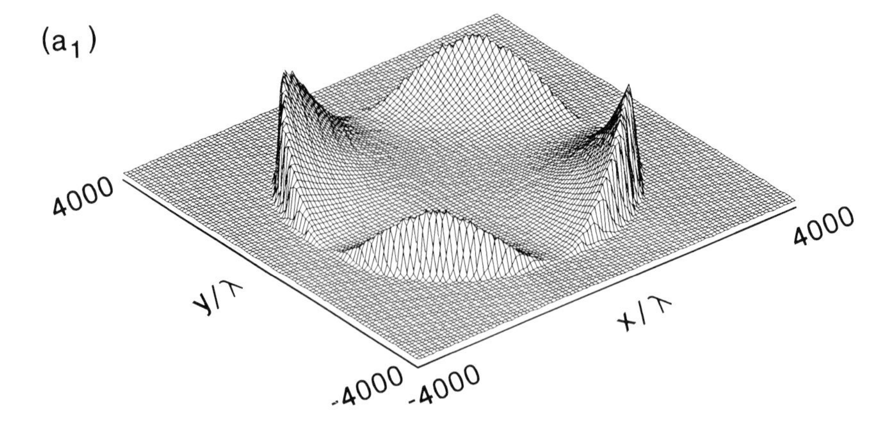
(a$_1$)
4000
y/λ
-4000
-4000
x/λ
4000

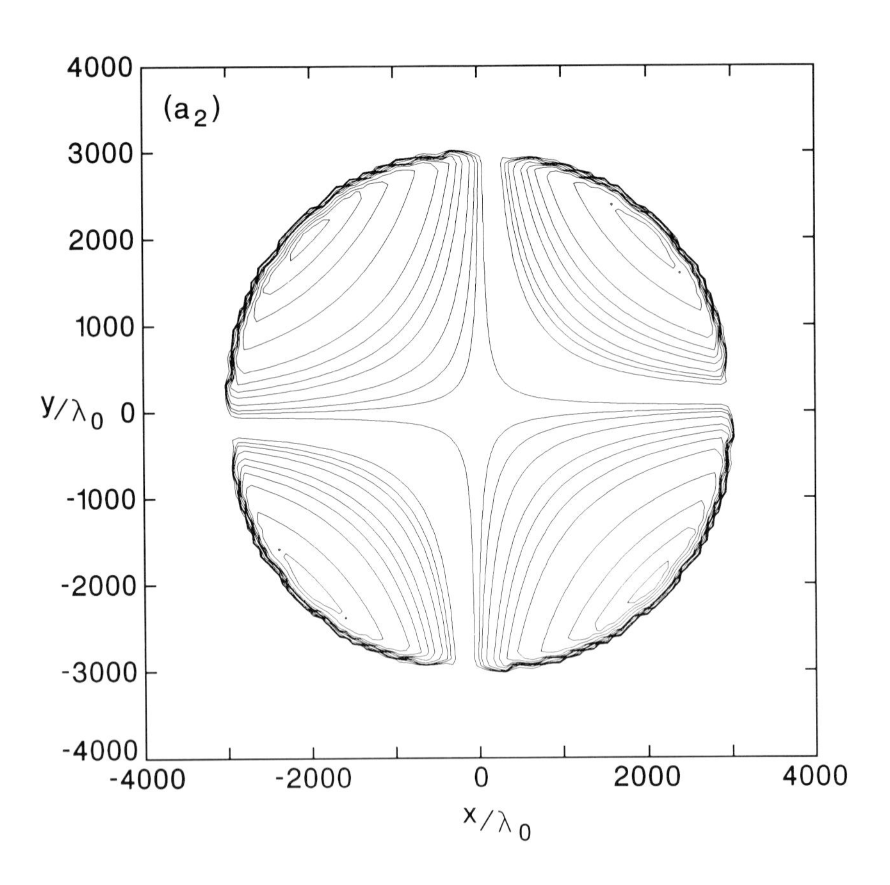
(a$_2$)
4000
3000
2000
1000
y/λ$_0$ 0
-1000
-2000
-3000
-4000
-4000
-2000
0
2000
4000
x/λ$_0$

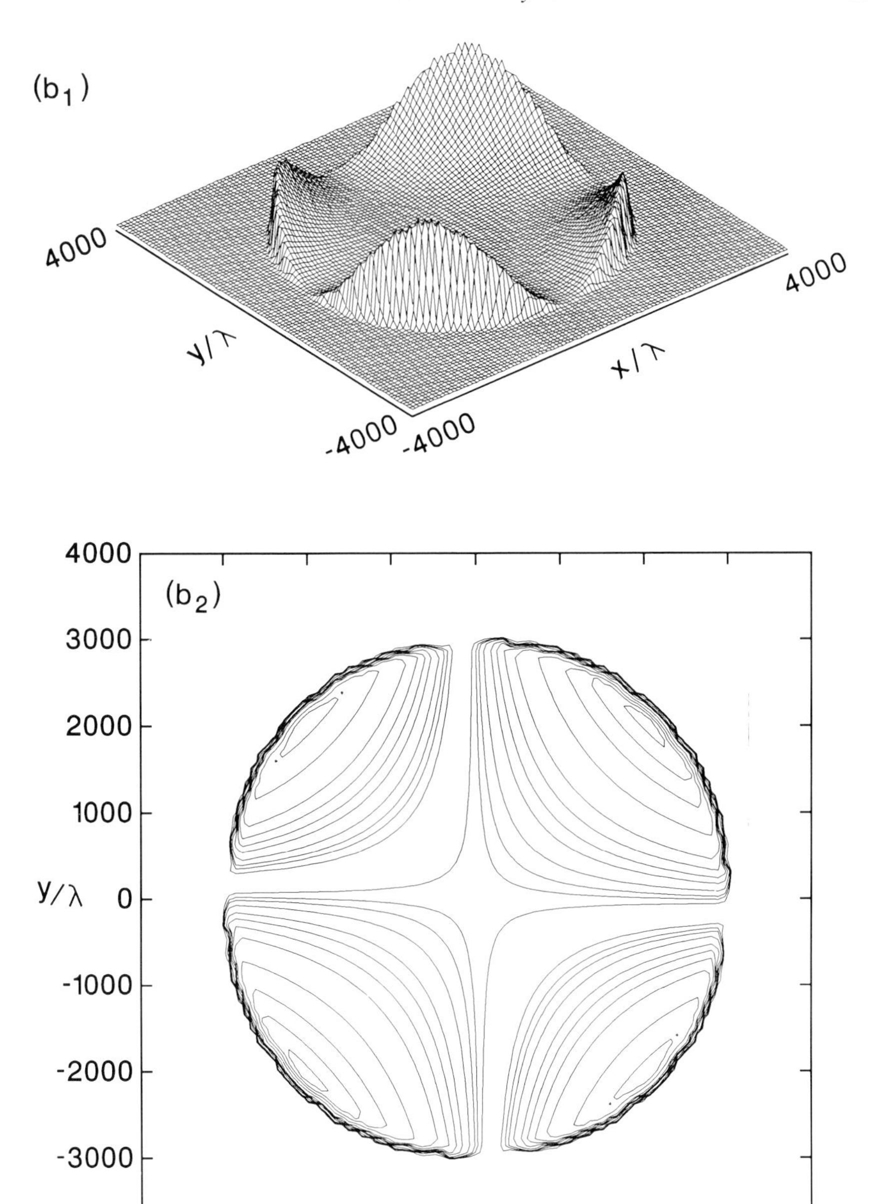

Figure 7.13. Distribution of I_y at the exit pupil for the beam reflected from the quadrilayer of Fig. 7.11; the off-diagonal elements of the tensor are now restored. (a_1), (a_2), Magnetization up; (b_1), (b_2), magnetization down. In both cases $(I_y)_{\max} = 0.15 \times 10^{-9}$.

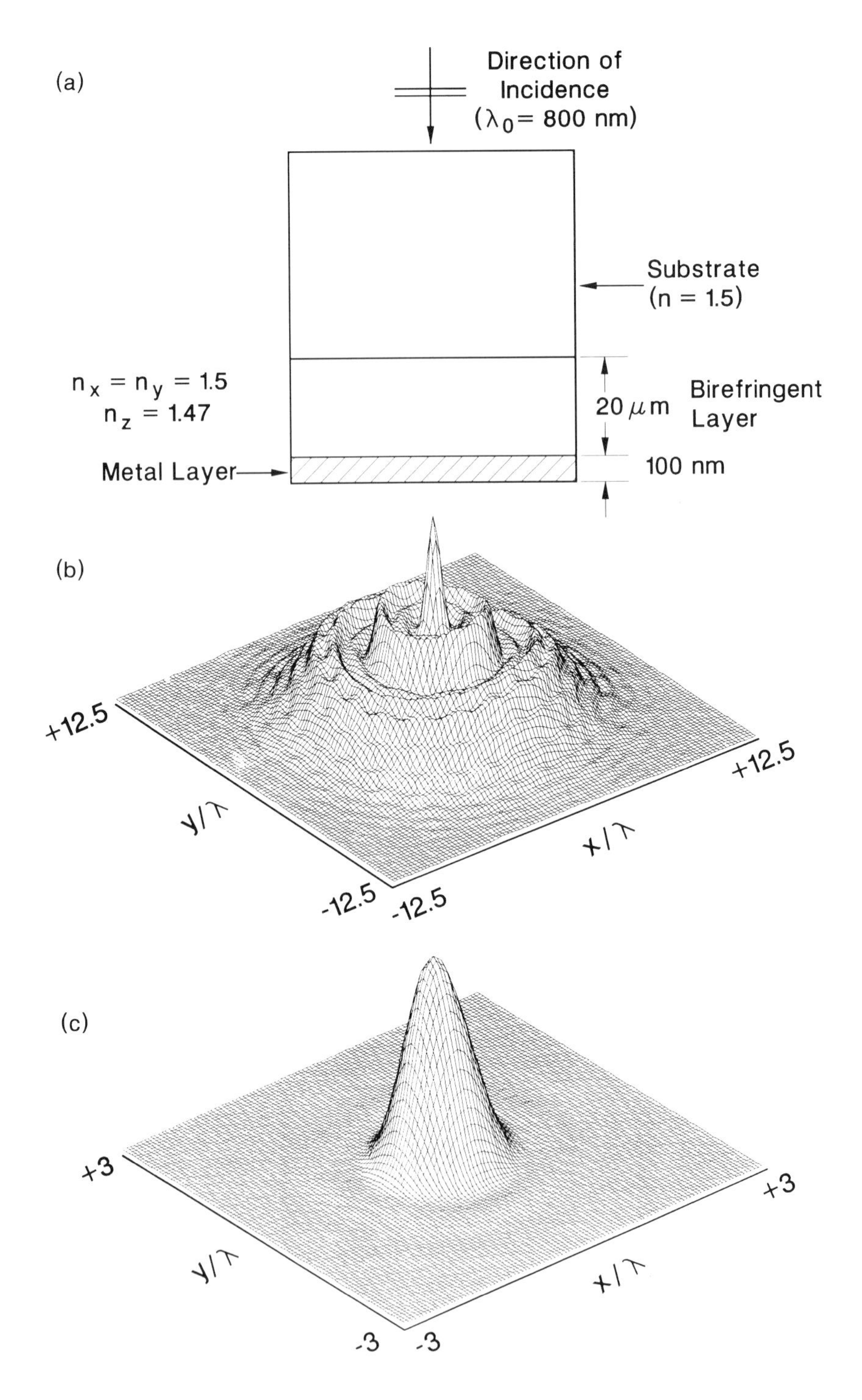

Figure 7.14. (a) Substrate birefringence is concentrated in the 20 μm-thick layer between the metallic film and the isotropic part of the substrate. The assumed vertical birefringence, n_r - n_z, is 60 times greater than that of typical polycarbonate substrates, since the layer's thickness is 60 times smaller than the standard thickness. (b) Distribution of light intensity at the interface between the substrate and the birefringent layer. The objective lens has NA = 0.55 and f = 3.76 mm. (c) Intensity distribution at the interface of the metallic and birefringent layers. The spherical aberrations induced by the 20 μm-thick layer have been compensated by the addition of defocus. The optimum amount of defocus is found by maximizing the peak intensity at the metallic layer.

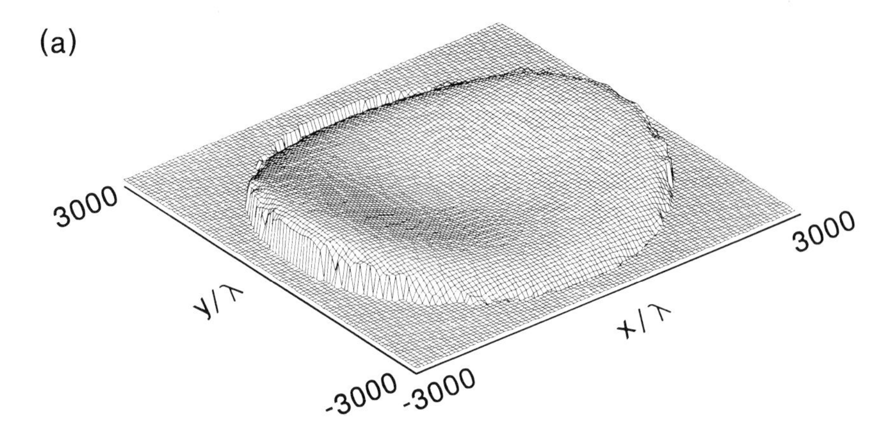

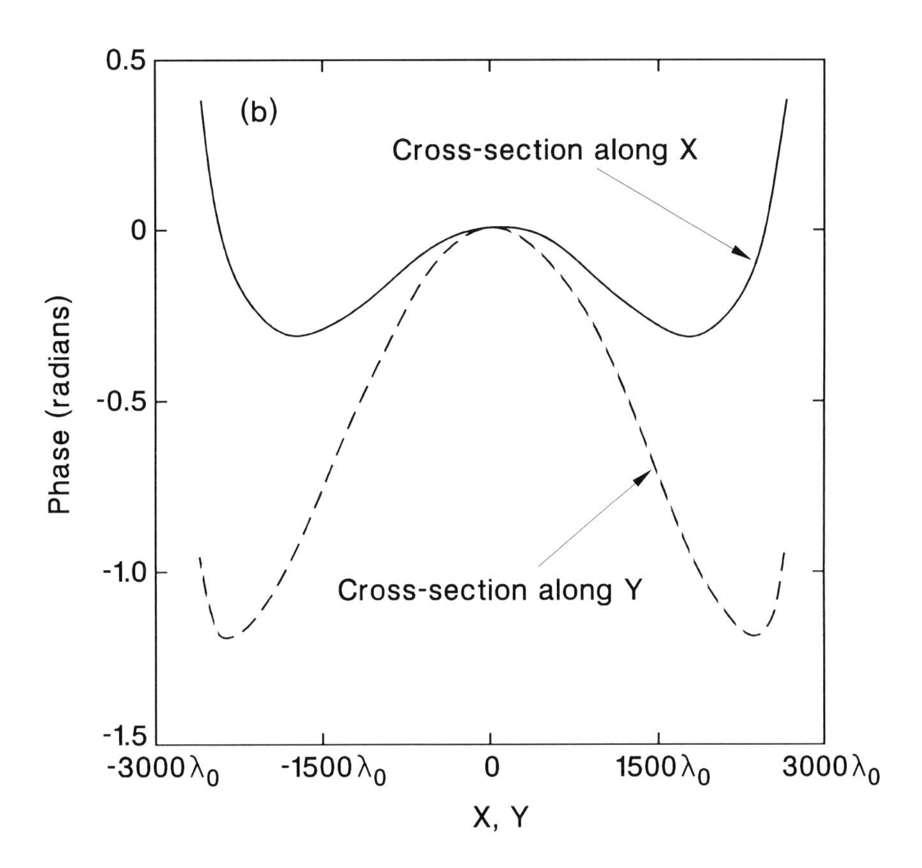

Figure 7.15. (a) Distribution of phase for the X-component of polarization at the exit pupil of the objective. The assumed disk structure is that of Fig. 7.14(a), with the metal layer acting as a perfect reflector. (b) Cross-sections of the phase plot in (a) along the X-direction and the Y-direction. The phase difference of 1.3 radians at the edge of the pupil corresponds to $0.2\lambda_0$ of astigmatism.

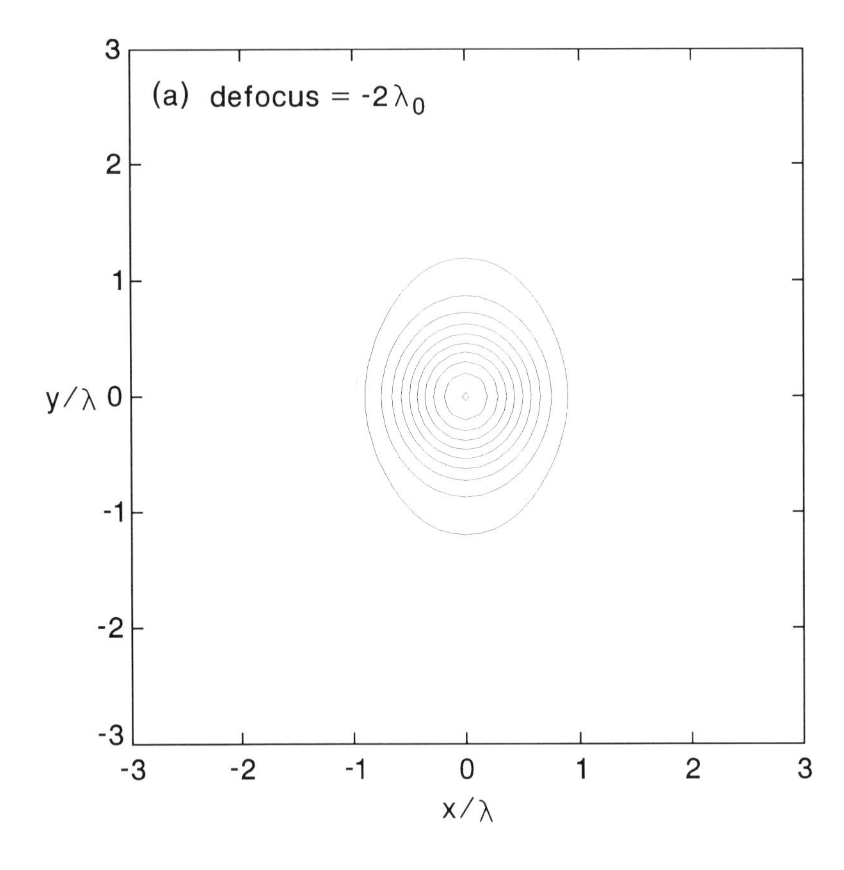

Figure 7.16. Contour plots of intensity, I_x, at the metallic layer for the disk structure of Fig. 7.14(a). The plots in (a)–(c) correspond, respectively, to a defocus of $-2\lambda_0$, best focus, and a defocus of $+2\lambda_0$. In each case the distribution is normalized to unity at the peak, and the individual contours range from 0.1 to 1 in steps of 0.1. Peak intensities are in the ratio of 0.67 : 1 : 0.74.

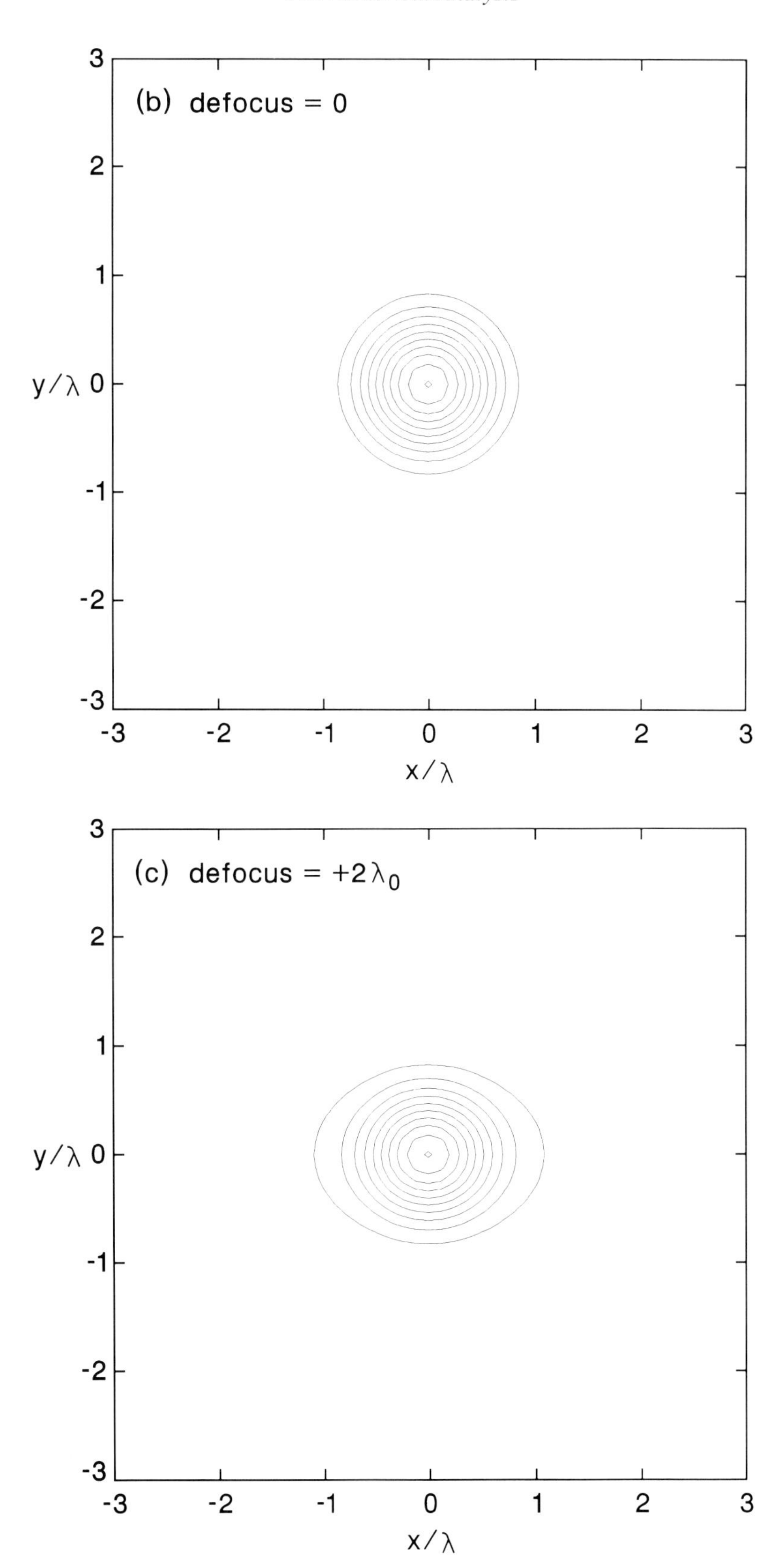
(b) defocus = 0
y/λ
x/λ
3
2
1
0
-1
-2
-3
(c) defocus = +2λ0
y/λ
x/λ

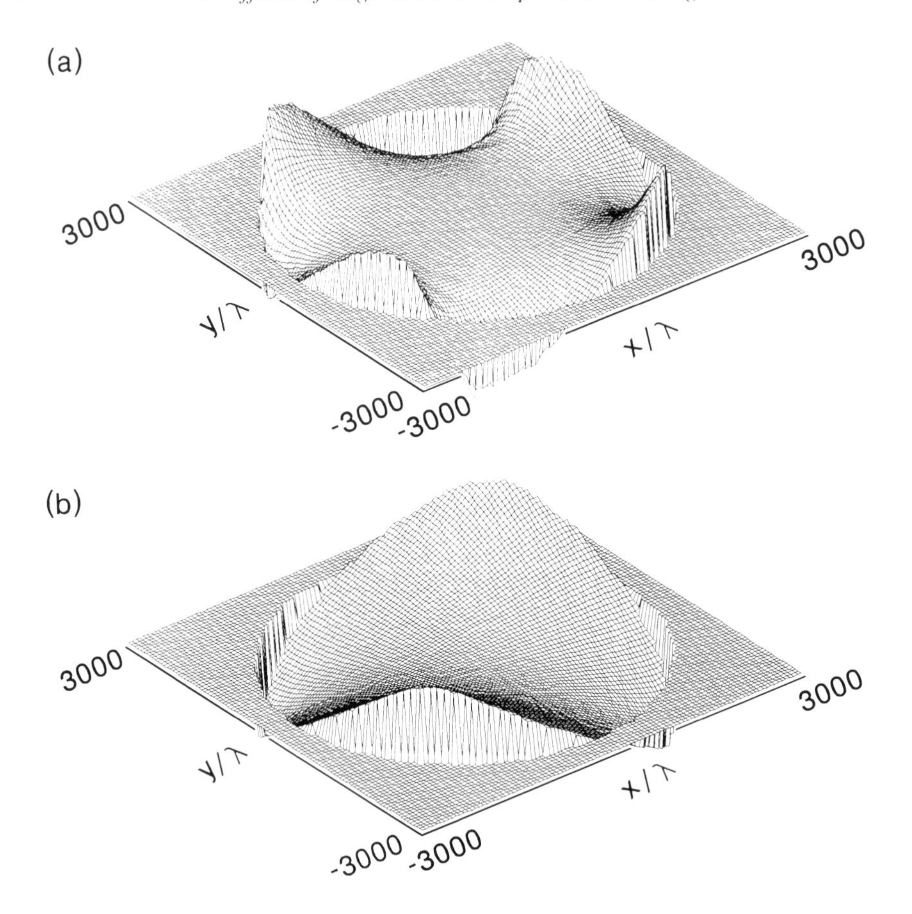

Figure 7.17. Distributions of the polarization rotation angle Θ and ellipticity ϵ at the exit pupil of the objective, corresponding to the disk structure of Fig. 7.14(a) with a magneto-optical metallic layer ($\epsilon_{xx} = -5 + 20i$, $\epsilon_{xy} = 0.5 + 0.005i$ at $\lambda_0 = 800$ nm). In (a) $\Theta_{min} = -18.2°$, $\Theta_{max} = +19.7°$. In (b) $\epsilon_{min} = -36.3°$, $\epsilon_{max} = +35.9°$.

obtained is $S_1 - S_2 \simeq \pm 0.0059$, while, in the absence of the birefringent layer, the maximum signal is found to be $S_1 - S_2 \simeq \pm 0.0063$. (In both cases a $\lambda/4$ plate is placed before the differential detector, its orientation adjusted to yield the maximum read signal.)

7.3. Summary and Conclusions

In this chapter the effects on the polarization state of the beam of high-NA focusing were analyzed. It was found that, upon reflection from the disk, the polarization rotation angle Θ and the ellipticity ϵ vary rather significantly over the objective's aperture. However, the average values of Θ and ϵ, which are the quantities picked up as the information signal by the differential readout system, show little dependence on the numerical aperture. In the worst cases studied (see the second example in subsection 7.2.5 and also the example in subsection 7.2.6) the effect of high-NA focusing was to convert a certain amount of rotation into ellipticity, but the signal could eventually be recovered with the aid of a $\lambda/4$ plate.

Plastic substrates of optical recording are often birefringent, with the largest refractive-index differential Δn found between the in-plane and vertical directions. A focused beam of low or moderate NA could travel through such substrates without "noticing" the birefringence. On the other hand, at high numerical apertures, oblique rays interact with the material in the vertical direction, giving rise to such problems as astigmatism and enhanced polarization ellipticity. Some of these problems may be alleviated by precompensation for astigmatism, and by the addition of phase plates for removal of ellipticity. At the same time, variations of birefringence across the disk surface, or differences between various manufacturers' substrates, could reduce the effectiveness of such measures.

Future generations of optical storage devices are expected to operate at shorter wavelengths, to spin faster, and to utilize higher numerical apertures. Under these conditions the readout signal and servo-channel performance could be seriously degraded by the inhomogeneous contributions of the various oblique rays. New, sensitive methods of readout and servo-signal generation that operate within manufacturing tolerances and that are immune to various defects (such as substrate birefringence) are therefore highly desirable.

Problems

(7.1) A lens with NA = 0.55 is corrected for operation through a glass slide of n = 1.5, and its diffraction-limited spot is allowed to form at a distance of 13 μm beneath the slide (see Fig. 7.18(a)). Let a second slide of the same material be placed in contact with the bottom surface of the first slide, as shown in Fig. 7.18(b). Assuming that the second slide has sufficient thickness, determine the new position of "best focus".

(7.2) In the schematic diagram shown in Fig. 7.19 the substrate thickness is 1.2 mm, its average refractive index $\langle n \rangle$ = 1.58, and its lateral (i.e., in-plane) birefringence is $n_r - n_t = 2 \times 10^{-5}$. Assuming that the MO film has diagonal and off-diagonal tensor elements $\epsilon = -5 + 20\mathrm{i}$ and $\epsilon' = 0.5 + 0.005\mathrm{i}$ at λ_0 = 800 nm, determine the Kerr rotation angle Θ and the ellipticity ϵ of the normally incident beam upon reflection from the disk. Compare the result with the values obtained if the substrate birefringence is neglected.

(7.3) A typical polycarbonate substrate is 1.2 mm thick and, at the operating wavelength of 800 nm, has different refractive indices along the radial, tangential, and vertical directions, as follows: $n_r = 1.58$, $n_r - n_t = 2 \times 10^{-5}$, $n_r - n_z = 5 \times 10^{-4}$.

(a) An oblique ray is incident on this substrate at an angle ϕ, as in Fig. 7.20. Determine the phase shift between the ray's p- and s-components of polarization after a single path through the thickness.

(b) A disk with the above substrate has been used in an MO drive where the objective NA is 0.55 and the incident polarization is along the tangential direction. Estimate the amount of astigmatism placed on the focused beam during its passage through the substrate.

(7.4) A typical optical disk substrate is 1.2 mm thick and has a refractive index of 1.58. The manufacturing tolerance on substrate thickness is ± 50 μm. With an objective lens having NA = 0.55 (corrected for the nominal substrate) and at the operating wavelength of 800 nm, what is the range of defocus needed to keep the MO layer at "best focus" as the substrate thickness varies within its tolerance range?

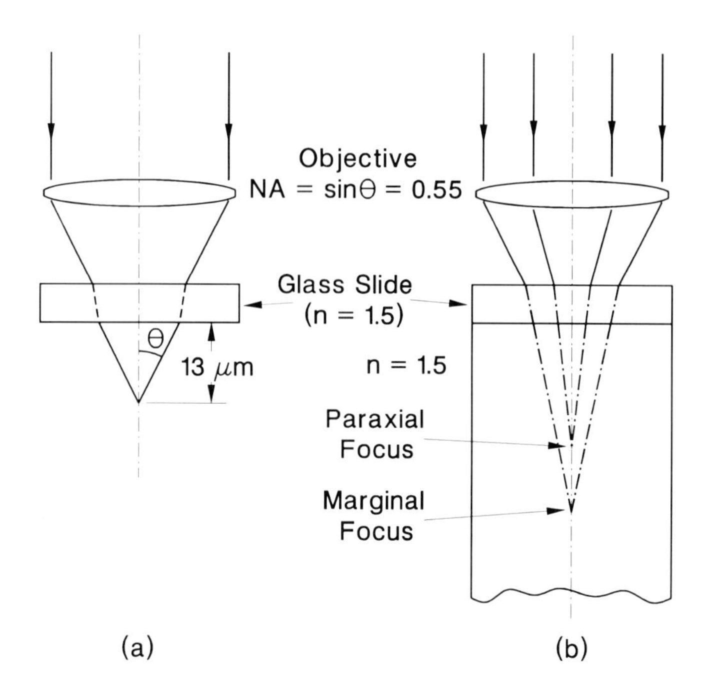

Figure 7.18.

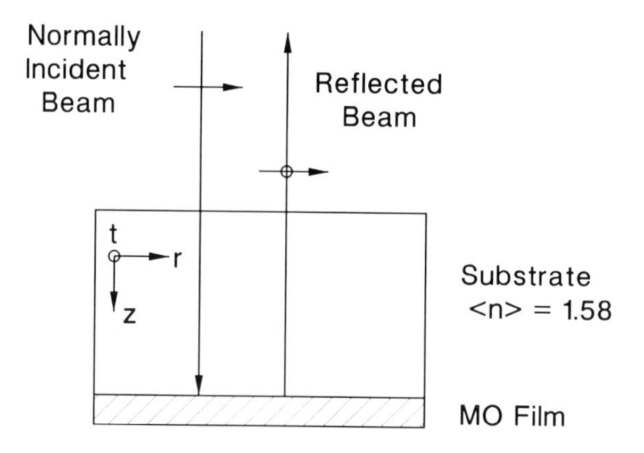

Figure 7.19.

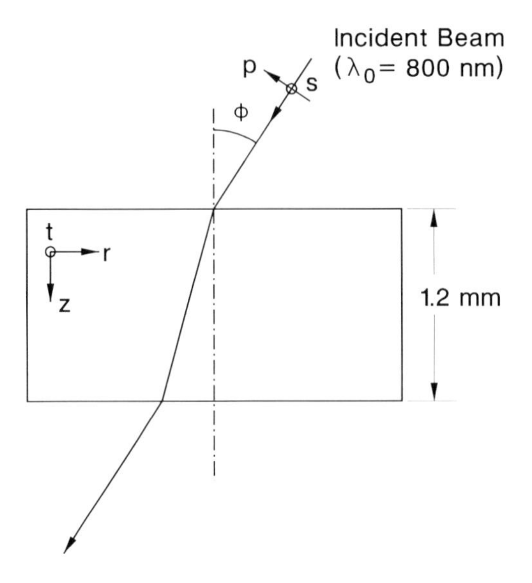

Figure 7.20.

8

Computer Modeling of the Optical Path

Introduction

The optical theory of laser disk recording and readout as well as that of the various focus-error and track-error detection schemes have been developed over the past several years.[1-25] Generally speaking, these theories describe the propagation of the laser beam in the optical head, its interaction with the storage medium, and its return to the photodetectors for final analysis and signal extraction. The work in this area has been based primarily on geometrical optics and the scalar theory of diffraction, an approach that has proven successful in describing a wide variety of observed phenomena.

The lasers for future generations of optical disk drives are expected to operate at shorter wavelengths; at the same time, the numerical aperture of the objective lens is likely to increase and the track-pitch will decrease. These developments will result in very small focused spots and shallow depths of focus. However, under these conditions the interaction between the light and the disk surface roughness increases, polarization-dependent effects (especially those for the marginal rays) gain significance, maintaining tight focus and accurate track position becomes exceedingly difficult, and, finally, small aberrations and/or misalignments deteriorate the quality of the readout and servo signals. Thus, attaining acceptable levels of performance and reliability in practice requires a thorough understanding of the details of operation of the system. In this respect, accurate modeling of the optical path is indispensable.

Ideally, one would like to have a computer program for vector diffraction calculations that allows the beam to propagate through the various optical elements in the head, interact with the storage medium, and propagate back to the detectors for final analysis. One important ingredient of such a program is obviously the subprogram that provides accurate solutions of Maxwell's equations at the disk surface, where the light interacts with the small features inscribed on the storage medium. At present this particular subprogram is far less developed than the rest of the program. Certain preliminary results regarding the interaction of polarized light with grooves and pits in simple disk structures exist,[1,24,25] but more work is needed. The examples presented in this chapter avoid the difficult question of interaction at the disk surface by treating grooves and marks as phase objects. In other respects, however, the problem is treated from the standpoint of diffraction theory and due attention is paid to the vector nature of light, when necessary. The term "quasi-vector" will be used when referring to this mixed scheme of diffraction calculations.

The optical path in a laser disk system begins at the laser. Typically, a GaAs semiconductor laser diode produces at its front facet a monochromatic beam of light with a small elliptical cross section (beam dimensions $\simeq$ 1 μm x 2 μm). Due to diffraction effects, this beam rapidly expands as it moves away from the laser. The expansion of the laser beam is arrested by a lens, resulting in a collimated beam a few millimeters in diameter, but with an elliptical cross section. This elliptical profile is usually corrected by a pair of anamorphic prisms[26] and the resulting beam, with circular cross-section, propagates towards the objective lens. Assuming that the incident wave-front and the objective lens are both free from aberrations, a small, diffraction-limited spot is subsequently formed at the focal plane of the objective. Essential features of this propagation process (from laser to disk) are covered in section 8.1. In general, the beam at the entrance pupil of the objective has a Gaussian profile. The lens, of course, does not allow the entire beam to pass through, but truncates it at the aperture's rim; certain compromises involving the truncation radius are explored in subsection 8.1.1.

Section 8.2 describes the action of various diffraction gratings on the beam profile, and gives the intensity distributions obtained from the gratings in both the near field and the far field. Aside from occasional use in optical storage systems as beam-splitters, diffraction gratings have many features in common with the grooved surface of an optical disk; their diffraction patterns presented in section 8.2, therefore, resemble those that will appear in later sections in conjunction with the pregrooved media of optical recording.

Focus- and track-control servomechanisms comprise two essential subsystems of any optical disk drive.[13-19] To achieve high-density recording and readout, one must use an objective lens with a large numerical aperture (NA), since the spot diameter at focus is inversely proportional to NA. Such a choice of NA, however, results in a small depth of focus, approximately equal to $\lambda/(\mathrm{NA})^2$. At a typical wavelength of $\lambda = 0.8$ μm and with NA = 0.5, the depth of focus is $\simeq$ 3 μm, thus necessitating stringent control of the position of the lens. At the same time, the requirement of high track density translates into the need for accurate positioning of the focused spot in the radial direction. In practice, with a typical track-pitch of $\simeq$ 1–2 μm, one cannot tolerate radial positioning errors of more than 0.1 μm.†

One of the more popular techniques for generating the focus-error signal (FES) is the so-called astigmatic method. This method employs a

†A method of controlling the spot position, used in many optical disk drives in use today, is to mount the objective in a pair of voice coils that support two independent movements along the directions of interest. Using one of several possible techniques for position-error detection, the deviation of the spot from the desired location on the disk is sensed, and the resulting bipolar signals are fed back to the voice coils. The closed-loop system thus obtained can maintain focus and track errors below acceptable levels at all times.

mildly astigmatic lens in the path of the reflected beam, with a quad detector placed between the two focal planes of the lens. In section 8.3, using at first the methods of geometrical optics, we derive an expression for the FES, and discuss the choice of parameters. A specific instance of the problem is treated in subsection 8.3.2 using the methods of scalar diffraction theory. Calculated intensity patterns at various cross-sections of the system are presented, and the extent of validity of the geometrical results is examined. The track-error signal (TES) from a pregrooved disk is most easily derived using the push–pull method. The optimum groove depth for this technique is $\lambda/8$, where λ is the laser wavelength; no additional optical elements are required and the difference signal obtained from a split detector in the far field contains the necessary information for control of the tracking servo. In magneto-optical drives, lack of interference from the MO domains simplifies the analysis of the push–pull method. The diffraction analysis of the TES on pregrooved MO disks is the subject of 8.3.3.†

Section 8.4 provides numerical simulation results for another method of focus-error detection based on a "ring-toric" lens. The basic property of a ring-toric is that it brings a collimated incident beam into sharp focus on a ring within its focal plane. When such a lens is placed in the path of the reflected light, its focal-plane intensity pattern contains the necessary information for generating the FES. Like the astigmatic method, the ring-toric technique is compatible with the push–pull scheme, and the TES can be obtained simultaneously from its focal-plane distribution. The ring-toric lens has not yet been adopted by commercial drive manufacturers, but it has the advantages of being simple, using the light optimally, producing a high sensitivity (i.e., fast slope) FES, and being fairly immune to false focus-error signals during track crossings. These features merit closer examination as they offer potential advantages for future generations of optical storage devices.

The last section is concerned with the process of MO readout. The currently standard technique of Kerr differential detection (which uses linearly polarized incident light) is analyzed first. Quasi-vector diffraction calculations in section 8.5 produce interesting plots of polarization distribution along the path of the beam. An example of readout of the crescent-shaped domains written by magnetic-field modulation is given in 8.5.1. Finally, a novel readout scheme based on edge detection with circularly polarized light that was described earlier (see section 6.6) will be further analyzed in 8.5.2 using quasi-vector numerical calculations.

†While in read-only and write-once devices the recorded data modulate the amplitude and/or phase of the readout beam, in MO systems it is the polarization that is being modulated and, as such, both FES and TES are rendered independent of the recorded information pattern.

8.1. Collimation and Focusing of the Laser Diode Beam

We follow the propagation of the beam from its emergence at the front facet of the semiconductor laser diode to the collimating lens and on to the objective lens. The schematic diagram of the system under consideration is shown in Fig. 8.1(a), where the waist of the Gaussian beam emerging from the laser is placed in the front focal plane of the collimating lens. The initial cross-section of the beam is elliptical with $1/e$ (amplitude) radii $r_{x0} = 4\lambda$ and $r_{y0} = 2\lambda$.† Thus the intensity pattern just in front of the laser is as shown in Fig. 8.1(b). Since at the waist the beam has a flat wavefront, its curvature is correspondingly set to zero.

Using scalar diffraction theory for a propagation distance of $10\,000\lambda$, this rapidly expanding beam is brought into the far field (i.e., the Fraunhofer regime). The lens L_1 with NA = 0.3 and focal length $f_1 = 10\,000\lambda$, arrests this expansion and produces a collimated beam. Figure 8.1(c) shows the computed intensity pattern at a distance of $20\,000\lambda$ beyond L_1, which is at the entrance pupil of the objective lens L_2. Note that in the X-direction (i.e., parallel to the laser's active layer), the beam has undergone a less rapid expansion than in the perpendicular Y-direction. This, of course, is consistent with one's expectations based on elementary diffraction considerations.

The aberration-free objective L_2 (with NA = 0.4 and $f_2 = 4000\lambda$) focuses the beam into a diffraction-limited spot; the intensity distribution at the focal plane of L_2 is shown in Fig. 8.1(d). Because of the slight truncation by the objective's aperture, the calculated power in the focused spot is 95% of that emitted by the laser. Also, there are faint intensity oscillations along the Y-axis, where truncation is most severe. These oscillations may be seen more clearly in the logarithmic plot of the focal plane intensity in Fig. 8.1(e). Here the vertical scale spans four orders of magnitude, from 10^{-4} at the bottom of the plot to unity at the top.

Certain assumptions were made in this example in order to simplify the analysis. We briefly mention some of the factors that complicate the situation in real life. First, the wave-front at the laser's front facet is usually distorted by astigmatism and possibly other aberrations; the collimator is therefore designed to correct these wave-front defects. Second, the elliptical cross section of the beam, if not corrected, will cause either under-utilization or overfilling of the objective's numerical aperture. This problem is solved in practice by the insertion of an anamorphic prism pair between the two lenses.[26] Third, polarization-related phenomena, which have been ignored in these scalar calculations, begin to play a role at high numerical apertures. Finally, residual aberrations, tilt, and decentering of the various elements cause deterioration of the quality of the focused

†Modern heterojunction laser diodes with buried stripe geometry have very thin and narrow active layers, resulting in initial beam dimensions even smaller than those assumed here. The results of analyses that take these smaller dimensions into account, however, are only slightly different from those presented here.

spot. A more realistic calculation (particularly one aimed at tolerancing the various components of the system) must take these nuances into account.

8.1.1. *Effect of beam profile on the focused spot*

We examine the computed distribution of intensity at the focal plane of a well-corrected lens for two different incident beams. The first beam has uniform amplitude and uniform phase across the aperture of the lens, while the second beam has a circularly symmetric Gaussian intensity profile. We will assume that the lens has NA = 0.55 and $f = 5000\lambda$, and that the Gaussian beam has a 1/e (amplitude) radius $r_0 = 2750\lambda$. The lens aperture thus truncates the beam at its 1/e point. Figure 8.2(a) is the result of a scalar-diffraction calculation of the intensity at the focal plane for an incident plane wave (the so-called Airy pattern); a similar calculation for the Gaussian beam yields the pattern of Fig. 8.2(b). (The logarithmic scale used in these plots enhances the weak rings relative to the strong central peak.) Clearly, the rings obtained for the truncated Gaussian beam are less intense than their counterparts in Fig. 8.2(a). This is understandable in view of the fact that the Gaussian beam, when not truncated, yields a focused spot with no rings at all.

A quantitative comparison between the above patterns is made in Fig. 8.2(c), which shows the dependence of intensity on radial distance from the focal point. The exact positions of peaks and valleys for the Airy pattern are marked in this figure. Note that the weaker rings of the truncated Gaussian beam are achieved at the expense of a larger diameter for the central bright spot. Truncation at or around the 1/e point is a reasonable compromise that allows the lens to collect a good fraction of the laser output while maintaining a small focused spot.

8.2. Diffraction Gratings and Grooved Optical Disks

Diffraction gratings may be used to split an incident beam into several reflected and/or transmitted beams (i.e., multiple diffraction orders). This property of gratings has found application in audio and video disk systems, where the original beam is split into three beams to produce tracking signals along with the information signal. In this so-called "twin spot" tracking scheme,[1] the central spot follows the main data track, while the satellite spots straddle the track. Upon reflection from the disk, the separately monitored twin spots provide the servo mechanism with the information necessary to guide the beam along the proper path.

Aside from being useful in their own right, diffraction gratings have features similar to grooved optical disks. These similarities have been used in the past to explain the functioning of the push–pull track-error detection scheme. We shall see in subsection 8.2.2 below, and also in 8.3.3, how the focused beam, in its interaction with the grooves, generates a push–pull TES. In 8.2.1 we examine a few simple gratings and show their

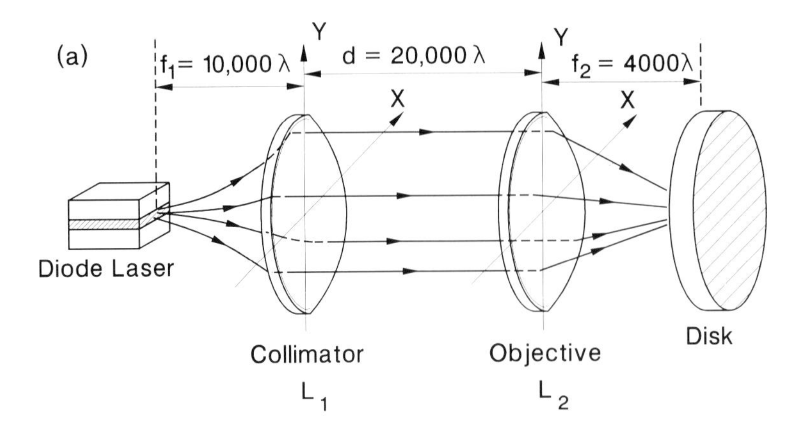

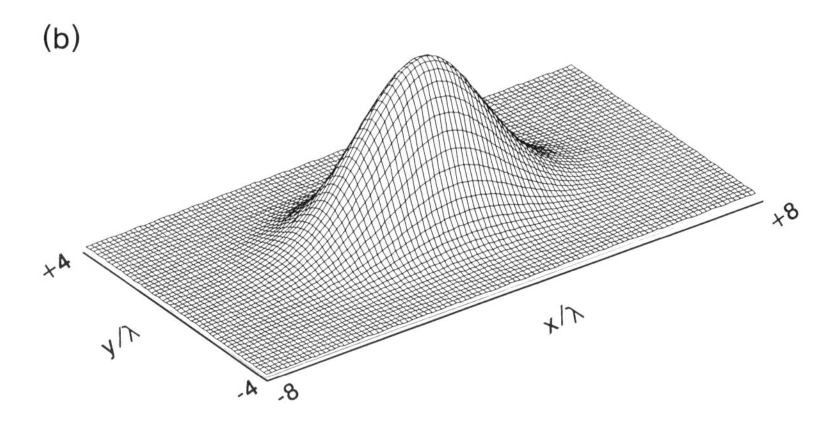

Figure 8.1. Propagation of a Gaussian beam from a laser to the focal plane of the objective. (a) Schematic diagram showing the semiconductor laser diode, collimating lens, and the focusing objective. (b) Intensity pattern at the front facet of the laser. The beam has an elliptical cross-section with an ellipticity ratio of 2 : 1. (c) Intensity distribution at the entrance pupil of the objective; (d) intensity pattern of the focused spot; (e) same as (d) but on a logarithmic scale with a span of four orders of magnitude.

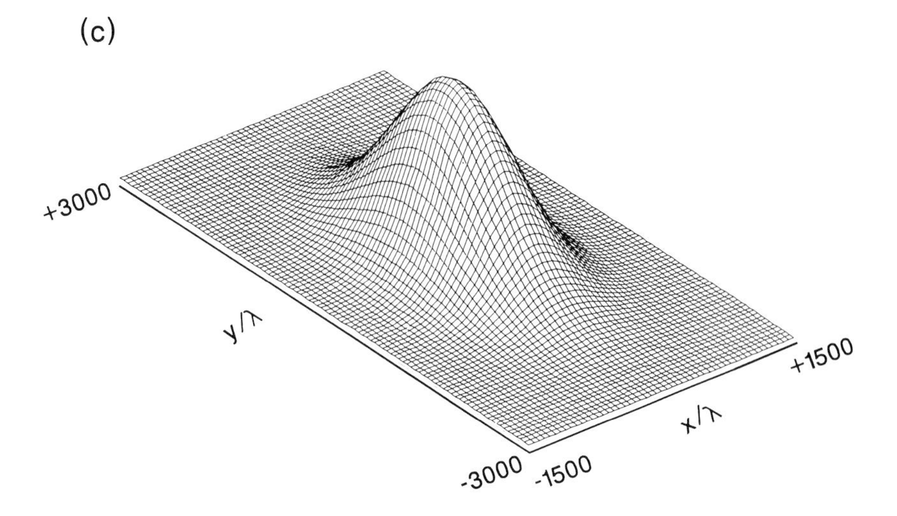
(c)
+3000
y/λ
-3000
-1500
x/λ
+1500

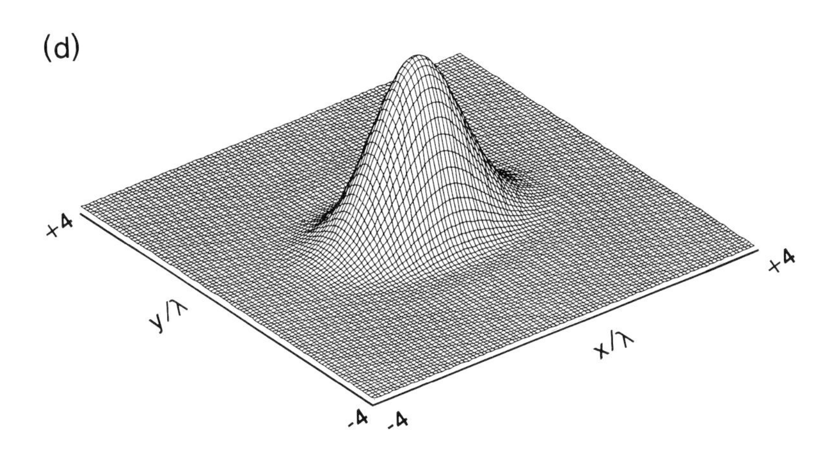
(d)
+4
y/λ
-4
-4
x/λ
+4

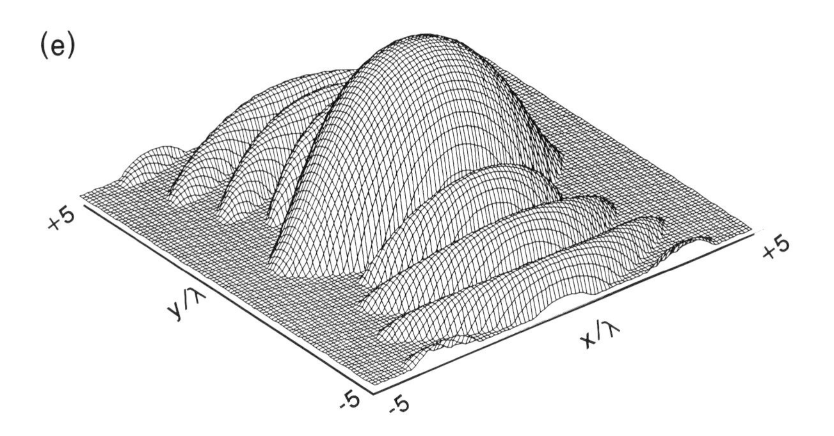
(e)
+5
y/λ
-5
-5
x/λ
+5

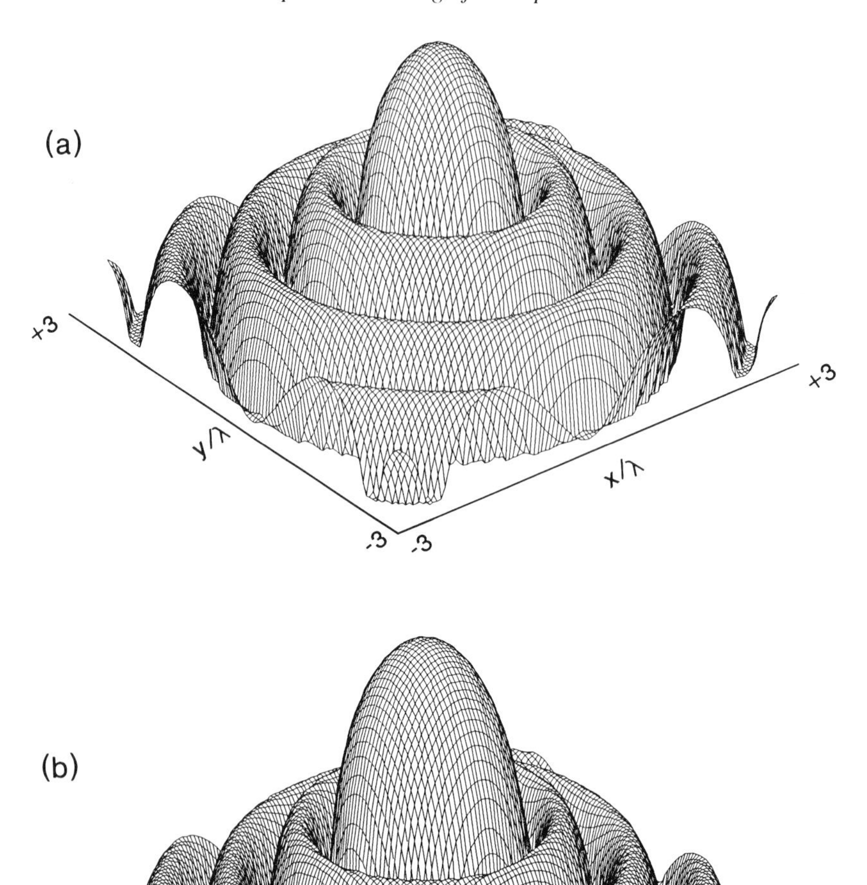
(a)
+3
y/λ
-3
-3
x/λ
+3
(b)
+3
y/λ
-3
-3
x/λ
+3

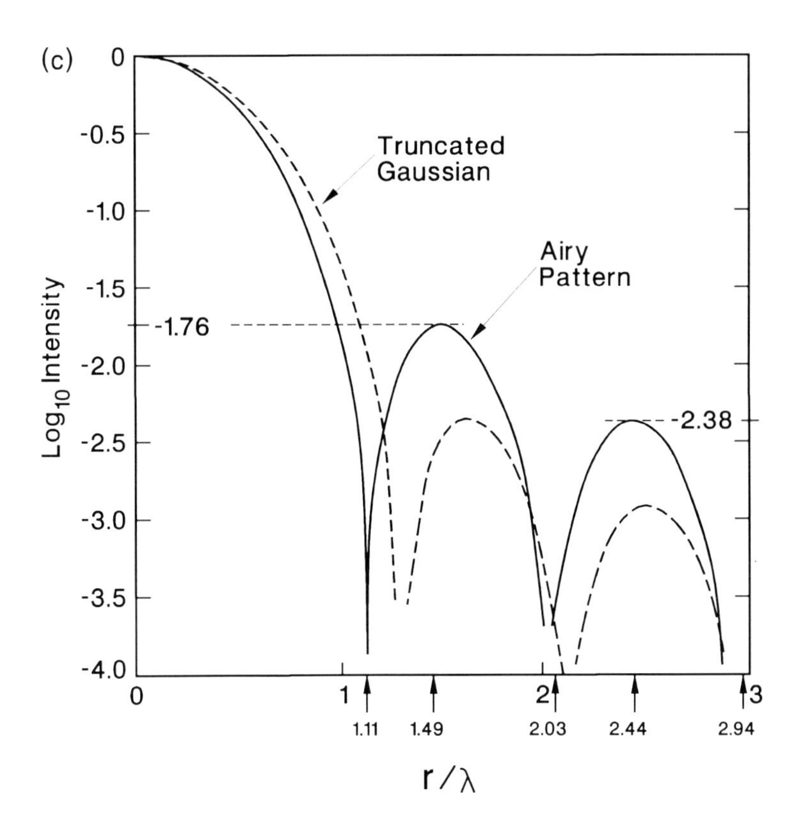

Figure 8.2. Logarithmic plots of the intensity distribution at the focal plane of an aberration-free objective with NA = 0.55 and $f = 5000\lambda$. The vertical scale spans four orders of magnitude (from 10^{-4} at the bottom of each plot to unity at the top). (a) With an incident plane-wave; (b) with an incident Gaussian beam, truncated at the $1/e$ (amplitude) point; (c) radial dependence of the focal-plane intensity for both uniform and Gaussian incident beams.

computed near-field and far-field patterns when the incident beam is large compared with the grating period. The far-field patterns vividly describe the beam-splitting property of the gratings. The near-field results should be compared with the patterns obtained in 8.3.3, where the concern is track-error detection in grooved optical disk systems.

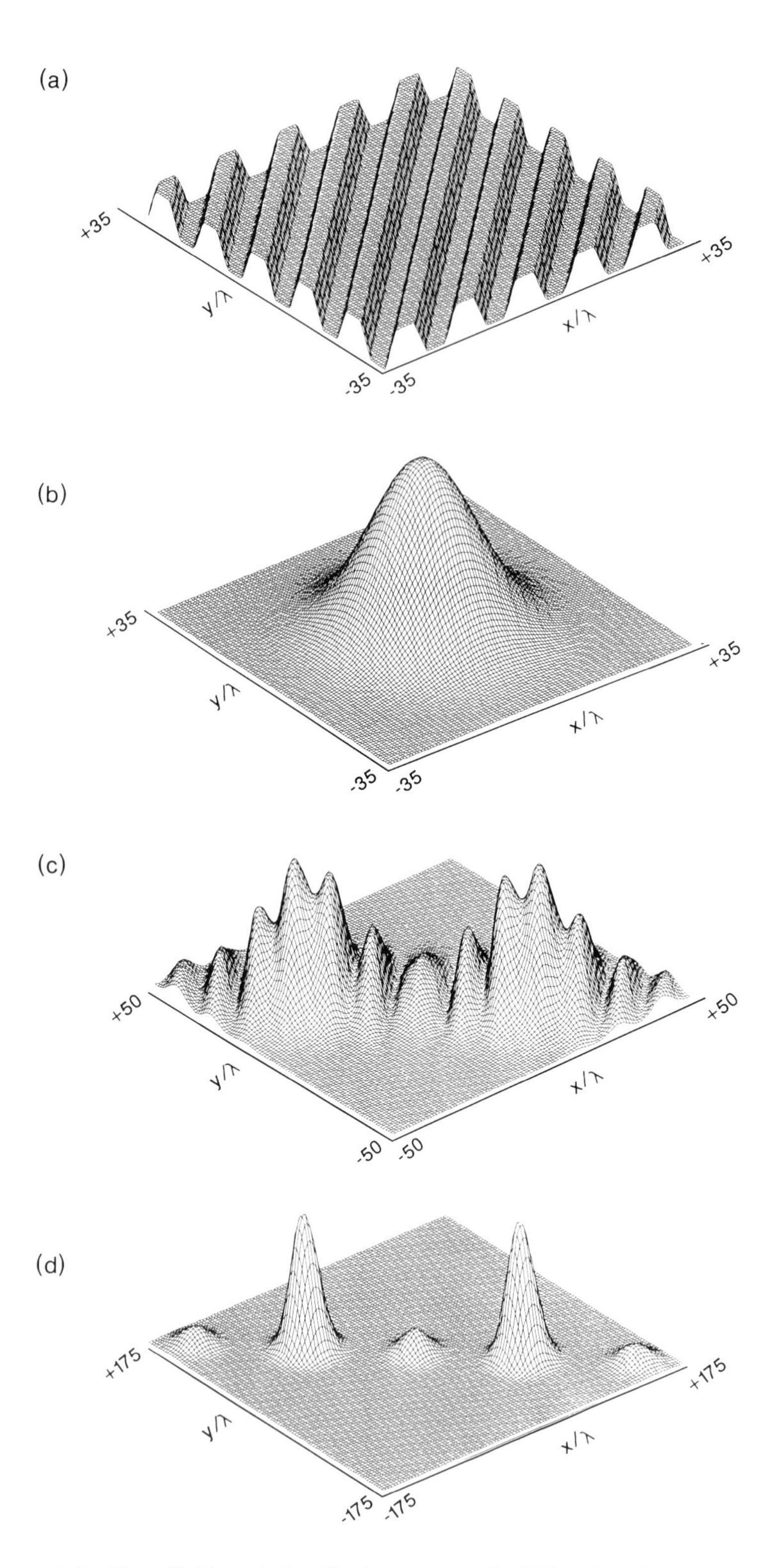

Figure 8.3. Near-field and far-field patterns of diffraction from a transmission grating with trapezoidal cross section. (a) Surface relief pattern of the grating. Groove depth = 0.5λ, groove width (at the bottom) = 3λ, land width = 3λ, slanted-wall width = 2λ. (b) Intensity plot showing the incident beam at the surface of the grating. The normally incident beam has a Gaussian profile with $1/e$ radius of 20λ. (c) Near-field intensity pattern at a distance of 300λ behind the grating. (d) Far-field intensity pattern at a distance of 1000λ behind the grating.

8.2.1. Near-field and far-field patterns of gratings

Figure 8.3(a) shows a simple transmission diffraction grating with trapezoidal cross-section. The groove depth is 0.5λ, both the lands and the grooves have width equal to 3λ, and the slanted walls each have a width of 2λ. (The numbers add up to a grating period of 10λ.) The normally incident beam whose intensity profile is shown in Fig. 8.3(b) is Gaussian with 1/e (amplitude) radius of 20λ. Figure 8.3(c) shows the computed near-field pattern at a distance of 300λ from the grating: the zero-order beam is weak and the main contribution to this distribution comes from the ±1 orders. Figure 8.3(d) shows the computed far-field pattern at a distance of 1000λ from the grating; the various orders are clearly separated and are readily identifiable.

The incident beam and grating parameters used for the calculations pertaining to Fig. 8.4 are the same as those used for Fig. 8.3, the exception being the groove depth which is now 0.25λ. The near-field and far-field patterns are shown in Figs. 8.4(a) and 8.4(b), respectively. The zero-order beam is strong in this case and the near-field pattern is less spread out.

Figure 8.5 corresponds to a grating with triangular cross-section (see frame (a)). The grating period is 10λ and the depth of the groove is 0.5λ. The near-field distribution in frame (b) is essentially that of the 0 order and -1 order beams superimposed. The lopsidedness of the diffraction pattern reflects the underlying asymmetry of the groove shape, which also presents itself in the corresponding far-field distribution in frame (c).

8.2.2. Reading preformat and track information from a grooved disk

In the simplified read system of Fig. 8.6(a) the objective has a numerical aperture of 0.65 and a focal length of 1.23 mm. The incident beam (λ = 780 nm) is Gaussian, truncated at the 1/e (amplitude) radius by the objective's aperture, and normalized to unit power within that aperture. A simple split detector is placed in the return path and oriented for push–pull detection of the track-error signal. Figure 8.6(b) shows a section of the disk containing grooves and preformat marks. The land is 0.75 μm wide and the grooves, with trapezoidal cross-section, are 0.35 μm wide at the top, 0.1 μm wide at the bottom, and 130 nm deep. The pits are centered on the land region and are 0.545 μm long, 0.3 μm wide, and 150 nm deep, with a center-to-center spacing of 1.09 μm. For simplicity's sake, the amplitude reflectivity of the disk is assumed to be unity, and the pits and grooves are treated as phase objects. Figure 8.6(c) shows the computed result of a scan along the track where the three preformat marks appear as dips in the reflected optical power $S_1 + S_2$ collected by the detectors. Away from the pits the sum signal is 83% of the incident power, the remaining 17% being diffracted out of the aperture. The peak-to-valley modulation of the sum signal produced by the pits is $\simeq 0.18$.

The computed track-crossing signal is depicted in Fig. 8.6(d). Three

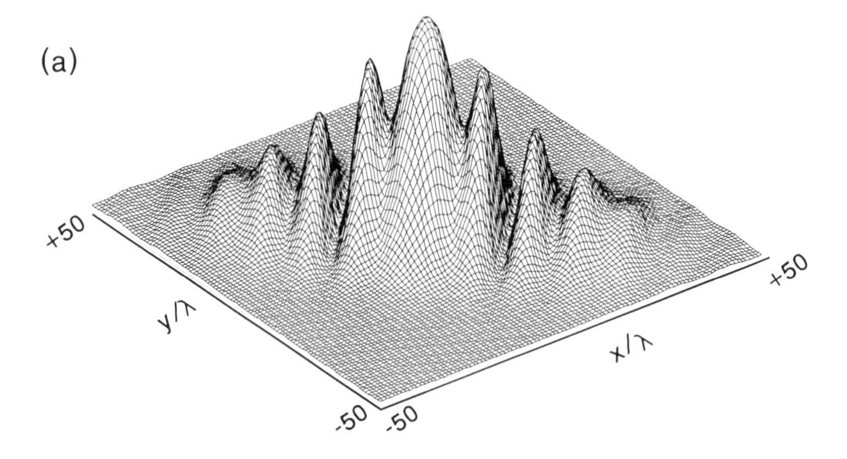

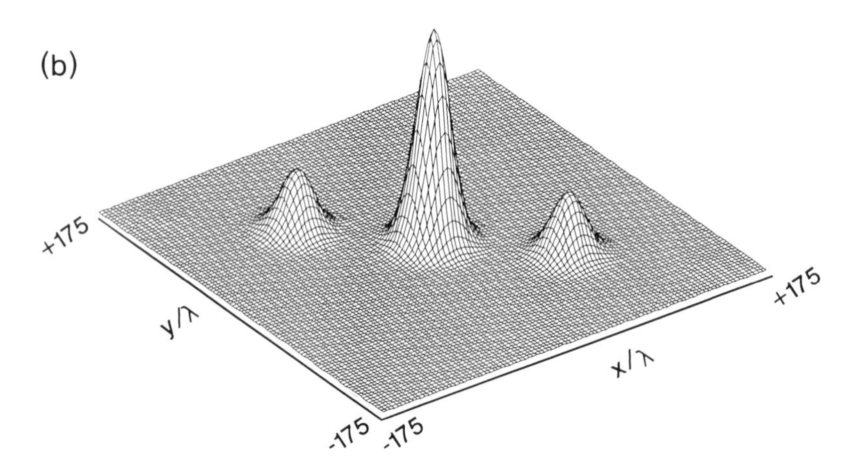

Figure 8.4. Diffraction patterns from a trapezoidal grating with a groove depth of 0.25λ; the grating structure and the incident beam are the same as those in Fig. 8.3. (a) Near-field intensity pattern at $z = 300\lambda$. (b) Far-field intensity pattern at $z = 1000\lambda$.

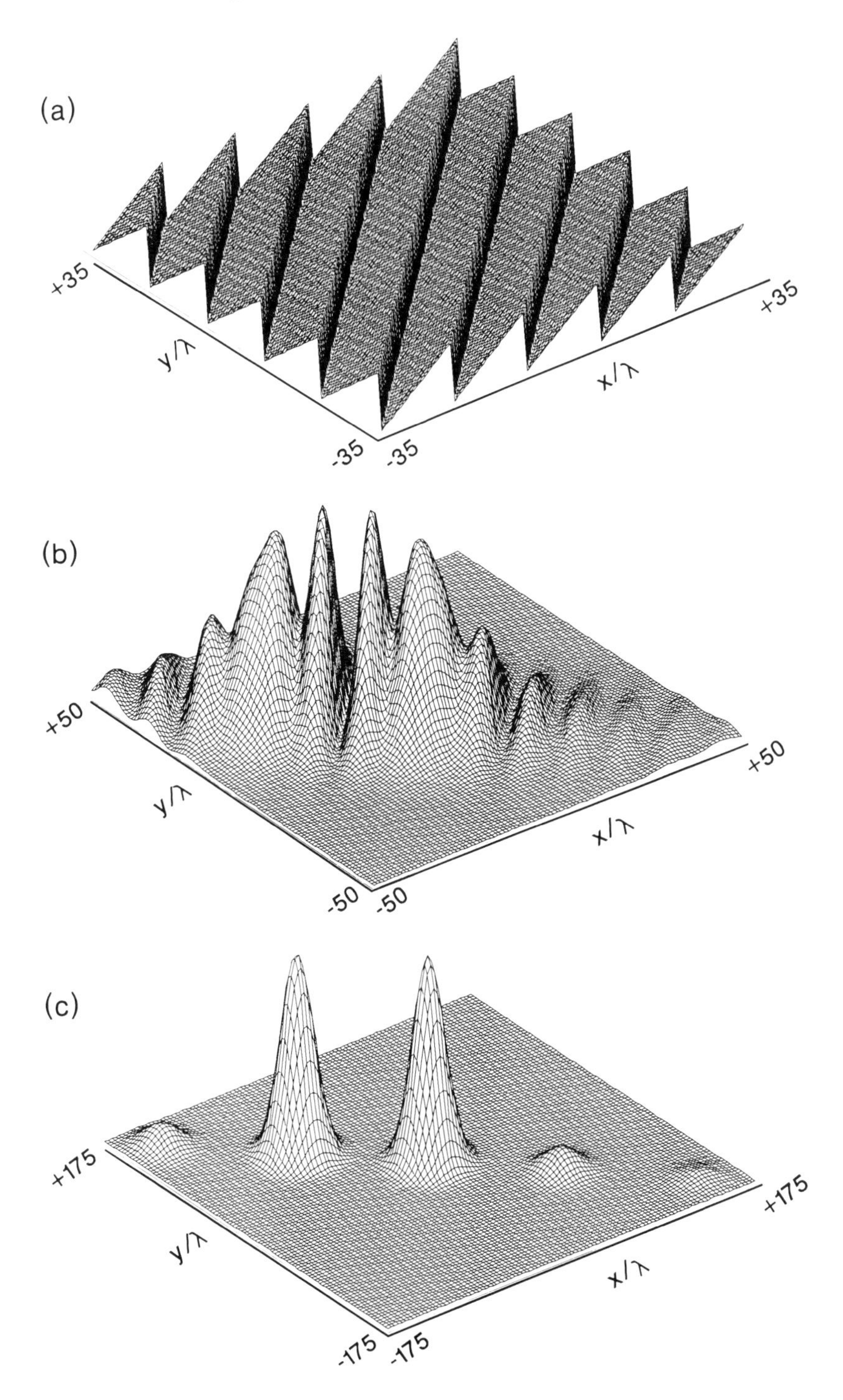

Figure 8.5. Diffraction from transmission grating with triangular cross-section. (a) Relief pattern of the grating, which has a period of 10λ and a groove depth of 0.5λ. (b) Near-field intensity pattern at $z = 300\lambda$. (c) Far-field intensity pattern at $z = 1000\lambda$.

(a)

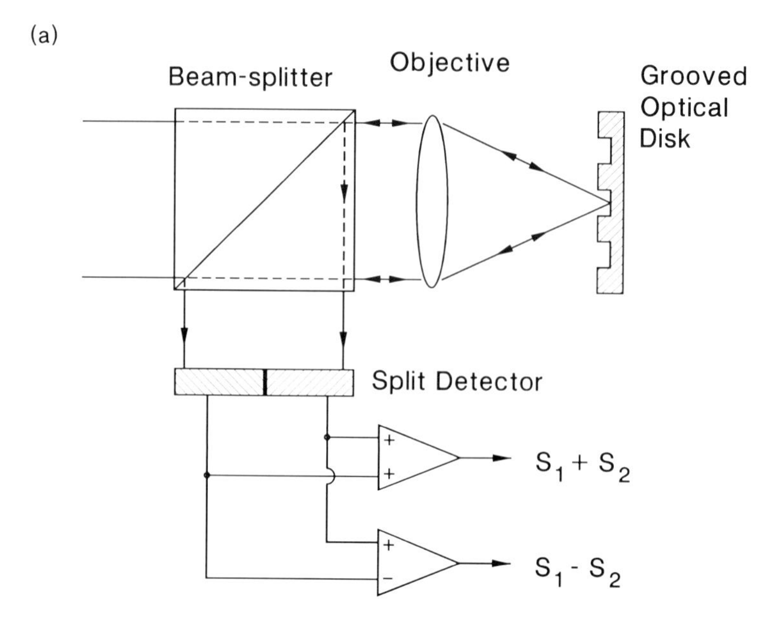

(b)

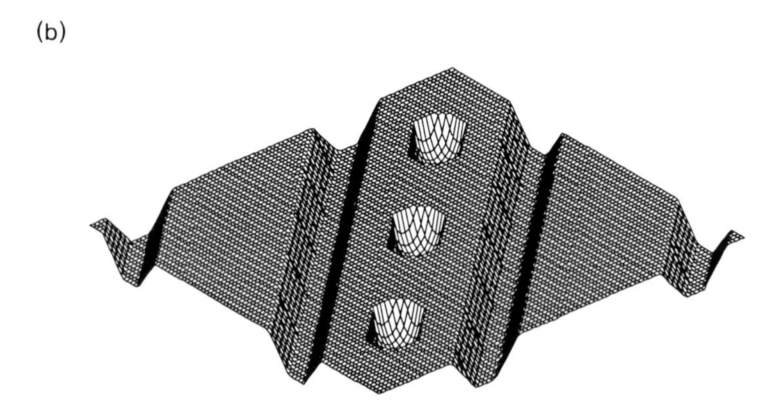

Figure 8.6. (a) Schematic diagram of a simplified readout system. (b) Section of grooved disk surface containing preformat marks (pits). (c) Sum signal of the split detector when the beam scans the preformat marks along the track. (d) Sum and difference signals of the detectors obtained when the beam scans in the radial direction, perpendicular to the tracks. During seek operations the swing of the push–pull signal from tracks 1 and 3 is strong enough to provide a count, but the signal from track 2, which contains the format marks, may be too weak for the purpose.

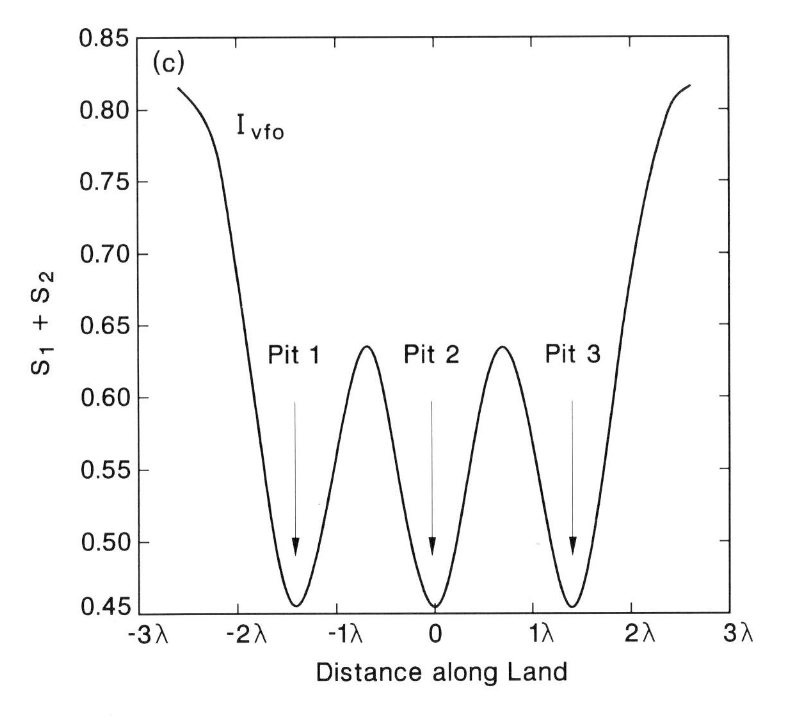
(c)
I_{vfo}
Pit 1
Pit 2
Pit 3
$S_1 + S_2$
0.85
0.80
0.75
0.70
0.65
0.60
0.55
0.50
0.45
-3λ
-2λ
-1λ
0
1λ
2λ
3λ
Distance along Land

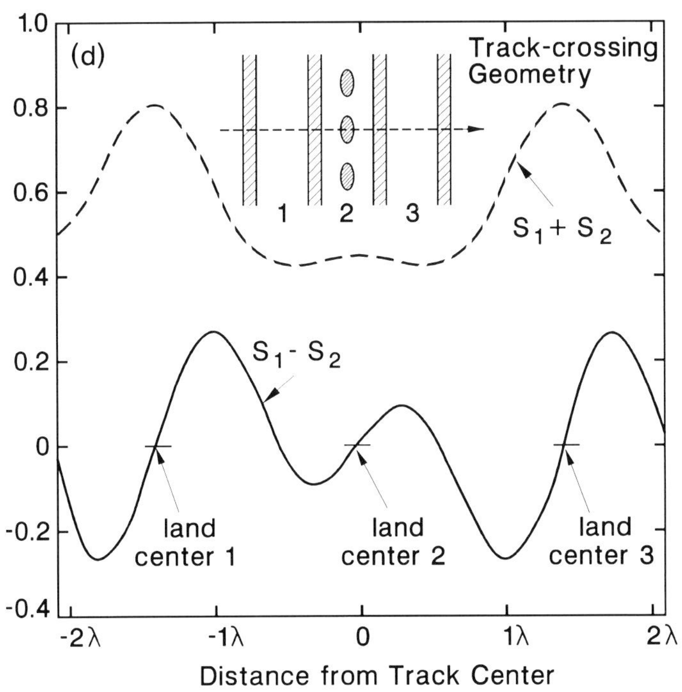
(d)
Track-crossing
Geometry
1
2
3
$S_1 + S_2$
$S_1 - S_2$
land
center 1
land
center 2
land
center 3
1.0
0.8
0.6
0.4
0.2
0
-0.2
-0.4
-2λ
-1λ
0
1λ
2λ
Distance from Track Center

tracks were crossed in this case, of which the middle track had the preformat marks. Both the push–pull signal $S_1 - S_2$ and the sum signal $S_1 + S_2$ are shown in the figure. The peak-to-valley modulation of the push–pull signal drops from a high of 0.55 away from the formatted region to a low of 0.2 upon crossing the format marks. Also, the sum signal, which normally reaches a value of 0.83 at the land center, stays low on the format marks.

8.3. Analysis of Focus-error Detection by the Astigmatic Method

We begin by presenting a geometric-optical analysis of the astigmatic method. The relevant path of the reflected light is depicted in Fig. 8.7, where a point source S, separated from the objective's focus by a distance δ, is shown to the left of the objective L_1. Assuming an infinity-corrected objective, S is the image of the original light source formed by L_1 and the reflecting surface of the disk. δ is thus twice the actual focus-error distance which, by definition, is the distance between the disk and the focal plane of L_1. The following sign convention is adopted: δ is positive when the source S moves away from L_1, and negative when S moves toward L_1. The remaining system parameters are as follows:

f_0, objective's focal length;
D, distance between objective L_1 and astigmatic lens L_2;
f_1, f_2, focal lengths of the astigmatic lens ($f_1 > f_2$);
d, separation between detector and astigmatic lens

Roughly speaking, the spot formed on the detector has an elliptical shape with major and minor axes h and w. When S is at the focus of L_1 we require that h be equal to w. Without imposing restrictions on other parameters, this requirement fixes the position d of the detector as

$$d = \frac{2 f_1 f_2}{f_1 + f_2} \tag{8.1}$$

Defining the focus-error signal (FES) as

$$\text{FES} = \frac{h^2 - w^2}{h^2 + w^2} \tag{8.2}$$

and carrying out straightforward, although tedious, geometric-optical calculations we find that

$$\boxed{\text{FES} = \frac{2\alpha(1 - \beta\Delta)\Delta}{1 - 2\beta\Delta + (\alpha^2 + \beta^2)\Delta^2}} \tag{8.3}$$

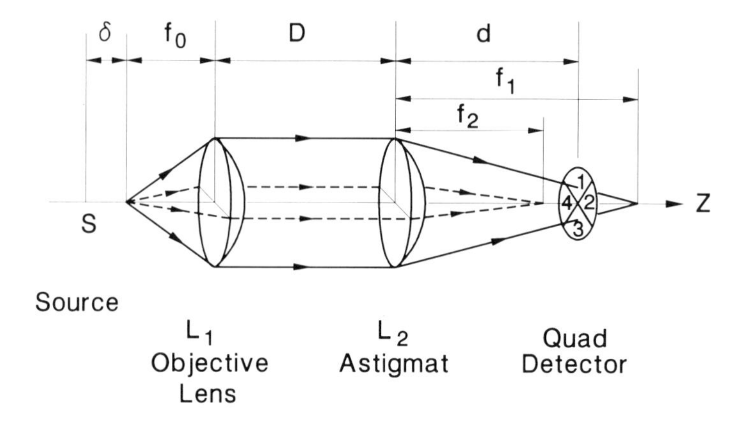

Figure 8.7. Schematic diagram of the astigmatic focus-error detection system. With a grooved disk the system can also generate a push–pull tracking-error signal. Note that δ, the distance between S and the focus of L_1, is twice the separation between the disk and the objective's focal plane.

Here $\alpha = 2f_1f_2/(f_1 - f_2)f_0$ is a normalized measure of the astigmatism of L_2, $\beta = (D/f_0) - 1$ is the normalized distance between L_1 and L_2, and $\Delta = \delta/f_0$ is the normalized separation of S from the objective's focal point. (Normalization is by f_0 in all cases.)

A typical plot of FES(Δ) for fixed values of α and β is shown in Fig. 8.8(a). Note that the maximum value of the function (+1) is attained at $\Delta_{max} = 1/(\alpha+\beta)$, while its minimum (-1) is achieved at $\Delta_{min} = -1/(\alpha-\beta)$. The first zero-crossing occurs at $\Delta = 0$, and the slope of the curve at this point is 2α. The second zero-crossing occurs at $\Delta = 1/\beta$ independently of α. In the limit when Δ goes to $\pm\infty$ the function approaches $-2\alpha\beta/(\alpha^2 + \beta^2)$. The first derivative of the FES with respect to Δ is found to be

$$\mathrm{FES}'(\Delta) = \frac{2\alpha\left[1 - 2\beta\Delta - (\alpha^2 - \beta^2)\Delta^2\right]}{\left[1 - 2\beta\Delta + (\alpha^2 + \beta^2)\Delta^2\right]^2} \tag{8.4}$$

Figure 8.8(b) shows a plot of FES′(Δ) for fixed values of α and β. The extrema of FES′(Δ), which are the inflection points of FES(Δ), occur at

$$\Delta_n = \frac{1}{\beta^2 - \alpha^2}\left\{\beta + 2\alpha\cos\left[\frac{2\pi n}{3} + \frac{1}{3}\arccos\left(\frac{2\alpha\beta}{\alpha^2+\beta^2}\right)\right]\right\}; \quad n = 0,1,2 \tag{8.5}$$

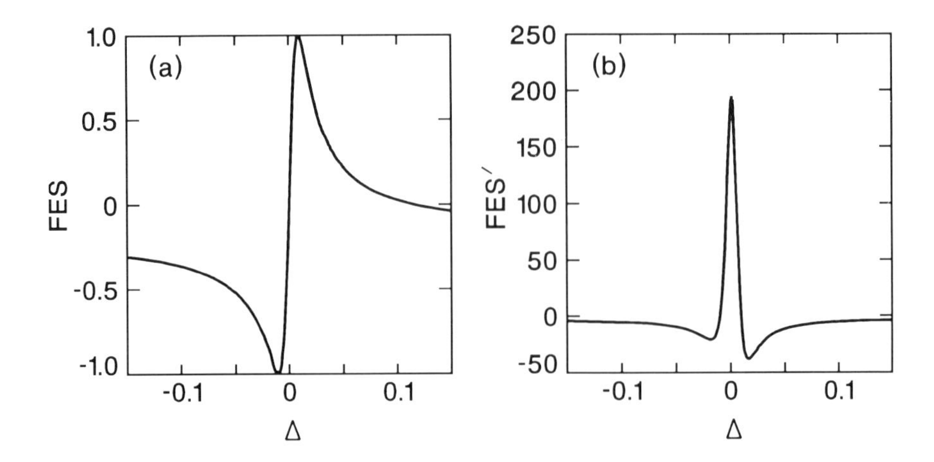

Figure 8.8. (a) FES(Δ) as obtained from Eq. (8.3). Although the plot is for the specific values $\alpha = 100$ and $\beta = 10$, it is typical of all other values of α and β. The maximum FES occurs at $\Delta_{max} = 1/(\alpha + \beta)$, the minimum at $\Delta_{min} = -1/(\alpha - \beta)$. The zero-crossings are at $\Delta = 0$ and $\Delta = 1/\beta$. (b) Derivative of the FES with respect to Δ (again $\alpha = 100$, $\beta = 10$). The extrema of this curve occur at Δ_0, Δ_1, and Δ_2 which are related to α and β through Eq. (8.5).

Note, in particular, that the maximum slope of FES(Δ) does not occur at $\Delta = 0$ but rather at $\Delta = \Delta_2$; however, as long as β is much less than α (and this is almost always the case in practice) one can verify that the two slopes are nearly identical.

To put the matters in perspective, consider the typical set of parameters listed in Table 8.1. For these parameters $\alpha = 425.5$ and $\beta = 4.32$; thus the range of focus error (i.e., disk displacement relative to the focal plane) over which the above FES is suitable for use in a closed feedback loop is

$$\left[\tfrac{1}{2}\, \Delta_{min} f_0 \,,\ \tfrac{1}{2}\, \Delta_{max} f_0 \right] = (-4.46\ \mu\text{m}, +4.37\ \mu\text{m})$$

For comparison, note that the depth of focus for this particular objective is $\simeq \lambda/(\text{NA})^2 = 2.64\ \mu\text{m}$. The FES of Eq. (8.3) is somewhat reduced by the effects of diffraction, as will be seen in 8.3.2.

8.3.1. The acquisition range

In addition to providing a bipolar, more or less linear signal to drive the servo loop in the neighborhood of focus, a focus-error detector must be able to inform the system about the position of the disk (relative to the focal plane) over a somewhat larger range of distances. This information is

Table 8.1. System parameters used in the examples in sections 8.3 and 8.4

Parameter	Definition	Numerical value
λ	wavelength	0.8 μm
f_0	objective focal length	3.76 mm
D	distance between L_1 and L_2	20.0 mm
f_1, f_2	focal lengths of L_2	20.25 mm, 19.75 mm
d	distance between L_2 and detectors	20.0 mm
NA_0	NA of objective	0.55
NA_1	NA of astigmatic lens	0.15
r_0	1/e radius at L_1 of incident Gaussian beam	3.2 mm
Ω	track-pitch	1.6 μm
ℓ	land width	1.1 μm
γ_1	groove width at the top	0.5 μm
γ_2	groove width at the bottom	0.3 μm
ς	groove depth	100 nm

necessary when the servo temporarily loses its lock on focus, or when the drive is first turned on, or when a new disk is inserted into the drive. In such situations, in order to avoid a head crash, the objective is automatically pulled away as far as possible. It is then moved slowly forward while the system monitors the output of the focus-error detector. Once the linear range of the FES curve has been attained, control is handed over to the servo and the servo's lock on focus is re-established. For a given focus-error detector one might therefore define an "acquisition range" as follows. The acquisition range is the maximum distance a disk can be out of focus (away from the objective) without compromising the ability of the system to move in the correct direction and establish a lock on focus. If the second zero-crossing of the FES curve in Fig. 8.8(a) is considered a measure of the acquisition range, then $f_0/2\beta = 435\ \mu$m is the acquisition range of the above system. However, because of the signal-to-noise limitations and also because of diffraction effects, the practical acquisition range is expected to be somewhat lower than this analysis has indicated.

8.3.2. Diffraction analysis

We investigate the effects of diffraction on the astigmatic focus-error signal described above. For concreteness, these effects are obtained numerically using the scalar diffraction theory of Chapter 3 for the system

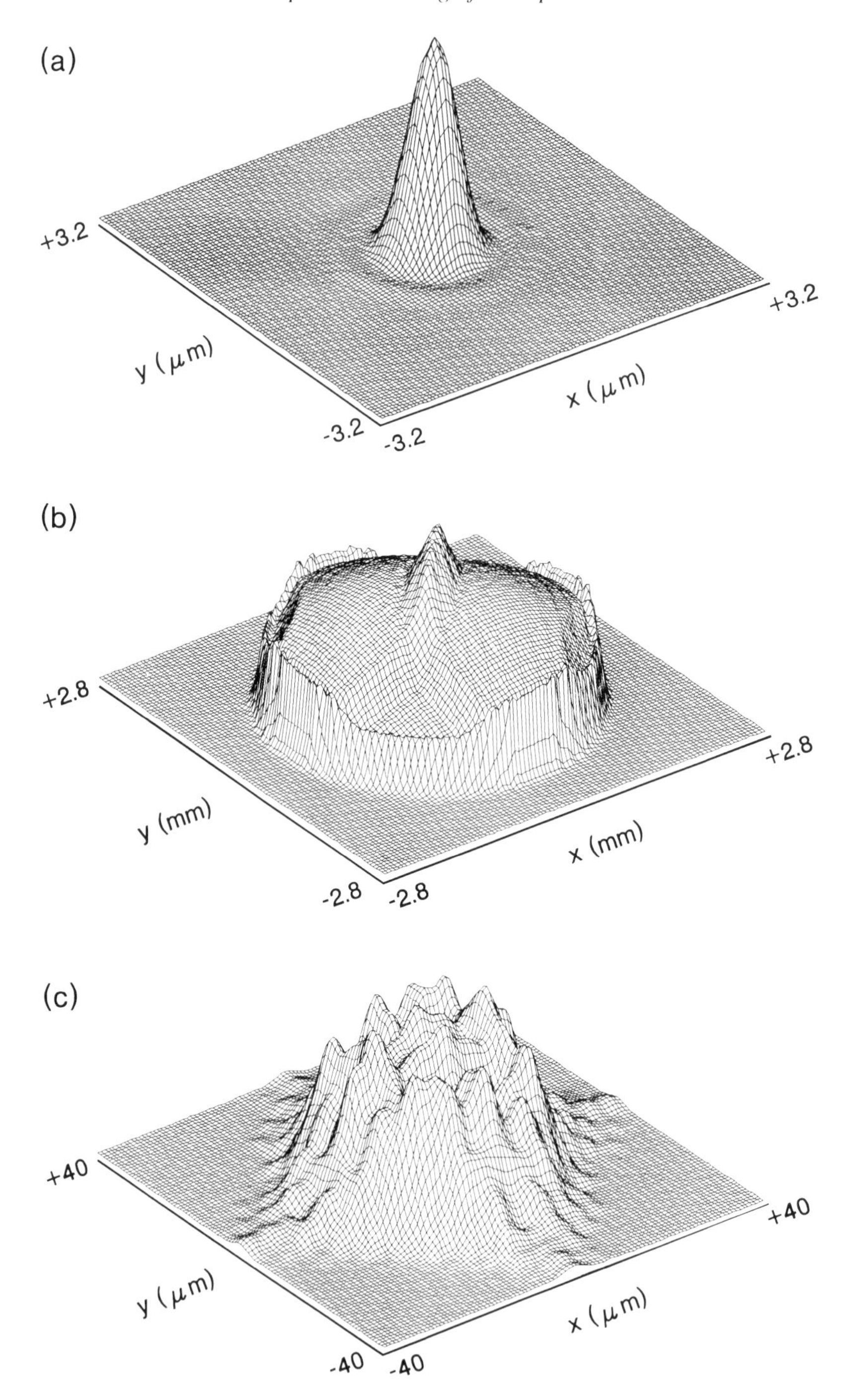

Figure 8.9. Calculated intensity patterns at various cross-sections of the system of Fig. 8.7 and Table 8.1. The disk is in focus and the spot is on-track. Intensity distribution (a) at the disk surface, (b) at the entrance pupil of the astigmatic lens L_2, (c) at the quad detector.

of Fig. 8.7 with parameter values of Table 8.1. The incident amplitude distribution at the objective is Gaussian with 1/e radius $r_0 = 3.2$ mm. The disk is parallel to the focal plane of L_1, has trapezoidal grooves that are 100 nm deep, and has equal reflectivities over the land and the groove. The land and the groove are 1.1 μm and 0.5 μm wide, respectively, and the tracks are oriented at 45° to the axes of the astigmat.

Figure 8.9(a) is a plot of the intensity distribution at the focal plane of the objective, which, as expected, shows a faint ring around a bright central spot. The diameter of the bright spot is $\simeq$ 1.8 μm, close to the value of $1.22\lambda/\text{NA}$ for the Airy pattern. When the disk is in focus and the spot is on-track (i.e., centered on the land) the reflected intensity profile at the entrance pupil of the astigmat is that of Fig. 8.9(b). The deviation of this pattern from the incident Gaussian shape is due to the presence of grooves on the disk. The finite aperture of the objective, of course, blocks some of the diffracted light and prevents it from reaching the astigmat. The distance from L_1 to L_2 is such that propagation between them is best described by the near-field (Fresnel) diffraction formulas; in fact, this distance is so short that no significant changes occur in the intensity pattern between L_1 and L_2. Figure 8.9(c) shows the intensity distribution at the quad detector, which is located half-way between the foci of the astigmat. The four segments of the detector in this case receive equal amounts of light. If their output signals are denoted by S_1, S_2, S_3, and S_4, and if the focus-error signal is defined as

$$\text{FES} = \frac{(S_1 + S_3) - (S_2 + S_4)}{S_1 + S_2 + S_3 + S_4} \tag{8.6}$$

then, for the distribution of Fig. 8.9(c), the FES is very nearly zero.

Figures 8.10 and 8.11 are similar to Fig. 8.9 except that they correspond to different positions of the disk relative to the focal plane; they represent out-of-focus distances of $+3\lambda$ and -3λ respectively. (The sign convention is that positive distances are farther away from the objective.) In both cases the beam is centered on the land and thus the track error is zero. It is clear from frame (c) in each case that two of the detectors receive more light than the others, and that the sign of the resulting FES corresponds to the sign of defocus.

Results of diffraction calculations for the focus-error signal are summarized in Fig. 8.12. In (a) the FES of Eq. (8.6) is plotted versus defocus distance (solid curve) and compared with the geometric-optical result of Eq. (8.3). These computations indicate that geometrical optics can fairly accurately predict the linear range of the FES curve, but that it overestimates the slope. Also shown in Fig. 8.12(a) is the FES curve corresponding to an off-track position of the spot (broken curve); here the spot is centered on a groove edge. Apparently tracking error has but a minor effect on the slope of the FES curve. Similar results are obtained when small amounts of coma are introduced on the objective. Coma, being an odd aberration, cancels itself out when the beam returns through the

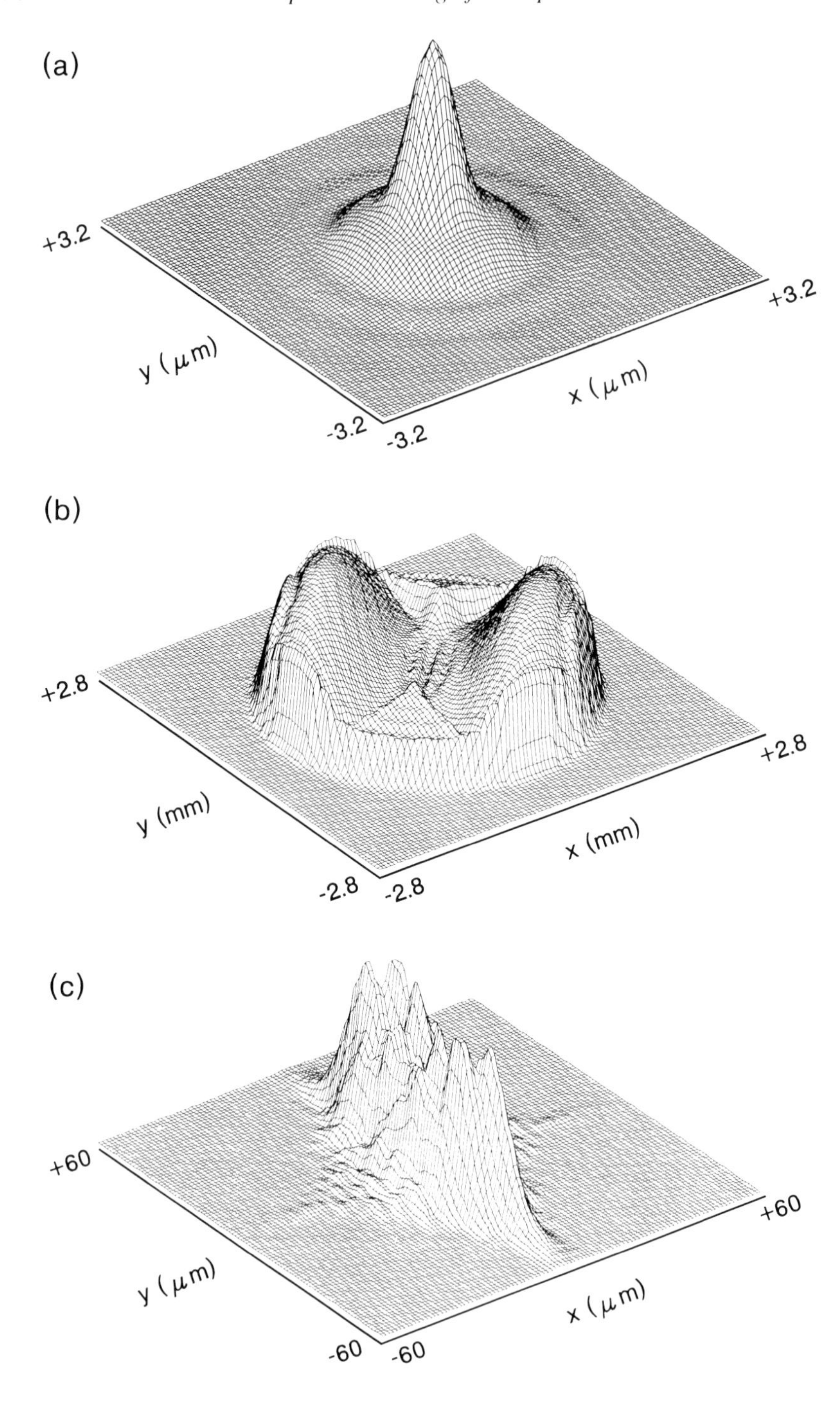

Figure 8.10. Same as Fig. 8.9 but for a disk defocus of $+3\lambda$.

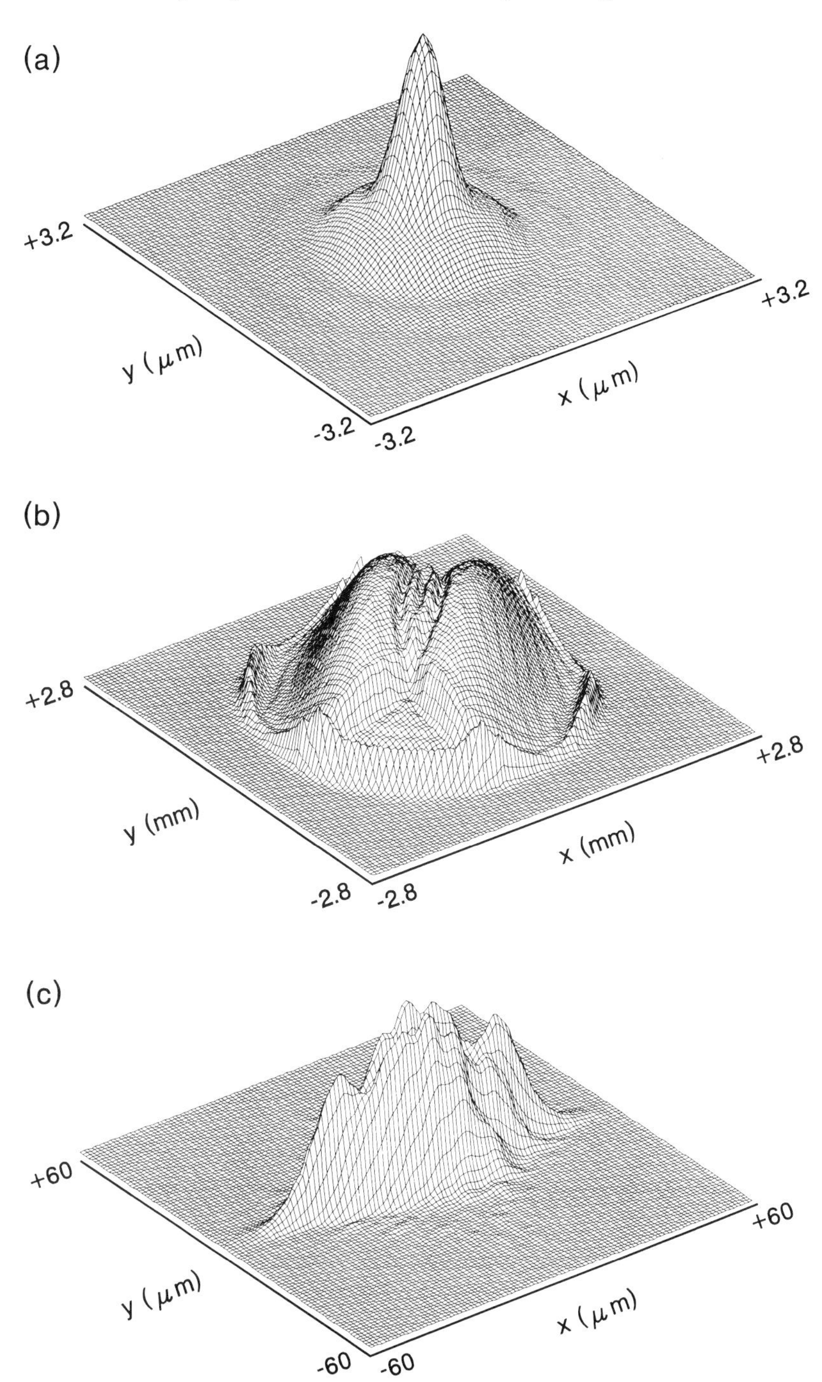

Figure 8.11. Same as Fig. 8.9 but for a disk defocus of -3λ.

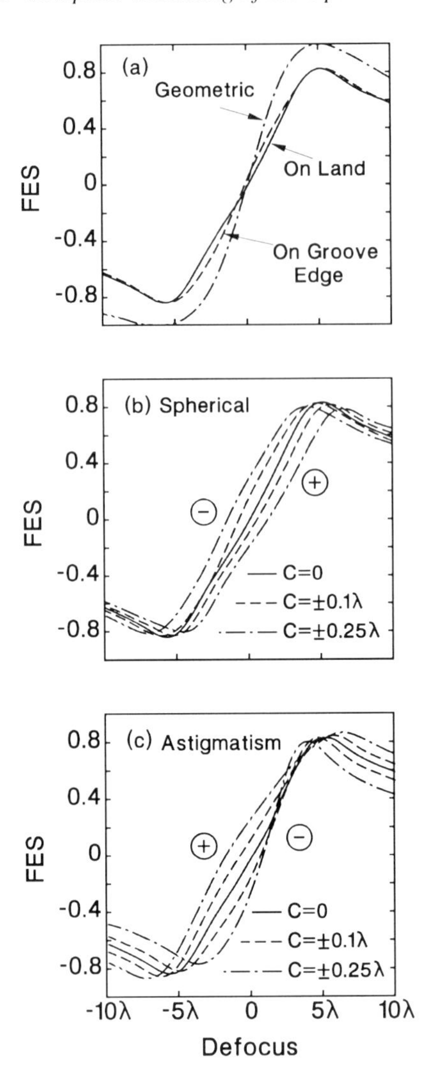

Figure 8.12. FES versus disk defocus (i.e., distance away from the focal plane of the objective) for the system of Fig. 8.7 and Table 8.1. Apart from the geometric-optical curve in (a) all curves are obtained from diffraction calculations. The beam is centered on the land unless specified otherwise. For a definition of the various aberrations see section 3.4.1. (a) The case of a well-corrected objective. The solid curve is the on-track FES; the dashed curve corresponds to a track error of 0.55 μm. Also shown for comparison is the geometric-optical FES curve from Eq. (8.3). (b) FES curves for objectives with different amounts of spherical aberration. (c) FES curves for objectives with different amounts of astigmatism. The orientation of the astigmatism is such that the two line-foci are at 45° to the track direction.

objective; the calculated FES (not shown) is rather insensitive to this type of wave-front defect.

Astigmatism and spherical aberration (SA) tend to affect the performance of the focus servo in different ways. Figure 8.12(b) shows computed FES curves for various amounts of SA associated with the objective. The observed shift of the FES curve will bring about a systematic focus error, which, fortuitously, corrects for the effects of SA to a large extent. As a result of this automatic adjustment, the system performance does not suffer excessively from the presence of moderate amounts of spherical aberration.† Astigmatism of the objective has similar effects on the FES curve, as can be seen from the computed plots in Fig. 8.12(c). In the case of astigmatism, however, automatic focus adjustment tends to degrade the quality of the spot. For this and other reasons (to be discussed in the next subsection) astigmatic aberrations in optical disk drives must be controlled and kept within tight tolerances.

8.3.3. Push–pull tracking, track-crossing signal, and feedthrough

We shall describe the results of a diffraction analysis of the track-error signal (TES) for the system of Fig. 8.7 and Table 8.1. The track-error signal generated by the push–pull method not only provides the necessary feedback for the tracking servo, but also contains sufficient information to enable the counting of tracks during seek operations. In terms of the detector signals S_1, S_2, S_3, and S_4, the push–pull TES is written

$$\text{TES} = \frac{(S_1 + S_2) - (S_3 + S_4)}{S_1 + S_2 + S_3 + S_4} \tag{8.7}$$

Under ideal circumstances, when the beam is centered on either the land or the groove, the above TES will be zero. Once the beam moves away from the track-center, the imbalance of $S_1 + S_2$ relative to $S_3 + S_4$ generates a bipolar error signal. The distributions of Fig. 8.13 correspond to situations in which the disk is in focus but the spot is off-track either by $+0.55\ \mu$m, as shown in frames (a_1), (a_2), or $-0.55\ \mu$m, as shown in frames (b_1), (b_2). Frames (a_1) and (b_1) are the intensity patterns at the entrance pupil of the astigmat, while frames (a_2) and (b_2) show the distribution at the quad detector. Clearly, diffraction from the groove edge has caused constructive interference on one side of the aperture and destructive interference on the other side. Similar effects occur for smaller amounts of track error, but the resulting imbalance of the detector signals is correspondingly smaller.

†The ability of the system to compensate for spherical aberrations with defocus is crucial for the proper operation of systems that read and write through the substrate, since in these systems deviations of the substrate from nominal thickness (or nominal refractive index) produce substantial amounts of spherical aberration.

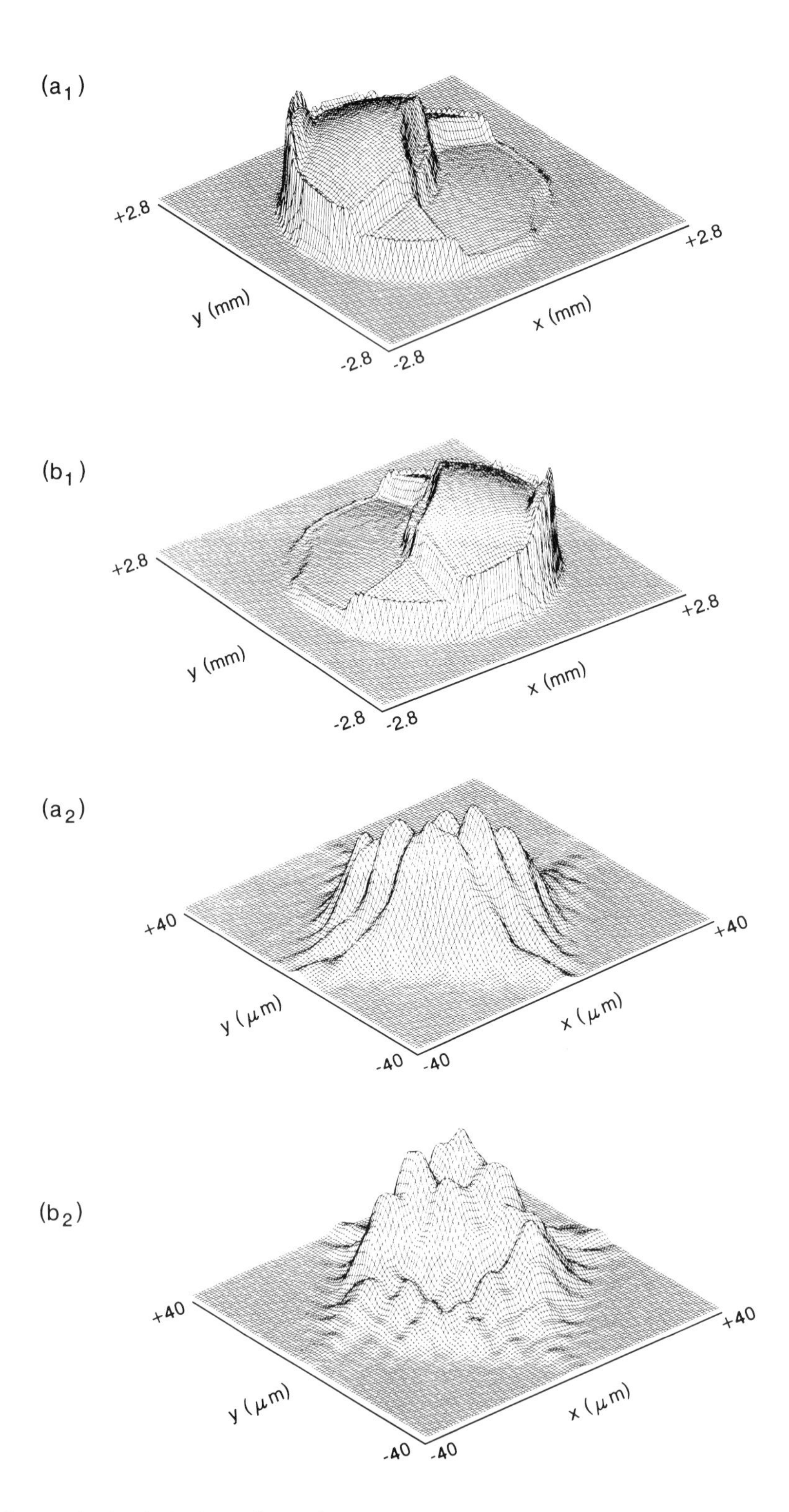

Figure 8.13. Calculated intensity patterns at two cross-sections of the system of Fig. 8.7 and Table 8.1. The disk is in focus but the spot is off-track by +0.55 μm in (a) and by -0.55 μm in (b). (a_1) and (b_1) are at the entrance pupil of the astigmat while (a_2) and (b_2) are at the quad detector.

Plots of track-error signal versus spot position in the cross-track direction are shown in Fig. 8.14. In (a) the solid curve corresponds to the in-focus position of the disk, and the other curves represent different amounts of defocus. While it is the linear part of these curves (near the center) that is used for automatic control of the track-follower, the zero-crossings may be used to count the tracks during seek operations. Observe that while deviations of one wavelength or so from focus leave the TES relatively undisturbed, three wavelengths of defocus can jeopardize both track-following and track-counting operations. One concludes from these observations that in order to achieve accurate counting of tracks during the seek operation, it is necessary to keep the disk tightly in focus.

The presence of coma on the objective lens shifts the TES curves, as shown in Fig. 8.14(b). The curves represent 0, 0.1λ and 0.25λ of primary coma, with the comatic tail oriented perpendicular to the tracks (which is the worst case as far as the TES is concerned). Coma, therefore, causes systematic track-following errors by displacing the TES curve; according to the plots in Fig. 8.14(b), 0.1λ of coma can cause as much as 0.07 μm of track decentering, while 0.25λ of coma could raise this error to about 0.17 μm. However, since the magnitude of the TES is not affected by moderate amounts of coma, the track-counting process should remain unscathed.

The consequences of astigmatic aberration of the objective for the tracking signal are shown in Fig. 8.14(c). These curves, which correspond to 0, 0.1λ, and 0.25λ of astigmatism, clearly show the progressive degradation of the TES with this type of aberration. Again, the orientation of wave-front distortion has been chosen to correspond to the worst possible scenario, with the astigmatic line-foci at 45° to the tracks. In accordance with the operating principles of disk drives, each curve in Fig. 8.14(c) has been computed for a disk that has been appropriately displaced from the nominal focal plane in order to achieve zero focus-error signal (see Fig. 8.12(c)).

A problem with all practical schemes for focus-error and track-error detection in use today is cross-talk between the two servo channels, a problem sometimes referred to as "feedthrough". During seek operations, as the beam scans the disk in search of a targetted track, diffraction from the grooves creates a false FES. When the head velocity in the cross-track direction is high enough, the false FES falls outside the bandwidth of the focus servo and, therefore, will be neglected. Feedthrough becomes a problem only at the beginning and at the end of a seek process, where the scan velocities are low and the focus servo tends to respond to the fluctuating FES. Figure 8.15 shows the magnitude of actual focus error produced over the period of one track-crossing for several cases of practical interest. The solid curve corresponds to the ideal case of an aberration-free lens, showing a maximum feedthrough of only about 0.3λ. Spherical aberration and coma do not seem to produce much feedthrough either. Astigmatism, when oriented with line-foci parallel and perpendicular to the tracks, creates about 0.75λ of focus error during track crossings, which should be tolerable. However, when the astigmatism is at

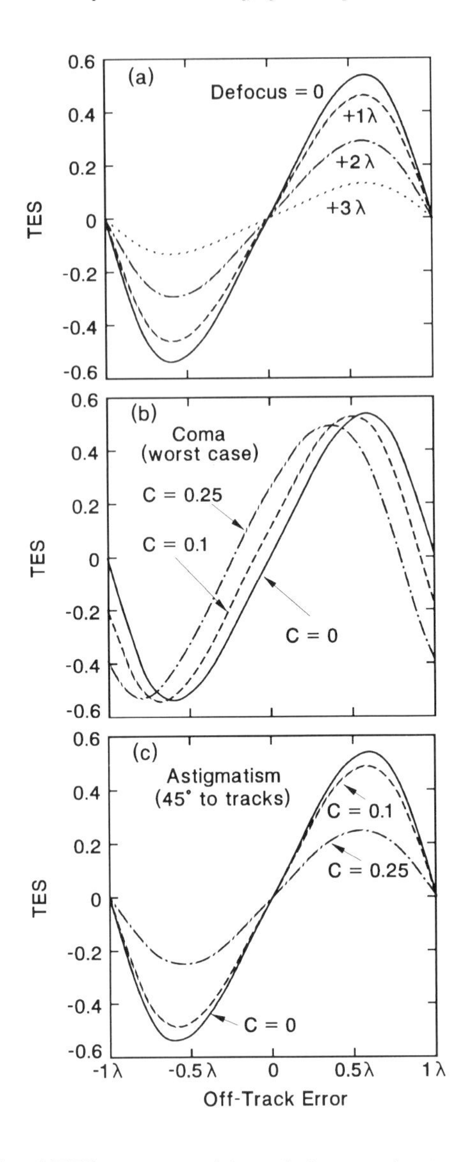

Figure 8.14. Calculated TES versus position of the spot in the cross-track direction. When the value of off-track error on the horizontal axis is zero the spot is centered on the land; off-track errors of $\pm 1\lambda$ place the spot on the adjacent grooves. For a definition of aberrations see section 3.4.1. (a) Plots of TES for different positions of the disk relative to the focal plane of the objective. (b) With the comatic tail oriented in the cross-track direction, coma of the objective shifts the TES curve away from the track-center. (Distortion aberration of equal magnitude and opposite sign has been added to the primary coma in order to assure that the TES shift is solely due to the comatic tail.) (c) Astigmatism of the objective can rapidly degrade the strength of the TES. These curves are obtained with the astigmatism oriented at 45° to the tracks, and the disk displaced from the nominal focal plane in order to nullify the FES.

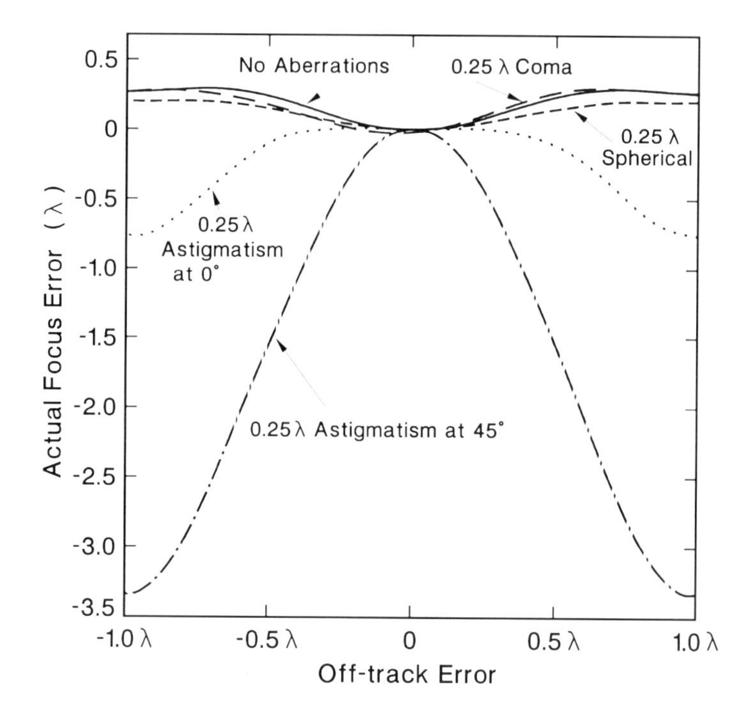

Figure 8.15. Residual focus error versus spot position in the cross-track direction. In each case the separation between the disk and nominal focus has been adjusted to yield zero FES at the land center; the disk is then fixed in that position and the spot scanned across one track period. The computed FES is converted into actual focus error (in units of wavelength) before being plotted. The solid curve is the feedthrough for an ideal system; the other curves show the effects of aberrations from the objective on the residual focus error. The worst-case feedthrough is observed when the objective has astigmatism at 45° to the tracks.

45° to the tracks, fluctuations of the FES pose a threat to the stability of the focus servo. The engineering challenge of automatic focusing and tracking is to create robust servos that operate within the range of tolerances of the various components of the system.

8.4. Analysis of Focus-error Detection by a Ring-toric Lens

A focus- and track-error detection scheme based on a ring-toric lens and a four-segment "phi detector" (so-called because of its configuration) is shown schematically in Fig. 8.16(a). The cross-section of the ring-toric,

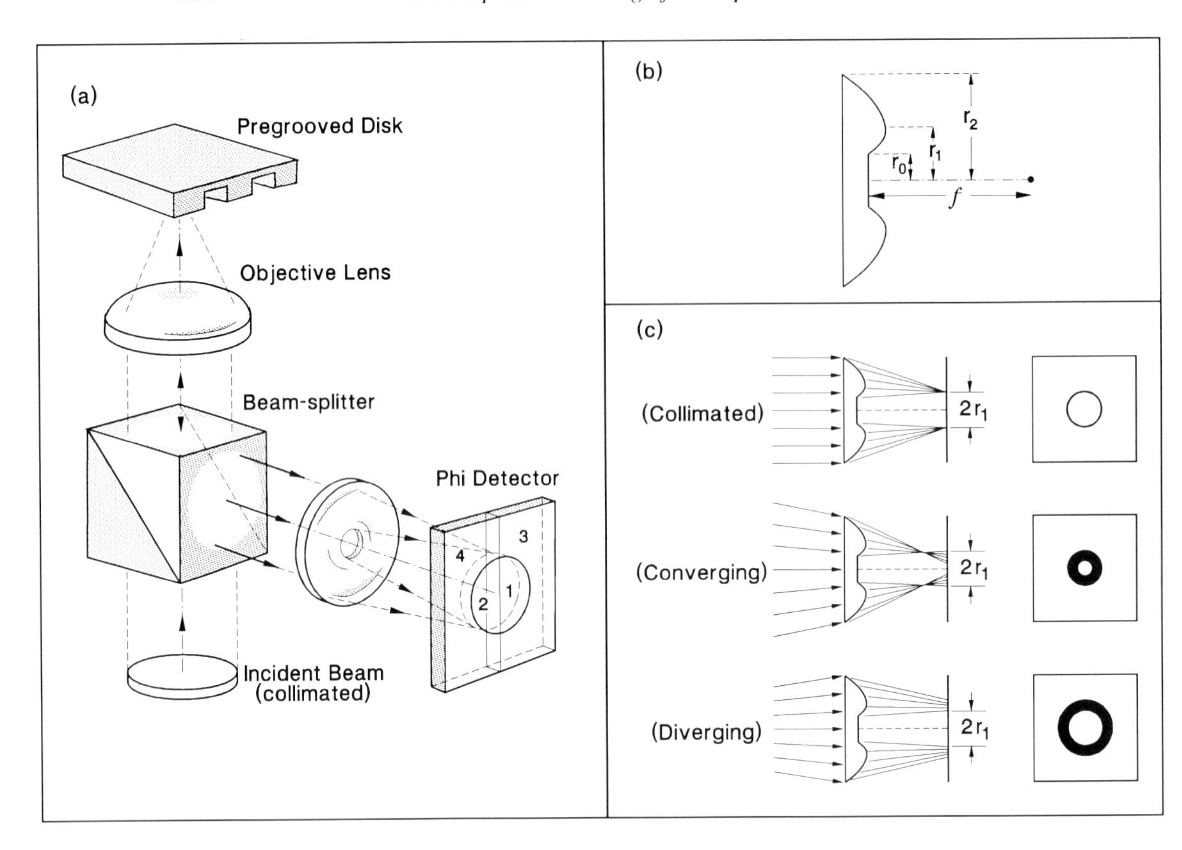

Figure 8.16. (a) Schematic diagram of a focus-error and track-error detection scheme based on a ring-toric lens and a phi detector. Note the resemblance of the detector to its namesake, the Greek letter ϕ. (b) Cross-section of the ring-toric lens, identifying the characteristic parameters r_0, r_1, r_2, f. (c) Ray diagrams showing various patterns of light distribution in the focal plane of the ring-toric lens.

shown in Fig. 8.16(b), defines its characteristic parameters r_0, r_1, r_2, and the focal length f. The ring-toric is designed to bring a collimated beam to focus on a sharp, uniform ring of radius r_1. For an aberration-free toric, the width of this ring is determined by diffraction. r_0 and r_2 are the aperture radii of the lens at its inner and outer rims. In the following discussion we shall refer to the ring of radius r_1 formed in the focal plane of the ring-toric lens simply as "the ring".

When the incident column of light is convergent, the ring-toric directs the rays towards the interior of the ring. A divergent beam, on the other hand, illuminates the exterior of the ring (see Fig. 8.16(c)). Therefore, the phi detector can generate the focus-error signal by a combination of its quadrant signals as follows:

$$\text{FES} = \frac{(S_1 + S_2) - (S_3 + S_4)}{S_1 + S_2 + S_3 + S_4} \quad (8.8)$$

The phi detector can also generate a push–pull track-error signal in conjunction with pregrooved disk structures. Since diffraction from the edges of grooves creates an imbalance of light intensity distribution on opposite halves of the detector, the TES may be obtained as follows:

$$\text{TES} = \frac{(S_1 + S_3) - (S_2 + S_4)}{S_1 + S_2 + S_3 + S_4} \quad (8.9)$$

The ring-toric method of focus-error detection is a generalization of the Fresnel biprism method,[1] which itself is an extension of the knife-edge technique (see Chapter 4). In contrast to the biprism, the circularly symmetric ring-toric utilizes all the available focus-error information carried by the various rays of the beam and, consequently, is more sensitive to focus errors (i.e., the slope of the FES curve is larger). Like the biprism, a useful property of the ring-toric method is its wide acquisition range. The circular symmetry of the ring-toric reduces the overall sensitivity of the signals to a number of rotational as well as translational misalignments. The slope of the ring-toric FES is diffraction limited, whereas in most other methods (notably astigmatic and wax–wane techniques) the slope is primarily limited by geometrical considerations. Also, the fact that the detector in this scheme is placed at the focal plane of the secondary lens helps to reduce the pattern noise, a major cause of feedthrough disturbances.

Diffraction calculations for the ring-toric scheme are somewhat more complicated than those for conventional lenses; the results presented below are based on the factorization method.[5] Shown in Fig. 8.17 are calculated intensity patterns for a ring-toric lens with $r_0 = 40\ \mu\text{m}$, $r_1 = 80\ \mu\text{m}$, $r_2 = 2.2$ mm, and $f = 20$ mm, the other system parameters being the same as those in Table 8.1. Figure 8.17(a) shows the intensity distribution on the phi detector when the disk is in focus and perfectly on-track. Here, as expected, the calculated values of both the FES and the TES are zero. When the beam is on-track but -3λ out of focus, the nonuniform ring of Fig. 8.17(b) appears at the detector. The nonuniformity here arises from the diffraction of a relatively large spot from the grooves. Figure 8.17(c) corresponds to an off-track position of the beam and zero defocus; the beam, which in this case is on the groove edge, forms a sharp half-ring at the focal plane of the ring-toric.

Several computed plots of FES versus actual disk defocus are shown in Fig. 8.18. These plots confirm our earlier assertions that the ring-toric scheme produces a fast FES (i.e., one with a large slope) and that its acquisition range is wide. Observe that the effect of objective-lens coma on the FES curve is minimal, and that spherical aberrations simply shift

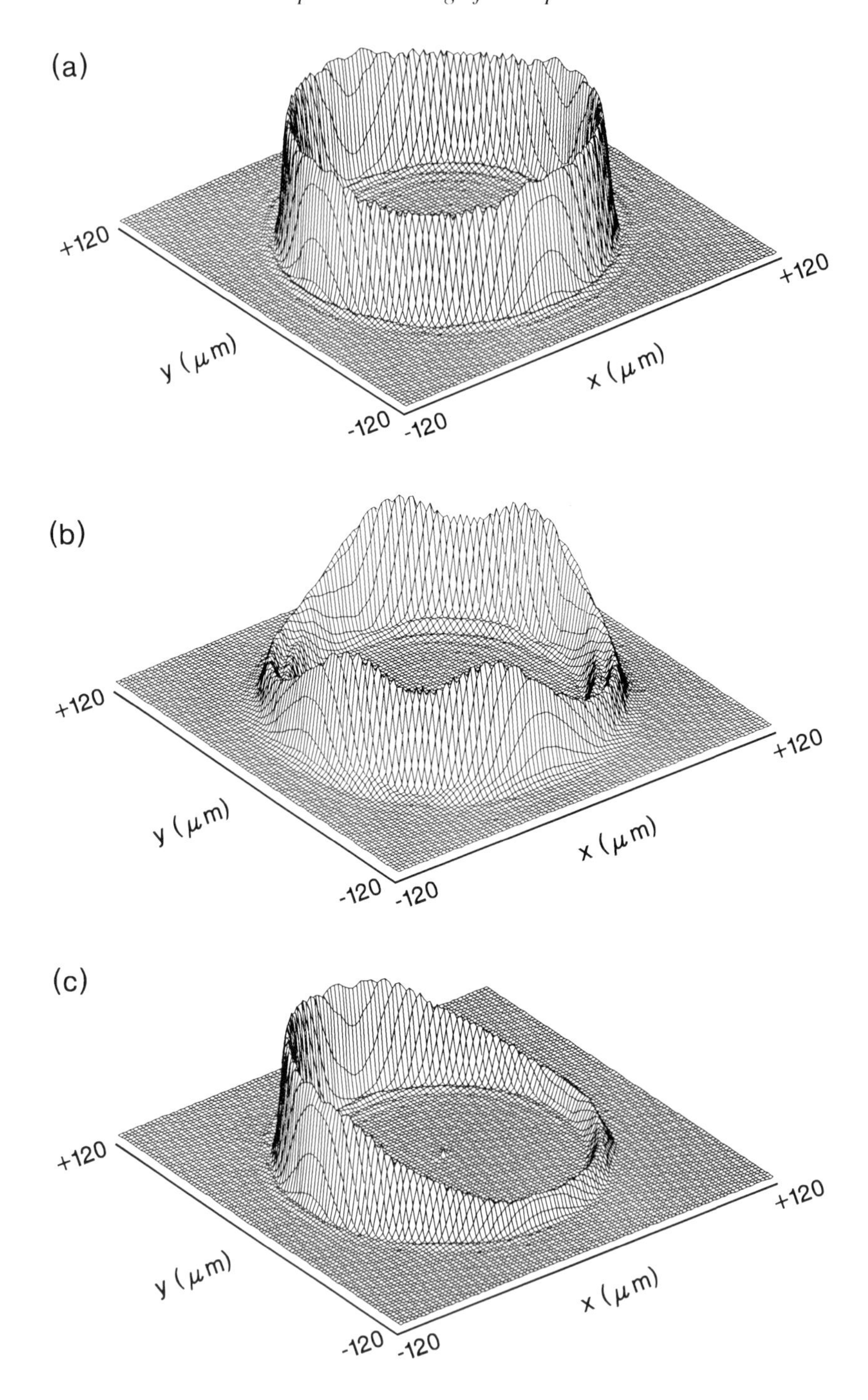

Figure 8.17. Calculated intensity distributions at the phi detector of Fig. 8.16 and Table 8.1 (a) when the disk is in focus and the spot is on-track; (b) when the spot is on-track but the disk is out of focus by -3λ; (c) when the disk is in focus but the spot is off-track by 0.55 μm.

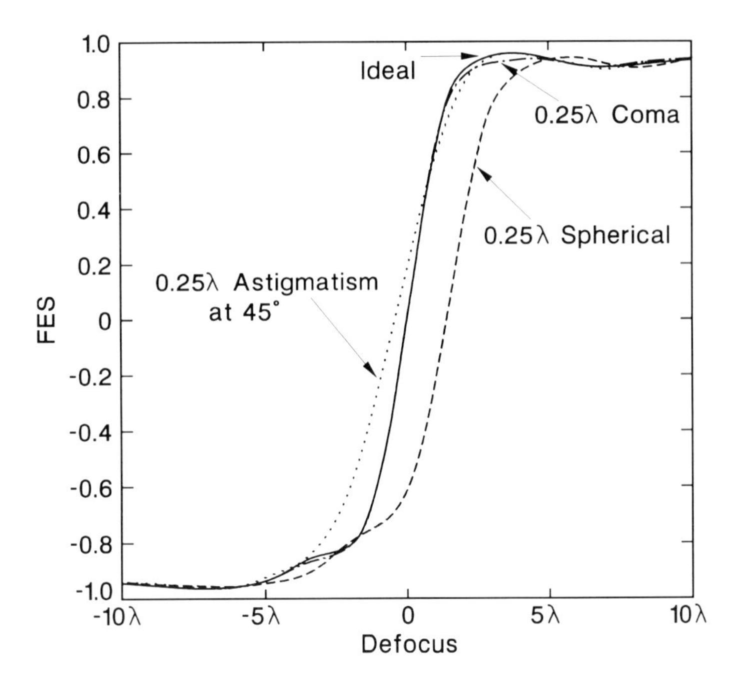

Figure 8.18. Calculated FES for the ring-toric scheme of Fig. 8.16 and Table 8.1, with ring-toric parameters r_0 = 40 μm, r_1 = 80 μm, r_2 = 2.2 mm, f = 20 mm. The solid curve is for an ideal objective; other curves represent objectives having 0.25λ of various aberrations. (The spot is centered on the land in all cases.)

the location of best focus.† The presence of astigmatism on the objective causes a small reduction of the slope of the FES curve, as well as a slight focus shift; none of these problems, however, could be considered a serious threat to the operation of the system. Nor should fluctuations of the FES during track-crossings (i.e., feedthrough) be cause for concern. Figure 8.19 shows plots of residual focus error versus position of the spot in the cross-track direction for various aberrations of the objective. In each case, the position of the disk relative to the nominal focal plane has been adjusted in order to reduce the FES to zero on the land center. In the worst case, when the objective has 0.25λ astigmatism at 45° to the tracks, feedthrough is less than 0.5λ, which is almost seven times better than feedthrough for the astigmatic focus detection scheme (see 8.3.3). Thus, by virtue of focusing

†As noted earlier, the shift of focus with spherical aberration is a highly desirable feature, since it counters the deteriorating effects of spherical aberration on the quality of the focused spot, thus correcting the aberration to a large extent.

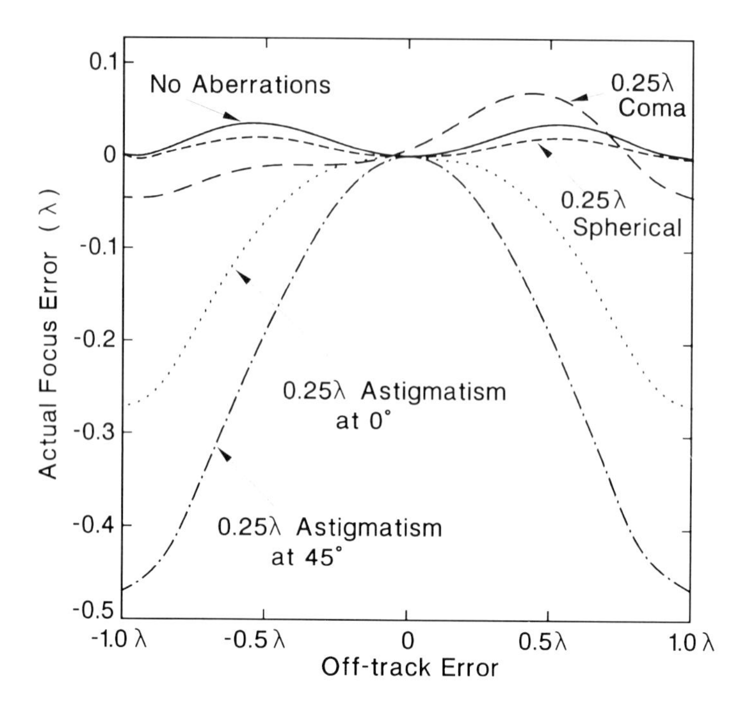

Figure 8.19. Residual focus error versus spot position in the cross-track direction. These feedthrough curves which belong to the ring-toric scheme may be compared directly with the curves of Fig. 8.15, which were obtained for the astigmatic method.

the return beam on the detector and reducing the pattern noise, the ring-toric scheme outperforms the astigmatic method in its lack of FES sensitivity to track-crossings.

8.5. Diffraction Effects in Magneto-optical Readout

A magneto-optical read system is depicted in Fig. 8.20(a). The incident beam from the laser diode ($\lambda = 800$ nm) is linearly polarized along X. At the objective's entrance pupil this beam has a Gaussian profile with $1/e$ radius of 2.92 mm, truncated by the rim of the lens ($r = 2.07$ mm). Figure 8.20(b) shows the intensity distribution at the entrance pupil of the objective, which has NA = 0.55 and $f = 3.76$ mm. Distributions of the intensity for X- and Y-polarized components of the focused spot are shown in Figs. 8.21(a), (b) respectively. Naturally, the Y-component is much weaker, with a peak intensity in the ratio of 1 : 3732 to that of the X-component.

The MO medium has reflection coefficients $r_{\parallel} = 0.7$, $r_{\perp} = 0.005$, and $\phi_{\parallel} = \phi_{\perp}$. A reverse domain simulated on this medium is shown in Fig. 8.21(c). (The three plots in Fig. 8.21 are on the same scale, so the spot dimensions can be readily compared with the domain size.) For the following calculations, ordinary reflection is assumed to have the coefficient $r_{\parallel}$, while magneto-optically generated polarization (perpendicular to the incident polarization) has the reflection coefficient $r_{\perp}$.

Figure 8.22 shows the intensity distribution after reflection from the MO surface, transmission through the objective lens, and propagation beyond the objective by a distance of d = 10 mm. The X-polarized component of this distribution is shown in frame (a). Note that this pattern is essentially the same as that of the incident beam in Fig. 8.20(b). (The X-component is not expected to change much upon reflection from the disk since it is unaffected by the magnetic domain.) Distribution of the Y-component, on the other hand, is a strong function of the position of the focused spot. When the spot is far from the domain, the Y-component distribution resembles that of the X-component, albeit with much less intensity (see Fig. 8.22(b)). When the spot center is on the edge of the domain (the semicircular edge on the left side of Fig. 8.21(c)) the Y-polarization pattern is that in Fig. 8.22(c). Here, in effect, diffraction from a 180° phase step has split the distribution in the middle, in much the same way as an analysis based on the methods of Chapter 4 would predict. With the beam centered on the domain, the Y-polarization pattern of Fig. 8.22(d) is obtained. (The difference between this pattern and that in Fig. 8.22(b) may be attributed to diffraction from the domain walls.) For the above positions of the spot, calculated outputs of the balanced differential detection module shown in Fig. 8.20(a) are tabulated below.

Focused spot position	Sum signal $(S_1 + S_2)$	Difference signal $(S_1 - S_2)$
far from domain	0.483	$+6.9 \times 10^{-3}$
on domain edge	0.483	$+1.8 \times 10^{-3}$
at domain center	0.483	-5.5×10^{-3}

8.5.1. Magneto-optical readout by differential detection

Figure 8.23(a) shows a section from a grooved MO disk containing three crescent-shaped magnetic domains; for better visualization the domains are shown slightly raised above the surface. The width of the land region is 0.75 μm while the trapezoidal grooves are 0.35 μm wide at the top, 0.1 μm wide at the bottom, and 130 nm deep. The reflection coefficients of the

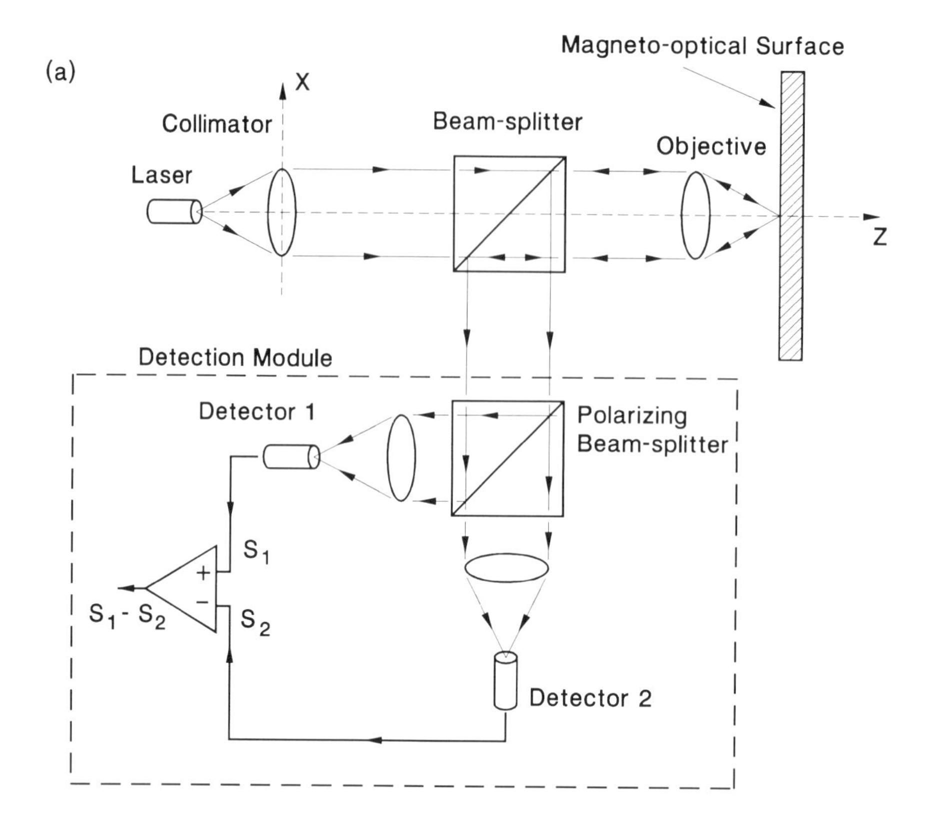

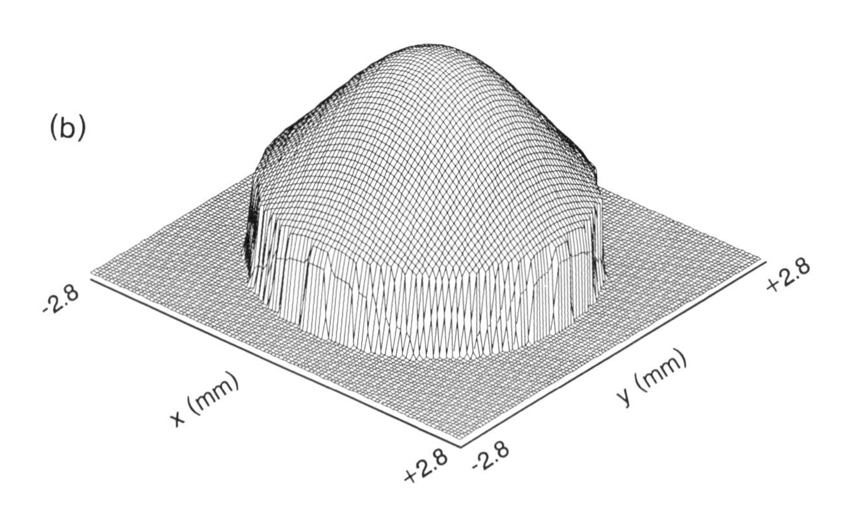

Figure 8.20. (a) Schematic diagram of an MO readout system using linearly polarized incident light. (b) Intensity profile of the truncated Gaussian beam at the objective's entrance pupil.

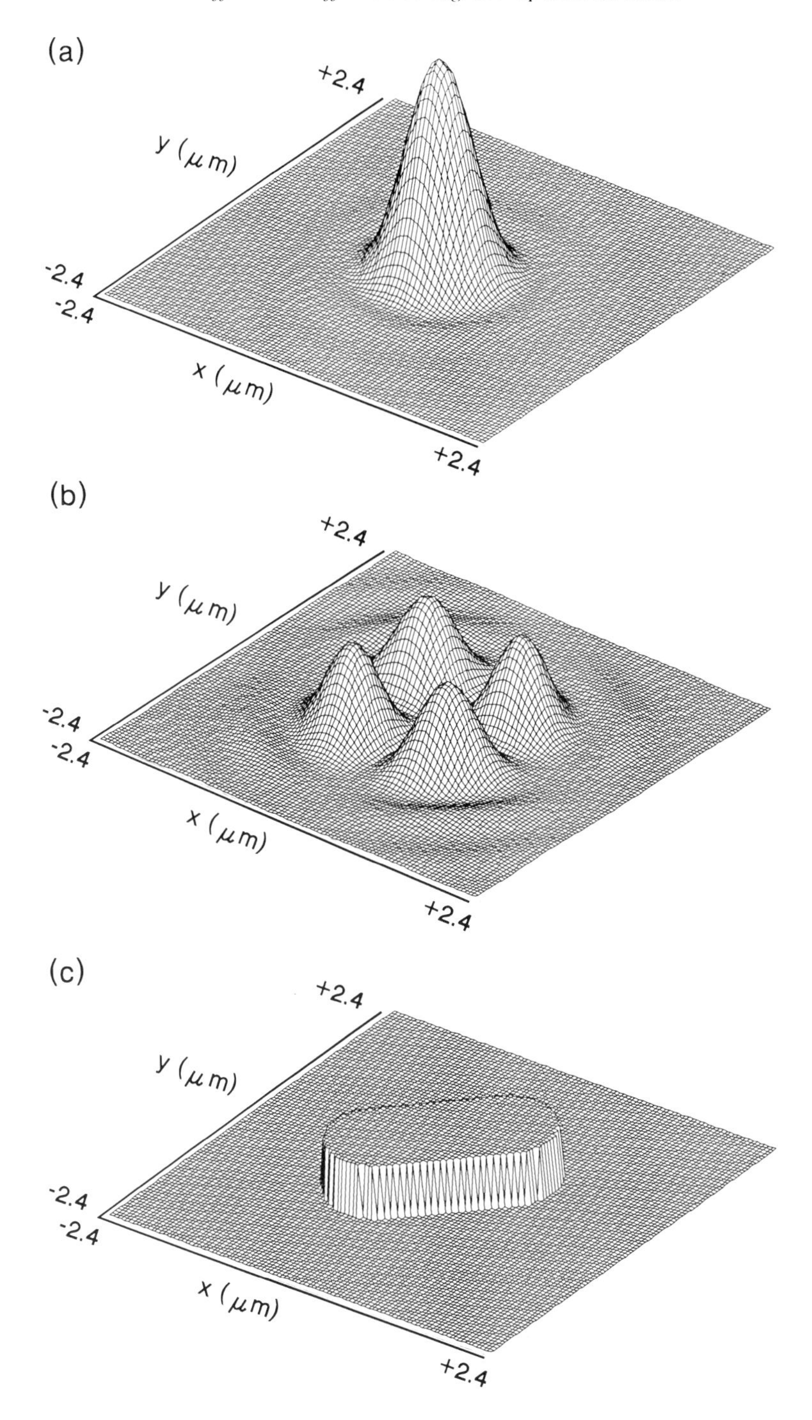

Figure 8.21. (a) Intensity pattern of the X-component of polarization at the disk surface. (b) Intensity pattern of the Y-component of polarization at the disk surface. (c) Magnetic domain with partly straight and partly semicircular edges. The beam moves parallel to (and centered on) the long axis of this domain.

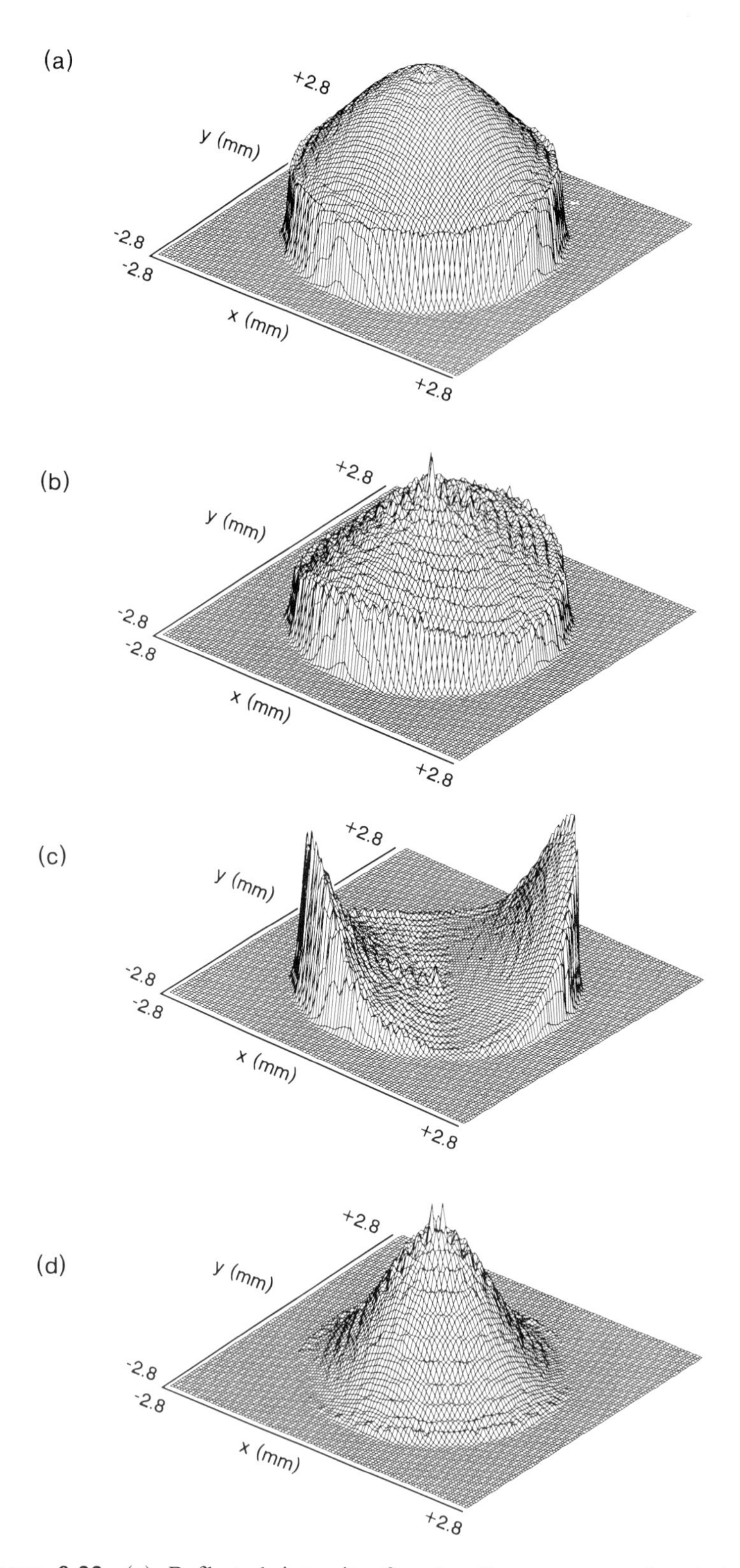

Figure 8.22. (a) Reflected intensity for the X-component of polarization upon entering the detection module (see Fig. 8.20). This pattern is the same irrespective of the relative position of the focused spot and the domain. (b) Intensity of the Y-component at the detection module, with the spot far from the domain. (c) Same as (b) when the spot is on the left edge of the domain. (d) Same as (b) when the spot is centered on the domain.

(a)

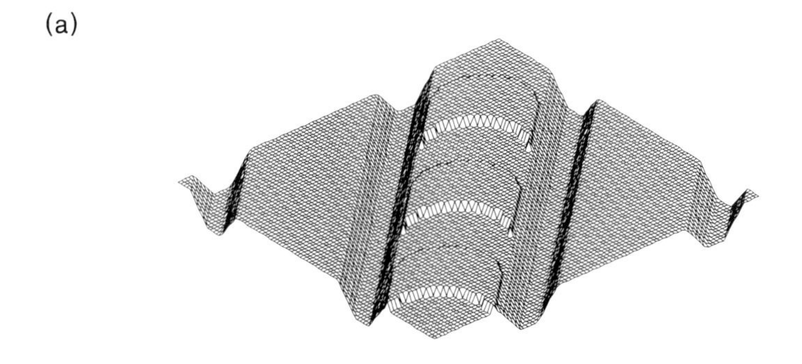

(b)

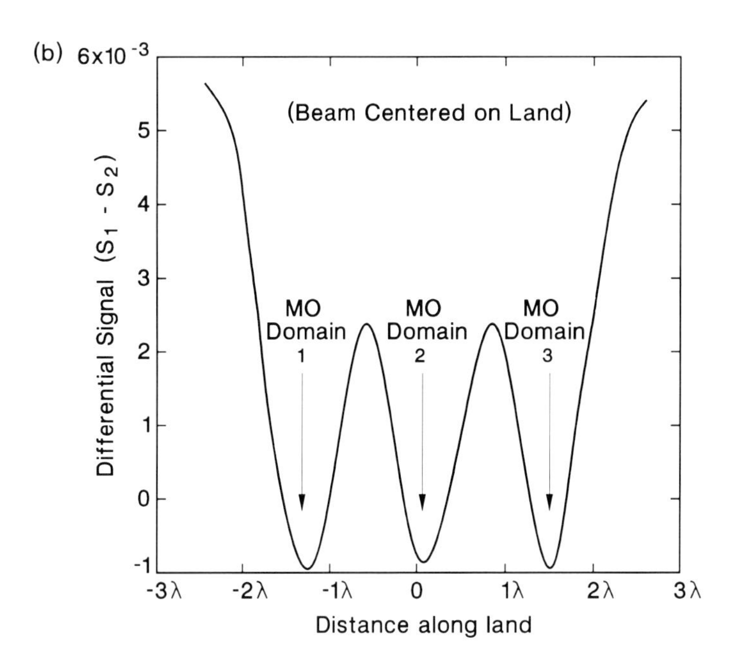

Figure 8.23. (a) Section of pregrooved disk showing three domains recorded with the magnetic-field modulation technique. (b) Computed MO readout signal with the differential detection method.

MO film are $r_{\|} = 0.7$ and $r_{\perp} = 0.005$, with zero phase difference between them (i.e., no ellipticity). The domains are 0.545 μm long and 0.7 μm wide, with a center-to-center spacing of 1.09 μm. The objective has NA = 0.65 and f = 1.23 mm, and the Gaussian read beam (λ = 780 nm) is truncated at the 1/e (amplitude) radius. A plot of the computed differential signal obtained upon scanning the recorded marks is shown in Fig. 8.23(b). If the domains were large and grooves were absent, the differential signal would have had a swing of $\pm 7 \times 10^{-3}$. The peak-to-valley signal in the present case, however, is merely 3.4×10^{-3}. The fact that the focused spot is somewhat greater than the domains can explain this reduction of the signal by a factor of nearly four.

8.5.2. Magneto-optical readout by diffraction from domain walls

In section 6.6 we described an MO read system that utilized diffraction from the domain walls to generate the readout signal; Fig. 6.12 is a schematic diagram of the system. The incident beam in this scheme is circularly polarized and the reflected beam, after going through the $\lambda/4$ plate and the PBS, is fully deflected towards the detection arm where a quad detector generates the data signal as well as the focus- and track-error signals. In the present section we further analyze this edge-detection scheme using numerical quasi-vector diffraction calculations.

Figure 8.24(a) shows a typical section from a disk containing a pattern of grooves, magnetic domains, and preformat marks. The V-shaped grooves are one wavelength wide (λ = 0.8 μm) and 128 nm deep, and the width of the land region is 1.6 μm. The disk has reflectivity R = 49% and a Kerr rotation angle of 0.4°, with no ellipticity. Of the three marks shown on the middle track, the one on the right-hand side is a sphero-cylindrical pit with length = 1.6 μm, width = 1.2 μm, and depth = 80 nm. The other two marks are magnetic domains which, for better visualization, are shown raised above the surface.

The objective has NA = 0.55 and f = 3.76 mm. It focuses a circularly polarized, truncated Gaussian beam onto the track and collects the reflected light, which then propagates a distance of 20 mm to the astigmatic lens. The astigmat has focal lengths f_1 = 20.25 mm and f_2 = 19.75 mm. Half-way between the two foci, the quad detector collects the light and generates the readout signal from the mark edges as well as the FES and the TES. Vector diffraction theory is used in these calculations throughout the optical path except at the disk surface, where marks and grooves are treated as phase objects.

Figure 8.24(b) shows the computed readout signal, the solid curve representing the case of best focus and the broken curve corresponding to a defocus of one wavelength; in both cases the focused spot was centered on the track. Note that the magnitude of the signal from the pit has been scaled down by a factor of 40 in order to fit on the same plot as the MO signal. The tail of the signal from the left edge of the pit interferes with

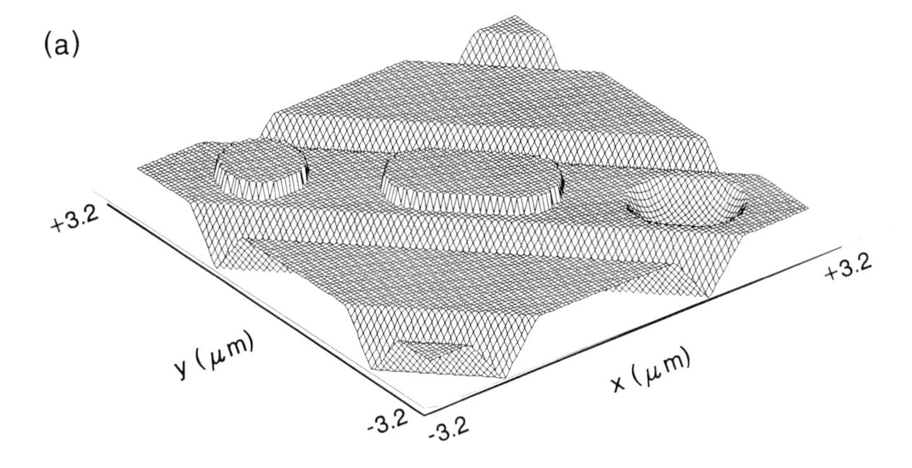

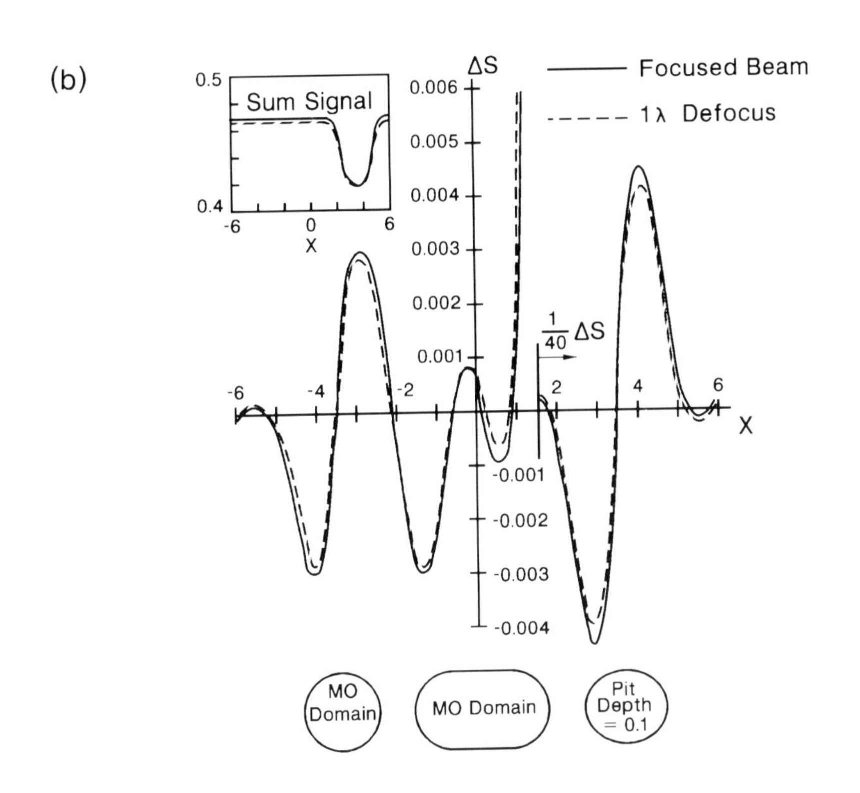

Figure 8.24. (a) Section of disk with V-shaped grooves and a pattern of marks (both magnetic and preformat). The focused spot is centered on the middle track and scans the marks from left to right. (b) Readout signal produced with an edge-detection scheme using circularly polarized light. The solid curve is obtained with the disk in focus, while the broken curve corresponds to a disk defocus of 1λ. The inset shows the sum signal of the quad detector.

the signal from the right edge of the large MO domain, rendering the latter undetectable. However, the left edge of the large MO domain as well as both edges of the smaller domain produce good signals and can readily be detected. The sum signal is shown in the inset, indicating a small drop at the position of the pit in the total amount of reflected light.

The magnitude of the signal obtained from the domain walls in this example compares favorably with that obtained from the domains in standard differential detection.[8,9] Unfortunately, the above edge-detection scheme has been found in practice to suffer from sensitivity to surface roughness. In hindsight, this should not be surprising, considering the effect of the pit on the MO signal in Fig. 8.24(b). There exist variations of the above scheme that do not suffer from surface roughness noise yet maintain many of the advantages of readout by diffraction from domain walls. The interested reader may consult the published literature for more information.[10]

Problems

(8.1) Let a Gaussian beam with 1/e (amplitude) radius r_0 be truncated by the objective aperture of radius ρ. Assuming that the optical axes of the beam and the lens are properly aligned, what fraction of the beam's power is transmitted through the lens? Find the aperture radius at which the value of transmitted power is most sensitive to the variations of r_0.

(8.2) Consider a shallow reflection grating with rectangular grooves of width W and depth d. Let the grating period be P and denote by a_1 and a_2 the amplitude reflection coefficients over the land and the groove, respectively. (a_1 and a_2 are real constants in the interval [0, 1].) A plane wave of unit amplitude propagating with wave vector $\mathbf{k} = (2\pi/\lambda_0)\boldsymbol{\sigma}$ is incident on the grating along the arbitrary direction $\boldsymbol{\sigma} = (\sigma_x, \sigma_y, \sigma_z)$. Determine the directions of propagation for the various diffracted orders, and also the corresponding diffraction efficiencies. (Note: treat the grating as a phase/amplitude object.)

(8.3) The blazed transmission grating in Fig. 8.25 has triangular grooves parallel to the Y-axis with period P and phase depth Δ. Upon transmission through the grating the complex amplitude $T(x)$ imparted to the normally incident plane wave (wavelength = λ_0) may be written as follows:

$$T(x) = \exp\left[\,\mathrm{i}\,\frac{2\pi\Delta}{\lambda_0}\,(x/P)\right]; \qquad 0 < x < P$$

Determine the efficiencies of the various diffracted orders. Under what circumstances does the grating behave like a prism in that it deflects the entire incident beam into a single diffracted order?

(8.4) Using the notation and definitions of section 8.3 derive the geometric-optical expression for the focus-error signal given in Eq. (8.3). Show that the inflection points of FES(Δ) are indeed Δ_0, Δ_1, and Δ_2 as given by Eq. (8.5). Finally, using the plot of the geometric-optical FES in Fig. 8.12(a) (corresponding to the parameter-set in Table 8.1) determine the slope of the FES curve in the vicinity of focus. Confirm that this slope is indeed the same as that predicted by Eq. (8.4) at $\Delta = 0$.

(8.5) The ring-toric lens depicted in Fig. 8.16(b) is characterized by the four parameters r_0, r_1, r_2, and f. Incorporate these parameters into an expression for the complex amplitude of a collimated incident beam immediately after transmission through the lens.

(8.6) Optical disk systems that operate in the substrate-incident configuration require a high degree of parallelism between the disk and the objective lens. Any tilt of the disk (or the lens) relative to the optic axis will place aberrations on the

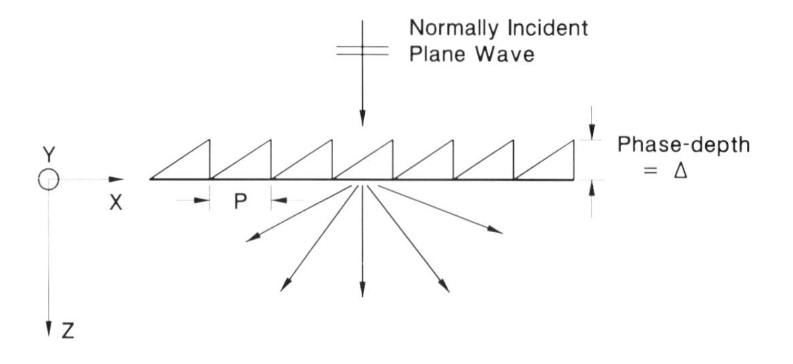

Figure 8.25.

focused spot and consequently deteriorate the performance of the system. Coma, in particular, is a major contributor to the overall wave-front distortion in such cases; in systems of high NA, coma increases rapidly as the amount of tilt increases. With reference to Fig. 8.26, estimate the magnitude of coma induced in a focused beam as a function of the objective NA, tilt angle δ, substrate thickness τ, and substrate refractive index n. You may assume that δ is small and thus omit $O(\delta^2)$ terms.

(8.7) Consider a 3.5" diameter disk, spinning at 3600 rpm in a drive that operates at the wavelength of 780 nm with a 0.55 NA objective . The track-pitch is 1.6 μm, the bandwidth of the focus servo is 10 kHz, the head weighs 70 grams, and the maximum acceleration and deceleration rates of the seek actuator are ± 50 m/s^2. For the sake of simplicity let us assume that during a seek operation the head moves with its maximum acceleration to the mid-point of the seek along the radial direction, then decelerates (again at maximum rate) and comes to a stop on the target track.

(a) In a 10 mm seek (i.e., a jump over 6250 tracks) what is the maximum velocity of the head? What is the duration of the seek? How much kinetic energy does the head acquire? How much power is, on the average, supplied to the head if friction losses are ignored?

(b) During a 20 mm seek from the inner to the outer disk radius, the head velocity reaches a maximum of 1 m/s. How long does the focused spot dwell on a given track at this speed? What is the length of the track that passes under the beam at this point, before the beam moves on to the next track?

(c) The track-crossings in the beginning and at the end of a seek interfere with the focus servo by creating a feedthrough signal within the pass-band of the focus actuator. What is the duration of this interference at either end of a seek operation?

(d) It has been suggested that the feedthrough problem described in (c) may be solved by opening the focus servo loop in the beginning and at the end of the seek operation. Let us make the reasonable assumption that, in order to follow the vertical movement of the disk surface under normal operating conditions, the focus actuator must move the objective lens at an average rate of 20 μm per millisecond. Evaluate the proposed solution to the feedthrough problem under the stated conditions and comment on its feasibility.

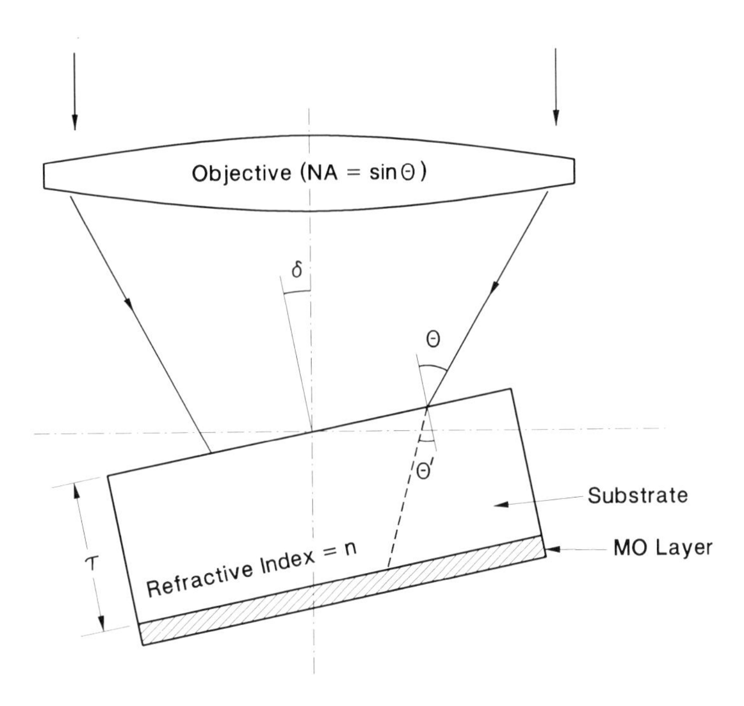

Figure 8.26.

(8.8) Figure 8.27 shows an optical disk readout system with an objective lens of numerical aperture NA operating at the wavelength of λ_0. The beam reflected from the disk is collected by the objective and sent to a photodetector; the read signal $S(t)$ is therefore proportional to the total optical power collected at the aperture of the objective lens. Consider a simple periodic pattern along a given track, having the following amplitude reflectivity function:

$$r(x) = r_0 + r_1 \sin(2\pi x/\Delta)$$

where x is the coordinate along the track and Δ is the period of the pattern. We assume that the track is sufficiently wide that a one-dimensional analysis suffices. The disk rotates at a constant angular velocity and, at the position of the track being considered, its linear velocity is V.

(a) Draw diagrams of the reflected intensity pattern at the exit pupil of the objective for the following four cases: $\Delta > \lambda_0/\text{NA}$, $\Delta = \lambda_0/\text{NA}$, $\lambda_0/(2\text{NA}) < \Delta < \lambda_0/\text{NA}$, and $\Delta = \lambda_0/(2\text{NA})$. These diagrams should show the relative positions and overlap of the 0 order, +1 order, and −1 order diffracted beams.

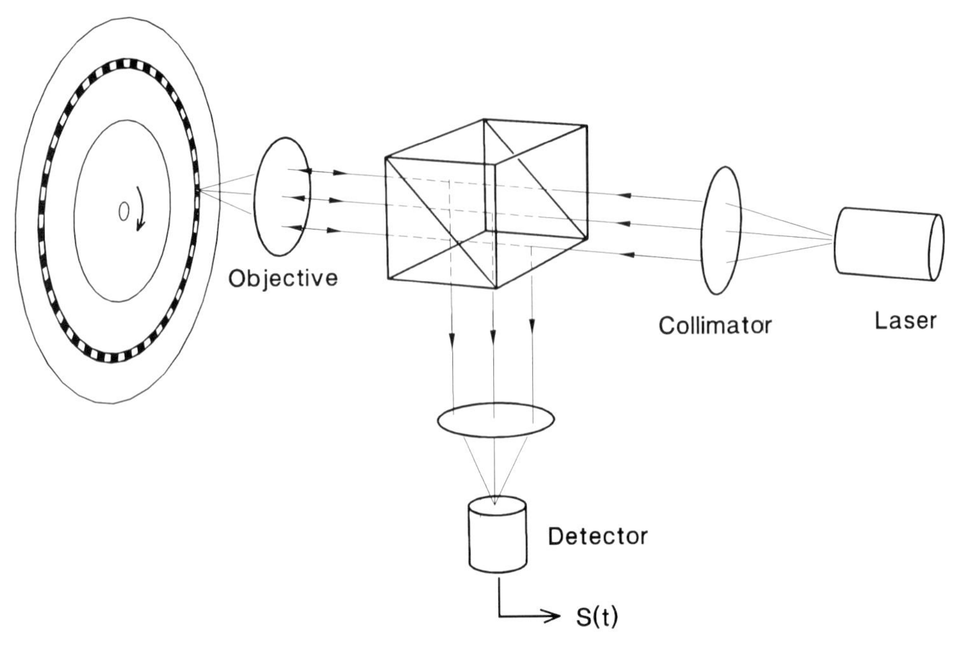

Figure 8.27.

(b) As the disk rotates, the detector signal $S(t)$ shows a periodic behavior. For the four cases considered in (a), what are the frequencies contained in $S(t)$?

(c) Now assume that the track is written with a periodic pattern of rectangular marks and spaces, each having a length of $\frac{1}{2}\Delta$. (You may assume that the mark and space reflection coefficients are $r_0 + r_1$ and $r_0 - r_1$ respectively, although the exact values of these reflectivities do not concern us here.) In terms of λ_0 and NA, what is the minimum mark length that can produce a signal at the detector?

(d) How does the read signal from the periodic mark patterns considered in (c) change if the period of the pattern is kept fixed, but the duty cycle is reduced (in other words, if the marks are made shorter than $\frac{1}{2}\Delta$ while the spaces are made longer, keeping the sum of their lengths fixed at Δ)?

(e) Why is it that a single mark can always produce a detectable signal at the detector but a periodic pattern of marks and spaces that has a period shorter than some critical value is impossible to detect?

9

Noise in Magneto-optical Readout

Introduction

The present chapter is devoted to the analysis of the various sources of noise commonly encountered in magneto-optical readout. Although the focus will be on magneto-optics, most of these sources are known to occur in other optical recording systems as well; the arguments of the present discussion, therefore, may be adapted and applied to various other media and systems.

The MO readout scheme considered in this chapter is briefly reviewed below, followed by a description of the standard methods of signal and noise measurement, and the predominant terminology in this area. After these preliminary considerations, the characteristics of thermal noise, which originates in the electronic circuitry of the readout, are described in section 9.1. Section 9.2 considers a fundamental type of noise, shot noise, and gives a detailed account of its statistical properties. Shot noise, which in semi-classical terms arises from the random fluctuations in photon arrival times, is an ever-present noise in optical detection. Since the performance of MO media and systems is approaching the limit imposed by shot noise, it is imperative to have a good grasp of this particular source of noise. In section 9.3 we describe a model for laser noise; here we present measurement results that yield numerical values for the strength of laser power fluctuations. Spatial variations of disk reflectivity and random depolarization phenomena also contribute to the overall level of noise in readout; these and related issues are treated in section 9.4. The final section is devoted to numerical simulation results describing some frequently encountered sources of noise that accompany the recorded waveform itself, namely, jitter and signal-amplitude-fluctuation noise.

The MO readout system. Throughout this chapter, frequent reference will be made to the readout system depicted in Fig. 9.1. The light source in this system is a semiconductor laser diode operating at a constant output power level, a mode of operation also known as continuous-wave or CW.†

† In practice, for noise reduction purposes, the laser's drive current is modulated at a frequency of several hundred MHz. This modulation frequency is well beyond the bandwidth of the detection system and is ignored by the detectors. Signal and noise analysis, therefore, may proceed by ignoring the modulation of the laser and treating its beam as CW.

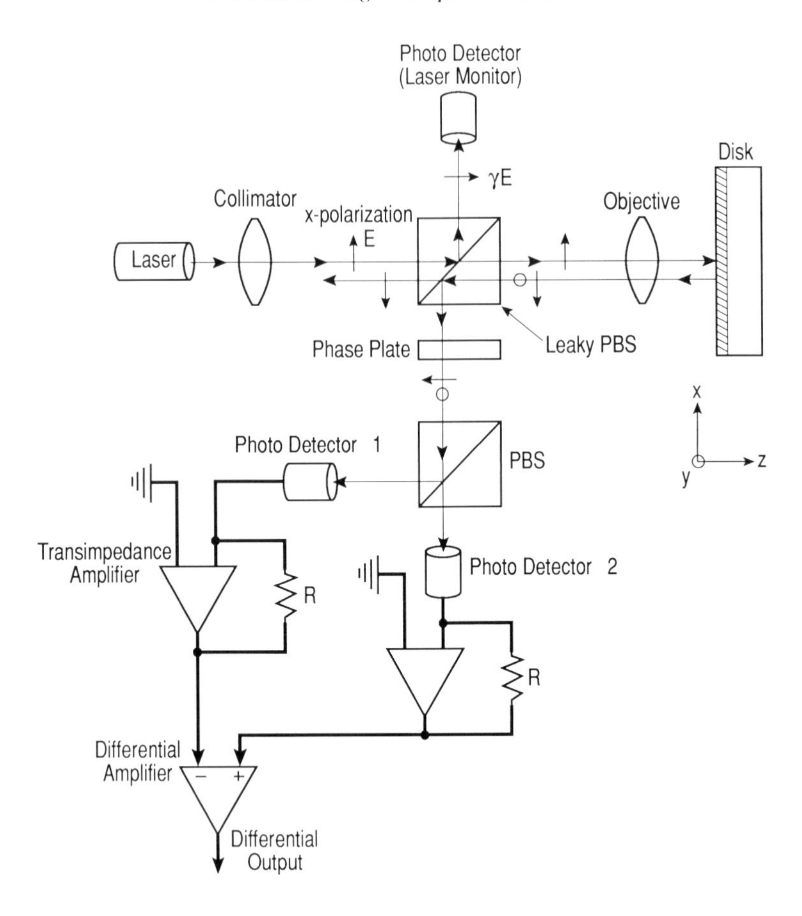

Figure 9.1. Schematic diagram of a magneto-optical readout system. In the forward path the leaky PBS provides a fraction γ^2 of the X-polarized light to the monitor of laser power. In the return path, 100% of the magneto-optically generated Y-polarization plus the fraction γ^2 of the returning X-polarization are deflected towards the differential detection module.

The light reflected from the MO disk consists of a polarization component parallel to the direction of incident polarization X, and a perpendicular component (parallel to Y); the latter carries information about the pattern of recorded data to the detection module.

For the purpose of monitoring laser power fluctuations, the leaky polarizing beam-splitter directs a small fraction γ^2 of the emitted power to a photodetector, the output of which is then fed back to the laser in an attempt to stabilize the level of output radiation. The light transmitted by the leaky PBS is linearly polarized along X but, upon reflection from the disk, it acquires a Y-component as well. In the return path, the leaky PBS provides the detection module with a fraction γ of the X-component plus the entire Y-component. The latter is the magneto-optically generated signal, carrying the desired information. The X-component, although barren of such information, is needed in the scheme of differential detection for the extraction of the data.

To eliminate the ellipticity of the beam, a phase compensator (either quarter-wave plate or adjustable Soleil–Babinet retarder) is employed. A regular PBS and detectors 1 and 2 comprise a balanced differential detection module. These detectors are low-noise photodiode-preamplifier integrated circuits whose outputs are fed to a differential amplifier for subtraction and further amplification.

Terminology of signal and noise analysis. We shall describe the units of signal and noise commonly used in practice. A constant (DC) voltage V or current I is specified in terms of the power P that it delivers to a fixed resistor R, that is, $P = V^2/R$ or $P = RI^2$. The standard value of R is 50 Ω and the power thus measured may be stated either in watts or in milliwatts. On the logarithmic scale (decibel or dB) the strength of the signal is defined as $10 \log_{10} P$; if P is in watts the logarithmic unit is called dBW, and if P is in milliwatts it is known as dBm. For example, the power of a 1 volt DC signal is 13 dBm, while that of a 1 amp DC current is 47 dBm. The same rules and definitions apply to single-frequency AC signals, provided of course that one remembers that the effective (r.m.s.) amplitude of a sinusoid is $1/\sqrt{2}$ times its maximum amplitude.

When the dimensionless ratio of two quantities such as P_1/P_2, V_1/V_2, or I_1/I_2 is expressed on the logarithmic scale, the units are specified as dB. It must be emphasized that the ratio of powers in dB is defined as $10 \log (P_1/P_2)$, whereas for currents and voltages the correct definitions are $20 \log (I_1/I_2)$ and $20 \log (V_1/V_2)$ respectively.

If a signal is amplified prior to the measurement, then the gain G must also be taken into account. Generally, when one uses a logarithmic scale, one must subtract the value of $20 \log_{10} G$ from the signal power in order to eliminate the gain factor. It is important here that the gain be dimensionless, that is, voltage be amplified to voltage and current to current. If it so happens that current is amplified to voltage (or vice versa) then the gain must be adjusted to take account of the relevant resistance R. For instance, if the initial noise current has a power of RI^2, after amplification (and conversion to voltage) its power will become G^2I^2/R. Therefore, the proper gain coefficient in this instance is not G but G/R (where R is typically 50 Ω).

Unlike the signal, noise has a broad spectrum and its strength (or power) depends on the bandwidth within which it is being measured. The spectral density of the noise is generally defined as the average noise power delivered to a 50 Ω resistor within a bandwidth of 1 Hz, and is denoted by $R\langle i_n{}^2\rangle$ or $\langle v_n{}^2\rangle/R$, depending on whether the noise is due to fluctuations of current or voltage. We shall denote the r.m.s. noise current by i_n, although formally the appropriate notation is $\sqrt{\langle i_n{}^2\rangle}$. Similarly the r.m.s. noise voltage will be denoted by v_n. It is customary in practice to measure the noise power not in a 1 Hz bandwidth, but in a 30 kHz bandwidth, and express the result in dBW or dBm. Thus, a noise current of 10 $\text{pA}/\sqrt{\text{Hz}}$

will have a power of

$$10 \log_{10}(50 \times 10^{-22} \times 30000) = -158.24 \text{ dBW} = -128.24 \text{ dBm}$$

Similarly, the measured power of a noise voltage with a spectral density of 1 $\text{nV}/\sqrt{\text{Hz}}$ will be

$$10 \log_{10}(10^{-18} \times 30000/50) = -152.22 \text{ dBW} = -122.22 \text{ dBm}$$

In the above example, if one assumes that an amplifier with a gain $G = 100$ V/A is used to convert current to voltage, then the normalized gain G/R will be equal to two. This means that the noise current must be 6 dB below the level of the noise voltage, as is indeed the case.

9.1. Noise in the Electronic Circuitry

The fluctuating conduction electrons within the resistors and transistors of an electronic circuit create a random voltage or current at the circuit's output terminal. When these fluctuations are caused by thermal motion of the electrons the resulting noise is referred to as Johnson noise. A resistor R at an equilibrium temperature T exhibits a noise whose spectrum is flat for practically all frequencies of interest. If the current noise of this resistor is measured within a bandwidth B, the resulting magnitude i_{th} will be given by

$$\langle i_{th}^2 \rangle = 4 k_B T B / R \tag{9.1}$$

In this equation the brackets $\langle \cdot \rangle$ indicate statistical averaging or averaging over time, k_B is the Boltzmann constant (1.38×10^{-23} J/K), T is the absolute temperature (in kelvins), B is the bandwidth (in hertz), and R is the resistance (in ohms). For the sake of simplicity we shall denote the root-mean-square value of the noise current by i_{th} instead of $\sqrt{\langle i_{th}^2 \rangle}$. Since the frequency spectrum of thermal noise is flat, one usually specifies its strength in a 1 Hz bandwidth. Thus the noise amplitude of a 1 kΩ resistor at room temperature is about 4 $\text{pA}/\sqrt{\text{Hz}}$. If the bandwidth of the measuring system happens to be 1 MHz, then the measured noise current will be 4 nA. The corresponding r.m.s. noise voltage is then given by $v_{th} = R i_{th} = 4\ \mu\text{V}$.

An amplifier, in addition to multiplying its input noise amplitude by a gain factor, introduces a noise of its own. This occurs because in transistors the random generation and recombination of electrons and holes, and the random arrival of these carriers at the collector, create current and voltage fluctuations.† The excess noise of an amplifier is characterized by the noise

†The terms shot noise and flicker noise (or $1/f$ noise) are often used in referring to these fluctuations. The $1/f$ noise is limited to the low-frequency range of the spectrum and, for all practical cases considered in this chapter, makes a negligible contribution to the overall noise level.

factor F, which is the ratio of the available output noise (including contributions by the amplifier) to that part of the latter arising from the input alone.[1] Although every stage of amplification contributes to the overall noise level, it is usually the case that the preamplifier (i.e., the first stage) makes the most contribution, simply because the noise from this stage is amplified more than that from the following stages. In any event, since the electronic noise is independent of the readout signal, there is no need to analyze its various components separately; instead, it will be treated in its entirety as one source of noise. By measuring the spectrum of noise at the differential output while the beam is being blocked (see Fig. 9.1), one obtains an accurate measure of the total electronic noise.

Another manifestation of thermal noise is the dark-current noise of photodetectors. Within the semiconductor crystal of a photodiode, some valence electrons are thermally excited into the conduction band, giving rise to a current similar to that generated by photo-induced electrons. Unlike Johnson noise, however, the average value of the dark current is nonzero. Fluctuations of the dark current around its average value produce dark-current noise. These fluctuations are very similar to shot noise, which will be described in section 9.2. For the purpose of analysis, dark current and its associated noise may be treated by adding an equivalent (constant) light power level to the incident power on the photodetector. We shall visit this topic again at the end of section 9.3.

9.2. Shot Noise in Photodetection

The direct detection of optical signals usually entails their conversion into electrical signals by a photodetector. Photodetection is a quantum process whereby incident photons cause the release of electrons, which then go on to produce an electric current. These photons do not arrive at well-defined instants of time; rather, their arrival times are random. The random arrival of photons at the detector creates fluctuations in the resulting photocurrent, known as shot noise.[2] In this section we study properties of the shot noise caused by coherent light. The basic property of coherent light is that the number of photons arriving during any time interval, say $[t, t+\tau]$, is independent of that arriving during any other (non-overlapping) interval. If the interval is sufficiently short, then there is either one photon arriving in that interval or none at all. Assuming that the incident beam is monochromatic with frequency ν, and that its power at time t is $P_0(t)$, the probability of a single photon arriving during the short interval Δt is

$$p\left\{n = 1, [t, t+\Delta t]\right\} = \frac{P_0(t)\,\Delta t}{h\nu} \tag{9.2}$$

where $h\nu$ is the energy of individual photons of frequency ν. Not every photon of course creates one photoelectron. The quantum efficiency η of a photodetector is defined as the probability of a free electron being generated by an incident photon. Thus the chance that one photoelectron is released during the short interval Δt is given by

$$p\left\{n = 1, [t, t + \Delta t]\right\} = (\eta/h\nu)\,P_0(t)\,\Delta t \tag{9.3}$$

Next we derive an expression for the probability that a number n of photoelectrons is produced in the interval $[0, t]$ by the incident laser power $P_0(t)$. The probability of having $n + 1$ electrons in $[0, t + \Delta t]$ is related to the course of events in $[0, t]$ by the fact that either there are n electrons in $[0, t]$, in which case the extra electron is created in $[t, t + \Delta t]$, or there are $n + 1$ electrons in $[0, t]$ and none in $[t, t + \Delta t]$. (Since Δt is very short, the chances of two or more photoelectrons being generated in $[t, t + \Delta t]$ are negligible.) From Eq. (9.3) we know the probability for the generation of one electron in a short interval; the probability of having no electrons at all in the same interval is one minus the probability of having one electron. Combining these facts, we arrive at the following equation:

$$p\left\{n + 1, [0, t + \Delta t]\right\}$$

$$= p\left\{n, [0, t]\right\} p\left\{1, [t, t + \Delta t]\right\} + p\left\{n + 1, [0, t]\right\} p\left\{0, [t, t + \Delta t]\right\}$$

$$= p\left\{n, [0, t]\right\}\left[(\eta/h\nu)\,P_0(t)\,\Delta t\right] + p\left\{n + 1, [0, t]\right\}\left[1 - (\eta/h\nu)\,P_0(t)\,\Delta t\right]$$

After some rearrangements, we obtain

$$\frac{\mathrm{d}}{\mathrm{d}t}\,p\left\{n + 1, [0, t]\right\} = (\eta/h\nu)\,P_0(t)\left\{p\left\{n, [0, t]\right\} - p\left\{n + 1, [0, t]\right\}\right\} \tag{9.4}$$

When there are no electrons in $[0, t]$, Eq. (9.4) simplifies to

$$\frac{\mathrm{d}}{\mathrm{d}t}\,p\left\{n = 0, [0, t]\right\} = -(\eta/h\nu)\,P_0(t)\;p\left\{n = 0, [0, t]\right\} \tag{9.5}$$

whose solution is readily obtained as follows:

$$p\left\{n = 0, [0, t]\right\} = A \exp\left[-(\eta/h\nu)\int_0^t P_0(t)\,dt\right]$$

The integration constant A must be equal to one, since at $t = 0$ the probability of having no electrons must be unity. Therefore

$$p\left\{n = 0, [0, t]\right\} = \exp(-\Lambda) \tag{9.6}$$

where

$$\Lambda = (\eta/h\nu)\int_0^t P_0(t)\,dt \tag{9.7}$$

We now substitute Eq. (9.6) in Eq. (9.4) and calculate $p\{n = 1, [0, t]\}$. Afterwards we calculate the probability for $n = 2$, then for $n = 3$, and so on. The general solution is found to be

$$p\left\{n, [0, t]\right\} = \exp(-\Lambda)\,\frac{\Lambda^n}{n!} \tag{9.8}$$

This probability distribution is known as the Poisson distribution. Several important properties of shot noise may be derived from Eq. (9.8). For instance, the average number of photoelectrons generated in $[0, t]$ is

$$\langle n\rangle = \sum_{n=0}^{\infty} n\,p\left\{n, [0, t]\right\} = \Lambda \exp(-\Lambda)\sum_{n=0}^{\infty}\frac{\Lambda^n}{n!} = \Lambda$$

That is,

$$\boxed{\langle n\rangle = (\eta/h\nu)\int_0^t P_0(t)\,dt} \tag{9.9}$$

Thus the average number of electrons released in $[0, t]$ is proportional to the total optical energy collected in the same time interval. Similarly, the variance of the number of electrons is

$$\sigma_n^{\,2} = \langle n^2 \rangle - \langle n \rangle^2 = \sum_{n=0}^{\infty} n^2 p \left\{ n, [0, t] \right\} - \langle n \rangle^2$$

$$= \sum_{n=0}^{\infty} n^2 \frac{\Lambda^n}{n!} \exp(-\Lambda) - \Lambda^2 = \exp(-\Lambda) \sum_{n=0}^{\infty} \frac{n\,\Lambda^n}{(n-1)!} - \Lambda^2$$

$$= \exp(-\Lambda) \left\{ \Lambda \sum_{n=0}^{\infty} \frac{\Lambda^n}{n!} + \Lambda^2 \sum_{n=0}^{\infty} \frac{\Lambda^n}{n!} \right\} - \Lambda^2 = \Lambda$$

The standard deviation σ_n of the number of photoelectrons released in $[0, t]$ is thus given by

$$\boxed{\sigma_n = \sqrt{(\eta/h\nu) \int_0^t P_0(t)\, \mathrm{d}t}} \tag{9.10}$$

Equations (9.9) and (9.10) are very important in the analysis of shot noise insofar as they relate the incident laser power $P_0(t)$ to the strength of the photocurrent signal and the inherent shot noise that accompanies it.

The sensitivity η_s of a photodiode is defined as the average current (in amperes) produced for 1 watt of incident power. From Eq. (9.9),

$$\eta_s = \frac{\eta e}{h\nu} \tag{9.11}$$

where $e = 1.6 \times 10^{-19}$ C is the electronic charge, $h = 6.62 \times 10^{-34}$ J s is Planck's constant, and ν is the frequency of the incident light. For the typical values $\eta = 0.85$, $\lambda = 800$ nm, the photodiode sensitivity is found from Eq. (9.11) to be $\eta_s = 0.55$ A/W. Now, if the signal S and the noise N are defined in terms of the integrated current (instead of the integrated number of electrons), and if the detector sensitivity η_s is used in place of the quantum efficiency η, we find

$$S = e\,\langle n \rangle = \eta_s \int_0^t P_0(t)\, \mathrm{d}t \tag{9.12}$$

$$N = e\sigma_n = \sqrt{e\eta_s \int_0^t P_0(t)\,\mathrm{d}t} \tag{9.13}$$

Example 1. Let a 1 mW, 100 ns pulse of laser (λ = 800 nm) be incident on a photodiode having quantum efficiency η = 0.85. If the pulse is uniformly integrated, the ratio of signal to noise (averaged over many repetitions of the experiment) will be

$$\frac{S}{N} = \frac{\langle n\rangle}{\sigma_n} = \sqrt{\frac{\eta}{h\nu}\int_0^t P_0(t)\,\mathrm{d}t} = 1.85 \times 10^4 = 85.34 \text{ dB.}$$

This is an excellent signal-to-noise ratio for most practical applications. □

9.2.1. Spectral analysis of shot noise

Another characteristic of shot noise is its power spectrum, which we are about to describe. Let each photoelectron generated within the detector give rise to an output electric current or voltage waveform $h(t)$.† The collective output $\psi(t)$ is thus written

$$\psi(t) = \sum_i eh(t - t_i) \tag{9.14}$$

where e is the electronic charge and t_i is the instant of time at which the ith electron has been released. Let us confine our attention temporarily to that part of $\psi(t)$ belonging to the finite time interval $[-T, T]$; later, T will be allowed to approach infinity in order to eliminate the consequences of this truncation. The Fourier transform of the truncated $\psi(t)$ is given by

†In general, $h(t)$ describes the response of a cascade of linear elements from the detector to the amplifier to the filter to the signal processing circuitry.

$$\Psi_T(f) = \int_{-T}^{T} \psi(t) \exp(-i2\pi ft)\, dt$$

$$= e \sum_i \exp(-i2\pi f t_i) \int_{-\infty}^{\infty} h(t) \exp(-i2\pi ft)\, dt$$

$$= e H(f) \sum_i \exp(-i2\pi f t_i) \tag{9.15}$$

$H(f)$ is the Fourier transform of $h(t)$ and is known as the detection system's transfer function. The power spectral density of $\psi(t)$ is defined as

$$\mathcal{S}_\psi(f) = \lim_{T\to\infty} \langle \frac{1}{2T} |\Psi_T(f)|^2 \rangle \tag{9.16}$$

where the angle brackets signify statistical averaging over all possible output functions $\psi(t)$. In Eq. (9.16) normalization by $2T$ ensures that the integral of the spectral density function $\mathcal{S}_\psi(f)$ over the entire range of frequencies remains finite. Now, substituting for $\Psi_T(f)$ from Eq. (9.15) yields

$$\mathcal{S}_\psi(f) = e^2 |H(f)|^2 \lim_{T\to\infty} \langle \frac{1}{2T} \sum_i \sum_j \exp[-i2\pi f(t_i - t_j)] \rangle \tag{9.17}$$

Upon separating the terms with $i = j$ from those with $i \neq j$, Eq. (9.17) may be rewritten as follows:

$$\mathcal{S}_\psi(f) = e^2 |H(f)|^2 \lim_{T\to\infty} \left\{ \langle \frac{1}{2T} \sum_i 1 \rangle + \langle \frac{1}{2T} \sum_{\substack{i \\ i \neq j}} \sum_j \exp[-i2\pi f(t_i - t_j)] \rangle \right\} \tag{9.18}$$

The first bracketed term in this equation is the average number of photoelectrons released in $[-T, T]$. As for the second term, it may be decomposed and expressed as the product of terms containing i and j only, since different electrons are released independently. Thus

$$\mathcal{S}_{\psi}(f) = e^2 \left|H(f)\right|^2 \lim_{T\to\infty} \left\{ \frac{1}{2T} \langle n \rangle + \frac{1}{2T} \langle n(n-1) \rangle \left| \int_{-T}^{T} P_0(t) \exp(-i2\pi ft)\, dt \Big/ \int_{-T}^{T} P_0(t)\, dt \right|^2 \right\} \tag{9.19}$$

Substituting for $\langle n \rangle$ and $\langle n^2 \rangle$ from Eqs. (9.9) and (9.10), we obtain

$$\mathcal{S}_{\psi}(f) = \left|H(f)\right|^2 \lim_{T\to\infty} \left\{ \frac{e\eta_s}{2T} \int_{-T}^{T} P_0(t)\, dt + \frac{\eta_s^2}{2T} \left| \int_{-T}^{T} P_0(t) \exp(-i2\pi ft)\, dt \right|^2 \right\} \tag{9.20}$$

Equation (9.20) is the main result of this section and contains several important pieces of information. First, it shows that the transfer function of the system, $|H(f)|^2$, simply multiplies the total spectrum. Second, the shot-noise spectral density (the first term on the right-hand side) is independent of frequency but proportional to the average incident power. Third, the spectrum of the optical signal $P_0(t)$, apart from addition of the shot noise and multiplication by $|H(f)|^2$, is fully reproduced within the spectrum of the signal $\psi(t)$.

The power spectral density of the shot noise is flat (i.e., frequency-independent) and is given by

$$N_{\text{shot}}(f) = e\eta_s \langle P_0(t) \rangle \tag{9.21}$$

The brackets in Eq. (9.21) indicate time averaging (as opposed to ensemble averaging). Assuming that the overall gain is unity, the system's bandwidth B may be defined as follows:

$$B = \tfrac{1}{2} \int_{-\infty}^{\infty} \left|H(f)\right|^2 df \tag{9.22}$$

The factor of one-half in this equation arises because $H(f)$ has both positive- and negative-frequency components. The total (integrated) shot-noise power within a given bandwidth then becomes

$$\langle i_{\text{shot}}^2 \rangle = \int_{-B}^{B} N_{\text{shot}}(f)\, \mathrm{d}f = 2e\eta_s \langle P_0(t) \rangle B \tag{9.23}$$

The total signal power is the integral of its spectral density over the entire frequency range. Assuming that the system transfer function $H(f)$ does not enhance, attenuate, or otherwise modify the spectrum of $P_0(t)$, we write

$$\text{signal power} = \eta_s^2 \lim_{T\to\infty} \int_{-\infty}^{\infty} \frac{1}{2T} \left| \int_{-T}^{T} P_0(t) \exp(-\mathrm{i}2\pi f t)\, \mathrm{d}t \right|^2 \mathrm{d}f$$

$$= \eta_s^2 \lim_{T\to\infty} \frac{1}{2T} \int_{-T}^{T} P_0^2(t)\, \mathrm{d}t = \eta_s^2 \langle P_0^2(t) \rangle \tag{9.24}$$

In writing the second equality above we have invoked Parseval's theorem,[2] which states that the integral of the squared modulus of a given function is equal to the integral of the squared modulus of the Fourier transform of that function. Using Eqs. (9.23) and (9.24), one can now write the signal-to-shot-noise ratio (SNR) as follows:

$$\boxed{\text{SNR} = 10 \log_{10} \left(\frac{\eta_s^2 \langle P_0^2(t) \rangle}{2e\eta_s \langle P_0(t) \rangle B} \right)} \tag{9.25}$$

Example 2. The noise equivalent power (NEP) for a photodetector is usually defined as the incident light power that produces a signal amplitude equal to the noise amplitude. For shot noise, the NEP may be calculated from Eq. (9.25) as $2eB/\eta_s$. Thus for the typical value $\eta_s = 0.5$ A/W, the NEP for a shot-noise-limited detector is 6.4×10^{-19} W/Hz. □

9.2.2. Dark-current noise

As mentioned earlier, the dark-current noise of photodiodes is closely related to the photon shot noise. Their similarity arises from the fact that dark-current electrons, which are thermally generated, are produced randomly and independently of each other. If the photodiode is kept at a constant temperature, then the rate of generation of the dark-current electrons is constant. Denoting the average dark current by I_d, the corresponding noise spectral density (in A/$\sqrt{\text{Hz}}$) will be

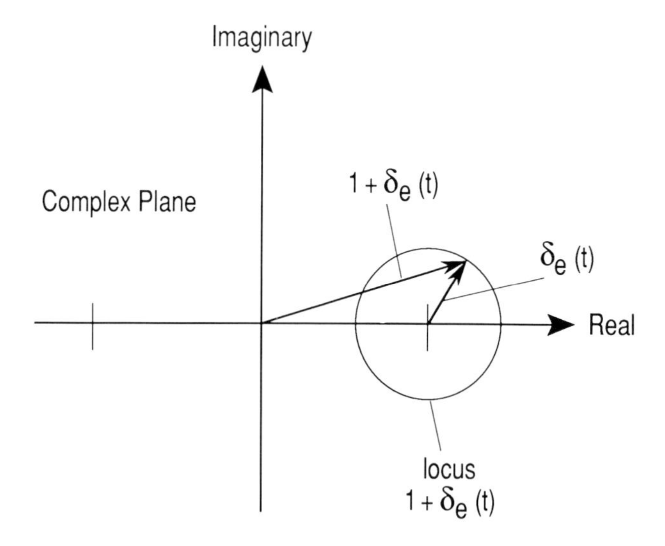

Figure 9.2. Complex-plane diagram showing the locus of $1+\delta_e(t)$, where $\delta_e(t)$ is a small complex quantity whose phase can assume arbitrary values in the interval $[0, 2\pi]$. The radius of the circle, which forms the upper bound on the magnitude of $\delta_e(t)$, is exaggerated.

$$i_{\text{dark}} = \sqrt{2eI_d} \tag{9.26}$$

In practice, the noises of the entire electronic circuitry (including dark-current noise of the photodiodes) are lumped together and treated as a single component of the total system noise.

9.3. Laser Noise

The laser light amplitude E is not constant, but varies randomly with time. Its fluctuations are rooted in the instabilities of the laser cavity; such instabilities result in mode-hopping and mode-competition.[3-5] One can write the following expression for the beam amplitude:

$$E(t) = E_0\left[1 + \delta_e(t)\right]\exp(-i\omega t) \tag{9.27}$$

where $\delta_e(t)$ is a dimensionless, complex quantity with $|\delta_e(t)| << 1$ for all times t. Figure 9.2 shows $1 + \delta_e(t)$ as the sum of two vectors in the complex plane. Assuming that the complex phase of $\delta_e(t)$ can assume all values between 0 and 2π, it is observed that $1 + \delta_e(t)$ is confined to a small disk centered at $(1,0)$ in the complex plane.

The laser power is proportional to the electric field intensity and is therefore given by

$$P(t) \propto |E_0|^2 \left[1 + \delta_e(t) + \delta_e^*(t) + |\delta_e(t)|^2 \right] \simeq P_0 \left\{ 1 + 2 \operatorname{Re}\left[\delta_e(t) \right] \right\} \qquad (9.28)$$

The r.m.s. fluctuations of the laser power are thus equal to $2P_0\Delta_e$ where

$$\Delta_e = \sqrt{\langle \operatorname{Re}^2 [\delta_e(t)] \rangle} \qquad (9.29)$$

Unlike the spectra of shot noise and Johnson noise, the spectrum of $\delta_e(t)$ is not necessarily flat. In general, the frequency response of any measurement system has a finite range and a nonuniform magnitude, thus requiring that the spectral content of $\delta_e(t)$ be properly trimmed and adjusted before using it in Eq. (9.29) to calculate the r.m.s. noise value.

When one measures the spectrum of the laser power fluctuations using a photodetector (i.e., intensity detection), one obtains a trace on the spectrum analyzer that, in addition to the desired spectrum, contains contributions from the shot noise and the thermal noise of the detection circuitry. The thermal noise, however, may be measured independently by observing the detector output when the beam is blocked, and the shot noise may be estimated from Eq. (9.23) using the average incident light power P_0. It may also happen that the fluctuations of $\delta_e(t)$ have a limited frequency content, in which case the noise outside the range of frequencies of $\delta_e(t)$ is identified as the sum of the shot noise and the electronic noise. Since both the shot noise and the thermal noise have flat spectra, one can then proceed to subtract their contributions from the total spectrum and obtain the spectrum of the laser power fluctuations.†

Example 3. Figure 9.3 shows the measured noise spectra at the output of the laser monitor in a typical experiment (see Fig. 9.1). The lower trace corresponds to the electronic noise, which is monitored in the absence of the light beam, and the upper trace shows the total noise with an incident light power of $P_0 = 380\ \mu$W. The p-i-n photodiode-amplifier used for these measurements had a sensitivity of $\eta_s = 0.42$ A/W at the laser wavelength of 680 nm, and a gain (i.e., current-to-voltage conversion factor) of $G = 10^4$ V/A. The electronic noise level is read from Fig. 9.3 as -84dBm. This is the power delivered to a 50 Ω resistor in a bandwidth of 30 kHz; thus the r.m.s. thermal-noise voltage at the amplifier output is $v_{th} = 80\ \text{nV}/\sqrt{\text{Hz}}$, and the corresponding noise current at the output of the

†Since the shot-noise spectral level is proportional to P_0, whereas that of the laser power fluctuations is proportional to P_0^2, it is tempting to suggest that a plot of the total noise power versus P_0 should behave as $a + bP_0 + cP_0^2$, from which one can sort out the individual contributions. The problem with this approach, however, is that $\delta_e(t)$ itself may depend in an unknown manner on P_0, causing the argument to fail.

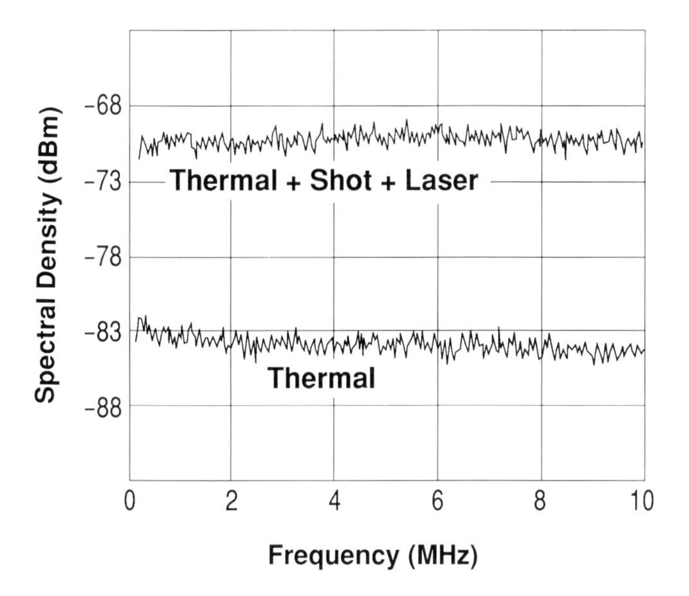

Figure 9.3. Spectrum of the combined electronic + shot + laser noise, as well as that of the electronic noise alone, at the output of a laser power monitor.[6] The operating wavelength is 680 nm and the incident light power is 380 μW.

photodiode prior to amplification is $i_{th} = 8$ pA/$\sqrt{\text{Hz}}$. The shot-noise current density is calculated from Eq. (9.23) as follows:

$$i_{sh} = \sqrt{2e\eta_s P_0} = \sqrt{2 \times 1.6 \times 10^{-19} \times 0.42 \times 380 \times 10^{-6}} = 7.2 \ \ \text{pA}/\sqrt{\text{Hz}}$$

Since the total noise power in Fig. 9.3 is about 14 dB (i.e., a factor of 25) above the thermal noise level, we have

$$\frac{i_{th}^2 + i_{sh}^2 + i_{laser}^2}{i_{th}^2} = 25$$

from which we find $i_{laser} = 38.5$ pA/$\sqrt{\text{Hz}}$. This is the r.m.s. current fluctuation caused by laser noise at the photodiode output (prior to amplification). The laser noise may be normalized by the average laser power and stated as a relative-intensity noise (RIN), that is,

$$\text{RIN} = 20 \log \left(\frac{i_{laser}}{\eta_s P_0} \right) = 20 \log \frac{38.5 \times 10^{-12}}{0.42 \times 380 \times 10^{-6}} \simeq -132.5 \text{ dB}$$

Since the laser-noise spectrum in Fig. 9.3 is relatively flat in the frequency

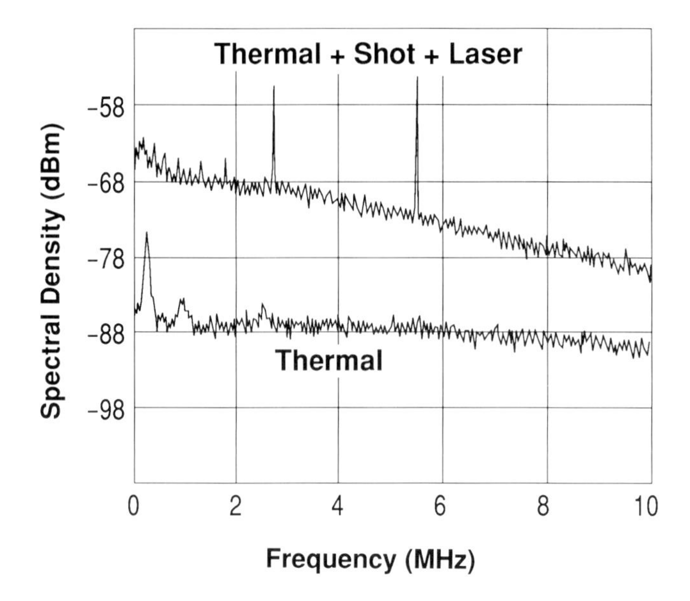

Figure 9.4. Spectrum of the combined electronic + shot + laser noise, as well as that of the electronic noise alone, at the output of a laser power monitor.[7] The operating wavelength is 830 nm and the incident light power is 39 μW. Since the spikes on the upper trace were generated by the electronic circuitry within the laser power supply, they were ignored in the noise analysis.

range from $f = 0$ to $f = 10$ MHz, one can readily integrate the RIN over the 10 MHz bandwidth of the system and obtain a total RIN of -62.5 dB. Comparison with Eqs. (9.28) and (9.29) then shows that the r.m.s. value of $\text{Re}[\delta_e(t)]$, confined to a 10 MHz bandwidth, is $\Delta_e \simeq 4 \times 10^{-4}$. □

Example 4. The spectra shown in Fig. 9.4 correspond to the output of a laser monitor, which is a low-noise, hybrid p-i-n photodiode-amplifier with a sensitivity of $\eta_s = 0.5$ A/W at $\lambda = 830$ nm.[7] The detector's current-to-voltage conversion factor $G = 3.64 \times 10^4$ V/A. The lower trace shows the spectrum of the detector output in the absence of light. The -88 dBm electronic noise level corresponds to $v_{\text{th}} = 51$ nV/$\sqrt{\text{Hz}}$ at the amplifier output and to $i_{\text{th}} = 1.4$ pA/$\sqrt{\text{Hz}}$ at its input. The upper trace was obtained when 39 μW of laser light was incident on the detector. From Eq. (9.23) the shot-noise current density (prior to amplification) is found to be $i_{\text{sh}} =$ 2.5 pA/$\sqrt{\text{Hz}}$, thus

$$\frac{i_{\text{th}}^2 + i_{\text{sh}}^2}{i_{\text{th}}^2} \simeq 4.2 = 6.2 \text{ dB}$$

Therefore, the thermal-plus-shot noise level must be about 6.2 dB above the thermal noise level in Fig. 9.4. The remaining noise is due to fluctuations of the laser power, which may now be separated out. Let us approximate the frequency-dependence of the total noise density in Fig. 9.4 with a linear function as follows:

$$10 \log \left(\frac{i_{th}^2 + i_{sh}^2 + i_{laser}^2}{i_{th}^2} \right) \simeq 23 - 1.3f \qquad (f \text{ in MHz})$$

This leads to the following frequency-dependence for the laser noise:

$$\left(\frac{i_{laser}}{i_{th}} \right)^2 \simeq -4.2 + 200 \exp(-0.3f) \qquad (f \text{ in MHz})$$

The above function is now averaged over the frequency range from zero to 10 MHz, yielding an average value of 59 for $(i_{laser}/i_{th})^2$. Thus the average noise current arising from laser power fluctuations (within a 10 MHz bandwidth) is $i_{laser} \simeq 10.75\ \text{pA}/\sqrt{\text{Hz}}$. Normalization by $\eta_s P_0$ yields an average RIN of -125.2 dB, and the r.m.s. value of the **E**-field fluctuations over the 10 MHz bandwidth of interest is $\Delta_e \simeq 8.7 \times 10^{-4}$. □

9.4. Noise due to Disk Reflectivity Fluctuations and Depolarization

In the following analysis it will be assumed that the disk spins at a constant velocity, and that the noise is measured on an erased track, that is, one with no reverse-magnetized domains. The effective light power incident on the disk surface is denoted by P_0, and the losses due to reflections from or transmissions through the various optical elements in the system are ignored. (If need be, however, such losses can easily be incorporated into the results later.) The three sources of noise considered here are: (i) amplitude and phase variations of the X-polarized reflected light; (ii) random depolarization of the incident X-polarized beam; and (iii) amplitude and phase fluctuations of the magneto-optically generated Y-polarization.[8–11] The physical mechanisms responsible for these various noises are described below.

For various reasons, there will be fluctuations in the effective reflectivity r_x of the X-component of polarization. First, the material composition and structure vary across the disk. Second, if there are grooves on the disk, their edge roughness is likely to scatter the light in a random fashion. Third, there are residual amounts of off-track and defocus errors that vary with time and, therefore, cause the reflectivity to fluctuate. Some of these noise sources are fixed on the disk and, consequently, their time-dependences scale with the disk velocity; others, such as those due to mechanical vibrations, defocus, and tracking error, are caused by fluctuations elsewhere in the system and do not scale with the velocity. In any event, assuming that these fluctuations are small, we may write

$$r_x(t) = r_{x0}\left[1 + \delta_x(t)\right]; \qquad |\delta_x| << 1 \tag{9.30}$$

where $\delta_x(t)$ is a small, dimensionless, complex coefficient, representing the fractional variations of r_x.

There exists the possibility of conversion of some of the X-polarized incident light into Y-polarized reflected light due to depolarization, as distinct from MO conversion. This phenomenon is partly due to substrate birefringence, but is also caused by scattering at the disk surface. Perhaps the best way to characterize the depolarization contribution to noise is by defining a disk reflection coefficient r_{xy} as follows:

$$r_{xy}(t) = r_{x0}\,\delta_{xy}(t); \qquad |\delta_{xy}| << 1 \tag{9.31}$$

where, as before, $\delta_{xy}(t)$ is a small, dimensionless, complex coefficient.

Finally, there is the MO contribution to the Y-polarized reflected light. For a completely erased track, one may write the Y-component of reflectivity as follows:

$$r_y(t) = r_{y0}\left[1 + \delta_y(t)\right]; \qquad |\delta_y| << 1 \tag{9.32}$$

Some of these fluctuations in r_y originate from the structural/magnetic variations of the disk along the track, but there is also a correlation between the fluctuations of r_y and r_x. For instance, if more X-polarized light is reflected, then less light will be available to interact with the magnetization of the material and produce the Y-component. For the sake of simplicity, however, we shall ignore all possible correlations among δ_x, δ_{xy}, and δ_y, and treat these parameters as independent random variables.

The phase plate in the system of Fig. 9.1 eliminates the nominal ellipticity of the beam by bringing r_{x0} and r_{y0} in phase with each other. Consequently, we assume at the outset that r_{y0}/r_{x0} is real and, to emphasize its realness, we shall write it as $|r_{y0}/r_{x0}|$. The compensator will have no influence whatsoever on the noise coefficients δ_x and δ_y, and will modify δ_{xy} only by the addition of an irrelevant constant phase. The power incident on individual detectors is thus written

$$P_1, P_2 \propto \left| \frac{E_0}{\sqrt{2}}(1 + \delta_e)\left[\gamma r_{x0}(1 + \delta_x) \pm r_{x0}\delta_{xy} \pm r_{y0}(1 + \delta_y)\right] \right|^2$$

where the plus (minus) sign applies to the incident power P_1 (P_2) on detector 1 (2). Expanding the above expression and ignoring the second-order terms in the noise variables, we find the photocurrents S_1 and S_2 of the detectors as follows:

$$S_1, S_2 \simeq \frac{1}{2}\eta_s P_0 \left[\gamma\,|r_{x0}| \pm |r_{y0}|\right]^2$$

$$+ \eta_s P_0 |r_{x0}|^2 \left(\gamma \pm \left|\frac{r_{y0}}{r_{x0}}\right|\right) \mathrm{Re}\left(\gamma(\delta_e + \delta_x) \pm \delta_{xy} \pm \left|\frac{r_{y0}}{r_{x0}}\right| (\delta_e + \delta_y)\right) \tag{9.33}$$

The sum of the two detector signals is thus given by

$$S_1 + S_2 \simeq \eta_s P_0 \gamma^2 \,|r_{x0}|^2 \left(1 + |r_{y0}/\gamma r_{x0}|^2\right)$$

$$+ \left\{2\eta_s P_0 \,|r_{x0}|^2 \left[\gamma^2\,\mathrm{Re}(\delta_e + \delta_x) + |r_{y0}/r_{x0}|\,\mathrm{Re}(\delta_{xy})\right.\right.$$

$$\left.\left. + |r_{y0}/r_{x0}|^2\,\mathrm{Re}(\delta_e + \delta_y)\right]\right\} \tag{9.34}$$

The first term on the right-hand side of Eq. (9.34) is the average photocurrent of the two detectors. The second term represents the fluctuations of this signal, and consists of three contributions, as follows. The contribution due to the laser noise δ_e and the noise in the X-component of reflectivity δ_x is proportional to γ^2. The second contribution comes from the depolarization noise δ_{xy} and is proportional to $|r_{y0}/r_{x0}|$. Note that individual channels have a much larger noise component arising from depolarization, but in adding the two channels, much of this noise cancels out. The third contribution is a combination of the laser noise δ_e and the noise in the MO component of the reflected light, δ_y; it is proportional to $|r_{y0}/r_{x0}|^2$. In practice, $|r_{y0}/r_{x0}|$ is usually much smaller than γ^2 and, therefore, the second and third contributions to noise may be negligible. Under these circumstances the total fluctuations of the sum signal arise from electronic noise, shot noise, laser noise, and the r_x noise. Since all but the last of these fluctuations can be measured when the disk is stationary, the excess noise observed upon spinning the disk is attributed to the r_x contribution.

The output of the differential detection system is proportional to the difference of the two detector signals, namely,

$$S_1 - S_2 \simeq 2\eta_s P_0 \gamma\,|r_{x0}|\,|r_{y0}| \left[1 + \mathrm{Re}(2\delta_e + \delta_x + \delta_y) + |r_{x0}/r_{y0}|\,\mathrm{Re}(\delta_{xy})\right] \tag{9.35}$$

The first term on the right-hand side of this equation is the MO signal, observed in the erased state of the disk. The second term is the combined contribution of the laser noise δ_e and the reflectivity noises δ_x and δ_y; depolarization noise δ_{xy} constitutes the third term. Note that the depolarization term has a larger coefficient than the other noise terms, and, therefore, it may be the dominant noise in MO readout.

One observes that in going from the sum signal of Eq. (9.34) to the difference signal of Eq. (9.35), the contribution of $(\delta_e + \delta_x)$ is attenuated by a factor of $|r_{y0}/\gamma r_{x0}|$. The reason is that the laser noise and the r_x noise are identical in the two channels and, for a balanced system, they cancel out. The residual noise observed in this instance is due to interference with the MO component, r_{y0}. The same thing, of course, happens to the r_y noise, but there interference with r_x creates a term larger than the original r_y noise; that is why the coefficient of $(\delta_e + \delta_y)$ in the difference signal is larger than that in the sum signal by a factor of $|\gamma r_{x0}/r_{y0}|$. Similar considerations apply to the depolarization noise δ_{xy}.

The signals S_1 and S_2 of the individual detectors may themselves be written in terms of the sum and difference signals in the following way:

$$S_1, S_2 = \tfrac{1}{2}(S_1 + S_2) \pm \tfrac{1}{2}(S_1 - S_2) \tag{9.36}$$

It thus becomes clear that the sum signal is the common-mode signal, shared by the two detectors and rejected by the differential amplifier. Practical differential amplifiers, unfortunately, have an only finite common-mode rejection-ratio (CMRR), and the noise accompanying the sum signal in Eq. (9.34) also appears at the output, albeit after a substantial attenuation. In the remainder of this section we shall assume that the differential amplifier has a large CMRR, and thus proceed to ignore the common-mode noise.

The r.m.s. values of the various fluctuating parameters (within the bandwidth of the system) are now defined as follows:

$$\Delta_x = \sqrt{\langle \mathrm{Re}^2\,[\delta_x(t)]\,\rangle} \tag{9.37a}$$

$$\Delta_y = \sqrt{\langle \mathrm{Re}^2\,[\delta_y(t)]\,\rangle} \tag{9.37b}$$

$$\Delta_{xy} = \sqrt{\langle \mathrm{Re}^2\,[\delta_{xy}(t)]\,\rangle} \tag{9.37c}$$

The total signal and the r.m.s. noise (including thermal and shot noises but excluding the residual common-mode noise) at the differential output are thus written as

$$\boxed{i_{\text{signal}} \simeq 2\eta_s P_0 \gamma\,|r_{x0}|\,|r_{y0}|} \tag{9.38}$$

$$(i_{\text{noise}})^2 \simeq \left\{2i_{\text{th}}^2 + 2e\eta_s P_0 \gamma^2 |r_{x0}|^2 \left[1 + |r_{y0}/\gamma r_{x0}|^2\right]\right\} B + \left(2\eta_s P_0 \gamma |r_{x0}|\, |r_{y0}|\right)^2 \left(4\Delta_e^2 + \Delta_x^2 + \Delta_y^2 + |r_{x0}/r_{y0}|^2 \Delta_{xy}^2\right) \tag{9.39}$$

From a practical design standpoint, there is an advantage in choosing γ as small as possible, so that the maximum amount of light can get through to the disk during the writing process. However, Eqs. (9.38) and (9.39) indicate that γ cannot be made too small either, since in that eventuality the thermal noise becomes dominant and the overall signal-to-noise ratio (SNR) suffers. γ must therefore be large enough to bring the total noise level well above the level of the thermal noise. In the ideal case, when γ is sufficiently large to make the thermal noise negligible, the SNR becomes independent of γ and can be expressed as follows:

$$\text{SNR}_{\text{ideal}} \simeq \frac{2\eta_s P_0 |r_{y0}|^2}{eB + 2\eta_s P_0 |r_{y0}|^2 \left(4\Delta_e^2 + \Delta_x^2 + \Delta_y^2 + \left|\dfrac{r_{x0}}{r_{y0}}\right|^2 \Delta_{xy}^2\right)} \tag{9.40}$$

Of course, $\text{SNR}_{\text{ideal}}$ is never achieved in practice, but can be approached closely. From Eq. (9.39) the critical value γ_c of γ, at which the strength of thermal-noise (plus a small amount of shot noise whose strength is independent of γ) becomes equal to the combined strength of all other noises, is given by

$$\gamma_c^2 = \frac{\dfrac{i_{\text{th}}^2}{e\eta_s P_0 |r_{x0}|^2} + \left|\dfrac{r_{y0}}{r_{x0}}\right|^2}{1 + 2(\eta_s/eB) P_0 |r_{y0}|^2 \left(4\Delta_e^2 + \Delta_x^2 + \Delta_y^2 + \left|\dfrac{r_{x0}}{r_{y0}}\right|^2 \Delta_{xy}^2\right)} \tag{9.41}$$

When $\gamma = \gamma_c$, the SNR is only 3 dB below its ideal value; when $\gamma = 2\gamma_c$, the gap shrinks to 1 dB, and with $\gamma = 3\gamma_c$, the ideal SNR is only 0.5 dB away. Further increases of γ beyond these values are hardly justified; in fact, for large values of γ, common-mode noise begins to appear in the differential output, causing the SNR to deteriorate.

According to Eq. (9.40), beyond the shot-noise limit one is in a regime where the dominant noise is proportional to the signal; therefore, increasing the laser power P_0 does not enhance the SNR. In this regime, if the depolarization noise δ_{xy} happens to be an important contributor, then increasing r_{y0} will diminish the noise. On the other hand, if the laser noise δ_e and/or the reflectivity noises δ_x and δ_y are dominant, increases in r_{y0} will have no effect on the SNR. These are some of the issues to contemplate when optimizing the readout system.

Example 5. Consider the readout system of Fig. 9.1, operating under the following conditions: laser wavelength $\lambda = 680$ nm, effective laser power incident on the disk $P_0 = 2$ mW, disk reflectivity $r_x = 0.5$, disk MO reflection coefficient $r_y = 0.005$, leakage parameter of the beam-splitter $\gamma = 0.4$, photodetector sensitivity $\eta_s = 0.42$ A/W, preamplifier conversion factor $G = 3.3 \times 10^4$ V/A, and differential amplifier gain = 13. The laser noise for this system was analyzed in Example 3 (section 9.3), and the r.m.s. fluctuations of the **E**-field were found to be $\Delta_e \simeq 4 \times 10^{-4}$.

Figure 9.5(a) shows the spectra of the differential amplifier's output voltage, measured with the light blocked from both detectors (lower trace), and with the light reaching detectors while the disk is stationary (upper trace). The electronic noise level at -65 dBm is the sum of contributions from the two detectors and the differential amplifier. Referred to each photodiode current (prior to amplification), this thermal noise has a density of $i_{th} \simeq 1.2$ pA/$\sqrt{\text{Hz}}$. The shot-noise current density for the 40 μW of incident power on each photodiode is calculated as $i_{sh} \simeq 2.3$ pA/$\sqrt{\text{Hz}}$. This means that the thermal-plus-shot noise level is about 6.7 dB above the thermal noise, as is indeed the case in Fig. 9.5(a). Apparently, laser noise does not contribute much to the upper trace in Fig. 9.5(a), in agreement with Eq. (9.39) which predicts a laser-noise current density of 0.4 pA/$\sqrt{\text{Hz}}$ at the output of the differential amplifier. (The common-mode contribution of the laser noise, estimated from Eq. (9.34), is also negligible. Assuming the reasonable value of CMRR = 100, the common-mode laser noise appearing at the differential output is only about 0.04 pA/$\sqrt{\text{Hz}}$.)

Figure 9.5(b) shows two spectra of the differential output with the disk spinning. The lower trace corresponds to an erased track, while the upper trace contains a recorded 8 MHz carrier. The erased track is about as noisy as the recorded one, leading one to believe that jitter and other signal fluctuations do not contribute much to the noise in this case. Integrating the erased-state noise over the frequency range from zero to 10 MHz, we obtain the figure of -30 dBm. This is the total noise power delivered to a 50 Ω resistor at the amplifier's output, and corresponds to an r.m.s. noise voltage of 7 mV. Dividing by the net gain of the system, the equivalent r.m.s. noise current is found to be 16.3 nA. After subtracting the thermal and shot noise contributions, we find the remaining r.m.s. noise to be 11.5 nA, which, according to Eq. (9.39), can arise from a small amount of depolarization noise ($\Delta_{xy} \simeq 0.7 \times 10^{-4}$). Of course, reflectivity fluctuations could just as well be responsible for the observed noise, but in that case the values of Δ_x or Δ_y would be larger than Δ_{xy} by a factor of $|r_{x0}/r_{y0}| = 100$.

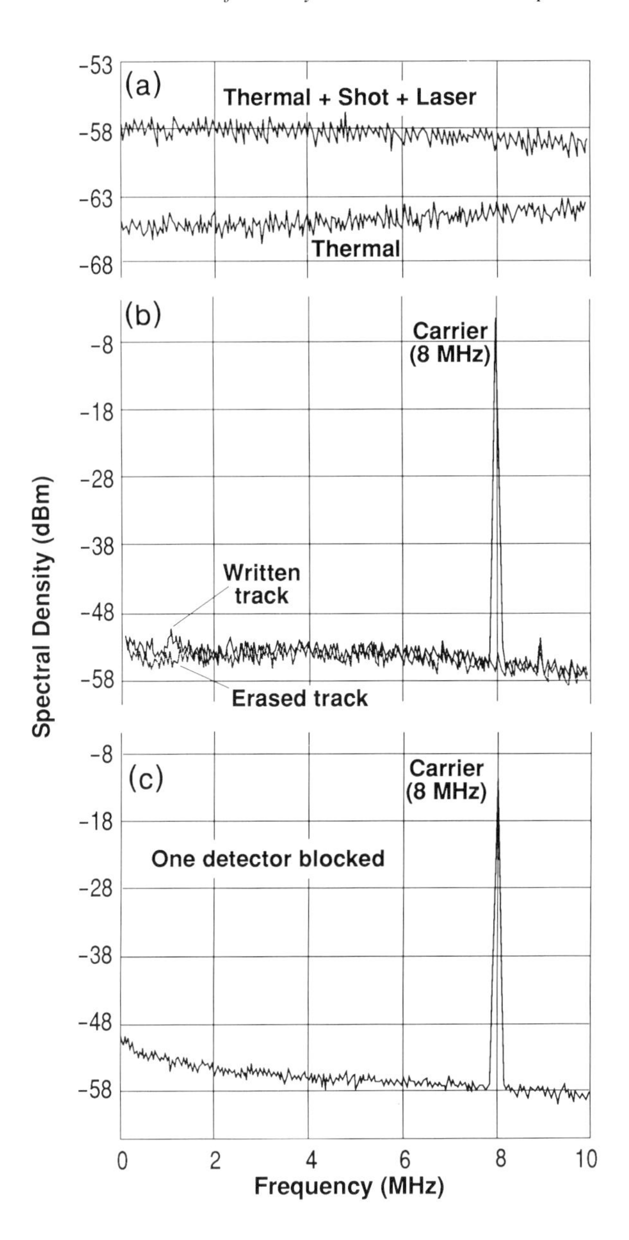

Figure 9.5. Measured noise spectra at the differential output of a magneto-optical readout system.[6] The operating wavelength and the light power incident on the disk are $\lambda = 680$ nm and $P_0 = 2$ mW, respectively. Other system parameters are given in the text (see Example 5). In (a) the disk is stationary. In (b) the spectrum of an erased track shows little deviation from that of a recorded track (carrier frequency = 8 MHz). The spectrum in (c) is obtained with the light blocked from one of the detectors.

Figure 9.5(c) shows the spectrum of the differential output with one of the detectors blocked. Although the erased-track spectrum is not shown here, it is reasonable to neglect the noise associated with the signal (i.e., jitter and signal amplitude fluctuation noise). The integrated noise power within the 10 MHz bandwidth of the system is -30 dBm, corresponding to a preamplification r.m.s. noise current of 16.3 nA. Subtracting the thermal noise (of two detectors) and the shot noise (of one detector) from this figure, we find the remaining noise current to be 13.6 nA. According to Eq. (9.33), this noise is almost exclusively due to laser power fluctuations, thus ruling out the possibility of a large r_x noise. One concludes, therefore, that the differential output noise (with neither detector blocked) is due either to a small Δ_{xy} or a large Δ_y, but not to a large Δ_x. □

Example 6. Consider the readout system of Fig. 9.1, operating under the following conditions: laser wavelength $\lambda = 830$ nm, effective laser power incident on the disk $P_0 = 2$ mW, disk reflectivity $r_x = 0.4$, disk MO reflection coefficient $r_y = 0.004$, leakage parameter of the beam-splitter $\gamma = 0.35$, photodetector sensitivity $\eta_s = 0.5$ A/W, preamplifier conversion factor $G = 3.3 \times 10^4$ V/A, and differential amplifier gain = 18. The laser noise for this system was analyzed in Example 4 (section 9.3), and the r.m.s. fluctuations of the **E**-field were found to be $\Delta_e \simeq 8.7 \times 10^{-4}$.

Figure 9.6 shows the measured spectra of the differential output voltage under various circumstances. With the light blocked from both detectors the electronic noise is -62 dBm; see the lowest trace in Fig. 9.6(a). Referred to individual photodiode currents (prior to amplification), this thermal noise has a density of $i_{th} \simeq 1.2$ pA/$\sqrt{\text{Hz}}$. The shot-noise current density for the 20 μW of power incident on each photodiode is calculated as $i_{sh} \simeq 1.8$ pA/$\sqrt{\text{Hz}}$. This means that the thermal-plus-shot noise level is about 5 dB above the thermal noise. (See the middle trace in Fig. 9.6(a), which shows the output noise spectrum with light reaching both detectors while the disk is stationary.) The laser noise contributes only 0.5 dB in this case since, according to Eq. (9.39), its current density is 0.6 pA/$\sqrt{\text{Hz}}$. (The common-mode contribution of the laser noise, estimated from Eq. (9.34), is negligible: if CMRR = 100, the common-mode laser noise at the differential output is only about 0.1 pA/$\sqrt{\text{Hz}}$.)

The top trace in Fig. 9.6(a) shows the spectrum of the differential output with the disk spinning and the track containing a 4 MHz carrier signal. Assuming the erased track has the same level of noise as the recorded one (i.e., ignoring jitter noise and other signal fluctuations) we integrate the noise from $f = 0$ to $f = 10$ MHz and obtain the figure of -27 dBm. This is the noise power delivered to a 50 Ω resistor at the amplifier's output, and corresponds to an r.m.s. noise voltage of 10 mV. Dividing by the net gain of the system, the equivalent r.m.s. noise current is found to be 16.8 nA. After subtracting the contributions of the thermal noise and the shot noise, we find the remaining r.m.s. noise to be 13.7 nA

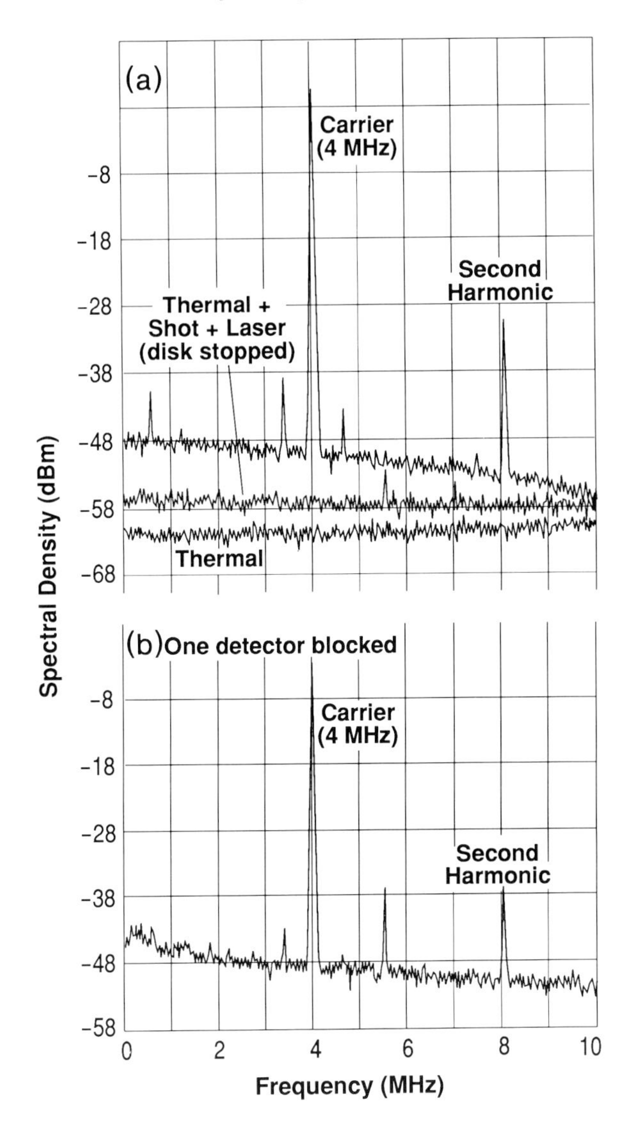

Figure 9.6. Measured noise spectra at the differential output of a magneto-optical readout system.[7] The operating wavelength and the light power incident on the disk are $\lambda = 830$ nm and $P_0 = 2$ mW respectively. Other system parameters are given in the text (see Example 6). The traces in (a) were obtained with the light blocked from both detectors (lower trace), with the disk stationary and the light allowed to reach the detectors (middle trace), and with the disk spinning and the light reaching both detectors (upper trace). The track under consideration was recorded with a 4 MHz carrier. The spectrum in (b) is similar to the upper trace in (a), but the light in this case was blocked from one of the detectors.

which, according to Eq. (9.39), can arise from a small amount of depolarization noise ($\Delta_{xy} \simeq 1.2 \times 10^{-4}$). Of course, reflectivity fluctuations could just as well be responsible for the observed noise, but in that case the values of Δ_x or Δ_y would be larger than Δ_{xy} by a factor of $|r_{x0}/r_{y0}| = 100$.

Figure 9.6(b) shows the spectrum of the differential output with one of the two detectors blocked. Again, the erased-track spectrum is not shown here, but it is reasonable to neglect the noise associated with the signal. The integrated noise power within the 10 MHz bandwidth of the system is -24 dBm, corresponding to a preamplification r.m.s. noise current of 23.7 nA. Subtracting the thermal noise (of two detectors) and the shot noise (of one detector) from this figure, we find the remaining noise current to be 22.4 nA. According to Eq. (9.33), this noise arises from laser power fluctuations Δ_e, reflectivity fluctuations Δ_x, and depolarization Δ_{xy}. Using the known values of Δ_e and Δ_{xy} in Eq. (9.33), we find $\Delta_x = 6.6 \times 10^{-4}$. One concludes, therefore, that the output noise when neither detector is blocked is due either to a small Δ_{xy} or a large Δ_y, but not to a large Δ_x. □

9.5. Jitter and Signal-amplitude Fluctuations

In this section we present computer simulation results pertaining to the noise arising from fluctuations of the information-carrying signal. A distinction will be made between jitter, which is caused by random variations of the zero-crossings of the signal waveform, and signal-amplitude-fluctuation noise. The basic waveform considered here is a 4 MHz, 50% duty-cycle carrier whose amplitude alternates periodically between the values of +1 and -1, with transitions between these levels being infinitely sharp. The power spectrum of this waveform has been computed numerically and shown in Fig. 9.7. The (approximate) sampling intervals in the time and frequency domains were 0.5 ns and 30 kHz respectively, corresponding to a total number of samples $N_{max} = 65536$. The spectrum contains the first and third harmonics of the waveform at 4 MHz and 12 MHz. With unit power of the square-wave signal at 0 dB, the fundamental and third harmonic power levels are at -0.912 dB and -10.455 dB. In Fig. 9.7 the spikes at frequencies other than 4 MHz and 12 MHz are numerical noise due to truncation errors, their very small magnitudes attesting to the basic accuracy of the numerical routine.

Next, we introduced jitter in the transition times of the signal. A random number generator selected values in the interval $[-\Delta t, \Delta t]$ with uniform distribution. The random numbers were then added independently to individual transition times of the ideal signal, creating a waveform with jitter magnitude of Δt. As an example, Fig. 9.8 shows a section of the signal between $t = 0$ and $t = 2$ μs, with $\Delta t = 25$ ns of jitter. The computed power spectra of the resulting noisy signals are shown in Figs. 9.9(a)–(d) which correspond, respectively, to jitter magnitudes of $\Delta t = 3$, 5, 15, and 25 ns. (The spectral power densities are computed in a 30 kHz bandwidth.)

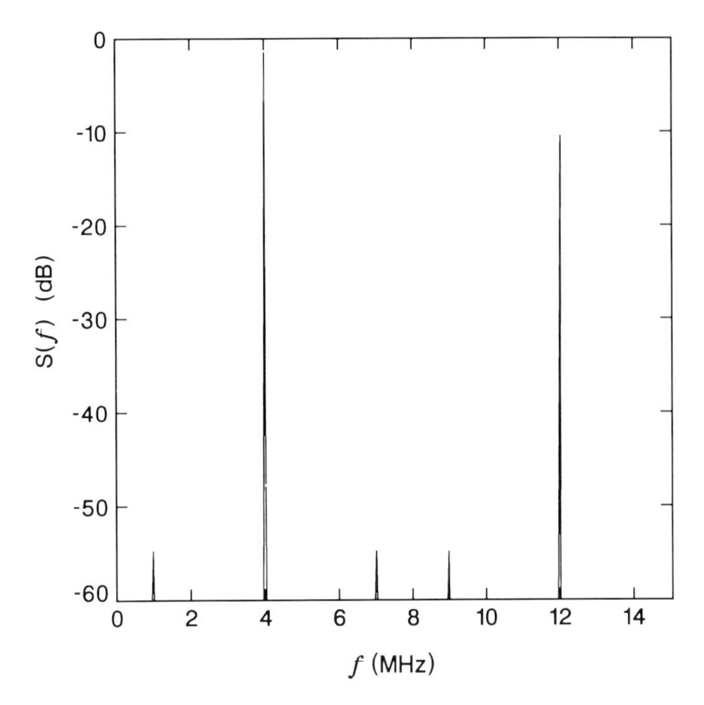

Figure 9.7. Computed power spectrum $\mathcal{S}$ of a 4 MHz, 50% duty-cycle, square-wave signal in the absence of jitter and other signal fluctuations. A total of 65536 samples were used in this fast Fourier transform (FFT) computation. The exact duration of the signal waveform was 33.25 μs, which is an integer multiple of the carrier period of 0.25 μs. The approximate values of the sampling intervals in time and frequency are, therefore, 0.5 ns and 30 kHz, respectively. (The exact values of the sampling intervals are slightly different from those quoted above; they are adjusted to fulfill two requirements: (i) the duration of the signal in time is an integer multiple of the carrier period, (ii) the total number of samples is an integer power of two.)

Note that within the bandwidth of interest the spectra are relatively flat and that, understandably, their levels rise with increasing jitter.

The effects of amplitude fluctuations on the signal waveform and its power spectrum are shown in Fig. 9.10. Figure 9.10(a) is a section of a jitter-free waveform, exhibiting random amplitude variations in its upper half; the values of these signal amplitudes were chosen randomly and independently from the interval [0.85, 1.0] with a uniform probability distribution. The corresponding power spectrum in Fig. 9.10(b) shows a significant noise contribution only at low frequencies. A similar conclusion is reached upon inspecting the spectrum of Fig. 9.10(c), which corresponds

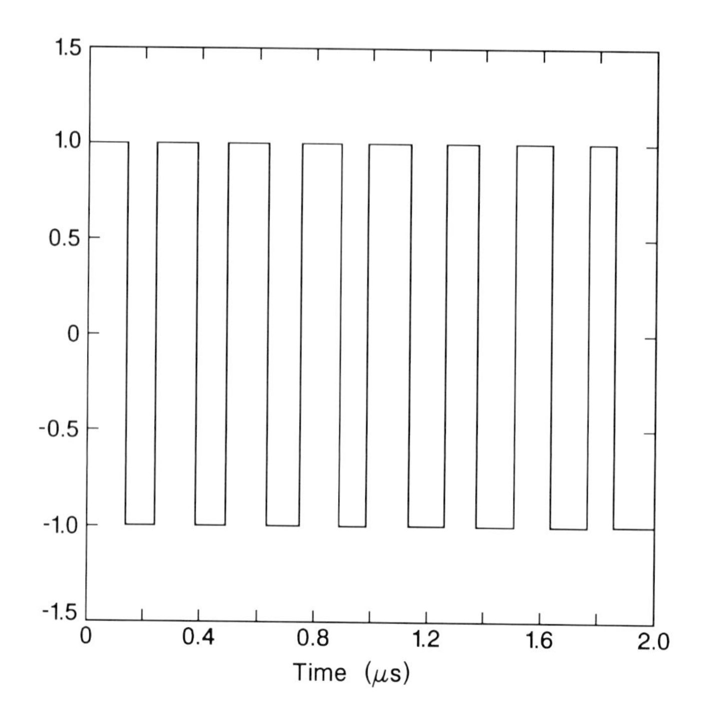

Figure 9.8. Section of a square-wave signal in the time interval [0, 2 μs], with jitter magnitude Δt = 25 ns. A random number generator is used to select the deviation of each zero-crossing from its nominal position. The deviations are uniformly distributed in the interval [$-\Delta t$, Δt], and the random number corresponding to any given point is independent of that for any other point.

to a signal whose positive amplitude fluctuates within [0.70, 1.0] and has, in addition, a jitter noise component with Δt = 3 ns. Comparison with Fig. 9.9(a) indicates that the high-frequency noise is solely due to jitter, whereas at low frequencies the noise may be attributed to amplitude fluctuations.

Finally, Fig. 9.11(a) shows a section of a signal waveform with positive amplitude fluctuations within [0.70, 1.0], negative amplitude fluctuations within [-1.0, -0.70], and a jitter magnitude of Δt = 3 ns. The corresponding spectrum in Fig. 9.11(b) shares its general features with that in Fig. 9.10(c), but it has a larger noise content in the low-frequency regime which is, of course, expected. The spectrum in Fig. 9.11(c) is obtained when the jitter magnitude is increased to Δt = 15 ns while amplitude fluctuations are maintained at their previous level. The noise in this instance is almost exclusively due to jitter, as a comparison with Fig. 9.9(c) indicates.

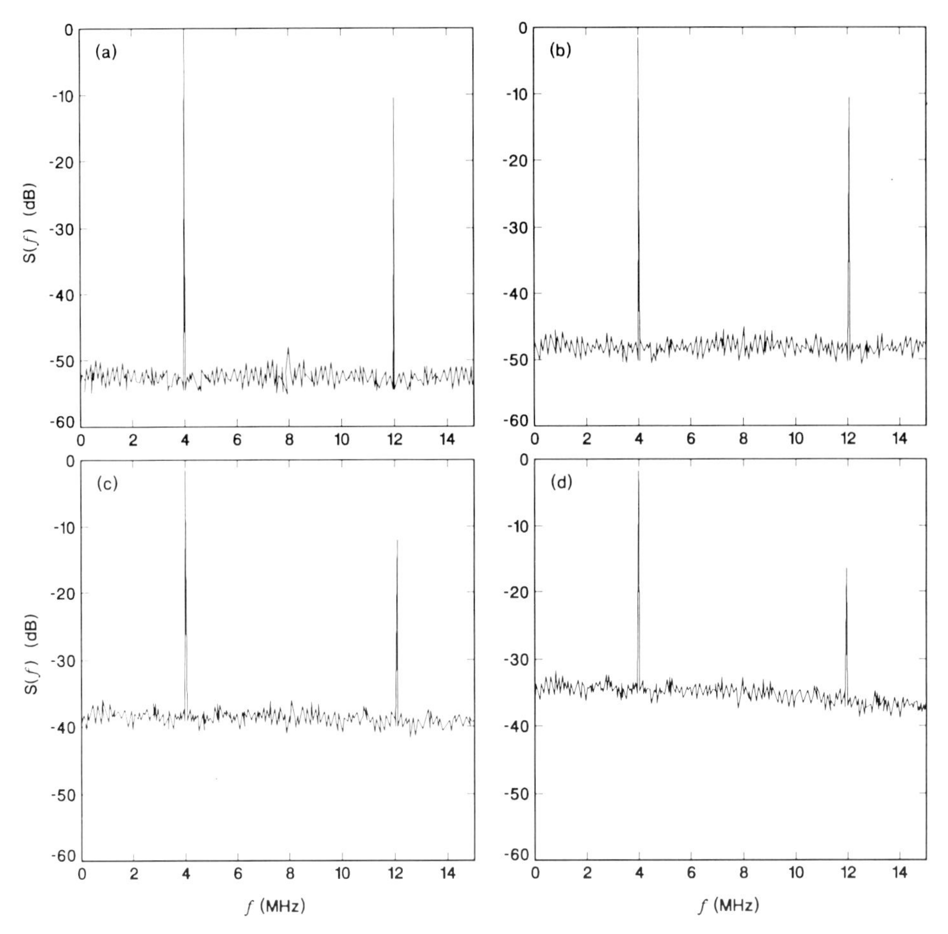

Figure 9.9. Spectra of square-wave signals (such as the one shown in Fig. 9.8) with different amounts of random jitter. The noise density shown here is computed within a 30 kHz bandwidth. (a) Δt = 3 ns, (b) Δt = 5 ns, (c) Δt = 15 ns, (d) Δt = 25 ns.

9.5.1. Effects of finite beam size on signal and noise spectra

Up to this point we have ignored the fact that the focused spot has a finite diameter. The effects of finite spot size are readily incorporated into the preceding results if one ignores diffraction effects. The signal at the output of the readout system is then obtained by convolution of the intensity profile of the spot with the recorded pattern of information on the disk. For the sake of simplicity, let us assume a Gaussian intensity profile for the spot as follows:

$$I(x) = \frac{1}{\sqrt{\pi\rho^2}} \exp\left(-x^2/\rho^2\right) \tag{9.42}$$

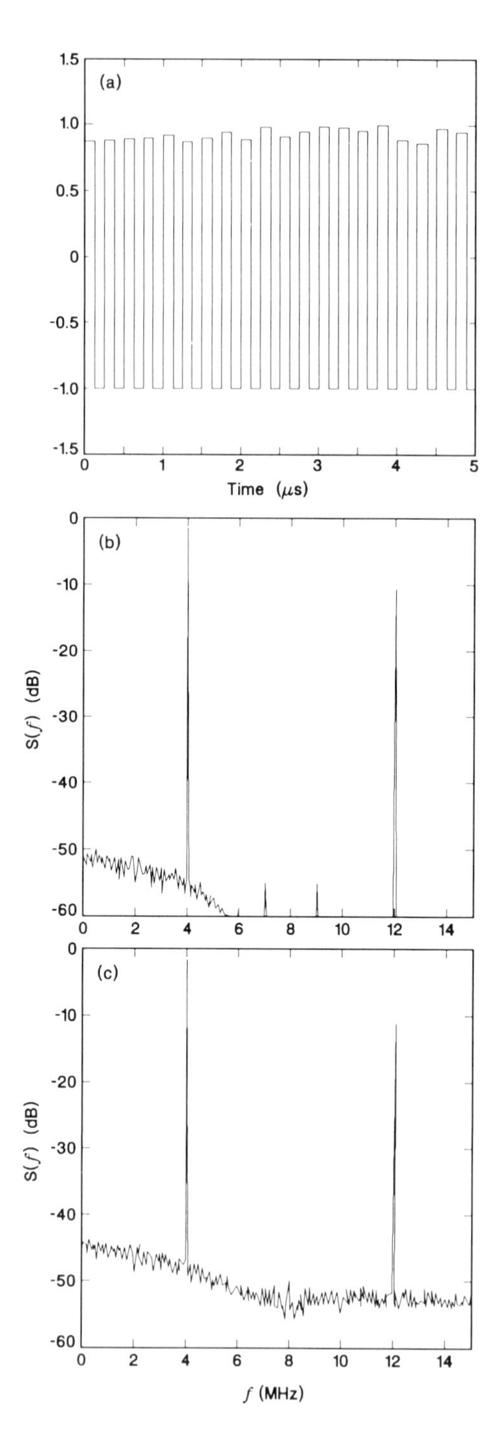

Figure 9.10. (a) Section of a square-wave signal in the time interval [0, $5\,\mu s$]. The waveform is free from jitter, but its positive amplitude fluctuates randomly within the interval [0.85, 1.0]. (b) Spectrum of the signal in (a), showing that the noise appears only at low frequencies. (c) Spectrum of a signal waveform with $\Delta t = 3$ ns of jitter in addition to positive amplitude fluctuations in the interval [0.70, 1.0].

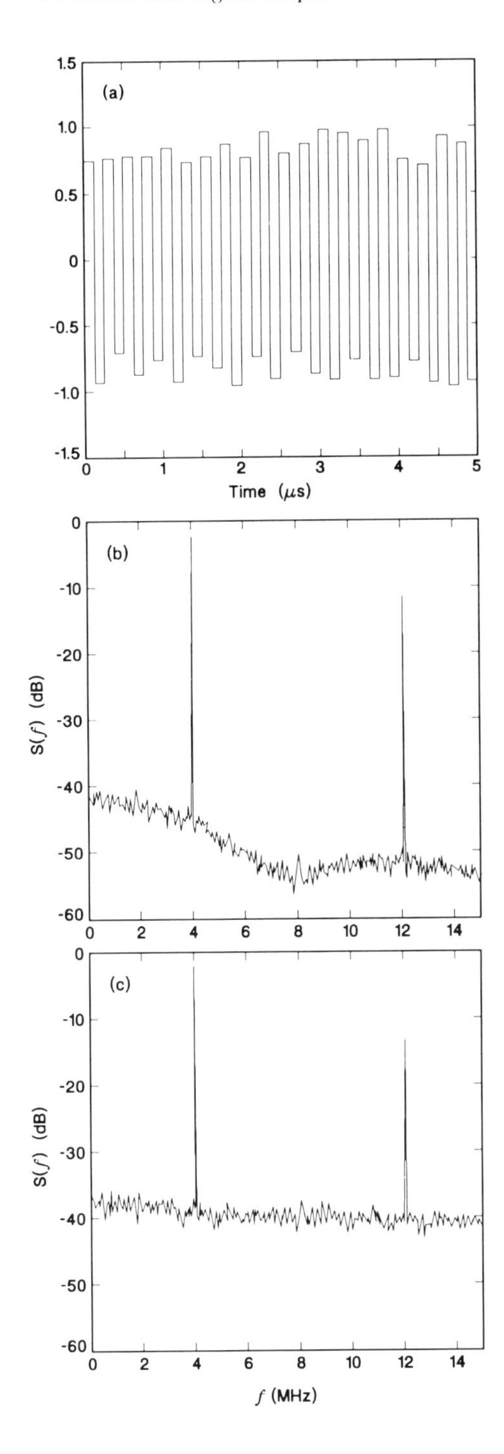

Figure 9.11. (a) Section of a square-wave signal in the time interval [0, 5 μs]. The waveform has Δt = 3 ns of jitter, its positive amplitude fluctuates randomly within the interval [0.70, 1.0], and its negative amplitude fluctuations are confined to the interval [- 1.0, - 0.70]. (b) Spectrum of the signal in (a). (c) Spectrum of the same signal when the jitter is increased to Δt = 15 ns.

Here x is the spatial coordinate on the disk surface along the track, and ρ is a measure of the spot size (FWHM = 1.66ρ). The function in Eq. (9.42) is properly normalized so that the integrated spot intensity is unity. If the actual pattern of recorded data along the track is denoted by $\psi_0(x)$, and if v_0 is the track velocity, then the output signal will be

$$\psi(v_0 t) = \int_{-\infty}^{\infty} I(x)\,\psi_0(x - v_0 t)\,\mathrm{d}x \tag{9.43}$$

The function $\psi(x)$, the result of convolution between $I(x)$ and $\psi_0(x)$, is scaled along the horizontal axis by the factor v_0 in order to yield the time dependence of the observed signal. Upon Fourier transformation, the convolution turns into the product of the Fourier transforms of the individual functions, that is,

$$\mathscr{F}\left\{\psi(v_0 t)\right\} = v_0\,\mathscr{F}\left\{I(v_0 t)\right\}\mathscr{F}\left\{\psi_0(v_0 t)\right\} \tag{9.44}$$

The transform of the Gaussian function is readily calculated as follows:

$$v_0\,\mathscr{F}\left\{I(v_0 t)\right\} = \exp\left[-(\pi\rho/v_0)^2 f^2\right] \tag{9.45}$$

If the spectra of $\psi_0(v_0 t)$ and $\psi(v_0 t)$ are denoted by $\mathscr{S}_{\psi_0}(f)$ and $\mathscr{S}_{\psi}(f)$ respectively, we will have

$$\boxed{\mathscr{S}_{\psi}(f) = \exp\left[-2(\pi\rho/v_0)^2 f^2\right]\,\mathscr{S}_{\psi_0}(f)} \tag{9.46}$$

Equation (9.46) shows that the spectrum of the signal (including jitter and amplitude noise) is attenuated by a multiplicative factor that is the Fourier transform of the spot intensity distribution. In the literature, this effect is usually attributed to the modulation transfer function (MTF) of the optical readout system. The frequency at which the spectrum is attenuated by a factor of two is a measure of the width of the MTF. According to this definition, the width of the Gaussian function in Eq. (9.46) $\simeq v_0/5\rho$.

Example 7. In Fig. 9.5(b) the recorded carrier is read from the disk at a constant velocity $v_0 = 14$ m/s. Assuming a Gaussian profile for the focused spot with FWHM = 1 μm (i.e., $\rho = 0.6$ μm), we find an MTF attenuation of 10 dB at $f = 8$ MHz. Similarly, in Fig. 9.6 where $v_0 = 10$ m/s and $\rho = 0.7$ μm, the attenuation of the 4 MHz carrier is found from Eq. (9.46) to be 6.7 dB.

□

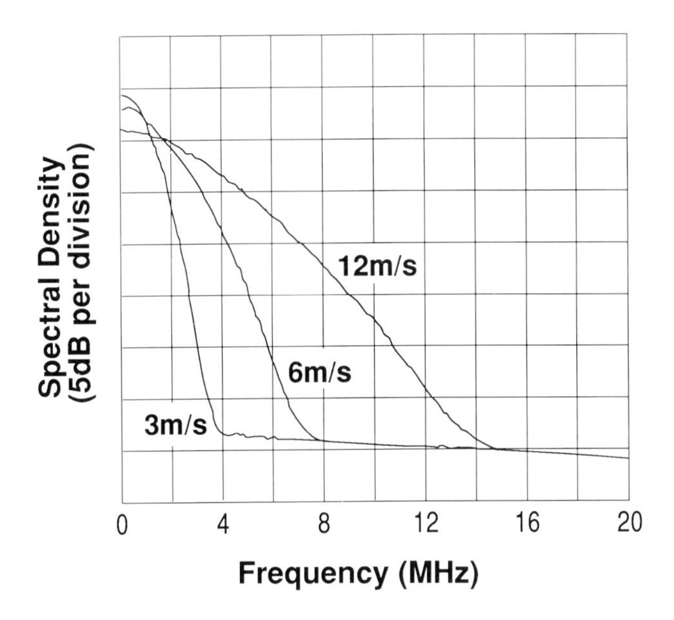

Figure 9.12. Measured noise spectra from an erased track on a magneto-optical disk.[12] The traces are obtained from the same track under identical conditions with the exception that the disk velocity is different for different traces. The velocities v_0 corresponding to these measurements are 3 m/s, 6 m/s, and 12 m/s.

Of the various noises discussed in the present chapter, the spectra of the electronic, shot, and laser noise are obviously unaffected by the MTF. The spectrum of the disk noise, on the other hand, has an attenuation factor similar to that of the information signal described by Eq. (9.46). An example of this phenomenon may be seen in Fig. 9.12, where the spectrum of an erased track is shown for three different disk velocities.[12] In each case, the tail of the trace corresponds to thermal + shot + laser noise, which is velocity-independent. The low-frequency magnitudes of the spectral density functions drop by 3 dB each time the velocity is doubled. This reflects the fact that the noise power is constant; in other words, when the velocity is doubled (causing the spectrum to stretch over a range of frequencies twice as large) the magnitudes of the spectral density drop by a factor of two (i.e., 3 dB) in order to maintain the total (integrated) power. If one assumes that the underlying sources of disk noise have flat spectra, then the curves of Fig. 9.12 are simply plots of the MTF.

Example 8. In Fig. 9.12 the curve corresponding to $v_0 = 6\,\text{m/s}$ drops by about 12 dB between $f = 0$ and $f = 4$ MHz. Assuming that the curve represents the actual MTF of the system and that the MTF has the Gaussian form of Eq. (9.46), we find $\rho = 0.56\ \mu\text{m}$, corresponding to FWHM = 0.93 μm. □

Problems

(9.1) In deriving Eq. (9.19) from Eq. (9.18) several steps were left out. Describe the missing steps and fill the gaps in the derivation.

(9.2) Photomultiplier tubes (or avalanche photodiodes) multiply photo-induced electrons (or electron–hole pairs), and produce, say, G electrons for every captured photon. In general, G is not the same for all photons; instead, it should be considered a random variable with average $\langle G \rangle$ and average-square $\langle G^2 \rangle$. The multiplication process is rapid enough to be assumed instantaneous, and the value of G for a given photon is independent of that for any other photon. Assuming an incident laser beam with power $P_0(t)$, determine the spectra of signal and shot noise at the photomultiplier output.

(9.3) Equation (9.33) was derived from the equation immediately preceding it, but details of the derivation were left out. Examine the approximations involved in this derivation and the conditions under which the second-order terms in δ_e, δ_x, δ_y, and δ_{xy} may be ignored. Pay special attention to those remaining terms that contain r_{y0}, since in practice the magnitudes of $|r_{y0}/r_{x0}|$ and $|r_{y0}/\gamma r_{x0}|$ are likely to be comparable to δ_e, δ_x, etc.

(9.4) Obtain the first few terms in the Fourier series representation of a periodic square-wave function such as the one whose spectrum is shown in Fig. 9.7. Assume that the function has frequency f_0 with a 50% duty cycle, and that its amplitude is confined to the values of +1 and -1. Show that the first and third harmonics of the waveform have power levels of -0.912 dB and -10.455 dB respectively, as the numerically computed spectrum in Fig. 9.7 indicates. (The numerical results are based on the formula

$$\mathcal{S}_\psi(m) = 2\left|\frac{1}{N_{max}}\sum_{n=0}^{N_{max}-1}\psi(n)\exp\left(-\frac{i2\pi mn}{N_{max}}\right)\right|^2$$

where the function $\psi(t)$ is uniformly sampled at times $t = n\Delta t$, its power spectrum $\mathcal{S}_\psi(f)$ is sampled (also uniformly) at frequencies $f = m\Delta f$, the total number of samples is N_{max}, and the various parameters are related by $\Delta t\,\Delta f = 1/N_{max}$.)

(9.5) Figure 9.12 shows three traces of the noise spectral density, obtained from the same erased track at three different disk velocities. According to Eq. (9.46) the modulation transfer function (MTF) should not affect the low-frequency end of the spectrum, yet the DC noise level in the figure drops by about 3 dB for each doubling of the velocity. Clearly, the spectrum $\mathcal{S}_{\psi 0}(f)$ of the signal itself is responsible for the observed behavior. Explain this phenomenon on the basis of the definition of the spectral density function in Eq. (9.16).

10

Modulation Coding and Error Correction

Introduction

In a digital system, the so-called "user-data" is typically an unconstrained sequence of binary digits 0 and 1. In such devices it is the responsibility of the information storage subsystem to record the data and to reproduce it faithfully and reliably upon request. To achieve high densities and fast data rates, storage systems are usually pushed to the limit at which the strength of the readout signal is of the same order of magnitude as that of the noise. Operating under these circumstances, it should come as no surprise that errors of misinterpretation do indeed occur at the time of retrieval. Also, because individual bits are recorded on microscopic areas, the presence of small media imperfections and defects, dust particles, fingerprints, scratches, and the like, results in imperfect reconstruction of the recorded binary sequences. For these and other reasons to be described below, the stream of user-data typically undergoes an encoding process before it finally arrives on the storage medium. The encoding not only adds some measure of protection against noise and other sources of error, but also introduces certain useful features in the recorded bit-pattern that help in signal processing and future data recovery operations. These features might be designed to allow the generation of a clocking signal from the readout waveform, or to maintain the balance of charge in the electronic circuitry, or to enable more efficient packing of data on the storage medium, or to provide some degree of control over the spectral content of the recorded waveform, and so on.

In today's practice, the encoding process consists of one or more error-correction-coding (ECC) steps designed to protect the data against random and burst errors, followed by a modulation coding step where desirable features are incorporated into the bit-pattern. In the ECC step additional bits, known as check bits, are generated and added to the stream of user-data; these check bits create an appropriate level of redundancy in the overall bit-sequence. The modulation step entails mapping of small blocks from the error-correction-coded sequence into somewhat larger blocks known as modulation code-words; this mapping, which is one-to-one, is typically done by looking up the code-words in a code table. The modulated bit pattern is thus a cascade of code-words from a code table. Given that the code table is designed with the channel characteristics in mind, the modulated sequence naturally conforms to the channel constraints.

The data retrieval process begins by reading the modulated bit sequence from the storage medium and chopping it into its constituent modulation code-words. The code table is then used to associate these code-words with the corresponding blocks of data. If there are readout errors in the sense that some of the recorded 1's are identified as 0's and vice versa, the code-words in which the erroneous bits are embedded may be incorrectly identified, thus giving rise to one or more incorrect blocks of data. These errors, however, are corrected in the next step using the built-in error-correction capability. Since the ECC protects both the check bits and the user-data bits, there is no need to distinguish between the two during the decoding process. (Once the decoding is over, however, the check bits have served their purpose and are discarded.)

It is not our intention here to describe the conventional methods of modulation and error-correction coding, since these are well known and abundantly expounded in the literature of information theory and coding. Instead, our primary goal is to describe a powerful and general technique for modulation encoding and decoding. This technique, which is based on the method of enumeration, can handle practically any set of constraints that the channel might impose on the recording bit-pattern. Modulation constraints of practically any degree of complexity can be described by a state-transition table with a finite number Ω of states. From the state-transition table we show how to construct a trellis diagram for code-words of arbitrary length L_0. A method of enumeration that assigns a number to each code-word according to its lexicographic order will then be described. All the necessary information for the enumerative encoding and decoding of binary data will be subsequently stored in an array of size $L_0 \times \Omega$; both encoding and decoding can be achieved with a few simple operations using this array. Since no additional constraints are imposed, the code rate approaches Shannon's noiseless channel capacity in the limit of long sequences.

In contrast to conventional modulation schemes, the enumerative technique works best when the code-words are long. Long code-words, however, make the enumerative technique quite sensitive to random errors. To remedy this shortcoming, error-correction coding for random errors is done *after* modulation, with the result that recovery from such errors occurs *prior* to demodulation. These and related ECC issues will be discussed in some detail in this chapter. Also introduced here is a simple, efficient algorithm for burst-error correction that may be applied either before or after modulation.

10.1. Preliminary Remarks

The use of modulation codes in optical and magnetic data storage is widespread and the advantages of modulation coding are well known.[1-8] Usually a 1 is represented by a transition between the up and down states of the recording waveform, while a 0 is represented by no change at all (see subsection 1.2.2). An unconstrained sequence of 1's and 0's, however, is not acceptable, nor is it desirable in practice. For instance, a long sequence of uninterrupted 0's results in loss of synchronization when the data is self-clocking. Or an imbalance between the up and down states of the waveform may result in significant charge accumulation in the electronic circuitry, thus forcing the system beyond acceptable levels of operation. The objective of modulation coding is to create a one-to-one correspondence between sequences of user-data (usually streams of random binary digits) and constrained binary sequences. The nature of the constraints imposed on the modulation waveform is determined by the system designer, and is dependent on the particular characteristics of the system under consideration. A typical set of restrictions imposed on the recording sequence is the so-called $(d,k;c)$ constraint. Here d is the minimum allowed run-length of zeros, k is the maximum allowed run-length of zeros, and c is the maximum charge that the recording waveform is permitted to accumulate. In addition, certain sequences of bits may be reserved for special purposes (e.g., synchronization, signaling the beginning of a block, etc.) and thus banned from appearing in the modulated waveform.

The minimum run-length constraint d is somewhat more subtle than the other constraints and needs further elaboration. The physical separation between successive transitions on the recording medium cannot be made arbitrarily small. The minimum distance between transitions is a function of the structural/magnetic properties of the medium, as well as the characteristics of the read/write head. The wavelength of light in optical recording and the head gap-width in magnetic recording, together with the structural/magnetic features of the recording medium, determine the minimum distance Δ between successive transitions. It is useful to measure distance along the track in units of this length Δ, so in the following discussion, density should be understood as the number of bits contained within the length Δ. Now, if there are no constraints on the minimum run-length of zeros (i.e., if $d = 0$), each modulation bit will occupy one interval of length Δ. Since the number of modulation bits is always greater than the number of data bits, the overall recording density will become less than one data bit per Δ. On the other hand, if $d \neq 0$, one can pack $d + 1$ modulation bits into each Δ. When the code parameters are chosen properly, the higher density of modulation bits translates into a higher density of recorded data bits, so that the overall density becomes greater than one data bit per Δ. The price of this increase in capacity is, of course, the reduced period of time available to each modulation bit. If τ is defined as the time that the read/write head dwells on an interval of length Δ, then the time window available to each modulation bit will be $\tau/(d + 1)$.

In this chapter we introduce a general algorithm for modulation coding with arbitrary constraints. This algorithm, which uses enumeration to encode blocks of data into constrained sequences, is much more powerful than other existing schemes, which typically apply only to (d,k) or $(d,k;c)$ constraints. The enumerative algorithm can handle any set of constraints that can be expressed in terms of a state-transition table; in addition, it solves the cascading problem, which has complicated all previous schemes and requires the addition of "merging" bits between successive blocks.

In section 10.2 several examples of (d,k) and $(d,k;c)$ constraints, together with the corresponding state-transition tables, are given. We also give an example of a (d,k) code that excludes a specific pattern of bits, for synchronization purposes. Although the examples in this section are for certain representative values of parameters, the techniques used for constructing the state-transition tables are quite general, and the reader should be able to apply the methods to any set of parameters.

After the state-transition table corresponding to a given set of constraints has been constructed, one must choose a parameter L_0 for the block length of the modulation code-words. Any choice of L_0 can be handled by the algorithm, but longer code-words are preferred since they give a capacity that is arbitrarily close to Shannon's noiseless-channel capacity[9] for the given set of constraints. The encoding and decoding routines are based on enumeration,[10,11] and map blocks of user-data of fixed length N_0 to modulation code-words of fixed length L_0. Both encoding and decoding routines utilize an array A of pre-calculated numbers. The array A has dimensions $L_0 \times \Omega$ where Ω is the total number of states in the state-transition table. The mappings are of the fixed-to-fixed-length-block type, and the beginning state is always the same as the final state, so that errors cannot propagate beyond the block boundaries. The trellis diagram, discussed in section 10.3, helps in understanding this point as well as in constructing the array A. The encoding and decoding algorithms are then described in section 10.4.

In conjunction with the enumerative modulation technique, two error-correction schemes will also be described. The first, discussed in section 10.5, deals with burst-error correction. Here one takes advantage of the fact that the position of a block of data affected by a long burst of errors is known to the decoder, simply because the modulation constraints within that block are seriously violated. A simple construction based on parity check bits and capable of restoring the lost block will be presented. The second type of error correction deals with random errors that occur rather infrequently and that corrupt modulation code-words in a few, a priori unknown, locations. Some of these errors may be readily detectable (and perhaps even correctable), because they violate the modulation rules. Depending on the severity of the constraints, however, this kind of error correction may or may not have the desired power in practice. In any event, a Viterbi decoder[12–14] can be used to find the closest code-word (in the Hamming sense†) to the readout sequence. In this context, the applicability of Viterbi's algorithm for minimum-distance decoding arises from the fact that modulation coding is based on state-transition tables. This point will be further elaborated in section 10.6. The more powerful

† The Hamming distance between two binary sequences of equal length is the number of positions in which the sequences differ from each other. For example, the distance between 1101001 and 1001101 is 2, since the second and the fifth bits of these sequences are different.

alternative for random-error correction within a modulated block is the use of an appropriate block error-correcting code such as a Reed–Solomon code.[15,16] Here, depending on the probable number of errors within a block of length L_0, the ECC scheme generates a certain number of check bits, say N_1. These check bits do not necessarily satisfy the modulation constraints and, therefore, must themselves be modulated. Let us denote by L_1 the length of the block of N_1 check bits after being modulated. Since encoded blocks always begin and end in the same state, no problems will arise during recording and readout if the block of length L_1 immediately follows the block of length L_0. In certain situations it might be desirable to apply another level of error correction to the block of length L_1, and to record the new set of check bits after modulating them. Indeed the process may be repeated any number of times. After the modulated block of data, followed by one or more blocks of modulated check bits, has been formed, one might add a special block of length L_s to signal the end of the stream. All in all, the sequence of user data of length N_0 has been mapped onto a sequence that satisfies the modulation constraints and has a total length $L = L_0 + L_1 + L_2 + \cdots + L_s$. In readout, this block of length L is separated into its sub-blocks, its various levels of error correction are demodulated, and the corresponding errors, if any, are corrected. Finally, the block of length L_0 is demodulated and the original N_0 bits of user-data are recovered. The details of this procedure are described in section 10.7.

Section 10.8 gives several examples of codes with various constraints and discusses their performance. The final section contains some closing remarks, including a suggestion for experimental determination of the appropriate set of modulation constraints for any given data storage system.

10.2. The State-transition Table

In this section we demonstrate, by way of examples, the construction of state-transition tables from the rules and constraints that a modulation code must obey.

Example 1. A (d,k) code is one in which the run-length of zeros has a minimum equal to d and a maximum equal to k. Figure 10.1 shows the state-transition diagram for a (1,3) code. (Extension to arbitrary values of d and k is straightforward.) A circle in this diagram represents a state and the identity of that state is written within the circle. The states in Fig. 10.1 are S_1, S_2, S_3, and S_4. Each state leads to at least one and at most two states. If the transition from one state to another is indicated by a solid line then, during that transition, the encoder's output will be 1. If, on the other hand, the transition is indicated by a broken line, the output will be 0. The system thus begins in an arbitrary state, say S_1, and with each clock pulse moves to another state, generating a binary digit in the process. In the diagram of Fig. 10.1 the only way out of S_1 is through a broken line to S_2; thus at least one 0 must follow a 1, which is always the output of the system on its way to S_1. Similarly, S_4 can only lead to S_1 with a solid line,

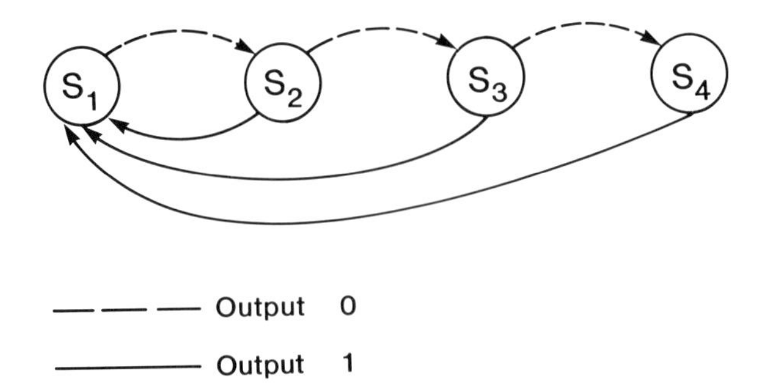

Figure 10.1. State-transition diagram for the (1, 3) code.

which indicates that no more than three zeros can follow each other without interruption. □

Example 2. The state-transition diagram in Fig. 10.2 corresponds to a (2, 7) modulation code with one additional constraint: the sequence 1001001001001 is reserved for synchronization and thus cannot appear in the output of the modulation encoder. States S_9 through S_{17} are added to the standard diagram for the (2, 7) code in order to accommodate this additional constraint. □

Example 3. A $(d,k;c)$ code is a (d,k) code with charge constraint c, namely, the accumulated charge in the circuitry during read/write operations must remain in the interval $[-c, +c]$. The occurrence of a 1 in the sequence does not change the charge but reverses the direction of its accumulation. For example, if there are n units of charge in the system at a given instant of time and the charge is *on the rise*, then immediately after a transition (corresponding to the appearance of a 1 in the modulated sequence) the charge will still be n units but *on the fall*. Each 0 in the modulated sequence changes the accumulated charge by one unit in a direction determined by the preceding 1's.

Let us consider the case of a (1, 3; 2) code for which the charge is allowed to be either +2, +1, 0, −1, or −2 units. The state diagram for this code is shown in Fig. 10.3 where $S_m{}^{n}\!\uparrow$ is defined as the state corresponding to a sequence of m successive 0's after the latest 1, having n units of charge with the direction of charge accumulation on the rise (↑). If the system begins in $S_0{}^{0}\!\uparrow$ or $S_0{}^{0}\!\downarrow$, some states, such as $S_0{}^{2}\!\uparrow$, $S_1{}^{2}\!\downarrow$, etc., will never be reached. These states are shown as broken circles in Fig. 10.3. □

10.3. The Trellis Diagram

The trellis diagram is obtained by replicating the state diagram in time. Figure 10.4 shows the trellis for the state diagram of the (1, 3) code

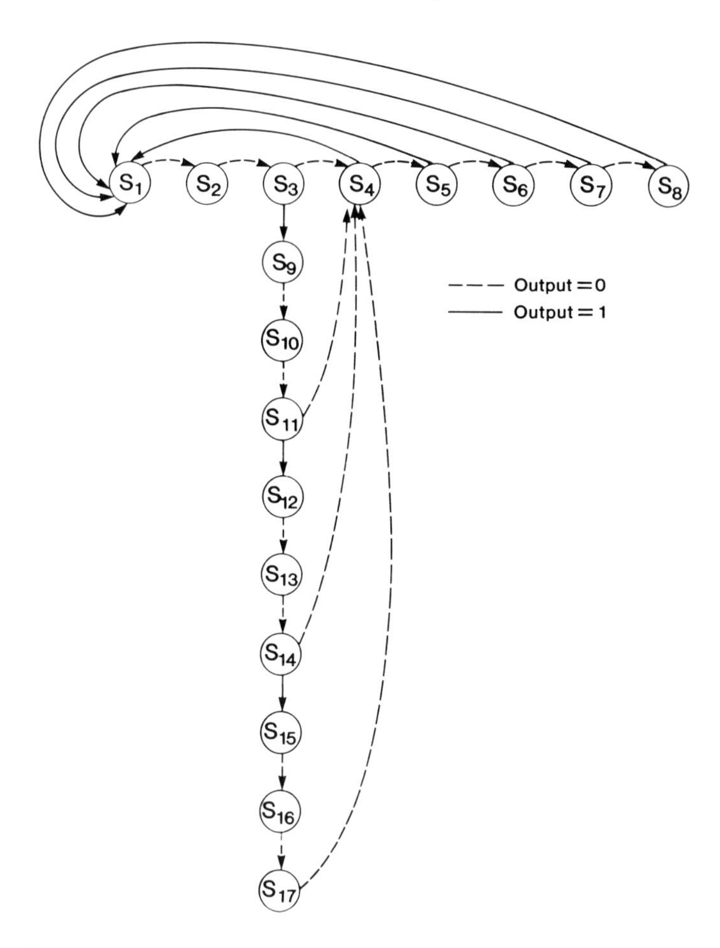

Figure 10.2. State-transition diagram for a (2, 7) code that excludes the synchronization sequence 1001001001001.

discussed in Example 1. Only seven time steps are shown in the figure; thus this particular diagram corresponds to a modulation code of length $L_0 = 7$. A transition from a state at time t to a state at time $t + 1$ is indicated with a solid line if the output is 1, and a broken line if the output is 0. The initial state being chosen as S_1, this is the only state at $t = 0$ that can connect to states at $t = 1$. Like the initial state, the final state can be selected arbitrarily. In fact, one may choose to end in any subset of states. In this chapter we shall assume that the initial and final states are one and the same in all cases. Accordingly, the final state in Fig. 10.4 is also S_1. This assumption will help simplify the discussion without seriously restricting the scope. It should be emphasized, however, that our particular choice of the final state is by no means essential for this encoding and decoding scheme.

Once the trellis has been constructed, various code-words of the modulation code may be obtained by simply following a connected path on

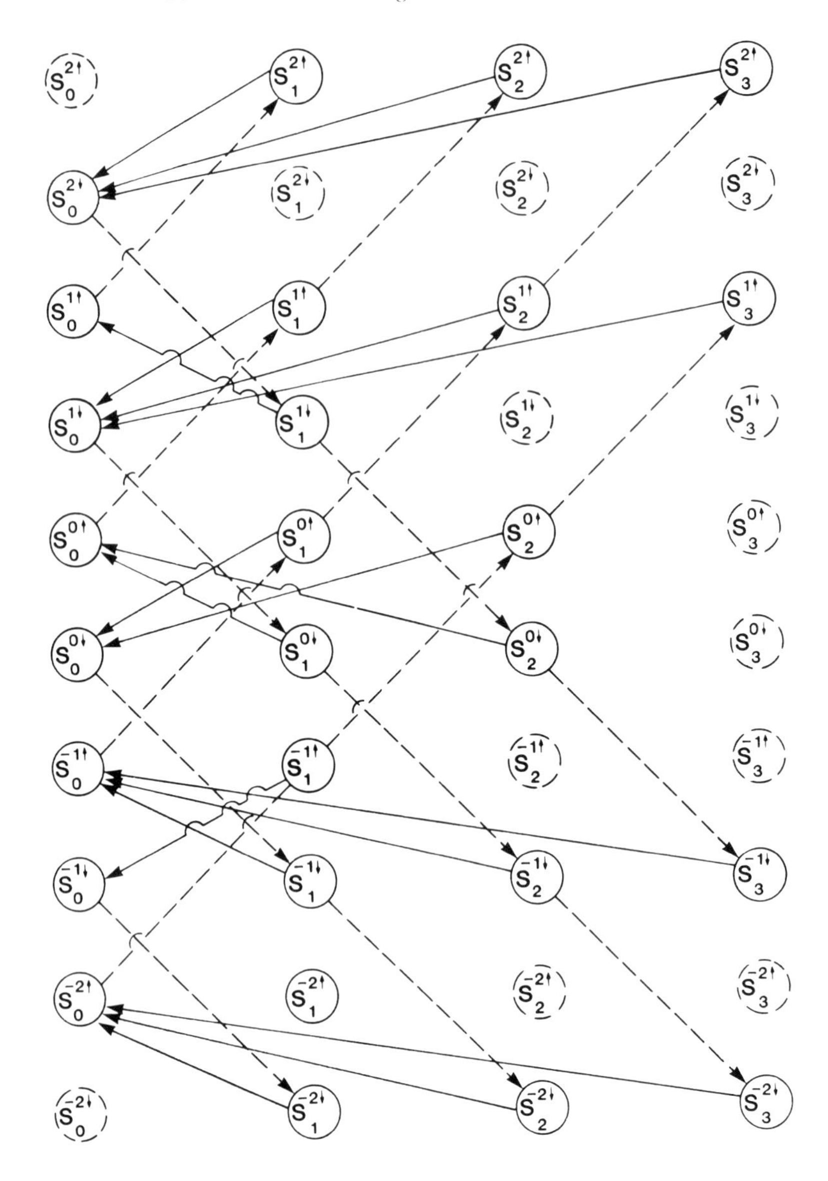

Figure 10.3. State-transition diagram for the (1, 3;2) code.

the trellis between the initial state and the final state. There is a one-to-one correspondence between the connected paths and the acceptable modulation code-words. To convert a path into a code-word, we simply replace each segment of the path with a 1 or a 0, depending on whether that segment is a solid or broken line. To find the path corresponding to a given code-word, we start at the initial state and move to the next state either along a solid or broken line, depending on whether the next digit of the code-word is 1 or 0. In Fig. 10.4 there are five complete paths,

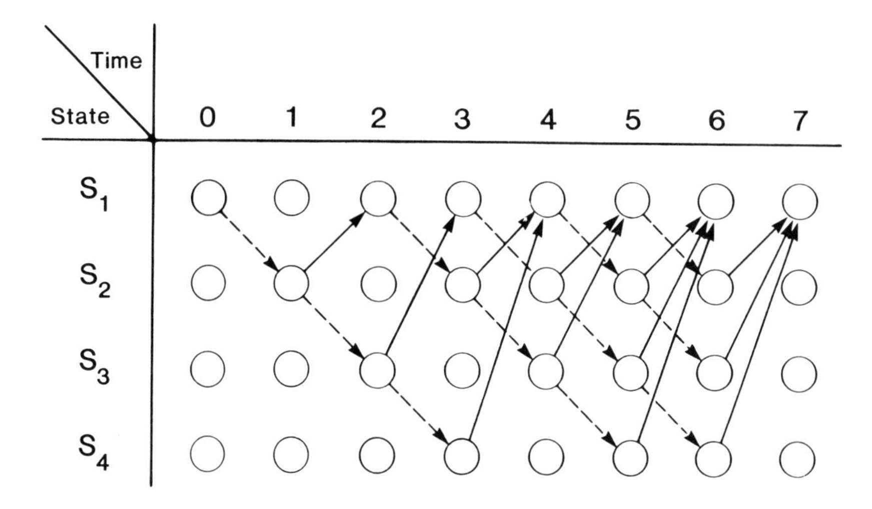

Figure 10.4. Trellis diagram for the (1, 3) code with block length $L_0 = 7$. The initial and final states are both chosen to be S_1.

indicating that there are only five code-words of length $L_0 = 7$ that satisfy the (1, 3) constraint while beginning and ending in S_1.

We now assign a number to each state S in the trellis. This number will represent the total number of connected paths between S and the final state. We begin at the last (rightmost) column of states in the trellis and assign the number 1 to those states that are acceptable as final states. Unacceptable states in this column will receive the number 0. In our example the only acceptable final state is S_1 which, as shown in Fig. 10.5, is the recipient of the number 1 in the last column. Next we move one column to the left and, to each state S in this new column, assign the sum of the numbers already assigned to the states immediately to the right of S and directly connected to it. By moving to the left one column at a time and repeating the above procedure, we obtain the desired number for every state in the trellis. Note, in particular, that the number associated in this manner with the initial state is the total number of acceptable modulation code-words. Figure 10.5 shows the end result of these assignments for the trellis depicted in Fig. 10.4.

Next we assign a second number to each state in the trellis. This second set of numbers, which is derived from the first set, will be used in encoding and decoding. In fact the first set will no longer be needed once the second set has been constructed. To obtain the second set we start with the leftmost column of the trellis and assign a number to each state S in that column. The number assigned to S will be 0 if there are less than two lines connecting S to its adjacent states on the right. Thus if a solid line only, or a broken line only, or no lines at all lead from S to the next state, the number assigned to S will be 0. If, on the other hand, S moves to the next state via *both* a solid line and a broken line, we follow the *broken line*

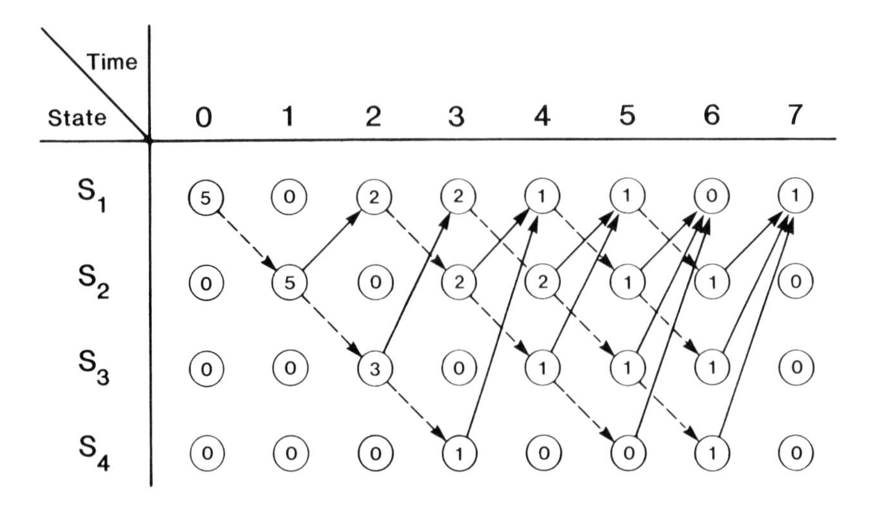

Figure 10.5. Trellis diagram of Fig. 10.4 with a number assigned to each one of its states. The number assigned to any state S is the total number of continuous paths between S and the final state.

to the corresponding next state, say S', and use the number previously assigned to S' (that is, the total number of connected paths between S' and the final state) as the new number for S. By moving to the right one column at a time and repeating the above procedure for all the states in each column, we obtain a complete trellis with new numbers assigned to each state. Figure 10.6 shows this trellis with new numbers derived from the trellis of Fig. 10.5.

A zero assigned to a state S in the new trellis indicates either that S does not belong to any path that connects the initial and final states, or that there is only one path that leads out of S. When a number other than zero is assigned to a state S, it represents the total number of paths that begin at S *with a broken line* and reach the final state. The set of numbers assigned to the states in this trellis can be stored in an array A of dimensions $L_0 \times \Omega$, where L_0 is the block-length of the modulation code-words and Ω is the number of states in the state-transition table. Array A and the transition table are all that is needed for modulation encoding and decoding.

10.4. Encoding and Decoding Algorithms

In this section we describe enumerative encoding and decoding algorithms based on the methods described in section 10.3. Consider all possible modulation code-words of length L_0 as obtained, for instance, by identifying all the continuous paths of a given trellis diagram. These code-words may be arranged according to their lexicographic order, just as

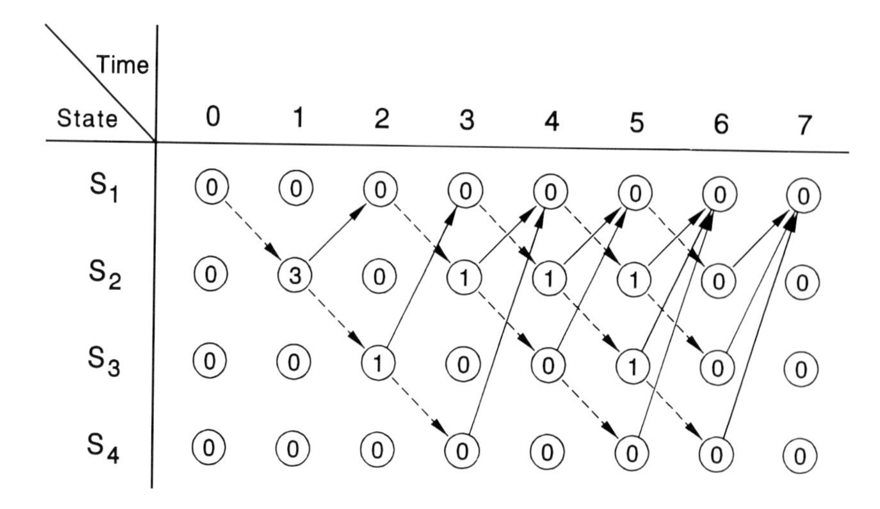

Figure 10.6. Trellis diagram of Fig. 10.5 with a different number assigned to each one of its states. See section 10.3 for a description of the assigned number.

words in a dictionary are arranged alphabetically.† We denote the total number of code-words by M_0, and assign to each code-word a unique integer (between 0 and $M_0 - 1$) that represents its position in the lexicographic order. This is called enumeration and in the following we shall describe a systematic method of obtaining this unique integer for a given code-word (i.e., decoding), as well as identifying the code-word from the knowledge of its corresponding integer (i.e., reverse enumeration or encoding).

Let us first consider the decoding procedure, since it is somewhat easier to describe. With reference to the last trellis discussed in the previous section (see Fig. 10.6), a given code-word is simply a connected path between the initial state and the final state. Starting at the initial state and moving along this path, we visit a total of $L_0 + 1$ states. Those states that lead to the next with a broken line must be ignored, while for the remaining states, (i.e., those that lead to the next state with a solid line), the corresponding numbers on the trellis must be added. The reason is that each time we move along a solid line (which corresponds to a 1 in the code-word), we leap over all the code-words that are the same until then but that have a zero at that junction. Therefore the lexicographic number of the code-word that follows the solid line must increase by the number of code-words that take the broken line at the junction. The sum thus obtained is the lexicographic number of the code-word under consideration

† In fact, alphabetical ordering of code-words is easier than ordering the words of a language, since the code alphabet consists only of two letters, 0 and 1; what is more, all the words have the same length L_0.

Table 10.1. Code-words corresponding to the trellis of Fig. 10.6 and their respective number in the lexicographic order

Code-word	Number
0101001	3 + 1 + 0 = 4
0100101	3 + 0 + 0 = 3
0001001	0 + 0 = 0
0010101	1 + 1 + 0 = 2
0010001	1 + 0 = 1

and the decoding is therefore complete. Table 10.1 illustrates the decoding process for all the code-words contained in the trellis diagram of Fig. 10.6.

As for the encoding process, in which the lexicographic order number m of a code-word ($0 \leq m \leq M_0 - 1$) must be translated into the code-word itself, once again we refer to the trellis of Fig. 10.6, on which we identify the path corresponding to the desired code-word as follows. We define a dummy variable n and set $n = m$ at the outset. Then, starting at the initial state of the trellis, we follow either a broken line or a solid line to the next state. When there is only one line leading to the next state, there is obviously no choice and that line must be followed. However, when there is a choice, we compare the current value of n with the number n_S assigned to the current state S of the trellis. If $n \geq n_S$ we follow the solid line and update the value of n to $n - n_S$. If, on the other hand, $n < n_S$ we follow the broken line without modifying n. The process continues until the final state is reached, at which point the value of n is zero and the path taken corresponds to the desired code-word.

In using these algorithms for the modulation and demodulation of user-data, it is usually desirable to have a fixed-length block of data mapped onto a fixed-length modulation code-word. If the block-length of the user-data is denoted by N_0, the total number of code-words M_0 must be greater than or equal to 2^{N_0}. Thus N_0 must be chosen as the largest integer less than or equal to $\log_2 M_0$. Although the remaining $M_0 - 2^{N_0}$ legitimate code-words are never used in this modulation scheme, their neglect has but a minor effect on the code rate R (which is defined as N_0/L_0). The effect of this truncation becomes less significant with increasing code-word length, and tends to zero in the limit of large L_0.

10.5. Burst-error Correction

There are at least two ways to recognize that a block is affected by a burst of errors. First, the read signal corresponding to all or a portion of the block may have an unusual waveform. Second, there may be significant and frequent violation of the modulation constraints within the block.† In any event, if one or more blocks within a sector are identified as erroneous, it will be possible to restore them with a simple scheme and with a relatively small penalty in overhead. The following example is based on realistic numerical values and explains the proposed method of burst-error correction.

Suppose that the 512 user-bytes that typically comprise a sector are divided into 64 blocks of 64 bits each. Subsequently, each block is modulated by some appropriate modulation code and, after additional bits for random-error correction and synchronization have been added, we may assume that its length L, in units of modulation bits, has become 150. Suppose now that the longest possible burst that could occur within this sector has length 1000 (in units of modulation bits), and that the probability of two or more bursts occurring within a given sector is negligible. The maximum number of consecutive blocks affected by the burst is therefore eight, which corresponds to 512 successive user-bits. The proposed burst-error-correction scheme divides the 512 user-bytes (prior to modulation) into eight segments of length 512 bits each, and generates a ninth segment (also 512 bits long), consisting solely of parity check-bits. The nth parity bit ($1 \leq n \leq 512$) is chosen such that the nth bits from all nine segments satisfy the condition of even (or odd) parity. Figure 10.7 shows the schematic of a circuit that may be used to generate the parity bits in the general case. The total length of the block in this figure is N, the length of each sub-block is K, and CBG stands for check-bit generator. In our example, therefore, $N = 4096$ and $K = 512$, resulting in an encoded total length of 4608 bits (user-data plus parity bits). It is not difficult now to see that any contiguous group of 512 bits or less (out of this 4608-bit-long block) that is lost to a single burst of errors can be recovered (irrespective of the location of the burst along the block) by virtue of the fact that only one out of every nine bits that originally satisfied the parity conditions is now missing. Thus by adding 512 parity bits to the original 4096 bits of user-data, dividing the entire block into 72 sub-blocks of 64 bits each (these sub-blocks are then independently modulated into 150-bit-long blocks), and storing the resulting 72 modulated blocks all in one sector, we have achieved immunity against single bursts that could be as long as 9.72% of the total length of the sector.

In this example, the parity bits were generated *before* modulation. A more efficient method of implementing the same burst-error-correction technique would generate the parity bits *after* modulation. In this case, because they do not satisfy the modulation constraints, the parity bits must be modulated separately before being recorded (see section 10.7 for more on this issue). Also, intelligence must be built into the decoder to enable it to recognize the boundaries of the burst.

†These are examples of the so-called erasure channel, where, by virtue of the knowledge of the position of an error, one can achieve significant gains in error-correction capability of the codes.

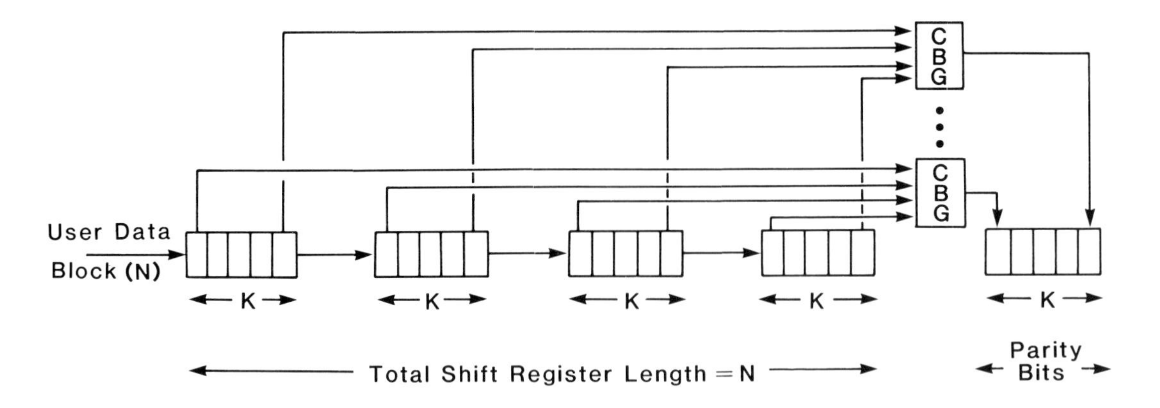

Figure 10.7. Implementation of a burst-error-correction encoder. The block of user-data has length N and fills a shift register of the same length. For clarity of presentation, this shift register is shown here divided into subsections of length K (naturally, N must be divisible by K). Check-bits are generated by the check-bit generators (CBGs) and stored in another shift register of length K. The input to the nth CBG comes from the nth bit of each subsection, and the output comprises the nth bit of the parity-check register. The resulting $(N + K)$-bit-long block (user-data followed by the check-bits) is subsequently sent to the next stage for modulation, where no distinctions are made between the actual user-data and the check-bits. In decoding, a contiguous group of erroneous bits (of length K or less), occuring anywhere within this $(N + K)$-bit-long block, can be corrected, provided that the decoder is informed of the location of the burst.

10.6. Viterbi Decoding

Since modulation constraints can be expressed in the form of state-transition tables, one should not be surprised to find that strong similarities exist between the modulation coding technique of this chapter and the methods of convolutional coding employed in communication systems.[12,13] In fact it is rather straightforward to apply Viterbi's algorithm,[12-14] which was originally developed for optimum decoding of convolutional codes, to identify our modulation code-words in the early stages of readout. Viterbi decoding of the read signal identifies a legitimate code-word that not only satisfies all the modulation constraints, but also is closest to the read-back waveform in the Hamming sense (or in some other appropriate sense). Thus the fact that certain paths along the trellis are disqualified by the rules of modulation guides the read system in its initial code-word identification process. Obviously, this initial screening would have the potential to be useful if there were no more than a few isolated, random errors in the waveform. Even then, though, it turns out that only some such errors could be recovered by the application of Viterbi's algorithm to simple modulation codes. (The more restrictive the modulation constraints, the better the correction capability of Viterbi's algorithm.) Numerical simulations show that ordinary $(d,k;c)$ constraints are not powerful enough to correct random errors in general. The exception perhaps is the case of very small c, which, unfortunately, results in small values of the rate R.

A fruitful subject for future research is to look for modulation constraints that also allow a significant degree of error correction without imposing severe penalties on the rate of the code. Some work along these lines has already been reported whereby it is sought to combine the advantages of convolutional coding with those of modulation codes.[17]

10.7. Random-error Correction

The modulation/demodulation algorithms described in section 10.4 are capable of encoding and decoding arbitrarily long blocks of data. The problem, from a practical standpoint, is that a single-bit error in a given block is usually sufficient to cause incorrect demodulation of an entire block. This problem is related to the well-known problem of error propagation in modulation coding,[1] which in the past has been "overcome" by choosing codes with short code-words, and by preventing the errors from propagating beyond the boundaries of the words within which they occur. A general solution to the problem of error propagation is to apply the error-correcting code to the modulated sequence itself and not, as is common practice today, to the data sequence prior to modulation. In other words, post-modulation error-correction eliminates the error-propagation problem of long-block codes. The following example should help clarify this concept.

Let a modulation code-word of length L_0 be error-correction coded for a certain number of errors, say ν. If the Reed–Solomon algorithm[15,16] is used for the purpose, it will generate 2ν check bits (to be appended to the original block of length L_0), but the original block itself will not be modified. These check bits, of course, do not satisfy the modulation constraints and will have to be modulated before being appended.† There are no problems in appending one modulated block to another in this manner, however, since individual blocks always begin and end in the same state. Of course, it is always possible to apply yet another ECC to the block of modulated check bits, and to repeat the process until the integrity of the data is assured. Figure 10.8 shows a two-level post-modulation error-correction encoder. The decoder corresponding to this encoder must follow the same steps in reverse order.

† It is even possible to use a different modulation scheme for the check bits, one which is less susceptible to random errors than the scheme used for the user-data. In this case the check bits may not be packed quite as efficiently as the data bits, but they occupy only a small fraction of the storage area and, therefore, the sacrifice will be relatively insignificant.

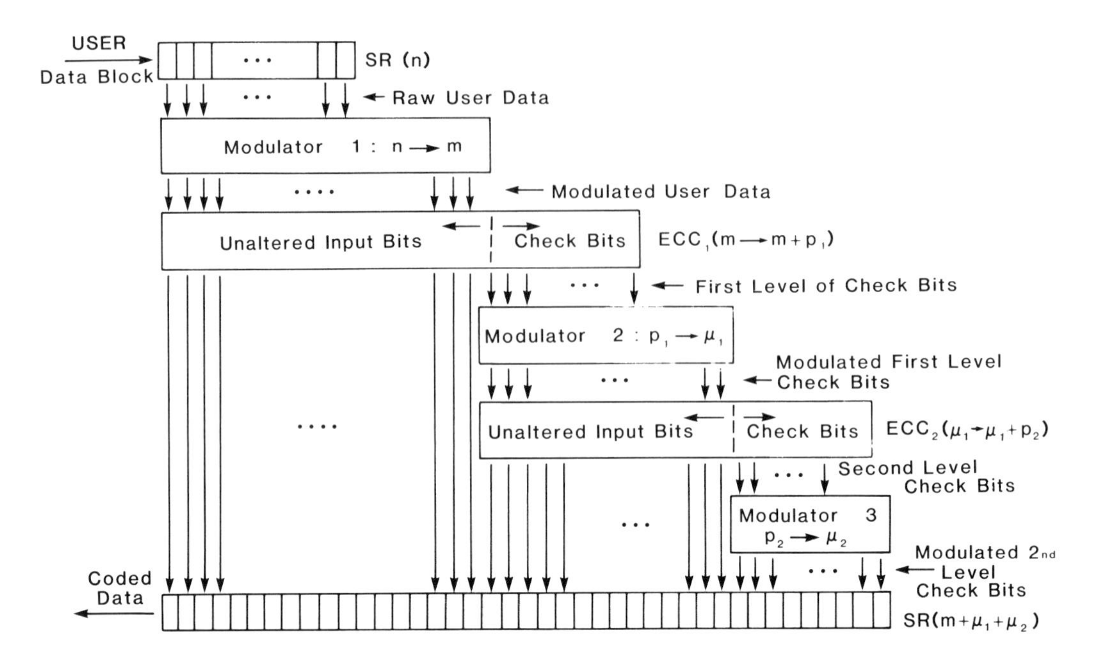

Figure 10.8. Two-level post-modulation error-correction encoder. The block of user-data of length n, placed in a shift register of the same length, SR(n), is sent to modulator 1. Subsequently, the m modulated bits are error-correction coded (at ECC_1) and p_1 additional bits (the first-level check bits) are generated. These check bits are then modulated by modulator 2 and converted into μ_1 bits. Next, the modulated check-bits are error-correction coded, giving rise to yet another p_2 check-bits at the output of ECC_2. Finally, these second-level check-bits are modulated and yield μ_2 bits at the output of modulator 3. The final code-word in this example consists of the m modulated user-bits, followed by μ_1 modulated first-level check-bits, followed by μ_2 modulated second-level check-bits.

10.8. Numerical Results and Discussion

A computer program has been developed to generate state-transition tables for arbitrary (d,k), $(d,k;c)$, and $(d,k;c,c')$ codes,†† and to implement the enumerative encoding and decoding algorithms described in the preceding sections. The results obtained with this program are summarized in Tables 10.2–10.4.

Table 10.2 corresponds to three codes each with $d = 2$ and $k = 7$. It shows possible block-lengths N_0 of the user-data for several lengths L_0 of

††A $(d,k;c,c')$ code is a $(d,k;c)$ code with the additional constraint that the cumulative integral of its accumulated charge over the length of any given code-word must remain confined between $-c'$ and $+c'$. Placing such constraints on the code reduces the low-frequency content of the signal waveform, and helps shape the low end of the spectrum.

Table 10.2. Comparison of three codes with $d = 2$ and $k = 7$ for different code-word lengths L_0. Code I is the (2,7) code with no restrictions other than those imposed by the initial and final conditions. Code II is the same as Code I except that the string 1001001001001 is prohibited from occurring in the code-words. Code III is the (2,7;8) code with each code-word beginning and ending with zero charge. The initial and final conditions for all three codes are such that each code-word ends in a 1 and starts as though the previous block ended with a 1

Modulation block-length (L_0)	User-data block-length (N_0)		
	Code I	Code II	Code III
50	23	22	19
100	49	48	44
150	75	74	70
200	101	100	95
250	127	125	120
300	153	151	145

the code-words. The initial and final conditions for all three codes in this table are such that each code-word ends in a 1, and begins as though the previous block ended with a 1. Code I is the (2,7) code with no restrictions other than those imposed by the initial and final conditions. The total number of states for this code is eight, and its theoretical maximum rate (Shannon's noiseless capacity) is 0.517.[1] Notice in Table 10.2 that the achievable rate $R = N_0/L_0$ increases with increasing L_0, and at $L_0 = 300$, R is already equal to 0.510, which is somewhat better than 0.5 for the currently standard (2,7) code. Code II is similar to Code I except that the string 1001001001001 is prohibited from occurring in its code-words. The state-transition table for this code, described in Example 2, section 10.2, consists of 17 states. Note that the extra constraint has reduced the code rate R by only a small amount. Code III is the (2,7;8) code with terminal states chosen such that, in addition to satisfying the previous boundary conditions, each code-word begins and ends with zero charge. The total number of states for this code is 210, and its theoretical maximum rate is 0.501.[1] As can be seen in Table 10.2, the achievable rate R increases with increasing L_0 and, for $L_0 = 300$, $R = 0.483$.

Table 10.3 corresponds to two codes, both with $d = 1$ and $k = 6$. Code IV is the (1,6) code with no restrictions other than those imposed by the initial and final conditions. The total number of states for this code is 7, and its theoretical maximum rate is 0.669. Notice that at $L_0 = 200$ achievable rate R is already equal to 0.660. Code V is the (1,6;8) code which, like Code III, begins and ends in a state with zero charge. The total number of states for this code is 194. Clearly, R increases with increasing L_0 and, for $L_0 = 300$, $R = 0.637$.

Table 10.4 corresponds to two codes each with $d=1$, $k=3$, and $c=4$.

Table 10.3. Comparison of two codes with $d = 1$ and $k = 6$ for different code-word lengths L_0. Code IV is the (1, 6) code with no restrictions other than those imposed by the initial and final conditions. Code V is the (1, 6;8) code with each code-word beginning and ending with zero charge. The initial and final conditions for both codes are such that each code-word ends in a 1 and starts as though the previous block ended with a 1

Modulation block-length (L_0)	User-data block-length (N_0)	
	Code IV	Code V
50	31	27
100	65	60
150	98	93
200	132	125
250	165	158
300	198	191

Code VI is the (1, 3; 4) code with no restrictions other than those imposed by the initial and final conditions; this code has a total of 58 states and, at $L_0 = 200$, its achievable rate R is 0.51. Code VII is the (1, 3; 4, 20) code that begins and ends in a state with zero charge *and* zero integrated charge. The total number of states for this code is 1362. The achievable rate increases with increasing L_0 and, for $L_0 = 200$, $R = 0.43$.

10.9. Concluding Remarks

We have described methods and algorithms for the modulation and error-correction coding of unconstrained streams of data. In the course of this exposition we have avoided abstractions and rigorous mathematical statements in order to convey the power and beauty of the concepts to a broad audience. Although the immediate concern of these ideas is the area of optical and magnetic recording, they might also be applicable in such areas as digital communication and digital signal processing.

The examples given so far have concentrated on known classes of modulation codes, but the enumerative technique is very powerful and should be applied to new problems and in new directions. For instance, a possible approach to the optimal design of constrained sequences for a given system might entail the experimental determination of correlation patterns similar to those shown in Fig. 10.9. Given a particular storage device and for a given clock cycle T, one must experiment with pulses whose durations are integer multiples of T. Recording and reading these pulses under realistic conditions, one obtains, after a fair number of trials, upper and lower bounds for the length of the readout pulse, as shown in

Table 10.4. Comparison of two codes with $d = 1$, $k = 3$, and $c = 4$ for different code-word lengths L_0. Code VI is the (1,3;4) code with no restrictions other than those imposed by the initial and final conditions. Code VII is the (1,3;4,20) code with each code-word beginning and ending with zero charge and zero integrated charge. The initial and final conditions for both codes are such that each code-word ends in a 1 and starts as though the previous block ended with a 1

Modulation block-length (L_0)	User-data block-length (N_0)	
	Code VI	Code VII
50	22	16
100	49	39
150	75	63
200	102	86

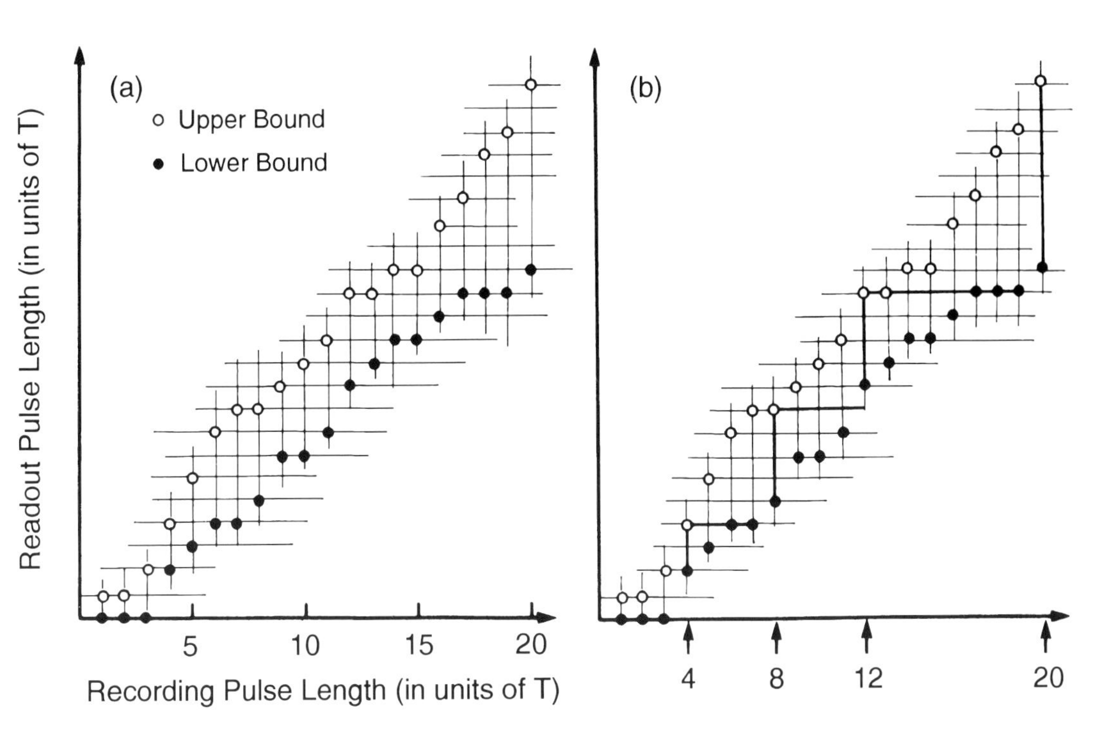

Figure 10.9. (a) Upper and lower bounds on the lengths of various recorded pulses after read-back, deduced after a number of trials. The horizontal axis represents the length of the pulse before recording. The durations of both read and write pulses are multiples of the clock cycle T. (Curves like these may be obtained for a given system by repeatedly recording marks of various lengths and monitoring the corresponding signal in readout.) (b) The method of selecting run-lengths for modulation. The first accepted run-length is the shortest pulse for which the lower bound is nonzero. The other run-lengths have the shortest possible length that allows them to avoid overlap.

Fig. 10.9(a). Run-lengths that do not overlap each other after being recorded and then retrieved are readily identifiable from these plots (see Fig. 10.9(b)). The correct set of constraints for such a system is now given by the collection of these non-overlapping run-lengths. For the imaginary system characterized by the curves of Fig. 10.9, therefore, the acceptable run-lengths are 4, 8, 12, 20, ... Once an acceptable set of run-lengths has been identified, the state-transition table can be constructed and the enumerative encoder and decoder implemented. The choice of maximum run-length is dictated by the acceptable level of complexity for the encoder and decoder. In fact the choice of the clock cycle T is not arbitrary either, and one might optimize it by trial and error, keeping in mind that although smaller values of T result in higher densities, the complexity of implementation grows with decreasing clock cycle.

Problems

(10.1) Investigate the problem of error propagation in the case of the (2,7) code described in subsection 1.2.2, and show that if one bit of the modulation code is in error then, upon demodulation, up to five erroneous bits could result.

(10.2) Show that the maximum theoretical rate (i.e., Shannon's noiseless capacity) for the (2,7) code is 0.517. Similarly, show that the maximum rate for the (2,7;8) code is 0.501.

(10.3) In the example of burst-error correction in section 10.5, verify the claim that the proposed method will accommodate bursts as long as 9.72% of the total length of the block.

(10.4) If the burst-error-correction method of section 10.5 were to be used *after* modulation, what sort of precautions would be required?

(10.5) If the scheme of section 10.5 were to be adapted to correct multiple bursts of error, what additional features would it have to incorporate?

11

Thermal Aspects of Magneto-optical Recording

Introduction

All media in use today for laser-assisted recording rely on the associated rise in temperature to achieve a local change in some physical property of the material.[1-3] Therefore the first step in analyzing the write and erase processes is the computation of temperature profiles. Except in very simple situations, these calculations are done numerically using either the finite difference method[4,5] or the finite element technique.[6] Generally speaking, the two methods produce similar results in comparable CPU times. Most often one assumes a flat surface for the disk and circular symmetry for the beam, which then allows one to proceed with solving the heat diffusion equation in two spatial dimensions (r and z in cylindrical coordinates). For more realistic calculations when the disk is grooved and preformatted, or when the beam has asymmetry due to aberrations or otherwise, the heat absorption and diffusion equations must be solved in three-dimensional space.

A problem with the existing thermal models is that the optical and thermal parameters of the media are assumed to be independent of the local temperature. In practice, as the temperature changes, these parameters vary (some appreciably).† While it is rather straightforward to incorporate such variations in the numerical models, at the present time it is difficult to obtain reliable data on the values of thermal parameters at a fixed temperature, let alone their temperature dependences. It is hoped that some effort in the future will be directed towards the accurate thermal characterization of thin-film media. Another difficulty in some thermal modeling efforts is caused by the changing geometry of the media during the heating cycle. This means that the thermal process cannot be completely decoupled from the mark formation process. (Some elementary methods for analyzing the thermal cycle in ablative and bubble-forming media have been reported.[8]) Fortunately, no such problems arise in MO recording, where the structure of the media remains intact while the magnetic properties change.

This chapter is devoted to the study of light absorption and heat

† The situation is perhaps most critical for erasable phase-change media, where the rates of heating and cooling in the vicinity of the melting point and in the neighborhood of the amorphous-to-crystalline phase transition determine the success or failure of the write and erase operations.[7]

diffusion in the multilayer disk structures of MO recording. Analytical solutions of the heat diffusion equation exist only for simple cases; these solutions are important, nonetheless, since they provide benchmarks for comparison with numerical results and also convey certain universal aspects of the thermal diffusion process. In section 11.1 we analyze the heat diffusion equation in detail and provide analytical solutions for simple problems in one-, two-, and three-dimensional geometries. Numerical techniques for solving the equations are the subject of section 11.2, where extension of the results to moving media is also discussed. Examples of numerical computations for problems of practical interest are presented in section 11.3. Here, for a multilayer illuminated by a focused beam, the absorption of optical energy in the absorptive layers and the subsequent diffusion of the resultant heat throughout the structure are studied and temperature profiles during recording and readout are displayed.

11.1. The Heat Diffusion Equation

The diffusion process can be described in terms of two material parameters, the specific heat C and the thermal conductivity K.[9] The specific heat is the thermal energy required to raise the temperature of a unit volume of material by one kelvin. The conductivity is the proportionality constant between the rate of flow of thermal energy and the local temperature gradient. Using simple phenomenological arguments, one can write the diffusion equation in isotropic media as a second-order partial differential equation. (The ratio K/C appears in this equation, and the symbol D, for diffusivity, is usually used to refer to it.) The diffusion equation can be solved analytically in simple cases, and certain general features of the diffusion process can be studied with the help of these solutions. In the following subsections we determine temperature profiles by solving the diffusion equation for problems with simple geometries, and we analyze the interesting features of these solutions.

11.1.1. Heat diffusion in one-dimensional space

Consider an infinite rod of uniform cross-sectional area A, specific heat C and thermal conductivity K as shown in Fig. 11.1. In the absence of sinks and sources of heat, the balance of energy in an element of length Δx at the center of which the temperature is T may be written

$$C\left[T(x, t+\Delta t) - T(x,t)\right] A\,\Delta x =$$

$$K\frac{\partial}{\partial x} T(x+\tfrac{1}{2}\Delta x, t)\, A\,\Delta t - K\frac{\partial}{\partial x} T(x-\tfrac{1}{2}\Delta x, t)\, A\,\Delta t \qquad (11.1)$$

In the limit of small Δx and Δt this equation reduces to the one-dimensional heat diffusion equation:

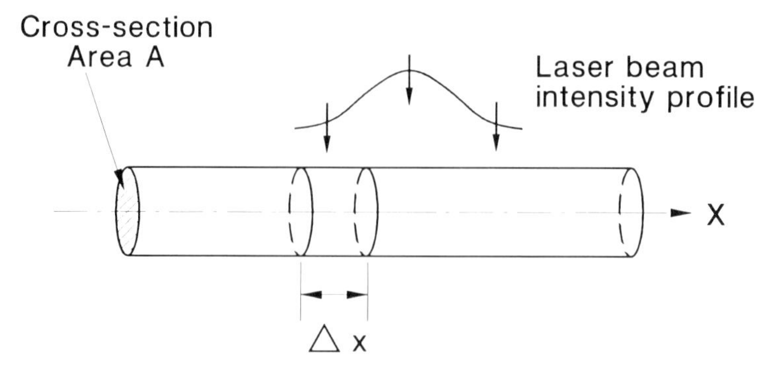

Figure 11.1. Uniform rod of cross-sectional area A, specific heat C, and thermal conductivity K. The thermal diffusion equations are written for an infinitesimal element of the rod of length Δx. The heat source is assumed to be a laser beam with arbitrary intensity distribution along X.

$$\boxed{\frac{\partial}{\partial t} T(x,t) = D \frac{\partial^2}{\partial x^2} T(x,t)} \tag{11.2}$$

where $D = K/C$ is the thermal diffusivity of the material. The solution to this equation is obtained by separating $T(x,t)$ into the product of a function of x and a function of t, namely, $f(x)\,g(t)$. Replacing in Eq. (11.2) then yields

$$\frac{g'(t)}{g(t)} = D \frac{f''(x)}{f(x)} \tag{11.3}$$

Since the left-hand side of Eq. (11.3) is a function of time only, while the right-hand side is a function of x, both sides must be equal to a constant, say $-\alpha$. (We shall see shortly why this constant should be negative.) Thus

$$g'(t) = -\alpha\, g(t) \qquad\qquad f''(x) = -\frac{\alpha}{D} f(x) \tag{11.4}$$

The solution to the first of these equations is an exponential function:

$$g(t) = \exp(-\alpha t) \tag{11.5a}$$

Since in the absence of external sources of heat the temperature profile is expected to decay, the only acceptable values for the constant α are positive. A positive α, when placed in the second of Eqs. (11.4), yields

$$f(x) = \exp(\mathrm{i}\sqrt{\alpha/D}\, x) \tag{11.5b}$$

The solution to Eq. (11.2) is thus written

$$T(x,t) = T_0 \exp(\mathrm{i}\sqrt{\alpha/D}\,x - \alpha t) \tag{11.6}$$

where T_0 and α are arbitrary constants. Note that the distribution at $t = 0$ is

$$T(x,\, t = 0) = T_0 \exp(\mathrm{i}\sqrt{\alpha/D}\,x) \tag{11.7}$$

which is a sinusoid with amplitude T_0 and spatial frequency $\sqrt{\alpha/D}$. As time elapses, the amplitude decays exponentially while the spatial frequency remains unchanged. The time constant for this decay is $1/\alpha$, indicating that a high-frequency sinusoid decays faster than a low-frequency one. This is certainly expected, since high-frequency profiles have large gradients, which causes them to decay rapidly through heat conduction.

Next, let the initial temperature profile be $T(x)$, and proceed to express it as the superposition of complex exponentials using the Fourier integral

$$T(x) = \int_{-\infty}^{\infty} \hat{T}(\sigma) \exp(\mathrm{i}2\pi\sigma x)\, \mathrm{d}\sigma \tag{11.8a}$$

Here $\hat{T}(\sigma)$ is the Fourier transform of $T(x)$, namely,

$$\hat{T}(\sigma) = \int_{-\infty}^{\infty} T(x) \exp(-\mathrm{i}2\pi\sigma x)\, \mathrm{d}x \tag{11.8b}$$

According to Eq. (11.8a) the initial profile is the superposition of sinusoidal functions of various frequencies σ. On the other hand, Eq. (11.6) describes the decay of any sinusoidal distribution with time. The diffusion equation (11.2) being linear, one can combine these observations to obtain

$$\boxed{T(x,t) = \int_{-\infty}^{\infty} \hat{T}(\sigma) \exp(\mathrm{i}2\pi\sigma x - 4\pi^2\sigma^2 D t)\, \mathrm{d}\sigma} \tag{11.9}$$

Equation (11.9) is the general solution to Eq. (11.2) for an arbitrary initial distribution $T(x)$ in the absence of sinks or sources of thermal energy. Before proceeding further, however, we provide an explicit solution for the case of Gaussian initial distributions.

Example 1. Let the initial temperature profile of the rod be Gaussian, i.e.,

$$T(x) = \frac{1}{\sqrt{\pi}\, x_0} \exp\left[-(x/x_0)^2\right] \tag{11.10}$$

Here x_0 is the $1/e$ point of the profile, and the normalization makes the integrated temperature over the entire rod equal to unity. (Since the stored thermal energy is proportional to the temperature, the integrated temperature may be used as a measure of total energy; in the absence of sinks and sources of heat, this thermal energy is preserved.) The Fourier transform of the above initial distribution is readily obtained as follows:

$$\hat{T}(\sigma) = \exp(-\pi^2 x_0^2 \sigma^2) \tag{11.11}$$

From Eq. (11.9) we now have

$$T(x,t) = \int_{-\infty}^{\infty} \exp\left[-\pi^2 (x_0^2 + 4Dt)\sigma^2\right] \exp(\mathrm{i}2\pi\sigma x)\, \mathrm{d}\sigma \tag{11.12}$$

This is simply the expression for the inverse Fourier transform of another Gaussian function. The new function has its $1/e$ point at $x_0(t)$, where

$$\boxed{x_0(t) = \sqrt{x_0^2 + 4Dt}} \tag{11.13a}$$

Therefore,

$$\boxed{T(x,t) = \frac{1}{\sqrt{\pi}\, x_0(t)} \exp\left[-\frac{x^2}{x_0^2(t)}\right]} \tag{11.13b}$$

Notice that the temperature profile remains Gaussian at all times, but its width, which is related to $x_0(t)$, increases with time according to Eq. (11.13a). The time is scaled by the diffusivity D, such that if one material has twice the diffusivity of another, the spread of heat will take half as much time in the first material compared with that in the second. Also notice that if the initial distribution is an impulse (i.e., $x_0(0) = 0$), the width of the distribution would increase as the square root of time. □

Next we investigate the effect of a heat source. Let a one-dimensional laser beam be focused on the rod with intensity profile $I(x,t)$, and assume that all the focused energy is absorbed. The absorbed energy is converted locally into heat before diffusing along the X-axis. If Δt is chosen to be small enough compared to the time constant of diffusion, then the contribution of the beam to the right-hand side of Eq. (11.1) will be

$I(x,t)\,\Delta x\,\Delta t$. Consequently, Eq. (11.2) will have to be augmented to include the heat source, as follows:

$$\frac{\partial}{\partial t} T(x,t) = D \frac{\partial^2}{\partial x^2} T(x,t) + \frac{1}{C'} I(x,t) \tag{11.14}$$

$C' = CA$ is the specific heat per unit length of the rod. To understand the effect of the laser beam on the temperature profile it is useful to consider the following scenario. Let the laser be turned on for a short period Δt (short enough that diffusion during Δt may be ignored). Then all the energy delivered to the rod is used to raise its temperature, and the temperature profile immediately after the pulse is the sum of the profile before the pulse and the normalized laser intensity profile, namely,

$$T(x,\, t_0 + \Delta t) = T(x,t_0) + \frac{\Delta t}{C'} I(x,t_0) \tag{11.15}$$

The laser thus changes the initial profile, and, after it is turned off, the new profile decays as discussed earlier. If the laser is on for more than just a short period, then time may be divided into short intervals, in each of which small parcels of energy will be delivered to the rod, causing the local temperature to rise instantaneously. This approach is quite general and may be applied to any arbitrary distribution of intensity in time and space. In particular, it may be applied to a situation where the beam moves along the rod. We shall return to this point shortly.

In the special case when the beam has a fixed spatial profile $I(x)$, is stationary relative to the rod, and has a time dependence $h(t)$, we have

$$I(x,t) = I(x)\,h(t) \tag{11.16}$$

In this case, assuming that the rod is initially at a uniform temperature, the temperature profile created by the external source may be written as a superposition integral:

$$\boxed{T(x,t) = \int_{\tau=-\infty}^{t} \frac{1}{C'} h(\tau)\,\mathrm{d}\tau \int_{\sigma=-\infty}^{\infty} \hat{I}(\sigma) \exp\left[\mathrm{i}2\pi\sigma x - 4\pi^2\sigma^2 D(t-\tau)\right] \mathrm{d}\sigma} \tag{11.17}$$

Here $\hat{I}(\sigma)$ is the Fourier transform of the beam's spatial intensity distribution $I(x)$. Equation (11.17) represents a convolution between the laser pulse shape $h(t)$ and the temperature profile $T(x,t)$ at a fixed point x along the rod due to an initial temperature profile $I(x)/C'$.

Example 2. When the profile $I(x)$ of the beam is Gaussian, the inner integral in Eq. (11.17) can be evaluated analytically, in which case the temperature distribution becomes

$$T(x,t) = \frac{1}{C'} \int_{-\infty}^{t} \frac{h(\tau)}{\sqrt{\pi\,[x_0^2 + 4D(t-\tau)]}} \exp\left(-\frac{x^2}{x_0^2 + 4D(t-\tau)}\right) \mathrm{d}\tau \tag{11.18}$$

□

The case of a moving heat source can also be investigated using similar arguments. Consider the following intensity profile for a moving beam:

$$I(x,t) = I_0(x - Vt)\,h(t) \tag{11.19}$$

where V is the velocity of the beam along X, $I_0(x)$ is the intensity profile at $t = 0$, and $h(t)$ is an arbitrary pulse shape. At $t = \tau$ the Fourier transform of the intensity profile is written

$$\mathscr{F}\left\{I_0(x - V\tau)\right\} = \exp(-\mathrm{i}2\pi\sigma V\tau)\,\mathscr{F}\left\{I_0(x)\right\} = \hat{I}_0(\sigma)\exp(-\mathrm{i}2\pi\sigma V\tau) \tag{11.20}$$

This instantaneous intensity profile gives rise to the same profile of temperature, albeit with a proportionality coefficient $1/C'$. The thermal profile decays at later times $t > \tau$, with each Fourier component decaying with its own characteristic time constant. Thus

$$\boxed{T(x,t) = \frac{1}{C'} \int_{\tau=-\infty}^{t} h(\tau)\,\mathrm{d}\tau \int_{\sigma=-\infty}^{\infty} \hat{I}_0(\sigma)\exp(-\mathrm{i}2\pi\sigma V\tau) \times \exp\left[\mathrm{i}2\pi\sigma x - 4\pi^2\sigma^2 D(t-\tau)\right] \mathrm{d}\sigma} \tag{11.21}$$

Example 3. Again, for a Gaussian beam profile the inner integral in Eq. (11.21) can be evaluated analytically, yielding

$$T(x,t) = \frac{1}{C'} \int_{-\infty}^{t} \frac{h(\tau)}{\sqrt{\pi\,[x_0^2 + 4D(t-\tau)]}} \exp\left(-\frac{(x - V\tau)^2}{x_0^2 + 4D(t-\tau)}\right) \mathrm{d}\tau \tag{11.22}$$

The above distribution is the superposition of many Gaussian terms, each widened (by diffusion) for a time period of $t - \tau$ from an initial Gaussian profile of width x_0, centered at $x = V\tau$. The weighting factor for each of these Gaussians is $h(\tau)$, the laser power at time τ.

□

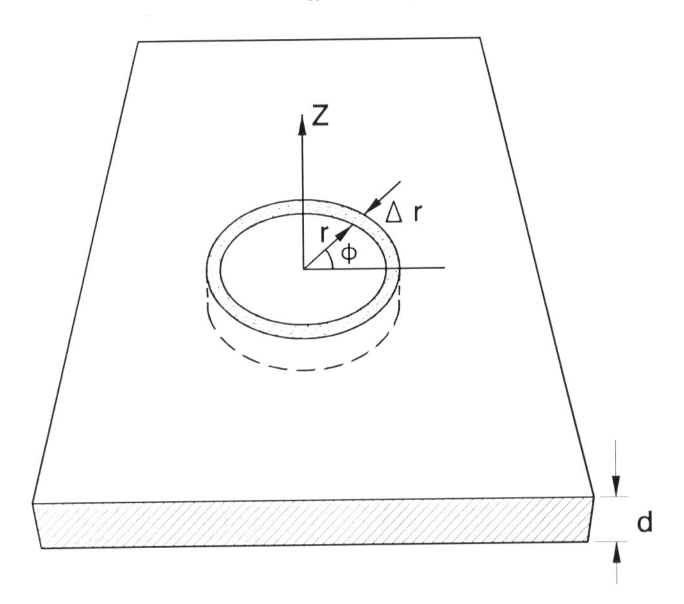

Figure 11.2. Geometry of a two-dimensional heat diffusion problem. An annulus of radius r and width Δr is used to derive equations that describe the balance of heat in this element. The temperature distribution is uniform throughout the thickness.

11.1.2. Heat diffusion in two-dimensional problems with circular symmetry

The schematic diagram in Fig. 11.2 shows a film of uniform thickness d, specific heat C and thermal conductivity K, at an initial temperature distribution $T(r, t = 0)$. The temperature is assumed uniform through the film thickness and in the angular coordinate ϕ. In the absence of sources and sinks the initial profile will spread by diffusion and smooth itself out. To obtain the diffusion equation we write the balance of energy in the annular region of radius r and width Δr, as follows:

$$2\pi r \Delta r C d \left[T(r, t + \Delta t) - T(r, t) \right] =$$

$$2\pi (r + \Delta r) K d \frac{\partial}{\partial r} T(r + \Delta r, t) \Delta t - 2\pi r K d \frac{\partial}{\partial r} T(r, t) \Delta t \qquad (11.23)$$

In the limit when $\Delta r \to 0$ and $\Delta t \to 0$ the above equation becomes

$$\boxed{\frac{\partial}{\partial t} T(r, t) = \frac{D}{r} \frac{\partial}{\partial r} \left[r \frac{\partial}{\partial r} T(r, t) \right]} \qquad (11.24)$$

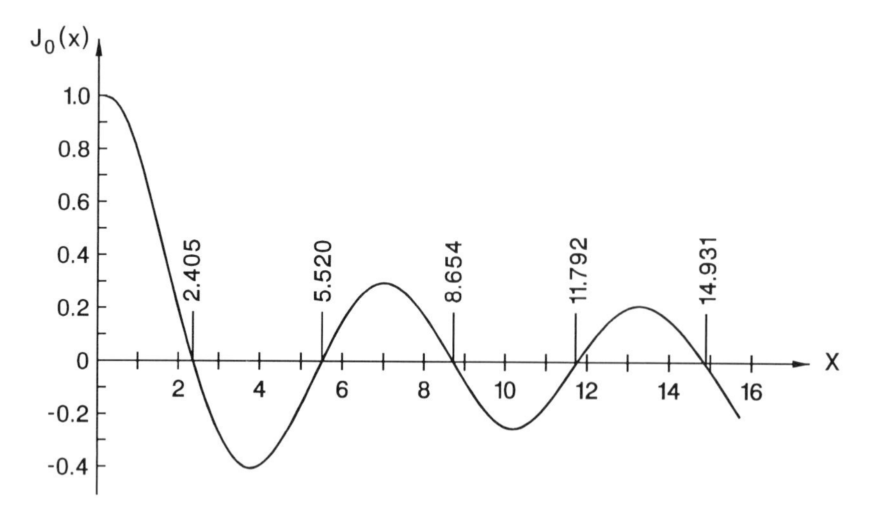

Figure 11.3. Plot of the zero-order Bessel function of the first kind, $J_0(x)$. The zero-crossings are identified on the graph.

where D is the thermal diffusivity. To solve the above equation we separate the variables, i.e., $T(r,t) = f(r)g(t)$, then substitute this in Eq. (11.24) to obtain

$$g'(t) = -\alpha g(t) \qquad f''(r) + \frac{1}{r}f'(r) + \frac{\alpha}{D}f(r) = 0 \tag{11.25}$$

As before, we have chosen $-\alpha$, a negative constant, to separate the two equations. The first of Eqs. (11.25) is a first-order ordinary differential equation whose solution is readily found to be an exponential function. The second is a Bessel equation whose solution is a Bessel function of zero order. Thus

$$\boxed{g(t) = \exp(-\alpha t)} \qquad \boxed{f(r) = T_0 J_0(\sqrt{\alpha/D}\,r)} \tag{11.26}$$

(A plot of $J_0(x)$ is shown in Fig. 11.3.) Thus, when the initial distribution is a zero-order Bessel function, the temperature continues to maintain its profile, while its magnitude decays exponentially with time. If we define an effective width parameter, r_0, for the Bessel function of Eq. (11.26), by

$$r_0 = \sqrt{D/\alpha} \tag{11.27}$$

the function $g(t)$ is then given by

$$g(t) = \exp\left(-\frac{Dt}{r_0^2}\right) \tag{11.28}$$

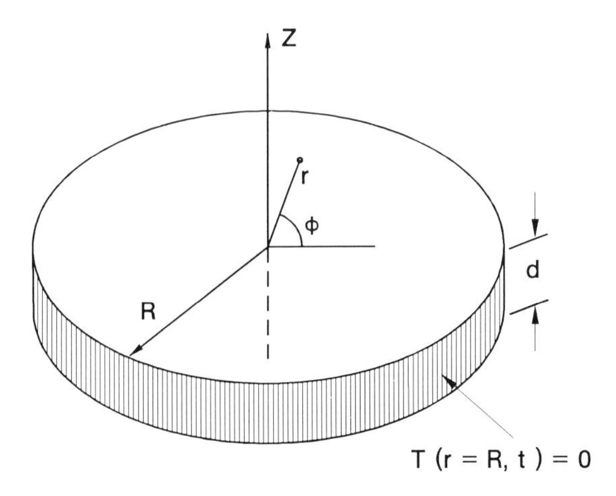

Figure 11.4. A uniform, flat disk of radius R and thickness d. The periphery of the disk at $r = R$ is held at the constant ambient temperature.

According to the above equation, time is scaled by the diffusivity D. Also, the wider a Bessel function, the longer it takes for the corresponding temperature to decay. These observations are consistent with one's expectations based on the nature of thermal diffusion in uniform media.

In general, an arbitrary initial distribution with circular symmetry may be decomposed into a number of zero-order Bessel functions, each with its own width parameter. The subsequent decay of each such term according to Eq. (11.28), and the superposition of all terms, then yields the temperature profile at later times.

Example 4. With reference to Fig. 11.4, consider an initial temperature distribution with the following Gaussian profile:

$$T(r, t = 0) = \frac{1}{\pi r_0^2} \exp\left(-\frac{r^2}{r_0^2}\right) \tag{11.29}$$

The disk-shaped slab in Fig. 11.4 has thickness d, thermal diffusivity D, and large (but finite) radius R. Furthermore, the boundary at $r = R$ is held at zero temperature, that is, $T(r = R, t) = 0$. The solution in Eq. (11.26) must now be slightly modified to reflect this new boundary condition: $J_0(\sqrt{\alpha/D}\,R)$ must be set equal to zero. This leads to

$$\alpha = \frac{D}{R^2} \lambda_n^2 \tag{11.30}$$

where λ_n is any one of the infinite number of roots of $J_0(x)$. Under these circumstances we expand the Gaussian profile of Eq. (11.29) as follows:

$$\frac{1}{\pi r_0^2} \exp\left(-\frac{r^2}{r_0^2}\right) = \sum_{n=1}^{\infty} T_n J_0\left(\frac{\lambda_n r}{R}\right) \tag{11.31}$$

The coefficients T_n are yet to be determined. Multiplying both sides of the above equation with $r J_0(\lambda_m r/R)$ and integrating from 0 to R yields

$$\frac{1}{\pi r_0^2} \int_0^R r \exp\left(-\frac{r^2}{r_0^2}\right) J_0\left(\frac{\lambda_m r}{R}\right) \mathrm{d}r = \sum_{n=1}^{\infty} T_n \int_0^R r J_0\left(\frac{\lambda_n r}{R}\right) J_0\left(\frac{\lambda_m r}{R}\right) \mathrm{d}r \tag{11.32}$$

Orthogonality of the Bessel functions ensures that the right-hand side terms of Eq. (11.32) are all zero except when $\lambda_n = \lambda_m$. As for the left-hand side, we assume $R >> r_0$, set the upper limit of integration to ∞, and obtain

$$\frac{1}{2\pi} \exp\left(-\frac{\lambda_m^2 r_0^2}{4R^2}\right) = \tfrac{1}{2} T_m R^2 J_1^2(\lambda_m)\,; \qquad R >> r_0 \tag{11.33}$$

Equation (11.33) yields T_m for $m = 1, 2, 3, \ldots$. Now, the time-dependence of each term in Eq. (11.31) is given by the first of Eqs. (11.26); therefore

$$T(r,t) = \sum_{n=1}^{\infty} T_n \exp\left(-\frac{\lambda_n^2}{R^2} Dt\right) J_0\left(\frac{\lambda_n r}{R}\right) \tag{11.34}$$

Substituting for T_n in Eq. (11.34) from Eq. (11.33) and combining the exponential terms, we find that the profile for $t > 0$ remains Gaussian but its size r_0 increases with time:

$$\boxed{r_0(t) = \sqrt{r_0^2 + 4Dt}} \tag{11.35}$$

It is thus seen that radial diffusion has characteristics very similar to those of diffusion in one-dimensional space described in subsection 11.1.1. □

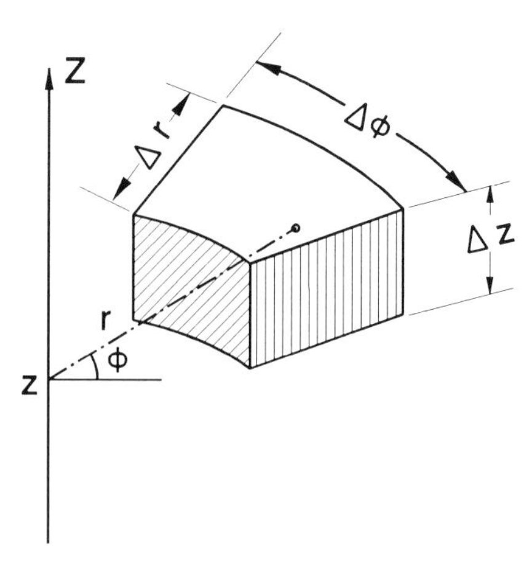

Figure 11.5. Infinitesimal volume element in cylindrical coordinates.

11.1.3. A three-dimensional heat diffusion problem

With reference to the cylindrical geometry of Fig. 11.5, the equation of heat diffusion in three-dimensional space in the absence of sinks and sources of heat can be written as follows:

$$C\left[T(r,\phi,z,t+\Delta t)-T(r,\phi,z,t)\right]r\Delta\phi\Delta r\Delta z =$$

$$K\left\{\left[\frac{\partial}{\partial r}T(r+\Delta r,\phi,z,t)\right](r+\Delta r)\Delta\phi\Delta z\Delta t-\left[\frac{\partial}{\partial r}T(r,\phi,z,t)\right]r\Delta\phi\Delta z\Delta t\right.$$

$$+\left[\frac{1}{r}\frac{\partial}{\partial\phi}T(r,\phi+\Delta\phi,z,t)\right]\Delta r\Delta z\Delta t-\left[\frac{1}{r}\frac{\partial}{\partial\phi}T(r,\phi,z,t)\right]\Delta r\Delta z\Delta t$$

$$\left.+\left[\frac{\partial}{\partial z}T(r,\phi,z+\Delta z,t)\right]r\Delta\phi\Delta r\Delta t-\left[\frac{\partial}{\partial z}T(r,\phi,z,t)\right]r\Delta\phi\Delta r\Delta t\right\} \quad (11.36)$$

Normalizing the above equation by $r\Delta r\Delta\phi\Delta z\Delta t$, and allowing the various Δ's to approach zero, we obtain

$$\boxed{\frac{\partial T}{\partial t}=D\left[\frac{1}{r}\frac{\partial}{\partial r}\left(r\frac{\partial T}{\partial r}\right)+\frac{1}{r^2}\frac{\partial^2 T}{\partial\phi^2}+\frac{\partial^2 T}{\partial z^2}\right]} \quad (11.37)$$

The general solution to this equation may be found by the method of separation of variables. Let $T(r,\phi,z,t) = f(r,\phi)\,p(z)\,g(t)$, where $f(\cdot)$, $p(\cdot)$, and $g(\cdot)$ are arbitrary functions of their respective variables. Substituting for $T(r,\phi,z,t)$ in Eq. (11.37) and rearranging the terms yields

$$\frac{g'(t)}{g(t)} = D\left[\frac{r\dfrac{\partial}{\partial r}\left(r\dfrac{\partial f}{\partial r}\right) + \dfrac{\partial^2 f}{\partial \phi^2}}{r^2 f(r,\phi)} + \frac{p''(z)}{p(z)}\right] \tag{11.38}$$

As before, we find that each term in Eq. (11.38) must be a constant. Thus,

$$p''(z) + \beta^2 p(z) = 0 \tag{11.39a}$$

$$r\frac{\partial}{\partial r}\left(r\frac{\partial f}{\partial r}\right) + \frac{\partial^2 f}{\partial \phi^2} + \gamma^2 r^2 f(r,\phi) = 0 \tag{11.39b}$$

$$g'(t) + D(\beta^2 + \gamma^2)\,g(t) = 0 \tag{11.39c}$$

The second equation can further be resolved into two equations in r and ϕ by another separation of variables:

$$W''(\phi) + n^2 W(\phi) = 0 \tag{11.40a}$$

$$r^2 q''(r) + r\,q'(r) - (n^2 - \gamma^2 r^2)\,q(r) = 0 \tag{11.40b}$$

These equations can be solved straightforwardly. The azimuthal profile $W(\cdot)$ is found from Eq. (11.40a) to be a sinusoid:

$$W(\phi) = \sin(n\phi + \phi_n) \tag{11.41a}$$

where ϕ_n is an arbitrary constant, but n must be an integer since $W(\phi)$ must repeat itself when multiples of 2π are added to ϕ. (One possible solution, of course, is $n = 0$ which yields a circularly symmetric profile around Z.) $q(r)$ will satisfy Eq. (11.40b) if it is a Bessel function of order n, namely,

$$q(r) = J_n(\gamma r) \tag{11.41b}$$

Figure 11.6 shows plots of $J_n(r)$ for several values of n. Note that, with the exception of $J_0(\cdot)$, all these functions have a null at the origin. Also, the range of values of r over which $J_n(r)$ is appreciable moves further away from the origin as n increases.

From Eq. (11.39a) the dependence of temperature on z is seen to be the same as that for the one-dimensional case discussed in 11.1.1. Thus

$$p(z) = \exp(i\beta z) \tag{11.41c}$$

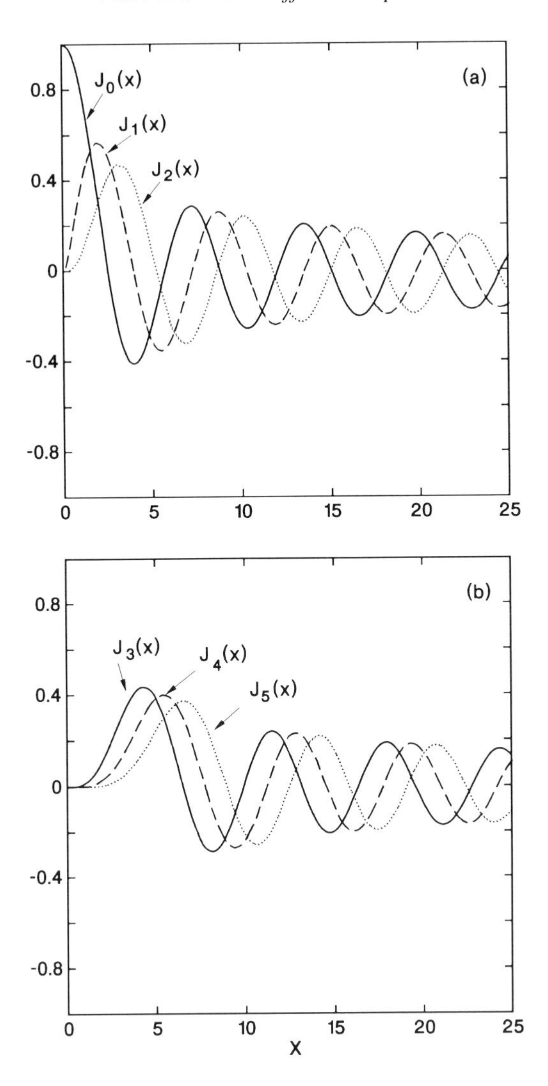

Figure 11.6. Plots of several Bessel functions of the first kind and order n, $J_n(x)$. Functions of order 0 through 5 are shown here. Aside from $J_0(x)$, all functions have a zero at the origin, and their peaks shift to the right as n increases.

Finally, the evolution of the temperature profile in time according to Eq. (11.39c) is as follows:

$$g(t) = \exp\left[-D(\beta^2 + \gamma^2)\, t\right] \tag{11.41d}$$

Notice that T at a given point in space decays with two time constants: one related to β (diffusion along Z) and another related to γ (radial diffusion). At first, it might seem that the angular distribution of T does not affect the time rate of this decay. This is deceiving, however, since the value of n

in Eq. (11.41a) determines the radial pattern through Eq. (11.41b); larger values of n, therefore, correspond to radial distributions that are confined to regions further away from the origin. Thus, as n becomes larger and causes more oscillations in the azimuthal profile, the lobe of the Bessel function acquires a larger radius, keeping the gradient of temperature more or less constant.

Example 5. The following thermal profile has been established in a uniform material of diffusivity D that covers the entire space:

$$T(r,\phi,z,t=0) = \frac{1}{\sqrt{\pi}\,z_0} \exp\left(-\frac{z^2}{z_0^2}\right) J_n(\gamma r) \sin(n\phi + \phi_n) \tag{11.42}$$

Given that the time dependence of the radial and azimuthal parts of the above distribution is readily known from Eq. (11.41d) to be $\exp(-D\gamma^2 t)$, Fourier analysis of the z-dependent part reveals the general pattern of time evolution as follows:

$$T(r,\phi,z,t) = \frac{1}{\sqrt{\pi}\,z_0(t)} \exp\left[-\frac{z^2}{z_0^2(t)}\right] J_n(\gamma r) \sin(n\phi + \phi_n) \exp(-D\gamma^2 t) \tag{11.43a}$$

Here $z_0(t)$, the width of the profile along Z, is given by

$$\boxed{z_0(t) = \sqrt{z_0^2 + 4Dt}} \tag{11.43b}$$

Thus the profile along Z widens (in proportion to the square root of Dt) as time progresses. The width along r, however, does not change since the initial radial profile was an eigenfunction of the diffusion equation.

However, it is not necessary for the radial profile to be an eigenfunction. One may replace the factor carrying the radial and azimuthal dependence in Eq. (11.42) by a superposition of such terms having different values of n and γ, and the Gaussian profile along Z will continue to obey Eq. (11.43b). Thus, if a thin, hot layer of some material happens to be embedded within a block of the same material, as in Fig. 11.7(a), the width of the heated region along Z evolves in time as $z_0 = 2\sqrt{Dt}$. For instance, if $D = 0.1$ cm²/s, and the initial width of the hot layer is negligible, then after 100 ns the width will become 2 μm. The same argument can be applied to the case of a thin, hot layer on top of a substrate, as in Fig. 11.7(b). Although we deal here with one half of a Gaussian profile, symmetry assures us that the same results apply to this case as well. □

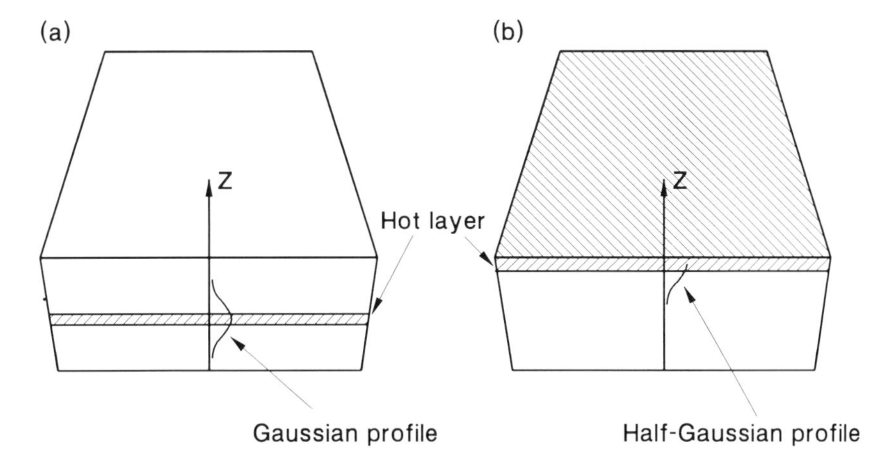

Figure 11.7. Hypothetical embedding of a thin layer of heated material within a block of identical material. In (a) the layer is inside, while in (b) it is on the surface of the block.

11.2. Numerical Solution of the Heat Diffusion Equation

Analytical solutions to the diffusion equation are rare and exist only for simple situations. For practical cases involving complex geometries such as multilayer media, one usually has to resort to numerical techniques for determining the temperature profiles. Both finite element and finite difference methods have been applied to the problem of thermal profiling in optical recording media.[4-6] With today's powerful personal computers there is essentially no limitation to numerical calculations, and highly accurate results may be obtained with either method. In this section we describe a finite difference method, the alternating-direction implicit technique, for solving partial differential equations;[10,11] this technique has been used in all the numerical computations presented in the next section. We outline the implicit method in subsection 11.2.1 for a problem in one-dimensional space, and explain in 11.2.2 how the method may be extended to problems involving two or more spatial dimensions. Subsection 11.2.3 considers the application of these results to the case of moving media.

11.2.1. The implicit method of solving linear partial differential equations

Consider the problem of heat absorption and diffusion in a uniform, one-dimensional rod of cross-sectional area A, similar to that described by Eq. (11.14). We confine our attention to a finite length of the rod between $x = 0$ and $x = L$, and assume that one end of the rod is kept at the ambient temperature, while the other end loses heat at a rate proportional to its temperature differential with the ambient (see Fig. 11.8). We also allow for an initial temperature profile $T_0(x)$. Under these circumstances the temperature above ambient, $T(x,t)$, may be described by the following set of equations:

$$T(x,\ t=0) = T_0(x) \qquad 0 \le x \le L \tag{11.44a}$$

$$T(x=L,\ t) = 0 \qquad t \ge 0 \tag{11.44b}$$

$$\frac{\partial}{\partial x} T(x=0,\ t) = \mu T(x=0,\ t) \qquad \mu \ge 0 \text{ is a constant} \tag{11.44c}$$

$$\frac{\partial}{\partial t} T(x,t) = D \frac{\partial^2}{\partial x^2} T(x,t) + \frac{1}{CA} I(x,t) \qquad 0 \le x \le L \quad t \ge 0 \tag{11.44d}$$

To convert the above into a set of finite difference equations, we define the time step Δt, and select the integer N as the number of sub-intervals of length Δx that cover $[0, L]$. The discrete form of Eqs. (11.44) will then be

$$T(n\Delta x,\ 0) = T_0(n\Delta x) \qquad 0 \le n \le N \tag{11.45a}$$

$$T(N\Delta x,\ m\Delta t) = 0 \qquad m \ge 1 \tag{11.45b}$$

$$T(0,\ m\Delta t) - T(0,\ (m-1)\,\Delta t) = \frac{2D\Delta t}{(\Delta x)^2}\left[T(\Delta x,\ m\Delta t) - (1+\mu\Delta x)\,T(0,\ m\Delta t)\right] + \frac{2}{CA\Delta x}\int_0^{\frac{1}{2}\Delta x}\int_{(m-1)\Delta t}^{m\Delta t} I(x,t)\,\mathrm{d}x\,\mathrm{d}t \qquad m \ge 1 \tag{11.45c}$$

$$T(n\Delta x,\ m\Delta t) - T(n\Delta x,\ (m-1)\Delta t) = \frac{D\Delta t}{(\Delta x)^2}\left\{T((n+1)\,\Delta x,\ m\,\Delta t) - 2T(n\Delta x,\ m\Delta t) + T((n-1)\,\Delta x,\ m\,\Delta t)\right\} + \frac{1}{CA\Delta x}\int_{(n-\frac{1}{2})\Delta x}^{(n+\frac{1}{2})\Delta x}\int_{(m-1)\Delta t}^{m\Delta t} I(x,t)\,\mathrm{d}x\,\mathrm{d}t \qquad 1 \le n \le N-1,\ \ m \ge 1 \tag{11.45d}$$

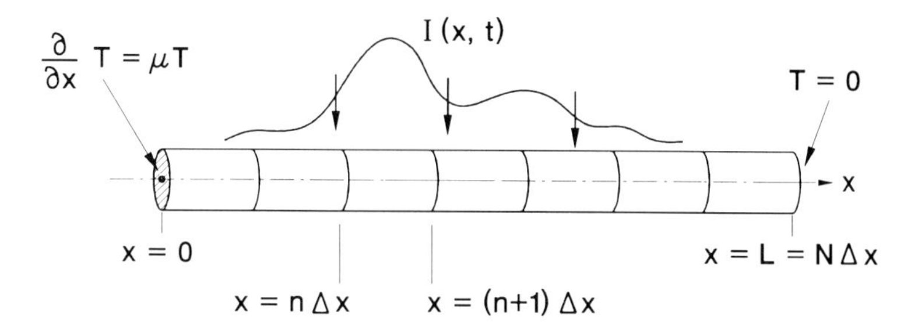

Figure 11.8. Discretization of the one-dimensional heat flow problem. The rod of length L and cross-sectional area A is subjected to a heat source in the form of a laser beam with intensity profile $I(x,t)$. One end of the rod is held at the constant ambient temperature, while the other end loses heat at a rate proportional to its temperature differential with the ambient.

Note that in these formulas the derivatives of $T(x,t)$ with respect to x have been evaluated at $t = m\Delta t$. This is the essence of the implicit technique. Had we opted for the explicit method and evaluated the derivatives at $t = (m-1)\Delta t$, the algorithm would have been easier to implement but the solution would have become less stable. The problem posed by Eqs. (11.45) may be solved by first rearranging the equations as follows:

$$T(n\Delta x,\, 0) = T_0(n\Delta x) \qquad 0 \le n \le N \tag{11.46a}$$

$$T(N\Delta x,\, m\Delta t) = 0 \qquad m \ge 1 \tag{11.46b}$$

$$T(0,\, m\Delta t) = Q_1^{(1)}\, T(\Delta x,\, m\Delta t) + Q_2^{(1,m)} \qquad m \ge 1 \tag{11.46c}$$

$$T((n-1)\Delta x,\, m\Delta t) - \beta T(n\Delta x,\, m\Delta t) + T((n+1)\Delta x,\, m\Delta t) + \gamma^{(n,m)} = 0$$

$$1 \le n \le N-1, \quad m \ge 1 \tag{11.46d}$$

In these equations,

$$Q_1^{(1)} = \frac{1}{1 + \mu\Delta x + \dfrac{(\Delta x)^2}{2D\Delta t}} \tag{11.46e}$$

$$Q_2^{(1,m)} = \frac{\dfrac{(\Delta x)^2}{2D\Delta t} T(0,\,(m-1)\Delta t) + \dfrac{\Delta x}{KA\Delta t} \displaystyle\int_0^{\frac{1}{2}\Delta x} \int_{(m-1)\Delta t}^{m\Delta t} I(x,t)\, \mathrm{d}x\mathrm{d}t}{1 + \mu\Delta x + \dfrac{(\Delta x)^2}{2D\Delta t}} \tag{11.46f}$$

$$\beta = 2 + \frac{(\Delta x)^2}{D\Delta t} \tag{11.46g}$$

$$\gamma^{(n,m)} = \frac{(\Delta x)^2}{D\Delta t} T(n\Delta x, (m-1)\Delta t) + \frac{\Delta x}{KA\Delta t} \int_{(n-\frac{1}{2})\Delta x}^{(n+\frac{1}{2})\Delta x} \int_{(m-1)\Delta t}^{m\Delta t} I(x,t)\, \mathrm{d}x\,\mathrm{d}t \tag{11.46h}$$

Next, taking our cue from Eq. (11.46c), we postulate the existence of the following relation between successive values of T along the X-axis:

$$\boxed{T((n-1)\Delta x, m\Delta t) = Q_1^{(n)} T(n\Delta x, m\Delta t) + Q_2^{(n,m)} \qquad 1 \le n \le N, \quad m \ge 1} \tag{11.47}$$

Substituting in Eq. (11.46d) for $T((n-1)\Delta x, m\Delta t)$ from Eq. (11.47) and rearranging the terms, we find the following recursive relations for Q_1, Q_2:

$$\boxed{Q_1^{(n+1)} = \frac{1}{\beta - Q_1^{(n)}}} \tag{11.48a}$$

$$\boxed{Q_2^{(n+1,m)} = \frac{\gamma^{(n,m)} + Q_2^{(n,m)}}{\beta - Q_1^{(n)}}} \tag{11.48b}$$

The implicit algorithm for solving Eq. (11.44) may now be summarized in the following steps.

(i) Set $m = 1$.

(ii) Calculate β from Eq. (11.46g) and $\gamma^{(n,m)}$ for $1 \le n \le N-1$ from Eq. (11.46h).

(iii) Calculate $Q_1^{(n)}$ and $Q_2^{(n,m)}$ for $1 \le n \le N$ from Eqs. (11.46e,f) and Eqs. (11.48).

(iv) Starting with the end-point given by Eq. (11.46b), compute $T(n\Delta x, m\Delta t)$ for $0 \le n \le N-1$ using Eq. (11.47).

(v) Set $m = m+1$ and repeat the above steps.

11.2.2. The alternating-direction implicit technique

The implicit method described in 11.2.1 for a problem with one spatial dimension may be extended to PDEs in two or more dimensions. For example, in heat diffusion problems with symmetry in the azimuthal coordinate ϕ, the temperature may be expressed as a function of the spatial coordinates r and z, as well as the time t. In such problems the time step Δt is divided into two half-steps. In one half-step the z-derivatives are treated explicitly and the r-derivatives implicitly; in the other half-step z becomes implicit and r explicit. Thus in each half-step the problem is similar to the one-dimensional case, and is treated likewise. Stable numerical solutions to PDEs with multiple variables can be obtained with this so-called alternating-direction implicit method.

11.2.3. Extension to moving media

Consider a cylindrical coordinate system centered on the axis of a beam of light, with the origin at the surface of the medium, as in Fig. 11.9. The positive Z-axis enters the medium, so that (x_0, y_0, z_0) is at depth z_0 below the surface and at radial distance $\sqrt{(x_0^2 + y_0^2)}$ from the beam axis. Assuming that the medium moves in the positive X-direction with constant velocity V, we approximate its motion by a discrete set of movements in which the medium jumps a distance $V\Delta t$ at the end of each time-step Δt. During each Δt, however, the medium is stationary and the static technique described in the preceding subsections applies. Moreover, for the specific case of temperature-independent optical and thermal parameters, the absorption and diffusion equations become linear, and the moving-medium problem may be solved by superposition of the static solutions at progressive time intervals.

Using the static technique we calculate $T(r, z, t)$ corresponding to a narrow stationary pulse with the following Gaussian profile:

$$I(r,t) = \begin{cases} \dfrac{1}{\pi r_0^2} \exp\left[-(r/r_0)^2\right] & 0 \le t \le \Delta t \\ 0 & \text{otherwise} \end{cases} \tag{11.49}$$

To determine the temperature $T(x_0, y_0, z_0,\ t_0 = M\Delta t)$ that develops in response to the laser pulse $P(t)$, we compute the set $\{P_m\}$ where

$$P_m = \frac{1}{\Delta t} \int_{(m-1)\Delta t}^{m\Delta t} P(t)\,\mathrm{d}t \qquad 1 \le m \le M \tag{11.50}$$

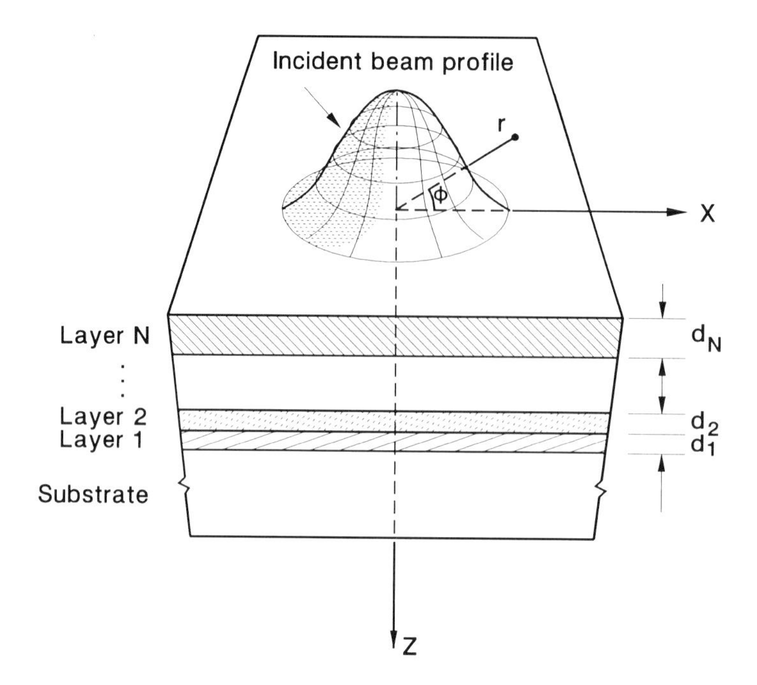

Figure 11.9. Multilayer with a cylindrical coordinate system (r,ϕ,z). The number of layers is N, and the surface is at $z = 0$. A circularly symmetric Gaussian beam propagating in the positive Z-direction illuminates the surface. The axis of symmetry of the beam is at $r = 0$.

P_m represents the average pulse power in the mth time interval. Next, we compute the sequence $\{r_m\}$ where

$$r_m^2 = \left[x_0 - (M - m + \tfrac{1}{2})\, V\Delta t\right]^2 + y_0^2 \qquad 1 \le m \le M \qquad (11.51)$$

r_m is the radial distance between (x_0, y_0, z_0) and the beam axis during the mth time interval. Finally, we add the individual responses and obtain the total response, as follows:

$$T(x_0, y_0, z_0, t_0) = \sum_{m=1}^{M} P_m\, T(r_m,\, z_0,\, (M - m + 1)\Delta t) \qquad (11.52)$$

Since $T(r,z,t)$ is originally calculated on a discrete mesh, it is necessary to interpolate in order to obtain its intermediate values. One may use linear interpolation in the r-direction, but interpolation in z and in t can be avoided by appropriate choices of z_0 and Δt at the outset.

11.3. Light Absorption and Heat Diffusion in Multilayers

We consider again the N-layer structure illuminated by a focused beam of light shown in Fig. 11.9. The coordinate system is cylindrical, with the surface of the multilayer at $z = 0$; the structure extends to infinity in all other directions. Starting at the substrate with $k = 0$, the layers are numbered in increasing order. The kth layer has thickness d_k, refractive index $n_k = \mathrm{Re}(n_k) + \mathrm{i}\,\mathrm{Im}(n_k)$, specific heat C_k, and thermal conductivity K_k. The incident beam (vacuum wavelength = λ_0) has a Gaussian intensity profile with radius r_0 at the $1/e$ point; it moves with constant velocity V along $+X$, and its time-dependent power will be denoted by $P_0(t)$. It is assumed that the entire multilayer structure stays within the depth of focus and, as such, variations of r_0 along the Z-axis will be ignored.

The temperature rise above the ambient at point (r, z) and time t is $T(r, z, t)$, and the rate of heat flow from the surface is assumed to be proportional to the local temperature, that is,

$$\frac{\partial}{\partial z} T(r,\ z = 0,\ t) = \mu T(r,\ z = 0,\ t) \tag{11.53}$$

where $\mu \geq 0$ is the proportionality constant. For a unit-intensity incident plane wave, the Poynting vector $Y(z)$ gives the average rate of flow of optical energy through a plane perpendicular to Z at z; $Y(z)$, which may be obtained from the Maxwell equations, has been described in detail in Chapter 5. The slope of $Y(z)$ determines the rate of absorption of energy per unit volume at $Z = z$. As for the diffusion of the heat thus generated, numerical methods (such as the alternating-direction implicit technique) yield reliable solutions for all cases of practical interest. In the examples that follow several such cases will be described in detail.

In the first example we assume that the medium is stationary (i.e, $V = 0$), and pay particular attention to the effects of light absorption and heat diffusion. In later examples we consider the thermal profiles of moving media, and examine the laser power requirements for raising the temperature of the recording layer to certain values. Two situations will be considered: (i) the laser pulse length is r_0/V, and (ii) the pulse is long enough for thermal equilibrium to be established. The two cases, therefore, relate to that of writing at fixed resolution and reading under CW illumination, respectively.

Example 6: A stationary, surface-incident, multilayer disk. Consider the quadrilayer structure of Fig. 11.10(a), the material parameters for which are listed in Table 11.1. This is a surface-incident structure designed for operation in non-removable environments where there is no need for illumination through the substrate. Figure 11.10(b) shows the calculated Poynting vector $Y(z)$ for this multilayer; $Y(z)$ is the rate of flow of energy at $Z = z$ for a unit-intensity incident beam. Note in particular that almost all the energy is absorbed by the magnetic film (100 nm $\leq z \leq$ 112 nm),

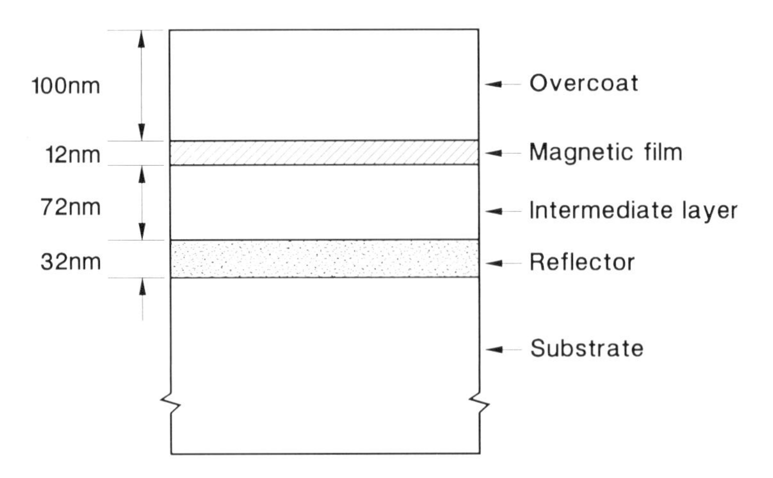
(a)
100nm
12nm
72nm
32nm
Overcoat
Magnetic film
Intermediate layer
Reflector
Substrate

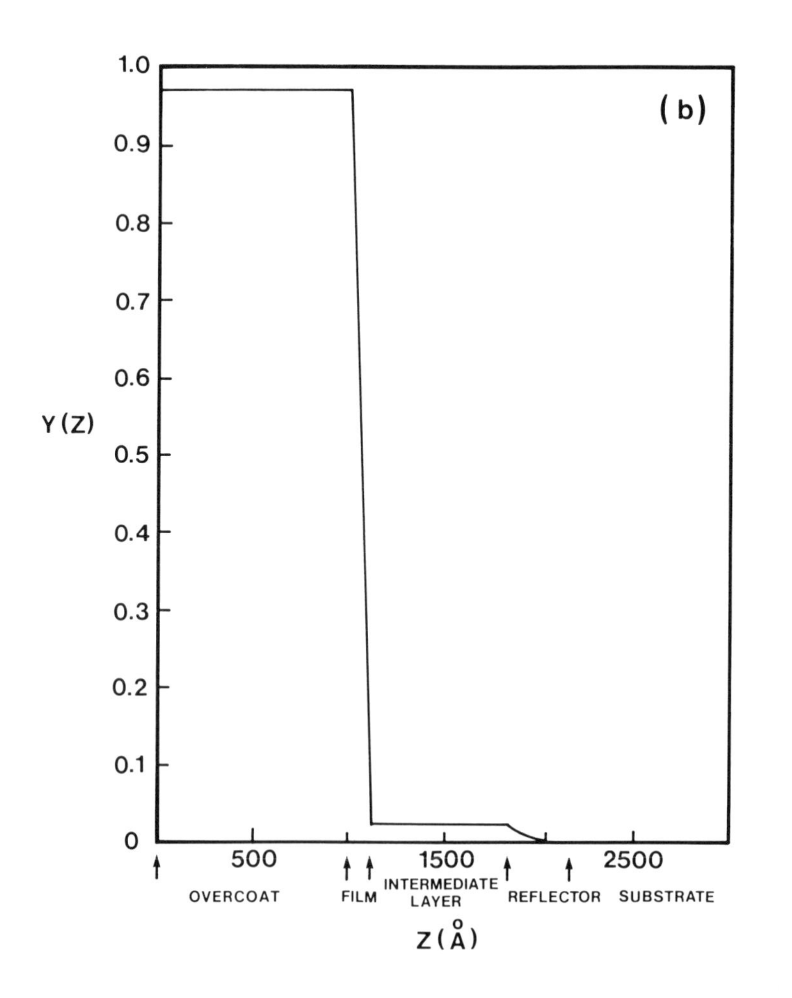
(b)
Y(Z)
1.0
0.9
0.8
0.7
0.6
0.5
0.4
0.3
0.2
0.1
0
500
1500
2500
OVERCOAT
FILM
INTERMEDIATE LAYER
REFLECTOR
SUBSTRATE
Z(Å)

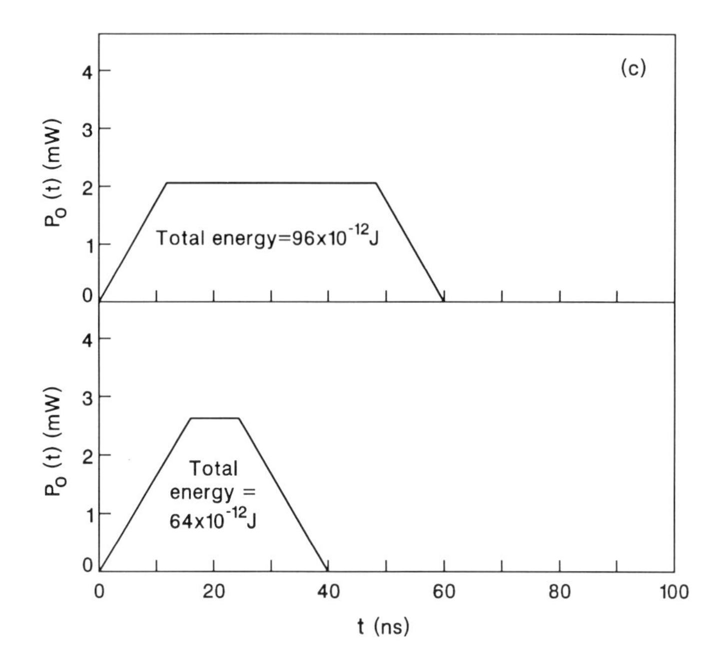

Figure 11.10. (a) Structure of a quadrilayer magneto-optical disk. This device is used in a surface-incident configuration. (The aluminum layer blocks illumination through the substrate.) (b) Plot of $Y(z)$, the average rate of flow of optical energy through the quadrilayer for a plane wave with unit intensity. (c) Two laser pulses used in the numerical calculations of temperature distribution. The 60 ns pulse with peak power of 2 mW and total energy of 96 pJ is used in calculations pertaining to Figs. 11.11 and 11.12. The 40 ns pulse with peak power of 2.67 mW and total energy of 64 pJ is used in calculations pertaining to Fig. 11.13.

Table 11.1. Optical and thermal parameters of the materials in Example 6. The focused beam (λ_0 = 633 nm, r_0 = 350 nm) is incident from the surface side

Layer	Thickness d (nm)	Refractive index n	Specific heat C (J/cm³ K)	Thermal conductivity K (J/cm K s)
substrate (PMMA)	∞	1.46	1.7	0.002
1. reflector (aluminum)	32	1.2 + 6.9 i	2.7	2.4
2. intermediate layer (SiO_2)	72	1.5	2.0	0.015
3. magnetic film	12	3.1 + 3.5 i	2.6	0.16
4. overcoat layer (SiO_2)	100	1.5	2.0	0.015

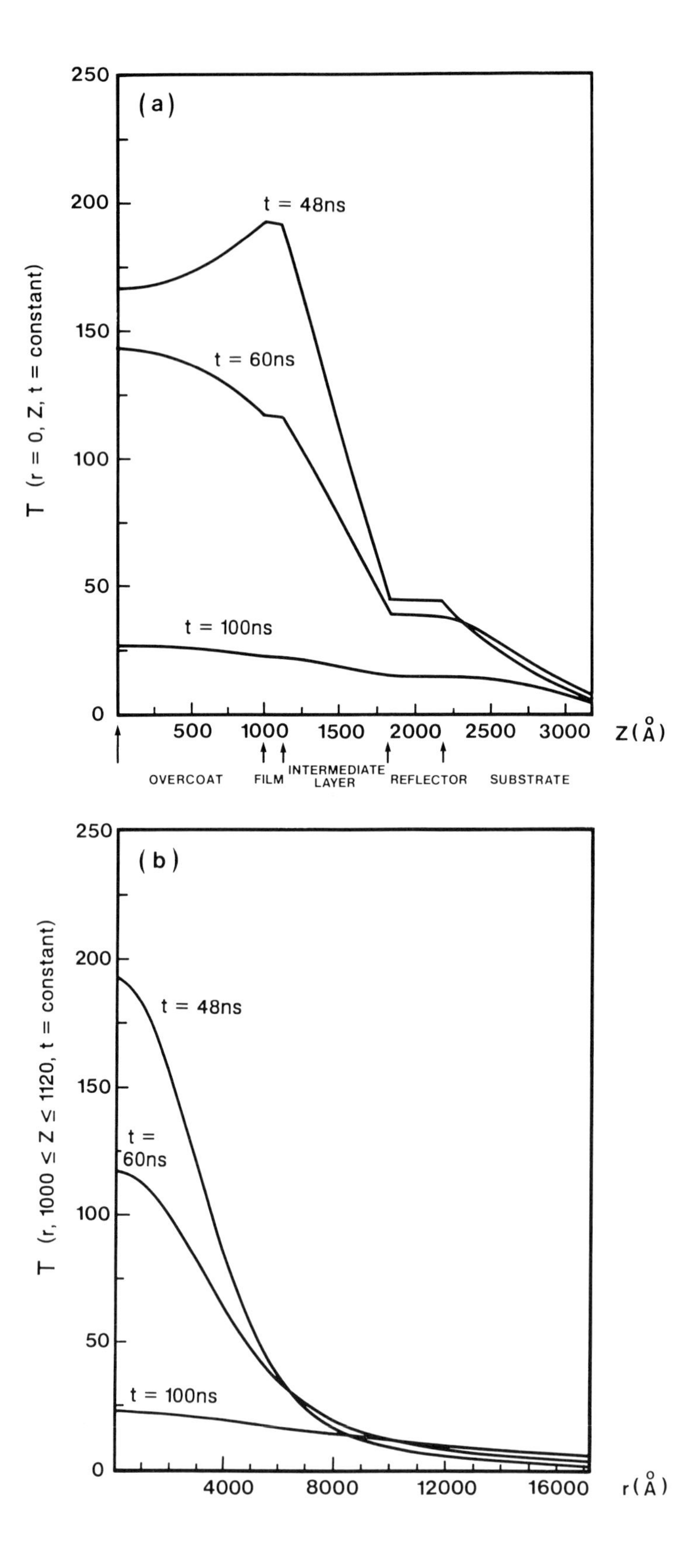

(a)
T (r = 0, Z, t = constant)
250
200
150
100
50
0
t = 48ns
t = 60ns
t = 100ns
500
1000
1500
2000
2500
3000
Z(Å)
OVERCOAT
FILM
INTERMEDIATE LAYER
REFLECTOR
SUBSTRATE
(b)
T (r, 1000 ≤ Z ≤ 1120, t = constant)
250
200
150
100
50
0
t = 48ns
t = 60ns
t = 100ns
4000
8000
12000
16000
r(Å)

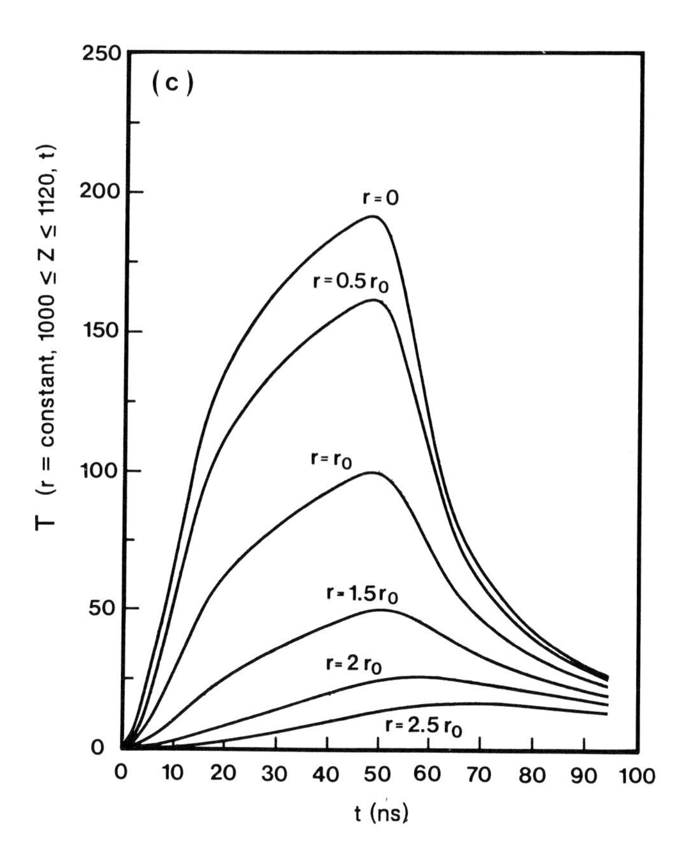

Figure 11.11. Distribution of temperature rise (above ambient) for the quadrilayer of Fig. 11.10(a). At $t = 0$ the quadrilayer is at the ambient temperature, the laser pulse is the longer pulse of Fig. 11.10(c), and the radius of the beam at the $1/e$ point is 350 nm. Heat flow from the surface is neglected here ($\mu = 0$). (a) Temperature versus z at the beam center for several instants of time. (b) Radial temperature profile in the MO film for several instants of time. (c) Time-dependence of the MO film's temperature for several radii.

and that within the film the absorption is fairly uniform. Two trapezoidal laser pulses will be used in this analysis, a 2 mW 60 ns pulse and a 2.67 mW 40 ns pulse; both are shown in Fig. 11.10(c).

The heat generated in the absorbing layers diffuses in all directions and creates a complex pattern of temperature distribution. Let us first assume that the heat flow from the surface is negligible (i.e., $\mu = 0$). For the longer laser pulse shown in Fig. 11.10(c), temperature profiles of the quadrilayer are plotted in Fig. 11.11. Frame (a) shows the temperature above ambient versus z along the optical axis (i.e., at $r = 0$); the curves for three instants of time are shown. At $t = 48$ ns, while the laser is still on, the film has a high temperature; some of its energy is escaping to the overcoat, but most of it is flowing rapidly to the highly conductive aluminum layer. At t = 60 ns, with the laser extinguished completely, the film is cooling down. Again the aluminum layer and the substrate are absorbing most of the film's thermal energy. Finally at t = 100 ns, only 40 ns after the laser is turned off, the temperature has become fairly uniform throughout the structure, and the situation is essentially back to normal. Note that at all instances the film maintains a uniform temperature throughout its thickness. Radial profiles of temperature above ambient in the MO film for the same three instants of time are shown in Fig. 11.11(b). Figure 11.11(c) shows the time-dependence of temperature above ambient in the MO layer for several values of r. Note that the temperature decay around the beam center begins with the decline of the laser power, but is delayed by the effects of diffusion at larger radii.

In reality, especially when the disk is rotating, convection losses at the front surface become important. To see what effect these may have on the thermal profiles, we set $\mu = 10^5\ \mathrm{cm}^{-1}$ in the numerical calculations. This value of μ causes the convection losses at the surface to be of the same order of magnitude as the losses to the substrate. The results, which are the parallels of those in Fig. 11.11, are shown in Fig. 11.12. As expected, the temperatures attained in this case are lower than those obtained previously, but the general features of the curves have remained intact. In the plots of Fig. 11.12(a) the nonzero gradients at $z = 0$ are indicative of heat loss at the front surface.

Roughly speaking, in thermomagnetic recording the region of the magnetic film whose temperature rises above a critical temperature, T_c, becomes reverse-magnetized. For a medium with $T_c = 130\ °\mathrm{C}$, Fig. 11.11(c) shows that a 60 ns pulse with 2 mW of power is capable of writing a spot of diameter 1 μm when convection losses from the surface are ignored. When heat loss is taken into account the same achievement obviously requires more laser power and, according to Fig. 11.12(c) (where $\mu = 10^5\ \mathrm{cm}^{-1}$), the required power increases to 2.5 mW.

Finally, let us consider the short laser pulse of Fig. 11.10(c), for which the computed time-dependence of temperature above ambient in the MO film is given in Fig. 11.13. Comparison with Fig. 11.12(c), which is the corresponding plot for the longer pulse, shows that though less energy is being consumed, higher temperatures are attained with a shorter pulse. This apparent discrepancy may be explained by noting that in the case of the narrow pulse the local temperature buildup occurs before the accumulated heat diffuses away.

□

Example 7: Thermal effects during readout. The readout of MO disks is achieved by means of a CW laser beam focused to a small spot, comparable in size with the recorded magnetic domains. The polar Kerr effect used in readout is proportional to magnetization of the MO layer, which, in general, decreases with the increasing temperature. The read signal is thus expected to increase sublinearly with laser power as a result

Table 11.2. Optical and thermal parameters of the materials in Example 7. The focused beam (λ_0 = 840 nm, r_0 = 500 nm) is incident from the surface side

Material	Refractive index n	Specific heat C (J/cm³ K)	Thermal conductivity K (J/cm K s)
glass substrate	1.5	2.0	0.015
dielectric	2.0	2.0	0.015
aluminum	2.0 + 7.1 i	2.7	2.4
magnetic film	3.67 + 3.85 i	3.2	0.4

of heating of the medium by the read beam. We examine the temperature distribution during readout of the quadrilayer MO device shown in Fig. 11.14(a), with parameters given in Table 11.2. A constant linear disk velocity of V = 20 m/s is assumed, and the effects of heat loss from the surface are ignored ($\mu = 0$). Since the temperature above ambient is proportional to the beam power, the rise in temperature will be given for unit incident power, in °C/mW.

Figure 11.14(b) shows the steady-state contours of constant temperature in the magnetic film. The central contour represents the hottest spot on the film at 40.6 °C/mW, the second contour corresponds to 90% of the maximum temperature, the third to 80%, and so on. The laser beam is centered at the origin of the coordinate system, and the disk moves in the positive X-direction. Note that the maximum temperature does not occur at the beam center, and that, due to radial diffusion, the temperature distribution is somewhat wider than the focused beam's intensity profile.

For comparison, we have shown similar isotherms for the surface of the multilayer and for the aluminum layer in frames (c) and (d) respectively of Fig. 11.14. The maximum temperature above ambient on the surface is 37.9 °C/mW, which is close to that of the magnetic film itself. This is partly due to the fact that we have ignored effects of heat loss from the surface, and partly due to the low thermal conductivity of the overcoat material. The latter is also responsible for the lack of appreciable broadening of the surface profile as compared with the MO film. In contrast, the temperature profile in the aluminum layer (whose thermal conductivity is large) is broadened substantially, resulting in a maximum temperature rise of only 12.9 °C/mW. □

Example 8: A multilayer disk illuminated through the substrate. The quadrilayer MO device shown in Fig. 11.15(a) is designed for use in substrate-incident configurations. The optical and thermal parameters listed in Table 11.3 are used in the computations described below. Figure 11.15(b) shows the temperature distribution at the beam center along the

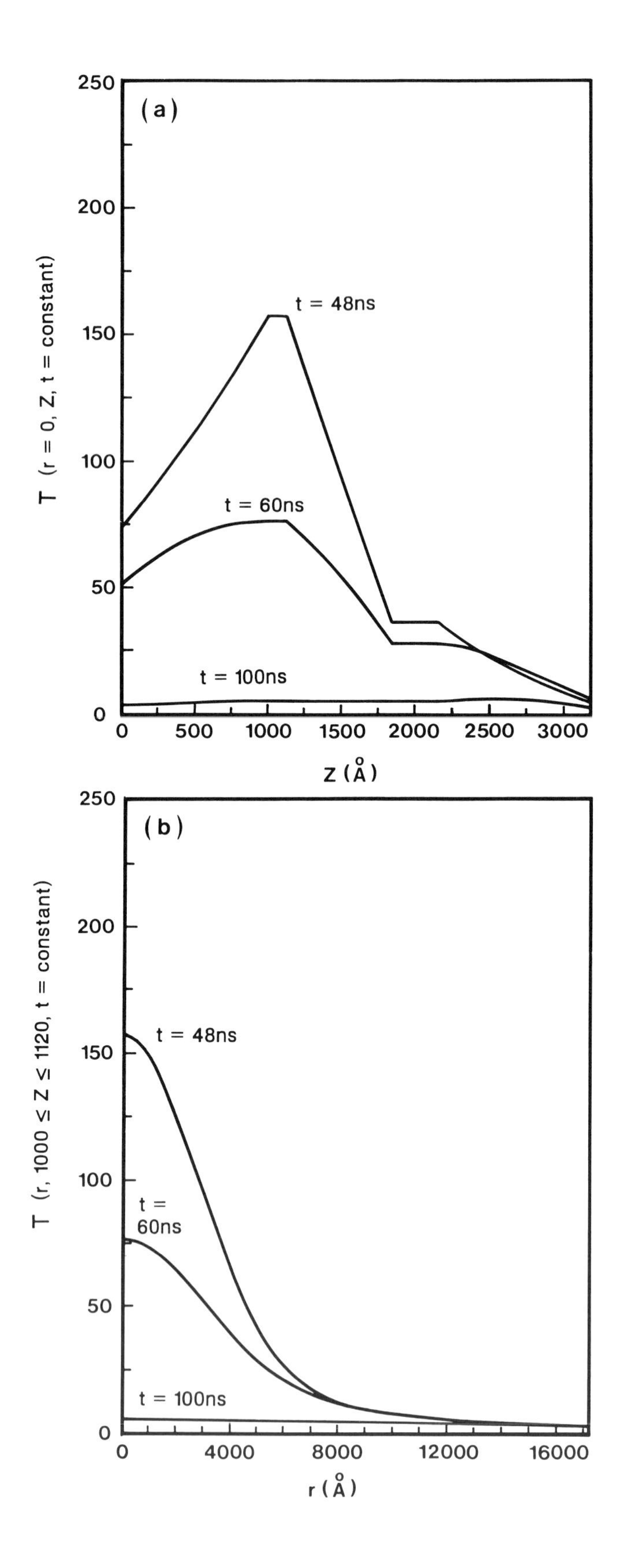

(a)
T (r = 0, Z, t = constant)
Z (Å)
t = 48ns
t = 60ns
t = 100ns
0
50
100
150
200
250
0
500
1000
1500
2000
2500
3000
(b)
T (r, 1000 ≤ Z ≤ 1120, t = constant)
r (Å)
t = 48ns
t = 60ns
t = 100ns
0
50
100
150
200
250
0
4000
8000
12000
16000

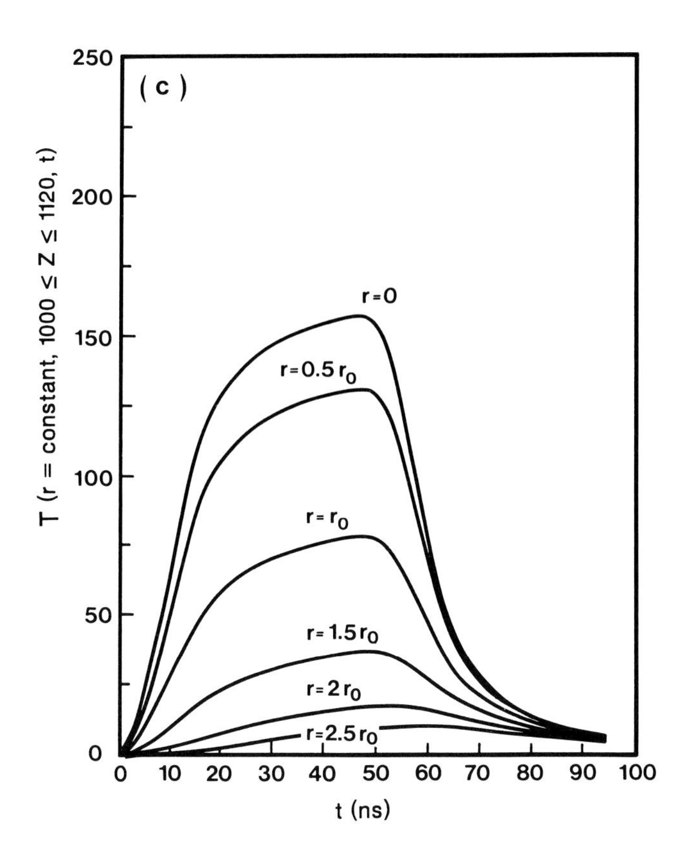

Figure 11.12. Distribution of temperature rise (above ambient) for the quadrilayer of Fig. 11.10(a) when convection losses are included. The quadrilayer is at the ambient temperature at $t = 0$, the laser pulse is the longer pulse of Fig. 11.10(c), and the radius of the beam at the $1/e$ point is 350 nm. Heat flow from the surface is taken into account by setting $\mu = 10^5\ cm^{-1}$. (a) Temperature versus z at the beam center for several instants of time. (b) Radial temperature profile in the MO film for several instants of time. (c) Time-dependence of the MO film's temperature for several radii.

Table 11.3. Optical and thermal parameters of the quadrilayer in Example 8. The focused beam (λ_0 = 830 nm, r_0 = 560 nm) is incident through the substrate

Layer	Thickness d (nm)	Refractive index n	Specific heat C (J/cm^3 K)	Thermal conductivity K (J/cm K s)
substrate	∞	1.53	2.6	0.01
1. dielectric layer 1	75	2.0	2.5	0.02
2. magnetic film	25	3.67 + 3.86i	3.5	0.5
3. dielectric layer 2	60	2.0	2.5	0.02
4. reflector	40	2.7 + 8.3i	2.4	2.4

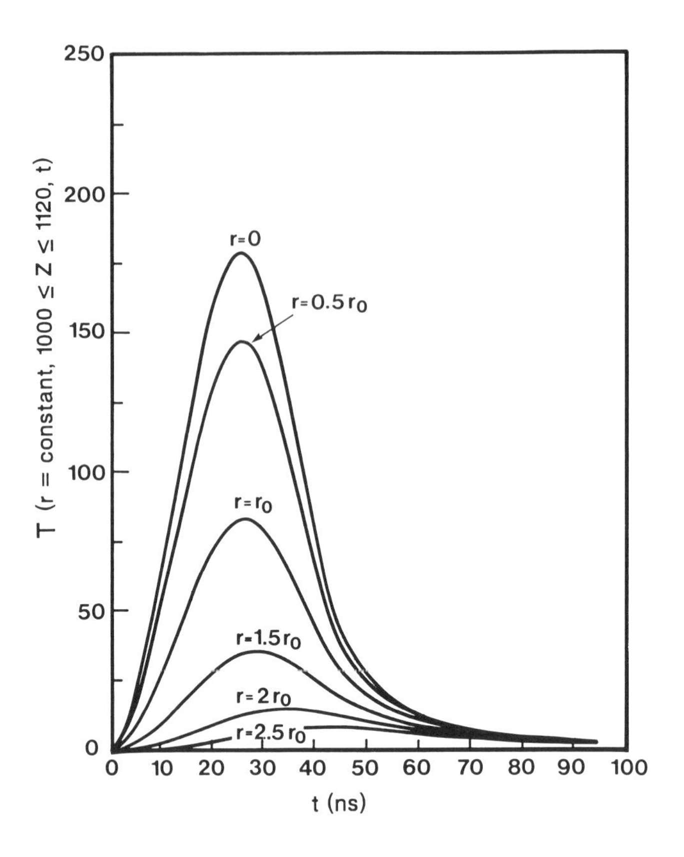

Figure 11.13. Time-dependence of the MO film's temperature rise above ambient at several radii. The conditions for these calculations are identical to those of Fig. 11.12(c), except that the laser pulse is the shorter pulse in Fig. 11.10(c).

Z-axis for three instants of time. The 10 mW 50 ns laser pulse used in this case had a rectangular shape, and the disk was stationary ($V = 0$). The maximum temperature reached in the MO layer at the end of the pulse is 230 °C. Observe that during the cooling process the heat from the MO layer is transferred to both the substrate and the aluminum reflecting layer. Note also that both metallic layers, thanks to their high thermal conductivity, maintain an essentially uniform temperature through their thicknesses.

For a 12 mW 100 ns laser pulse and a disk velocity $V = 10$ m/s (in the negative X-direction) the MO layer isotherms corresponding to $T = 125$ °C (above ambient) are shown in Fig. 11.15(c_1); these isotherms are shown at 10 ns intervals. The maximum temperature of the MO layer reached in the process is 295 °C (above ambient). The envelope of these isotherms is the likely shape of the recorded domain. In applications, such "tear-drop"

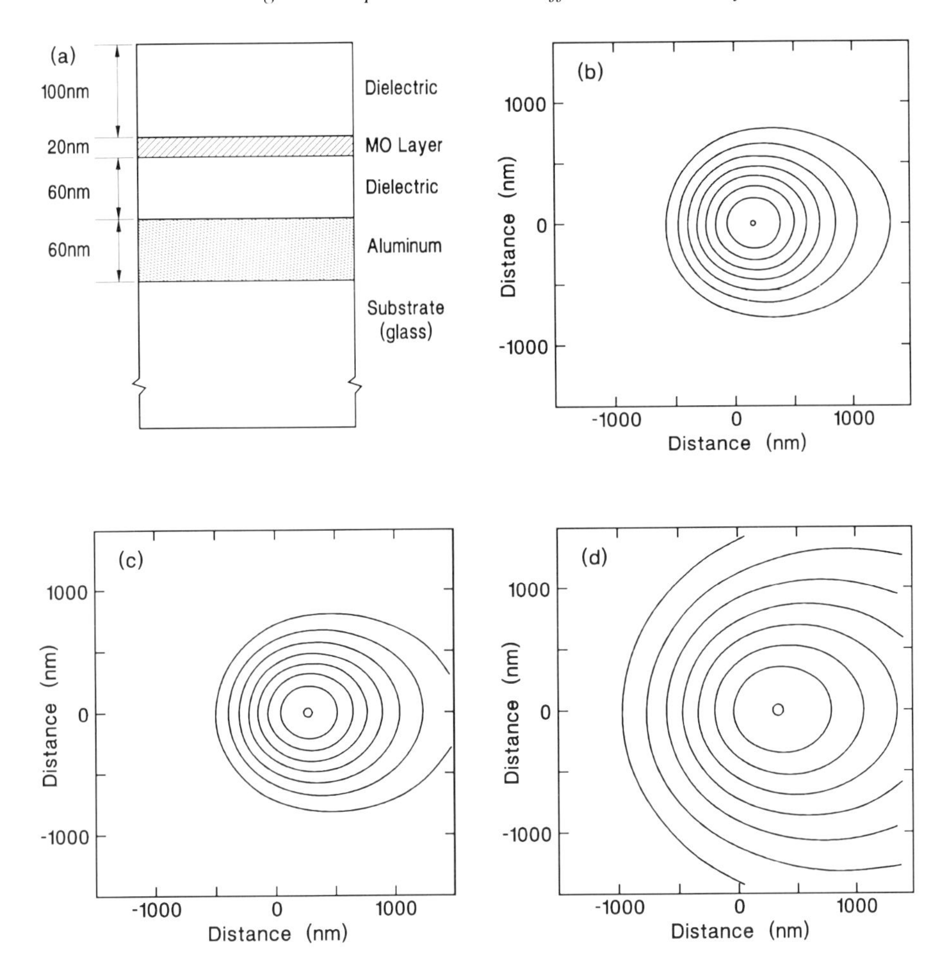

Figure 11.14. Steady-state isotherms at various planes within the quadrilayer MO device shown in (a). The focused beam (λ_0 = 840 nm, r_0 = 500 nm) is incident from the surface side on a disk that has velocity V = 20 m/s in the positive X-direction. The central contour in each case shows the hottest spot in the layer; each successive contour has a temperature lower than the preceding contour by 10% of this maximum. (b) Isotherms in the MO layer with peak temperature rise above ambient, per unit incident power, at 40.6 °C/mW. (c) Isotherms at the top surface of the quadrilayer; peak temperature rise is 37.9 °C/mW. (d) Isotherms in the aluminum layer with a peak temperature rise of 12.9 °C/mW.

shaped domains are not desirable, since they cause differences in readout between the leading edge and the trailing edge of the marks. (As will be discussed shortly, one can eliminate this problem by pre-emphasizing the write pulse.) Plots of temperature versus time at several points along the track are shown in Fig. 11.15(c_2). These curves, which are closely related to the isotherms in (c_1), describe the rise and fall of temperature as fixed

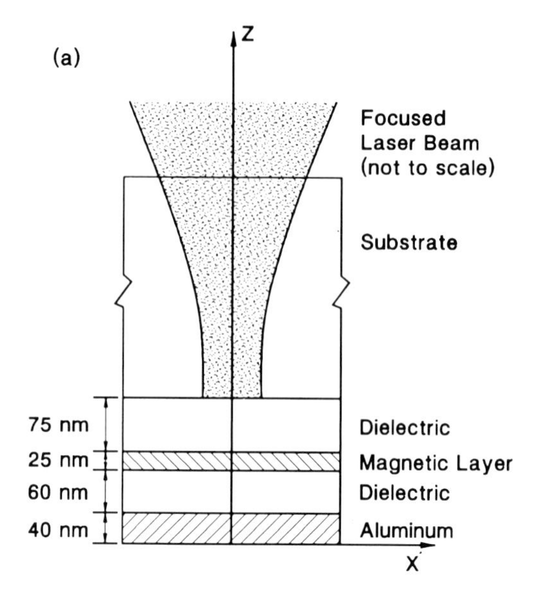
(a)
Z
Focused
Laser Beam
(not to scale)
Substrate
75 nm
25 nm
60 nm
40 nm
Dielectric
Magnetic Layer
Dielectric
Aluminum
X

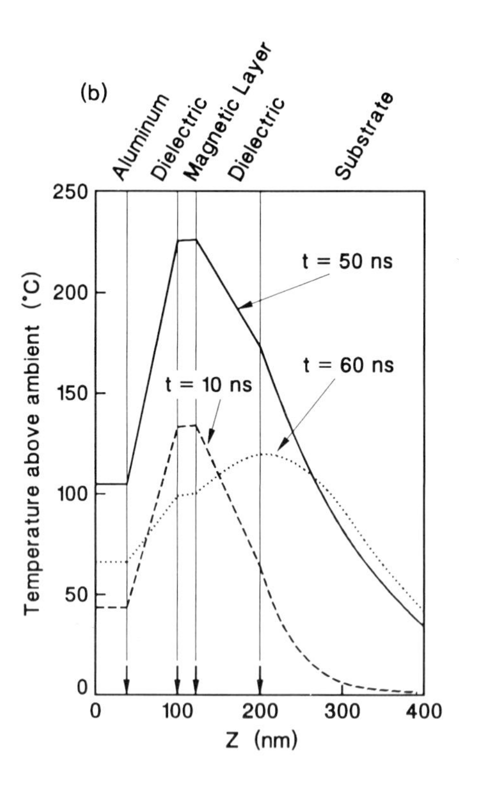
(b)
Aluminum
Dielectric
Magnetic Layer
Dielectric
Substrate
Temperature above ambient (°C)
t = 50 ns
t = 60 ns
t = 10 ns
0
50
100
150
200
250
0
100
200
300
400
Z (nm)

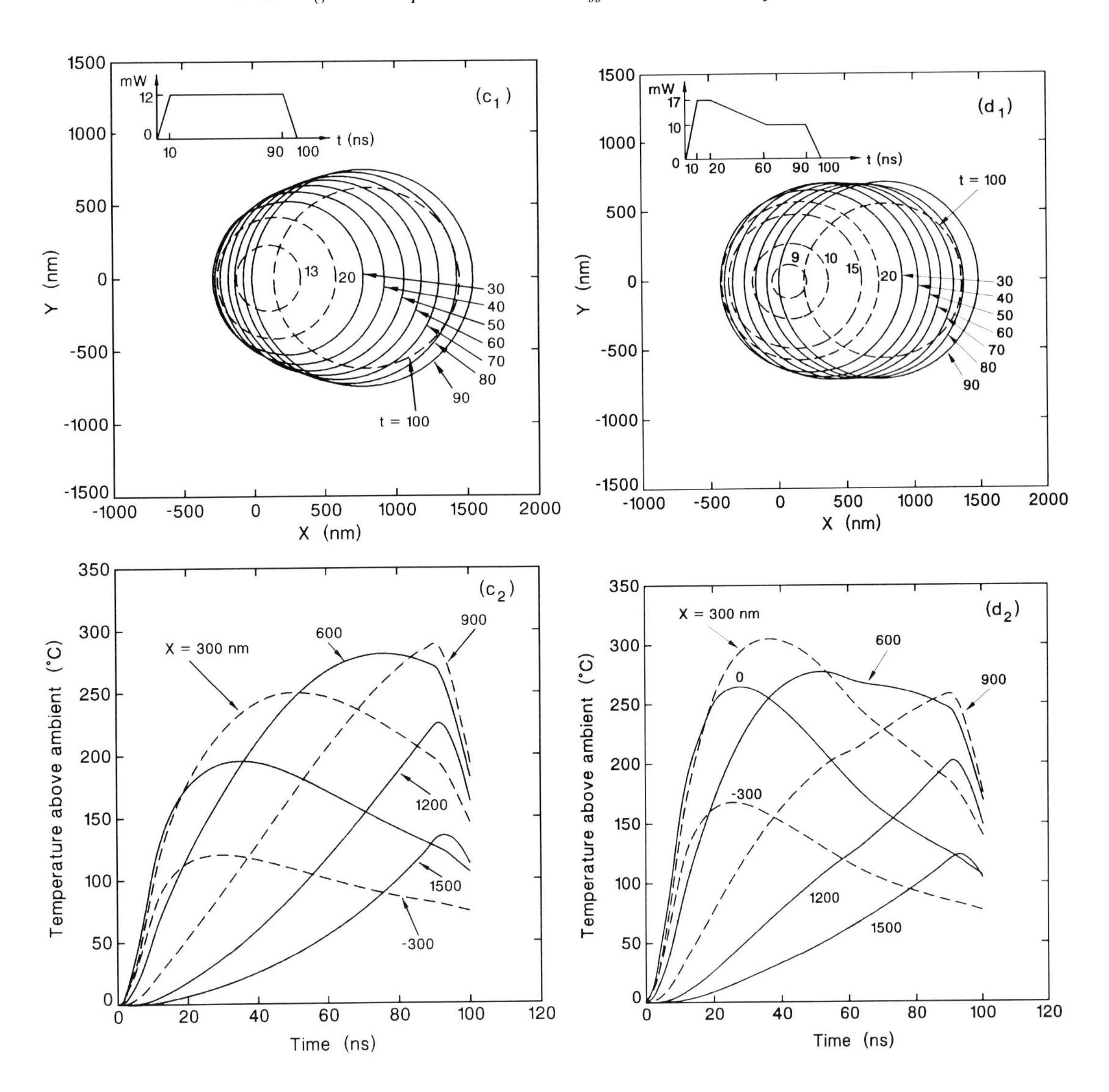

Figure 11.15. Calculated temperature profiles in the quadrilayer MO device shown in (a). The focused beam (λ_0 = 830 nm, r_0 = 560 nm) is incident from the substrate side. (b) Temperature profiles at the beam center along the Z-axis for three instants of time. The laser pulse is a rectangular 10 mW 50 ns pulse, and the disk is stationary. (c_1) Isotherms in the magnetic layer at T = 125 °C (above ambient). The trapezoidal laser pulse shown in the inset has 12 mW of power and 100 ns duration. The disk moves at V = 10 m/s in the negative X-direction, and the maximum temperature reached in the film is 295 °C (above ambient). (c_2) Plots of temperature above ambient versus time corresponding to the case in (c_1). The seven curves shown here represent seven different points on the disk. At t = 0, the beginning of the pulse, the beam is centered at (x = 0, y = 0); the point (x = - 300, y = 0) is 300 nm to the left of this point, while (x = 1200, y = 0) is 1200 nm to the right. Other points are identified in similar fashion. (d_1), (d_2) Similar to (c_1) and (c_2) except that these plots correspond to the pre-emphasized laser pulse shown in the inset to (d_1).

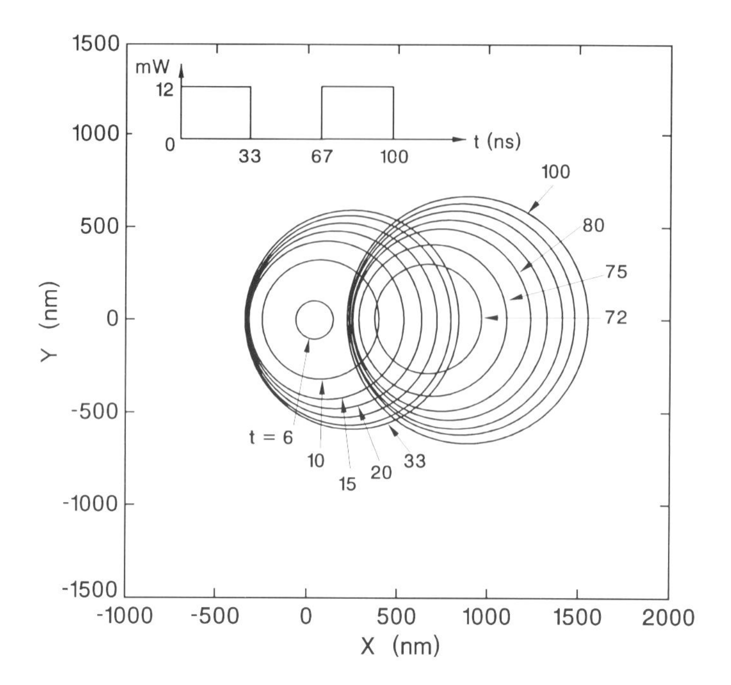

Figure 11.16. Contours of constant temperature (T = 125 °C above ambient) in the MO layer of the structure of Fig. 11.15(a). The disk velocity is 10 m/s along the negative X-axis, and the laser pulse, shown in the inset, is composed of two 12 mW 33 ns short pulses separated by 34 ns. The envelopes of the two sets of contours enclose the area whose temperature rises to more than 125 °C during each pulse.

points on the medium approach the beam, receive energy from it, and then recede into the distance.

When a pre-emphasized pulse is used to write on the medium, the plots in frames (d_1) and (d_2) of Fig. 11.15 are obtained (the write pulse is shown in the inset to (d_1)). These are the counterparts to frames (c_1) and (c_2). The envelope of the isotherms is now more symmetric, and there should be little, if any, difference between the leading and trailing edges of the recorded marks.

Finally, Fig. 11.16 shows contour plots at the fixed temperature of 125 °C (above ambient) for the pair of 12 mW 33 ns pulses shown in the inset. As before, the multilayer structure is that of Fig. 11.15(a), and the disk velocity is 10 m/s. The first set of isotherms corresponds to the first pulse, during which the temperature reaches a maximum of 242 °C. The second set occurs during the second pulse and, due to preheating by the first, reaches the slightly higher maximum of 261 °C. If the envelopes of the two sets of isotherms represent regions whose magnetization will be switched by the applied field, then thermal cross-talk between successive marks will be appreciable. This cross-talk (or intersymbol interference) may be minimized either by pulse shaping, as described earlier, or by proper media design, which is the subject of the next subsection. □

11.3.1. Thermal engineering of the media

We saw in Fig. 11.11 the results of computations that simulate the writing process for a stationary quadrilayer medium. Note that in Fig. 11.11(c) the writing pulse is terminated prior to the establishment of thermal equilibrium. In contrast, readout is done in steady state, and our example therefore implies that readout must be carried out with even less power than might straightforwardly be deduced from the write power alone. In the present subsection we are concerned with the selection of media structures that provide the best overall system operation when these differences between reading and writing are properly taken into account. In the following example we will discuss the read and write performances of optically similar but thermally different quadrilayer media.

When the read power is sufficiently low for thermal effects to be negligible, an "optimum" quadrilayer provides the maximum signal-to-noise-ratio (SNR) during readout at a fixed value of laser power.[12] The structure can be adjusted to make best use of the laser power available for read and write processes by detuning the optimum device in such a way that all the laser power available is used for recording. The effects of such detuning on the SNR are then more than compensated by the extra power that can be used during readout. The following example puts this result on a quantitative basis.

Example 9: Quadrilayer design with desired thermal characteristics. The media designs considered here and the parameters used in the calculations are given in Tables 11.4 and 11.5 respectively. The quadrilayers considered fall into two categories: the first comprises optimum designs, in the sense of providing maximum SNR at fixed laser power when heating during readout is ignored, and the second comprises sub-optimum designs that have, as we shall see, attractive thermal characteristics. The SNR at the same fixed, small laser power, however, is only slightly different in the two cases (the difference in SNR is less than 0.5 dB) and therefore any major differences in performance will derive directly from differing thermal characteristics. It therefore follows that the power required for a fixed temperature rise under continuous illumination will be a direct relative measure of the readout SNR for a given structure. Similarly, the power required for a fixed temperature rise under pulsed illumination will be a direct relative measure of the writing sensitivity at fixed applied magnetic field.

Table 11.4. Quadrilayer designs with varying thermal characteristics

Quadrilayer label		Layer thickness (nm)			
		overlayer	magnetic layer	intermediate layer	reflecting layer
optimum	*a*	100	20	60	thick (dielectric)
	b	100	20	60	50 (aluminum)
	c	310	20	60	50 (aluminum)
	d	100	20	60	thick (aluminum)
detuned	*e*	100	20	30	50 (aluminum)
	f	100	20	30	thick (aluminum)

Table 11.5. Optical and thermal parameters of the materials in Example 9. The focused beam ($\lambda_0 = 840$ nm, $r_0 = 500$ nm) is incident from the surface side

Material		Refractive index n	Specific heat C (J/cm^3 K)	Thermal conductivity K (J/cm K s)
glass substrate		1.5	2.0	0.015
MO layer		3.67 + 3.85 i	3.2	0.4
overlayer		2.0	2.0	0.015
intermediate layer		2.0	2.0	0.015
reflector	aluminum	2.0 + 7.1 i	2.7	2.4
	dielectric†	2.0 + 7.1 i	2.0	0.015

For the structures in Table 11.4 the power required to raise the temperature of the magnetic medium by 100 °C during readout and writing, versus the velocity of the medium, is shown in Fig. 11.17. At the simplest level, the behavior for each structure is as expected: more power is needed for readout at fixed maximum temperature, and for writing at fixed resolution, as the velocity is increased. However, at the next level, an interesting effect can be seen, namely that the rate of power increase is larger for writing than for reading, a result of considerable value when selecting a medium for a particular system. Moreover, different structures

† As far as the optics of the quadrilayer is concerned it is sufficient to use an effective refractive index for the dielectric mirror, which makes it resemble the aluminum reflector. The thermal constants of this dielectric mirror, however, are chosen to be identical with those of the substrate.

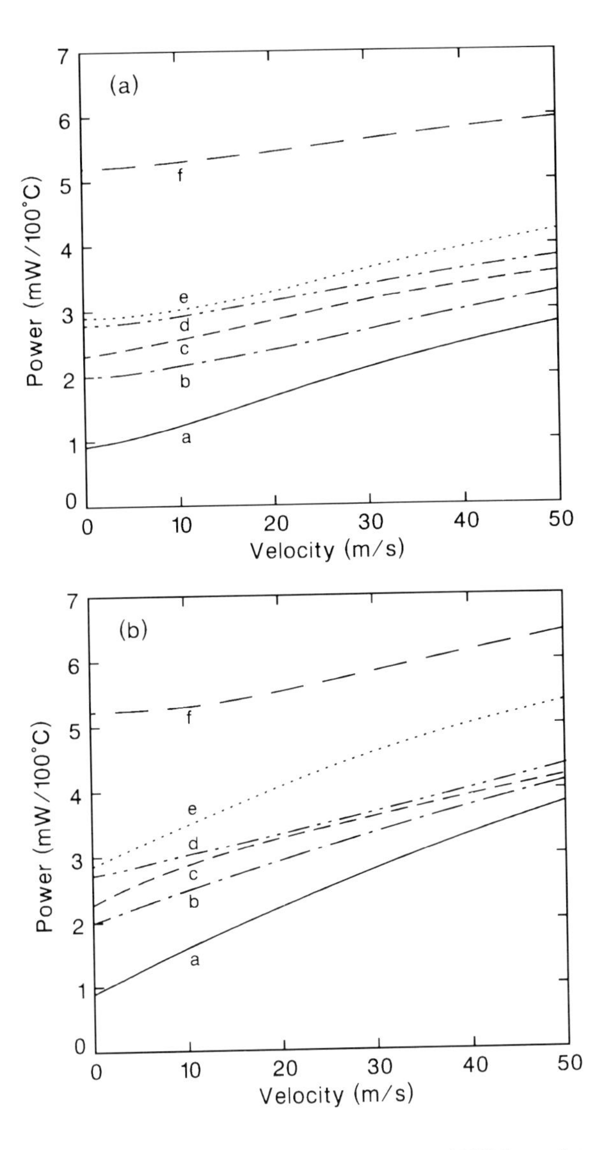

Figure 11.17. (a) The power required during readout (CW beam) to raise the MO film to a peak temperature of 100 °C (above ambient) as a function of the linear velocity of the medium, V. The curves labeled $a-f$ represent the various quadrilayer designs in Table 11.4. (b) The power required during writing (with rectangular pulse of duration r_0/V) to raise the MO film to a peak temperature of 100 °C as a function of the linear velocity of the medium, V.

are seen to require very different power levels. In the remainder of this example, we will trace the origin of these effects and establish the conditions under which a particular structure would provide the best performance.

The structure with the greatest writing sensitivity (structure *a*) is an optimum quadrilayer with dielectric reflector. This reflector is chosen to have the optical parameters of aluminum and the thermal properties of the glass substrate. The structure thus has the readout SNR of an optimum quadrilayer, but thermally behaves like a simple bilayer. Of all these structures it tolerates the least power in readout, and therefore its selection as the medium of choice in a recording system would be appropriate only in the case of an insensitive magnetic material or of severely limited laser power. For the magnetic materials of current interest and in view of the present availability of power from semiconductor lasers, neither of these conditions will obtain.

When the dielectric reflector is replaced by an aluminum one, the power levels for the readout and writing curves increase progressively with aluminum thickness to provide a satisfactory way of adjusting the quadrilayer structure to the writing constraints. For example, at a medium velocity of 20 m/s, a decrease of writing sensitivity by a factor of about 1.5 leads to an increase of about 3 dB in readout SNR at the read power now allowed when the dielectric mirror is replaced by thick aluminum. These effects, of course, are caused by increased heat flow to the aluminum sink, and the isotherms for the magnetic layer shown in Fig. 11.18 show this very clearly. When heat flow to the substrate is minimized by using the dielectric reflector, a significant part of the heat generated in the magnetic film must be lost by lateral flow. Thus in Fig. 11.18(a) the maximum temperature in the MO film is displaced downstream from the center of the laser spot, and a long tail is apparent in the temperature distribution. However, as heat flow toward the substrate is increased by replacing the dielectric mirror with an aluminum layer, Fig. 11.18(b), or by increasing the aluminum layer thickness, Fig. 11.18(c), the temperature distributions tend to move toward the original Gaussian intensity distribution of the laser spot. Note in particular the disappearance of the downstream tail and the movement of the temperature maximum toward the center of the beam. This is an important result because it demonstrates that thermal coupling to the substrate can be made large enough to overcome the enlarging effects of lateral heat flow on bit size that may occur during writing.

Further control of the quadrilayer's thermal characteristics may be obtained by adjusting the overlayer thickness. If this is increased discretely in half-wave steps, the quadrilayer maintains its optical response but now requires more power for writing. Figure 11.17 shows the effects of a half-wave increase in the overlayer thickness when a 50 nm-thick aluminum layer is used as reflector (structure *c*). Fortuitously, the writing and reading results are now very similar to those for the structure having a 100 nm-thick overlayer and a relatively thick aluminum reflector (structure *d*). A combination of overcoat and reflector thickness adjustments can therefore yield the desired result under a variety of circumstances.

Finally, structures requiring the highest write powers (and therefore providing the largest SNR in readout) are obtained by decreasing the

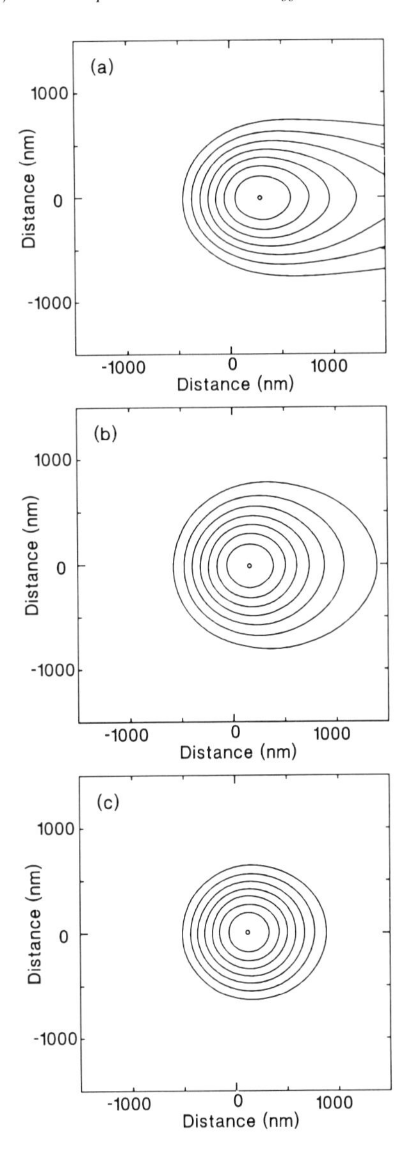

Figure 11.18. Steady-state isotherms in the MO layer of some of the quadrilayers in Table 11.4. The focused beam ($\lambda_0 = 840$ nm, $r_0 = 500$ nm) is incident from the surface side on a disk with a velocity of 20 m/s in the positive X-direction. The central contour in each case shows the hottest spot in the layer; each successive contour has a temperature lower than the contour preceding it by 10% of the maximum. (a) Isotherms for the quadrilayer a, which has a dielectric reflector; peak temperature rise is 57.6 °C/mW. (b) Isotherms for the quadrilayer b, which has a 50 nm aluminum reflector; peak temperature rise is 41.5 °C/mW. (c) Isotherms for the quadrilayer d, which has a thick aluminum reflector; peak temperature rise is 31.4 °C/mW.

intermediate layer thickness to create a detuned design. Two examples using an intermediate layer thickness of 30 nm and different aluminum layer thicknesses are shown in Fig. 11.17 as curves *e* and *f*. As before, increases in write power and readout signal are evident when the aluminum layer thickness is increased, but now the media and system can be matched over a much wider range. For example, if sufficient laser power is available for writing at V = 20 m/s, the SNR attainable with the increased read power now allowed for the detuned structure with thick reflector is about 2 dB greater than that for the equivalent "optimum" structure. In this case, the write power required at fixed applied magnetic field is about 1.7 times greater.

There is a common link in all of these trends. As the thermal coupling to the aluminum heat sink is improved, the time constant for attaining the steady-state decreases. The temporal difference between the writing and reading processes therefore gets smaller and indeed has almost disappeared in the most detuned structure (see curve *f* in Figs. 11.17(a),(b)). If thermomagnetic writing can still be achieved satisfactorily in this condition, then both the read SNR and the domain resolution are simultaneously maximized. □

Problems

(11.1) Verify that Eq. (11.33) can indeed be derived from Eq. (11.32).

(11.2) Consider the Gaussian temperature profile of Eq. (11.29) with r_0 having the time-dependence of Eq. (11.35):

$$T(r,t) = \frac{1}{\pi(r_0^2 + 4Dt)} \exp\left(-\frac{r^2}{r_0^2 + 4Dt}\right)$$

Show directly that the above profile satisfies the appropriate diffusion equation in two-dimensional space, namely, Eq. (11.24).

(11.3) Write the discrete form of Eqs. (11.44) using the explicit technique; that is, in evaluating the derivatives with respect to x, use the values of the temperature T at $t = (m - 1)\Delta t$. Show that the set of discrete equations thus obtained is trivial to solve.

(11.4) Substitute Eq. (11.47) in Eq. (11.46d) and verify the recursive relations for $Q_1^{(n)}$ and $Q_2^{(n,m)}$ given in Eqs. (11.48).

12

Fundamentals of Magnetism and Magnetic Materials

Introduction

Magneto-optical recording is an important mode of optical data storage; it is also the most viable technique for erasable optical recording at the present time. The MO read and write processes are both dependent on the interaction between the laser beam and the magnetic medium. In the preceding chapters we described the readout process with the aid of the dielectric tensor of the storage medium, without paying much attention to the underlying magnetism. Understanding the write and erase processes, on the other hand, requires a certain degree of familiarity with the concepts of magnetism in general, and with the micromagnetics of thin-film media in particular. The purpose of the present chapter is to give an elementary account of the basic magnetic phenomena, and to introduce the reader to certain aspects of the theory of magnetism and magnetic materials that will be encountered throughout the rest of the book.

After defining the various magnetic fields (**H**, **B**, and **A**) in section 12.1, we turn our attention in section 12.2 to small current loops, and show the equivalence between the properties of these loops and those of magnetic dipoles. The magnetization **M** of magnetic materials is also introduced in this section. Larmor diamagnetism is the subject of section 12.3. In section 12.4 the magnetic ground state of free atoms (ions) is described, and Hund's rules (which apply to this ground state) are presented. The paramagnetism of a collection of identical atoms (ions) is described in the first half of section 12.5, followed by an analysis of Pauli paramagnetism and its consequences for the conduction electrons in metals. The exchange interaction is the basis of magnetic ordering in solids; this topic is taken up in section 12.6 where we use qualitative arguments to clarify the nature and origins of this quantum mechanical phenomenon. An analysis of magnetic ordering in ferromagnets based on the mean-field theory is the subject of section 12.7; here we examine the critical temperature T_c of the order–disorder transition, and derive the Curie–Weiss law, which pertains to the temperature-dependence of the susceptibility in the paramagnetic state. Section 12.8 describes some of the interesting magnetic properties of the rare-earth-based solids. The transition-metal-based solids in the iron group are briefly described in section 12.9. A discussion of magnetic anisotropy, its physical origins and macroscopic manifestations is the subject of section 12.10.

In the literature of magnetism there are two widely used systems of units: the CGS (centimeter-gram-second) system, also known as the Gaussian system, preferred by physicists, and the MKSA (meter-kilogram-second-ampere) system, which is predominant in the engineering literature. Unfortunately, there is some confusion surrounding these systems, and at times there is even some mixing of the two in the same context. To set matters straight, the important equations encountered in this chapter will be described in both systems of units; the Gaussian formulas will be written on the left and the corresponding MKSA formulas on the right. (This rule is adhered to only when the formulas are different; when the same formula applies in both systems it will be written only once.) Occasionally, when a quantity is expressed in Gaussian units, its equivalent in MKSA will follow in parentheses. Derivations of formulas and proofs of theorems, however, will be given exclusively in Gaussian units.

12.1. Magnetic Fields in Free Space

In magnetostatics there are two sources of magnetic fields: electrical currents (i.e., moving charges) and permanently magnetized materials. In this section we review the fields produced by electrical currents in free space, and pay particular attention to the units commonly used for the quantities involved.

12.1.1. Units of electric charge

The confusion regarding units in magnetism is rooted in the fact that electric charge is treated fundamentally differently in the two systems. Consider the unit of charge in these systems as defined through the Coulomb law of force between two point charges q_1 and q_2, separated by a distance r:

Gaussian	MKSA	
$F = \dfrac{q_1 q_2}{r^2}$ unit of charge $= \text{cm}\sqrt{\text{dyne}}$ = statcoulomb	$F = \dfrac{1}{4\pi\epsilon_0}\dfrac{q_1 q_2}{r^2}$ unit of charge $= \text{m}\sqrt{\text{newton} \times \text{farad/m}}$ = coulomb	(12.1)

Numerically, $4\pi\epsilon_0 = 10^7/c^2$ where $c \simeq 3 \times 10^8$ m/s is the speed of light in vacuo; the units of ϵ_0, however, are farads/m which are not the same as those of c^{-2}. Now, if $q_1 = q_2 = 1$ coulomb, and if $r = 1$ m, the resulting force in MKSA will be $10^{-7}c^2 \simeq 9 \times 10^9$ newtons; this is equal to 9×10^{14} dynes in CGS. One coulomb of charge, therefore, must be equivalent to 3×10^9 statcoulombs, if two such charges are to exert a net force of 9×10^{14} dynes on each other at a distance of $r = 100$ cm. Thus we see that, numerically, 1 coulomb = 3×10^9 statcoulombs, but if the dimensions of force and distance are taken to be more "fundamental" than that of the electric charge, then coulomb and statcoulomb will have different dimensions.

12.1.2. Electric current

The electric current density **J** is defined as a vector in the direction of the current whose magnitude is given by the amount of charge crossing unit normal area per unit time. Thus the units of **J** are statcoulombs/s cm^2 in CGS. In MKSA, current density has units of amperes/m^2, where 1 ampere = 1 coulomb/s. The conservation of charge in Maxwell's theory of electromagnetism is expressed as follows:

$$\boxed{\nabla \cdot \mathbf{J} + \frac{\partial \rho}{\partial t} = 0} \tag{12.2}$$

Here ρ is the density of electrical charge. This equation is valid in both systems of units. In magnetostatics, since the charge density does not vary with time, $\nabla \cdot \mathbf{J} = 0$ everywhere in space at all times.

In MKSA, the unit of current is the ampere (A), which is equivalent to 3×10^9 statamperes in CGS. (Again, in terms of the more basic units, the ampere and the statampere are not the same.) In both systems, the current flow I out of a surface is the integral of the current density **J** over the area of that surface, that is,

$$I = \int_{\text{surface area}} \mathbf{J} \cdot d\mathbf{s} \tag{12.3}$$

12.1.3. H-field and B-field

The second of Maxwell's equations, relating the magnetic field **H**, the current density **J**, and the electric displacement **D**, is expressed differently in the two systems:

$$\boxed{\nabla \times \mathbf{H} = \frac{4\pi}{c}\mathbf{J} + \frac{1}{c}\frac{\partial \mathbf{D}}{\partial t}} \qquad \boxed{\nabla \times \mathbf{H} = \mathbf{J} + \frac{\partial \mathbf{D}}{\partial t}} \tag{12.4}$$

In magnetostatics $\partial \mathbf{D}/\partial t = 0$, and the above equation simply relates the magnetic field to the current density. Let a long, straight wire carry a constant current I_0, as shown in Fig. 12.1. At a distance r_0 from the wire, the strength of the **H**-field is found by integrating Eq. (12.4) and using Stokes' theorem.[1] Thus:

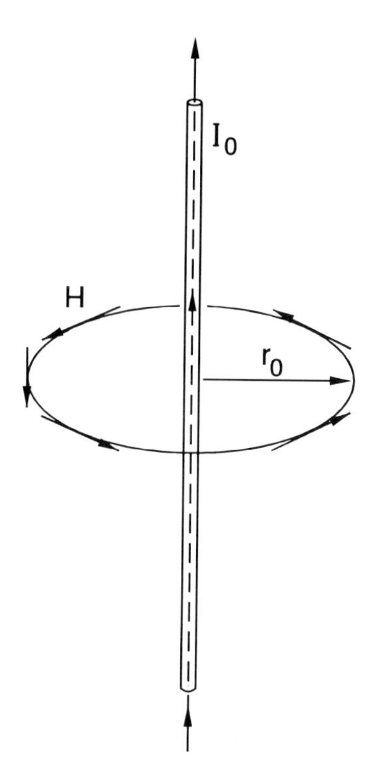

Figure 12.1. A long, straight wire carrying a constant current I_0. The magnetic field **H** at a distance r_0 from the wire is given by Eq. (12.5).

$2\pi r_0 H = \frac{4\pi}{c} I_0$ unit of H = oersted	$2\pi r_0 H = I_0$ unit of H = A/m

(12.5)

Let $I_0 = 1$ A and $r_0 = 1$ m. Then $H = 1/2\pi$ A/m in MKSA. Since $I_0 = 3 \times 10^9$ statamperes, $r_0 = 100$ cm, and $c = 3 \times 10^{10}$ cm/s, the same field strength in CGS units is 0.002 Oe. Therefore

$$1 \text{ oersted} = \frac{1000}{4\pi} \simeq 79.6 \text{ A/m}$$

(The field of the Earth at ground level is around 0.5 Oe.) Because 1 A/m is such a small field, the **H**-field is sometimes expressed in kiloamperes/m (kA/m) where 1 kA/m = 4π Oe.

Another complicating factor in these systems of units is the permeability μ_0 of free space:

$\mu_0 = 1$ (dimensionless)	$\mu_0 = 4\pi \times 10^{-7}$ henry/m

(12.6)

In free space, the magnetic induction **B** is defined as $\mu_0\mathbf{H}$. Since both μ_0 and **H** have different dimensions in the two systems of units, **B** turns out to be different as well.

$\mathbf{B} = \mathbf{H}$ (in free space)	$\mathbf{B} = 4\pi \times 10^{-7}\,\mathbf{H}$ (in free space)
unit of B = gauss	unit of B = tesla

(12.7)

One can thus say, for instance, that a certain magnetic field (in free space) is 1 oersted or 1 gauss, as long as it is understood that the former statement refers to **H** while the latter refers to **B**. In MKSA, however, the same field is either $1000/4\pi \simeq 79.6$ A/m, if **H** is being referred to, or 0.0001 tesla, if the **B**-field is under consideration. Consequently,

$$1 \text{ tesla} = 10\,000 \text{ gauss}$$

Once again, in addition to the numerical factor of 10 000, the **B**-field in MKSA differs in dimensionality from that in CGS, i.e., they cannot be expressed identically in terms of the same fundamental quantities (length, mass, time, etc.). Despite these differences, the fourth Maxwell equation has the same form in both systems of units:

$$\nabla \cdot \mathbf{B} = 0 \tag{12.8}$$

The above equation simply states that there are no free magnetic charges (monopoles) in nature.

12.1.4. Vector potential A

Another field of interest in the theory of electromagnetism is the vector potential **A**. In magnetostatics **A** is the field whose curl is the magnetic induction **B**, that is,

$$\mathbf{B} = \nabla \times \mathbf{A} \tag{12.9}$$

Since the divergence of the curl of any vector field is always zero, the above definition of **A** is consistent with Eq. (12.8). The definition of **A** in Eq. (12.9) does not define it in a unique way, however. In fact, if one picks an arbitrary scalar field ψ, and adds its gradient to **A**, then $\mathbf{A} + \nabla\psi$

will also satisfy Eq. (12.9).† To specify **A** uniquely, one must therefore constrain it in some other way. Usually, one identifies the divergence of **A** in addition to its curl, which is fixed by Eq. (12.9). The specification of divergence is referred to as "choosing the gauge"; for instance, the Coulomb gauge is defined as follows:

$$\nabla \cdot \mathbf{A} = 0 \tag{12.10}$$

With the choice of gauge, Eq. (12.9) yields a unique solution for the vector potential field **A**.††

Let us now find the vector potential field for an arbitrary distribution of currents $\mathbf{J}(\mathbf{r})$ in free space. In the magnetostatic regime, Eq. (12.9) can be combined with Eqs. (12.4) and (12.7) to yield

$$\nabla \times \nabla \times \mathbf{A} = \frac{4\pi}{c}\mathbf{J} \tag{12.11}$$

But $\nabla \times \nabla \times \mathbf{A} = \nabla(\nabla \cdot \mathbf{A}) - \nabla^2\mathbf{A}$, where $\nabla^2\mathbf{A} = \nabla^2 A_x + \nabla^2 A_y + \nabla^2 A_z$ in Cartesian coordinates. (See Digression 1 below.) With the Coulomb gauge $\nabla \cdot \mathbf{A} = 0$, Eq. (12.11) is written thus:

$$\nabla^2 A_x = -\frac{4\pi}{c} J_x \tag{12.12a}$$

$$\nabla^2 A_y = -\frac{4\pi}{c} J_y \tag{12.12b}$$

$$\nabla^2 A_z = -\frac{4\pi}{c} J_z \tag{12.12c}$$

The solution to the above equations is readily obtained as follows:†††

† Recall that $\nabla \times \nabla\psi = 0$.

†† Since $\nabla \cdot (\mathbf{A} + \nabla\psi) = \nabla \cdot \mathbf{A} + \nabla^2\psi$, one can always find the scalar field ψ whose Laplacian cancels the divergence of any **A** that satisfies Eq. (12.9).

††† Each of Eqs. (12.12) is similar to the Poisson equation that relates the electrostatic potential ϕ to the charge density distribution ρ in free space, namely, $\nabla^2\phi = -4\pi\rho$. The solution to this equation is

$$\phi(\mathbf{r}) = \int_{\text{all space}} \frac{\rho(\mathbf{r}')}{|\mathbf{r} - \mathbf{r}'|}\, d^3\mathbf{r}'$$

By analogy, Eq. (12.13) is the solution to Eqs. (12.12).

$\mathbf{A} = \dfrac{1}{c}\displaystyle\int_{\text{all space}} \dfrac{\mathbf{J}(\mathbf{r}')}{\lvert\mathbf{r} - \mathbf{r}'\rvert}\, d^3\mathbf{r}'$	$\mathbf{A} = \dfrac{\mu_0}{4\pi}\displaystyle\int_{\text{all space}} \dfrac{\mathbf{J}(\mathbf{r}')}{\lvert\mathbf{r} - \mathbf{r}'\rvert}\, d^3\mathbf{r}'$
units of **A** = gauss cm	units of **A** = tesla m

(12.13)

Since this equation already satisfies the Coulomb gauge (see Digression 2 below), there is no need to augment it by an appropriate $\nabla\psi$.

Digression 1. To prove that $\nabla \times \nabla \times \mathbf{A} = \nabla(\nabla \cdot \mathbf{A}) - \nabla^2\mathbf{A}$, one may use the matrix notation of Chapter 5, as follows:

$$\nabla \times \nabla \times \mathbf{A} = \begin{pmatrix} 0 & -\partial/\partial z & \partial/\partial y \\ \partial/\partial z & 0 & -\partial/\partial x \\ -\partial/\partial y & \partial/\partial x & 0 \end{pmatrix} \begin{pmatrix} 0 & -\partial/\partial z & \partial/\partial y \\ \partial/\partial z & 0 & -\partial/\partial x \\ -\partial/\partial y & \partial/\partial x & 0 \end{pmatrix} \begin{pmatrix} A_x \\ A_y \\ A_z \end{pmatrix}$$

$$= \begin{pmatrix} -\dfrac{\partial^2}{\partial z^2} - \dfrac{\partial^2}{\partial y^2} & \dfrac{\partial}{\partial y}\dfrac{\partial}{\partial x} & \dfrac{\partial}{\partial z}\dfrac{\partial}{\partial x} \\ \dfrac{\partial}{\partial x}\dfrac{\partial}{\partial y} & -\dfrac{\partial^2}{\partial z^2} - \dfrac{\partial^2}{\partial x^2} & \dfrac{\partial}{\partial z}\dfrac{\partial}{\partial y} \\ \dfrac{\partial}{\partial x}\dfrac{\partial}{\partial z} & \dfrac{\partial}{\partial y}\dfrac{\partial}{\partial z} & -\dfrac{\partial^2}{\partial y^2} - \dfrac{\partial^2}{\partial x^2} \end{pmatrix} \begin{pmatrix} A_x \\ A_y \\ A_z \end{pmatrix}$$

$$= \left\{ -\left(\frac{\partial^2}{\partial x^2} + \frac{\partial^2}{\partial y^2} + \frac{\partial^2}{\partial z^2}\right) \begin{pmatrix} 1 & 0 & 0 \\ 0 & 1 & 0 \\ 0 & 0 & 1 \end{pmatrix} \right.$$

$$\left. + \begin{pmatrix} \partial/\partial x \\ \partial/\partial y \\ \partial/\partial z \end{pmatrix} \begin{pmatrix} \dfrac{\partial}{\partial x} & \dfrac{\partial}{\partial y} & \dfrac{\partial}{\partial z} \end{pmatrix} \right\} \begin{pmatrix} A_x \\ A_y \\ A_z \end{pmatrix}$$

$$= -\nabla^2\mathbf{A} + \nabla(\nabla \cdot \mathbf{A})$$

□

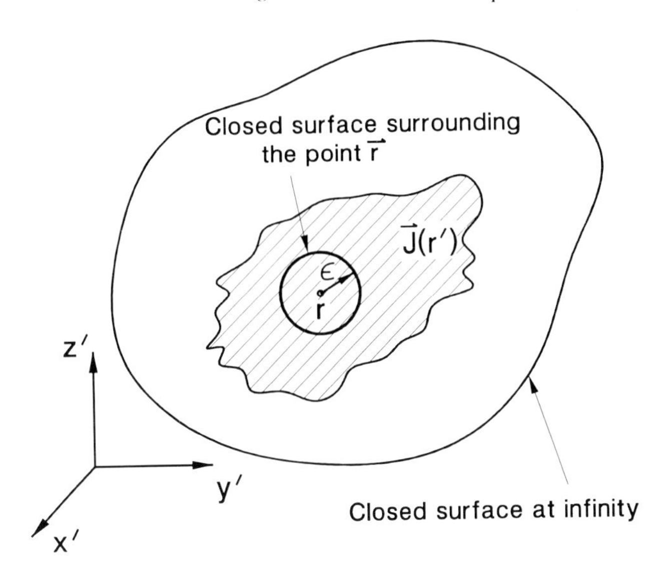

Figure 12.2. Application of Gauss's theorem to convert a volume integral to a surface integral. The outer surface is far enough from the current distribution $\mathbf{J}(\mathbf{r})$ to be considered at infinity. The inner surface is a sphere of radius ϵ centered at $\mathbf{r}$, so that $|\mathbf{r} - \mathbf{r}'|$ is equal to ϵ over the surface of this sphere.

Digression 2. We shall prove that the vector potential field $\mathbf{A}$ in Eq. (12.13) satisfies the Coulomb gauge $\nabla \cdot \mathbf{A} = 0$:

$$\nabla \cdot \int \frac{\mathbf{J}(\mathbf{r}')}{|\mathbf{r} - \mathbf{r}'|} d^3\mathbf{r}' = \int \nabla \cdot \frac{J_x(\mathbf{r}')\,\hat{\mathbf{x}} + J_y(\mathbf{r}')\,\hat{\mathbf{y}} + J_z(\mathbf{r}')\,\hat{\mathbf{z}}}{|\mathbf{r} - \mathbf{r}'|} d^3\mathbf{r}'$$

$$= -\int \frac{(\mathbf{r} - \mathbf{r}') \cdot \mathbf{J}(\mathbf{r}')}{|\mathbf{r} - \mathbf{r}'|^3} d^3\mathbf{r}' = -\int \nabla' \left(\frac{1}{|\mathbf{r} - \mathbf{r}'|} \right) \cdot \mathbf{J}(\mathbf{r}')\, d^3\mathbf{r}'$$

Using the relation $\nabla \cdot [f(\mathbf{r})\mathbf{J}(\mathbf{r})] = f(\mathbf{r})\nabla \cdot \mathbf{J}(\mathbf{r}) + [\nabla f(\mathbf{r})] \cdot \mathbf{J}(\mathbf{r})$, and the fact that in magnetostatics $\nabla \cdot \mathbf{J} = 0$, we write the preceding equation as follows:

$$\nabla \cdot \int \frac{\mathbf{J}(\mathbf{r}')}{|\mathbf{r} - \mathbf{r}'|} d^3\mathbf{r}' = -\int_{\text{all space}} \nabla' \cdot \left(\frac{\mathbf{J}(\mathbf{r}')}{|\mathbf{r} - \mathbf{r}'|} \right) d^3\mathbf{r}'$$

Using Gauss's theorem,[1] the above volume integral can be expressed as a surface integral, as shown in Fig. 12.2:

$$\nabla \cdot \int \frac{\mathbf{J}(\mathbf{r}')}{|\mathbf{r} - \mathbf{r}'|} \, d^3\mathbf{r}' = - \int_{\text{closed surface}} \frac{\mathbf{J}(\mathbf{r}')}{|\mathbf{r} - \mathbf{r}'|} \cdot d\mathbf{s}'$$

The surface at infinity makes no contribution to the integral, since at infinity $\mathbf{J}(\mathbf{r}') = 0$. The surface at $|\mathbf{r} - \mathbf{r}'| = \epsilon$ makes no contribution either, since the integral of $\mathbf{J}(\mathbf{r}')$ over any closed surface is zero, i.e., $\nabla' \cdot \mathbf{J} = 0$. The total integral over the closed surface is therefore zero. □

12.2. Current Loops and the Magnetic Dipole Moment

In electrostatics the source of the **E**-field is electric charge. Both positive and negative charge carriers (protons and electrons) exist abundantly in nature, and give rise to electric fields around them. In magnetostatics, on the other hand, there are no carriers of magnetic charge (monopoles). The elementary sources of magnetic field are the electrons; their orbital motion around the nuclei of atoms as well as their intrinsic spins create the ordinarily observed magnetic fields. An orbiting electron resembles a small current loop, and the magnetic field produced by such a loop is essentially that of a dipole placed parallel to the axis at the loop center. Every electron in its tiny orbit thus behaves as a small bar magnet. In this section we analyze the current loop and relate the strength of the dipole moment to the loop characteristics.

Consider the current loop of Fig. 12.3, which is in the XY-plane and carries a counterclockwise current I. Let the observation point $\mathbf{r}$ be remote from the loop, so that the following approximation can be made:

$$\frac{1}{|\mathbf{r} - \mathbf{r}'|} = \left[(\mathbf{r} - \mathbf{r}') \cdot (\mathbf{r} - \mathbf{r}')\right]^{-1/2} = (r^2 + r'^2 - 2\mathbf{r} \cdot \mathbf{r}')^{-1/2}$$

$$= \frac{1}{|\mathbf{r}|} \left(1 + \frac{r'^2}{r^2} - 2\frac{\mathbf{r} \cdot \mathbf{r}'}{r^2}\right)^{-1/2} \simeq \frac{1}{|\mathbf{r}|} \left(1 + \frac{\mathbf{r} \cdot \mathbf{r}'}{|\mathbf{r}|^2}\right)$$

This is a Taylor series expansion in which we have retained only the first-order term in r'/r. The integral in Eq. (12.13) is thus written

$$\int_{\text{all space}} \frac{\mathbf{J}(\mathbf{r}')}{|\mathbf{r} - \mathbf{r}'|} \, d^3\mathbf{r}' \simeq \frac{I}{|\mathbf{r}|} \int_{\text{loop}} \left(1 + \frac{\mathbf{r} \cdot \mathbf{r}'}{|\mathbf{r}|^2}\right) d\boldsymbol{\ell}$$

where $d\boldsymbol{\ell}$ is an infinitesimal vector along the direction of the current. Around the loop, the vectors $d\boldsymbol{\ell}$ add up to zero, that is, $\int_{\text{loop}} d\boldsymbol{\ell} = 0$. Thus

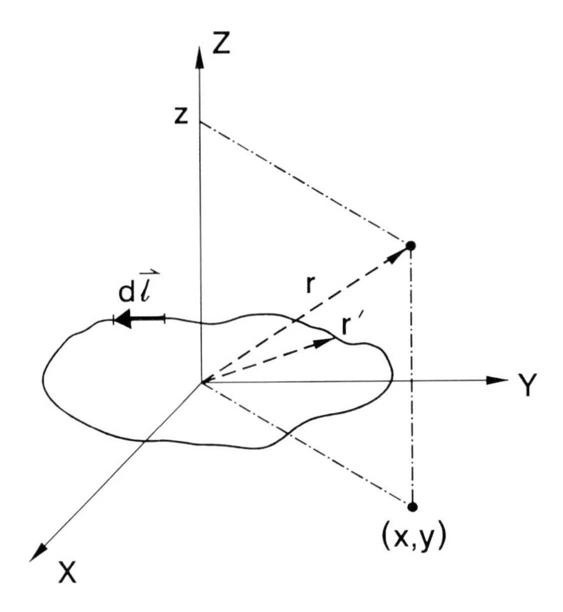

Figure 12.3. A current loop in the XY-plane carries the counterclockwise current I. The observation point $\mathbf{r}$ is sufficiently removed from the loop for the approximate expression for $|\mathbf{r} - \mathbf{r}'|^{-1}$ given in the text to be valid.

we find

$$\int \frac{\mathbf{J}(\mathbf{r}')}{|\mathbf{r} - \mathbf{r}'|} \mathrm{d}^3\mathbf{r}' \simeq \frac{I}{|\mathbf{r}|^3} \int_{\text{loop}} (xx' + yy')(\mathrm{d}x'\,\hat{\mathbf{x}} + \mathrm{d}y'\,\hat{\mathbf{y}})$$

$$= \frac{I}{|\mathbf{r}|^3} \left\{ \left[x \int_{\text{loop}} x'\mathrm{d}x' + y \int_{\text{loop}} y'\mathrm{d}x' \right] \hat{\mathbf{x}} \right.$$

$$\left. + \left[x \int_{\text{loop}} x'\mathrm{d}y' + y \int_{\text{loop}} y'\mathrm{d}y' \right] \hat{\mathbf{y}} \right\}$$

$\hat{\mathbf{x}}$ and $\hat{\mathbf{y}}$ are unit vectors along X and Y. Now, $\int x'\mathrm{d}x' = 0$, because for each point on the loop there is always another whose contribution to the integral exactly cancels that of the first (see Fig. 12.4). Similarly $\int y'\mathrm{d}y' = 0$. On the other hand,

$$\int_{\text{loop}} x'\mathrm{d}y' = -\int_{\text{loop}} y'\mathrm{d}x' = \mathcal{A}$$

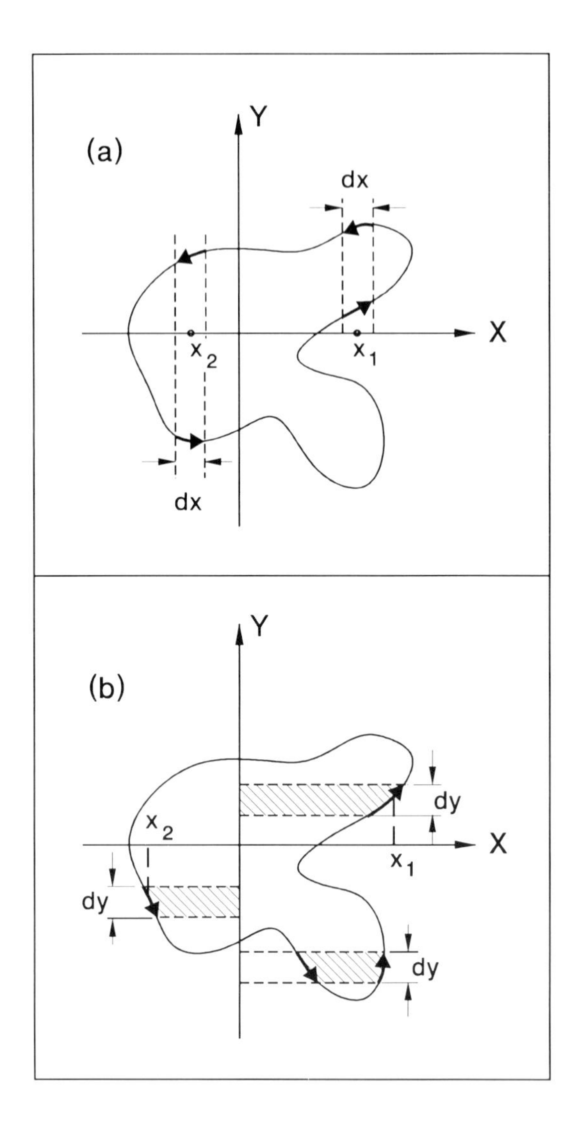

Figure 12.4. (a) For any point such as x_1 or x_2 there are two segments on the loop with equal but opposite values of $\mathrm{d}x$. The contributions of these segments to $\int x\,\mathrm{d}x$ cancel out, leading to a net value of zero for the total integral. (b) Each point on the loop contributes the (infinitesimal) area $x\,\mathrm{d}y$ of the corresponding strip to $\int x\,\mathrm{d}y$. Note that depending on the signs of $\mathrm{d}x$ and $\mathrm{d}y$ the contribution may be positive or negative, but the net result of integration is the total area enclosed by the loop.

where $\mathcal{A}$ is the area enclosed by the loop. We conclude that

$$\int \frac{\mathbf{J}(\mathbf{r}')}{|\mathbf{r} - \mathbf{r}'|}\,\mathrm{d}^3\mathbf{r}' \simeq \frac{I\mathcal{A}}{|\mathbf{r}|^3}(-y\hat{\mathbf{x}} + x\hat{\mathbf{y}})$$

Next, we define the magnetic dipole moment **m** of the loop as follows:

CGS	MKSA	
$\mathbf{m} = \dfrac{I\mathcal{A}}{c}\,\hat{\mathbf{z}}$ units of **m** = statcoulomb cm = emu	$\mathbf{m} = I\mathcal{A}\,\hat{\mathbf{z}}$ units of **m** = ampere m^2	(12.14)

Combining the above results, one obtains the vector potential **A**(**r**) of an infinitesimal current loop:

CGS	MKSA	
$\mathbf{A}(\mathbf{r}) = \dfrac{\mathbf{m} \times \mathbf{r}}{\lvert\mathbf{r}\rvert^3}$	$\mathbf{A}(\mathbf{r}) = \dfrac{\mu_0}{4\pi}\,\dfrac{\mathbf{m} \times \mathbf{r}}{\lvert\mathbf{r}\rvert^3}$	(12.15)

Equation (12.15) indicates that the magnetic field produced by a current loop is independent of the shape of the loop, depending only on the product of its current and its area, so long as the point of observation is not too close to the loop. The magnetic induction **B** can be readily derived from **A**(**r**) using Eq. (12.9):

CGS	MKSA	
$\mathbf{B}(\mathbf{r}) = \dfrac{3\hat{\mathbf{r}}\,(\mathbf{m} \cdot \hat{\mathbf{r}}) - \mathbf{m}}{\lvert\mathbf{r}\rvert^3}$	$\mathbf{B}(\mathbf{r}) = \dfrac{\mu_0}{4\pi}\,\dfrac{3\hat{\mathbf{r}}\,(\mathbf{m} \cdot \hat{\mathbf{r}}) - \mathbf{m}}{\lvert\mathbf{r}\rvert^3}$	(12.16)

In Eq. (12.16) $\hat{\mathbf{r}}$ is the unit vector in the direction of **r**. Figure 12.5 shows the fields **A** and **B** in the neighborhood of a magnetic dipole **m**. The similarity between the **B**-field of a current loop, given by Eq. (12.16), and the **E**-field of an electrostatic dipole is the reason why a current loop is called a magnetic dipole.

Magnetic dipole moments in nature primarily arise from the motion of electrons in their orbits, as well as from their intrinsic spins. When there is a net moment in a given material, the material is said to have magnetization $\mathbf{M} = \mathcal{N}\mathbf{m}$ where $\mathcal{N}$ is the number density (per unit volume) of the atomic dipoles, and **m** is the strength of individual dipole moments. In MKSA the units of **M** are amperes/m, the same as for the magnetic field **H**. Similarly, in CGS the units of both **M** and **H** are statcoulombs/cm^2, even though traditionally the units of the former are referred to as emu/cm^3 while the units of the latter are referred to as oersteds.

Example 1. A free electron (charge $e = 4.803 \times 10^{-10}$ statcoulombs) has a spin magnetic moment of one Bohr magneton, $\mu_B = 0.927 \times 10^{-20}$ emu in CGS. Following Eq. (12.14), in order to find the value of μ_B in MKSA one

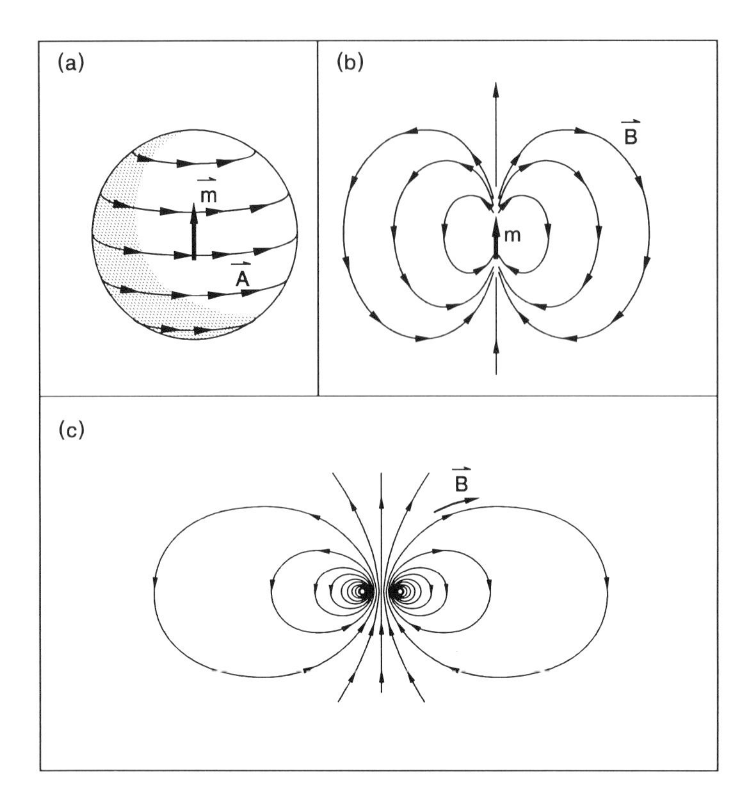

Figure 12.5. (a) Vector potential (**A**-field) of a magnetic dipole moment **m**. (b) Magnetic induction (**B**-field) of a magnetic dipole moment **m**. (c) **B**-field of a charge moving in a small circular loop. Note the similarity between the field distributions in (b) and (c).

must first multiply the above value by $c = 3 \times 10^{10}$ cm/s, then divide by 3×10^9 to convert statamperes to amperes, then divide by 10^4 to convert cm^2 to m^2. The result is $\mu_B = 0.927 \times 10^{-23}$ A m^2. □

Example 2. The magnetization **M** has units of emu/cm^3 in CGS, and A/m in MKSA. Since 1 emu = 10^{-3} A m^2, then 1 emu/cm^3, which is equivalent to 10^6 emu/m^3, is equal to 1000 A/m. On the other hand, 1 oersted of **H** in CGS is equivalent to $1000/4\pi$ A/m in MKSA. Thus one must be careful in converting quantities between the two systems of units. In both systems **M** has the dimensionality of **H**; however, when converting **H** from oersted to A/m, one multiplies by $1000/4\pi$, whereas conversion of **M** from emu/cm^3 to A/m requires multiplication by 1000. □

12.2.1. Angular momentum of a current loop

Let the loop in Fig. 12.3 carry an electric current I. In a small segment $\Delta\boldsymbol{\ell}$ of the loop, the mass and the charge of the current carrier are Δm and Δq, respectively. The charge moves with velocity $\mathbf{v}$, taking Δt seconds for Δq to leave the segment $\Delta\boldsymbol{\ell}$. The angular momentum $\mathbf{L}$ of this moving mass is given by

$$\mathbf{L} = -\int_{\text{loop}} (\Delta m)\,\mathbf{r}' \times \mathbf{v} = -\int_{\text{loop}} \Delta q \,\frac{\Delta m}{\Delta q}\,\mathbf{r}' \times \frac{\Delta\boldsymbol{\ell}}{\Delta t} = -\frac{\Delta q}{\Delta t}\,\frac{\Delta m}{\Delta q}\int_{\text{loop}} \mathbf{r}' \times \Delta\boldsymbol{\ell}$$

The minus sign in front of the above expression signifies that electrons actually move clockwise when the current flow is counterclockwise. Clearly $I = \Delta q/\Delta t$ and $\Delta m/\Delta q = m_e/e$, which is the mass-to-charge ratio for electrons. The cross product $\mathbf{r}' \times \Delta\boldsymbol{\ell}$ is twice the area swept by $\mathbf{r}'$ during the time interval Δt. Therefore

$$\boxed{\mathbf{L} = -\frac{2m_e I \mathcal{A}}{e}\,\hat{\mathbf{z}}} \tag{12.17}$$

Comparing Eq. (12.17) with Eq. (12.14), we conclude that

$\mathbf{m} = -\dfrac{e}{2m_e c}\mathbf{L} = -\gamma\,\mathbf{L} = -\mu_B\,\dfrac{\mathbf{L}}{\hbar}$	$\mathbf{m} = -\dfrac{e}{2m_e}\mathbf{L} = -\gamma\,\mathbf{L} = -\mu_B\,\dfrac{\mathbf{L}}{\hbar}$
$\gamma = 0.88 \times 10^{7}$ Hz/gauss	$\gamma = 0.88 \times 10^{11}$ Hz/tesla
$\mu_B = 0.927 \times 10^{-20}$ emu	$\mu_B = 0.927 \times 10^{-23}$ A m^2

(12.18)

γ is known as the gyromagnetic ratio. Later we shall see why γ is expressed in units of Hz/gauss (or Hz/tesla); for the moment, however, the reader should verify that these are indeed consistent with the definition of γ in terms of e, m_e, and c. The reason for introducing Planck's constant in the above equation is that in quantum mechanics $\mathbf{L}$ is quantized in units of $\hbar$. The magnetic dipole moment of the loop is thus proportional to its orbital angular momentum quantum number $L/\hbar$, the proportionality constant being the Bohr magneton μ_B.

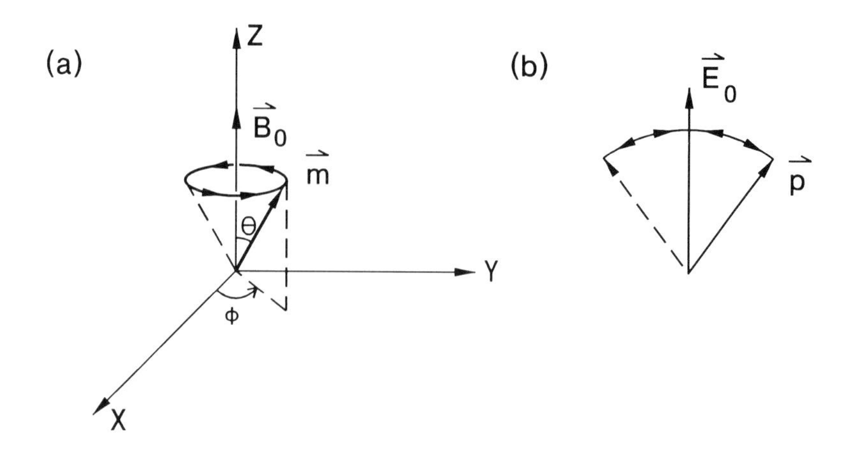

Figure 12.6. (a) Magnetic dipole $\mathbf{m}$ in a uniform magnetic field $\mathbf{B}_0$. In the absence of damping the dipole gyrates around the field, sweeping out the surface of a cone. (b) Electric dipole $\mathbf{p}$ in uniform electric field $\mathbf{E}_0$. In the absence of damping the dipole oscillates back and forth indefinitely, but stays within the plane of $\mathbf{p}$ and $\mathbf{E}_0$.

constant γ. Equation (12.22) also shows that the units of γ must be Hz/gauss (CGS) or Hz/tesla (MKSA) as stated earlier (see Eq. (12.18)).

The undamped precession of $\mathbf{m}$ around $\mathbf{B}_0$ is known as Larmor precession. This gyrational motion is one of the major differences between electric and magnetic dipole moments. In a uniform electric field $\mathbf{E}_0$ the electric dipole $\mathbf{p}$ does not precess since it does not have an inherent angular momentum. In the absence of friction, $\mathbf{p}$ simply oscillates in the plane of $\mathbf{p}$ and $\mathbf{E}_0$, as shown in Fig. 12.6(b).

12.2.4. Force on a current loop in a magnetic field

In a uniform field $\mathbf{B}_0$ the net force on a current loop is zero. We thus investigate the case of a nonuniform field $\mathbf{B}(\mathbf{r})$, and assume that variations of $\mathbf{B}$ over the loop surface are small enough to justify the discarding of higher-order terms in the following Taylor series expansion (see Fig. 12.7):

$$\mathbf{B}(\mathbf{r}) \simeq \mathbf{B}(0) + (\nabla B_x \cdot \mathbf{r})\,\hat{\mathbf{x}} + (\nabla B_y \cdot \mathbf{r})\,\hat{\mathbf{y}} + (\nabla B_z \cdot \mathbf{r})\,\hat{\mathbf{z}}$$

The force expression will be derived in CGS, although the final result applies in MKSA as well. Denoting by $\mathrm{d}q$ the charge contained within a small segment $\mathrm{d}\boldsymbol{\ell}$ of the loop, we write:

$$\mathbf{F} = \int_{\text{loop}} \mathrm{d}q\, \frac{\mathbf{v}}{c} \times \mathbf{B}(\mathbf{r}) = \int_{\text{loop}} \mathrm{d}q\, \frac{\mathrm{d}\boldsymbol{\ell}}{c\,\mathrm{d}t} \times \mathbf{B}(\mathbf{r})$$

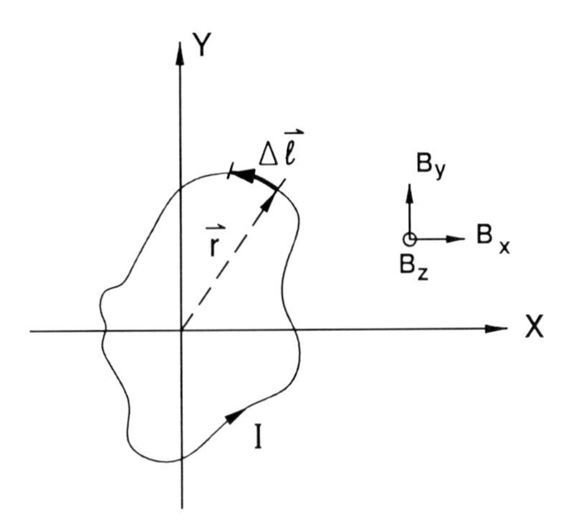

Figure 12.7. Current loop in a nonuniform magnetic field $\mathbf{B}(\mathbf{r})$. Although the components of the field, B_x, B_y, and B_z, are shown at one point only, it is assumed that the field is present everywhere within and around the loop. The net force on the loop may be calculated from Eq. (12.23); for it to be nonzero, the magnitude and/or direction of the field must vary with position over the loop surface since a uniform field does not exert any force.

$$= \frac{I}{c}\left\{\int_{\text{loop}} d\boldsymbol{\ell} \times \mathbf{B}(0) + \int_{\text{loop}} (\nabla B_x \cdot \mathbf{r})\, d\boldsymbol{\ell} \times \hat{\mathbf{x}}\right.$$

$$\left. + \int_{\text{loop}} (\nabla B_y \cdot \mathbf{r})\, d\boldsymbol{\ell} \times \hat{\mathbf{y}} + \int_{\text{loop}} (\nabla B_z \cdot \mathbf{r})\, d\boldsymbol{\ell} \times \hat{\mathbf{z}}\right\}$$

Since $\int d\boldsymbol{\ell} = 0$ the first term in the above expression vanishes. To evaluate the remaining terms, we write $d\boldsymbol{\ell} = dx\,\hat{\mathbf{x}} + dy\,\hat{\mathbf{y}}$ and use the previously established identities $\int x dx = \int y dy = 0$ and $\int x dy = -\int y dx = \mathcal{A}$. We find

$$\int (\nabla B_x \cdot \mathbf{r})\, d\boldsymbol{\ell} \times \hat{\mathbf{x}} = -\int_{\text{loop}} \left(\frac{\partial B_x}{\partial x} x + \frac{\partial B_y}{\partial y} y\right) \hat{\mathbf{z}}\, dy = -\mathcal{A}\, \frac{\partial B_x}{\partial x}\, \hat{\mathbf{z}}$$

$$\int (\nabla B_y \cdot \mathbf{r})\, d\boldsymbol{\ell} \times \hat{\mathbf{y}} = \int_{\text{loop}} \left(\frac{\partial B_y}{\partial x} x + \frac{\partial B_y}{\partial y} y\right) \hat{\mathbf{z}}\, dx = -\mathcal{A}\, \frac{\partial B_y}{\partial y}\, \hat{\mathbf{z}}$$

$$\int(\nabla B_z \cdot \mathbf{r})\, d\boldsymbol{\ell} \times \hat{\mathbf{z}} = \int_{\text{loop}} \left[\frac{\partial B_z}{\partial x} x + \frac{\partial B_z}{\partial y} y\right](\hat{\mathbf{x}}\, dy - \hat{\mathbf{y}}\, dx)$$

$$= \mathcal{A} \frac{\partial B_z}{\partial x} \hat{\mathbf{x}} + \mathcal{A} \frac{\partial B_z}{\partial y} \hat{\mathbf{y}}$$

We also know from Maxwell's fourth equation (12.8) that

$$\frac{\partial B_x}{\partial x} + \frac{\partial B_y}{\partial y} + \frac{\partial B_z}{\partial z} = 0$$

Combining the above results, we find

$$\mathbf{F} = \frac{I\mathcal{A}}{c}\left[\frac{\partial B_z}{\partial x}\hat{\mathbf{x}} + \frac{\partial B_z}{\partial y}\hat{\mathbf{y}} + \frac{\partial B_z}{\partial z}\hat{\mathbf{z}}\right] = \frac{I\mathcal{A}}{c}\nabla(\mathbf{B}\cdot\hat{\mathbf{z}})$$

Finally, we substitute for **m** from Eq. (12.14), and write the above equation in compact form:

$$\boxed{\mathbf{F} = \nabla(\mathbf{m} \cdot \mathbf{B}(\mathbf{r}))} \tag{12.23}$$

This is the general expression for the force on a magnetic dipole **m**, exerted by the field $\mathbf{B}(\mathbf{r})$. On this basis one can define the potential energy U of a dipole **m** in a magnetic field **B** as follows:

$$\boxed{U = -\mathbf{m} \cdot \mathbf{B}(\mathbf{r})} \tag{12.24}$$

The usefulness of the potential U defined by Eq. (12.24) is not limited to the derivation of the force expression, Eq. (12.23). In fact, in a uniform field $\mathbf{B}_0$, the same potential when regarded as function of the orientation of **m** can yield the torque expression of Eq. (12.20). The potential U of a dipole in an external field given by Eq. (12.24) is fairly general;† it is often used in conjunction with energy minimization schemes in micromagnetic calculations aimed at identifying the equilibrium state of a system.

†The limitations imposed on $\mathbf{B}(\mathbf{r})$ when deriving Eqs. (12.20) and (12.23) restrict the validity of Eq. (12.24) to localized loops, such as those formed by intrinsic spin and by orbiting electrons in atoms. When the current loops are not localized, Eq. (12.24) must be augmented by higher-order terms.

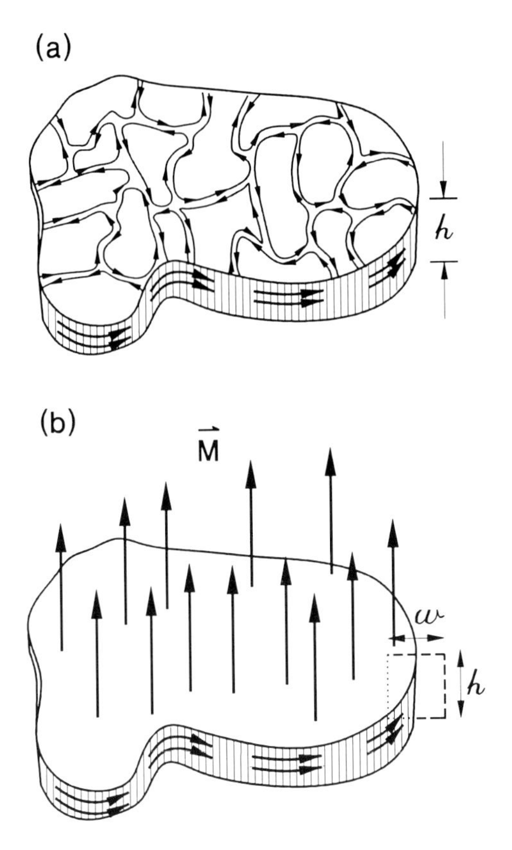

Figure 12.8. (a) Ribbon of current forming a planar closed loop. If small loops are imagined to cover the interior of the ribbon, then a dipole moment can be assigned to each of these little loops. In the limit of infinitesimally small loops the ribbon is filled with uniform magnetization **M** where M is proportional to the surface current density. (b) Slab of uniform magnetization **M** in free space. The magnetization vector is perpendicular to the flat surface. The current density in the ribbon surrounding the slab is $\mathbf{J} = \nabla \times \mathbf{M}$, where **J** and **M** are treated as functions over the entire space. The rectangular closed path with width w and height h is used to calculate $\nabla \times \mathbf{M}$ at the slab boundary.

12.2.5. Equivalence of current loops and slabs of magnetic material; relation between B, H, and M

Consider a flat, arbitrarily shaped loop with finite depth h, as shown in Fig. 12.8(a). The current I, confined to a thin ribbon, is uniformly distributed in the depth direction. This loop is equivalent to a collection of small, planar loops of arbitrary shapes and sizes, each carrying the same current I. (The equivalence is due to the fact that the currents on the adjacent inside walls cancel out, leaving only the current on the external wall.) Each tiny loop with surface area $\mathcal{A}$ is equivalent to a magnetic dipole moment $\mathbf{m} = (I\mathcal{A}/c)\hat{\mathbf{z}}$ (in MKSA, $I\mathcal{A}\hat{\mathbf{z}}$). The magnetization $\mathbf{M}$ at any given point is now given by the local value of $\mathbf{m}$ divided by the volume $\mathcal{A}h$ of the corresponding small loop, namely, $\mathbf{M} = (I/ch)\hat{\mathbf{z}}$ (in MKSA, $\mathbf{M} = (I/h)\hat{\mathbf{z}}$). The current loop in Fig. 12.8(a) and the slab having uniform magnetization in Fig. 12.8(b) are therefore equivalent.

Now, consider the slab of magnetic material in Fig. 12.8(b), with uniform magnetization $\mathbf{M}$ perpendicular to the flat face of the slab. Assuming the slab is surrounded by free space, we show that the equivalent loop current density $\mathbf{J}(\mathbf{r})$ around the slab is given by

$\mathbf{J}(\mathbf{r}) = c\,\nabla \times \mathbf{M}(\mathbf{r})$	$\mathbf{J}(\mathbf{r}) = \nabla \times \mathbf{M}(\mathbf{r})$	(12.25)

Since the curl of $\mathbf{M}$ is obviously zero both within the interior and in the external region of the slab, the validity of Eq. (12.25) need only be verified at the boundary. Thus consider the closed integration path in Fig. 12.8(b), and write $\nabla \times \mathbf{M}$ as the integral of $\mathbf{M}$ divided by the surface area of the path:

$$\nabla \times \mathbf{M} = \lim_{w \to 0} \frac{Mh}{wh}\,\mathbf{n} = \frac{1}{c} \lim_{w \to 0} \frac{I}{wh}\,\mathbf{n} = \frac{1}{c}\,\mathbf{J}(\mathbf{r})$$

where $\mathbf{n}$ is a unit vector perpendicular to the integration path. Any arbitrary distribution of magnetization can now be broken up into small slabs of uniform $\mathbf{M}$, similar to the one in Fig. 12.8(b), and the total magnetization can be regarded as the sum of all the slab magnetizations. Since "curl" is a linear operation, the curl of $\mathbf{M}$ will then be the superposition of the curls of all the individual slabs. In other words, the current distribution $\mathbf{J}(\mathbf{r})$ that gives rise to an arbitrary distribution of the magnetization $\mathbf{M}(\mathbf{r})$ can *always* be derived from Eq. (12.25).

The second Maxwell equation, Eq. (12.4), is written in terms of $\mathbf{H}$ rather than $\mathbf{B}$. The reason is that the current density $\mathbf{J}$ refers to the current of free charge carriers (such as conduction electrons) rather than the bound currents (orbital and spin motion of the electrons) that give rise to magnetic dipole moments through Eq. (12.25). If the bound currents were included in $\mathbf{J}$ of Eq. (12.4), then $\mathbf{H}$ on the left-hand side would have become $\mathbf{B}$ (or in MKSA, $\mathbf{B}/\mu_0$). The following relationship thus exists between $\mathbf{B}$, $\mathbf{H}$, and $\mathbf{M}$:

$\mathbf{B} = \mathbf{H} + 4\pi\mathbf{M}$	$\mathbf{B} = \mu_0(\mathbf{H} + \mathbf{M})$	(12.26)

In MKSA the units of both $\mathbf{H}$ and $\mathbf{M}$ are ampere/m, and Eq. (12.26) shows that these two entities may be combined without any constraints. In

CGS, on the other hand, **H** is in oersteds while **M** is in emu/cm^3; both have the same physical dimensions, but a numerical factor of 4π is needed to convert one to the other. (This is the same 4π that appeared earlier in Example 2.) Other than this peculiarity, the relationship between **B**, **H**, and **M** in Eq. (12.26) is straightforward and easy to remember.

In conjunction with Eq. (12.26), the fourth Maxwell equation (12.8) yields

$\nabla \cdot \mathbf{H} = -4\pi \nabla \cdot \mathbf{M} = 4\pi\rho_m$	$\nabla \cdot \mathbf{H} = -\nabla \cdot \mathbf{M} = \rho_m$	(12.27)

where the "magnetic charge density" ρ_m is defined as follows:

$$\rho_m = -\nabla \cdot \mathbf{M} \tag{12.28}$$

In magnetostatics, therefore, there are two sources of the **H**-field: the free-carrier current density $\mathbf{J}(\mathbf{r})$, and the magnetic "charge" density $\rho_m(\mathbf{r})$. Equations (12.4) and (12.27) show the dependence of **H** on these sources, respectively.†

Example 3. Figure 12.9(a) shows a thin film with uniform, perpendicular magnetization **M**. Using Eq. (12.28), the values of the surface magnetic charge density σ_m on the upper and lower surfaces of this film are found to be $+M$ and $-M$ respectively. This gives rise to an **H**-field of magnitude $-4\pi\mathbf{M}$ (in MKSA, $-\mathbf{M}$) inside the film.†† Outside the film the magnetic field is zero. According to Eq. (12.26), therefore, the **B**-field everywhere (both inside the film and outside) must be zero. The case of a uniformly magnetized cylinder is more interesting and is explored in some detail in frames (b)–(d) of Fig. 12.9. □

12.3. Larmor Diamagnetism

An electron of charge $-e$ in a circular orbit of radius r_0 is shown in Fig. 12.10. Let a magnetic field $\mathbf{B}_0$, uniform over the orbit of the electron, be applied in the Z-direction. Since the **B**-field has risen from an initial value of zero to its final value of $\mathbf{B}_0$ as some function of time, say $\mathbf{B}(t)$, an **E**-field must have been generated in the process by $\partial\mathbf{B}/\partial t$, as required by Maxwell's third equation,

†Compare Eq. (12.27) with Maxwell's first equation, $\nabla \cdot \mathbf{D} = 4\pi\rho_e$ (in MKSA, $\nabla \cdot \mathbf{D} = \rho_e$) where ρ_e is the electric charge density, and **D** is the displacement vector.

††This is similar to a parallel-plate capacitor with surface charge density σ, which produces a **D**-field of magnitude $-4\pi\sigma$ between the plates.

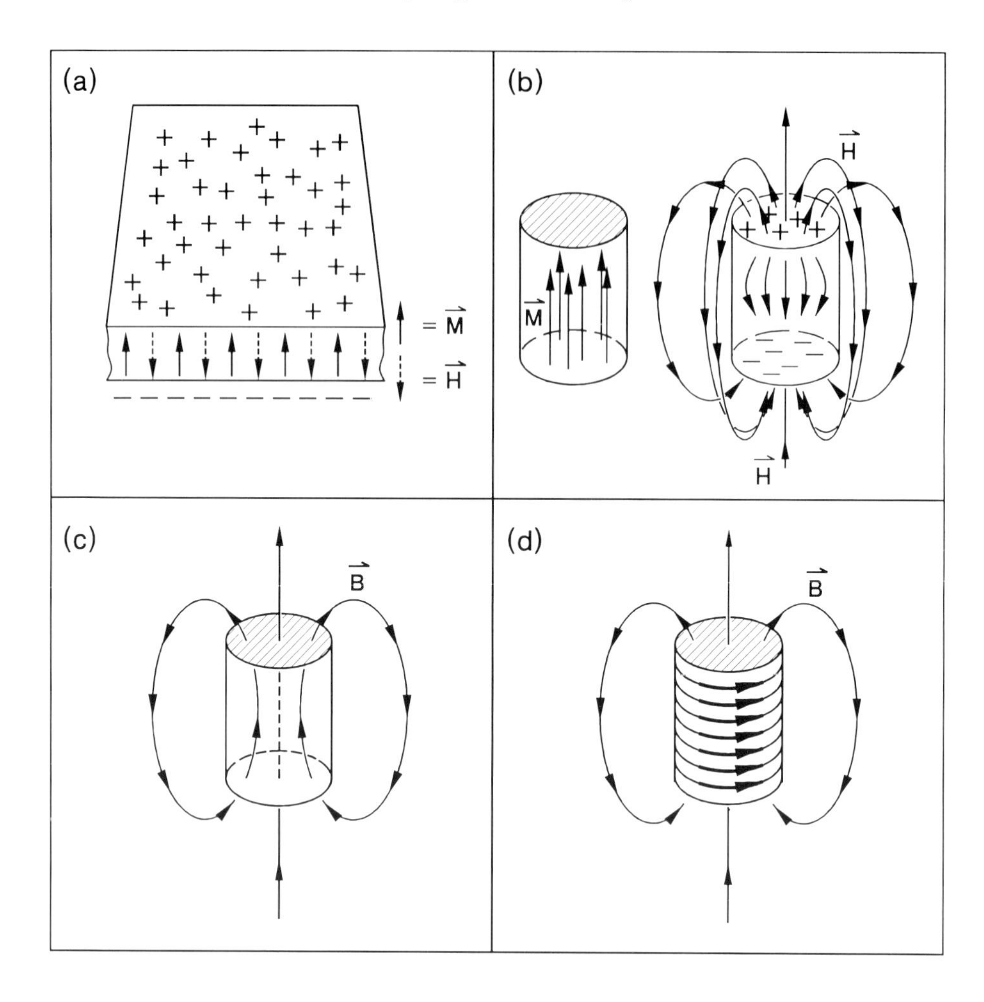

Figure 12.9. (a) Cross-section of magnetic film with infinite lateral extent. The magnetization **M** gives rise to surface magnetic charges with density $\sigma_m = \pm \mathrm{M}$. These charges in turn create a uniform **H**-field within the film, equal to $-4\pi\mathbf{M}$. The **B**-field is zero throughout the entire space. (b) Uniformly magnetized cylinder with magnetization **M** and surface magnetic charge density $\sigma_m = \pm \mathrm{M}$. The lines of the **H**-field originate on positive magnetic charges and terminate on negative magnetic charges. (c) In contrast to the **H**-field, the lines of **B** are continuous; they differ from those of **H** inside the cylinder, but in the outside region **B** = **H**. (d) A solenoidal current sheet wrapped around the cylinder produces the same **B**-field as the uniformly magnetized cylinder.

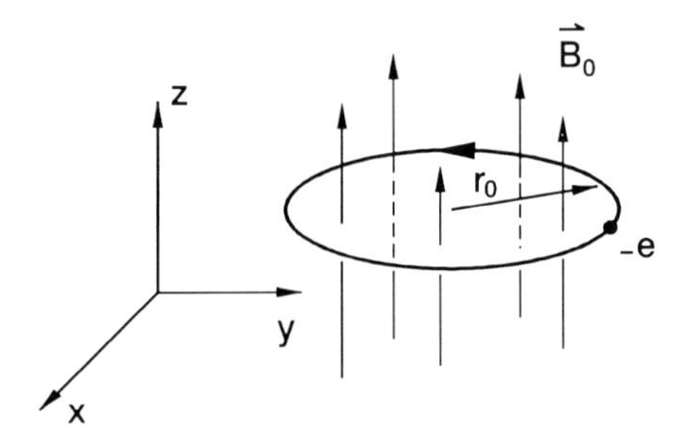

Figure 12.10. Electron in a circular orbit of radius r_0 subject to a uniform magnetic field $\mathbf{B}_0$. The plane of the orbit is parallel to XY and the field is along the Z-axis.

$\nabla \times \mathbf{E} = -\frac{1}{c}\frac{\partial \mathbf{B}}{\partial t}$	$\nabla \times \mathbf{E} = -\frac{\partial \mathbf{B}}{\partial t}$

(12.29)

Within a small segment $\Delta\boldsymbol{\ell}$ of the electron orbit, the charge Δq is

$$\Delta q = -\frac{e}{2\pi r_0}\Delta\ell$$

The force exerted on this segment by the $\mathbf{E}$-field is $\Delta q\mathbf{E}$, and the corresponding torque $\Delta\mathbf{T}$ is given by

$$\Delta\mathbf{T} = -r_0\left(\frac{e}{2\pi r_0}\right)\mathbf{E}\cdot\Delta\boldsymbol{\ell}\,\hat{\mathbf{z}}$$

The total instantaneous torque on the loop is obtained by integrating the above equation around the loop, which yields

$$\mathbf{T} = \frac{e r_0^2}{2c}\frac{\partial \mathbf{B}}{\partial t}$$

But $\mathbf{T} = \mathrm{d}\mathbf{L}/\mathrm{d}t$ and $\mathbf{m} = -\gamma\mathbf{L}$ where $\mathbf{L}$ is the angular momentum of the electron and γ, its gyromagnetic ratio, is given by Eq. (12.18). Therefore,

$$\Delta\mathbf{m} = -\frac{e^2 r_0^2}{4m_e c^2}\Delta\mathbf{B}$$

The susceptibility χ is defined as $|\Delta\mathbf{m}|/|\Delta\mathbf{B}|$. For diamagnetic substances χ is negative. Accordingly, an increasing $\mathbf{B}$-field causes the magnetization to decrease, and vice versa, as if a magnetic moment opposing the field

Table 12.1. Measured diamagnetic susceptibilities for noble gases[2]

Element	Atomic number, Z	Susceptibility, χ (cm^3/mole)
He	2	-1.9×10^{-6}
Ne	10	-7.2×10^{-6}
Ar	18	-19.4×10^{-6}
Kr	36	-28.0×10^{-6}
Xe	54	-43.0×10^{-6}

were being produced. In the case of Larmor diamagnetism, which arises from the orbital motion of electrons around nuclei, the susceptibility per electron according to the preceding equation is

$\chi = -\dfrac{e^2 r_0^2}{4m_e c^2}$ unit of χ = cm^3	$\chi = -\dfrac{e^2 r_0^2}{4m_e}$ unit of χ = m^3

(12.30)

A typical numerical value for χ may be obtained using $r_0 \simeq 0.5$ Å, $e = 4.8 \times 10^{-10}$ statcoulombs, $m_e = 9.1 \times 10^{-28}$ g, and $c = 3 \times 10^{10}$ cm/s, resulting in $\chi \simeq 2 \times 10^{-30}$ cm^3. If an atom of a given material has Z electrons, its susceptibility per mole will be $ZN_a\chi$, where N_a, Avogadro's number = 6.023×10^{23}, is the number of atoms per mole. Typical values of χ for noble gases are listed in Table 12.1; the order-of-magnitude agreement between the measured values and those obtained from Eq. (12.30) is quite satisfactory. Because χ is relatively small compared to paramagnetic and ferromagnetic susceptibilities, diamagnetism is usually obscured by these other phenomena. Only in cases where para- and ferromagnetic effects are absent (such as for noble gases) does one observe an unambiguous diamagnetic effect.

12.4. Ground State of Atoms with Partially Filled Shells; Hund's Rules

The total angular momentum quantum number J of an atom (ion) is always a good quantum number. The orbital (L) and spin (S) angular momentum quantum numbers, on the other hand, are good quantum numbers only insofar as the spin–orbit coupling can be ignored. In the absence of spin–orbit coupling the states of the atom, described by quantum numbers L, L_z, S, S_z, J, and J_z, are eigenstates of the operators $\mathbf{L}^2$, $\mathbf{L}_z$, $\mathbf{S}^2$, $\mathbf{S}_z$, $\mathbf{J}^2$, and $\mathbf{J}_z$, whose eigenvalues are $L(L + 1)$, L_z, $S(S + 1)$, S_z, $J(J + 1)$, and J_z respectively. For filled shells $J = L = S = 0$, so that only partially filled shells contribute to the angular momentum (and therefore to the magnetic moment) of free atoms.[2,3] The intra-atomic interaction amongst the various electrons of a given shell (the so-called Russell–Saunders coupling) lifts some of the degeneracy associated with the electronic state of the atom. It thus results in particular arrangements of the spins and orbital momenta in the ground state, which are best described by Hund's rules.[4-6]

Hund's first rule: in the ground state, the spin angular momentum quantum number S of an atomic subshell† will have the largest possible value consistent with the exclusion principle. Thus, with one electron in the subshell, $S = \frac{1}{2}$, with two electrons, $S = 1$, and so forth, until the subshell is half-filled. At this point there are $2\ell + 1$ electrons in the subshell, and the net spin quantum number is $S = \ell + \frac{1}{2}$. Adding electrons after this will reduce the spin back to zero. As is generally the case with angular momenta, the values of S_z for a given S range from $-S$ to $+S$ in integer steps.

Hund's second rule: the orbital angular momentum quantum number L of an atomic subshell has the largest value consistent with the first rule and with the exclusion principle. Thus the first electron in the subshell goes to a state with $|\ell_z| = \ell$, the second to $|\ell_z| = \ell - 1$, and so forth. In this way, the orbital angular momentum quantum number L assumes the values ℓ, $\ell + (\ell - 1)$, $\ell + (\ell - 1) + (\ell - 2)$, and so on. By the time the subshell is half-filled, the net value of L is zero. After this the opposite spins begin to come in and the same set of values for L is visited again. Now, S_z can be anywhere from $-S$ to $+S$, and L_z can be anywhere from $-L$ to $+L$. The spin–orbit coupling determines the relative energy of these $(2L + 1) \times (2S + 1)$ states. That is where Hund's third rule comes into play.

Hund's third rule:†† if the subshell is less than half-filled, the spin and

† A shell, identified by the principal quantum number n, has n subshells. Each subshell is characterized by an orbital angular momentum quantum number ℓ, which is an integer between 0 and $n - 1$. (For historical reasons, the subshells are referred to as s, p, d, f, g, h corresponding to $\ell = 0, 1, 2, 3, 4, 5$.) A subshell in turn consists of $2\ell + 1$ orbitals, each capable of holding a pair of electrons, one with spin up, the other with spin down. All electrons within the subshell ℓ have the same orbital angular momentum, $\sqrt{\ell(\ell + 1)}\,\hbar$, but the projection $\ell_z\hbar$ of this angular momentum along the Z-axis differs for different orbitals, ranging from $-\ell\hbar$ to $+\ell\hbar$ in steps of $\hbar$.

†† The basis of the third rule is the coupling between the two components of an electron's magnetic moment, arising from its spin and its orbital motion. In the rest frame of a given electron, the charged atomic nucleus revolves around the electron, producing a magnetic field that interacts with the electron's spin magnetic moment. This interaction couples $\mathbf{L}$ and $\mathbf{S}$ of the electron, making them align either parallel or antiparallel to each other.

Table 12.2. Ground states of ions with partially filled d- or f-shells as constructed from Hund's rules (↑ = spin $\frac{1}{2}$; ↓ = spin $-\frac{1}{2}$). Source: N.W. Ashcroft and N.D. Mermin, *Solid State Physics*, copyright © 1976 by Holt, Rinehart & Winston, reproduced by permission of Saunders College Publishing

d-shell ($\ell = 2$)

ν	$\ell_z = 2$	1	0	−1	−2	S	$L = \lvert\Sigma\ell_z\rvert$	J	Symbol
1	↓					1/2	2	3/2	$^2D_{3/2}$
2	↓	↓				1	3	2	$^3F_2$
3	↓	↓	↓			3/2	3	3/2	$^4F_{3/2}$
4	↓	↓	↓	↓		2	2	0	$^5D_0$
5	↓	↓	↓	↓	↓	5/2	0	5/2	$^6S_{5/2}$
6	↓↑	↑	↑	↑	↑	2	2	4	$^5D_4$
7	↓↑	↓↑	↑	↑	↑	3/2	3	9/2	$^4F_{9/2}$
8	↓↑	↓↑	↓↑	↑	↑	1	3	4	$^3F_4$
9	↓↑	↓↑	↓↑	↓↑	↑	1/2	2	5/2	$^2D_{5/2}$
10	↓↑	↓↑	↓↑	↓↑	↓↑	0	0	0	$^1S_0$

f-shell ($\ell = 3$)

ν	$\ell_z = 3$	2	1	0	−1	−2	−3	S	$L = \lvert\Sigma\ell_z\rvert$	J	Symbol
1	↓							1/2	3	5/2	$^2F_{5/2}$
2	↓	↓						1	5	4	$^3H_4$
3	↓	↓	↓					3/2	6	9/2	$^4I_{9/2}$
4	↓	↓	↓	↓				2	6	4	$^5I_4$
5	↓	↓	↓	↓	↓			5/2	5	5/2	$^6H_{5/2}$
6	↓	↓	↓	↓	↓	↓		3	3	0	$^7F_0$
7	↓	↓	↓	↓	↓	↓	↓	7/2	0	7/2	$^8S_{7/2}$
8	↓↑	↑	↑	↑	↑	↑	↑	3	3	6	$^7F_6$
9	↓↑	↓↑	↑	↑	↑	↑	↑	5/2	5	15/2	$^6H_{15/2}$
10	↓↑	↓↑	↓↑	↑	↑	↑	↑	2	6	8	$^5I_8$
11	↓↑	↓↑	↓↑	↓↑	↑	↑	↑	3/2	6	15/2	$^4I_{15/2}$
12	↓↑	↓↑	↓↑	↓↑	↓↑	↑	↑	1	5	6	$^3H_6$
13	↓↑	↓↑	↓↑	↓↑	↓↑	↓↑	↑	1/2	3	7/2	$^2F_{7/2}$
14	↓↑	↓↑	↓↑	↓↑	↓↑	↓↑	↓↑	0	0	0	$^1S_0$

orbital angular momenta couple in an antiparallel fashion (in the low-lying energy states, of course). Under these circumstances the total angular momentum quantum number J is given by

$$J = |L - S| \qquad \text{(number of electrons in the subshell} \le 2\ell + 1) \qquad (12.31a)$$

On the other hand, if the subshell is more than half-filled, the spin and orbital momenta couple parallel to each other, and the net angular momentum quantum number will be

$$J = L + S \qquad \text{(number of electrons in the subshell} \ge 2\ell + 1) \qquad (12.31b)$$

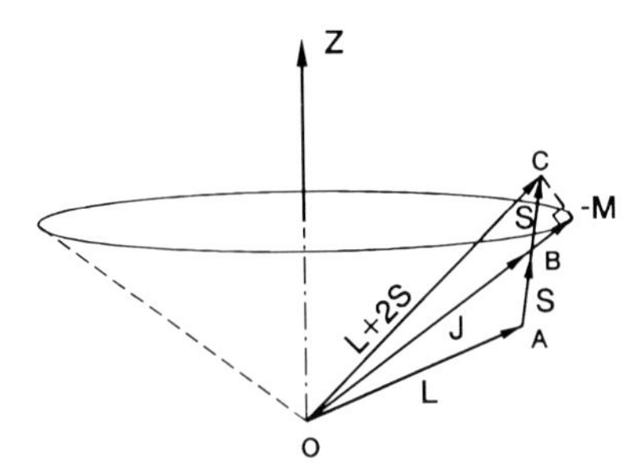

Figure 12.11. Schematic diagram showing the various components of the atomic magnetic moment **M**. The magnetic field $\mathbf{B}_0$ is applied along the Z-axis. The vectors identified as **L** and **S** are the orbital and spin components, respectively, of the total angular momentum **J**. The net magnetic moment is proportional to **L** + 2**S**, whose projection along **J** is equal to $g(JLS)\mathbf{J}$. Because of their precessing motion around the field, only the projections of these vectors along Z are observable. Source: S. Chikazumi and S.H. Charap, *Physics of Magnetism*, copyright © 1964 by Krieger Publishing Company, reproduced by permission of the publisher.

The projection on Z of the total angular momentum is degenerate, with values between $-J\hbar$ and $+J\hbar$ in steps of $\hbar$. In magnetic systems of practical interest, one usually deals only with the set of $(2L + 1) \times (2S + 1)$ states determined by the first two rules, all others lying so much higher in energy as to be of no interest. Furthermore, it is often enough to consider only the $2J+1$ lowest-lying of these states, specified by the third rule. Thus the JLS triplet of numbers identifies the ground state of any free atom (ion). For historical reasons, the value of L is identified by the letters S, P, D, F, G, H, I instead of the numbers 0, 1, 2, 3, 4, 5, 6. The spin is specified by the value of $2S + 1$ as a superprefix, and the total angular momentum quantum number J is given as a subscript. Thus ${}^{2S+1}\mathrm{X}_J$, where X is the letter identifying the L-value, provides complete identification for a given ground state. Table 12.2 lists the ground states of ions with partially filled d or f shells as constructed from the Hund rules.[2]

12.4.1. Spectroscopic splitting factor

We learned in section 12.2.1 that the magnetic moment **m** of an electron in orbital motion is proportional to its orbital momentum quantum number L, with proportionality constant of one Bohr magneton μ_B. The situation is somewhat different for the electron spin **S** where, according to relativistic quantum electrodynamics, $\mathbf{m} = g_0\mu_B\mathbf{S}$. Here $g_0 = 2.0023$ is the so-called spectroscopic splitting factor (the g-factor) for the free electron. ($g_0 \simeq 2$ is usually a good approximation.)

When the splitting between the zero-field atomic ground state multiplet and the first excited multiplet is large compared with k_BT (as is frequently the case), only the $2J + 1$ states in the ground-state multiplet contribute appreciably to the free energy, in which case it can be shown that:[7]

$$\boxed{\mathbf{L} + g_0 \mathbf{S} = g(JLS)\,\mathbf{J}} \tag{12.32a}$$

and

$$\boxed{\mathbf{m} = g(JLS)\,\mu_B\,\mathbf{J}} \tag{12.32b}$$

The situation is depicted schematically in Fig. 12.11, but proper treatment of the problem requires quantum mechanical calculations well beyond the scope intended here. For free ions, the g-factor appearing in Eq. (12.32) can be shown to obey the Landé formula:[2]

$$\boxed{g(JLS) = \frac{3}{2} + \frac{1}{2}\,\frac{S(S+1) - L(L+1)}{J(J+1)}} \tag{12.32c}$$

The Hund rules, together with the Landé formula for the g-factor, provide an accurate picture of the magnetic state of free atoms (ions) for most cases of practical interest. Exceptions occur when the separation between the low-lying and excited levels is small as, for example, in the case of Sm^{+++} and Eu^{+++} ions, or when the spin–orbit coupling is too strong for the assumptions that lead to the above results to hold.

12.5. Paramagnetism

Consider a collection of N non-interacting particles confined within a volume V. The particles, each carrying a magnetic moment $\mathbf{m}$, are in thermal equilibrium at a finite temperature T. In the absence of an external magnetic field the randomly oriented dipoles cancel each other out along any given direction in space; thus the collection will have no net magnetic moment. However, when a field $\mathbf{B}_0$ is applied, the individual dipoles tend to align themselves with the field and, as a result, a net magnetic moment parallel to the field will develop. This behavior is known as paramagnetism and, unless the field is too strong or the temperature too low, the net magnetization of the volume will be proportional to the applied field, as follows:

$$\mathbf{M} = \frac{N}{V}\langle \mathbf{m} \rangle = \chi \mathbf{B}_0$$

The paramagnetic susceptibility χ is generally a function of the temperature T. At high fields and/or low temperatures, the magnetization approaches saturation and all the individual dipole moments become fully aligned with the field; under these circumstances there will be no further increases of $\mathbf{M}$ with an increasing magnetic field. Now, depending on whether the particles follow the Maxwell–Boltzmann distribution law or obey Fermi–Dirac statistics, two different types of paramagnetic behavior may be observed. In the former case, the susceptibility depends strongly on the temperature, following a $1/T$ law; this is the so-called Langevin paramagnetism, which will be covered in the following subsection. The latter case, which corresponds to a collection of indistinguishable fermions (such as the conduction electrons in metals) gives rise to a weak paramagnetic χ that is more or less independent of temperature; this is the case of Pauli paramagnetism and will be considered in subsection 12.5.2.

12.5.1. Langevin paramagnetism of a collection of identical atoms

We derive the temperature-dependence of the magnetic moment for a collection of ions subject to an external magnetic field $B_0\hat{z}$. Let the individual ions have a total angular momentum quantum number J and a spectroscopic splitting factor g. Each ion can exhibit a Z-component of angular momentum $J_z\hbar$ where J_z is between $-J$ and $+J$ (in integer steps). The corresponding energy of the ion in the field is thus $U = -g\mu_B J_z B_0$. At a finite temperature T, the moments are distributed according to the Boltzmann distribution; therefore, the average magnetic moment $\langle m \rangle$ of individual ions is given by

$$\langle m \rangle = \frac{\sum_{j=-J}^{J} -g\mu_B j \exp\left(-\frac{g\mu_B jB_0}{k_B T}\right)}{\sum_{j=-J}^{J} \exp\left(-\frac{g\mu_B jB_0}{k_B T}\right)} \tag{12.33}$$

A closed form for this equation may be derived by considering the identity

$$\sum_{j=-J}^{J} \exp(-\alpha j) = \frac{\exp[\alpha(J+1)] - \exp(-\alpha J)}{\exp(\alpha) - 1} = \frac{\sinh[\alpha(J+\frac{1}{2})]}{\sinh(\frac{1}{2}\alpha)}$$

which leads to

$$\frac{d}{d\alpha}\left\{\ln\left[\sum_{j=-J}^{J} \exp(-\alpha j)\right]\right\} = \frac{\Sigma[-j\exp(-\alpha j)]}{\Sigma \exp(-\alpha j)}$$

$$= \frac{d}{d\alpha}\left\{\ln\{\sinh[\alpha(J+\tfrac{1}{2})]\} - \ln[\sinh(\tfrac{1}{2}\alpha)]\right\}$$

$$= (J+\tfrac{1}{2})\coth[\alpha(J+\tfrac{1}{2})] - \tfrac{1}{2}\coth(\tfrac{1}{2}\alpha)$$

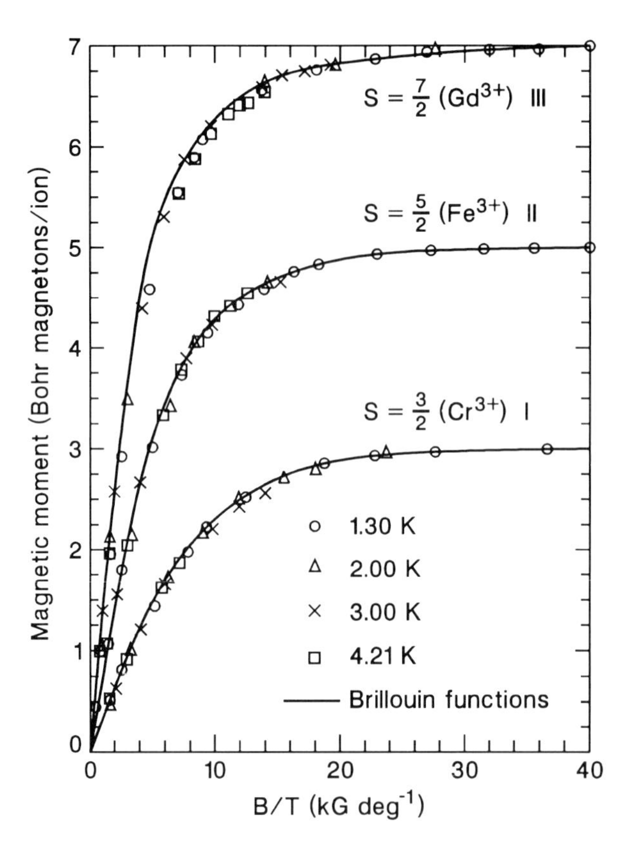

Figure 12.12. Magnetization curves of paramagnetic salts: potassium chromium alum (I), ferric ammonium alum (II), and gadolinium sulfate octahydrate (III). After W. E. Henry, *Phys. Rev.* **88**, 559 (1952).

This yields the following closed form for Eq. (12.33):

$$\langle m \rangle = g\mu_B \left\{ (J + \tfrac{1}{2}) \coth \left[(J + \tfrac{1}{2}) \frac{g\mu_B B_0}{k_B T} \right] - \tfrac{1}{2} \coth \left[\frac{g\mu_B B_0}{2k_B T} \right] \right\}$$

In terms of the Brillouin function $\mathscr{B}_J(x)$, defined as[6]

$$\mathscr{B}_J(x) = \frac{2J+1}{2J} \coth \left(\frac{2J+1}{2J} x \right) - \frac{1}{2J} \coth \left(\frac{1}{2J} x \right) \tag{12.34}$$

the average moment $\langle m \rangle$ of a given ion is written as

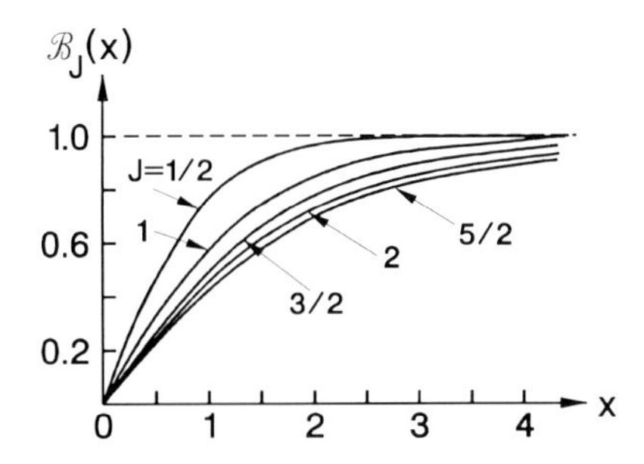

Figure 12.13. Plots of the Brillouin function $\mathscr{B}_J(x)$ for various values of J. The function is defined in Eq. (12.34).

$$\boxed{\langle m \rangle \;=\; g\mu_B\, J\mathscr{B}_J\left(\frac{g\mu_B J}{k_B T}\, B_0\right)} \qquad (12.35)$$

To obtain the magnetization M we multiply the average moment per ion, $\langle m \rangle$, by N/V, the number of ions per unit volume.

Figure 12.12 shows the excellent agreement between the above theory and experiment for three paramagnetic salts. The solid curves are plots of $\langle m \rangle/\mu_0$ versus B_0/T as given by Eq. (12.35). Note that in the case of Cr^{+++} (which has three electrons left in the d-shell) the orbital angular momentum is "quenched" by the crystal electric field, resulting in $L = 0$. (We shall say more about quenching in section 12.9). Consequently, the values of $J = S = 3/2$ and $g = g_0 = 2$ are used in this case. Fe^{+++} with five electrons in the d-shell and Gd^{+++} with seven electrons in the f-shell are both S-state ions (see Table 12.2); therefore, $L = 0$, $J = S$, and $g = 2$ are used in both these cases.

Figure 12.13 shows plots of $\mathscr{B}_J(x)$ for several values of J. At low temperatures (or high fields) the argument of $\mathscr{B}_J(\cdot)$ in Eq. (12.35) is large; the function is thus close to unity and $\langle m \rangle \simeq g\mu_B J$, the largest possible moment for the given ion. In the other extreme of large T or small B_0,

$$\mathscr{B}_J(x) \simeq \frac{J+1}{3J}\, x + O(x^3) \qquad (12.36)$$

Consequently, the paramagnetic susceptibility χ per unit volume becomes

$$\boxed{\chi = \frac{N}{V}\frac{\langle m \rangle}{B_0} \;\simeq\; \frac{N}{V}\, g^2 {\mu_B}^2\, \frac{J(J+1)}{3k_B T}\,; \qquad k_B T >> g\mu_B J B_0} \qquad (12.37)$$

This inverse temperature-dependence of χ is known as the Curie law. The paramagnetic susceptibility is generally larger than the diamagnetic susceptibility by a factor of about 10^2-10^3, so that in comparison diamagnetic effects may be ignored. From the coefficient of $1/T$ in the experimentally obtained plots of susceptibility one usually derives an "effective Bohr magneton number" $\mathscr{P}$, which is defined as follows:[2]

$$\mathscr{P} = g(JLS)\sqrt{J(J+1)} \tag{12.38}$$

Later, in our discussion of rare earths (section 12.8) and transition metals (section 12.9), we shall have occasion to compare the experimentally determined values of $\mathscr{P}$ with those obtained from Eq. (12.38) in conjunction with Hund's rules and the Landé formula.

12.5.2. Conduction electron (Pauli) paramagnetism

Before the advent of the quantum theory of solids, it was thought that the conduction electrons within a solid, which carry a spin magnetic moment of $g_0\mu_B S \simeq \mu_B$, must exhibit a paramagnetic susceptibility as predicted by Eq. (12.37). Failure to observe such behavior was one of the outstanding problems of the theory of solids in the early days. Pauli's discovery of the exclusion principle and the recognition that the occupancy of energy levels in a free-electron gas is governed by Fermi–Dirac statistics brought about a satisfactory resolution of this controversy. In this section we shall derive the paramagnetic susceptibility of the conduction electrons based on an elementary treatment of the free-electron gas. But first we must digress to discuss Fermi–Dirac statistics and derive the density of states for a gas of free electrons.

Digression 3: the Fermi–Dirac distribution. Consider a system of N identical particles having a number of states indexed by i, and let the energy of the ith state be dentoed by E_i. At the equilibrium temperature T the Helmholtz free energy F_N of this system satisfies the relation

$$\text{partition function} = \sum_i \exp\left(-\frac{E_i}{k_B T}\right) = \exp\left(-\frac{F_N}{k_B T}\right) \tag{12.39}$$

In accordance with classical statistical mechanics, the probability of the system being in a particular state j with energy E_j is

$$P_N(j) = \frac{\exp\left(-\frac{E_j}{k_B T}\right)}{\sum_i \exp\left(-\frac{E_i}{k_B T}\right)} = \exp\left(-\frac{E_j - F_N}{k_B T}\right) \qquad (12.40)$$

Now, consider all states of this N-particle system in which a certain single-particle state σ is occupied. The probability $p(\sigma)$ of this event can be written in one of two possible ways. The first is

$$p(\sigma) = \sum_i P_N(i) \qquad (12.41a)$$

where the sum is over all N-particle states i in which the single-particle state σ is occupied. The second is

$$p(\sigma) = 1 - \sum_i P_N(i) \qquad (12.41b)$$

where the sum is over all N-particle states i in which the single-particle state σ is *not* occupied. For every N-particle state i in which the state σ is empty, there exists a $(N + 1)$-particle state i' where σ is occupied while everything else is the same. We thus have

$$P_N(i) = \exp\left(-\frac{E_i - F_N}{k_B T}\right)$$

$$P_{N+1}(i') = \exp\left(-\frac{E_{i'} - F_{N+1}}{k_B T}\right)$$

Let $E_{i'} = E_i + \mathcal{E}_\sigma$, where $\mathcal{E}_\sigma$ is the energy associated with the single-particle state σ. Also, we define the chemical potential μ for the N-particle system as $\mu = F_{N+1} - F_N$. The preceding equations then yield

$$\frac{P_N(i)}{P_{N+1}(i')} = \exp\left(\frac{\mathcal{E}_\sigma - \mu}{k_B T}\right) \qquad (12.42)$$

Therefore, from Eq. (12.41b) we have

$$p(\sigma) = 1 - \exp\left(\frac{\mathcal{E}_\sigma - \mu}{k_B T}\right) \sum_{i'} P_{N+1}(i') \tag{12.43}$$

Here the sum is over all $(N + 1)$-particle states i' in which the single-particle state σ is occupied. Since N is typically very large (of the order of Avogadro's number) the addition of one particle to the system cannot alter the probabilities significantly; the summation on the right-hand side of Eq. (12.43) is thus equal to $p(\sigma)$, in accordance with Eq. (12.41a). Therefore

$$p(\sigma) = \frac{1}{1 + \exp\left(\frac{\mathcal{E}_\sigma - \mu}{k_B T}\right)} \tag{12.44}$$

This is the well-known Fermi–Dirac distribution function. We have derived it here for a system of N particles, and the chemical potential μ carries this dependence on N implicitly. At a fixed temperature T and for a fixed system of known energy levels, the probability of a given level being occupied is solely a function of the energy $\mathcal{E}$ associated with that level. The Fermi function $\mathcal{f}(\mathcal{E})$ is thus defined as a universal function to represent this probability distribution as follows:

$$\boxed{\mathcal{f}(\mathcal{E}) = \frac{1}{1 + \exp\left(\frac{\mathcal{E} - \mu}{k_B T}\right)}} \tag{12.45}$$

Strictly speaking, the chemical potential μ for a given system is a function of the temperature T. In practice, however, this temperature-dependence is often insignificant and one can take μ to be a constant. The value of μ at $T = 0$ K is known as the Fermi energy $\mathcal{E}_F$. If the energy $\mathcal{E}$ of a single-particle state is less than $\mathcal{E}_F$ then, at $T = 0$ K, that state is occupied and $\mathcal{f}(\mathcal{E}) = 1$. On the other hand, if $\mathcal{E} > \mathcal{E}_F$ then $\mathcal{f}(\mathcal{E}) = 0$ and the state is unoccupied (see Fig. 12.14). The Fermi energy, therefore, is the boundary between occupied and unoccupied states at $T = 0$ K. To obtain the chemical potential μ, we note that at $T = 0$ K there are N particles for which $\mathcal{f}(\mathcal{E}) = 1$. Generalizing this notion to $T \neq 0$ K, we have

$$\sum_\sigma \frac{1}{1 + \exp\left(\frac{\mathcal{E}_\sigma - \mu}{k_B T}\right)} = N \tag{12.46}$$

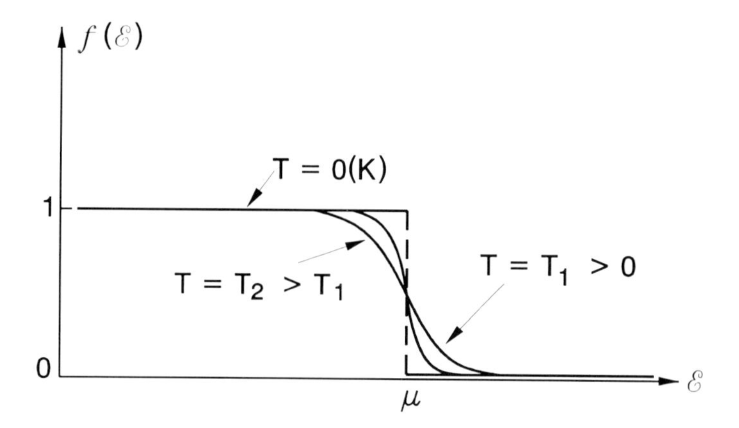

Figure 12.14. The Fermi function $f(\mathcal{E})$. At $T = 0$ K the states with $\mathcal{E} < \mu$ are occupied and those with $\mathcal{E} > \mu$ are empty. As the temperature rises, the sharp cutoff near the Fermi surface disappears. Note that μ for a given system is not constant but varies with temperature. The value of μ at $T = 0$ K is the Fermi energy $\mathcal{E}_F$.

where the sum is over all single-particle states σ with energy $\mathcal{E}_\sigma$. Since the total number of particles N and the energy levels $\mathcal{E}_\sigma$ are known, the only unknown in Eq. (12.46), μ, can be determined as a function of T. □

Digression 4: a gas of non-interacting free electrons. Consider N electrons occupying an empty $L \times L \times L$ cube. In the absence of electron–electron interactions, the wave-function ψ for each electron will be

$$\psi(\mathbf{r}) = \exp(i\mathbf{k}\cdot\mathbf{r}) \tag{12.47a}$$

Of course, the wave-vector $\mathbf{k}$ is not arbitrary; confinement to the box imposes restrictions on the acceptable values of $\mathbf{k}$. The Born–Von Karman boundary conditions[2] require periodicity at the walls of the box, namely,

$$\mathbf{k} = \frac{2\pi}{L}(n_x\hat{\mathbf{x}} + n_y\hat{\mathbf{y}} + n_z\hat{\mathbf{z}}) \tag{12.47b}$$

where n_x, n_y, and n_z are arbitrary integers. The momentum $\mathbf{p}$ of the electron is related to its wave-vector $\mathbf{k}$ through De Broglie's relation, $\mathbf{p} = \hbar\mathbf{k}$. The total energy of an electron in the present situation where potential energies are absent and interactions amongst electrons are

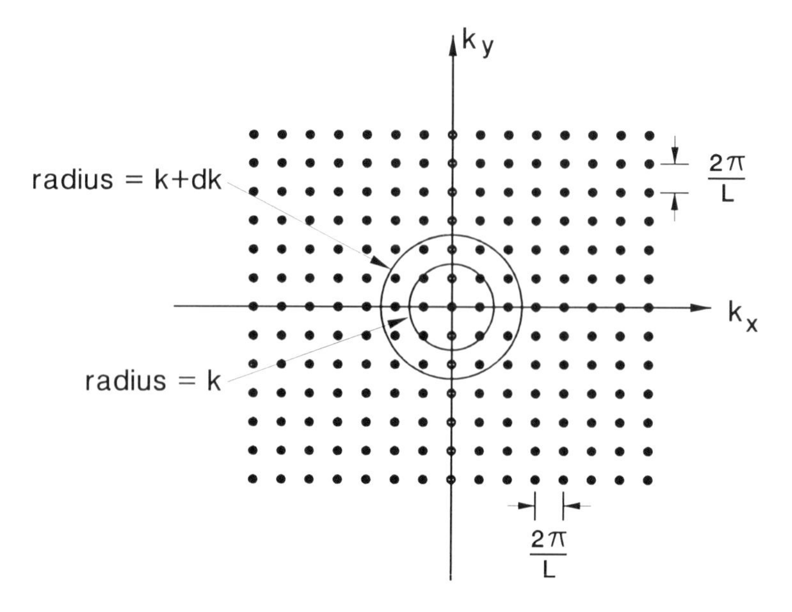

Figure 12.15. Allowed values of **k** for free electrons confined to a cube of side L. The annulus is the cross-section with the $k_x k_y$-plane of the region between two spheres with radii k and $k + \mathrm{d}k$.

neglected, is simply the kinetic energy of the electron,† that is,

$$\mathscr{E}(\mathbf{k}) = \frac{p^2}{2m} = \frac{\hbar^2 k^2}{2m} \tag{12.48}$$

Since L, the side of the cube, can be taken to be very large compared with k^{-1}, the **k**-space will be densely populated with acceptable wave-vectors defined by Eq. (12.47b), as shown in Fig. 12.15. We calculate the number of allowed states whose energy is between $\mathscr{E}_0$ and $\mathscr{E}_0 + \mathrm{d}\mathscr{E}$. Here $\mathrm{d}\mathscr{E}$ is an infinitesimal interval which, according to Eq. (12.48), is given by $(\hbar^2/m)k\mathrm{d}k$. The volume of **k**-space occupied by states within the above range of energies is

$$4\pi k^2 \mathrm{d}k = 4\pi k(k\mathrm{d}k) = \frac{4\pi m}{\hbar^2}\sqrt{\frac{2m\mathscr{E}(k)}{\hbar^2}}\,\mathrm{d}\mathscr{E} \tag{12.49}$$

†The wave-function in Eq. (12.47a) and its corresponding energy in Eq. (12.48) satisfy the time-independent Schrödinger equation:

$$-\frac{\hbar^2}{2m}\left(\frac{\partial^2}{\partial x^2} + \frac{\partial^2}{\partial y^2} + \frac{\partial^2}{\partial z^2}\right)\psi(\mathbf{r}) + V(\mathbf{r})\psi(\mathbf{r}) = \mathscr{E}\psi(\mathbf{r})$$

For free electrons the potential $V(\mathbf{r})$ within the $L \times L \times L$ cube is zero. Periodicity at the cube boundaries limits the acceptable wave vectors to those that satisfy Eq. (12.47b).

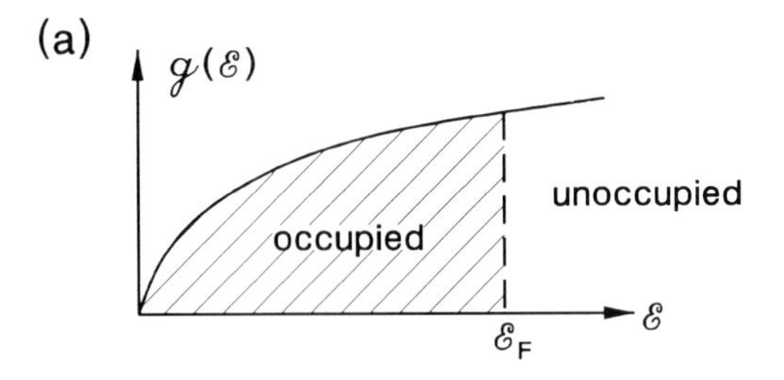

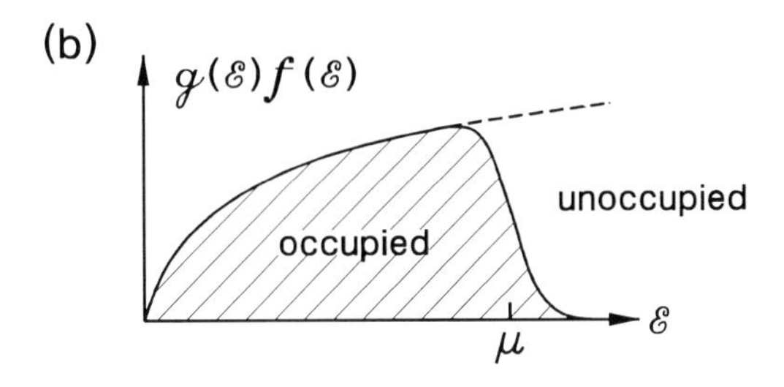

Figure 12.16. Density of states for a gas of free electrons. (a) At $T = 0$ K all the states below the Fermi level are occupied. (b) At $T \neq 0$ K some of the electrons near the Fermi surface spill over to higher energy states.

From Eq. (12.47b) it must be clear that each acceptable $\mathbf{k}$ occupies a volume $\Delta V = (2\pi/L)^3$ within $\mathbf{k}$-space. The number of acceptable wave-vectors corresponding to energies between $\mathcal{E}_0$ and $\mathcal{E}_0+\mathrm{d}\mathcal{E}$ is thus obtained by dividing the total volume in Eq. (12.49) by ΔV, namely,

$$\frac{4\pi k^2 \mathrm{d}k}{(2\pi/L)^3} = \frac{mL^3}{2\pi^2\hbar^2}\sqrt{\frac{2m\mathcal{E}(k)}{\hbar^2}}\,\mathrm{d}\mathcal{E} \tag{12.50}$$

This number, when divided by the volume L^3 of the cube, gives the number density of states per unit volume of space. The number density must be multiplied by a factor of two to account for the fact that each state can be occupied by two electrons of opposite spin. The result, when divided by $\mathrm{d}\mathcal{E}$, is the number density per unit interval of energy, $g(\mathcal{E})$. This so-called electronic density-of-states function is given by

$$g(\mathcal{E}) = \frac{m}{\pi^2\hbar^2}\sqrt{\frac{2m\mathcal{E}(k)}{\hbar^2}} \tag{12.51}$$

A plot of $g(\mathcal{E})$ is shown in Fig. 12.16. At $T = 0$ K all states with $\mathcal{E} \leq \mathcal{E}_F$ are occupied, and all states with $\mathcal{E} > \mathcal{E}_F$ are empty. An equivalent statement is that at $T = 0$ K the sphere of radius $k_F = \sqrt{2m\mathscr{E}_F}/\hbar$ in **k**-space is fully occupied (see Fig. 12.15). The volume of this sphere, $(4\pi/3)k_F^3$, when divided by $(2\pi/L)^3$, multiplied by two to account for the two spin states, and divided by L^3 to normalize to unit volume, yields the number density $\mathscr{N}$ of the electrons. Thus $\mathscr{N} = k_F^3/3\pi^2$ and, using Eq. (12.48),

$$\boxed{\mathscr{E}_F = \frac{\hbar^2}{2m}\left(3\pi^2\mathscr{N}\right)^{2/3}} \tag{12.52}$$

Finally, substituting for $m/\hbar^2$ in Eq. (12.51) in terms of $\mathscr{E}_F$ and $\mathscr{N}$ from Eq. (12.52), we arrive at

$$\boxed{g(\mathscr{E}) = \frac{3}{2}\frac{\mathscr{N}}{\mathscr{E}_F}\sqrt{\frac{\mathscr{E}}{\mathscr{E}_F}}} \tag{12.53}$$

The consistency of this result may be verified by noting that

$$\int_0^{\mathscr{E}_F} g(\mathscr{E})\,\mathrm{d}\mathscr{E} = \mathscr{N}$$

For a typical value of $\mathscr{N}$, 10^{22} cm^{-3}, Eq. (12.52) with $m = 0.91 \times 10^{-27}$ g, $h = 6.62 \times 10^{-27}$ erg s, and $k_B = 1.38 \times 10^{-16}$ erg/K yields $\mathscr{E}_F/k_B \simeq$ 10 000 K. This value of the Fermi energy for conduction electrons is typical of ordinary conductors (such as copper and aluminum). Thus, at all temperatures of practical interest, the Fermi energy is well above k_BT.

With the density-of-states function $g(\mathscr{E})$ at hand, the chemical potential μ for a gas of free electrons at any temperature T can be obtained from the following identity (see Eq. (12.46)):

$$\boxed{\int_0^{\infty} f(\mathscr{E})\,g(\mathscr{E})\,\mathrm{d}\mathscr{E} = \mathscr{N}} \tag{12.54}$$

Here $f(\mathscr{E})$ is the Fermi function of Eq. (12.45) and $\mathscr{N}$ is the number density of the free electrons. At $T = 0$ K this relation yields $\mu = \mathscr{E}_F$. At other temperatures, μ turns out to be only slightly different from $\mathscr{E}_F$. In any event, since $\mathscr{E}_F$ is so much greater than the temperatures of interest in solid state physics, the distribution of electrons at $T \neq 0$ K is affected only in the vicinity of the Fermi energy, and left more or less intact far from $\mathscr{E}_F$. Figure 12.16 shows schematically the density of states $g(\mathscr{E})f(\mathscr{E})$ for a gas of free electrons both at $T = 0$ and at $T = T_0 \neq 0$. Note that the difference between the two distributions is confined to a region of width $\simeq k_B T_0$ in the vicinity of $\mathcal{E}_F$.

□

Free-electron gas in an external magnetic field. Let us now assume that the free-electron gas described above is subject to an external magnetic field $B_0\hat{\mathbf{z}}$. We ignore the effect of the field on the orbital motion of the electrons (i.e., as though the electron carried no charge) and concentrate on the interaction between the electron spin and the applied magnetic field. Those electrons with spin parallel to the field will have magnetic moment $\mathbf{m} = -\mu_B\hat{\mathbf{z}}$, and are therefore shifted up in energy by $\Delta\mathcal{E} = \mu_B B_0$. The density-of-states function for these electrons is therefore $\frac{1}{2}\mathcal{g}(\mathcal{E} - \mu_B B_0)$ where $\mathcal{g}(\cdot)$ is given by Eq. (12.51). The remaining electrons, with spin antiparallel to the field, are shifted down in energy and the corresponding density-of-states function for these electrons will be $\frac{1}{2}\mathcal{g}(\mathcal{E} + \mu_B B_0)$. We recognize that $\mu_B B_0$ is small compared to $\mathcal{E}_F$, and make the following approximation:†

$$\mathcal{g}(\mathcal{E} \pm \mu_B B_0) \simeq \mathcal{g}(\mathcal{E}) \pm \mu_B B_0 \mathcal{g}'(\mathcal{E}) \tag{12.55}$$

The split density-of-states function is shown in Fig. 12.17, with shaded areas corresponding to the occupied states at $T = 0$ K. The majority spins are antiparallel to the field (their magnetic moments are parallel); these are the electrons aligned favorably with the field and therefore have somewhat lower energies. The number density $\mathcal{N}^+$ of electrons with parallel magnetization is somewhat greater than that of electrons with antiparallel moments, $\mathcal{N}^-$. To determine the Fermi energy $\mathcal{E}_F$ note that

$$\mathcal{N} = \mathcal{N}^+ + \mathcal{N}^- \simeq \int_0^{\mathcal{E}_F} \tfrac{1}{2}\left[\mathcal{g}(\mathcal{E} - \mu_B B_0) + \mathcal{g}(\mathcal{E} + \mu_B B_0)\right] \, d\mathcal{E}$$

$$\simeq \int_0^{\mathcal{E}_F} \mathcal{g}(\mathcal{E}) \, d\mathcal{E} \tag{12.56}$$

Thus, to a first approximation, the Fermi energy retains its value in the absence of the field. As for the net moment of the electron gas, we have

$$\mathbf{M} = (\mathcal{N}^+ - \mathcal{N}^-)\mu_B\hat{\mathbf{z}} \simeq \left\{\int_0^{\mathcal{E}_F} \mathcal{g}'(\mathcal{E}) \, d\mathcal{E}\right\} \mu_B^2 B_0\hat{\mathbf{z}} = \mathcal{g}(\mathcal{E}_F)\,\mu_B^2 B_0\hat{\mathbf{z}} \tag{12.57}$$

†Even with $B_0 = 10^4$ gauss, $\mu_B B_0/k_B \simeq 1$ K whereas $\mathcal{E}_F/k_B \simeq 10^4$ K.

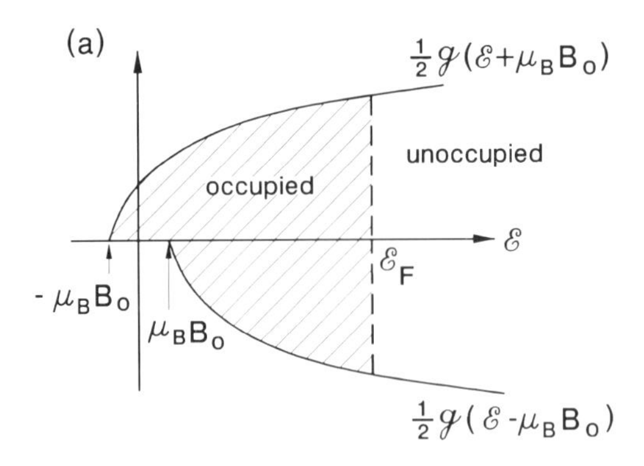

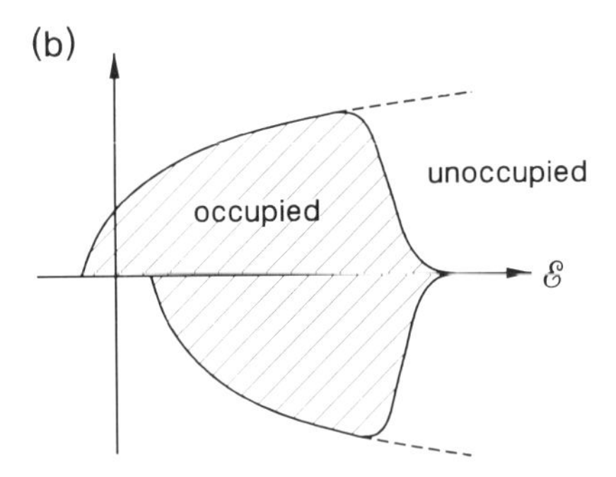

Figure 12.17. Split density of states for a free-electron gas in the presence of an external field $\mathbf{B} = B_0\hat{\mathbf{z}}$. The density-of-states function for electrons with parallel spin is shifted to the right, while for electrons with antiparallel spin it is shifted to the left. (a) At $T = 0$ K and (b) at $T \neq 0$ K.

$\mathcal{g}(\mathcal{E}_F)$ is the electronic density of states at the Fermi energy in the absence of the field. Substituting in Eq. (12.57) for $\mathcal{g}(\mathcal{E}_F)$ from Eq. (12.53), we derive Pauli's paramagnetic susceptibility:

$$\boxed{\chi_{\text{Pauli}} \simeq \mathcal{g}(\mathcal{E}_F)\mu_B^2 = \frac{3\mathcal{N}\mu_B^2}{2\mathcal{E}_F}} \tag{12.58}$$

With $\mathcal{N} \simeq 10^{22}$ cm^{-3}, $\mu_B = 0.927 \times 10^{-20}$ emu, and $\mathcal{E}_F \simeq 10^{-12}$ erg, we find $\chi_{\text{Pauli}} \simeq 10^{-6}$ (dimensionless units). This is comparable to the typical values obtained for Larmor diamagnetic susceptibility (albeit with a change of sign), and is much smaller than the paramagnetic susceptibility for the free-electron gas in the absence of the exclusion principle.

At $T > 0$ the density of states in the vicinity of $\mathcal{E}_F$ broadens somewhat, but the net magnetization of the gas remains more or less constant. The reason is that the difference between $\mathcal{N}^+$ and $\mathcal{N}^-$ is due to the shift of energy levels throughout the entire conduction band, and not just at the Fermi surface. With increasing temperature, therefore, the Pauli susceptibility remains close to its value at $T = 0$ K, disobeying Curie's law. From the foregoing analysis we see not only a very small susceptibility for the conduction electrons, but also one that, to a good approximation, is independent of temperature.

12.6. Exchange Interaction

The most common magnetic phenomenon, ferromagnetism, occurs because in certain solids individual atomic magnetic dipoles couple to each other and form magnetically ordered states. The coupling, which is quantum mechanical in nature, is known as the exchange interaction and is rooted in the overlap of electrons in conjunction with Pauli's exclusion principle. Whether it is a metal or a dielectric ferromagnet (Fe, Co, Gd, EuO, Cu_2MnAl, etc.), antiferromagnet (MnO, FeO, CoO, $KFeF_3$, etc.), or ferrimagnet (Fe_3O_4, $GdCo_5$, $Y_3Fe_5O_{12}$, etc.), the exchange interaction between neighboring magnetic ions will force the individual moments into parallel (ferromagnetic) or antiparallel (antiferromagnetic) alignment with their neighbors. The three types of exchange that are currently believed to exist are direct exchange, indirect exchange, and superexchange; these are depicted schematically in Fig. 12.18 and will be described qualitatively in the present section.

Direct exchange. The simplest example of direct-exchange coupling is found between the two electrons of the hydrogen molecule H_2. The Coulomb interaction between the electrons and the two nuclei is minimized when the electrons spend most of their time in-between the nuclei; this despite the Coulomb repulsion between the electrons themselves. Since the electrons are thus supposed to be at the same place in space at the same time, Pauli's exclusion principle requires that they possess opposite spins. Thus for purely electrostatic reasons (i.e., the attraction of nuclei for the electrons) the electrons of H_2 are forced into a state with oppositely oriented spins. It is as though each spin exerts a strong negative magnetic field on the other spin in order to keep it properly aligned. This fictitious magnetic field is the same "molecular field" whose existence Pierre Weiss[8] postulated in 1907 in order to account for the phenomenon of ferromagnetism. In the more precise language of quantum mechanics the

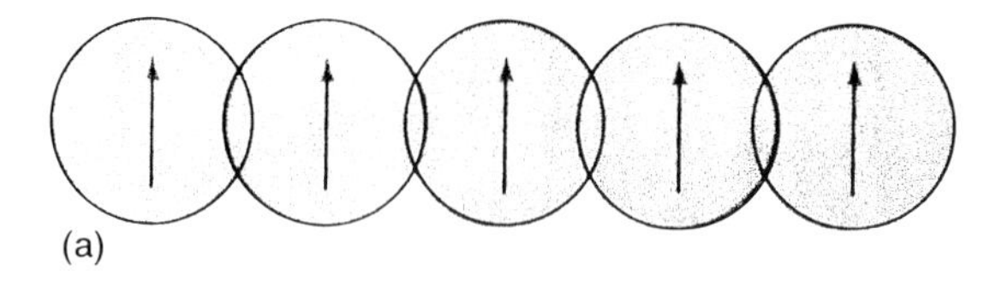

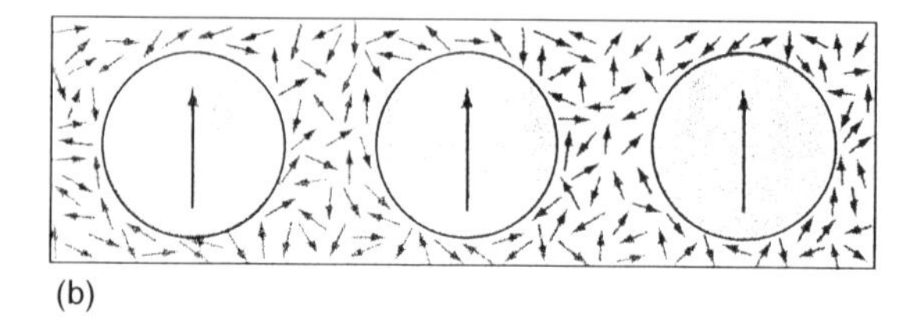

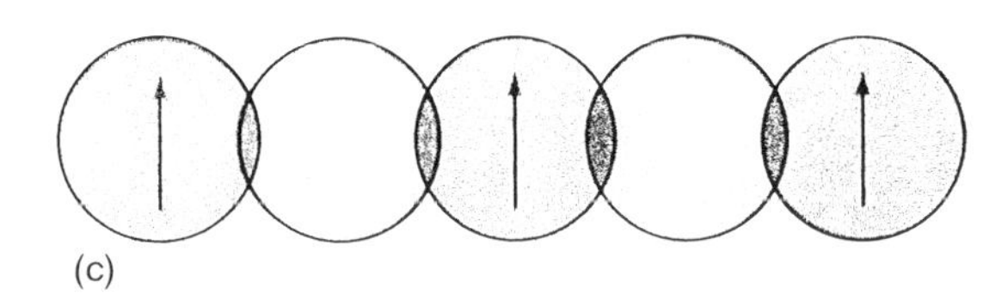

Figure 12.18. Schematic illustrations of (a) direct exchange, in which the magnetic ions interact because their charge distributions overlap; (b) indirect exchange, in which in the absence of overlap a magnetic interaction is mediated by interactions with the conduction electrons; (c) superexchange, in which two magnetic ions with non-overlapping charge distributions interact because both have overlap with the same non-magnetic ion. Source: N.W. Ashcroft and N.D. Mermin, *Solid State Physics*, copyright © 1976 by Holt, Rinehart & Winston, reproduced by permission of Saunders College Publishing.

above statements may be rephrased as follows.[2,9] *The spatial distribution of the electrons in a hydrogen molecule consists of symmetric (singlet) and antisymmetric (triplet) wave-functions. Because of the exclusion principle, the singlet state must have antisymmetric spin, whereas the spin state of the triplet is symmetric. It so happens that in the hydrogen molecule the singlet is the state of minimum electrostatic energy, resulting in an exchange coupling between the spins of the electrons that is antiferromagnetic.*

In more complex systems the distribution of electrons is very complicated and seldom is it possible to compute the exact distribution from first principles. When the charge distributions on neighboring magnetic ions have a significant overlap, it is usually the case that direct exchange is the dominant magnetic interaction between them. Bethe and Slater surmised that when the interatomic distance is small the electrons spend most of their time in-between neighboring atoms, as for the

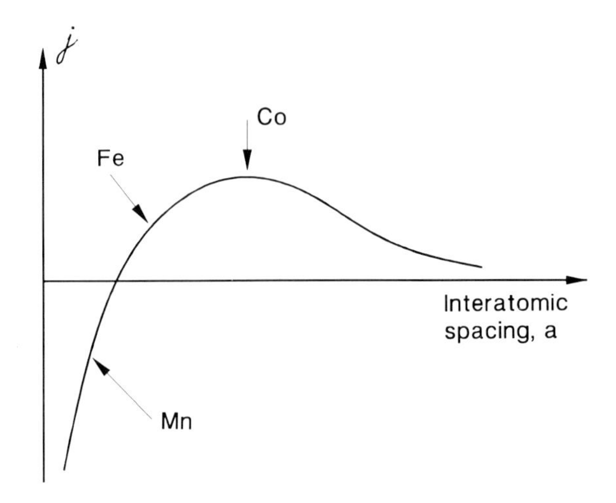

Figure 12.19. The Bethe–Slater curve qualitatively represents the magnitude of direct exchange as function of the interatomic separation.

hydrogen molecule. This gives rise to antiparallel alignment and, therefore, negative exchange. On the other hand, if the atoms are far apart, the respective electrons spend their time away from each other in an effort to minimize the electron–electron repulsion. This gives rise to parallel alignment or positive exchange. The Bethe–Slater curve shown in Fig. 12.19 depicts this dependence of the coefficient of direct exchange, $\mathcal{J}$, on interatomic spacing. (For a definition of $\mathcal{J}$ see Eq. (12.59) below.) Cobalt in its normal metallic state is believed to be situated near the peak of this curve, while chromium and manganese are on the side of negative exchange. Iron, with the sign of its exchange depending sensitively on the crystal structure, is probably near the zero-crossing of this curve.

Indirect exchange. The indirect or RKKY-type exchange is dominant in metals when there is little or no direct overlap between neighboring magnetic electrons, and the conduction electrons mediate the exchange.[5,6] (RKKY stands for Ruderman–Kittel–Kasuya–Yosida.) The RKKY exchange coefficient $\mathcal{J}$ is believed to have the damped oscillatory nature shown in Fig. 12.20. Thus, depending on the separation between a pair of ions, their magnetic coupling can be ferromagnetic or antiferromagnetic. In rare earth metals (Gd, Tb, Dy, etc.) whose magnetic electrons in the 4f shell are shielded by the 5s and 5p electrons, direct exchange is rather weak and insignificant; indirect exchange via the conduction electrons then gives rise to a variety of magnetic order in these materials. The magnetic structure of the rare earths will be described in some detail in section 12.8.

Superexchange. The third mechanism for the exchange coupling of magnetic ions in magnetically ordered solids is superexchange. Here two

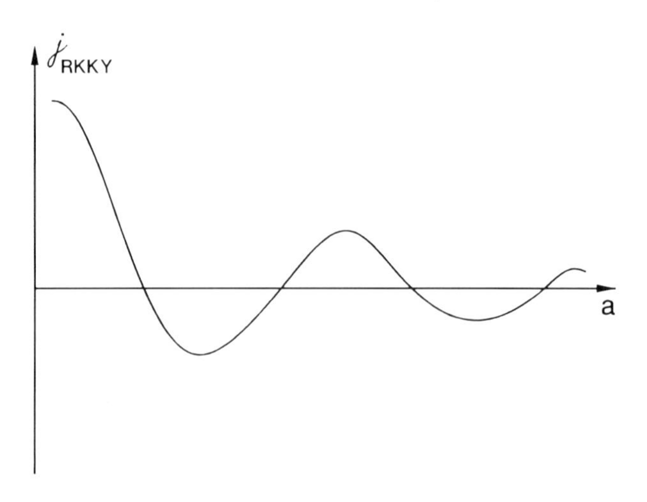

Figure 12.20. The coefficient of indirect (RKKY) exchange versus the interatomic spacing a. The magnetic ions interact via the conduction electrons, whose polarization could extend over a range of several lattice constants.

magnetic ions with non-overlapping charge distributions interact via their overlap with a third, non-magnetic ion. As an example, consider the antiferromagnetic dielectric MnO, which has the crystal structure of NaCl. The Mn ions in this oxide form a face-centered cubic (fcc) lattice, and their spins are aligned antiparallel to one another as shown in Fig. 12.21(a).† The direct exchange between Mn and Mn is relatively weak, since the manganese ions are separated by oxygens. On the other hand, there must be a strong exchange between these magnetic ions, as evidenced by a fairly large Néel temperature (T_N = 122 K). This type of magnetic coupling has been explained in terms of superexchange,[10,11] whereby the spins of the metal ions on opposite sides of an oxygen ion interact through the p-orbitals of oxygen. Consider the two ions Mn_1 and Mn_2 separated by the oxygen ion O, as shown in Fig. 12.21(b). The ground state of the oxygen ion is doubly charged, O^{--}, with the configuration $(2s)^2(2p)^6$. In this state there are no spin couplings with the metal ions. There is, however, the possibility of an excited state in which one of the two extra electrons of O^{--} is transferred to a neighbor, say Mn_1, where the strong intra-atomic exchange will orient the transferred electron's spin in such a way as to impart to Mn_1 a maximum spin magnetic moment (Hund's first rule). In like manner, the unpaired electron left in the p-orbital of O^{--} will be coupled with the other metal ion, Mn_2. Since, according to Pauli's

†The arrangement of atomic magnetic dipoles has been determined from neutron diffraction experiments. The neutron has no electric charge and accordingly is insensitive to the electric charge of ions of the crystal lattice. On the other hand, it can be scattered by the magnetic moments of the electron spins (and the nuclei) because it has a magnetic moment itself.

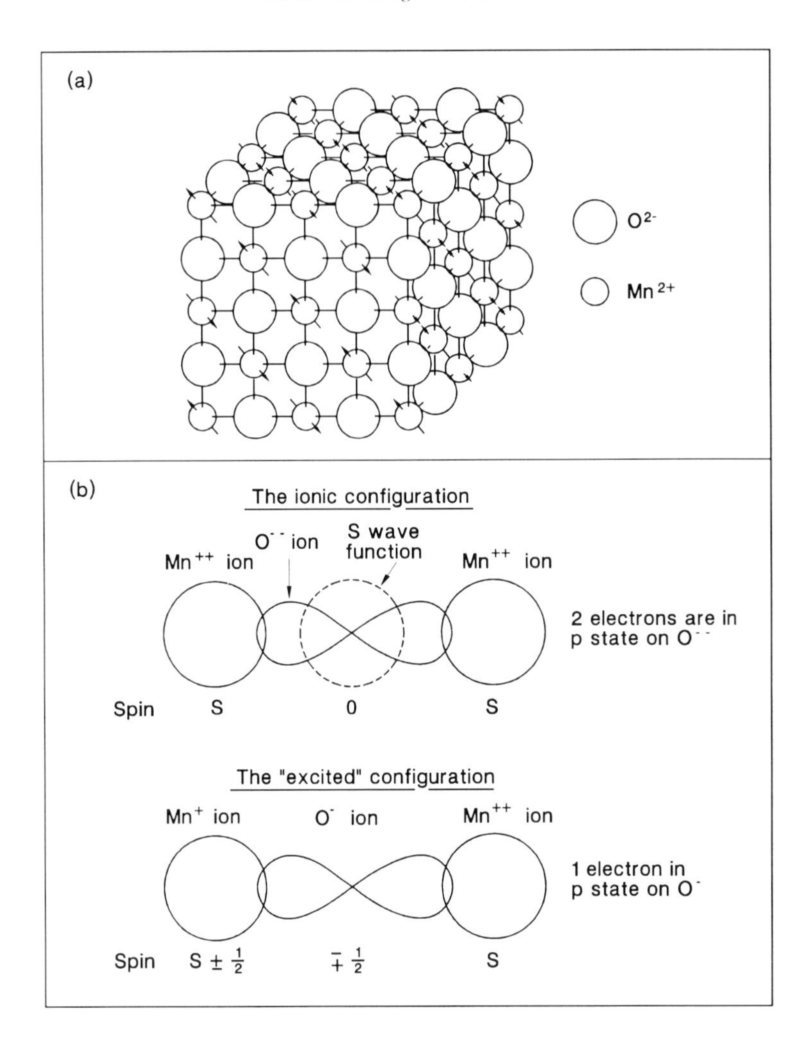

Figure 12.21. (a) Crystal and magnetic structure of MnO. There are planes of Mn^{++} ions magnetized in one direction, while their neighboring planes are magnetized in the opposite direction (net moment = 0). (b) Ground and excited states of the Mn–O–Mn bond in the superexchange process. The Mn^{++} are S-state ions with spherically symmetric charge distribution. The participating electrons of the oxygen ion are in the 2p orbital. Source: S. Chikazumi and S.H. Charap, *Physics of Magnetism*, copyright ©1964 by Krieger Publishing Company, reproduced by permission of the publisher.

exclusion principle, the two electrons must have opposite spins, Mn_1 and Mn_2 end up with antiparallel magnetic moments. Such a superexchange interaction is expected to be strongest when the bond Mn_1–O–Mn_2 is along a straight line, because the stretched p-orbital can maximally overlap the orbitals of both metal ions in this configuration. This is indeed the case for the MnO crystal where the oxygen ions are at the midpoint of the $\langle 100 \rangle$ bond between the manganese ions.

12.6.1. The Heisenberg model

In this highly simplified model of the exchange interaction, a pair of neighboring ions with spin magnetic moments $g_0\mu_B\mathbf{S}_i$ and $g_0\mu_B\mathbf{S}_j$ are assumed to have the mutual energy $\mathcal{E}_{ij}$ where

$$\mathcal{E}_{ij} = -2\mathcal{J}\ \mathbf{S}_i \cdot \mathbf{S}_j \tag{12.59}$$

Here $\mathcal{J}$ is the coefficient of exchange, and the pairwise interaction energy is assumed to be independent of all other spins in the system.[9] Typically, $\mathcal{J}$ is of the order of 10^{-14} ergs per pair of ions.[††] In a classical interpretation, one may assume that a given spin $\mathbf{S}_i$ exerts an effective magnetic field $\mathbf{B}_{\text{eff}}$ on its neighboring spins, where

$$\mathbf{B}_{\text{eff}} = (\mathcal{J}/g\mu_B)\ \mathbf{S}_i \tag{12.60}$$

If $\mathcal{J}$ is positive, the effective field will be parallel to $\mathbf{S}_i$ and the interaction is ferromagnetic. On the other hand, if $\mathcal{J}$ is negative, the field is antiparallel to $\mathbf{S}_i$ and the exchange will be antiferromagnetic.

12.6.2. Exchange stiffness coefficient

In micromagnetic theory, the exchange interaction is averaged over a large number of sites, and the exchange stiffness coefficient A_x is defined in association with a continuous magnetization distribution $\mathbf{M}(\mathbf{r})$. Let $|\mathbf{M}|$ be constant at all positions $\mathbf{r}$, and define the unit vector $\mathbf{m}(\mathbf{r}) = \mathbf{M}(\mathbf{r})/|\mathbf{M}|$. For a pair of nearest neighbor ions, the exchange energy is

$$\begin{aligned}
\mathcal{E}_{ij} &= -2\mathcal{J}\mathbf{S}_i \cdot \mathbf{S}_j = -2\mathcal{J}\ |\mathbf{S}|^2\ \mathbf{m}_i \cdot \mathbf{m}_j \qquad (12.61)\\
&= -2\mathcal{J}|\mathbf{S}|^2 (m_{ix}\, m_{jx} + m_{iy}\, m_{jy} + m_{iz}\, m_{jz})\\
&= -2\mathcal{J}|\mathbf{S}|^2 \Big\{(m_x + \tfrac{1}{2}\Delta m_x)(m_x - \tfrac{1}{2}\Delta m_x) + (m_y + \tfrac{1}{2}\Delta m_y)(m_y - \tfrac{1}{2}\Delta m_y)\\
&\qquad + (m_z + \tfrac{1}{2}\Delta m_z)(m_z - \tfrac{1}{2}\Delta m_z)\Big\}\\
&= -2\mathcal{J}|\mathbf{S}|^2 \Big\{(m_x^{\,2} + m_y^{\,2} + m_z^{\,2}) - \tfrac{1}{4}\left[(\Delta m_x)^2 + (\Delta m_y)^2 + (\Delta m_z)^2\right]\Big\}
\end{aligned}$$

[††] Compare this with the classical interaction energy between a pair of magnetic dipoles, which is of the order of $m_1 m_2/r^3$. With m_1, $m_2 \simeq \mu_B \simeq 10^{-20}$ emu and $r \simeq$ lattice spacing $\simeq$ 1 Å, this energy is about 10^{-16} erg, which is at least two orders of magnitude below the exchange energy.

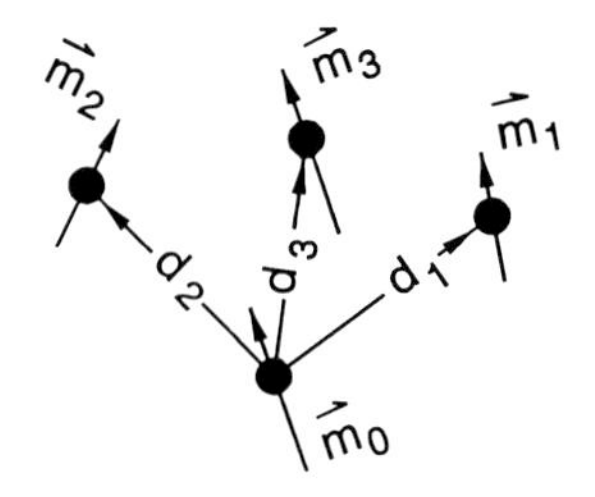

Figure 12.22. Three ions with magnetic dipole moments $\mathbf{m}_1$, $\mathbf{m}_2$, $\mathbf{m}_3$, whose distances $\mathbf{d}_1$, $\mathbf{d}_2$, $\mathbf{d}_3$ from the central ion, which has moment $\mathbf{m}_0$, are mutually orthogonal in three-dimensional space.

Now, since $m_x^2 + m_y^2 + m_z^2 = 1$, this term only adds a constant to $\mathcal{E}_{ij}$. As for the other term, let $\mathbf{d}$ be the separation vector between the two neighboring ions; then $\Delta m_x = (\nabla m_x) \cdot \mathbf{d}$. Equation (12.61) is now written:

$$\Delta\mathcal{E}_{ij} = \tfrac{1}{2}\mathcal{J}|\mathbf{S}|^2 \left[(\nabla m_x \cdot \mathbf{d})^2 + (\nabla m_y \cdot \mathbf{d})^2 + (\nabla m_z \cdot \mathbf{d})^2\right] \qquad (12.62)$$

Figure 12.22 shows that for a given ion the separation vectors $\mathbf{d}$ to its various neighbors may be considered in groups of three (such as $\mathbf{d}_1$, $\mathbf{d}_2$, $\mathbf{d}_3$), where, within each group, the three vectors are mutually orthogonal. Noting that $\nabla m_x \cdot \mathbf{d}$ is simply the projection of ∇m_x along $\mathbf{d}$, and assuming that $|\mathbf{d}_1| = |\mathbf{d}_2| = |\mathbf{d}_3| = d$, we have

$$(\nabla m_x \cdot \mathbf{d}_1)^2 + (\nabla m_x \cdot \mathbf{d}_2)^2 + (\nabla m_x \cdot \mathbf{d}_3)^2 = d^2 (\nabla m_x)^2 \qquad (12.63)$$

Therefore, on the average, for a pair of spins separated by d one can write

$$\langle (\nabla m_x \cdot \mathbf{d})^2 \rangle = \tfrac{1}{3} d^2 (\nabla m_x)^2 \qquad (12.64)$$

with similar expressions for m_y and m_z. Equation (12.62) for an average pair of spins now becomes

$$\Delta\mathcal{E}_{\text{ave}} = \tfrac{1}{6}\mathcal{J} d^2 |\mathbf{S}|^2 \left[(\nabla m_x)^2 + (\nabla m_y)^2 + (\nabla m_z)^2\right] \qquad (12.65)$$

The above energy must be divided by two to yield the energy associated with each member of the ion pair, then multiplied by the average number Z of nearest neighbors (i.e., the coordination number) in order to yield the total energy per ion, and finally divided by d^3 to give the exchange energy per unit volume. The resulting exchange energy density $\mathcal{E}_{\text{xhg}}$ will then be

$$\boxed{\mathcal{E}_{\text{xhg}} = A_x \left[(\nabla m_x)^2 + (\nabla m_y)^2 + (\nabla m_z)^2\right]} \qquad (12.66a)$$

where A_x, the exchange stiffness coefficient, is given by

$$\boxed{A_x = \frac{\mathscr{J} Z S^2}{12d}} \tag{12.66b}$$

A typical value of A_x, 10^{-6} erg/cm, may be obtained from the above expression taking $\mathscr{J} \simeq 10^{-14}$ erg, $Z \simeq 10$, $S \simeq 1$, and $d \simeq 10^{-8}$ cm. The exchange stiffness coefficient is an important parameter in micromagnetic theory; it will appear frequently in the following chapters in connection with the properties of magnetic domains and domain walls.

12.7. Magnetic Order

12.7.1. Ferromagnetism

Due to the exchange interaction, certain materials composed of magnetic ions exhibit a spontaneous magnetic moment below a certain critical temperature. The phenomenon is called ferromagnetism, and the critical (or Curie) temperature is associated with a second-order, ferromagnetic-to-paramagnetic phase transition. Table 12.3 lists several ferromagnetic materials along with their measured critical temperatures and saturation moments.

The simplest model on which to base a description of ferromagnetism is provided by the mean-field theory.[2,5,6] Consider a lattice of identical ions with individual magnetic moments $\mathbf{m} = g\mu_B \mathbf{J}$, coordination number Z, and nearest-neighbor exchange interaction coefficient $\mathscr{J}$, subject to an external magnetic field $\mathbf{B}_0$. At finite temperature T, the competition between exchange forces and thermal excitation will result in an average moment per ion, $\langle \mathbf{m} \rangle$, along the direction of the applied field. Taking into account the effective exchange field, Eq. (12.60), from all near-neighbor ions, the total effective field on a given ion will be

$$\mathbf{B}_{\text{eff}} = \mathbf{B}_0 + \frac{Z\mathscr{J}}{g^2 \mu_B^2} \langle \mathbf{m} \rangle \tag{12.67}$$

Now, as in subsection 12.5.1, we have a collection of magnetic ions, each subject to the same field. The average moment of individual ions, $\langle m \rangle$, will therefore be given by Eq. (12.35) provided that $\mathbf{B}_0$ in that equation is replaced by the above $\mathbf{B}_{\text{eff}}$, that is,

$$\boxed{\langle m \rangle = g\mu_B J \mathscr{B}_J \left\{ \frac{g\mu_B J}{k_B T} \left[B_0 + \frac{Z\mathscr{J}}{g^2 \mu_B^2} \langle m \rangle \right] \right\}} \tag{12.68}$$

Table 12.3. Selected ferromagnets with their critical temperatures T_c and saturation magnetization M_s. Source: F. Keffer, *Handbuch der Physik* **18**, Springer, New York (1966)

Material	T_c (K)	M_s (0) (emu/cm³)	Material	T_c (K)	M_s (0) (emu/cm³)
Fe	1043	1752	Cu_2MnAl	630	726
Co	1388	1446	Cu_2MnIn	500	613
Ni	627	510	EuO	77	1910
Gd	293	1980	EuS	16.5	1184
Dy	85	3000	MnAs	318	870
$CrBr_3$	37	270	MnBi	670?	675
CrO_2	386	540?	MnSb	587	750?
Au_2MnAl	200	323	$GdCl_3$	2.2	550

From this equation, using a graphical technique, we determine $\langle m \rangle$ self-consistently. Equating the argument of the Brillouin function in Eq. (12.68) with the dummy variable x, one obtains:

$$\langle m \rangle = g\mu_B J \mathscr{B}_J(x) \tag{12.69a}$$

$$\langle m \rangle = \frac{k_B T}{J} \frac{g\mu_B}{Z\mathscr{J}} x - \frac{g^2\mu_B^2}{Z\mathscr{J}} B_0 \tag{12.69b}$$

At fixed temperature T, the above equations represent $\langle m \rangle$ as two functions of x, which can both be plotted on the same set of axes. The crossing point of the two curves is the self-consistent solution of Eq. (12.68). We shall discuss the case of $B_0 = 0$ first, deriving the temperature-dependence of spontaneous magnetization and an expression for the critical temperature T_c, before we embark on a discussion of the case $B_0 \neq 0$.

(i) The case $\mathbf{B_0 = 0}$. Figure 12.23(a) shows plots of $\langle m \rangle$ versus x for the two functions given by Eqs. (12.69). At low temperatures the straight line has a small slope, and the crossing occurs near the tail of the Brillouin function; then $\langle m \rangle \simeq g\mu_B J$. The average moment $\langle m \rangle$ decreases with increasing temperature until, at a critical point T_c, the straight line becomes tangent to the Brillouin function, at which point $\langle m \rangle$ becomes zero and remains zero afterwards. To obtain the critical temperature, we use Eq. (12.36), which gives the slope of $\mathscr{B}_J(x)$ at small x, and equate that with the slope of the straight line:

$$g\mu_B J \left(\frac{J+1}{3J} \right) = \frac{k_B T_c}{J} \frac{g\mu_B}{Z\mathscr{J}}$$

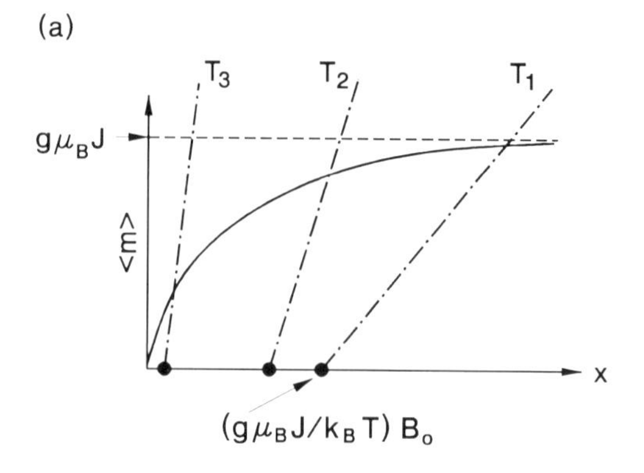

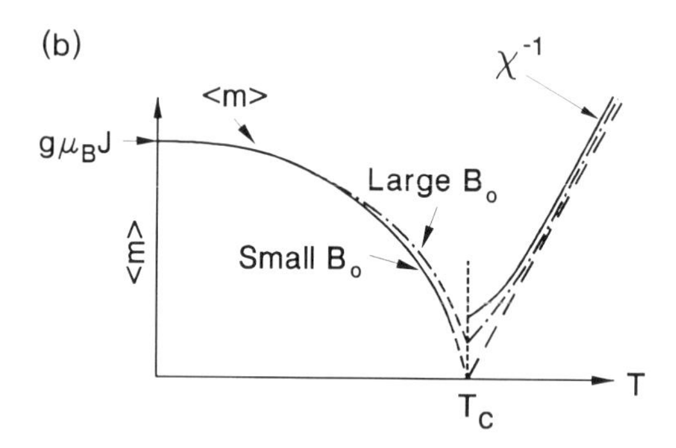

Figure 12.24. (a) Graphical solution of Eq. (12.68) with $B_0 \neq 0$. The straight-line function given by Eq. (12.69b) crosses the horizontal axis at $x = (g\mu_B J/k_B T)B_0$. As the temperature rises, the slope of the straight line increases and its intercept moves closer to the origin. (b) Magnetization versus temperature ($T < T_c$) and inverse susceptibility χ^{-1} versus temperature ($T > T_c$). The external field eliminates the discontinuity of slope of the magnetization curve at the Curie point and makes the transition smooth, but otherwise causes little change in the $\langle m \rangle$ versus T curve. The inverse susceptibility approaches a linear asymptote with increasing temperature.

$$\boxed{\chi = \frac{N}{V}\frac{\langle m \rangle}{B_0} \simeq \frac{N}{V} g^2 {\mu_B}^2 \frac{J(J+1)}{3k_B(T - T_c)}; \qquad T > T_c} \tag{12.71}$$

This result is identical to the Curie law, Eq. (12.37), except for the denominator, which is $(T - T_c)$ instead of T. The temperature-dependence of χ in Eq. (12.71) is known as the Curie–Weiss law.[2,3,5,6]

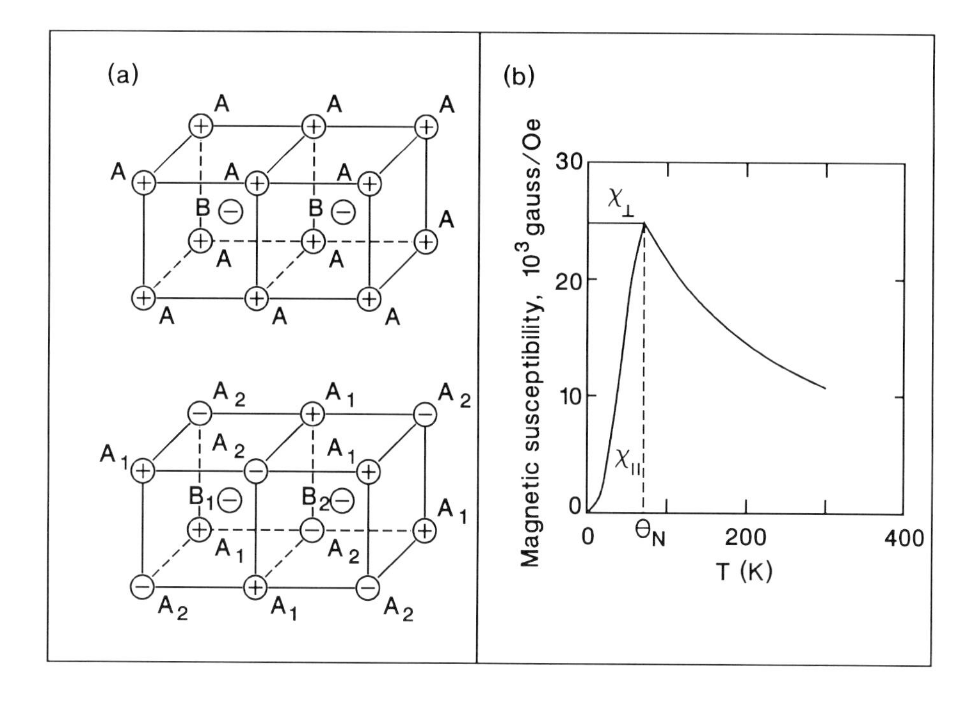

Figure 12.25. (a) Two possible antiferromagnetic orderings in a body-centered cubic (bcc) lattice. All sites are occupied by magnetically active ions. The + and - signs indicate the orientations of the atomic magnetic moments at the corresponding sites. (b) Temperature-dependence of the magnetic susceptibility (per mole) of antiferromagnetic MnF_2 single crystal. $\chi_{\|}$ is measured with the applied field parallel to the tetragonal axis of the crystal, which is the easy axis along which the moments of Mn^{++} ions are normally aligned. χ corresponds to an applied magnetic field perpendicular to the easy axis. Above the Néel temperature T_N the antiferromagnetic order disappears and the susceptibility follows the Curie–Weiss law. Source: S.V. Vonsovskii, *Magnetism*, Keter Publishing House, Jerusalem (1974).

12.7.2. Antiferromagnetism

In an antiferromagnet the exchange interaction coefficient $\mathcal{J}$ between neighboring ions is negative. The moments form two interpenetrating sublattices, each of which resembles a ferromagnetic lattice, but the relative orientation of the two is antiparallel (see Fig. 12.25); the net magnetization of the lattice as a whole is therefore zero. Nonetheless, there exists an order–disorder phase transition from antiferromagnetism to paramagnetism in this type of material; the transition temperature is referred to as the Néel temperature T_N. Table 12.4 lists several antiferromagnetic materials along with their critical temperatures.

Since an antiferromagnetic arrangement of spins has no net (spontaneous) magnetization, the substance does not show the kind of

Table 12.4. Selected antiferromagnets with critical temperatures T_N. Source: F. Keffer, *Handbuch der Physik* **18**, Springer, New York (1966)

Material	T_N (K)	Material	T_N (K)
MnO	122	$KCoF_3$	125
FeO	198	MnF_2	67.3
CoO	291	FeF_2	78.4
NiO	600	CoF_2	37.7
CuO	450	$CuBr_2$	193
$RbMnF_3$	54.5	$MnCl_2$	2
$KFeF_3$	115	VS	1040
$KMnF_3$	88.3	Cr	312
α-Mn	95	Cr_2O_3	308

behavior typically exhibited by ferromagnetic materials. The tendency to maintain antiferromagnetic order, however, opposes magnetization by an external field and, accordingly, gives rise to a characteristic temperature-dependence of the susceptibility.[5,6] Thus, contrary to the case of paramagnetism, where the susceptibility increases with the decrease of temperature, antiferromagnetic substances show a decrease of susceptibility below the transition temperature T_N. This is exemplified by the $\chi_{\|}$ curve in Fig. 12.25(b), where the magnetic field is applied parallel to the spin axis. When the field is applied perpendicular to the spin axis, magnetization takes place by the rotation of each spin away from this axis, in which case the susceptibility becomes nearly independent of the temperature, as shown by the $\chi_{\perp}$ curve in Fig. 12.25(b). Above T_N the susceptibility always decreases with increasing temperature, independently of the direction of the applied field; thus $\chi(T)$ shows a kink at the transition point. This behavior is one of the distinguishing features of antiferromagnetism.

12.7.3. Ferrimagnetism†

A ferrimagnet is rather similar to an antiferromagnet, the difference being that the sublattices of a ferrimagnet are not identical and their magnetization vectors do not cancel each other. It is usually the case that within each sublattice the ions are ferromagnetically coupled to their neighbors (i.e., $\mathcal{J}_{11} > 0$, $\mathcal{J}_{22} > 0$) but the intersublattice coupling is antiferromagnetic (i.e., $\mathcal{J}_{12} < 0$). An important class of magnetic oxides known as ferrites shows ferrimagnetic behavior; in fact, the term

†Adapted from C. Kittel, *Introduction to Solid State Physics*, 6E revised, copyright © 1995 by John Wiley & Sons, with permission of the publisher.

Table 12.5. Selected ferrimagnets with Curie temperatures T_c and saturation magnetization M_s. Source: F. Keffer, *Handbuch der Physik* **18**, Springer, New York (1966)

Material	T_c (K)	$M_s(0)$ (emu/cm³)	Material	T_c (K)	$M_s(0)$ (emu/cm³)
Fe_3O_4 (magnetite)	858	510	$CuFe_2O_4$	728	160
$CoFe_2O_4$	793	475	$MnFe_2O_4$	573	560
$NiFe_2O_4$	858	300	$Y_3Fe_5O_{12}$(YIG)	560	195

ferrimagnet was originally coined to describe the ferrite-type spin order. The usual chemical formula for a ferrite is $MO \cdot Fe_2O_3$, where M is a divalent cation, often Zn, Cd, Fe, Ni, Cu, Co, or Mg. The cubic ferrites have the spinel crystal structure shown in Fig. 12.26(a). There are eight occupied tetrahedral (or A) sites and 16 occupied octahedral (or B) sites in a unit cube; the lattice constant is about 8 Å. A remarkable feature of the spinels is that all exchange interactions (between AA, AB, and BB) favor antiparallel alignment of the spins connected by the interaction; all exchange integrals $\mathcal{J}_{AA}$, $\mathcal{J}_{AB}$, $\mathcal{J}_{BB}$ are thus believed to be negative. But the AB interaction is the strongest, so that the A-spins must be parallel to each other and the B-spins must be parallel to each other, in order that the A-spins may be antiparallel to the B-spins. As an example consider the case of magnetite $FeO \cdot Fe_2O_3$. Each ferric ion Fe^{+++} is in a state with spin $S = 5/2$ and $L = 0$. Thus each ion should contribute $5\mu_B$ to the saturation moment. The ferrous ions Fe^{++} have a spin of 2 and should contribute $4\mu_B$, apart from any residual orbital moment contribution. Thus the number of Bohr magnetons per Fe_3O_4 formula unit would be $\simeq 2 \times 5 + 4 = 14$ if all spins were parallel. However, the observed value is $4.1\,\mu_B$. The discrepancy is accounted for if the moments of Fe^{+++} ions are antiparallel to each other: then the observed moment arises only from the Fe^{++} ions, as in Fig. 12.26(b).

The two sublattices exhibit different magnetization-versus-temperature dependences, but they both have the same critical (Curie) temperature T_c. The net saturation moment $\mathbf{M}_s$ of a ferrimagnetic substance is the vector sum of the individual sublattice magnetizations. In ferrimagnets the behavior of M_s versus T depends on the material's constituent elements, its crystal structure, and the relative strengths of the various exchange parameters. Schematic plots of the observed $M_s(T)$ for several commonly encountered ferrimagnetic materials are shown in Fig. 12.26(c). If at some temperature T (below T_c) the magnetizations of the two sublattices become equal, thereby reducing the net magnetic moment to zero, the material is said to have reached the compensation point. Table 12.5 lists several ferrimagnetic materials along with their Curie temperatures T_c and saturation moments M_s at $T = 0$ K.

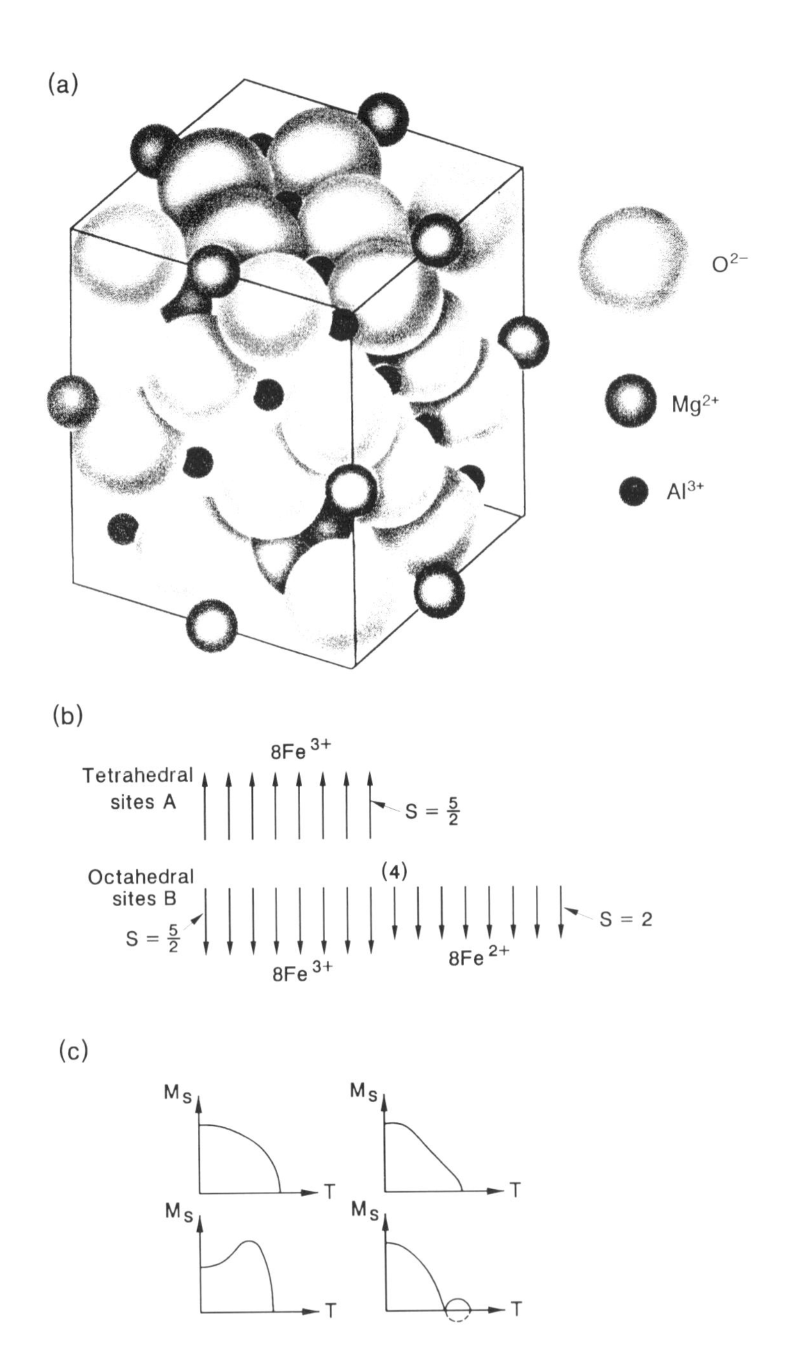

Figure 12.26. (a) Crystal structure of the mineral spinel $MgAl_2O_4$. The Mg^{++} ions occupy tetrahedral sites, each surrounded by four oxygen ions. The Al^{+++} occupy octahedral sites, each surrounded by six oxygen ions. This is a **normal spinel** arrangement. In the **inverse spinel** arrangement the tetrahedral sites are occupied by trivalent metal ions, while the octahedral sites are occupied half by divalent and half by trivalent metal ions. (b) Arrangement of atomic magnetic moments in magnetite $FeO{\cdot}Fe_2O_3$, which has the inverse spinel crystal structure. The moments of the Fe^{+++} ions cancel out, leaving only the moments of the Fe^{++} ions. (c) Different types of $M_s(T)$ behavior exhibited by ferrimagnetic materials. Source: C. Kittel, *Introduction to Solid State Physics*, 6E revised, copyright © 1995 by John Wiley & Sons, reproduced by permission of the publisher.

Iron garnets. The iron garnets are cubic ferrimagnetic insulators with the general formula $M_3Fe_5O_{12}$, where M is a trivalent metal ion and the iron is the trivalent ferric ion ($S = 5/2$, $L = 0$). An example is yttrium iron garnet $Y_3Fe_5O_{12}$, known as YIG. Here Y^{+++} is diamagnetic. The net magnetization of YIG is the resultant of two oppositely magnetized lattices of Fe^{+++} ions. At absolute zero each ferric ion contributes $\pm 5\mu_B$ to the magnetization, but in each formula unit the three Fe^{+++} ions on sites denoted as d sites are magnetized in one sense and the two Fe^{+++} ions on a sites are magnetized in the opposite sense, giving a resultant of $5\mu_B$ per formula unit, in good agreement with measurements. The only magnetic ions in YIG are the ferric ions. Because these are in an $L = 0$ state with a spherical charge distribution, their interaction with lattice deformations and phonons is weak. As a result YIG is characterized by very narrow line widths in ferromagnetic resonance experiments.

In the rare earth iron garnets the ions M^{+++} are paramagnetic trivalent rare earth (RE) ions. Magnetization curves are given in Fig. 12.27. The RE ions occupy sites labeled c; the magnetization of the ions on the c lattice is opposite to the net magnetization of the ferric ions on the $a + d$ sites. At low temperatures the combined moments of the three RE ions in a formula unit may dominate the net moment of the Fe^{+++} ions, but because of the weak $c-a$ and $c-d$ couplings the RE lattice rapidly loses its magnetization with increasing temperature. The total moment can pass through zero and then rise again as the Fe^{+++} moments begin to dominate.

12.8. Electronic Structure and Magnetic Properties of the Rare Earths†

The rare earth (RE) metals are the fifteen elements that range from lanthanum (La, $Z = 57$), to lutetium (Lu, $Z = 71$). They are usually arranged outside the regular array of the atomic table, because they all exhibit similar chemical characteristics. The reason is that the electronic structure of these metals is given by

$$(4f)^n (5s)^2(5p)^6(5d)^1(6s)^2 ,$$

where n increases from 0 to 14 as the atomic number increases from 57 to 71. Thus the outer-shell electronic structure, which determines the chemical properties of elements, is the same for all RE metals. Normally the three outer electrons, $(5d)^1(6s)^2$, are removed from these atoms, leaving trivalent ions. Because of this similarity in chemical character, it has been difficult to separate these elements from each other and, before the development of the ion-exchange method, it was not possible to obtain them in pure form.

† Adapted from S. Chikazumi and S.H. Charap, *Physics of Magnetism*, copyright © 1964 by Krieger Publishing Company, with permission of the publisher.

Table 12.6. Various properties of the rare earths

Element (Z)	Density (g/cm^3)	Crystal structure at 300 K	Melting point (° C)	Curie point T_C (K)	Néel point T_N (K)	Electronic state of 3^+ ion: no. of 4f electrons	S	L	J	Effective moment $\mathscr{P}$: Theory Hund	Theory VV-F	Experiment 3^+ion	Experiment metal	Saturation moment gJ: Theory	Experiment
La (57)	6.17	hcp	920			0	0	0	0			diamagnetic			
Ce (58)	6.77	fcc	795		12.5	1	1/2	3	$2\frac{1}{2}$	2.54	2.56	2.52	2.51	2.14	
Pr (59)	6.78	hex	935			2	1	5	4	3.58	3.62	3.60	3.56	3.20	
Nd (60)	7.00	hex	1024		7.5	3	3/2	6	$4\frac{1}{2}$	3.62	3.68	3.50	3.3-3.71	3.27	
Pm (61)	--	--	1035			4	2	6	4	2.68	2.83	--	--	2.40	
Sm (62)	7.54	rhomb.	1072		14.8	5	5/2	5	$2\frac{1}{2}$	0.85	1.55	1.5	1.74	0.72	
Eu (63)	5.53	bcc	826		(90)	6	3	3	0	0.00	3.40	3.4	8.3	0.0	
Gd (64)	7.89	hcp	1312	289		7	7/2	0	$3\frac{1}{2}$	7.94	7.94	7.80	7.93	7.0	7.12
Tb (65)	8.27	hcp	1356	218	230	8	3	3	6	9.72	9.70	9.74	9.62	9.0	9.25
Dy (66)	8.53	hcp	1407	90	179	9	5/2	5	$7\frac{1}{2}$	10.64	10.6	10.5	10.67	10.0	10.2
Ho (67)	8.80	hcp	1461	20	133	10	2	6	8	10.60	10.6	10.6	10.9	10.0	9.7
Er (68)	9.05	hcp	1497	20	80 (52)	11	3/2	6	$7\frac{1}{2}$	9.58	9.6	9.6	10.0	9.0	8.3
Tm (69)	9.33	hcp	1545	22	53	12	1	5	6	7.56	7.6	7.1	7.56	7.0	
Yb (70)	6.98	fcc	824			13	1/2	3	$3\frac{1}{2}$	4.53	4.5	4.4	0.0	4.0	
Lu (71)	9.84	hcp	1652			14	0	0	0			diamagnetic			

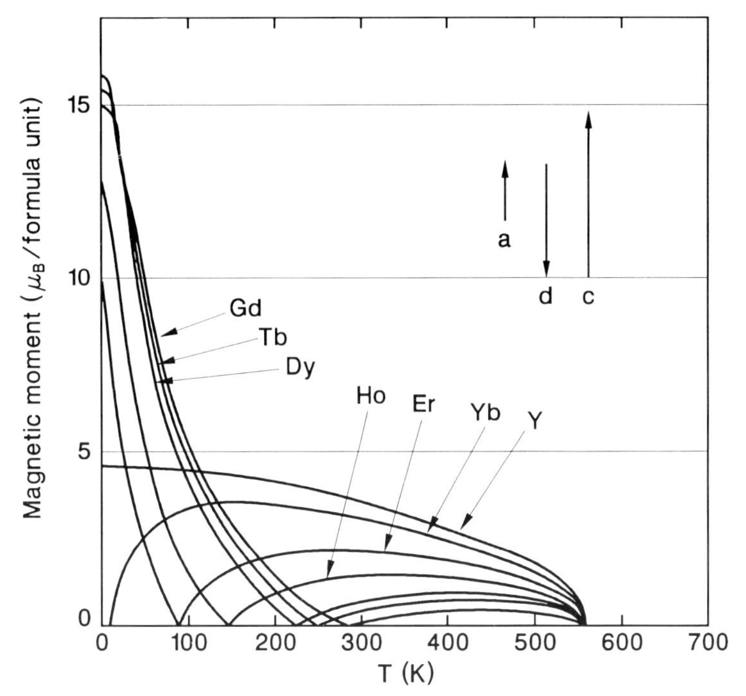

Figure 12.27. Experimental values of the saturation magnetization versus temperature of various iron garnets. The formula unit is $M_3Fe_5O_{12}$, where M is a trivalent metal ion. The temperature at which the magnetization crosses zero is called the compensation temperature; here the magnetization of the M sublattice is equal and opposite to the net magnetization of the ferric ion sublattices. Per formula unit there are three Fe^{+++} ions on tetrahedral sites *d*, two Fe^{+++} ions on octahedral sites *a*, and three M^{+++} ions on sites denoted by *c*. The ferric ions contribute $(3 - 2)5\mu_B = 5\mu_B$ per formula unit. The ferric ion coupling is strong and determines the Curie temperature. If the M^{+++} ions are rare earth ions they are magnetized opposite to the resultant of the Fe^{+++} ions. The M^{+++} contribution drops rapidly with increasing temperature because the M–Fe coupling is weak. Source: C. Kittel, *Introduction to Solid State Physics*, 6E revised, copyright © 1995 by John Wiley & Sons, reproduced by permission of the publisher.

Various physical constants of the rare earth metals are listed in Table 12.6. As seen in this table, most of the RE elements crystallize in the hexagonal close packed (hcp) structure, with a few exceptions. The atomic radius remains almost the same for all RE elements, decreasing from about 1.8 Å to 1.74 Å as the atomic number increases from 57 to 71. Exceptional are Eu and Yb, which are divalent in the metallic state.

The incomplete 4f shell has a close relation to the magnetic properties of this series of elements, as does the 3d shell for the iron-group transition elements. The difference is that in these RE metals, even when the atoms are ionized by the loss of three electrons, the 4f shell remains enclosed by an outer shell $(5s)^2(5p)^6$. Therefore, the orbital angular momentum of the 4f electrons remains unquenched by the crystal field of the neighboring ions. The 4f shell has seven orbitals with orbital quantum number $\ell = 3$, so that the magnetic quantum numbers of the orbitals are $\ell_z = -3, -2, -1, 0, 1, 2, 3$. According to the Pauli principle, each orbital can have two electrons with spins $\frac{1}{2}$ and $-\frac{1}{2}$. The possible electronic states can therefore be shown as the 14 boxes of Fig. 12.28(a), where, according to Hund's rules, the electrons are expected to fill up the boxes in the order indicated. The resultant values of L, S, and J for the trivalent ions are plotted in Fig. 12.28(b) as functions of the number of 4f electrons.

In order to calculate the magnetic moment of individual ions, one must first calculate the g-factor from the Landé formula, Eq. (12.32c). The effective magneton number $\mathscr{P}$, obtained from Eq. (12.38), is listed in Table 12.6 and also shown as a solid curve in Fig. 12.28(c); the experimental points shown as circles in the figure are determined from the temperature-dependence of the susceptibility of the RE salts. The agreement between theory and experiment is excellent except for Eu^{+++} and Sm^{+++}. Van Vleck[12] interpreted this discrepancy as a result of the fact that the energy separation between the ground state and the first excited state (the multiplet interval) is fairly small for these two elements, where L and S nearly compensate each other. Since such an excited state has a larger value of J than the ground state for the less than half-filled 4f shell, thermal excitation is expected to increase the magnetic moment above that calculated according to Hund's rules. The values of $\mathscr{P}$ calculated in this way by Van Vleck and Frank are listed in Table 12.6 and also plotted in Fig. 12.28(c) (broken line); these show excellent agreement with the measured values.

The experimental points shown as crosses in Fig. 12.28(c) are observed for RE elements in the metallic state.[13-15] These are also in good agreement with the theory for trivalent ions except for Eu and Yb, both of which are believed to have divalent ion cores. (The anomalously large ionic radii of these two elements can be interpreted along the same lines.) If we suppose that an extra electron goes into the 4f shell, the inner core structure of Eu^{+++} and Yb^{++} should be the same as that of Gd^{+++} and Lu^{+++} respectively. By this picture the magnetic moments of the metals in question can be adequately explained.

Thus, even in the metallic state, the 4f electrons are well localized in the inner core of each atom. Despite this confinement, some of these metals exhibit fairly strong exchange interactions, giving rise to ferromagnetism, antiferromagnetism, and helical spin structures. The exchange coupling between neighboring atomic magnetic moments in RE metals is believed to be of the indirect (RKKY) type.[16-18] Most of the RE metals with a more than half-filled 4f shell exhibit ferromagnetism, albeit at low temperatures. (In RE metals with a less than half-filled 4f shell ferromagnetism does not appear.) Gadolinium shows ferromagnetism below 289 K; its saturation moment gJ also agrees well with the theoretical value calculated for the trivalent ion. This is also true for Tb and Dy† (see Table

†Because of its large crystalline anisotropy, it has been hard to saturate Dy at low temperature. Nonetheless, the saturation moment measured along the easy direction of a single-crystal specimen is in good agreement with gJ.

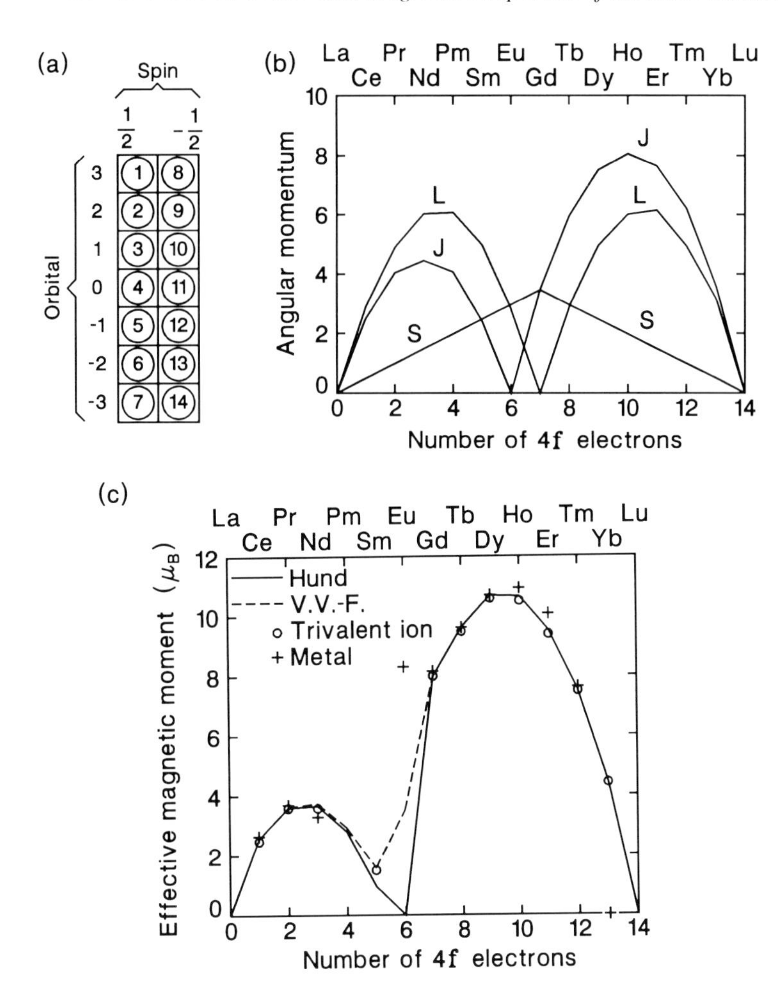

Figure 12.28. (a) Spin and orbital states of electrons in the 4f shell, and the order in which they are filled according to Hund's rules. (b) Plots of S, L, and J, the spin, orbital, and total angular momentum quantum numbers, as functions of the number of 4f electrons of trivalent RE ions. (c) Atomic magnetic moment as a function of the number of 4f electrons measured for trivalent RE ions in compounds (circles) and for RE metals (crosses). Comparison with the Hund (solid line) and Van Vleck-Frank (broken line) theories. Source: S. Chikazumi and S.H. Charap, *Physics of Magnetism*, copyright ©1964 by Krieger Publishing Company, reproduced by permission of the publisher.

12.6). Holmium has a very large crystalline anisotropy, so that its apparent magnetization, extrapolated to 0 K, gives a saturation moment of $8.55\mu_B$, which is somewhat lower than the theoretical value of $10\mu_B$. Neutron diffraction experiments, however, have revealed magnetic moments of about $9.7\mu_B$ arranged on a conical surface about the c-axis.[19] The large magnetic anisotropy of these metals originates in the interaction of their large orbital moment with the crystal **E**-field. (For a description of various anisotropy mechanisms see section 12.10.)

A striking feature of the magnetism of rare earth metals with a more than half-filled 4f shell is that above the Curie point T_c most of them show a helical arrangement of spins, which is again destroyed at the Néel point T_N. These temperatures T_c and T_N are listed in Table 12.6 and the various magnetic structures of RE metals are depicted in Fig. 12.29. Holmium appears to be ferromagnetic below 20 K, and antiferromagnetic between 20 K and 133 K. In the antiferromagnetic region it exhibits a helical spin structure with its moment parallel (or nearly parallel) to the c-plane, as shown in Fig. 12.29. The turn angle ϕ varies from 50° at 133 K to 36° at 35 K. In the ferromagnetic region there remains a helical spin structure ($\phi = 36°$) with a small component of moment ($2\mu_B$) parallel to the c-axis.

Erbium is ferromagnetic below 20 K and antiferromagnetic between 20 K and 80 K. The antiferromagnetic region is divided into two subregions. In the upper region, above 52 K, the c-axis is the easy axis of anisotropy and the spins oscillate parallel to the c-axis with a period of seven lattice constants. Between 52 K and 20 K there appears a component of moment parallel to the c-plane, which forms a helical spin structure. Below 20 K neutron diffraction lines are explained by a model wherein the spins have a ferromagnetic component parallel to the c-axis but show a helical spin arrangement in the c-plane. Thulium is antiferromagnetic below 53 K where an oscillation of the Z-component of spins occurs with a period of seven lattice constants. Terbium and dysprosium also exhibit a helical spin structure, although in the case of Tb this occurs within a narrow range of temperatures (218 K to 230 K).

The alloys of Gd with 3d transition metals have been studied extensively and the intermetallic compounds $GdCo_5$, $GdCo_2$, $GdFe_5$, and $GdFe_2$ are found to exist. According to measurements on single crystals, the magnetic moment of Co in $GdCo_5$ is aligned exactly antiparallel to that of Gd, while for $GdFe_5$ some of the Fe moments are antiparallel, and the others parallel, to the Gd moment.[20] Systematic studies of $RE-Co_5$ compounds have revealed almost complete antiparallel alignment of the RE and Co moments for RE = Ce, Sm, Gd, Tb, Dy, Ho, Er, and Tm. For Pr and Nd the saturation moments are considerably higher than those expected from antiparallel alignment. Intermetallic compounds between RE metals and 4d or 5d transition metals have also been investigated. Bozorth[21] measured the Curie points and the saturation moments for $RE-Ir_2$, $RE-Os_2$, and $RE-Ru_2$. The Curie point, when plotted as a function of the number of 4f electrons, was found to be highest (T_c = 80 K) for RE = Gd and to decrease on both sides of Gd. The saturation moments were lower than $gJ\mu_B$ but higher than $g_0S\mu_B$, indicating that the orbital moments in these materials are only partially quenched. $PrPt_2$, $NdPt_2$, and $GdPt_2$ are ferromagnetic below 7.9 K, 6.7 K and 77 K respectively. Also, the $RE-Al_2$ compounds are found to be ferromagnetic for all rare earths. Again the maximum Curie temperature, 175 K, occurs for RE = Gd, with the Curie point decreasing with either an increase or a decrease in the number of 4f electrons. Here again the saturation moments are found to be smaller than $gJ\mu_B$.

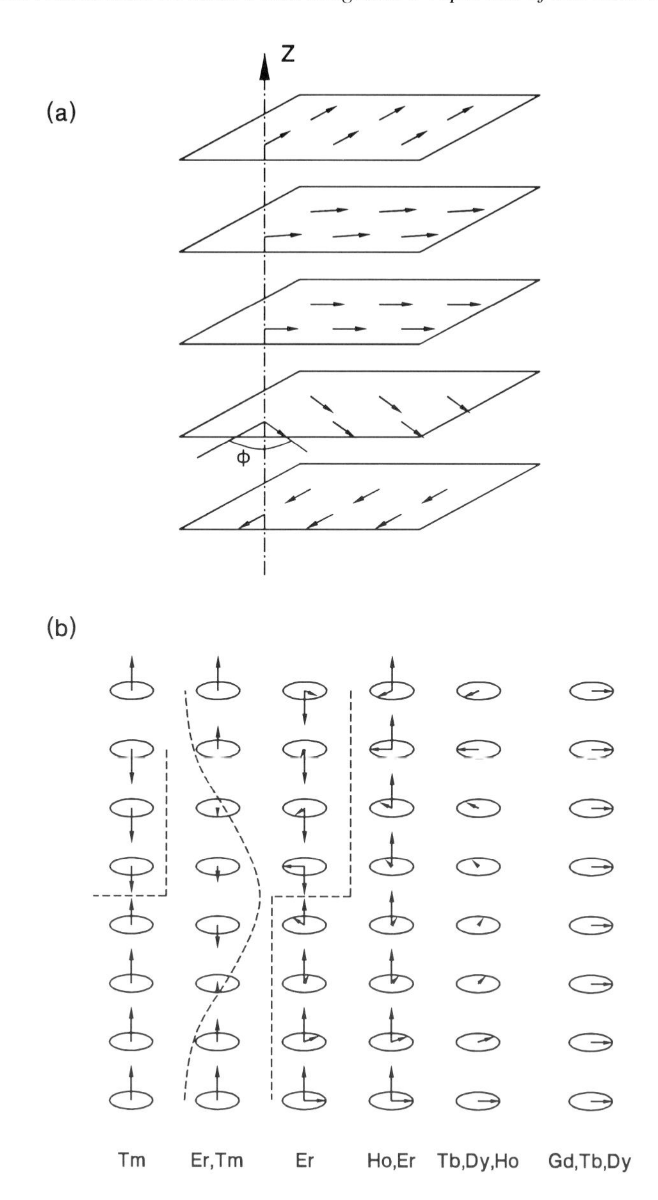

Figure 12.29. (a) Helical spin configuration in rare earth metals. The moments in the individual basal planes are ferromagnetically aligned, but they turn through an angle ϕ between neighboring planes. The Z-axis corresponds to the c-axis of the hexagonal crystal. Source: S. Chikazumi and S.H. Charap, *Physics of Magnetism*, copyright © 1964 by Krieger Publishing Company, reproduced by permission of the publisher. (b) Schematic diagram showing non-collinear atomic magnetic structures in heavy RE metals. Source: S.V. Vonsovskii, *Magnetism*, Keter Publishing House, Jerusalem (1974).

Table 12.7. Experimental values of the effective Bohr magneton number $\mathscr{P}$ for transition metal ions of the iron group

Ion	Configuration	State	$g\sqrt{J(J+1)}$	$2\sqrt{S(S+1)}$	$\mathscr{P}$
Ti^{+++}, V^{++++}	$3d^1$	$^2D_{3/2}$	1.55	1.73	1.8
V^{+++}	$3d^2$	$^3F_2$	1.63	2.83	2.8
Cr^{+++}, V^{++}	$3d^3$	$^4F_{3/2}$	0.77	3.87	3.8
Mn^{+++}, Cr^{++}	$3d^4$	$^5D_0$	0	4.90	4.9
Fe^{+++}, Mn^{++}	$3d^5$	$^6S_{5/2}$	5.92	5.92	5.9
Fe^{++}	$3d^6$	$^5D_4$	6.70	4.90	5.4
Co^{++}	$3d^7$	$^4F_{9/2}$	6.63	3.87	4.8
Ni^{++}	$3d^8$	$^3F_4$	5.59	2.83	3.2
Cu^{++}	$3d^9$	$^2D_{5/2}$	3.55	1.73	1.9

Finally, let us mention the remarkable magnetic properties of EuO. Below 69 K this oxide exhibits strong ferromagnetism in which the magnetic moments of Eu^{++} ions are aligned in parallel. The saturation magnetization at low temperature is as high as 1920 emu/cm^3, which is comparable with that of iron, gadolinium, or FeCo alloys.

12.9. Transition Metals of the Iron Group†

In solids containing transition metal ions from the iron group, the 3d shell responsible for magnetism is the outermost ionic shell, experiencing the intense inhomogeneous **E**-field of the neighboring ions. This so-called crystal field does not have spherical symmetry, but only the symmetry of the crystalline site at which the ion is located. As a result, the basis for Hund's rules is partially invalidated. The interaction of the paramagnetic ions with the crystal field has two major effects: (i) the coupling of the **L** and **S** vectors is largely broken up, so that the states are no longer specified by their J values; (ii) the $2L+1$ sublevels belonging to a given L, which are degenerate in the free ion may now be split by the crystal field, as shown in Fig. 12.30. This splitting diminishes the contribution of the orbital motion to the magnetic moment and is known as the quenching of the orbital angular momentum. Table 12.7 shows the experimental magneton numbers for salts of the iron transition group. One finds that although Curie's law is obeyed, the value of $\mathscr{P}$ determined from this law is consistent with Eq. (12.38) only if one assumes that although S is still given by Hund's rules, $L = 0$ and hence $J = S$.

†Adapted from C. Kittel, *Introduction to Solid State Physics*, 6E revised, copyright © 1995 by John Wiley & Sons, with permission of the publisher, and from N.W. Ashcroft and N.D. Mermin, *Solid State Physics*, copyright © 1976 by Holt, Rinehart & Winston, with permission of Saunders College Publishing.

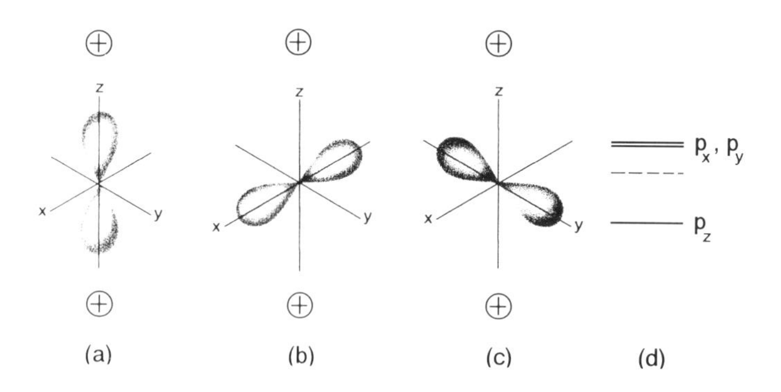

Figure 12.30. An atom with orbital angular momentum quantum number $L = 1$ placed in the uniaxial crystalline **E**-field of two positive ions along the Z-axis. In the free atom the states $L_z = -1, 0, +1$ have identical energies, i.e., they are degenerate. In the crystal the atom has a lower energy when the electron cloud is close to positive ions as in (a) than when it is oriented with the lobes transverse to the axis between them as in (b) and (c). The wave-functions that give rise to these charge densities are called the p_z, p_x, and p_y orbitals respectively. The new energy levels, in relation to that in the free atom (dotted line), are shown on the right. In this axially symmetric field the p_x and p_y orbitals are degenerate. Source: C. Kittel, *Introduction to Solid State Physics*, 6E revised, copyright © 1995 by John Wiley & Sons, reproduced by permission of the publisher.

In an electric field directed toward a fixed nucleus, the plane of a classical orbit is fixed in space, so that all the orbital angular momentum components $\mathbf{L}_x$, $\mathbf{L}_y$, $\mathbf{L}_z$ are constant. In quantum theory one angular momentum component, usually taken as $\mathbf{L}_z$, and the square of the total orbital angular momentum $\mathbf{L}^2$ are constant in a central field. In a non-central field the plane of the orbit will move about; the angular momentum components are no longer constant and may average to zero. In a crystal, $\mathbf{L}_z$ will no longer be a constant of motion, although to a good approximation $\mathbf{L}^2$ may continue to be constant. One can interpret this classically as arising from a precession of the orbital angular momentum in the crystal field, so that although its magnitude is unchanged, its projections along any particular direction in space average out to zero. When $\mathbf{L}_z$ averages to zero, the orbital angular momentum is said to be quenched. The magnetic moment of a state is given by the average value of the operator $\mu_B(\mathbf{L} + 2\mathbf{S})$. In a magnetic field along Z the orbital contribution to the magnetic moment is proportional to the quantum expectation value of $\mathbf{L}_z$; the orbital magnetic moment is thus quenched along with the mechanical moment $\mathbf{L}_z$.

It turns out that even in a crystalline environment the first two of Hund's rules (section 12.4) can be retained. The crystal field must, however, be introduced as a perturbation on the $(2S + 1) \times (2L + 1)$-fold set of states determined by the first two rules; this perturbation acts in addition to the spin–orbit coupling. The crystal field is very much stronger than the spin–orbit coupling, so that to a first approximation a new version of Hund's third rule can be constructed in which the spin–orbit coupling is ignored. The crystal field perturbation alone will not split the spin degeneracy, since it depends only on spatial variables and therefore commutes with **S**, but, if sufficiently low in symmetry, it can completely lift the degeneracy of the orbital L-multiplet.† The result will then be a ground-state multiplet in which the mean value of every component of **L** vanishes, even though $\mathbf{L}^2$ still has the mean value of $L(L+1)$.

When the spin–orbit interaction is introduced, the spin may drag some orbital moment along with it. If the sign of the interaction favors parallel orientation of the spin and orbital moments, the total magnetic moment will be enhanced, and the g-factor will be larger than 2. The experimental results are in agreement with the known variation of the sign of the spin–orbit interaction: $g > 2$ when 3d shell is more than half-filled, $g = 2$ when the shell is half-filled, and $g < 2$ when the shell is less than half-filled.

For simple magnetic materials the net magnetic moment per ion, $gJ\mu_B$, is determined by measuring the low-temperature saturation magnetization, $M_s(0)$, and dividing the result by N, where N is the number of ions per unit volume. Based on measurements of single-crystal samples of iron, cobalt, and nickel, it has been determined that the atomic magnetic moments of these metals are $2.22\,\mu_B$, $1.72\,\mu_B$, and $0.6\,\mu_B$ respectively. The observed values of gJ are often non-integral. One reason is the spin–orbit interaction, which as mentioned above, adds or subtracts some orbital magnetic moment. Another cause (in metals) is the conduction-electron magnetization induced locally about a paramagnetic ion core. Perhaps the most appropriate description of ferromagnetism in transition metals and alloys is given by the band model. To give an example we mention here the band model for single crystals of pure nickel. As a point of reference, Fig. 12.31(a) shows the relationship of the 4s and 3d bands in copper, which has one electron more than nickel but is not ferromagnetic. For nickel at absolute zero $gJ = 0.60$ per atom. After allowance for the magnetic moment contribution of the orbital motion, nickel has an excess of 0.54 of an electron per atom having spin preferentially oriented in one direction. In the band structure of nickel shown in Fig. 12.31(b) for $T > T_c$, $2 \times 0.27 = 0.54$ of an electron has been taken away from the 3d band and 0.46 from the 4s band of copper. The band occupancy of nickel at $T = 0$ K is shown in Fig. 12.31(c), where the splitting between the 3d↑ and 3d↓ bands provides the required excess magnetic moment of $0.54\,\mu_B$ in the "up" direction.

The situation for the higher transition metals, with partially filled 4d or 5d shells, is more complex, since in these heavier elements the spin-orbit coupling is stronger. In these cases the multiplet splitting due to spin–orbit coupling may be comparable to (or greater than) the crystal-field splitting.

†If one adds the spin–orbit coupling to the Hamiltonian as an additional perturbation on the crystal field, even the remaining $(2S+1)$-fold degeneracy of the ground state will be split. However this additional splitting may well be negligible compared with both k_BT and the splitting in an applied magnetic field.

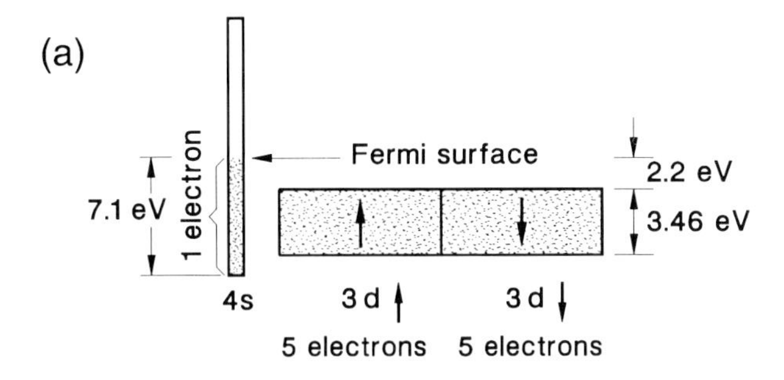

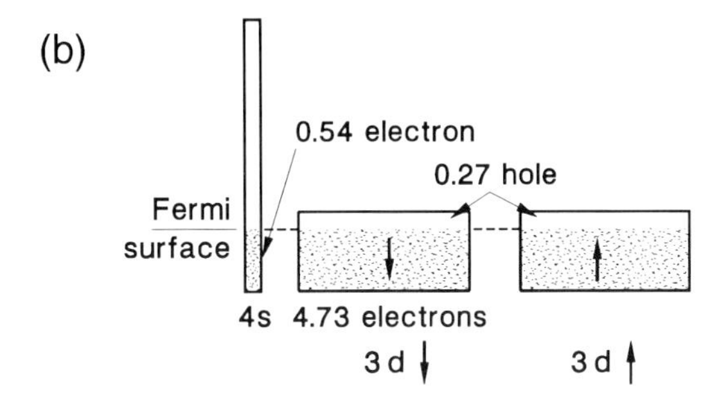

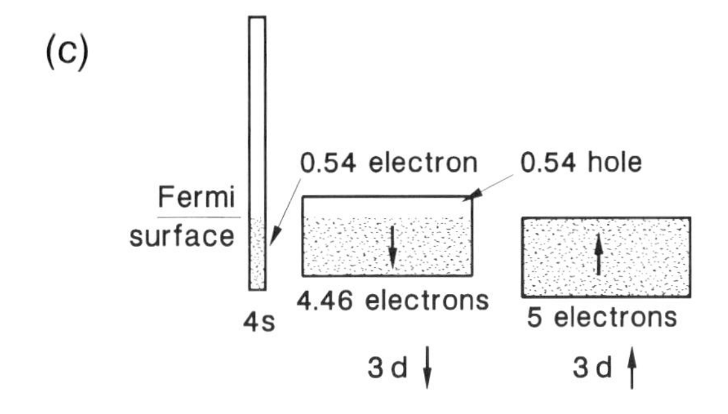

Figure 12.31. (a) Schematic relationship of 4s and 3d bands in metallic copper. The 3d band holds 10 electrons per atom and is filled in copper. The 4s band can hold two electrons per atom; it is shown half-filled, as copper has one valence electron outside the filled 3d shell. The filled 3d band is shown as two separate sub-bands of opposite electron spin orientation, each band holding five electrons. The 4s band is usually thought to contain approximately equal numbers of electrons of both spins, and so we have not divided it into sub-bands. With both sub-bands filled, the net spin (and hence the net magnetization) of the d band is zero. (b) Band occupancy in nickel above the Curie temperature. The net magnetic moment is zero, as there are equal numbers of holes in both 3d↑ and 3d↓ bands. (c) Schematic relationship of bands in nickel at absolute zero. The energies of the 3d↑ and 3d↓ sub-bands are separated by an exchange interaction. The 3d↑ band is filled, while the 3d↓ band contains 4.46 electrons and 0.54 holes. The net magnetic moment of $0.54\mu_B$ per atom arises from the excess population of the 3d↑ band over the 3d↓ band. Adapted from C. Kittel, *Introduction to Solid State Physics*, 6E revised, copyright © 1995 by John Wiley & Sons, by permission of the publisher.

12.10. Magnetic Anisotropy

In a crystalline material there are certain crystallographic directions along which the magnetization prefers to orient itself (easy axes) or tries to avoid (hard axes). The tendency of a magnetic crystal to direct the magnetization along certain directions is called magnetic anisotropy, and the energy associated with the various orientations of the magnetization vector is the anisotropy energy. For instance, in cobalt, which is a hexagonal crystal, the c-axis is the direction of easy magnetization at room temperature (see Fig. 12.32(a)). Denoting by Θ the angle between the magnetization direction and the c-axis, the anisotropy energy density $\mathcal{E}_a(\Theta)$ of cobalt may be written

$$\boxed{\mathcal{E}_a = K_1 \sin^2\Theta + K_2 \sin^4\Theta} \tag{12.72}$$

where, at $T = 300$ K, $K_1 = 4.1 \times 10^6$ erg/cm³, and $K_2 = 1.0 \times 10^6$ erg/cm³. Iron is a cubic crystal (bcc) and the cube edges are the directions of easy magnetization, as can be seen from the magnetization curves of Fig. 12.32(b). To represent the anisotropy energy of iron magnetized in an arbitrary direction with direction cosines α_1, α_2, α_3 referred to the cube edges, we are guided by cubic symmetry. The expression for the anisotropy energy must be an even power of each α_i, provided opposite ends of a crystal axis are equivalent magnetically, and it must be invariant under interchanges of the α_i among themselves. The lowest-order combination satisfying the symmetry requirements is $\alpha_1^2 + \alpha_2^2 + \alpha_3^2$, but this is identically equal to unity and does not describe anisotropic effects. The next combinations are of the fourth degree, $\alpha_1^2\alpha_2^2 + \alpha_1^2\alpha_3^2 + \alpha_3^2\alpha_2^2$, and then of the sixth degree, $\alpha_1^2\alpha_2^2\alpha_3^2$. Thus

$$\boxed{\mathcal{E}_a = K_1(\alpha_1^2\alpha_2^2 + \alpha_2^2\alpha_3^2 + \alpha_3^2\alpha_1^2) + K_2\, \alpha_1^2\alpha_2^2\alpha_3^2} \tag{12.73}$$

For pure iron at $T = 300$ K it is found that $K_1 = +4.2 \times 10^5$ erg/cm³ and $K_2 = +1.5 \times 10^5$ erg/cm³. Nickel is an fcc crystal with easy axis along the cube diagonal, i.e., in the [111] direction. The anisotropy energy of single-crystal nickel samples is also given by Eq. (12.73) and, at $T = 300$ K, the measured values of the anisotropy constants are $K_1 = -4.5 \times 10^4$ erg/cm³ and $K_2 = +2.3 \times 10^4$ erg/cm³. Figure 12.32(c) shows magnetization curves of nickel along the crystal axes.

The simplest kind of magnetic anisotropy is uniaxial anisotropy with the following dependence of energy on the angle between the magnetization vector $\mathbf{M}$ and the (unique) axis of anisotropy $\mathcal{N}$:

$$\boxed{\mathcal{E}_a = -K_u \left(\frac{\mathbf{M}}{|\mathbf{M}|} \cdot \mathcal{N}\right)^2} \tag{12.74}$$

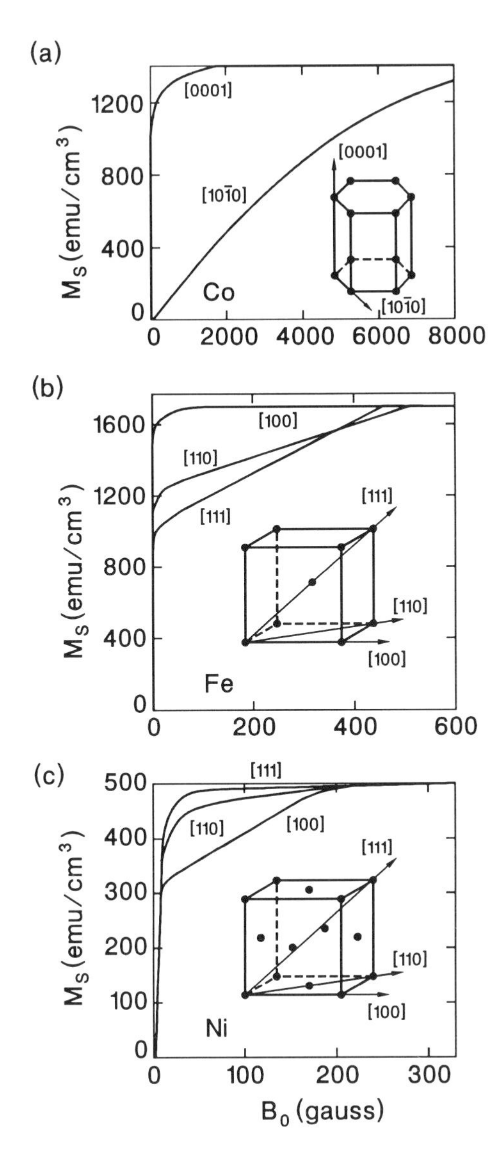

Figure 12.32. Magnetization curves for single crystals of cobalt, iron, and nickel at room temperature; B_0 is the applied field. In cobalt the c-axis of the hcp crystal is the direction of easy magnetization. From the curves for bcc iron we see that the [100] directions are easy directions whereas the [111] directions are hard. In fcc nickel the [111] direction is the easy axis. Source: K. Honda and S. Kaya, *Sci. Rep. Tohoku Univ.* **15**, 721 (1926).

If K_u happens to be positive, both parallel and antiparallel alignments of **M** relative to $\mathcal{N}$ are favored; under these circumstances **M** perpendicular to $\mathcal{N}$ is the state of highest energy, and $\mathcal{N}$ is referred to as the bidirectional easy axis. The opposite is true when $K_u < 0$; here the plane perpendicular to $\mathcal{N}$ is the easy plane for **M**.

12.10.1. Single-ion anisotropy

In certain magnetic solids the crystal **E**-field acts on magnetic ions and renders certain orientations of their orbital angular momentum **L** preferable over others. The crystal field thus lifts the degeneracy of the orbital momentum, and the occupied states tend to favor certain crystallographic directions for their orbital moments. The spin angular momentum **S** then couples to **L** via the spin–orbit interaction and assumes the same preferred orientation. Thus the anisotropy of magnetization direction relative to the crystal axes is established.

The rare earth metals Tb and Dy are good examples of materials with this type of anisotropy. Both metals have hexagonal close-packed (hcp) crystal structures, are ferromagnetic below their respective Curie temperatures (T_c = 218 K for Tb and 90 K for Dy), and have large unquenched orbital angular momenta ($L = 3$ for Tb; $L = 5$ for Dy). In both cases the spatial distributions of the 4f electrons interact with the crystal field and create a situation whereby the individual ions' orbital momenta prefer to lie in the basal plane, causing the c-axis of the crystal to become the (magnetic) hard axis.† Even in the antiferromagnetic state observed between T_c and T_N, the magnetic moments of both these metals continue to favor orientation in the basal plane, the helical magnetic structure being caused by the peculiarities of the RKKY exchange interaction between the RE ions and the conduction electrons.

The single-ion mechanism is often the main source of anisotropy in antiferromagnetic and ferrimagnetic dielectric crystals, where non-magnetic elements surround the moment-carrying ions, leaving in relative isolation the partially quenched orbital momenta of the latter to interact with the local crystal field. Examples of dielectric crystals with single-ion anisotropy are Fe_3O_4, $MnFe_2O_4$, $CoFe_2O_4$, $NiFe_2O_4$, FeF_2, $BaFe_{12}O_{19}$ and $Y_3Fe_5O_{12}$.

12.10.2. Anisotropy by pair-ordering

A major origin of magnetic anisotropy is illustrated in Fig. 12.33; here the magnetization sees the crystal lattice through orbital overlap of the electrons on neighboring sites. The spin interacts with the orbital motion through spin–orbit coupling and aligns itself relative to the crystal axes accordingly. This kind of magnetic anisotropy is not rooted in the purely isotropic exchange interaction that obtains when spin–orbit effects are ignored: without the coupling between **L** and **S**, the exchange between two ions with overlapping charge clouds would be isotropic in the total spin orientation, in the sense that the coupled spins could have arbitrary orientation with respect to the bond between the two ions. Inclusion of the

†Measured[5] uniaxial anisotropy constants for these metals at 77 K are $K_u(\text{Tb}) \simeq -2.7 \times 10^7$ erg/cm^3 and $K_u(\text{Dy}) \simeq -2.0 \times 10^7$ erg/cm^3.

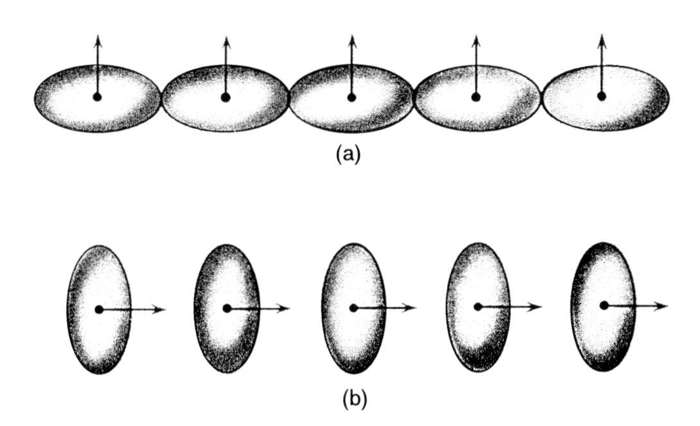

Figure 12.33. Asymmetry of the overlap of electron distributions on neighboring ions provides one mechanism of magneto-crystalline anisotropy. Because of the spin–orbit interaction the charge distribution is spheroidal and not spherical. The asymmetry is tied to the direction of the spin on an ion, so that a rotation of the spin directions relative to the crystal axes changes the exchange energy and also changes the electrostatic interaction energy of the charge distributions on pairs of atoms. Both effects give rise to an anisotropy energy. The energy of (a) is not the same as that of (b). After C. Kittel, *Introduction to Solid State Physics*, 6E revised, copyright © 1995 by John Wiley & Sons, reproduced by permission of the publisher.

$\mathbf{L} \cdot \mathbf{S}$ term in the Hamiltonian of the system lifts this degeneracy, and allows the magnetization to interact with the crystal lattice. Thus, if there are small amounts of orbital magnetic moment left unquenched by the crystal field, a part of the orbital will rotate in the wake of a rotation of the spin because of the interaction between the two; the rotation of the orbital will, in turn, change the overlap of the wave functions between the two atoms, giving rise to a change in the electrostatic energy and, consequently, the exchange energy. This type of interaction, which is sometimes referred to as anisotropic exchange,[6,9] is believed to be the main contributor to the magnetic anisotropy of the 3d transition metals (Cr, Mn, Fe, Co, Ni) as well as the alloys of these with other metals.

12.10.3. Shape anisotropy

This form of anisotropy is rooted in classical dipole–dipole interactions, and can be readily explained in terms of the tendency of a magnetized body to minimize its magnetostatic energy. For instance, a long, slender magnetic needle tends to orient its magnetization along the long axis, as shown in Fig. 12.34(a). By the same token a flat slab of ferromagnetic material resists magnetization in the perpendicular direction, preferring instead to be magnetized in the plane of the slab; see Fig. 12.34(b). Because of the long-range nature of dipole–dipole interactions, and also because magnetic specimens usually contain a multitude of domains and therefore a

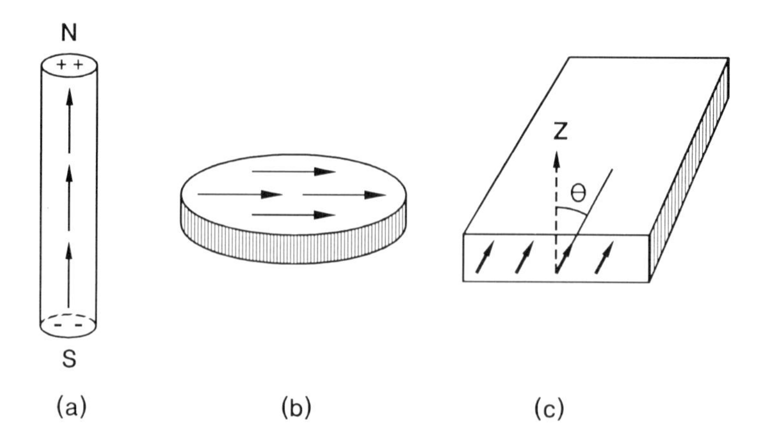

Figure 12.34. (a) In a long, slender needle the magnetization orients itself along the length of the needle. The positive and negative magnetic charges thus produced are well separated, thus lowering the magnetostatic energy. (b) A uniformly magnetized flat disk reduces the magnetostatic energy by pulling **M** into the plane of the disk. (c) Thin film with uniform magnetization **M**. When the angle between **M** and the normal to the plane is Θ, the demagnetizing energy is $2\pi|\mathbf{M}|^2\cos^2\Theta$. Shape anisotropy for thin films is therefore equivalent to uniaxial perpendicular anisotropy with an effective constant $K_u = -2\pi|\mathbf{M}|^2$.

complex magnetization distribution, the analysis of shape anisotropy can be very complicated. We shall describe shape anisotropy only for a thin film with uniform magnetization as shown in Fig. 12.34(c). Assuming that the magnetization of the film is uniform and denoting by Θ the angle between **M** and the Z-axis, the magnetic field within the film is found to be

$$\mathbf{H} = -4\pi\ |\mathbf{M}|\ \cos\Theta\ \hat{\mathbf{z}} \tag{12.75}$$

The magnetostatic energy density is therefore given by†

$$\mathcal{E}_{\text{dmag}} = -\tfrac{1}{2}\ \mathbf{M}\cdot\mathbf{H} = 2\pi|\mathbf{M}|^2\ \cos^2\Theta \tag{12.76}$$

This equation indicates that the magnetostatic energy of a thin-film sample with uniform magnetization has a minimum at $\Theta = 90°$ and a maximum at $\Theta = 0$. Therefore, barring any other source of anisotropy, the shape anisotropy brings the magnetization into the plane (i.e., the plane of the film is the easy plane). In the case of a thin film with magnetization **M** and uniaxial anisotropy (unrelated to shape) in the perpendicular direction with constant K_u, the shape anisotropy is overcome and the magnetization stands

†The factor $\frac{1}{2}$ in Eq. (12.76) accounts for the fact that, unlike an externally applied field, the **H**-field in the present situation is self-induced. We shall have more to say about the demagnetizing field and self energy in Chapter 13.

perpendicular to the film plane provided that $K_u > 2\pi|\mathbf{M}|^2$. In measurements of the anisotropy constants of thin films, it is necessary to add $2\pi M^2$ to the apparent value of K_u in order to obtain the intrinsic constant of magnetic anisotropy.

12.10.4. Anisotropy due to classical dipole–dipole interactions

This type of anisotropy is dominant in several antiferromagnetic crystals; examples include MnF_2, $MnTe_2$, $MnSe_2$, MnO, MnS_2, and NiO. It must be distinguished from shape anisotropy, which is also dependent on dipole–dipole interactions. Perhaps the best way to separate the two effects is to measure (or calculate) the anisotropy constants for spherical samples, where the shape anisotropy is absent.

Figure 12.35 shows the arrangement of Mn^{++} ions in the crystal lattice of MnF_2; in (a) the moments are aligned with the c-axis of the body-centered tetragonal lattice, which is also the easy axis of magnetization; in (b) the moments are parallel to the basal plane. According to Eqs. (12.24) and (12.16), the mutual energy of a pair of ions carrying magnetic dipole moments $\mathbf{m}_1$ and $\mathbf{m}_2$, and separated by a distance $\mathbf{r}_{12}$, is

$$U_{12} = \frac{\mathbf{m}_1 \cdot \mathbf{m}_2 - 3\,(\mathbf{m}_1 \cdot \hat{\mathbf{r}})\,(\mathbf{m}_2 \cdot \hat{\mathbf{r}})}{|\mathbf{r}_{12}|^3} \tag{12.77}$$

where $\hat{\mathbf{r}} = \mathbf{r}_{12}/|\mathbf{r}_{12}|$ is the unit vector joining the two ions. Clearly, two ions whose moments are antiferromagnetically coupled by exchange forces will have the lowest value of dipole–dipole energy, U_{12}, when their dipole moments are perpendicular to the bond ($U_{12} = -\mathbf{m}_1 \cdot \mathbf{m}_2/|\mathbf{r}_{12}|^3$). On the other hand, the dipole–dipole energy will be maximized when the antiparallel moments are aligned with the bond ($U_{12} = +2\,\mathbf{m}_1 \cdot \mathbf{m}_2/|\mathbf{r}_{12}|^3$). Based on this type of interaction alone, the energy of the two states shown in Fig. 12.35 have been computed[22] for MnF_2 at absolute zero and found to differ by as much as 4.91×10^6 erg/cm^3; this is indeed very close to the observed magnitude of the anisotropy energy in this material.

Similar effects, of course, can be found in other materials, not only antiferromagnets but ferromagnets and ferrimagnets as well. The symmetry of the lattice and the strength of the interactions, however, are such that the dipole–dipole contribution to magnetic anisotropy is usually insignificant.

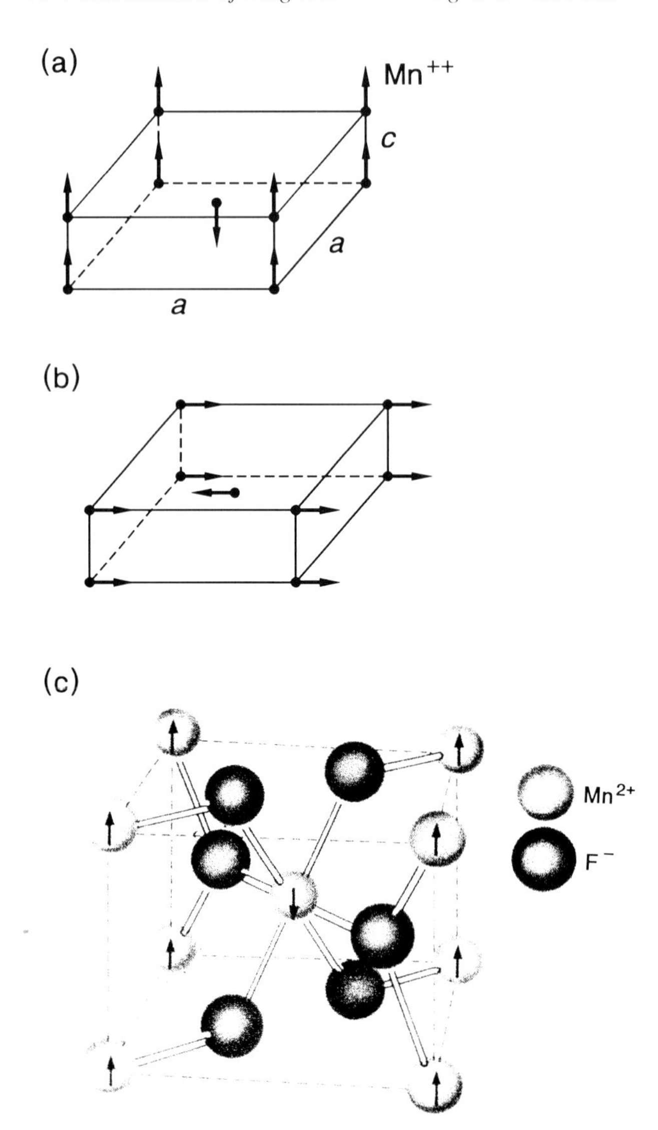

Figure 12.35. Body-centered trigonal lattice of the Mn^{++} ions in MnF_2 crystal, where $a > c$. This is an antiferromagnet with the moment at the body center aligned antiparallel to the moments at the cube corners. The classical dipole–dipole interactions prefer the magnetic orientation in (a) over that in (b), as explained in the text. The crystal therefore exhibits magnetic anisotropy with an easy axis along c. Note that since Mn^{++} is a S-state ion with zero orbital angular momentum ($L = 0$), its contribution to the single-ion anisotropy is usually negligible, thus allowing other mechanisms, such as in this case the dipole–dipole interaction, to dominate. The chemical structure of MnF_2, shown in (c), is reproduced with permission from C. Kittel, *Introduction to Solid State Physics*, 6E revised, copyright © 1995 by John Wiley & Sons.

Problems

(12.1) Prove that the solution to Eqs. (12.12) is given by Eq. (12.13).

(12.2) A wire of negligible diameter forms a planar loop of radius R. Assuming that the loop carries a uniform electric current I, determine the magnetic field distribution in the vicinity of the loop. (Figure 12.5(c) shows the field lines produced by one such current carrier.)

(12.3) Consider a spherical particle of some magnetic material with radius R and uniform magnetization $\mathbf{M}$. Determine the magnetic field $\mathbf{H}$ both inside and outside the particle.

(12.4) The diameters of an ellipsoidal particle along the principal axes are D_x, D_y, D_z and the magnitude of its saturation magnetization is M_s. With the particle magnetized uniformly along one of its principal axes, determine the strength of the magnetic field $\mathbf{H}$ within the ellipsoid.

(12.5) Magnetic films are generally deposited on a substrate such as a glass or plastic slide. When a small piece of a typical glass substrate was measured in a vibrating sample magnetometer (VSM), a plot of magnetic moment versus the applied field, shown in Fig. 12.36 was obtained. What kind of magnetic behavior does this material exhibit? Is the order of magnitude of susceptibility consistent with the estimates given in the text?

(12.6) Consider a magnetic dipole moment $\mathbf{m}$ in an applied field $\mathbf{B}_0$. Ignore the quantization of angular momentum and allow the dipole to assume all possible orientations in space. Determine the average moment $\langle m \rangle$ at a given temperature T. Compare the result with its quantum counterpart in Eq. (12.35), and show that in the limit when $J \to \infty$ the two are identical.

(12.7) Verify the approximate expression given in Eq. (12.36) for the Brillouin function $\mathscr{B}_J(x)$ in the limit of small x.

(12.8) A simple cubic lattice is occupied by the identical ions of a certain magnetic species. Denoting the lattice constant by a, each ion has six nearest neighbors at distance a, twelve second neighbors at distance $\sqrt{2}a$, and eight third neighbors at distance $\sqrt{3}a$. Let the exchange interaction between a given ion and each one of its neighbors be antiferromagnetic, and assume that the strength of exchange drops as the fourth power of distance, with no interaction beyond the third neighbor. Develop a mean-field theory for this lattice and derive an expression for its Néel temperature T_N similar to that given in Eq. (12.70) for the Curie temperature.

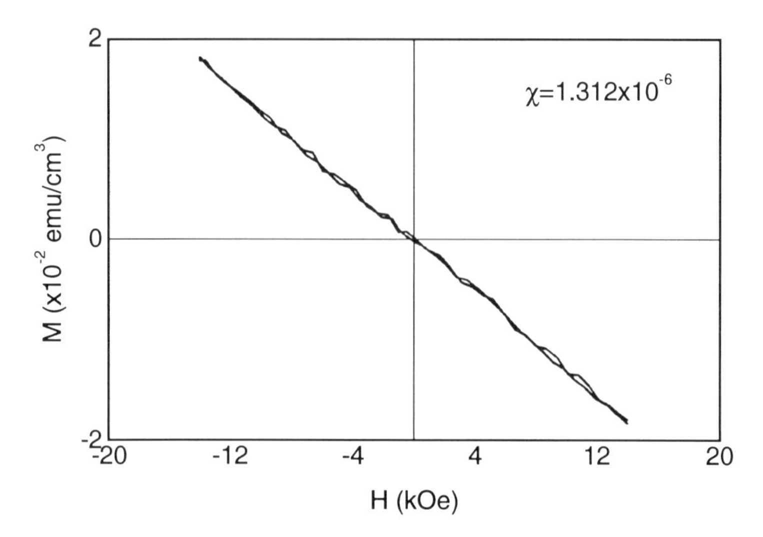

Figure 12.36.

(12.9) Use a mean-field model of antiferromagnetism to derive the temperature dependence of the magnetic susceptibility χ, and verify the behavior exemplified by the plots in Fig. 12.25(b).

(12.10) The amorphous ferrimagnet $A_x B_{1-x}$ is composed of two magnetic species, A and B, whose ions are of the same size and show the same affinity towards each other as each shows towards one of its own species. Thus if the coordination number is denoted by Z, then, on average, each atom of type A will be surrounded by $(Z + 1)(1 - x)$ atoms of type B and by $(Z + 1)x - 1$ atoms of type A. Similarly, each B-type atom will have $(Z + 1)x$ neighbors of type A and $(Z + 1)(1 - x) - 1$ neighbors of type B. Let the individual atomic magnetic moments be $m_a = g_a \mu_B J_a$ and $m_b = g_b \mu_B J_b$, and let the mutual exchange coefficients be $\mathcal{J}_{aa} > 0$, $\mathcal{J}_{bb} > 0$ and $\mathcal{J}_{ab} < 0$. Develop a mean-field theory of magnetization for this material and derive an expression for its critical temperature T_c.

(12.11) Figure 12.35 shows two arrangements of magnetic dipoles in MnF_2 crystal. Estimate the energy difference between these states due to the classical dipole–dipole interaction.

13

Magnetostatics of Thin-film Magneto-optical Media

Introduction

Magnetostatics is the study of stationary patterns of magnetization in magnetic media. The central role in these studies is played by domains and domain walls. Domains are regions of uniform magnetization that are large on the atomic scale, but may be small in comparison with the dimensions of the medium under consideration. The walls are narrow regions that separate adjacent domains. In uniform media, the primary source of the spontaneous breakdown of magnetization into domains is the long-range dipole–dipole interaction, commonly referred to as the demagnetizing effect. Breakdown into domains provides a means of lowering the energy of demagnetization at the expense of increased exchange and anisotropy energies. Since within the domains the magnetization is fairly uniform and often aligned with an easy axis, it is at the site of the domain walls that the excess energies of exchange and anisotropy accumulate, as though the walls were endowed with an energy of their own. This chapter is devoted to the study of domains and domain walls and their structure and energy, as well as their magnetostatic interactions.

The basic tenets of magnetism were reviewed in Chapter 12, and the fundamental notions of magnetic field, magnetization, exchange and anisotropy were introduced. In the present chapter we shall confine attention to homogeneous thin-film media, where the magnetization $\mathbf{M}$ is treated as a continuous function of position $\mathbf{r}$ over the volume of the material (see Fig. 13.1). The film will be taken as parallel to the XY-plane and its thickness τ as uniform. Moreover, we shall assume that the magnetization at any given point (x, y) does not vary throughout the film thickness, and is therefore independent of the z-coordinate. This is a reasonable assumption considering the fact that the media of MO recording are usually very thin ($\tau < 500$ Å is typical).

Because of exchange, which will be represented by the macroscopic stiffness coefficient A_x, the magnetization distribution $\mathbf{M}(\mathbf{r})$ can be taken as smooth, in the sense that the magnetization direction does not undergo rapid variations in space; this will allow us to treat $\mathbf{M}$ as a differentiable function of $\mathbf{r}$. Furthermore, since the materials of interest in MO recording have uniaxial anisotropy with easy axis perpendicular to the plane of the film, our analyses will be restricted to media with a perpendicular easy axis; the constant of uniaxial anisotropy K_u will therefore be positive, and deviations in $\mathbf{M}$ will be measured relative to the Z-axis. Under these

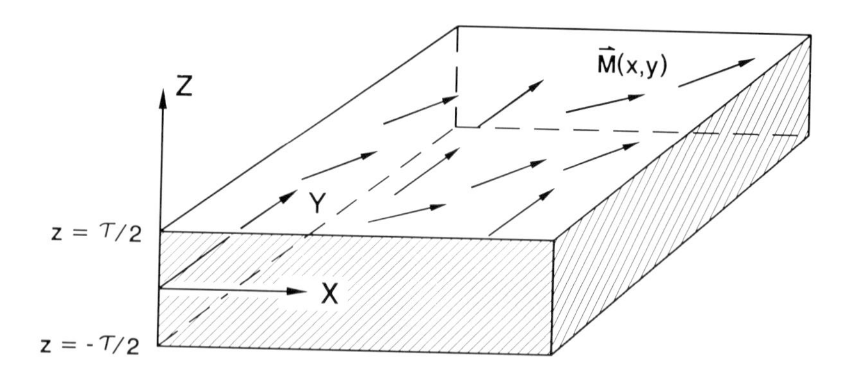

Figure 13.1. Thin magnetic film in the *XY*-plane of a Cartesian coordinate system. The film thickness τ is uniform, and the magnetization $\mathbf{M}(x, y)$ is constant through the thickness. The distribution of $\mathbf{M}$ is smooth in the sense that $\mathbf{M}(x, y)$ is a differentiable function of x and y.

circumstances the distribution of magnetization over the *XY*-plane of the film will be influenced by the competing forces of exchange, anisotropy, long-range dipole–dipole interactions, and the interaction between the magnetization and any external magnetic field that might be present.[1–7]

A major contributing factor to the formation process and stability of various patterns of magnetization in thin-film media is the coercive force. Coercivity is rooted in defects and inhomogeneities as well as in the nonuniform distribution of magnetic properties across the film. Later we shall devote an entire chapter (Chapter 16) to this subject; for the time being, however, we ignore coercivity and concentrate on the simpler case of defect-free, homogeneous, uniform media. These are media that may be assumed to possess a single constant of uniaxial anisotropy K_u, a single exchange coefficient A_x, and a single saturation moment M_s over their entire volume; the only spatially varying parameter then is the direction of the magnetization vector $\mathbf{M}$.

Within thin films of magnetic media it is often the case that fairly large regions (large on the atomic scale) are found uniformly magnetized along certain directions. Such regions, or domains, are surrounded by other domains of differing magnetic orientations. The transition between a domain and its neighbor is not atomically sharp; rather, the intermediate region, which is known as the domain wall, assumes a finite width. (In other words, $\mathbf{M}$ needs a finite space within which to rotate "continuously" from one direction to another.) Figure 13.2 is a schematic diagram of the magnetization within a simple domain wall.

When the film has strong uniaxial anisotropy, there can be only two kinds of domains: those having $\mathbf{M}$ either parallel or antiparallel to the easy axis (i.e., "up" and "down" domains). Figure 13.3 shows two micrographs of thin magnetic plates in the demagnetized state, exhibiting several up and down domains. In (a) the domain structure has been revealed in polarized

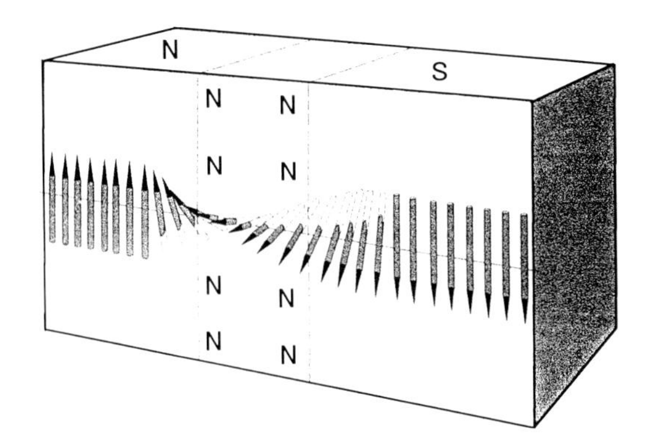

Figure 13.2. Magnetic structure of a wall separating "up" and "down" domains. The north and south poles of the dipole magnets are identified as N and S, respectively. Source: C. Kittel, *Introduction to Solid State Physics*, 6E revised, copyright © 1995 by John Wiley & Sons, reproduced by permission of the publisher.

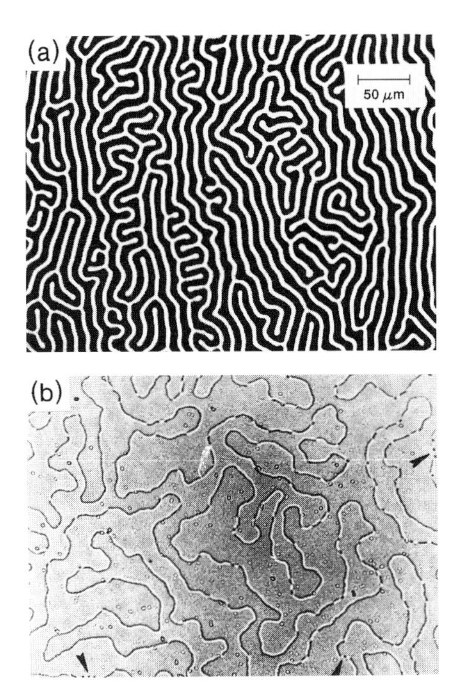

Figure 13.3. (a) Polarized-light microphotograph of domain pattern in a 3 μm-thick plate of $BaFe_{12}O_{19}$ at zero field after cooling from the Curie point. The black and white stripes correspond to domains of opposite polarity. After C. Kooy and U. Enz, Philips Res. Repts. No. 15, 7 (1960). (b) Lorentz electron micrograph of domains and domain walls in a 100 nm-thick amorphous film of GdCoAu. The thin lines are domain walls that separate domains of opposite magnetization. The small circles are defects, but the breaks in domain walls (see black pointers) indicate the presence of vertical Bloch lines. Source: P. Chaudhari and S. R. Herd, *IBM J. Res. Dev.* **20**, 102 (1976).

light with the aid of Faraday effect; the contrast is caused by domains of opposite magnetization. In (b) the electron beam of a transmission electron microscope is used to enhance (via the Lorentz force) domain walls against the background of the domains themselves. The walls in this case are seen as thin black and white pairs of lines on a gray background.†

In general, exchange, anisotropy, and magnetostatic fields (the latter being caused by classical dipole–dipole interactions) determine the magnetic structure of a domain wall. In the media of MO recording it so happens that exchange and anisotropy outweigh the dipole–dipole interactions, thus dominating the basic characteristics of the wall. To a good approximation, in these media one may ignore dipole–dipole effects and obtain expressions for the shape and energy of the wall in terms of A_x and K_u ; this will be done in section 13.1. Also considered briefly in this section is the distinction between the Bloch and Néel walls, and the role played by the magnetostatic force in favoring one structure over the other.

Section 13.2 is devoted to a method of computing the magnetostatic fields that arise from arbitrary distributions of magnetization. The **H**-field arising from the magnetic charge density $\nabla \cdot \mathbf{M}$ is referred to as the stray field when it is outside the volume occupied by the film; within the body of the film the field is known as the demagnetizing field. A uniformly saturated film with perpendicular magnetization **M** has stray field equal to zero†† and demagnetizing field equal to $-4\pi\mathbf{M}$. When the magnetization distribution is nonuniform, these fields become difficult to compute, in part because of the long-range nature of the dipolar interaction. The method developed in section 13.2 is based on Fourier transforms, and offers a powerful tool for the numerical evaluation of both stray and demagnetizing fields. The technique is first developed for discretization on a square lattice, then extended to cover hexagonal lattices as well.

Circular domains are of particular interest in MO recording. Their significance arises from the fact that domains recorded on stationary media tend to have circular symmetry. A good deal of theoretical work in thermomagnetic recording and erasure has been based on the properties of circular domains.[8,9] In section 13.3 we analyze the energetics of circular domains and pay particular attention to the demagnetizing energy. Later, in Chapter 17, the results of section 13.3 will be used to analyze domains in the presence of temperature profiles imposed by the recording laser beam. An interesting application of the results of section 13.3 occurs in the measurement of domain wall energy density σ_w; a technique for measuring σ_w is described in section 13.4.

† Transmission electron microscopy (TEM) is now used routinely to reveal the micromagnetic structure of thin films. The technique is known as Lorentz electron microscopy, and is a valued tool for high-resolution observations of magnetic domains and domain walls. Section 18.3 gives a detailed account of this method of microscopy.

†† It is assumed that the film is infinite in its lateral extent. Real films, of course, are finite and exhibit fringing fields near the edges. As long as the region of interest is far from an edge, however, it is safe to assume that the stray field of a uniformly magnetized film is vanishingly small.

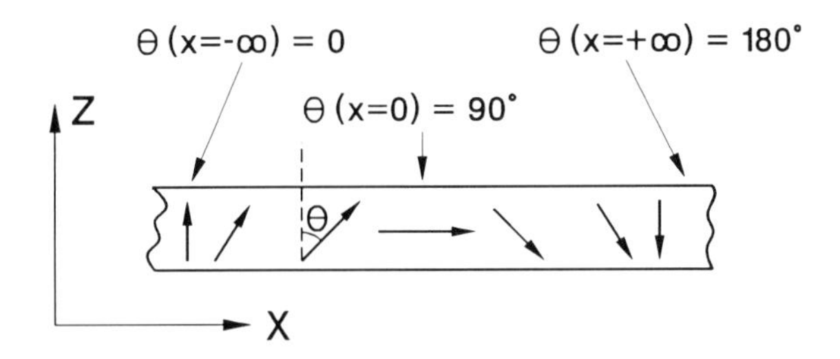

Figure 13.4. Cross section of thin magnetic film with a domain wall centered at $x = 0$. The magnetization is parallel to the Z-axis at $x = -\infty$, and antiparallel to it at $x = +\infty$. The deviation angle of **M** from Z, denoted by $\theta(x)$, is a continuous function of x.

13.1. Domain Walls in Perpendicular Films

Consider a uniform film with thickness τ, saturation moment M_s, anisotropy constant K_u and exchange coefficient A_x. Let the magnetization direction be up in one half of the film and down in the other half, as in Fig. 13.4. Assuming a smooth transition between these two regions, the exchange energy density will be

$$\mathcal{E}_{\text{xhg}} = A_x\left[(\nabla m_x)^2 + (\nabla m_y)^2 + (\nabla m_z)^2\right]$$

$$= A_x\left\{\left[\frac{\mathrm{d}}{\mathrm{d}x}\sin\theta(x)\right]^2 + \left[\frac{\mathrm{d}}{\mathrm{d}x}\cos\theta(x)\right]^2\right\} = A_x\left[\frac{\mathrm{d}\theta(x)}{\mathrm{d}x}\right]^2 \tag{13.1}$$

Ignoring the energy associated with the external field as well as that of dipole–dipole interactions, the total energy of the system arises from anisotropy and exchange, as follows: [1,10]

$$\mathcal{E}_{\text{tot}} = \mathcal{E}_{\text{xhg}} + \mathcal{E}_{\text{anis}} = \int_{-\infty}^{\infty}\left\{A_x\left[\mathrm{d}\theta(x)/\mathrm{d}x\right]^2 + K_u\sin^2\theta(x)\right\}\mathrm{d}x \tag{13.2}$$

To find the equilibrium magnetization distribution one must use a variational technique to minimize the energy. Let $\theta_0(x)$ be a minimum energy configuration, and assume that another configuration $\hat{\theta}(x)$ differs only slightly from $\theta_0(x)$, that is, $\delta\theta(x) = \hat{\theta}(x) - \theta_0(x)$ is small for all x. The energy differential $\delta\mathcal{E}$ between these two states must then be zero. Now

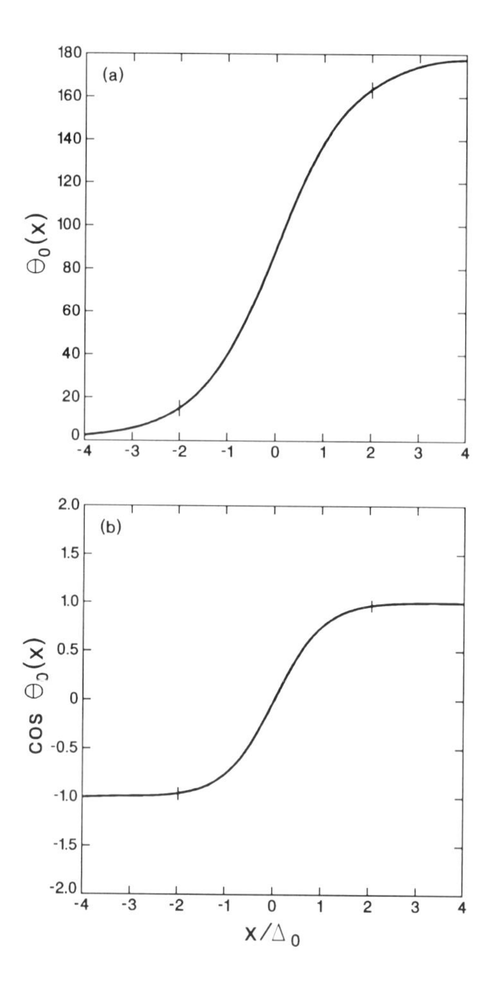

Figure 13.5. Deviation angle θ_0 of the magnetization vector from the Z-axis, over the width of a domain wall. The parameter by which the values of x on the abscissa are scaled is $\Delta_0 = \sqrt{A_x / K_u}$. Away from the wall center the magnetization approaches saturation asymptotically; for all practical purposes, however, the wall is confined within $\pm 2\Delta_0$ from the center. The plot in (a) is of $\theta_0(x)$, while in (b) $\cos\theta_0(x)$ is plotted.

$$\sigma_w = \int_{-\infty}^{\infty} \left\{ \frac{A_x}{\Delta_0^2} \sin^2\theta_0(x) + K_u \sin^2\theta_0(x) \right\} \mathrm{d}x = 2\int_{-\infty}^{\infty} K_u \sin^2\theta_0(x)\, \mathrm{d}x \tag{13.8}$$

Note that the exchange and anisotropy energies of the equilibrium wall structure are identical. Using Eq. (13.6f′) in the above expression, we find

$$\sigma_w = 8K_u \int_{-\infty}^{\infty} \frac{\exp(2x/\Delta_0)}{[1 + \exp(2x/\Delta_0)]^2} \, dx = 4K_u \Delta_0 \int_1^{\infty} \frac{dy}{y^2} = -4\sqrt{A_x K_u} \left. \frac{1}{y} \right|_1^{\infty} \tag{13.9}$$

Consequently,

$$\boxed{\sigma_w = 4\sqrt{A_x K_u}} \tag{13.10}$$

Again, assuming $A_x = 10^{-7}$ erg/cm and $K_u = 10^6$ erg/cm^3, we find $\sigma_w \simeq 1.26$ erg/cm^2 for typical films of amorphous TbFeCo at room temperature. Under the circumstances that led to Eq. (13.10), it is always the case that half the wall energy comes from anisotropy and half from exchange. The wall width is a result of a compromise between the opposing forces of exchange and anisotropy. When anisotropy is large the wall is narrrow, since anisotropy favors those directions of **M** that are either parallel or antiparallel to the easy axis, thereby encouraging the wall to avoid intermediate orientations of **M**. On the other hand, strong exchange tends to widen the wall, since the energy of exchange is proportional to $\int(d\theta_0/dx)^2\,dx$, and a wide wall reduces this by reducing $d\theta_0/dx$. These considerations, of course, are consistent with the expression derived earlier for the wall width, namely, $4\sqrt{A_x/K_u}$: increasing K_u will result in a narrow wall, whereas increasing A_x will make the wall wider.

13.1.2. Effect of demagnetizing field; Bloch and Néel walls

Thus far we have ignored the demagnetizing effects arising from classical dipole–dipole interactions. To justify this supposition we point to the fact that the effective field acting on a given dipole due to anisotropy is of the order of K_u/M_s, while that due to exchange is of the order of $A_x/(M_s d^2)$, d being the average interatomic distance. The demagnetizing field, on the other hand, is of the order of M_s. (A derivation of these effective fields will be given in Chapter 15.) For typical TbFeCo films of interest in MO recording $M_s \simeq 100$ emu/cm^3, $K_u \simeq 10^6$ erg/cm^3, $A_x \simeq 10^{-7}$ erg/cm, and $d \simeq 1$ Å. Consequently, the orders of magnitude of the effective anisotropy, exchange, and demagnetizing fields are 10^4 Oe, 10^7 Oe, and 10^2 Oe respectively. It is thus possible to ignore the demagnetizing field in a first approximation, and to leave it for later treatment as a perturbation.

Consider the walls depicted schematically in Fig. 13.6. In (a), the so-called Bloch wall, the in-plane component of magnetization is parallel to the wall itself, i.e., $\psi = 0$. There are no magnetic charges on the wall and the demagnetization energy, $\mathcal{E}_{dmag}$, is small. In (b), the Néel wall, the in-plane component is perpendicular to the wall ($\psi = 90°$) and the resulting magnetic charges produce an excess demagnetizing energy. As far as exchange and anisotropy are concerned the two walls are identical; however, if the wall is free to choose either structure, it always settles in the Bloch configuration in order to minimize its energy of demagnetization.†

It may happen that two sections of a wall are in the Bloch state with oppositely oriented in-plane components, i.e., $\psi = 0$ in one part and 180° in the other, as shown in Fig. 13.6(c). In this case there will be a twist in the in-plane component where these sections join. The twist, known as a vertical Bloch line (VBL), has an excess of exchange and demagnetizing energies; it is inherently unstable and exists only when trapped in some local potential well, or when the boundary conditions require the twist.

13.2. Mathematical Analysis of Stray and Demagnetizing Fields

In this section we show that the magnetic field distribution $\mathbf{H}(x,y,z)$ of a thin magnetic film can be accurately and efficiently computed with the aid of fast Fourier transforms.[11,12] We begin by computing the vector potential field $\mathbf{A}(x,y,z)$. Consider a magnetic film parallel to the XY-plane of a Cartesian coordinate system, as in Fig. 13.1, and denote its magnetization distribution by $\mathbf{M}(x,y)$. Assume that the film has thickness τ, with surfaces at $z = \pm\frac{1}{2}\tau$. Also assume that $\mathbf{M}(x,y)$ is periodic along both X and Y, with periods L_x and L_y, respectively. The Fourier series representation of the film's magnetization will then be

$$\mathbf{M}(x,y) = \sum_{m=-\infty}^{\infty} \sum_{n=-\infty}^{\infty} \mathcal{M}_{mn} \exp\left\{i2\pi\left(\frac{mx}{L_x} + \frac{ny}{L_y}\right)\right\} \tag{13.11a}$$

where

$$\mathcal{M}_{mn} = \frac{1}{L_x L_y} \int_0^{L_x} \int_0^{L_y} \mathbf{M}(x,y) \exp\left\{-i2\pi\left(\frac{mx}{L_x} + \frac{ny}{L_y}\right)\right\} dx dy \tag{13.11b}$$

In general, an arbitrary magnetization $\mathbf{M}(\mathbf{r})$ gives rise to the vector potential $\mathbf{A}(\mathbf{r})$ (see Eq. (12.15)) described by the convolution integral:

†There is nothing special about $\psi = 0$ or 90°; the in-plane component may in fact have any arbitrary orientation in the XY-plane. Simply stated, $\mathcal{E}_{dmag}$ is a minimum for the Bloch structure ($\psi = 0$ or 180°) and a maximum for the Néel structure ($\psi = \pm 90°$), but it may also assume an intermediate value in between the two configurations.

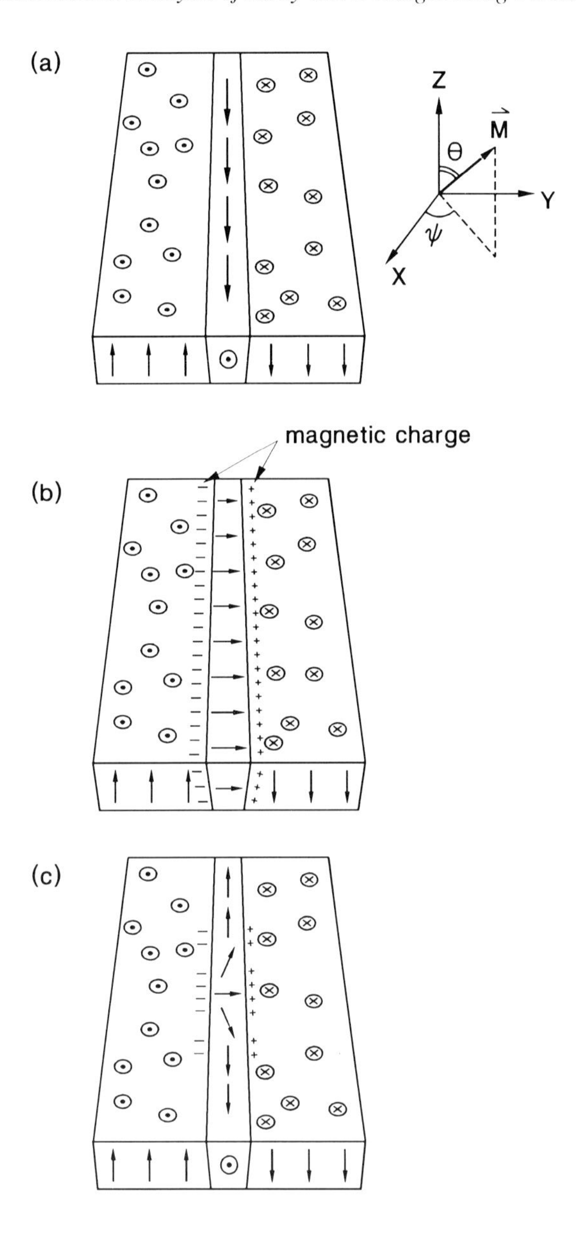

Figure 13.6. Domain walls in thin magnetic films with perpendicular anisotropy. The deviation of **M** from the Z-axis, $\theta(y)$, is the same in all cases. The walls differ in their distributions of the azimuthal angle $\psi(x, y)$. (a) In a Bloch wall the in-plane component of **M** is parallel to the wall (i.e., $\psi = 0$ or $180°$). (b) In a Néel wall the in-plane component is perpendicular to the wall (i.e., $\psi = \pm 90°$). This results in a charged wall and a significant increase of the demagnetization energy over the Bloch wall. (c) Vertical Bloch line within a domain wall. The two halves of the wall have the structure of a Bloch wall, with $\psi = 0$ and $\psi = 180°$, forcing a twist of the in-plane component in the middle where they meet.

13.2.3. Energy of demagnetization

The magnetostatic energy density for the thin film discussed in the preceding subsections is

$$\mathscr{E}_{\text{dmag}} = -\frac{1}{2L_x L_y}\int_0^{L_y}\int_0^{L_x} \mathbf{M}(x,y)\cdot\mathbf{H}^{(\text{ave})}(x,y)\,\mathrm{d}x\mathrm{d}y$$

Using Parseval's theorem,[14] the above equation may be written in terms of the Fourier coefficients $\mathscr{M}_{mn}$ and $\mathscr{H}_{mn}^{(\text{ave})}$ as follows:

$$\mathscr{E}_{\text{dmag}} = -\tfrac{1}{2}\sum\sum \mathscr{M}_{mn}^{*}\cdot\mathscr{H}_{mn}$$

$$= 2\pi\sum_{n=-\infty}^{\infty}\sum_{m=-\infty}^{\infty} \mathscr{G}(\tau s)\,\left|\mathscr{M}_{mn}\cdot\hat{\mathbf{z}}\right|^2 + \left[1-\mathscr{G}(\tau s)\right]\,\left|\mathscr{M}_{mn}\cdot\boldsymbol{\sigma}\right|^2 \qquad (13.24)$$

In this equation the function $\mathscr{G}(\cdot)$ is given by Eq. (13.23b), s is the magnitude of the frequency vector $\mathbf{s}$ defined in Eq. (13.21c), τ is the film thickness, and $\boldsymbol{\sigma}$ is the unit vector along $\mathbf{s}$ as defined in Eq. (13.21d). According to Eq. (13.24) the contribution of the perpendicular component of $\mathbf{M}$ to the demagnetizing energy comes mainly from the low-frequency content of M_z. On the other hand, since the contribution to $\mathscr{E}_{\text{dmag}}$ of the in-plane component of $\mathbf{M}$ passes through the high-pass filter $1-\mathscr{G}(\tau s)$, it comes mainly from the high-frequency end of the spectrum.

Digression. Another useful relation that may be derived by applying Parseval's theorem is the following relation between the average magnitude of $\mathbf{M}$ and its Fourier components $\mathscr{M}_{mn}$:

$$\langle\mathbf{M}^2\rangle = \frac{1}{L_x L_y}\int_0^{L_y}\int_0^{L_x}\mathbf{M}\cdot\mathbf{M}^*\,\mathrm{d}x\mathrm{d}y = \sum_{n=-\infty}^{\infty}\sum_{m=-\infty}^{\infty}\left|\mathscr{M}_{mn}\right|^2$$

For a homogeneous film kept at uniform temperature, the magnitude of $\mathbf{M}$ is the saturation moment M_s, in which case $\langle\mathbf{M}^2\rangle = M_s^2$. □

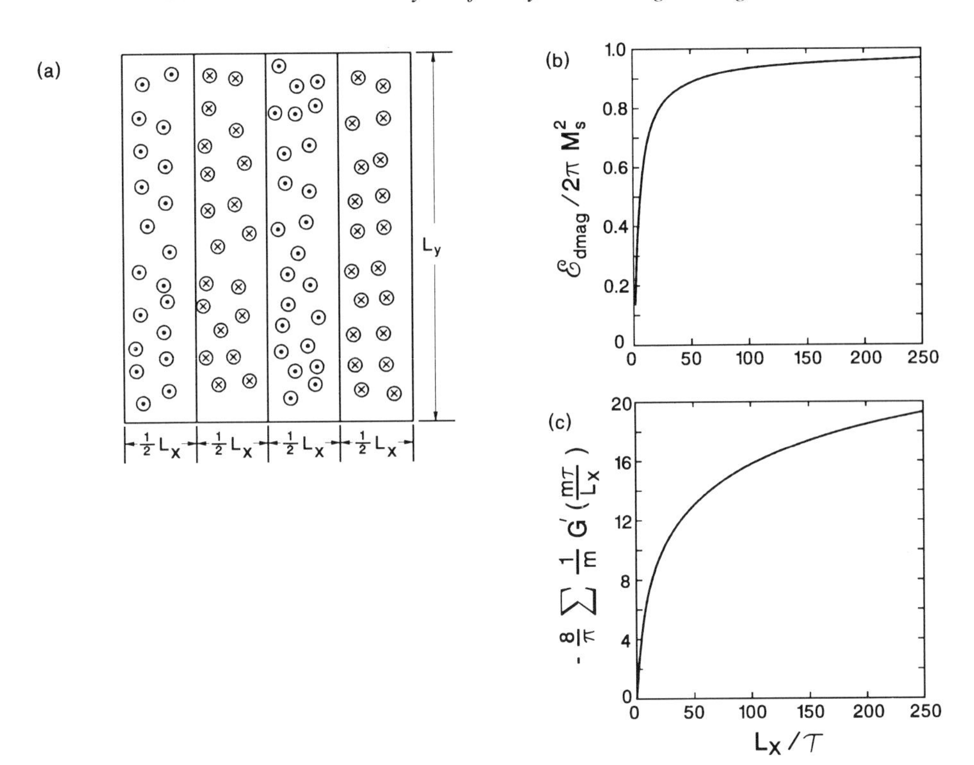

Figure 13.9. (a) Periodic array of stripe domains in a perpendicular medium. The "up" and "down" domains have equal widths, $\frac{1}{2}L_x$, and the walls are infinitely sharp. In computing the demagnetizing field by the Fourier method, only one pair of stripes (say, the region between $x = 0$ and $x = L_x$) need be considered; the periodic boundary condition inherent in the method automatically accounts for the periodicity. (b) Demagnetizing energy density $\mathcal{E}_{\text{dmag}}$ (normalized by $2\pi M_s^2$) versus stripe period L_x (normalized by film thickness τ). In the limit of large L_x the function approaches unity. (c) The left-hand side of Eq. (13.28), computed numerically and plotted versus L_x/τ. This is therefore a plot of $\sigma_w/\tau M_s^2$ versus the optimum stripe period L_x. The function does not reach an asymptotic value with increasing L_x, but continues to grow at a logarithmic rate. Thus the stripe pattern will always have an equilibrium period, irrespective of how large σ_w or how small M_s may be. In practice, of course, there is a limit to how large the period can become before deviations from the ideal stripe pattern occur.

Example 2. Figure 13.9(a) shows a periodic array of straight domain walls in a thin magnetic film. The walls have zero width, are parallel to the Y-axis, and the separation between adjacent walls is $\frac{1}{2}L_x$. Because of the periodic boundary conditions used in the Fourier method, one need consider only a section of the film with dimensions $L_x \times L_y$. Thus

$$\mathbf{M}(x,y) = M_s \mathcal{U}(x/L_x)\,\hat{\mathbf{z}} \tag{13.25a}$$

where

$$\mathcal{U}(x) = \begin{cases} +1 & 0 < x \le \frac{1}{2} \\ -1 & \frac{1}{2} < x \le 1 \end{cases} \tag{13.25b}$$

Since there are no variations of **M** along Y, the Fourier coefficients $\mathcal{M}_{mn}$ with $n \neq 0$ vanish. Moreover, the remaining coefficients $\mathcal{M}_{mo}$ have no components along $\hat{\mathbf{x}}$ and $\hat{\mathbf{y}}$, with projections along $\hat{\mathbf{z}}$ only. We find:

$$\mathcal{M}_{mo} = \frac{1}{L_x} \int_0^{L_x} M_s \mathcal{U}(x/L_x) \hat{\mathbf{z}} \exp\left(-\mathrm{i}2\pi \frac{mx}{L_x}\right) \mathrm{d}x$$

$$= M_s \hat{\mathbf{z}} \int_0^1 \mathcal{U}(x) \exp(-\mathrm{i}2\pi mx)\, \mathrm{d}x$$

$$= \begin{cases} -\dfrac{2\mathrm{i}}{\pi m} M_s \hat{\mathbf{z}} & m = \pm 1, \pm 3, \pm 5, \cdots \\ 0 & m = 0, \pm 2, \pm 4, \cdots \end{cases} \tag{13.26}$$

Substituting the above coefficients in Eq. (13.24) one obtains the following expression for the demagnetizing energy density:

$$\mathcal{E}_{\mathrm{dmag}} = 2\pi M_s^2 \left[\frac{8}{\pi^2} \sum_{m=1,3,5,\cdots} \frac{1}{m^2} \mathcal{G}(m\tau/L_x) \right] \tag{13.27}$$

In the limit where $L_x >> \tau$, for all significant values of m we have $\mathcal{G}(m\tau/L_x) \simeq 1$. Since $\Sigma(1/m^2) = \pi^2/8$, we find $\mathcal{E}_{\mathrm{dmag}} \simeq 2\pi M_s^2$, as expected. Figure 13.9(b) is a plot of $\mathcal{E}_{\mathrm{dmag}}$ versus L_x/τ obtained from Eq. (13.27), showing that the demagnetizing energy density asymptotically approaches $2\pi M_s^2$ as $L_x \to \infty$.

Let us now assume that each wall has surface energy density σ_w arising from exchange and anisotropy. The corresponding energy density per unit volume, $\mathcal{E}_{\mathrm{wall}}$, is $2\sigma_w/L_x$. To find the optimum period L_x, the total energy $\mathcal{E}_{\mathrm{wall}} + \mathcal{E}_{\mathrm{dmag}}$ must be minimized. Setting the derivative of energy with respect to L_x equal to zero, we obtain

$$-\frac{8}{\pi} \sum_{m=1,3,5,\cdots} \frac{1}{m} \mathcal{G}'(m\tau/L_x) = \frac{\sigma_w}{\tau M_s^2} \tag{13.28}$$

The function on the left-hand side of this equation is numerically evaluated and plotted versus L_x/τ in Fig. 13.9(c). To determine the optimum stripe period L_x for given values of M_s, σ_w and τ, one simply locates the point $\sigma_w/\tau M_s^2$ on the vertical axis and reads the corresponding value of L_x/τ on the horizontal axis. As L_x increases, the curve in Fig. 13.9(c) goes to infinity at a logarithmic rate. The period of the stripe pattern thus increases rapidly as $\sigma_w/\tau M_s^2$ assumes larger and larger values. It is interesting to observe that no matter how large σ_w or how small M_s become, there always exists a value of L_x for which the stripes are stable. This behavior is due to the long-range nature of demagnetization, and indicates that a single straight wall affects the demagnetizing energy in distant regions. In other words, there is no distance that is far enough from the wall for its demagnetizing effects to be negligible. □

13.2.4. Field computation on the hexagonal lattice

In subsection 13.2.1 we described a method of calculating the magnetic field distribution for thin magnetic films. The method was based on Fourier transforms and the corresponding numerical computations were performed on square lattices. This method is now extended to hexagonal lattices.

Consider a rectangular slab of magnetic material in the XY-plane. The slab's dimensions are $L_x \times L_y \times \tau$, where τ is the film thickness. With the imposed periodic boundary conditions, the magnetization pattern will be periodic in the XY-plane, L_x and L_y being the periods along X and Y. Figure 13.10 shows a regular two-dimensional hexagonal lattice; the original slab is the rectangular region of dimensions $L_x \times L_y$ having its lower left-hand corner at the origin of the XY-plane. To emphasize the periodicity, a few lattice cells beyond the slab are also shown.

Discrete Fourier transforms cannot be readily performed on the rectangular region, since it does not conform to the natural bases of the hexagonal lattice. The bases of the hexagonal lattice are vectors **a** and **b** having equal length, i.e., $|\mathbf{a}| = |\mathbf{b}| = d$. These are shown in the lower left-hand corner of Fig. 13.10. The shaded region in Fig. 13.10 over which the Fourier transform will be performed has N_1 cells along **a** and N_2 cells along **b**, where $L_x = N_1 d$ and $L_y = \frac{1}{2}\sqrt{3}\,N_2 d$. This new parallelepiped slab has the same number of cells as the original rectangular slab; either slab could be considered as a basic pattern of magnetization which, when replicated in the XY-plane, would reproduce the distribution $\mathbf{M}(x,y)$. Note, however, that although periodicity along **a** follows from the periodicity in the X-direction, periodicity along Y does not lead to the same along **b**. Nonetheless, if N_2 is chosen to be equal to $2N_1$ (or any integer multiple of $2N_1$), then periodicity along **b** will be ensured. The following restrictions thus apply to the lattice:

$$N_2 = 2N_1 \tag{13.29a}$$

$$L_y = \sqrt{3}\, L_x \tag{13.29b}$$

Under these circumstances, one can express the periodic nature of $\mathbf{M}(x, y)$ in the new coordinate system as follows:

$$\mathbf{M}\left[(\alpha + mN_1)\mathbf{a} + (\beta + nN_2)\mathbf{b}\right] = \mathbf{M}(\alpha\mathbf{a} + \beta\mathbf{b}) \tag{13.30}$$

Here α and β are arbitrary real numbers, while m and n are arbitrary integers. Next we define the reciprocal lattice vectors $\mathbf{A}$ and $\mathbf{B}$ in the frequency domain (i.e., the $s_x s_y$-plane). $\mathbf{A}$ and $\mathbf{B}$ must satisfy the following conditions:

$$\mathbf{a} \cdot \mathbf{B} = \mathbf{b} \cdot \mathbf{A} = 0 \tag{13.31a}$$

$$\mathbf{a} \cdot \mathbf{A} = \mathbf{b} \cdot \mathbf{B} = 1 \tag{13.31b}$$

$\mathbf{A}$ and $\mathbf{B}$ have equal lengths ($|\mathbf{A}| = |\mathbf{B}| = 2/(\sqrt{3}d)$) and are shown in the inset in Fig. 13.10. With the aid of these reciprocal lattice vectors, we write the following expression for sinusoidal patterns of magnetization that have the same periodicity as in Eq. (13.30):

$$\mathcal{M}_{mn} \exp\left\{ i2\pi \left(\frac{m}{N_1}\mathbf{A} + \frac{n}{N_2}\mathbf{B} \right) \cdot (\alpha\mathbf{a} + \beta\mathbf{b}) \right\}$$

$$= \mathcal{M}_{mn} \exp\left\{ i2\pi \left(\frac{m\alpha}{N_1} + \frac{n\beta}{N_2} \right) \right\} \tag{13.32}$$

The Fourier series expansion of $\mathbf{M}(\alpha\mathbf{a} + \beta\mathbf{b})$ is now written:

$$\mathbf{M}(\alpha\mathbf{a} + \beta\mathbf{b}) = \sum_{m=-\infty}^{\infty} \sum_{n=-\infty}^{\infty} \mathcal{M}_{mn} \exp\left\{ i2\pi \left(\frac{m\alpha}{N_1} + \frac{n\beta}{N_2} \right) \right\} \tag{13.33a}$$

where

$$\mathcal{M}_{mn} = \frac{1}{L_x L_y} \iint_{\substack{\text{parallelepiped} \\ \text{slab}}} \mathbf{M}(\alpha\mathbf{a} + \beta\mathbf{b}) \exp\left\{ -i2\pi \left(\frac{m\alpha}{N_1} + \frac{n\beta}{N_2} \right) \right\} \mathrm{d}x\,\mathrm{d}y \tag{13.33b}$$

In the above equation $L_x L_y$ is the area of the parallelepiped slab, over the surface of which the integral is being taken. Approximating the integral in Eq. (13.33b) with a finite sum over the hexagonal lattice, we obtain

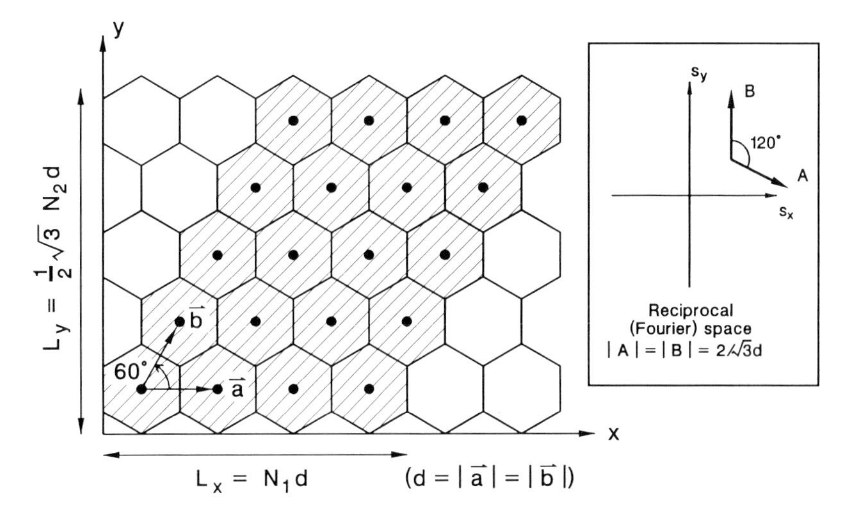

Figure 13.10. The hexagonal lattice and its basis vectors **a** and **b**. The inset shows the reciprocal lattice vectors **A** and **B**.

$$\mathscr{M}_{mn} \simeq \frac{1}{N_1 N_2} \sum_{I=0}^{N_1-1} \sum_{J=0}^{N_2-1} \mathbf{M}(I\mathbf{a} + J\mathbf{b}) \exp\left\{-\mathrm{i}2\pi\left(\frac{mI}{N_1} + \frac{nJ}{N_2}\right)\right\} \tag{13.34}$$

Thus a straightforward application of the discrete Fourier transform formula to the magnetization distribution, sampled over the parallelepiped slab, produces the coefficients $\mathscr{M}_{mn}$.

Next, we relate the **H**-field distribution to that of the magnetization. In analogy with Eq. (13.22a):

$$\mathbf{H}(\alpha\mathbf{a} + \beta\mathbf{b}, z) = \sum_{m=-\infty}^{\infty} \sum_{n=-\infty}^{\infty} \mathscr{H}_{mn}(z) \exp\left\{\mathrm{i}2\pi\left(\frac{m\alpha}{N_1} + \frac{n\beta}{N_2}\right)\right\} \tag{13.35}$$

where, as before, $\mathscr{H}_{mn}(z)$ is given by Eq. (13.22b), but now the frequency vector is

$$\mathbf{s} = \frac{m}{N_1}\mathbf{A} + \frac{n}{N_2}\mathbf{B} = \frac{m}{L_x}\hat{\mathbf{x}} + \left(\frac{n}{L_y} - \frac{m}{\sqrt{3}\,L_x}\right)\hat{\mathbf{y}} \tag{13.36}$$

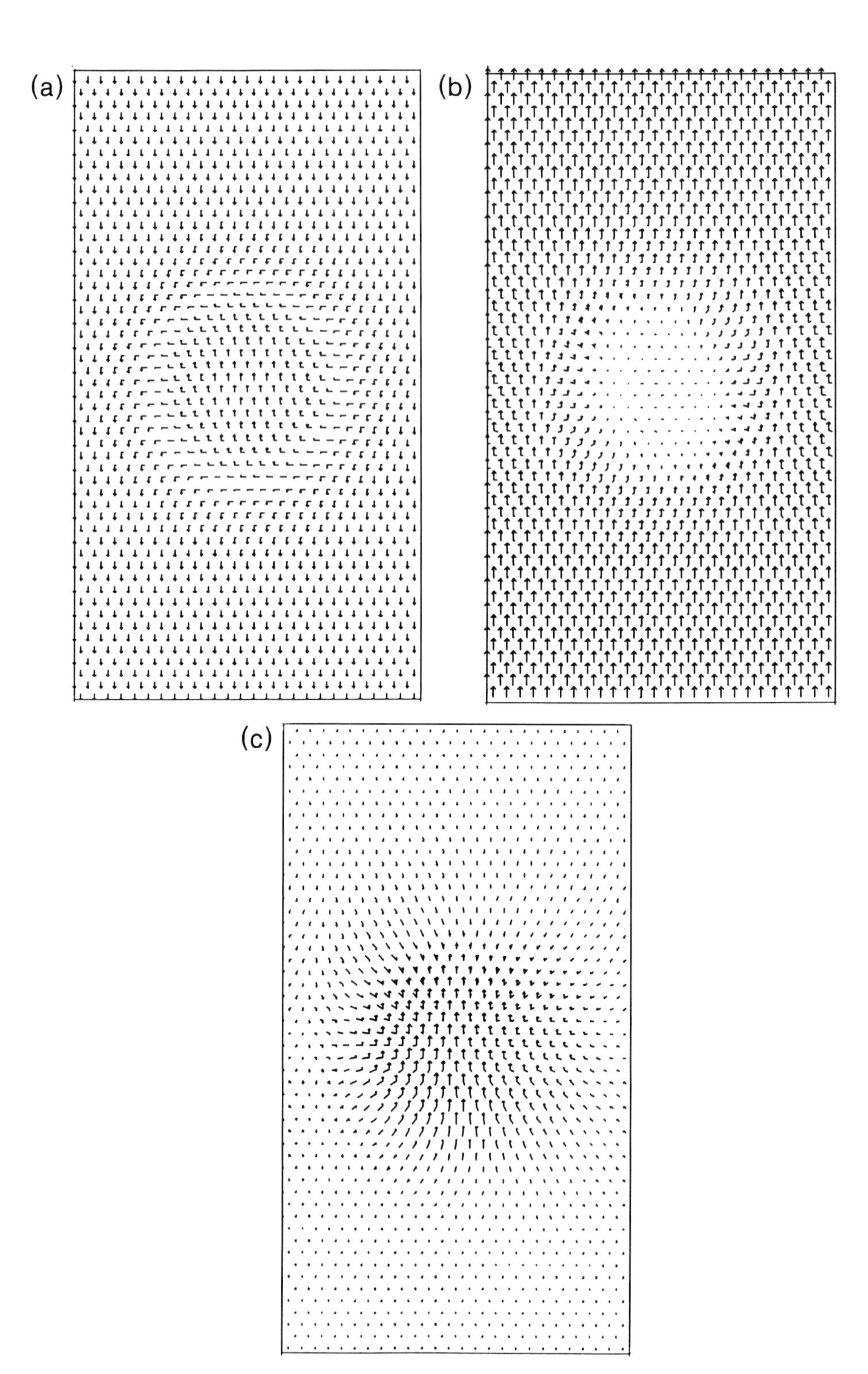

Figure 13.11. (a) Assumed magnetization distribution in a perpendicular film containing a circular domain. In units of the lattice constant d the domain radius $R_0 = 7d$ and the wall-width parameter $\Delta_0 = 2d$. The arrows represent the perpendicular component of magnetization, while the appendage to each arrow shows the corresponding in-plane component. (b) Thickness-averaged demagnetizing field for the magnetization pattern shown in (a). The assumed film thickness $\tau = 10d$. (c) **H**-field distribution outside the film whose magnetization pattern is shown in (a). The assumed film thickness $\tau = 50d$, and the distance above the surface at which the field is computed is d.

In summary, the results obtained previously for the square lattice can also be applied to the hexagonal lattice provided that

(i) The parallelepiped slab of Fig. 13.10, satisfying the constraints of Eq. (13.29), is taken as the pertinent region.
(ii) The frequency vector **s** in Eq. (13.36), its magnitude $s = |\mathbf{s}|$, and the unit vector $\boldsymbol{\sigma} = \mathbf{s}/|\mathbf{s}|$ are associated with the reciprocal lattice space.

Example 3. We compute the **H**-field pattern of a circular domain in a perpendicular thin film, shown in Fig. 13.11(a). The hexagonal lattice has $N_1 = 26$ pixels and $N_2 = 52$ pixels along the X-axis and the Y-axis respectively. In units of the lattice constant, d, the domain radius $R_0 = 7\text{d}$ and the wall-width parameter $\Delta_0 = 2\text{d}$. Figure 13.11(b) shows the thickness-averaged demagnetizing field $\mathbf{H}^{(\text{ave})}$ for this film assuming $\tau = 10d$. The magnetic field pattern outside the film is shown in Fig. 13.11(c). Here the assumed film thickness $\tau = 50d$ and the distance above the surface at which the field is computed is one lattice constant, i.e., $z = 26d$.

□

Example 4. Figure 13.12(a) shows an in-plane magnetization distribution with a pair of head-to-head walls. The hexagonal lattice has 26×52 pixels, and the wall-width parameter $\Delta_0 = 5d$, where d is the lattice constant; the assumed film thickness $\tau = 10d$. Figure 13.12(b) shows the computed demagnetizing field, $\mathbf{H}^{(\text{ave})}$, for this film; note that the field within the walls is rather weak, whereas in the central stripe between the two walls the field strength is appreciable. The wall on the left has a positive magnetic charge, while the charge on the right wall is negative. This charge pattern gives rise to a field in the central region that is oriented along the positive X-axis.

The field pattern at a distance of one lattice constant above the film surface (i.e., at $z = 6d$) is shown in Fig. 13.12(c). Here the arrow represents the component of the field along Z, while its appendage indicates the in-plane component. We observe that immediately above each wall the field is perpendicular to the plane of the lattice, being parallel to Z when the charge of the wall is positive, and antiparallel when the charge is negative.

□

13.3. Micromagnetics of Circular Domains

The process of thermomagnetic recording in thin films involves the formation of a reverse-magnetized domain in the region of the hot spot produced by a focused laser beam, followed by the growth or contraction of this domain as heating and then cooling proceeds. The behavior of the spin system under conditions of thermomagnetic writing is fairly complicated; nonetheless, the process lends itself to a phenomenological analysis by the consideration of magnetic energy and the notion of coercivity. The ingredients of such a study are calculation of different

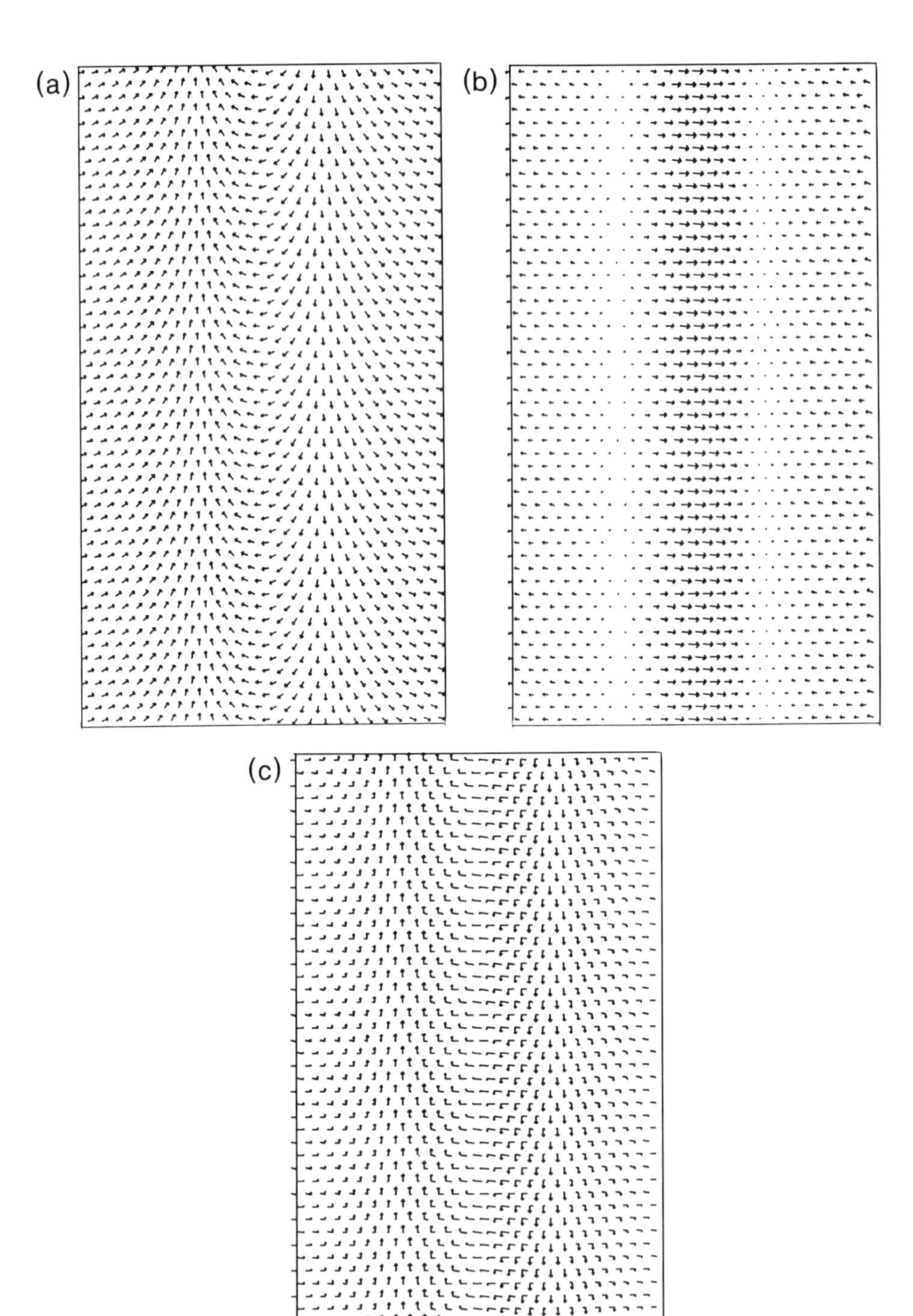

Figure 13.12. (a) In-plane magnetization distribution with a pair of head-to-head walls. Since there are no perpendicular components, the arrows represent the in-plane **M** only. The assumed film thickness τ in subsequent calculations is $10d$. (b) Thickness-averaged demagnetizing field for the magnetization pattern shown in (a). The arrows show the in-plane field, since there are no perpendicular components. (c) Distribution of the **H**-field outside the film shown in (a). The field is computed at a distance of d above the film surface, that is, $z = 6d$. The arrows represent the perpendicular component of the field, while the appendage to each arrow shows the corresponding in-plane component.

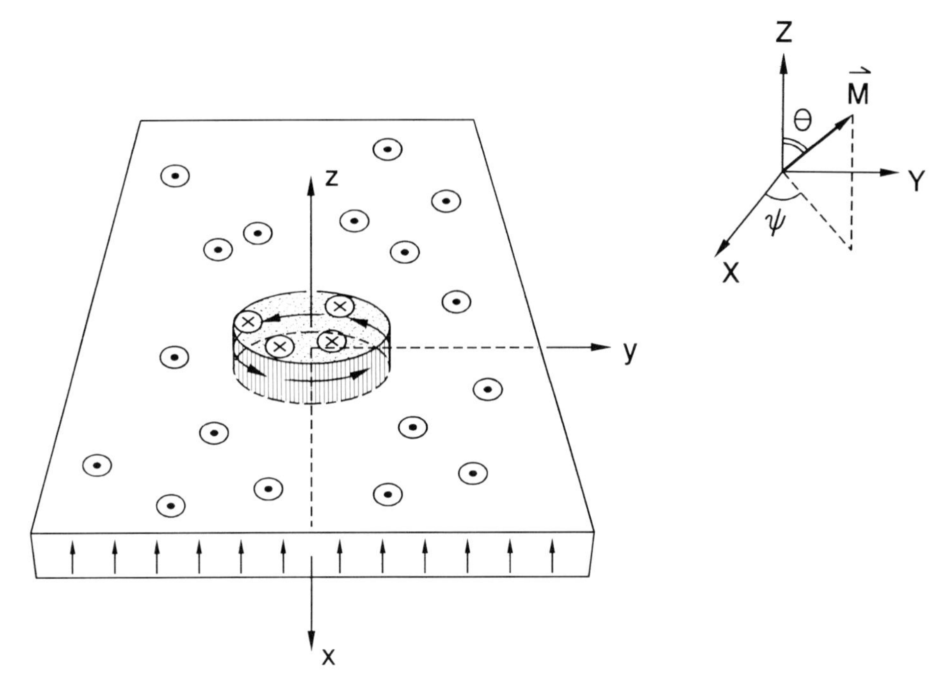

Figure 13.13. Schematic representation of a reverse-magnetized circular domain in a perpendicular thin film. Within the domain wall, the in-plane components of **M** form a closed loop in order to avoid the formation of magnetic charges. This type of structure is known as a Bloch wall.

contributions to energy, assessment of the relative significance of these contributions, and tracking of the net energy variations over a cycle of thermomagnetic recording.

In preparation for the investigation of thermomagnetic write/erase processes in Chapter 17, we derive in this section expressions for various contributions to the energy of a thin film containing a reverse-magnetized circular domain. As before, we shall postulate a homogeneous film of thickness τ and infinite lateral extent in the XY-plane of a Cartesian coordinate system. The laser beam, assumed circularly symmetric and centered on the origin of the XY-plane, creates a temperature profile that, at any instant of time, may be considered some function of the radial coordinate r, but independent of the azimuthal coordinate ϕ. At the moment, we are not interested in the details of the temperature profile; rather, the knowledge that it is circularly symmetric is sufficient to enable us to proceed with the derivation of expressions for energy. Throughout this section, therefore, it will be assumed that the saturation moment M_s, the anisotropy constant K_u, and the exchange stiffness coefficient A_x, are functions of the radial distance r from the Z-axis. We suppose that the circular domain as depicted in Fig. 13.13 has radius r_0. The magnetization through the thickness will be assumed uniform.

At first glance it might seem reasonable to assume for a circular domain the same kind of wall structure that was obtained for the straight wall in Eq. (13.6f′), namely,

$$\cos\theta(r) = \tanh\left(\frac{r - r_0}{\Delta_0}\right) \tag{13.37}$$

In particular, when $r_0 >> \Delta_0$ the curvature of the wall becomes negligible, and Eq. (13.37) will describe the magnetization pattern with sufficient accuracy. One problem with this assumption, however, is that the exchange energy corresponding to Eq. (13.37) is infinite at the center of the domain. Another problem is that Eq. (13.37) does not apply to small domains. Moreover, if A_x and K_u happen to have substantial variations with r in the vicinity of the wall, then the conditions under which Eq. (13.6f′) was derived will have been violated. To avoid these problems, we propose a somewhat different functional form for $\theta(r)$, namely,

$$\boxed{\cos\theta(r) = \tanh\left(\frac{r^3 - r_0^3}{3\Lambda r_0 r}\right)} \tag{13.38}$$

Λ is the wall-width parameter, much like Δ_0 in Eq. (13.37). The above function approximates Eq. (13.37) very closely when $r_0 >> \Lambda$, yet its asymptotic approach to $r = 0$ eliminates the singularity of exchange. Figure 13.14 shows plots of $\cos\theta(r)$ versus r/Λ for several values of r_0/Λ. The curves represent Eq. (13.38), while the solid circles correspond to Eq. (13.37). Notice that when $r_0 \geq 5\Lambda$, the two equations are nearly identical. For small values of r_0/Λ the results differ mainly in the vicinity of $r = 0$. It should be emphasized that Eq. (13.38) is *not* an analytical solution to the variational problem of wall-energy minimization, a problem that leads to an intractable nonlinear differential equation for circular domains; it is merely a simple function that has the correct asymptotic behavior and can be thought of as a parametric representation of a circular domain with radius r_0. The parameter Λ is no longer given by $\sqrt{A_x/K_u}$; instead, it must be determined numerically by energy minimization. In almost all cases of practical interest, however, Λ turns out to be very close to Δ_0 in value, namely, $\Lambda \simeq [A_x(r_0)/K_u(r_0)]^{\frac{1}{2}}$.

In addition to the deviation angle θ from Z, complete identification of the magnetization pattern requires specification of the azimuthal angle of **M**, which we shall denote by ψ (Fig. 13.13). For simplicity, we assume a Bloch structure for the wall; this has the added advantage of being the state of minimum demagnetizing energy. Thus

$$\boxed{\psi(r,\phi) = \frac{\pi}{2} + \phi} \tag{13.39}$$

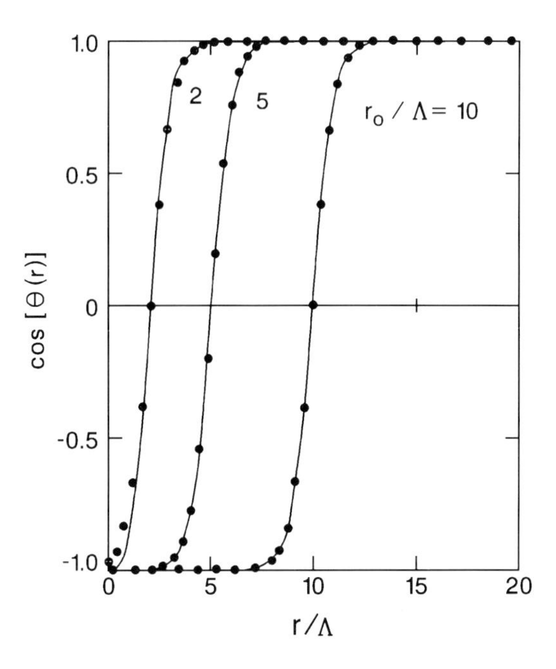

Figure 13.14. Plots of $\cos\theta(r)$ versus r/Λ for several values of r_0/Λ. The curves correspond to Eq. (13.38) while the solid circles represent Eq. (13.37). The two sets of results are nearly identical, except in the vicinity of $r = 0$.

ϕ, the azimuth of a point in the XY-plane, is measured counterclockwise from the positive X-axis.

There are four contributions to the energy of a reverse-magnetized domain: the external field energy, the anisotropy energy, the exchange energy, and the demagnetizing energy.[1] In what follows we calculate the contributions of these sources to the total energy when a domain of radius r_0 and wall parameter Λ satisfying Eqs. (13.38) and (13.39) is created. The initial magnetization is assumed uniform through the thickness, with perpendicular orientation and circular symmetry, namely, $\mathbf{M}(r,\phi,z) = M(r)\,\hat{\mathbf{z}}$. The presence of the domain will change the distribution, creating the following components of magnetization:

$$M_x(r,\phi) = -M(r)\sin\theta(r)\sin\phi = M(r)\,\xi_x(r,\phi) \tag{13.40a}$$

$$M_y(r,\phi) = M(r)\sin\theta(r)\cos\phi = M(r)\,\xi_y(r,\phi) \tag{13.40b}$$

$$M_z(r) = M(r)\cos\theta(r) = M(r)\,\xi_z(r,\phi) \tag{13.40c}$$

In these equations ξ_x, ξ_y, ξ_z are the direction cosines of $\mathbf{M}(r)$.

13.3.1. External field energy

Let a uniform magnetic field $\mathbf{B}_0$ be applied in the Z-direction. In the absence of domains the external field energy is

$$\mathcal{E}_{\text{ext}} = -\int_0^\infty 2\pi r\tau B_0 M(r)\,\mathrm{d}r$$

In contrast, when a domain is present, we have:

$$\mathcal{E}_{\text{ext}} = -\int_0^\infty 2\pi r\tau B_0 M(r)\cos\theta(r)\,\mathrm{d}r$$

The energy difference is therefore given by

$$\Delta\mathcal{E}_{\text{ext}} = 2\pi\Lambda^2\tau B_0 \int_0^\infty M(\Lambda\rho)\, f_{\text{ext}}(\rho;\rho_0)\,\mathrm{d}\rho \tag{13.41a}$$

where

$$f_{\text{ext}}(\rho;\rho_0) = \rho\left\{1 - \tanh\left[\frac{\rho^3 - \rho_0^{\,3}}{3\rho_0\rho}\right]\right\} \tag{13.41b}$$

and $\rho_0 = r_0/\Lambda$. Figure 13.15 shows plots of $f_{\text{ext}}(\rho;\rho_0)$ versus ρ for various values of ρ_0. When $M(r) = M_s$ and $r_0 >> \Lambda$, Eq. (13.41) closely approximates the relation $\Delta\mathcal{E}_{\text{ext}} = 2\pi r_0^{\,2}\tau B_0 M_s$.

13.3.2. Anisotropy energy

In the presence of uniaxial magnetic anisotropy with easy axis along Z, the deviation of $\mathbf{M}$ from the Z-axis gives rise to an excess magnetic energy. Assuming that the anisotropy energy density constant is $K_u(r)$, the increase in energy will be

$$\Delta\mathcal{E}_{\text{anis}} = \int_0^\infty 2\pi r\tau K_u(r)\sin^2\theta(r)\,\mathrm{d}r$$

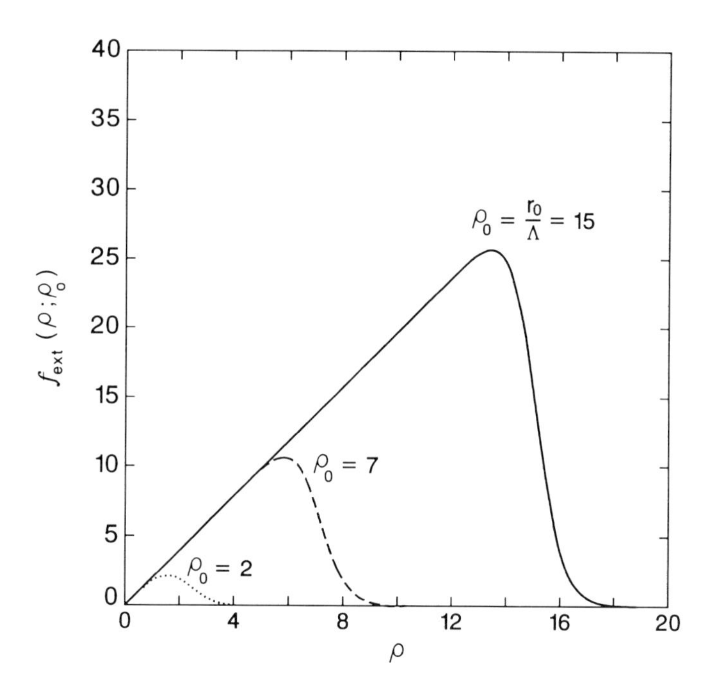

Figure 13.15. Plots of $f_{\text{ext}}(\rho;\ \rho_0)$ versus ρ for several values of $\rho_0 = r_0/\Lambda$. The function is described by Eq. (13.41b).

or, equivalently,

$$\Delta\mathscr{E}_{\text{anis}} = 2\pi r_0 \tau \Lambda \int_0^{\infty} K_u(\Lambda\rho)\, f_{\text{anis}}(\rho;\rho_0)\, \mathrm{d}\rho \tag{13.42a}$$

where

$$f_{\text{anis}}(\rho;\rho_0) = \frac{\rho}{\rho_0}\left\{1 - \tanh^2\left[\frac{\rho^3 - \rho_0{}^3}{3\rho_0\rho}\right]\right\} \tag{13.42b}$$

Plots of $f_{\text{anis}}(\rho;\rho_0)$ versus ρ for various values of $\rho_0 = r_0/\Lambda$ are shown in

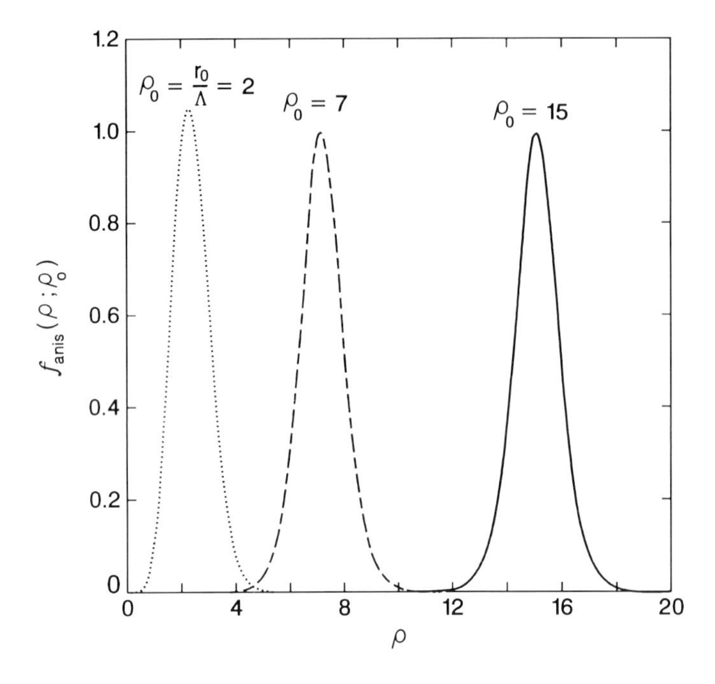

Figure 13.16. Plots of $f_{anis}(\rho;\ \rho_0)$ versus ρ for several values of $\rho_0 = r_0/\Lambda$. The function is described by Eq. (13.42b).

Fig. 13.16. When $K_u(r) = K_u$ and $r_0 >> \Lambda$, Eq. (13.42) closely approximates the relation $\Delta\mathscr{E}_{anis} = 4\pi r_0 \tau \Lambda K_u$, which would have obtained if the curvature of the wall had been neglected.

13.3.3. Exchange energy

The appearance of a domain in a magnetic medium is accompanied by an increase in the free energy associated with the variation of the direction of magnetization. The energy difference can be expressed as follows:

$$\Delta\mathscr{E}_{xhg} = \int_0^\infty 2\pi r \tau A_x(r) \left[(\nabla \xi_x)^2 + (\nabla \xi_y)^2 + (\nabla \xi_z)^2 \right] \mathrm{d}r \qquad (13.43)$$

in which $A_x(r)$ is the stiffness coefficient, ∇ is the gradient operator, and

ξ_x, ξ_y, ξ_z are the direction cosines of $\mathbf{M}(r)$ given in Eq. (13.40). After algebraic manipulations, Eq. (13.43) yields

$$\Delta\mathcal{E}_{\text{xhg}} = 2\pi r_0(\tau/\Lambda) \int_0^\infty A_x(\Lambda\rho)\, f_{\text{xhg}}(\rho;\rho_0)\, d\rho \tag{13.44a}$$

where

$$f_{\text{xhg}}(\rho;\rho_0) = \frac{1}{\rho_0\rho}\left\{1 + \left[\frac{2\rho^3 + \rho_0^3}{3\rho_0\rho}\right]^2\right\}\left\{1 - \tanh^2\left[\frac{\rho^3 - \rho_0^3}{3\rho_0\rho}\right]\right\} \tag{13.44b}$$

$f_{\text{xhg}}(\rho;\rho_0)$ is plotted for several values of $\rho_0 = r_0/\Lambda$ in Fig. 13.17. Notice that the function approaches zero as $\rho \to 0$ for all values of ρ_0. This is due to the asymptotic behavior of the wall profile in the neighborhood of the domain center, and is the primary reason why Eq. (13.38) was chosen over Eq. (13.37) to represent the domain. When $A_x(r) = A_x$ and $r_0 \gg \Lambda$, Eq. (13.44) closely approximates the relation $\Delta\mathcal{E}_{\text{xhg}} = 4\pi r_0(\tau/\Lambda)A_x$, which would have obtained if the curvature of the wall had been neglected.

13.3.4. Demagnetizing energy

The general formula for the demagnetizing energy is: [1]

$$\mathcal{E}_{\text{dmag}} = \frac{1}{2}\int_V\int_{V'} \frac{[\nabla\cdot\mathbf{M}(\mathbf{r})]\,[\nabla\cdot\mathbf{M}(\mathbf{r}')]}{|\mathbf{r}-\mathbf{r}'|}\, dV\, dV' \tag{13.45}$$

where V is the region containing the magnetic material, $\nabla\cdot$ is the divergence operator, and

$$|\mathbf{r}-\mathbf{r}'| = \sqrt{r^2 + r'^2 - 2rr'\cos(\phi-\phi') + (z-z')^2}$$

is the distance between two points located at $\mathbf{r}$ and $\mathbf{r}'$. Let the magnetization have the following (circularly symmetric) pattern:

$$\mathbf{M}(r,\phi,z) = M(r)\left[\sin\theta(r)\,\hat{\mathbf{a}}_\phi + \cos\theta(r)\,\hat{\mathbf{a}}_z\right] U(z) \tag{13.46}$$

Here $M(r)$ is the saturation moment at a distance r from the center, and $U(z)$ is defined in Eq. (13.13). Then:

$$\nabla \cdot \mathbf{M} = M(r) \cos \theta(r) \left[\delta(z + \tfrac{1}{2}\tau) - \delta(z - \tfrac{1}{2}\tau) \right] \tag{13.47}$$

$\delta(z)$ is Dirac's delta function. Equation (13.47) may now be placed in Eq. (13.45) to yield

$$\mathscr{E}_{\text{dmag}} = 16\pi \int_{r=0}^{\infty} \int_{r'=0}^{r} M(r)\, M(r') \cos \theta(r) \cos \theta(r') \times \left\{ r' \mathscr{K}(r'/r) - \frac{r r'}{\sqrt{(r + r')^2 + \tau^2}} \mathscr{K}\left[\sqrt{\frac{4 r r'}{(r + r')^2 + \tau^2}} \right] \right\} \mathrm{d}r\, \mathrm{d}r' \tag{13.48}$$

Here $\mathscr{K}(\cdot)$ is the complete elliptic integral of the first kind, namely,

$$\mathscr{K}(\chi) = \int_0^{\frac{1}{2}\pi} \frac{1}{\sqrt{1 - \chi^2 \sin^2 \theta}} \mathrm{d}\theta$$

To examine Eq. (13.48), we confine our attention to the case where the magnitude of $\mathbf{M}$ is constant, i.e., $M(r) = M_s$. Then:

$$\mathscr{E}_{\text{dmag}} = 2\pi M_s^2 \int_{r=0}^{\infty} 2\pi r \tau \int_{r'=0}^{r} \cos \theta(r) \cos \theta(r')\, g(r, r'; \tau)\, \mathrm{d}r'\, \mathrm{d}r \tag{13.49a}$$

where

$$g(r, r'; \tau) = \frac{4}{\pi \tau} \left\{ \frac{r'}{r} \mathscr{K}(r'/r) - \frac{r'}{\sqrt{(r + r')^2 + \tau^2}} \mathscr{K}\left(\sqrt{\frac{4 r r'}{(r + r')^2 + \tau^2}} \right) \right\} \tag{13.49b}$$

In the absence of domains, when $\cos \theta(r) = \cos \theta(r') = 1$, the above equation becomes

$$\mathscr{E}_{\text{dmag}} = 2\pi M_s^2 \int_{r=0}^{\infty} 2\pi r \tau\, f(r/\tau)\, \mathrm{d}r \tag{13.50a}$$

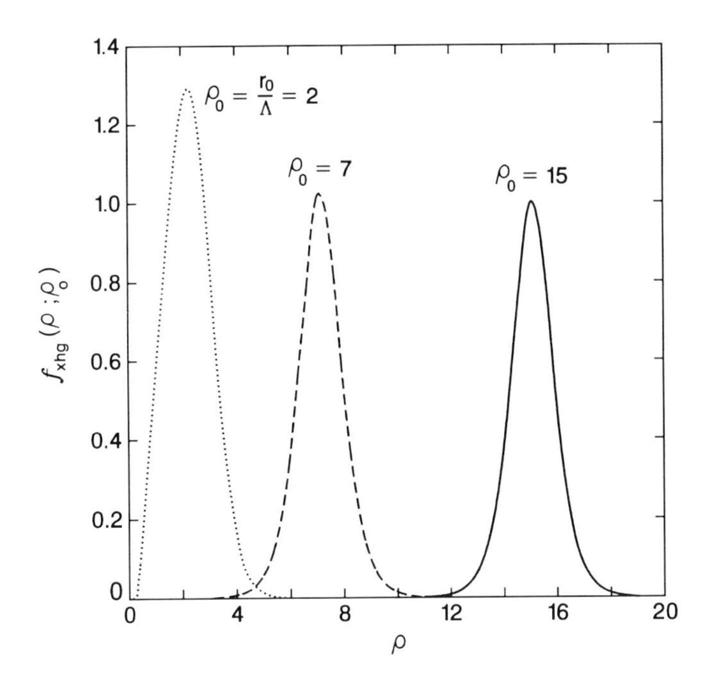

Figure 13.17. Plots of $f_{xhg}(\rho;\ \rho_0)$ versus ρ for several values of $\rho_0 = r_0/\Lambda$. The function is described by Eq. (13.44b).

where

$$\mathcal{f}(r/\tau) = \int_{r'=0}^{r} \mathcal{g}(r, r'; \tau)\ \mathrm{d}r'$$

$$= \frac{4r}{\pi\tau}\int_0^1 \left\{ \mathcal{K}(x) - \frac{\mathcal{K}\left(\sqrt{\dfrac{4x}{(1+x)^2 + (\tau/r)^2}}\right)}{\sqrt{(1+x)^2 + (\tau/r)^2}} \right\} x\mathrm{d}x \qquad (13.50\mathrm{b})$$

The function $\mathcal{f}(\cdot)$ may be computed numerically. The plot of $\mathcal{f}(r/\tau)$ in Fig. 13.18(a) shows that it approaches unity when $r >> \tau$.

With a reverse domain present, $\cos\theta(r)$ will be given by Eq. (13.38), and Eq. (13.49) may be written:

$$\mathscr{E}_{\text{dmag}} = 2\pi M_s^2 \int_{r=0}^{\infty} 2\pi r \tau F\left(\frac{r}{\tau}, \frac{r_0}{\tau}; \frac{\Lambda}{\tau}\right) \mathrm{d}r \tag{13.51a}$$

where

$$F(\rho, \rho_0; \hat{\Lambda}) = \frac{4\rho}{\pi} \tanh\left(\frac{\rho^3 - \rho_0^3}{3\hat{\Lambda}\rho_0\rho}\right) \int_0^1 \tanh\left(\frac{\rho^3 x^3 - \rho_0^3}{3\hat{\Lambda}\rho_0\rho x}\right)$$

$$\times \left\{ \mathscr{K}(x) - \frac{\mathscr{K}\left(\sqrt{\dfrac{4x}{(1+x)^2 + \rho^{-2}}}\right)}{\sqrt{(1+x)^2 + \rho^{-2}}} \right\} x \mathrm{d}x \tag{13.51b}$$

Figure 13.18(b) is a plot of $F(\rho, \rho_0; \hat{\Lambda})$ versus ρ for fixed values of ρ_0 and $\hat{\Lambda}$. It is observed that the presence of a wall at $r = r_0$ reduces the energy of demagnetization principally in the neighborhood of r_0; this is due to a reduced demagnetizing field in the region of the wall. For other values of ρ_0 and $\hat{\Lambda}$ the behavior of $F(\rho, \rho_0; \hat{\Lambda})$ remains qualitatively the same. The increasing of $\hat{\Lambda}$ makes the notch around ρ_0 wider, suggesting that a wide wall would reduce the demagnetizing energy more than a narrow wall.

The reduction of $\mathscr{E}_{\text{dmag}}$, caused by the creation of a circular domain of radius r_0 and wall-width parameter Λ, can now be obtained by subtracting Eq. (13.51) from Eq. (13.50). Thus

$$\boxed{\Delta\mathscr{E}_{\text{dmag}} = -2\pi M_s^2 (2\pi r_0 \tau^2)\, G\left(\frac{r_0}{\tau}; \frac{\Lambda}{\tau}\right)} \tag{13.52a}$$

where

$$\boxed{G(\rho_0; \hat{\Lambda}) = \int_0^{\infty} \frac{\rho}{\rho_0} \left[\mathscr{F}(\rho) - F(\rho, \rho_0; \hat{\Lambda})\right] \mathrm{d}\rho} \tag{13.52b}$$

A plot of $G(\rho_0; \hat{\Lambda})$ versus ρ_0 for a fixed value of $\hat{\Lambda}$ is shown in Fig. 13.19(a); the same general behavior can be expected for other values of $\hat{\Lambda}$ as well. One observes that the function increases logarithmically with ρ_0 when $\rho_0 >> 1$. This logarithmic increase, which is due to the long-range

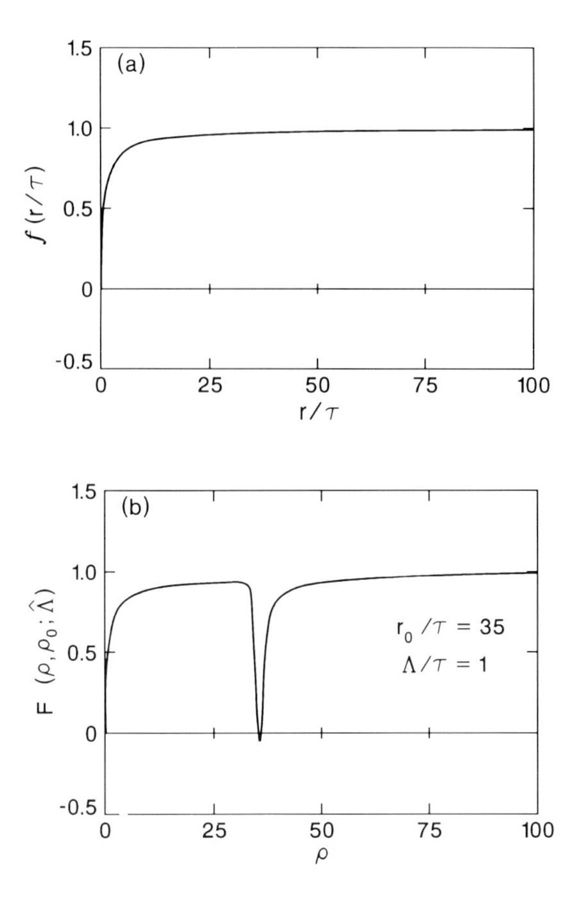

Figure 13.18. (a) Plot of $\mathcal{f}(r/\tau)$ versus r/τ. The function is described by Eq. (13.50b). (b) Plot of $F(\rho, \rho_0; \hat{\Lambda})$ versus ρ for $\rho_0 = 35$ and $\hat{\Lambda} = 1$. The function is described by Eq. (13.51b).

nature of dipole–dipole interactions, shows that demagnetizing effects cannot be confined to the vicinity of the wall.† Nonetheless, if one happens to be interested in a fairly narrow range of values of ρ_0 (say, $10 \leq \rho_0 \leq 30$) then $G(\rho_0; \hat{\Lambda})$ may be treated as a constant, independent of ρ_0. The value of this constant, of course, will depend on the normalized wall-width parameter Λ/τ. Figure 13.19(b) shows a plot of $G(\rho_0; \hat{\Lambda})$ versus $\hat{\Lambda}$ for $\rho_0 = 20$. Note, in particular, that as $\hat{\Lambda} \to 0$ the function approaches the

†We observed similar behavior previously, in conjunction with straight domain walls; see subsection 13.2.3, Example 2.

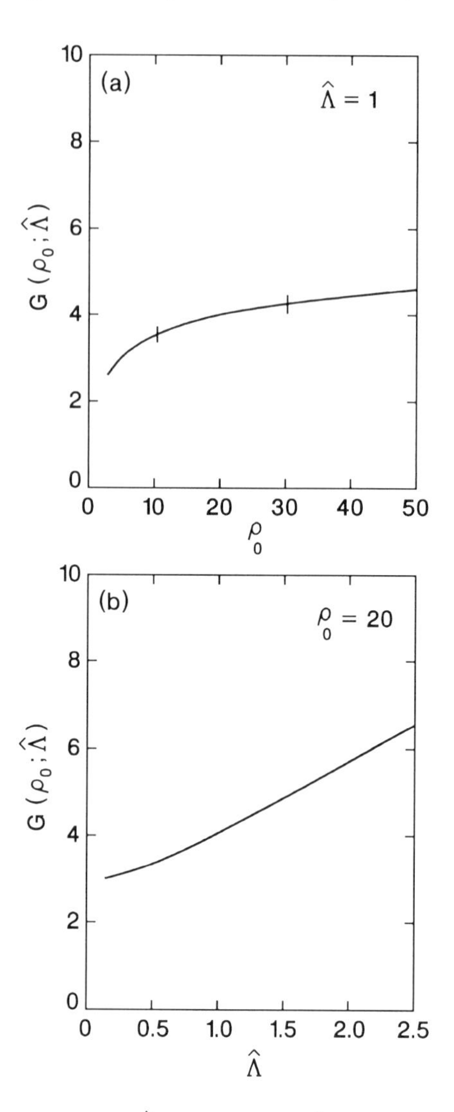

Figure 13.19. (a) Plot of $G(\rho_0;\ \hat{\Lambda})$ versus ρ_0 for $\hat{\Lambda} = 1$. (b) Plot of $G(\rho_0;\ \hat{\Lambda})$ versus $\hat{\Lambda}$ for $\rho_0 = 20$. The function is described by Eq. (13.52b).

constant value $\simeq 3$. Therefore, for a narrow domain wall (i.e., $\Lambda << \tau$) and a reasonable domain size (say, $10\tau \leq r_0 \leq 30\tau$), a good estimate of $\Delta\mathscr{E}_{\text{dmag}}$ *per unit wall area* may be obtained from Eq. (13.52a) as follows:

$$\sigma_w^{(\text{dmag})} = \frac{1}{2\pi r_0 \tau} \Delta\mathscr{E}_{\text{dmag}} \simeq -6\pi\tau M_s^2 \tag{13.53}$$

$\sigma_w^{(\text{dmag})}$ is the quantity one must compare with the wall energy density σ_w in Eq. (13.10) in order to determine whether the demagnetizing energy alone can compensate for the exchange and anisotropy energies of the wall.

13.4. A Technique for Measuring the Energy Density of Domain Walls

We shall describe a method for obtaining the wall energy density σ_w from experimental measurements.[15] The technique requires measurement of the expansion and collapse fields for small circular domains, as shown in Fig. 13.20. σ_w is an important characteristic of MO media; according to the micromagnetic theory of section 13.1 it is equal to $4\sqrt{A_x K_u}$, where A_x and K_u are the stiffness coefficient and the anisotropy constant of the material respectively. The micromagnetic model is based on the following assumptions. (i) The anisotropy is uniaxial and its energy density is $K_u \sin^2\theta$. (ii) The exchange mechanism is isotropic and confined to nearest neighbor spins. (iii) The wall is simple, containing no Bloch lines or Bloch points. None of these assumptions, of course, is strictly valid for MO media, especially those media that contain non-S-state rare earth elements such as terbium and dysprosium: the Tb and Dy are responsible for random-axis anisotropy, which helps create complex wall structures. In the absence of precise theoretical models for the wall, and in view of the fact that reliable measurements of A_x are usually lacking, direct methods of measurement of σ_w have gained significance.†

The measurements are performed on small, thermomagnetically recorded domains. Consider a circular domain of radius R in a uniformly magnetized film of thickness τ and saturation moment M_s. Using any one of a number of techniques for domain observation, it is possible to determine the value of an external field, B_1, that initiates the expansion of the domain. Similarly one can determine the value of the field, B_2, at the onset of collapse. The effective force†† per unit wall area due to a perpendicular external field B_0 is given by

$$F_{\text{ext}} \simeq \pm 2B_0 M_s \tag{13.54}$$

† A comparison with the media of magnetic bubble storage is worthwhile here. In thin-film bubble media, the domain wall energy is determined by allowing the sample to break into stripe domains. As discussed in Example 2, subsection 13.2.3, an estimate of σ_w may be obtained from the observed pattern of stripes. In the absence of an external field, the equilibrium state of the sample is one in which the reduction of energy by demagnetization compensates for the increased energy caused by the creation of domain walls. The reduction of demagnetizing energy, calculated for given saturation moment M_s, film thickness τ, and stripe width ℓ, is set equal to the increase in wall energy, from which the value of σ_w is determined. This approach cannot be applied to MO media: unlike bubble materials, which have negligible coercivity, the large coercivity of amorphous RE-TM films prevents the magnetization state from reaching equilibrium. Without perfect balance between the two kinds of energy, it is impossible to estimate one energy from the other.

†† The effective force on the wall is the derivative of energy with respect to the wall displacement, i.e., $\Delta\mathscr{E} = F\Delta R$. The force may be normalized by $2\pi R\tau$ to yield force per unit wall area. The force is "expansive" when the energy declines with increasing domain radius, otherwise it is "contractive".

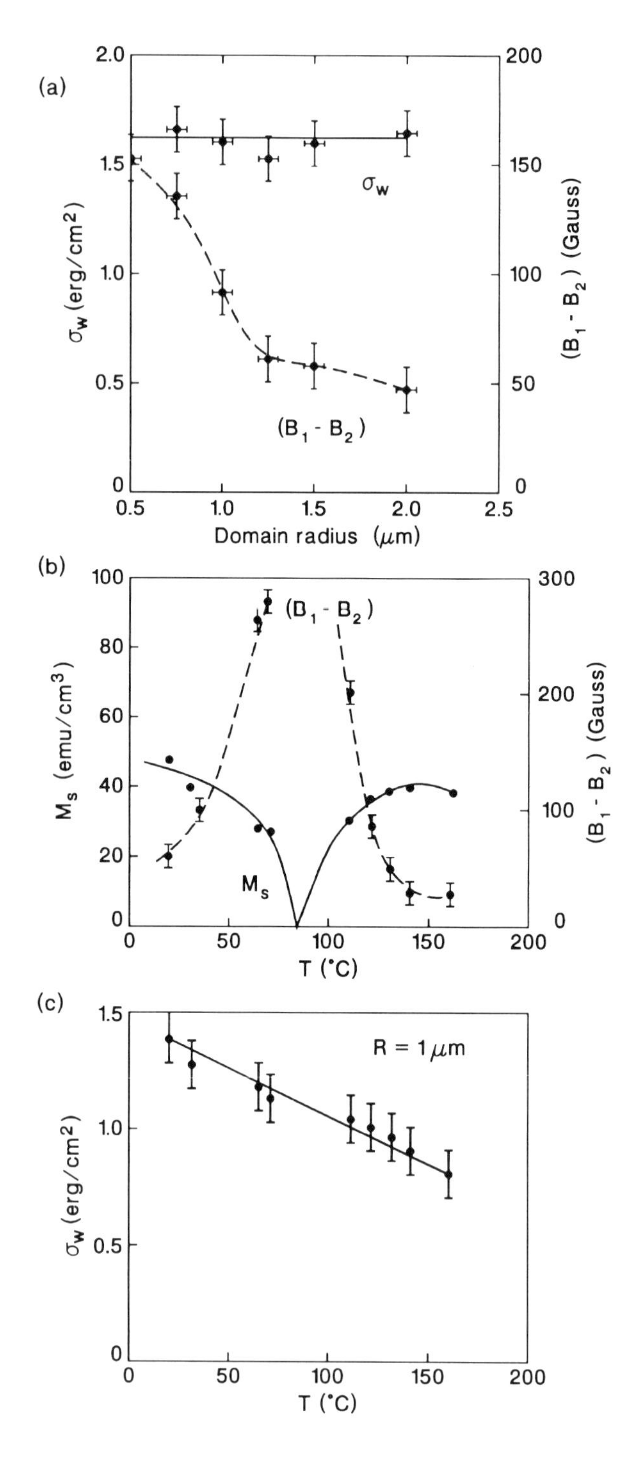

Figure 13.21. Measurement of σ_w in a $Gd_{13}Tb_{16}Fe_{49}Co_{18}Ar_4$ sample. (a) Measured values of $B_1 - B_2$ and the derived values of σ_w for domains of different radii at $T = 295$ K. (b) Temperature-dependence of M_s and $B_1 - B_2$; for the latter measurement the domain radius was 1 μm. (c) Temperature-dependence of σ_w, obtained from the data in (b) with the aid of Eq. (13.58).

Problems

(13.1) The equation of state for a straight wall in a thin magnetic film with perpendicular anisotropy, Eq. (13.5), is based on the assumption that the polar angle θ is a function of x only, while the azimuthal angle ϕ is constant (see Fig. 13.4). Generalize this equation by allowing ϕ to depend on y. Solve the resulting equation and determine the structure and energy density of walls whose azimuthal angle varies along the length of the wall.

(13.2) Figure 13.22 shows an exchange-coupled magnetic bilayer, where both layers have perpendicular magnetic anisotropy. The top layer has saturation magnetization M_1, exchange stiffness coefficient A_1, and uniaxial anisotropy constant K_1. The corresponding parameters for the bottom layer are M_2, A_2, and K_2. When the two layers are oppositely magnetized a domain wall forms at their interface, provided, of course, that they are thick enough to sustain the wall. Ignoring the demagnetizing effects and assuming a simple wall, determine the structure and energy density of this interfacial wall.

(13.3) Derive Eq. (13.15) from Eq. (13.12). (Note: The operator $\mathscr{F}_3$ is defined in Eq. (13.14b).)

(13.4) Verify the Fourier transform relation in Eq. (13.17).

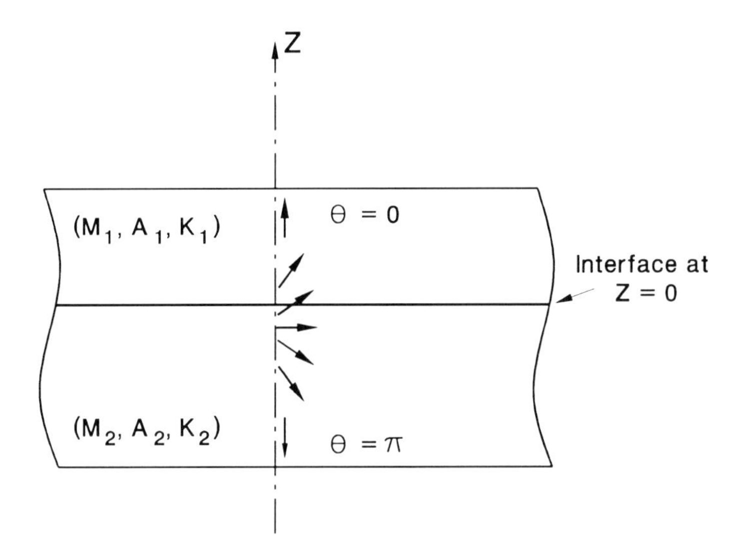

Figure 13.22.

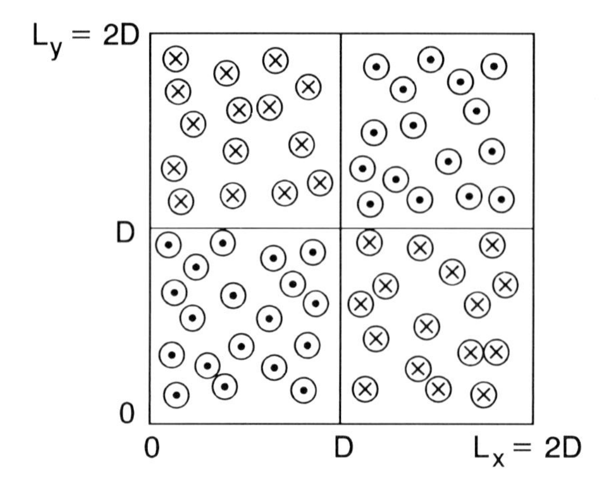

Figure 13.23.

(13.5) With reference to Fig. 13.9(a) let a uniform magnetic field H_{ext} be applied perpendicular to the plane of the film. Knowing that the external field encourages the expansion of stripes aligned with the field at the expense of those that are not, determine the optimum spacing of the stripes for a given strength of the applied field.

(13.6) Figure 13.13 shows a perpendicularly magnetized film of thickness τ and uniform magnetization M_s, containing a circular domain of radius r. Using the expression for $\mathscr{E}_{dmag}$ in Eq. (13.24) determine the demagnetizing energy of this domain. (You may ignore the width of the wall, assuming in effect that the transition from "down" magnetization inside the domain to "up" magnetization outside is infinitely sharp.)

(13.7) A perpendicular film with saturation moment M_s and thickness τ has a checker-board pattern of magnetization, that is, square regions of side D are alternatively "up" and "down" magnetized (see Fig. 13.23). Ignoring the width of the walls and using the expression for $\mathscr{E}_{dmag}$ in Eq. (13.24), determine the demagnetizing energy of this film.

14
Mean-field Analysis of Amorphous Rare Earth – Transition Metal Alloys

Introduction

In Chapter 12 we described the mean-field theory of magnetization for ferromagnetic materials, which consist of only one type of magnetic species. The atoms (or ions) comprising a simple ferromagnet are coupled via exchange interactions to their neighbors, and the sign of the exchange integral $\mathscr{J}$ is positive everywhere. In this chapter we develop the mean-field model of magnetization for amorphous ferrimagnetic materials, of which the rare earth–transition metal (RE–TM) alloys are the media of choice for thermomagnetic recording applications.[1–6] Ferrimagnets are composed of at least two types of magnetic species; while the exchange integral for some pairs of ions is positive, there are other pairs for which the integral is negative. This leads to the formation of two or more subnetworks† of magnetic species. When the ferrimagnet has a uniform temperature, each of its subnetworks will be uniformly magnetized, but the direction of magnetization will vary among the subnetworks. In simple ferrimagnets consisting of only two subnetworks, the magnetization directions of the two are antiparallel.

In thin-film form, amorphous RE–TM alloys exhibit perpendicular magnetic anisotropy, which makes them particularly useful for polar Kerr (or Faraday) effect readout. Being ferrimagnetic, they possess a compensation point temperature, T_{comp}, at which the net moment of the material becomes zero. T_{comp} can be brought to the vicinity of the ambient temperature by proper choice of composition. This feature preserves uniform magnetic alignment in the perpendicular direction by preventing the magnetization from breaking up into domains. Moreover, high coercivity around the compensation point protects the recorded data from stray magnetic fields and unwanted erasures. The amorphous nature of the films eliminates a significant source of noise previously encountered in polycrystalline media: surface roughness and grain-boundary noise are no longer limiting factors in the readout performance of RE–TM alloys.[7]

When the various subnetworks of a ferrimagnet are fairly strongly coupled to each other, they all will have the same Curie temperature T_c; that is, the phase transition between the magnetically-ordered and

†Strictly speaking, crystalline ferrimagnets consist of sublattices, whereas disordered or amorphous ferrimagnets contain subnetworks. Since the emphasis in this chapter is on amorphous media, we shall use the term "subnetwork" throughout.

paramagnetic states occurs at the same temperature for all the subnetworks. Despite the uniqueness of the Curie point, different subnetworks exhibit different magnetization-versus-temperature behavior, leading to complex behavior of the net magnetization M_s as T varies in the range from 0 K to T_c. The first step in the study of thermomagnetic recording and erasure is the development of a model that can explain $M_s(T)$ in RE–TM alloys.[8,9] Mean-field theory provides a simple solution to this problem, although its usefulness has been marred in the past by the existence of too many adjustable parameters.[10–15] The goal of the present chapter is to develop a mean-field model for amorphous RE–TM alloys that can explain the available data with as few adjustables as possible.

Typically, in a mean-field model there exist a number of adjustable parameters, notably the exchange integrals $\mathcal{J}_{ij}$ between pairs of atoms (or ions) belonging to subnetworks i and j. The numerical values of these parameters are determined by matching the predictions of the model to the experimental data. In the case of ferrimagnets, the proliferation of parameters with an increasing number of subnetworks can complicate the task of matching theory and experiment, and result in multiple values for each of the parameters. This undesirable state of affairs must be avoided, first by a judicious selection of the model and, second, through an effort to obtain sufficient experimental data points. One significant feature of the model presented here is the emphasis placed on reducing the number of parameters. The other feature is that we allow two separate iron subnetworks, one ferromagnetic and the other antiferromagnetic, to coexist. The model, therefore, takes into consideration the large variation of $\mathcal{J}_{\text{Fe-Fe}}$ with interatomic distance, a property that has long been the subject of speculation.[14] The presence of both ferromagnetic and antiferromagnetic Fe–Fe exchange gives rise to a spread in the magnetic moment of iron at $T \neq 0$ K, a feature that might be amenable to experimental verification.

Throughout the chapter examples are given in which the available experimental data on binary GdCo, GdFe, and TbFe alloys are compared with the mean-field calculations; with a limited number of adjustable parameters, good agreement has been obtained in all cases. Also derived in this chapter is an expression for the exchange stiffness coefficient A_x of amorphous ferrimagnets. A formula that relates the anisotropy constant K_u to the subnetwork magnetizations M_{RE} and M_{TM} is suggested, and, through comparisons with available data, its adjustable parameters are fixed. The temperature dependences of A_x and K_u thus obtained are used to study the characteristics of domain walls as functions of temperature.

The organization of the chapter is as follows. In section 14.1 we define the parameters and derive the equations of the mean-field theory of ferrimagnets. Section 14.2 is concerned with explaining the observed $M_s(T)$ behavior in several amorphous binary alloys. In section 14.3 the effect of uniaxial single-ion anisotropy on the mean-field model is explored. Sections 14.4 and 14.5 are devoted to the exchange stiffness coefficient A_x and the anisotropy energy constant K_u respectively. Comparison of the three binary alloy systems GdCo, GdFe, and TbFe in terms of their

Table 14.1.

N	total number of atoms (or ions) per unit volume ($1/cm^3$)
x_n	atomic percentage of the nth species in the alloy
r_n	atomic radius
Z_n	coordination number (i.e., average number of nearest neighbors)
S_n	spin angular momentum quantum number
L_n	orbital angular momentum quantum number
J_n	total angular momentum quantum number
g_n	spectroscopic splitting factor (g-factor)
$\mathcal{J}_{mn}$	exchange integral between ions of species m and n (erg)
M_n	saturation magnetization of the nth subnetwork (emu/cm^3)
M_s	total saturation magnetization (emu/cm^3)
T	absolute temperature (K)
T_c	Curie point temperature (K)
T_{comp}	compensation point temperature (K)
k_B	Boltzmann's constant (1.38×10^{-16} erg/K)
μ_B	Bohr magneton (9.27×10^{-21} emu).

domain wall characteristics is the subject of section 14.6. Final remarks and conclusions are contained in section 14.7.

14.1. The Mean-field Model

We describe a mean-field model of magnetization for an amorphous system with three magnetic subnetworks. Non-magnetic elements are also included in this model insofar as they affect the densities and the coordination numbers of the magnetic elements. The notation shown in Table 14.1 will be used throughout the chapter. In this notation the subscript n refers to the nth subnetwork. We shall generally assume that $n = 1$ for the rare earth subnetwork, $n = 2$ for the transition metal subnetwork, $n = 3$ for a third magnetic element, and $n = 4, 5$ for non-magnetic elements.

The atoms (ions) are assumed to be hard spheres and their radii r_n are calculated from a table of atomic concentrations[16] under the assumption that the atoms fill the entire space. The approximate values of r_n for materials of interest are shown in Table 14.2. Since amorphous materials are usually less dense than their crystalline counterparts, it is reasonable to assume that only 95% of the space is filled in the amorphous state.[15] The total number of atoms per unit volume, N, is thus given by

Table 14.2. Metallic radii, angular momenta, and g-factors of the elements used in the mean-field calculations. The values of angular momenta for the rare earth elements correspond to their free ions

	r (Å)	L	S	J	g
Gd	2.0	0	3.5	3.5	2.0
Tb	2.0	3.0	3.0	6.0	1.5
Co	1.4	0	adjustable		2.0
Fe	1.4	0	adjustable		2.0
Ar	2.0	--	--	--	--
B	1.2	--	--	--	--
Mo	1.5	--	--	--	--
Sn	2.0	--	--	--	--

$$N = \frac{0.95}{\sum_{n=1}^{5} \left(\frac{4}{3}\pi r_n{}^3\right) x_n} \tag{14.1}$$

The atomic density of the nth species is then equal to Nx_n.

In amorphous materials the coordination numbers Z_n are not constants but vary from site to site. For purposes of the mean-field theory, however, it suffices to have average values. Traditionally, researchers have assumed a model based on the random dense packing of hard spheres, and assigned Z the fixed value of 12.[10,11] This is inappropriate when the radii of the constituting elements differ substantially. To account for the dependence of Z_n on composition and on atomic radii, we consider a sphere of radius r_n in contact with another sphere of radius r_m, as shown in Fig. 14.1. Looking from the center of the first sphere, the spatial angle subtended by the second sphere is $4\pi \sin^2(\frac{1}{2}\phi_{mn})$, where $\phi_{mn} = \arcsin\,[r_m/(r_m + r_n)]$. If non-overlapping cones are assumed, the average number of atoms of species m that surround a given atom of species n will be equal to $Z_n x_m$, and together they will cover a fraction of space equal to $Z_n x_m \sin^2(\frac{1}{2}\phi_{mn})$. If it is further assumed that the entire space is filled with the nearest neighbor cones (an assumption only approximately valid in three-dimensional space), we obtain

$$\sum_{m=1}^{5} Z_n x_m \sin^2(\tfrac{1}{2}\phi_{mn}) = 1 \tag{14.2}$$

from which Z_n is readily calculated (see Problem 14.1 for a discussion of

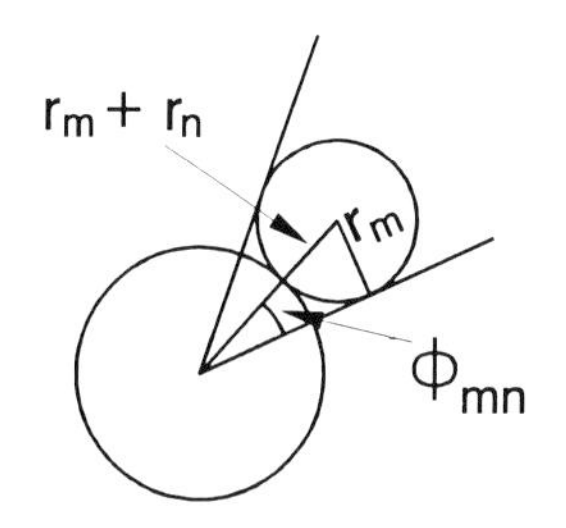

$$\phi_{mn} = \arcsin\left(\frac{r_m}{r_m + r_n}\right)$$

Figure 14.1. The conical half-angle ϕ_{mn} subtended by a sphere of radius r_m as viewed from the center of an adjacent sphere of radius r_n.

the approximations involved). In the special case where all atoms are identical, Z from Eq. (14.2) turns out to be about 15, which is somewhat greater than the average coordination number, 12, in the random dense packing of hard spheres. However, since only relative values of Z_n are important in the mean-field model, the type of approximation embodied by Eq. (14.2) should be acceptable.

The spin, orbital, and total angular momentum quantum numbers for the materials of interest are listed in Table 14.2. The angular momenta of the rare earth elements are identical with their free-ion values. This is a reasonable approximation, considering that the 4f electrons, responsible for the magnetization of the RE ion, are well shielded by the 5s and 5p shells and are therefore largely unaffected by the environment in which the ion is embedded. The values of S, L, and J for Tb and Gd ions as obtained from the Hund rules are consistent with the measured values.[16]

The situation with the transition metal elements is quite different. Here the magnetic electrons are in the 3d shell, whose structure is significantly affected by the local environment. The magnetic properties therefore vary with the composition and atomic structure of the alloy.[17] The orbital momentum is usually quenched in these materials and the assumption $L = 0$ is quite reasonable. The 3d electrons occupy a band of energies split between electrons with up and down spins (3d$\uparrow$ and 3d$\downarrow$ bands). The difference between the populations of these bands determines the net spin of the TM ion. The spin can thus assume non-integer values; moreover, the band structure and the number of electrons available to each band vary with composition. The TM spin is thus a complicated function of the composition and structure of the alloy and, to simplify matters, we have used it as an adjustable parameter in the following calculations.

The band structure of cobalt is believed to be of the form shown in Fig. 14.2(a), with the Fermi energy $\mathcal{E}_F$ beyond the upper edge of the 3d$\uparrow$

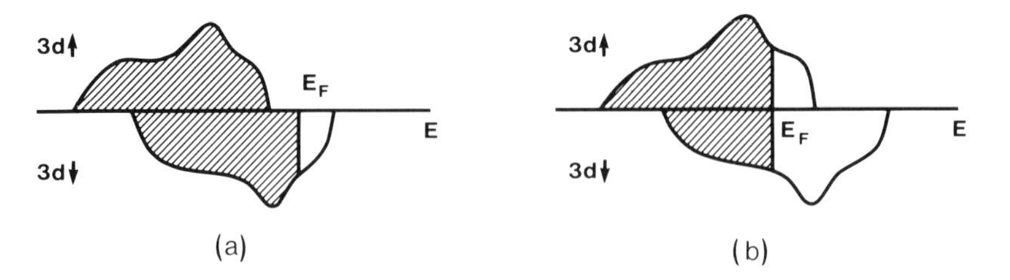

Figure 14.2. Schematic plots of the density of states versus energy for 3d electrons of the transition metal subnetwork. (a) Fermi energy above the upper edge of the 3d↑ band. (b) Fermi energy within both 3d↑ and 3d↓ bands.

band.[17] Assuming that alloying does not greatly modify this band structure, the addition of electrons can only fill the 3d↓ band and thereby reduce the net spin of individual cobalt ions. In contrast, the band structure of iron is believed to be of the form shown in Fig. 14.2(b). Here $\mathscr{E}_F$ is within both the 3d↑ and 3d↓ bands; depending on the exact structure of the bands, the addition of electrons could result in either an increase or a decrease of the net iron spin.[17] The band structure can be used qualitatively as a guide in adjusting the numerical value of the spin, although, in the absence of more elaborate information, its usefulness is quite limited.

The spectroscopic splitting factor g is a proportionality constant that relates the magnetic moment to the angular momentum. For pure orbital momenta $g = 1$, while for pure spin $g \simeq 2$. In general, the value of g is obtained from the Landé formula, Eq. (12.32c).

The exchange integral $\mathscr{J}$ is a quantum mechanical entity that arises from the overlap of electronic charge distributions. While in dielectric media the exchange interaction between neighboring atoms (ions) is direct, the interaction in metals is, to a certain extent, mediated by the conduction electrons (the RKKY interaction).[17] The magnitude and sign of $\mathscr{J}$ are, in general, functions of the electronic structure of the ions and the distance between them. For a pair of ions with angular momenta J_m and J_n, the mutual energy in the classical approximation is given by

$$E_{mn} = -2\mathscr{J}_{mn}\, j_m\, j_n$$

where j_m and j_n are the projections of the angular momenta along the axis of anisotropy. In the mean-field approximation, the exchange energy of ion m resulting from its interaction with ion n is

$$E_m = -\mathscr{J}_{mn}\, j_m \langle j_n \rangle$$

where $\langle j_n \rangle$ is the time average of j_n. The total exchange energy of ion m, arising from interactions with its nearest neighbors in our three-magnetic-subnetwork model, is thus given by

$$\mathcal{E}_m = -Z_m j_m \sum_{n=1}^{3} x_n \mathcal{J}_{mn} \langle j_n \rangle$$

Since j_m assumes the values $-J_m, -J_m + 1, \ldots, J_m - 1, J_m$ only, we have

$$\langle j_m \rangle = \frac{\displaystyle\sum_{j_m=-J_m}^{J_m} j_m \exp(-\mathcal{E}_m / k_B T)}{\displaystyle\sum_{j_m=-J_m}^{J_m} \exp(-\mathcal{E}_m / k_B T)}$$

or, equivalently,

$$\boxed{\langle j_m \rangle = \tfrac{1}{2}(2J_m + 1) \coth \left\{ \frac{Z_m (2J_m + 1)}{2k_B T} \sum_{n=1}^{3} x_n \mathcal{J}_{mn} \langle j_n \rangle \right\} - \tfrac{1}{2}\coth \left\{ \frac{Z_m}{2k_B T} \sum_{n=1}^{3} x_n \mathcal{J}_{mn} \langle j_n \rangle \right\}} \quad (14.3)$$

The three equations thus obtained for $m = 1, 2$, and 3 must be solved simultaneously to yield the values of $\langle j_n \rangle$ at any desired T. A simple way to tackle these equations is by way of a numerical algorithm that starts with reasonable initial estimates for $\langle j_n \rangle$ and iterates until a consistent solution is obtained. In practice, this numerical approach converges quickly and yields reliable solutions. The subnetwork magnetizations are then obtained from the following relation:

$$\boxed{M_n = N x_n \mu_B g_n \langle j_n \rangle} \quad (14.4)$$

14.1.1. Computing the Curie temperature

The Curie temperature T_c may be found analytically if one realizes that around T_c the values of $\langle j_n \rangle$ are small, thus allowing the following approximation to be used in Eq. (14.3):

$$\coth(x) \simeq \frac{1}{x} + \frac{x}{3} \qquad |x| << 1$$

In the vicinity of T_c, therefore, one writes

$$\langle j_m \rangle = \frac{Z_m J_m (J_m + 1)}{3k_B T} \sum_{n=1}^{3} x_n \mathscr{J}_{mn} \langle j_n \rangle$$

which, in matrix notation, is equivalent to

$$\begin{bmatrix} a_{11} & a_{12} & a_{13} \\ a_{21} & a_{22} & a_{23} \\ a_{31} & a_{32} & a_{33} \end{bmatrix} \begin{bmatrix} \langle j_1 \rangle \\ \langle j_2 \rangle \\ \langle j_3 \rangle \end{bmatrix} = T \begin{bmatrix} \langle j_1 \rangle \\ \langle j_2 \rangle \\ \langle j_3 \rangle \end{bmatrix}$$

where

$$a_{mn} = \frac{Z_m x_n \mathscr{J}_{mn} J_m (J_m + 1)}{3k_B}$$

Hence, T_c must be an eigenvalue of the matrix $[a_{mn}]$. It turns out, in fact, that T_c is always the largest real eigenvalue of the matrix. This method allows the calculation of T_c without the necessity of solving the mean-field equations for the entire range of temperatures.

14.2. Comparison with Experiment

The mean-field theory of section 14.1 is now employed to explain the experimentally observed behavior of $M_s(T)$ in several amorphous binary RE–TM alloys. Instead of trying to obtain a close match in every case by varying all the adjustable parameters, we shall try to obtain a reasonable match with as few adjustables as possible. This, we hope, will bring out the dominant trends and exclude the less significant factors.

We will study three classes of material for which experimental data are available in the published literature: GdCo, GdFe, and TbFe binary alloys. For each class, we use the fixed set of exchange integrals $\mathscr{J}_{mn}$ given in Table 14.3. It is true that the local environment and the interatomic distances play a role in determining the values of the exchange integrals, and it is also true that by changing the composition in a given material system, these factors are likely to change. We believe, however, that because of the nature of exchange in metallic alloys, such variations of the exchange parameters are of secondary importance. An exception is made for $\mathscr{J}_{\text{Fe-Fe}}$, which is apparently very sensitive to the interatomic distance. In fact, in certain compounds, the Fe–Fe exchange is known to be ferromagnetic for some iron pairs, and antiferromagnetic for others.[13] Thus for the RE–Fe alloys, we postulate the existence of two iron subnetworks,

Table 14.3. Numerical values of the exchange integrals used in the mean-field calculations

Material system	$\mathscr{J}_{\text{TM-TM}}$	$\mathscr{J}_{\text{RE-TM}}$	$\mathscr{J}_{\text{RE-RE}}$
GdCo	28.0×10^{-15}	-2.2×10^{-15}	0.5×10^{-15}
GdFe	$\pm 12.0 \times 10^{-15}$	-1.7×10^{-15}	0.5×10^{-15}
TbFe	$\pm 8.5 \times 10^{-15}$	-1.0×10^{-15}	0.2×10^{-15}

one with positive and the other with negative Fe–Fe exchange. (This is the meaning of the ± symbol in front of $\mathscr{J}_{\text{Fe-Fe}}$ in Table 14.3.) The coupling between the two subnetworks, however, remains ferromagnetic ($\mathscr{J}_{\text{Fe-Fe}}$ positive), and both subnetworks couple antiferromagnetically to the RE subnetwork with the same exchange parameter $\mathscr{J}_{\text{RE-Fe}}$. The only new parameter thus introduced is the fraction α of Fe in the antiferromagnetic subnetwork; α will be adjustable in these calculations.

As long as α is not too large, the effective local field on the antiferromagnetic iron subnetwork will remain parallel to, but smaller than, the local field on the ferromagnetic iron subnetwork. The two kinds of iron thus have parallel moments at all temperatures below T_c, but the moment of the antiferromagnetic kind quickly decays with rising temperature. This means that at $T \neq 0$ K the ferromagnetic Fe ions have a larger moment than the antiferromagnetic Fe ions. Figure 14.3 shows a typical arrangement of moments at temperatures below T_c.†

In Tables 14.4–14.6, we have compared the experimental data collected from the literature with the mean-field calculations. Information regarding the sources of data, sample compositions, preparation conditions, and measurement methods is also provided. In fitting the data, it was assumed that the nominal compositions were subject to a few per cent error; we thus searched the vicinity of the nominal composition for a good match. The best match was usually found within ±1% of the nominal composition. We also assumed a small amount of argon inclusion in the films in order to account for impurities, which are inevitably present in any sample. Since the measurement of M_s requires a precise knowledge of the sample thickness, systematic errors could be introduced if there is inaccuracy in the thickness measurement. Lack of instrument calibration is another source of systematic errors. In a few cases we allowed for the possibility of such systematic errors in the data. Figures 14.4 to 14.6 show typical plots of calculated $M_s(T)$ curves matched to the data points.

† At $T = 0$ K, the arrangement of the moments is the same as that in Fig. 14.3, but the two iron subnetworks have equal moments.

Table 14.4. Comparison between theoretical and experimental data for amorphous GdCo-based alloys. Argon is used in the theoretical (best match) compositions to represent impurities in the sample. The effect of non-magnetic impurities is only on the density and coordination numbers, and in that respect argon can be substituted for other contaminants. In reality, sputtered films contain a certain amount of argon while evaporated films are likely to be contaminated by other elements

	Nominal composition	Ref.	Deposition method	Composition analysis method	Magnetization measurement method	Theoretical (best match) composition	J_{Co}
1	$Gd_{21.5}Co_{78.5}$	1	sputtering	XRF	force balance	$(Gd_{22.5}Co_{77.5})_{95}Ar_5$	0.63
2	$Gd_{22.1}Co_{77.9}$	15	e-beam evap.	microprobe	VSM	$(Gd_{21.5}Co_{78.5})_{95}Ar_5$	0.63
3	$Gd_{18}Co_{75}B_6Ar_1$	13	sputtering	microprobe	force balance	$Gd_{15.5}Co_{77.5}B_6Ar_1$	0.53
4	$Gd_7Co_{74}Mo_{13}Ar_6$	19	sputtering	---	---	$Gd_8Co_{75}Mo_{12}Ar_5$	0.53
5	$Gd_{15}Co_{74}Mo_{11}$	21	sputtering	---	---	$(Gd_{15}Co_{74}Mo_{11})_{94}Ar_6$	0.49
6	$Gd_{11}Co_{59}B_{13}Ar_{17}$	13	sputtering	microprobe	force balance	$Gd_{11}Co_{59}B_{13}Ar_{17}$	0.45
7	$(Gd_{10}Co_{73}Mo_{17})_{98}Ar_2$	12	sputtering	microprobe	VSM	$(Gd_{10.5}Co_{72.5}Mo_{17})_{95}Ar_5$	0.40
8	$Gd_{11.3}Co_{67.2}Mo_{16}Ar_{5.5}$	19	sputtering	---	---	$Gd_{12.5}Co_{66.5}Mo_{16}Ar_5$	0.39

Table 14.5. Comparison between theoretical and experimental data for amorphous GdFe-based alloys

	Nominal composition	Ref.	Deposition method	Composition analysis method	Magnetization measurement method	Theoretical (best match) composition	J_{Fe}	α
1	$Gd_{19}Fe_{67}B_{12}Ar_2$	13	sputtering	microprobe	VSM	$Gd_{19}Fe_{66}B_{12}Ar_3$	0.92	0
2	$Gd_{20}Fe_{60}B_{18}Ar_2$	13	sputtering	microprobe	VSM	$Gd_{20}Fe_{60}B_{18}Ar_2$	0.95	0
3	$Gd_{24.5}Fe_{69.5}Sn_6$	6	e-beam evap.	microprobe	VSM	$(Gd_{23.5}Fe_{70.5}Sn_6)_{90}Ar_{10}$	0.95	0
4	$Gd_{26}Fe_{74}$	6	e-beam evap.	microprobe	VSM	$(Gd_{25}Fe_{75})_{95}Ar_5$	0.95	0.33
5	$Gd_{24}Fe_{76}$	2	sputtering	XRF	VSM	$(Gd_{25}Fe_{75})_{95}Ar_5$	0.95	0.40
6	$Gd_{26.3}Fe_{73.7}$	15	e-beam evap.	microprobe	VSM	$(Gd_{25.5}Fe_{74.5})_{95}Ar_5$	0.96	0.33
7	$Gd_{23}Fe_{77}$	2	sputtering	XRF	VSM	$(Gd_{23.5}Fe_{76.5})_{95}Ar_5$	0.97	0.40
8	$Gd_{24.9}Fe_{75.1}$	15	e-beam evap.	microprobe	VSM	$(Gd_{23.5}Fe_{76.5})_{95}Ar_5$	0.97	0.33
9	$Gd_{30.6}Fe_{69.4}$	15	e-beam evap.	microprobe	VSM	$(Gd_{31.5}Fe_{68.5})_{95}Ar_5$	1.12	0.48

Table 14.6. Comparison between theoretical and experimental data for amorphous TbFe-based alloys

	Nominal composition	Ref.	Deposition method	Composition analysis method	Magnetization measurement method	Theoretical (best match) composition	J_{Fe}	α
1	$Tb_{14}Fe_{86}$	22	sputtering	--	--	$(Tb_{13}Fe_{87})_{95}Ar_5$	1.02	0.50
2	$Tb_{19}Fe_{81}$	2	sputtering	XRF	VSM	$(Tb_{19}Fe_{81})_{95}Ar_5$	0.96	0.31
3	$Tb_{21}Fe_{79}$	2	sputtering	XRF	VSM	$(Tb_{20.5}Fe_{79.5})_{95}Ar_5$	0.97	0.25
4	$Tb_{22}Fe_{78}$	22	sputtering	--	--	$(Tb_{21.5}Fe_{78.5})_{95}Ar_5$	0.97	0.24
5	$Tb_{29}Fe_{71}$	22	sputtering	--	--	$(Tb_{29}Fe_{71})_{95}Ar_5$	1.14	0.30
6	$Tb_{33.3}Fe_{66.7}$	23	sputtering	--	--	$(Tb_{34}Fe_{66})_{95}Ar_5$	1.10	0.35

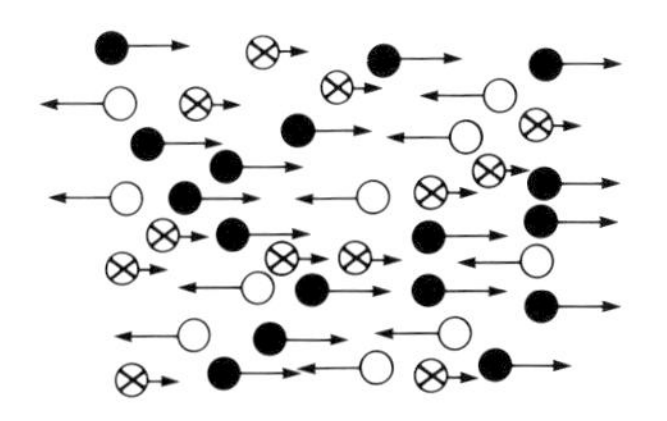

Figure 14.3. Typical arrangement of moments in an RE–Fe alloy. Open circles, RE; solid circles, Fe in the ferromagnetic subnetwork; circles with crosses, Fe in the antiferromagnetic subnetwork. The net exchange field at a given atom (or ion) is the sum of contributions from its nearest neighbors. The contribution of RE(↓) and ferromagnetic Fe(↑) to their Fe neighbors, whether in the ferromagnetic or in the antiferromagnetic subnetwork, is always positive. The contribution of antiferromagnetic Fe(↑) to a neighboring Fe in the ferromagnetic subnetwork is positive, while its contribution to a neighboring Fe in the antiferromagnetic subnetwork is negative. If the fraction of antiferromagnetic Fe in the alloy is not too large, then the net field on both types of Fe will be positive, while the net field on RE will be negative. This is why both Fe subnetworks have positive moments in this figure.

GdCo-based alloys. Table 14.4 corresponds to GdCo-based alloys. A two-magnetic-subnetwork model has been sufficient for explaining the data. The cobalt moment J_{Co} is seen in all cases to be below the value of 0.86 for pure hcp cobalt. This is consistent with the band model in which the 3d↓ band is successively filled with additional electrons. If we assume that Gd, B, and Mo atoms each contribute 1.5, 2.5 and 3 additional electrons respectively to the d band, then the values of J_{Co} obtained in these calculations can be explained. Let us emphasize, however, that charge-transfer arguments of this type are not very reliable, and, although one may use them as a guide in estimating the TM moment, one cannot entirely rely on the quantitative results.

GdFe-based alloys. Table 14.5 corresponds to GdFe-based alloys. A three-magnetic-subnetwork model has been used here to account for the antiferromagnetic coupling among a certain fraction of Fe ions. The adjustable parameters are the iron moment J_{Fe}, and the fraction α of iron in the antiferromagnetic subnetwork. It will be seen in Table 14.5 that with the addition of Gd the iron moment decreases from 1.11 (the value for pure bcc iron) to a minimum of 0.95 at around 25% Gd. Adding more gadolinium seems to increase the moment again. The fraction of antiferromagnetic iron, α, varies between 0.3 and 0.5 for pure GdFe alloys; this may represent the effect of the deposition environment on the structural characteristics of the alloy. The three alloys in Table 14.5 that contain Sn or B do not have an antiferromagnetic Fe subnetwork; also, their iron

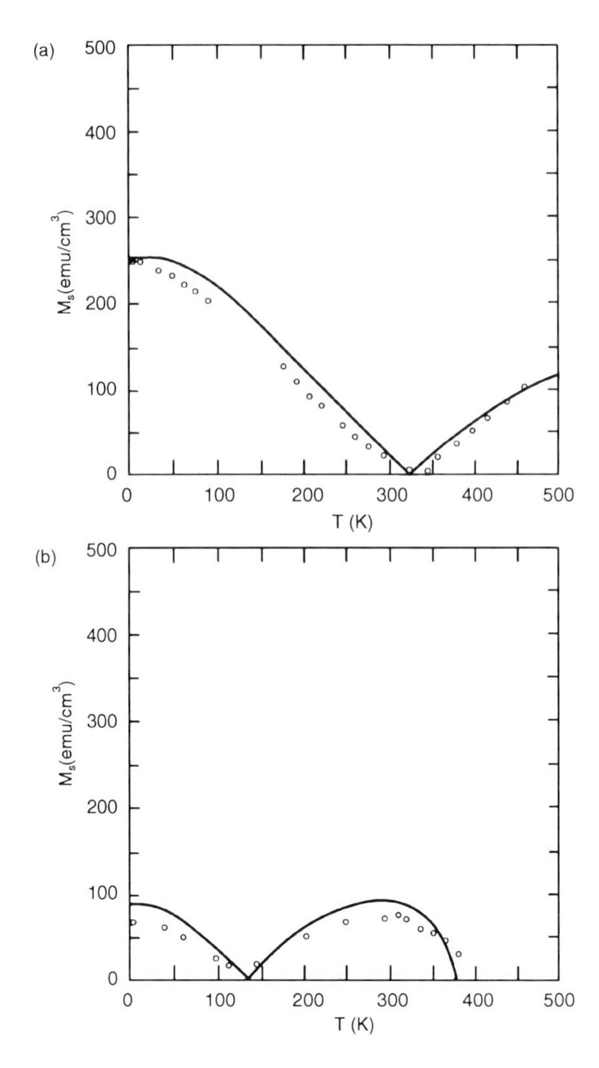

Figure 14.4. Experimental points and theoretical plots of $M_s(T)$ for two GdCo alloy films. (a) $Gd_{22.1}Co_{77.9}$ (reported composition); see second row in Table 14.4. (b) $(Gd_{10}Co_{73}Mo_{17})_{98}Ar_2$ (reported composition); see seventh row in Table 14.4.

moment J_{Fe} remains fixed at around 0.95. The absence of antiferromagnetic iron in this case may be a result of the reduced iron concentration in the alloy; without additional information, however, it is difficult to arrive at definite conclusions.

TbFe-based alloys. Table 14.6 corresponds to amorphous binary TbFe alloys. Again, the adjustable parameters are J_{Fe} and α. With the addition of Tb, the iron moment decreases to a minimum of 0.96 around 19% Tb, and then rises again. The similarity of this behavior of iron in TbFe to that in

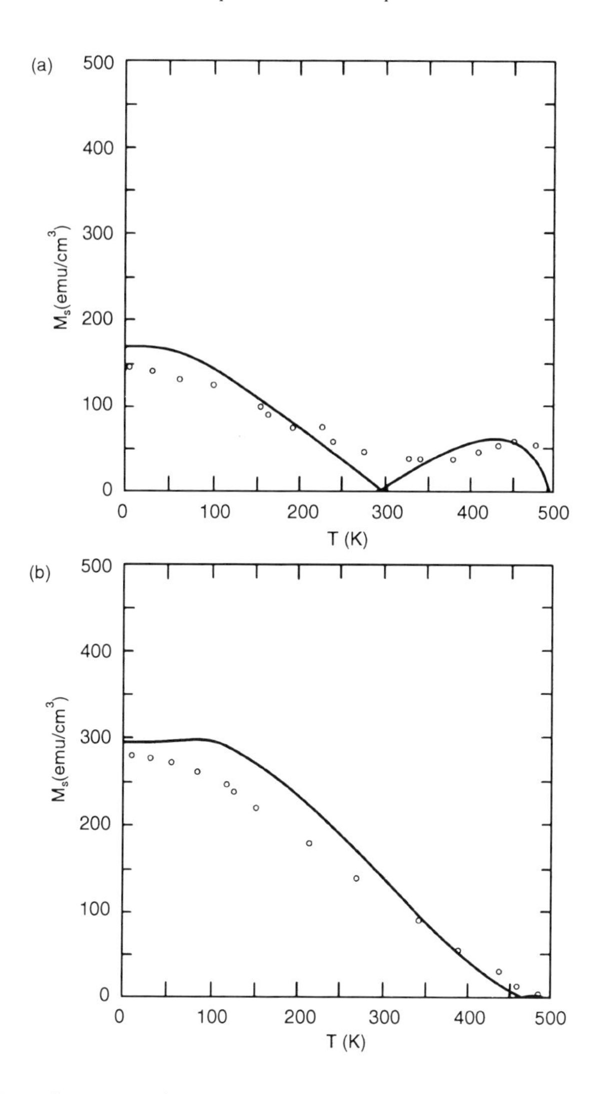

Figure 14.5. Experimental points and theoretical plots of $M_S(T)$ for two GdFe alloy films. (a) $Gd_{26.3}Fe_{73.7}$ (reported composition); see sixth row in Table 14.5. (b) $Gd_{30.6}Fe_{69.4}$ (reported composition); see ninth row in Table 14.5.

GdFe is encouraging, and may in fact suggest that a 3d band structure such as that shown in Fig. 14.7 is at work. This band structure has the property that additional electrons initially fill the 3d$\downarrow$ band, causing the net Fe moment to decrease. Further increases in the number of electrons, however, will strengthen the 3d$\uparrow$ band, and raise the Fe moment.

To appreciate the significance of the antiferromagnetic iron subnetwork in these calculations, we have plotted in Fig. 14.8(a) calculated Curie and compensation point temperatures versus α for a typical binary alloy. With increasing α, it is observed that T_c drops and T_{comp} increases, both by

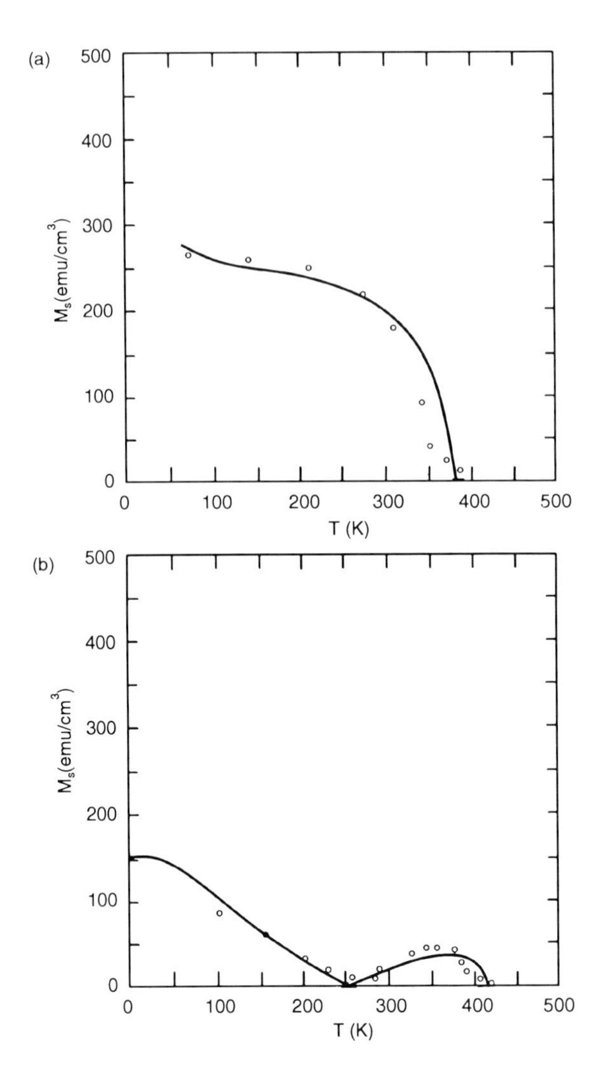

Figure 14.6. Experimental points and theoretical plots of $M_s(T)$ for two TbFe alloy films. (a) $Tb_{14}Fe_{86}$ (reported composition); see first row in Table 14.6. (b) $Tb_{21}Fe_{79}$ (reported composition); see third row in Table 14.6.

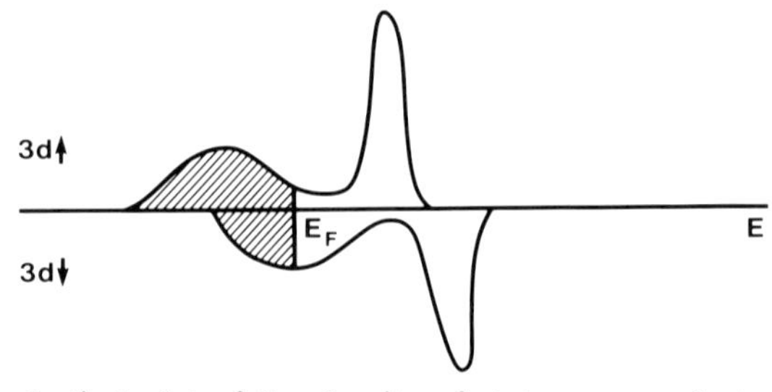

Figure 14.7. Hypothetical plot of the density of states versus electron energy for 3d electrons in amorphous RE–Fe alloys. Additional electrons will initially fill the 3d↓ band, thus reducing the net magnetic moment. After this initial phase, the added electrons will go to the 3d↑ band, and the magnetic moment begins to rise.

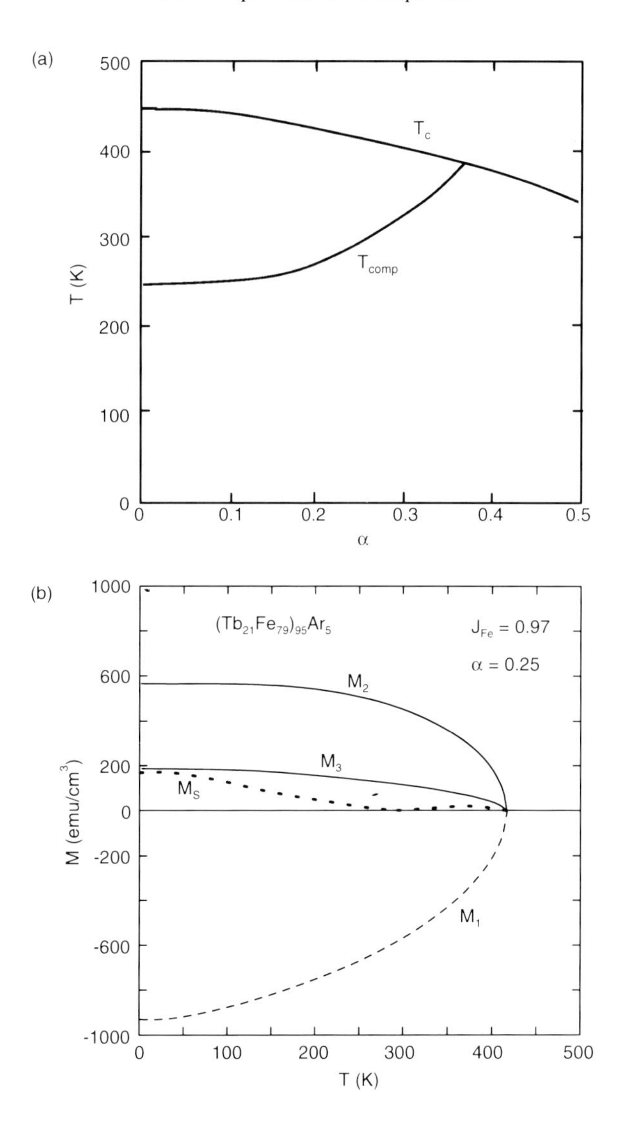

Figure 14.8. (a) Calculated plots of T_c and T_{comp} as functions of the fraction α of antiferromagnetic iron in $(Tb_{21}Fe_{79})_{95}Ar_5$, taking $J_{\mathbf{Fe}} = 0.97$. (b) Magnetizations of the three subnetworks as functions of temperature T, calculated with $\alpha = 0.25$. $M_1(T)$ is the magnetization of terbium, $M_2(T)$ is that of iron in the ferromagnetic subnetwork, and $M_3(T)$ corresponds to iron in the antiferromagnetic subnetwork. Also shown is the sum of the three curves (dotted line), which is the net moment $M_s(T)$ of the material.

significant amounts. The reason for this kind of dependence on α is that the effective exchange field on the antiferromagnetic iron subnetwork is generally small, and, as a consequence, its magnetization decays rather swiftly with rising temperature; witness Fig. 14.8(b), which shows plots of $M(T)$ for the three subnetworks in a film of $(Tb_{21}Fe_{79})_{95}Ar_5$, taking J_{Fe} = 0.97 and $\alpha = 0.25$. α, therefore, is an important parameter of the mean-field model, providing the extra degree of freedom needed to match the measured data with a fixed set of exchange integrals.

14.3. Single-ion Anisotropy and the Mean-field Model

In order to investigate the effect of single-ion anisotropy on the mean-field model, let us assume that the RE element is subject to uniaxial anisotropy of the simplest kind, and that its total energy is given by

$$\mathcal{E}_1 = -Dj_1^2 - Z_1 j_1 \sum_{n=1}^{3} x_n \mathcal{J}_{1n} \langle j_n \rangle \tag{14.5}$$

Here D is the constant of uniaxial anisotropy, with axis perpendicular to the plane of the film. It follows that

$$\langle j_1 \rangle = \frac{\sum_{j_1=-J_1}^{J_1} j_1 \exp(-\mathcal{E}_1/k_B T)}{\sum_{j_1=-J_1}^{J_1} \exp(-\mathcal{E}_1/k_B T)} \tag{14.6}$$

while $\langle j_2 \rangle$ and $\langle j_3 \rangle$ are still given by Eq. (14.3). Although a closed form does not exist for Eq. (14.6), the mean-field equations are still amenable to numerical solution. Figure 14.9 shows the computed plots of T_c and T_{comp} versus D for a typical amorphous alloy. It is seen that, for values of D up to 2×10^{-15} ergs, the increase of T_c with D is rather insignificant, but the effect on T_{comp} is dramatic.

In gadolinium-based alloys, single-ion anisotropy is negligible, since Gd is an S-state ion having little or no interaction with the ionic network's electric field. Terbium, on the other hand, couples strongly to local electric fields, giving rise to significant single-ion anisotropies. Thus, the best model for TbFe is one that includes both the single-ion anisotropy of Tb and the antiferromagnetic coupling amongst Fe ions. However, as will be seen in section 14.6, the measured values of the macroscopic anisotropy

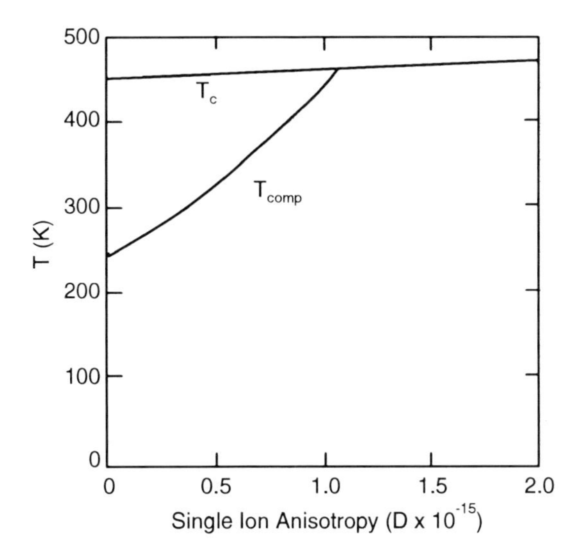

Figure 14.9. Calculated plots of T_c and T_{comp} as functions of the single-ion anisotropy constant D for $(Tb_{21}Fe_{79})_{95}Ar_5$, taking $J_{\text{Fe}} = 0.97$ and $\alpha = 0$.

constant K_u seem to indicate that $D \simeq 10^{-17}$ ergs, which is far too small to affect these mean-field calculations.†

14.4. Exchange Stiffness Coefficient

In the Heisenberg model (see subsection 12.6.1) the exchange energy density is written[12]

$$\mathscr{E} = -\sum_m \sum_n \mathscr{J}_{mn} \langle \mathbf{j}_m \rangle \cdot \langle \mathbf{j}_n \rangle \tag{14.7}$$

where the summations are over all sites in a unit volume, and the assumption $\langle \mathbf{j}_m \cdot \mathbf{j}_n \rangle = \langle \mathbf{j}_m \rangle \cdot \langle \mathbf{j}_n \rangle$ is implicit. When we ignore all but nearest-neighbor interactions, Eq. (14.7) becomes

$$\mathscr{E} = -\sum_m \sum_n \mathscr{J}_{mn} \langle j_m \rangle \langle j_n \rangle \cos \theta_{mn} \tag{14.8}$$

†It has been argued that the random-axis nature of single-ion anisotropy causes the effective value of D to be greater than that suggested by macroscopic measurements.[18] A meaningful discussion of this subject, however, is not possible until reliable data on local anisotropy is available.

where m and n are nearest neighbors, and θ_{mn} is the angle between $\langle \mathbf{j}_m \rangle$ and $\langle \mathbf{j}_n \rangle$. Now, $\cos\theta_{mn}$ must be replaced by its average over all orientations of m, n pairs. Let $\mathbf{d}$ be the displacement between a pair of unit vectors located at $\pm\frac{1}{2}\mathbf{d}$, with direction cosines $(\alpha \pm \nabla\alpha \cdot \frac{1}{2}\mathbf{d},\ \beta \pm \nabla\beta \cdot \frac{1}{2}\mathbf{d},\ \gamma \pm \nabla\gamma \cdot \frac{1}{2}\mathbf{d})$. The angle θ between the vectors is then given by

$$\cos\theta = 1 - \frac{1}{4}\left[(\nabla\alpha \cdot \mathbf{d})^2 + (\nabla\beta \cdot \mathbf{d})^2 + (\nabla\gamma \cdot \mathbf{d})^2\right]$$

which is obviously a function of $\mathbf{d}$. Now, for every vector $\mathbf{d}$ there are two other vectors $\mathbf{d}'$ and $\mathbf{d}''$, such that the three vectors are mutually orthogonal. Then $(\nabla\alpha \cdot \mathbf{d})^2 + (\nabla\alpha \cdot \mathbf{d}')^2 + (\nabla\alpha \cdot \mathbf{d}'')^2 = (\nabla\alpha)^2\,\mathbf{d}^2$; the same is true for β and γ. Consequently, the spatial average of $\cos\theta$ is given by

$$\langle \cos\theta \rangle = 1 - \frac{d^2}{12}\left[(\nabla\alpha)^2 + (\nabla\beta)^2 + (\nabla\gamma)^2\right]$$

The excess energy $\Delta\mathscr{E}$ relative to the state in which all moments are aligned is now written as

$$\Delta\mathscr{E} = \frac{1}{12}\sum_m \sum_n \mathscr{J}_{mn}\langle j_m \rangle \langle j_n \rangle\, d_{mn}^2 \left[(\nabla\alpha)^2 + (\nabla\beta)^2 + (\nabla\gamma)^2\right]$$

where m, n are nearest neighbors, and d_{mn} is the distance between nearest-neighbor atoms (i.e., the sum of the atomic radii r_m and r_n). The macroscopic exchange stiffness coefficient A_x is then given by

$$A_x = \frac{1}{12}\sum_{n=1}^{3} N x_n \sum_{m=1}^{3} Z_n x_m \mathscr{J}_{mn}\langle j_m \rangle \langle j_n \rangle\, d_{mn}^2 \qquad (14.9)$$

Figure 14.10 shows plots of $A_x(T)$, calculated from Eq. (14.9) for representative compositions from the three groups of alloys studied in section 14.2; the selected compositions all have compensation points in the vicinity of the usual ambient temperature (T_{amb} = 300 K). It is observed that the exchange stiffness coefficient is dominated by the transition metal subnetwork in these alloys. A_x decreases with increasing temperature until it reaches zero at the Curie point.

14.5. Macroscopic Anisotropy Energy Constant

It has been suggested that at least two sources of anisotropy are at work in amorphous RE–TM alloys (see section 12.10). The first is the so-called

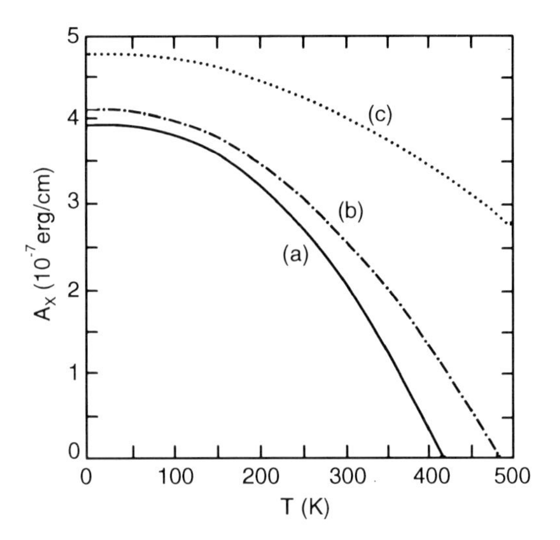

Figure 14.10. Calculated plots of the exchange stiffness coefficient $A_x(T)$. (a) $(Tb_{21}Fe_{79})_{95}Ar_5$, taking $J_{Fe} = 0.97$ and $\alpha = 0.25$. (b) $(Gd_{25}Fe_{75})_{95}Ar_5$, taking $J_{Fe} = 0.95$ and $\alpha = 0.33$. (c) $(Gd_{21}Co_{79})_{95}Ar_5$, taking $J_{Co} = 0.63$.

"pair-ordering" mechanism, due to the inhomogeneous distribution of atoms throughout the material, whereby the bond anisotropy (augmented perhaps by classical dipole–dipole interactions) creates a distinct axis of magnetic anisotropy.[19] A phenomenological expression for the anisotropy energy density of pair-ordering may be written as follows:

$$\mathscr{E} = -\sum_{n=1}^{3} N x_n \langle j_n \rangle \cos \psi \sum_{m=1}^{3} D_{mn} Z_n x_m \langle j_m \rangle \cos \psi \qquad (14.10)$$

where the D_{mn} are anisotropy coefficients for nearest neighbor pairs, and ψ signifies deviation from the easy axis. Assuming that the ions are already arranged in pairs such that both parallel and antiparallel configurations prefer their current orientations, D_{mn} must be positive for pairs of parallel moments and negative for pairs of antiparallel moments. From symmetry it is clear that $D_{mn} = D_{nm}$.

The second source of anisotropy is the interaction of the electronic charge distribution with the local electric field.[17] If the charge distribution is non-spherical, the **E**-field forces the distribution, and consequently the orbital angular momentum, into a preferred orientation; the spin then follows suit due to spin-orbit coupling. In RE–TM alloys the orbital moment of the TM is usually small, making its interaction with the **E**-field rather insignificant. The non-S-state RE ions, on the other hand, couple strongly to the **E**-field, and create random-axis anisotropy. We therefore conjecture that the following phenomenological expression will describe the

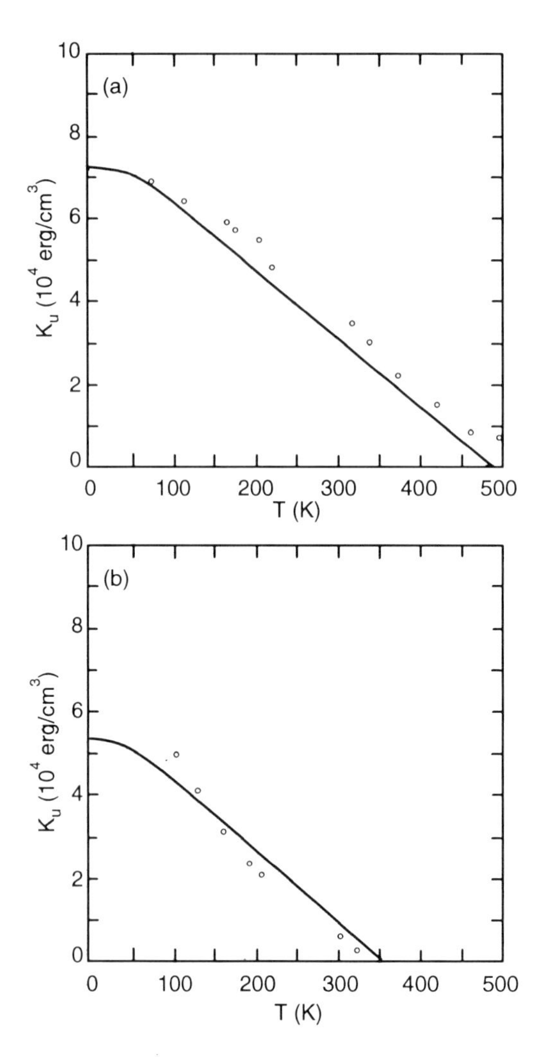

Figure 14.11. Measured points[19,21] and calculated curves showing the anisotropy constant K_u versus temperature. In the calculations $D_{mn} = \pm 10^{-19}$ erg and $D_n = 0$. (a) $Gd_{15}Co_{74}Mo_{11}$; the calculated curve is for $(Gd_{15}Co_{74}Mo_{11})_{94}Ar_6$ with $J_{Co} = 0.49$. (b) $Gd_{11.3}Co_{67.2}Mo_{16}Ar_{5.5}$; the calculated curve is for $Gd_{12.5}Co_{66.5}Mo_{16}Ar_5$ with $J_{Co} = 0.39$.

energy density of single-ion anisotropy:

$$\mathscr{E} = -N \sum_{n=1}^{3} x_n D_n \langle j_n \rangle^2 \cos^2\psi \tag{14.11}$$

Here D_n is the single-ion anisotropy constant. The total macroscopic anisotropy energy constant K_u is thus obtained by combining Eqs. (14.10) and (14.11), as follows:

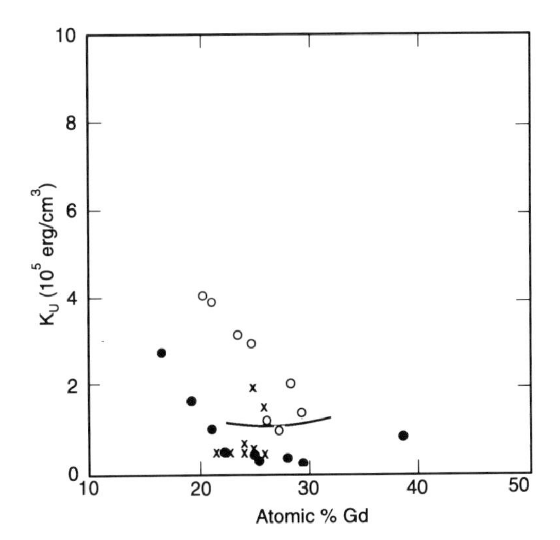

Figure 14.12. Measured and calculated anisotropy constants K_u as functions of the gadolinium content in GdFe alloys, at T = 300 K. The short solid curve is obtained from the available mean-field data with $D_{mn} = \pm 10^{-19}$ erg and $D_n = 0$. (Open circles from Ref. 2, solid circles from Ref. 24, and crosses from Ref. 25.)

$$K_u = N \left\{ \sum_{n=1}^{3} \sum_{m=1}^{3} D_{mn} Z_n x_n x_m \langle j_n \rangle \langle j_m \rangle + \sum_{n=1}^{3} D_n x_n \langle j_n \rangle^2 \right\} \tag{14.12}$$

The coefficients D_{mn} and D_n have complicated relations with the structure of the media, but, for our present purposes, it is sufficient to treat them as adjustable parameters.

Plots of $K_u(T)$ for two GdCo-based alloys in Fig. 14.11 show good agreement between experimental data and model calculations. In both cases the assumed parameter values are $D_n = 0$ and $D_{mn} = \pm 10^{-19}$ erg, with the plus sign applicable to Co–Co and Gd–Gd pairs, and the minus sign applicable to Gd–Co pairs.

Figure 14.12 shows K_u versus the atomic percentage of Gd in amorphous GdFe alloys, at $T \simeq 300$ K. A fairly large scatter in the reported experimental data is evident. Part of this scatter may be due to the fact that near T_{comp} the magnetization is small and measurements of K_u are subject to large errors. It is also well known that preparation conditions (such as argon pressure during sputtering, substrate treatment method, sputtering bias voltage, surface passivation layer, and so on) affect the magnetic anisotropy by changing the morphology and microstructure of the

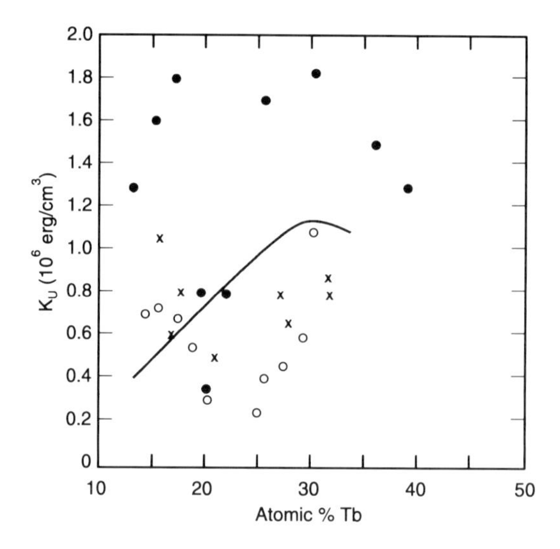

Figure 14.13. Measured and calculated anisotropy constants K_u as functions of the terbium content in TbFe alloys, at $T = 300$ K. The solid curve is obtained from the available mean-field data with $D_{mn} = \pm 10^{-19}$ erg, $D_1 = 0.4 \times 10^{-17}$, and $D_2 = D_3 = 0$. (Open circles from Ref. 2, solid circles from Ref. 24, and crosses from Ref. 3.)

film. The solid curve in Fig. 14.12 is calculated from the information available through the mean-field model. The anisotropy parameters used here are identical to those used in Fig. 14.11 for GdCo samples, namely $D_n = 0$ and $D_{mn} = \pm 10^{-19}$ erg. The order-of-magnitude agreement between this curve and the experimental data may be taken as evidence that the magnetic anisotropy of GdFe is controlled by pair-ordering and that single-ion anisotropy does not play a major role here. This observation is consistent with the fact that Gd^{+++} is an S-state ion.

Figure 14.13 shows K_u versus the atomic percentage of Tb in TbFe films, at $T \simeq 300$ K. The scatter in the experimental data probably arises from the sources mentioned earlier in conjunction with the GdFe data. The solid curve is based on the available mean-field data and corresponds to $D_{mn} = \pm 10^{-19}$ erg, $D_1 = 0.4 \times 10^{-17}$ erg, and $D_2 = D_3 = 0$. The explanation of this result is that Tb^{+++}, being a non-S-state ion, is subject to strong local uniaxial anisotropy, which dominates the behavior of K_u in terbium containing alloys. Notice that D_1, although much larger than the pair-ordering coefficients D_{mn}, is still too small to have any significant effects on the mean-field calculations (see Fig. 14.9).

14.6. Domain Wall Characteristics

An important characteristic of the MO media, connected with the formation and stability of domains in thermomagnetic recording, is the

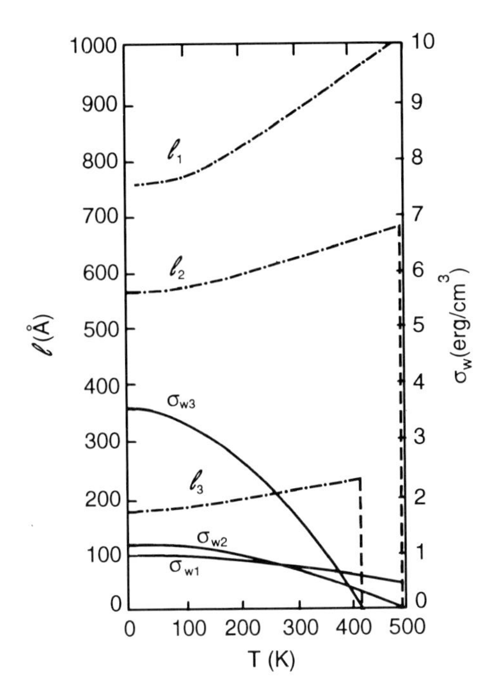

Figure 14.14. Calculated wall-width ℓ and wall energy density σ_w versus temperature. 1, $(Gd_{21}Co_{79})_{95}Ar_5$; 2, $(Gd_{25}Fe_{75})_{95}Ar_5$; 3, $(Tb_{21}Fe_{79})_{95}Ar_5$. All compositions have $T_{comp} \simeq 300$ K.

wall energy density σ_w. This entity is related to the exchange and anisotropy constants through the formula $\sigma_w = 4\sqrt{A_x K_u}$. Figure 14.14 shows calculated plots of $\sigma_w(T)$ for representatives of the three classes of amorphous materials studied in this chapter. Also shown in the figure are calculated plots of the wall width $\ell = 4\sqrt{A_x/K_u}$. The alloy compositions depicted here have $T_{comp} \simeq 300$ K. One observes that TbFe has a larger σ_w than either of the other two alloys. The narrow domain wall of TbFe is suitable for high-density recording applications, where dense packing of domains is of major concern.

14.7. Concluding Remarks

The mean-field model of magnetization for amorphous RE–TM alloys can explain the available data on these media with good accuracy. We have postulated the existence of an idealized antiferromagnetic subnetwork for the iron-based alloys, which results in two different values of the iron moment at $T \neq 0$ K. Realistically, however, the assumption of equal but opposite exchange coefficients for the two types of iron should be replaced by a distribution of exchange integrals among neighboring Fe ions, with the result that the iron moments will be distributed in a continuous fashion within a certain range.†

The mean-field model may be used to investigate other characteristics of amorphous thin-film alloys, such as the anisotropy energy constant and the domain-wall energy density. The anisotropy constant, in particular, is dependent on the microstructure of the film; this, in turn, depends on the environment in which the film has been fabricated. These dependences influence the adjustable parameters D_n and D_{mn} described in section 14.5. A systematic study of magnetic anisotropy in RE–TM films must be undertaken in order to clarify the nature of these relationships.

† It may be possible to obtain information concerning the distribution of iron moments from Mössbauer spectroscopy. Temperature-dependent measurements of the MO Kerr effect and the extraordinary Hall effect, which probe the TM subnetwork, may also provide useful information.

Problems

(14.1) Verify the validity of Eq. (14.2) under the circumstances stated in the text. This equation, of course, is only approximately valid since the cones do not fill the space in three dimensions, and also because the atom at the center is not taken into account when the population of the various species among its neighbors is estimated. To give an example, consider a dilute alloy of two elements, $A_x B_{1-x}$, with $x << 1$. Naturally, it is possible to arrange the elements in such a way that the A atoms are totally surrounded by the B atoms, in which case no two A atoms can be nearest neighbors. Equation (14.2), on the other hand, assumes that both A and B are present in the neighborhood of every A atom.

(a) Can you improve upon this approximation?

(b) If there happens to be a strong tendency for terbium ions to surround themselves exclusively with iron ions in an amorphous $Tb_x Fe_{1-x}$ alloy, what would be the largest possible value of x for which Tb ions remain isolated from each other?

(14.2) Verify the expression given for A_x in Eq. (14.9) by following the steps outlined in the text, filling in the gaps.

(14.3) A simple measurement of the anisotropy constant K_u uses vibrating-sample magnetometry. Here a magnetic field $\mathbf{H}_{\text{ext}}$ is applied along the X-axis, as in Fig. 14.15(a), and the component of M_s parallel to X is measured. M_x is then set equal to $M_s \sin \psi$, where ψ is the deviation of magnetization from its easy axis.

(a) Use energy minimization arguments to relate K_u to the measured plot of M_x versus H_{ext}.

(b) Measurements of K_u in ferrimagnetic films such as $Tb_x Fe_{1-x}$ show a dip in the plot of K_u versus x near the compensation composition. Show that finite exchange coupling between the RE and TM subnetworks is responsible for this phenomenon, by taking into account the bending of M_{RE} relative to M_{TM} under the influence of the applied field, as shown in Fig. 14.15(b).

(14.4)† Consider a monolayer film of $Tb_x Fe_{1-x}$, with x small enough to allow ions of terbium to be exclusively surrounded by iron ions. Antiferromagnetic exchange coupling between Tb and Fe aligns their respective dipole moments in opposite directions, but, since the exchange is isotropic, there is no preferred direction in space for the net moment. The classical dipole–dipole interaction, on the other hand, prefers an in-plane orientation for parallel moments and a perpendicular orientation for antiparallel moments. Estimate the energy difference, $\Delta\mathscr{E}_{\text{dmag}}$, between the following two states: (a) the state where all dipoles are in the plane of the film, (b) the state where all dipoles are perpendicular to the film.

†The phenomenon described in this problem could contribute to perpendicular magnetic anisotropy at the surface of ferrimagnetic films.[26]

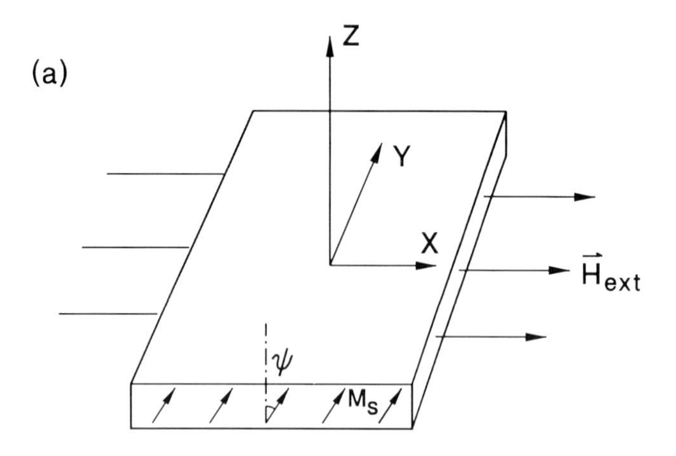

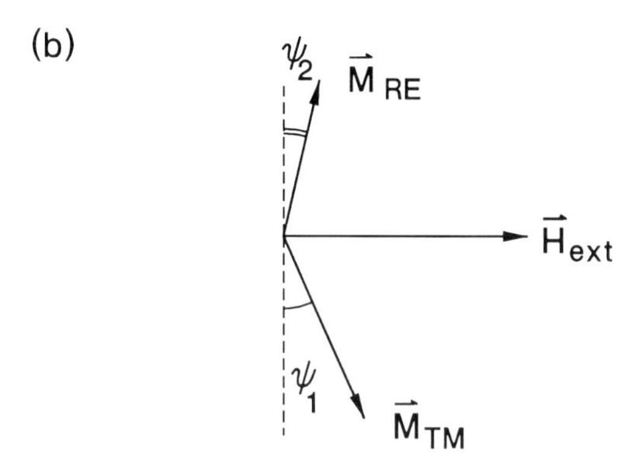

Figure 14.15.

15

Magnetization Dynamics

Introduction

Thermomagnetic recording involves the laser-assisted nucleation and growth of reverse-magnetized domains under the influence of external and/or internal magnetic fields. The nature of nucleation, the speed and uniformity of growth, and the local pinning of domain boundaries due to structural inhomogeneities of the material are among the factors that determine the final shape and size of recorded domains.[1-4] Since it is important to control domain size, while avoiding nonuniformities and jagged boundaries, and since it is important to create stable domains, a knowledge of magnetization-reversal dynamics, including the effects of the nanostructure within the "amorphous" material, is desirable.

The observed magnetization reversal in RE–TM films is a nucleation and growth process. During bulk reversal (at temperatures not too close to the Curie point T_c) the nucleation sites are mostly reproducible. Rather than being driven by thermal fluctuations, nucleation is believed to be rooted in the nonuniform spatial distribution of the structural and magnetic properties of the media.[5,6] The measured hysteresis loops generally have high squareness, which indicates that the coercivity of wall motion is less than the nucleation coercivity. In other words, once the nuclei are created at a certain applied field, they continue to grow without further hindrance. The recording and erasure processes rely on the creation and annihilation of reverse-magnetized domains. High densities are achieved when the recorded domains are small and uniform, have smooth boundaries, and are precisely positioned. Stringent requirements of this type can be fulfilled only when the process of domain formation is well understood and the material parameters that dominate this process are brought under control.

In this chapter we study magnetization dynamics as based on the classical equation of Landau, Lifshitz, and Gilbert (the LLG equation). The basic formulas are derived for a discrete lattice in section 15.1, with an eye towards their application in computer simulations. In section 15.2 we derive analytical results pertaining to the statics and dynamics of domain walls in simple or simplified situations. This study identifies the range of validity of some approximate solutions, and provides a basis for comparison with (and verification of) our simulation results. Section 15.3 describes an algorithm used for micromagnetic simulations on a Connection

Machine,† and presents several examples in which the behavior of thin-film MO media is numerically investigated. The subject of subsection 15.3.2 is the motion of straight walls in uniform media. Aside from significance in their own right, these results establish the reliability of simulations by showing close agreement with the analytic results of section 15.2. Subsection 15.3.3 contains examples of magnetization reversal by nucleation and growth under the influence of external and internal magnetic fields. These simulations incorporate certain attributes of amorphous magnetic materials having random spatial fluctuations.

15.1. Magnetization Dynamics for a Lattice of Interacting Dipoles

The Landau–Lifshitz–Gilbert (LLG) equation embodies the mechanism of magnetization dynamics.[7–11] This equation, which will be introduced in subsection 15.1.2, describes the motion of a dipole in the presence of an effective magnetic field. In this section we derive the effective field for a lattice of interacting dipoles, which is used to simulate a thin magnetic film; the field is then used in the LLG equation to yield suitable formulas for the numerical analysis of dynamic processes. We also show that the energy of the system driven by the LLG equation is a nonincreasing function of time.

To simulate the magnetization dynamics of thin-film media, we make several assumptions. First it is assumed that small volumes of the material act as single-domain particles, which is reasonable as long as the dimensions of the volume remain well below the wall width for the given material. Second, it is assumed that these single-domain "particles" are equal in size and shape, and are regularly distributed on a two-dimensional hexagonal lattice. The film thickness τ is constant and the magnetization through the thickness is uniform. Figure 15.1 shows the postulated geometry of the "particles" and their placement in the lattice. Each particle is a hexagonal prism of height τ and volume V where

$$V = \frac{\sqrt{3}}{2} d^2 \tau \tag{15.1}$$

(d is the lattice constant.) The saturation magnetization M_s of the material

†Such parallelism as inherent to the Connection Machine (CM) is indeed necessary if large arrays of interacting dipoles are to be considered for simulation. Each dipole in the lattice may be assigned to a single processor within the CM and, using the machine's vast internal communication network, information may be exchanged among the various processors. One can then evaluate the net torque on each dipole, and compute its new state accordingly. The architecture of the CM is particularly suited for this problem, since all dipoles follow the same dynamic law, albeit in different environments. Communication among the processors makes it possible for each dipole to know the state of its neighbors, which it needs for computation of the effective exchange field. The communication network also allows speedy calculation of the Fourier transforms needed for demagnetization computations.

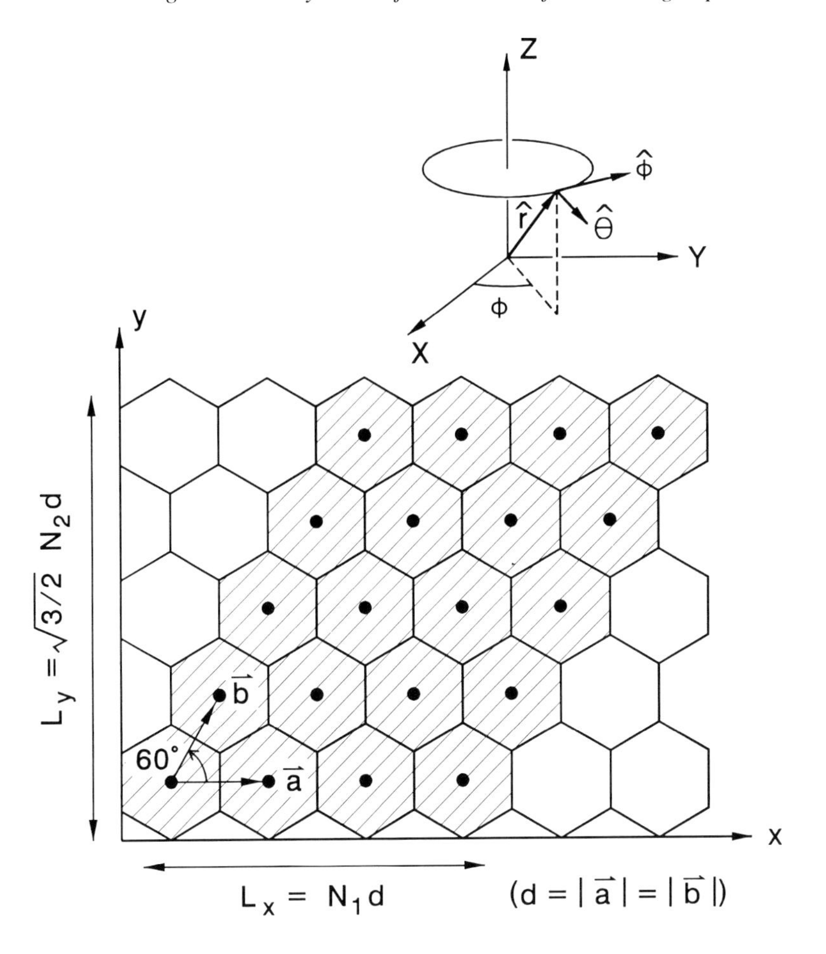

Figure 15.1. (a) Geometry of the hexagonal lattice in the XY-plane. Each cell represents a single-domain "particle" in the actual film. The basis vectors **a** and **b** have equal lengths, d, which is the lattice constant. The height of each cell in the Z-direction is the film thickness τ. The inset shows the unit vectors $\hat{\mathbf{r}}$, $\hat{\boldsymbol{\theta}}$, and $\hat{\boldsymbol{\phi}}$ of the spherical coordinate system used in the text; at each lattice site, the direction of the local dipole moment $\mathbf{m}_{ij}$ is designated as $\hat{\mathbf{r}}$.

is a model parameter. If the assumption is made that the particles are densely packed, then the magnitude of the dipole moment $\mathbf{m}$ assigned to each lattice cell will be $|\mathbf{m}| = M_s V$.

15.1.1. Effective magnetic field

The effective field $\mathbf{H}_{\text{eff}}$ acting on each dipole within the lattice has contributions from four different sources;[9] these sources are discussed separately in the following subsections. In addition to the externally applied field, the effective field arises from local anisotropy, nearest-neighbor exchange, and classical dipole–dipole interactions. The sum of these contributions constitutes the net field, which has components in the $\hat{\mathbf{r}}$, $\hat{\boldsymbol{\theta}}$, and $\hat{\boldsymbol{\phi}}$ directions (see the inset to Fig. 15.1). Since only the torque induced by the field is of interest here, and since the field component along the magnetization direction does not contribute to the torque, we shall define the $\hat{\mathbf{r}}$ direction to be that of $\mathbf{M}$ at the given site, and proceed to ignore the $\hat{\mathbf{r}}$ component of $\mathbf{H}_{\text{eff}}$; in the following analysis only the $\hat{\boldsymbol{\theta}}$ and $\hat{\boldsymbol{\phi}}$ components of $\mathbf{H}_{\text{eff}}$ will be evaluated.

Now, suppose that under the influence of $\mathbf{H}_{\text{eff}}$ the vector $\mathbf{M}$ rotates by the small amount $(\Delta\theta, \Delta\phi)$. The change in energy density $\Delta\mathcal{E}$ will then be

$$\Delta\mathcal{E} = -\mathbf{H} \cdot \Delta\mathbf{M} = -(H_r\hat{\mathbf{r}} + H_\theta\hat{\boldsymbol{\theta}} + H_\phi\hat{\boldsymbol{\phi}}) \cdot (|\mathbf{M}|\Delta\theta\,\hat{\boldsymbol{\theta}} + |\mathbf{M}|\sin\theta\,\Delta\phi\,\hat{\boldsymbol{\phi}})$$

$$= -|\mathbf{M}|H_\theta\,\Delta\theta - |\mathbf{M}|H_\phi\sin\theta\,\Delta\phi$$

Consequently, one may write

$$\boxed{\frac{\partial\mathcal{E}}{\partial\theta} = -|\mathbf{M}|H_\theta} \tag{15.2a}$$

$$\boxed{\frac{\partial\mathcal{E}}{\partial\phi} = -|\mathbf{M}|\sin\theta\,H_\phi} \tag{15.2b}$$

These equations give the effective-field components in terms of partial derivatives of the energy density. We shall use Eqs. (15.2) frequently to derive the various contributions that make up the effective field. Since we will have to switch back and forth between Cartesian and spherical coordinates, we list below the relations between the unit vectors of these coordinate systems:

$$\hat{\mathbf{r}} = \sin\theta\cos\phi\,\hat{\mathbf{x}} + \sin\theta\sin\phi\,\hat{\mathbf{y}} + \cos\theta\,\hat{\mathbf{z}} \tag{15.3a}$$

$$\hat{\boldsymbol{\theta}} = \cos\theta\cos\phi\,\hat{\mathbf{x}} + \cos\theta\sin\phi\,\hat{\mathbf{y}} - \sin\theta\,\hat{\mathbf{z}} \tag{15.3b}$$

$$\hat{\boldsymbol{\phi}} = -\sin\phi\,\hat{\mathbf{x}} + \cos\phi\,\hat{\mathbf{y}} \tag{15.3c}$$

External field. Suppose the external field $\mathbf{H}_{\text{ext}} = H_x\hat{\mathbf{x}} + H_y\hat{\mathbf{y}} + H_z\hat{\mathbf{z}}$ acts on the dipole moment $\mathbf{m}_{ij} = M_s V\hat{\mathbf{r}}$ at the lattice site (i, j). The θ and ϕ components of the field will then be

$$H_\theta^{(\text{ext})} = \mathbf{H}_{\text{ext}} \cdot \hat{\boldsymbol{\theta}} = H_x\cos\theta\cos\phi + H_y\cos\theta\sin\phi - H_z\sin\theta \tag{15.4a}$$

$$H_\phi^{(\text{ext})} = \mathbf{H}_{\text{ext}} \cdot \hat{\boldsymbol{\phi}} = -H_x\sin\phi + H_y\cos\phi \tag{15.4b}$$

The energy of interaction between $\mathbf{H}_{\text{ext}}$ and the magnetization at the given site may be written:

$$\mathcal{E}_{\text{ext}} = -\mathbf{H} \cdot \mathbf{M}_{ij} = -|\mathbf{M}_{ij}| \, (H_x \sin\theta \cos\phi + H_y \sin\theta \sin\phi + H_z \cos\theta) \qquad (15.5)$$

These relations are quite general, and allow any spatial or temporal distributions of $\mathbf{H}_{\text{ext}}$ to be incorporated into the computer simulations.

Anisotropy. Let the local easy axis have an arbitrary direction in space, specified in spherical coodinates by $\hat{\mathbf{r}}_0$ with angular coordinates (θ_0, ϕ_0). The dipole at this site is $\mathbf{m}_{ij} = M_s V \hat{\mathbf{r}}$, with angular coordinates (θ, ϕ). Denoting the constant of uniaxial anisotropy by K_u, the corresponding energy density will be

$$\mathcal{E}_{\text{anis}} = K_u \left[1 - (\hat{\mathbf{r}} \cdot \hat{\mathbf{r}}_0)^2 \right] \qquad (15.6a)$$

or, in terms of the angular coordinates,

$$\mathcal{E}_{\text{anis}} = K_u \left\{ 1 - \left[\cos\theta \cos\theta_0 + \sin\theta \sin\theta_0 \cos(\phi - \phi_0) \right]^2 \right\} \qquad (15.6b)$$

Application of Eq. (15.2) to $\mathcal{E}_{\text{anis}}$ yields the following components for the effective field:

$$H_\theta^{(\text{anis})} = -\frac{K_u}{|\mathbf{M}|} \left\{ \sin 2\theta \left[\cos^2\theta_0 - \sin^2\theta_0 \cos^2(\phi - \phi_0) \right] - \cos 2\theta \sin 2\theta_0 \cos(\phi - \phi_0) \right\} \qquad (15.7a)$$

$$H_\phi^{(\text{anis})} = -\frac{K_u}{|\mathbf{M}|} \left\{ \sin\theta \sin^2\theta_0 \sin 2(\phi - \phi_0) + \cos\theta \sin 2\theta_0 \sin(\phi - \phi_0) \right\} \qquad (15.7b)$$

An alternative but equivalent expression for $\mathbf{H}_{\text{anis}}$ may be derived from Eq. (15.6a), namely,

$$\boxed{\mathbf{H}_{\text{anis}} = \frac{2K_u}{|\mathbf{M}|} (\hat{\mathbf{r}} \cdot \hat{\mathbf{r}}_0) \, \hat{\mathbf{r}}_0} \qquad (15.8)$$

In the computer simulations of section 15.3, each lattice cell will be allowed to have its own axis of anisotropy. In some cases the local axes will be assigned randomly within a cone perpendicular to the plane of the lattice. Denoting by Θ the maximum deviation angle of the local easy axis from the normal direction, the angles θ_0 and ϕ_0 are chosen uniformly and independently within the intervals $[0, \Theta]$ and $[0, 2\pi]$ respectively. In such cases, specification of the cone's half-angle, Θ, should be sufficient for describing the distribution of the easy axis.

Exchange. Neighboring cells interact at their common boundaries through exchange forces. The effective exchange field on a given dipole moment $\mathbf{m}$ due to a neighboring moment $\mathbf{m}'$ will be denoted by $\mathbf{H}_{xhg}$. We show below that the magnitude of this field is related to the exchange stiffness coefficient A_x and its direction is that of $\mathbf{m}'$. With reference to Fig. 15.1, we note that since the magnitude of $\mathbf{M}(x, y)$ is the same across the lattice, its spatial variations can arise only from variations in its direction. Defining the unit-magnitude vector field $\boldsymbol{\mu}(x, y) = \mathbf{M}(x, y)/M_s$, we write the exchange energy of the system as follows:

$$\mathscr{E}_{xhg} = \tau A_x \iint \left[(\nabla\mu_x)^2 + (\nabla\mu_y)^2 + (\nabla\mu_z)^2 \right] \mathrm{d}x\mathrm{d}y \tag{15.9}$$

Next, we relate the above energy, expressed as a continuum integral, to the discrete distribution of dipoles on the hexagonal lattice. Let (x_0, y_0) be the center of an arbitrary lattice cell, and denote the dipole moment associated with this cell by $\mathbf{m}_0$ where

$$\mathbf{m}_0 = \frac{\sqrt{3}d^2\tau}{2} \mathbf{M}(x_0, y_0) \tag{15.10}$$

In a similar fashion, define the six nearest neighbors of $\mathbf{m}_0$ and denote them by $\mathbf{m}_n$, where $1 \le n \le 6$. $\boldsymbol{\mu}_n$ will be a unit vector parallel to $\mathbf{m}_n$. Let $\mathbf{d}_n$ be the vector connecting (x_0, y_0) to the center of its near-neighbor cell at (x_n, y_n). Clearly, the magnitude of $\mathbf{d}_n$ is equal to the lattice constant d. One then writes

$$\boldsymbol{\mu}_n - \boldsymbol{\mu}_0 \simeq (\nabla\mu_x \cdot \mathbf{d}_n)\,\hat{\mathbf{x}} + (\nabla\mu_y \cdot \mathbf{d}_n)\,\hat{\mathbf{y}} + (\nabla\mu_z \cdot \mathbf{d}_n)\,\hat{\mathbf{z}} \tag{15.11}$$

Forming the dot product of Eq. (15.11) with itself results in

$$2 - 2\boldsymbol{\mu}_n \cdot \boldsymbol{\mu}_0 \simeq (\nabla\mu_x)^2 d^2\cos^2\Theta_n + (\nabla\mu_y)^2 d^2\cos^2\Phi_n + (\nabla\mu_z)^2 d^2\cos^2\Omega_n \tag{15.12}$$

Here Θ_n, Φ_n, Ω_n are the angles between $\nabla\mu_x$, $\nabla\mu_y$, $\nabla\mu_z$ respectively and $\mathbf{d}_n$. Summing Eq. (15.12) over the six nearest neighbors of $\mathbf{m}_0$ yields

$$-2\sum_{n=1}^{6}\boldsymbol{\mu}_n\cdot\boldsymbol{\mu}_0 \simeq d^2\Bigg[(\nabla\mu_x)^2\sum_{n=1}^{6}\tfrac{1}{2}(1+\cos 2\Theta_n)$$

$$+(\nabla\mu_y)^2\sum_{n=1}^{6}\tfrac{1}{2}(1+\cos 2\Phi_n)+(\nabla\mu_z)^2\sum_{n=1}^{6}\tfrac{1}{2}(1+\cos 2\Omega_n)\Bigg] + \text{constant} \qquad (15.13)$$

Thanks to the symmetry of the problem, the sums of cosines in all cases turn out to be zero. Multiplying the remaining terms in Eq. (15.13) by the constant factor $2A_x/3d^2$ results in:

$$-\left(\frac{4A_x}{3d^2}\sum_{n=1}^{6}\boldsymbol{\mu}_n\right)\cdot\boldsymbol{\mu}_0 \simeq 2A_x\left[(\nabla\mu_x)^2+(\nabla\mu_y)^2+(\nabla\mu_z)^2\right] + \text{constant} \qquad (15.14)$$

The right-hand side of Eq. (15.14) is the exchange-energy density (i.e., energy per unit volume) for the interaction between $\mathbf{m}_0$ and its nearest neighbors; the factor of two has been included here to account for interactions in both directions. The effective exchange field acting on $\mathbf{m}_0$ is thus found to be

$$\boxed{\mathbf{H}_{\text{xhg}} = \frac{4A_x}{3M_s d^2}\sum_{n=1}^{6}\boldsymbol{\mu}_n} \qquad (15.15)$$

In terms of the spherical coordinates of two adjacent dipoles **m** and **m′**, the effective exchange field of **m′** on **m** is now written

$$H_\theta^{(\text{xhg})} = \frac{4A_x}{3M_s d^2}\left[\cos\theta\sin\theta'\cos(\phi'-\phi)-\sin\theta\cos\theta'\right] \qquad (15.16a)$$

$$H_\phi^{(\text{xhg})} = \frac{4A_x}{3M_s d^2}\sin\theta'\sin(\phi'-\phi) \qquad (15.16b)$$

From the above discussion it is clear that the mutual exchange-energy density between **m** and **m′** is

$$\mathcal{E}_{\text{xhg}} = \frac{4A_x}{3d^2}(1-\boldsymbol{\mu}\cdot\boldsymbol{\mu}') = \frac{4A_x}{3d^2}\left[1-\cos\theta\cos\theta'-\sin\theta\sin\theta'\cos(\phi-\phi')\right] \qquad (15.17)$$

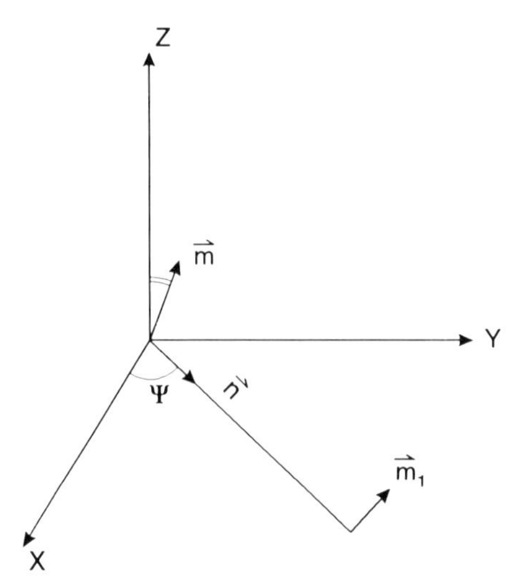

Figure 15.2. Assumed positions of two dipoles **m** and $\mathbf{m}_1$ in the XY-plane of the lattice for calculation of the demagnetizing field. **m** is at the origin and its orientation is specified by the spherical coordinates θ and ϕ. $\mathbf{m}_1$ is a distance ℓ from the origin and its orientation is specified by θ_1 and ϕ_1. The angular coordinate of the line joining the two dipoles is ψ. All three angles ϕ, ϕ_1, and ψ are measured from the positive X-axis.

This energy, of course, is shared between the two dipoles, and must be divided by two if the energy per dipole is needed.

Demagnetization. Classical dipole–dipole interactions give rise to demagnetization fields. These fields, which moderately influence the magnetic processes of thin-film RE–TM alloys, are the hardest of the effective fields to compute. The difficulty is rooted in the long-range nature of the interaction and the fact that every dipole in the system contributes to the effective field on all other dipoles. Fortunately, however, the FFT-based method described in Chapter 13 provides a fast and accurate algorithm for computation of demagnetizing fields.† Unlike direct dipole–dipole calculations, the FFT-based algorithm does not require that film thickness τ and lattice constant d be roughly equal. It also avoids the cumbersome integrations that must be carried out in order to go beyond the single-dipole approximation and average the interaction over the individual cell volumes.[12]

Digression. For completeness and for future reference, we shall derive expressions for the demagnetizing field based on direct dipole–dipole calculations. Figure 15.2 shows a dipole **m** at the origin ($x = 0$, $y = 0$), and a second diplole $\mathbf{m}_1$ at ($x_1 = \ell \cos \psi$, $y_1 = \ell \sin \psi$). The unit vector **n** points

†The FFT-based algorithm is particularly suited for the Connection Machine environment, where a vector of length N can be transformed in time $\log N$, as opposed to $N \log N$ on conventional computers.

from the origin to (x_1, y_1), i.e., $\mathbf{n} = \cos\psi\,\hat{\mathbf{x}} + \sin\psi\,\hat{\mathbf{y}}$. It was shown in Eq. (12.16) that the demagnetizing field of $\mathbf{m}_1$ at the site of $\mathbf{m}$ is

$$\mathbf{H}_{\text{dmag}} = \frac{3\mathbf{n}(\mathbf{n}\cdot\mathbf{m}_1) - \mathbf{m}_1}{\ell^3}$$

Using Eqs. (15.3) we find

$$H_\theta^{(\text{dmag})} = \mathbf{H}_{\text{dmag}}\cdot\hat{\boldsymbol{\theta}} = \frac{|\mathbf{m}_1|}{\ell^3}\left[3(\mathbf{n}\cdot\hat{\mathbf{r}}_1)(\mathbf{n}\cdot\hat{\boldsymbol{\theta}}) - (\hat{\mathbf{r}}_1\cdot\hat{\boldsymbol{\theta}})\right]$$

$$= \frac{|\mathbf{M}_1|V}{\ell^3}\Big\{\sin\theta\cos\theta_1 + \cos\theta\sin\theta_1$$

$$\times\left[3\cos(\phi-\psi)\cos(\phi_1-\psi) - \cos(\phi-\phi_1)\right]\Big\} \quad (15.18a)$$

$$H_\phi^{(\text{dmag})} = \mathbf{H}_{\text{dmag}}\cdot\hat{\boldsymbol{\phi}} = \frac{|\mathbf{m}_1|}{\ell^3}\left[3(\mathbf{n}\cdot\hat{\mathbf{r}}_1)(\mathbf{n}\cdot\hat{\boldsymbol{\phi}}) - (\hat{\mathbf{r}}_1\cdot\hat{\boldsymbol{\phi}})\right]$$

$$= \frac{|\mathbf{M}_1|V}{\ell^3}\sin\theta_1\left[\sin(\phi-\phi_1) - 3\sin(\phi-\psi)\cos(\phi_1-\psi)\right] \quad (15.18b)$$

V is the volume of a lattice cell, which, when multiplied by the magnetization $\mathbf{M}_1$, gives the strength of the dipole moment $\mathbf{m}_1$ at that lattice site. The field on any given dipole is the sum of the demagnetizing fields from all other dipoles calculated at the site of the given dipole. Since the strength of the dipole–dipole interaction drops with distance as $1/\ell^3$, whereas the number of dipoles at a distance ℓ is linear in ℓ, for two-dimensional lattices it is only necessary to include the "significant neighbors" in the computation of $\mathbf{H}_{\text{dmag}}$; typically, these are the dipoles within a radius of $\simeq 5\tau$ from the given site.

The interaction energy density of the dipole pair $(\mathbf{m}, \mathbf{m}_1)$ in Fig. 15.2 is readily found to be

$$\mathscr{E}_{\text{dmag}} = -\mathbf{H}_{\text{dmag}}\cdot\mathbf{M} = |\mathbf{M}|\,|\mathbf{M}_1|\Big\{\cos\theta\cos\theta_1 + \sin\theta\sin\theta_1$$

$$\times\left[\cos(\phi-\phi_1) - 3\cos(\phi-\psi)\cos(\phi_1-\psi)\right]\Big\}\frac{V}{\ell^3} \quad (15.19)$$

This energy is shared between the two dipoles; it must be divided by two if the energy associated with individual dipoles is desired. □

15.1.2. The Landau–Lifshitz–Gilbert equation

The fundamental law governing the dynamics of magnetization is the law of mechanics that relates the applied torque to the time rate of change of angular momentum. At a lattice site with dipole moment $\mathbf{m}$, gyromagnetic ratio γ, and magnetic field $\mathbf{H}$, the angular momentum is $\mathbf{m}/\gamma$ and the torque is $\mathbf{m} \times \mathbf{H}$; thus

$$\dot{\mathbf{m}} = \gamma \mathbf{m} \times \mathbf{H} \tag{15.20a}$$

Here $\dot{\mathbf{m}} = d\mathbf{m}/dt$. The field $\mathbf{H}$ consists of the external field and contributions from anisotropy, exchange, and demagnetization. In addition, it has a phenomenological term that represents the dissipative factors. The latter contribution is proportional to the time rate of change of $\boldsymbol{\mu} = \mathbf{m}/|\mathbf{m}|$. The proportionality constant is defined as α/γ, where α is the dimensionless viscous (Gilbert) damping parameter.[8] Therefore

$$\mathbf{H} = \mathbf{H}_{\text{ext}} + \mathbf{H}_{\text{anis}} + \left(\sum_{\text{n.n.}} \mathbf{H}_{\text{xhg}} \right) + \mathbf{H}_{\text{dmag}} + \frac{\alpha}{\gamma} \dot{\boldsymbol{\mu}} \tag{15.20b}$$

(n.n. stands for nearest neighbors.) In compact form, Eqs. (15.20) combine to relate the time rate of change of $\boldsymbol{\mu}$ to $\mathbf{H}_{\text{eff}}$ and α. Thus we have the Landau–Lifshitz–Gilbert equation:

$$\boxed{\dot{\boldsymbol{\mu}} = \gamma \boldsymbol{\mu} \times \mathbf{H}_{\text{eff}} + \alpha \boldsymbol{\mu} \times \dot{\boldsymbol{\mu}}} \tag{15.21}$$

The LLG equation forms the basis of nearly all studies of magnetization dynamics. Notice that in the absence of damping (i.e., when $\alpha = 0$) $\mathbf{m}$ gyrates around $\mathbf{H}_{\text{eff}}$, just like any object with angular momentum $\mathbf{m}/\gamma$, in response to the torque $\mathbf{m} \times \mathbf{H}_{\text{eff}}$. The effect of damping is to continuously reduce the radius of gyration until $\mathbf{m}$ aligns itself with $\mathbf{H}_{\text{eff}}$. Thus, according to Eq. (15.21), a magnetic dipole moment in a constant effective field within a dissipative medium follows a damped gyration path to equilibrium, as shown in Fig. 15.3. From Eq. (15.21) we obtain

$$\dot{\theta} = -\gamma H_\phi - \alpha \dot{\phi} \sin\theta \qquad \dot{\phi} \sin\theta = \gamma H_\theta + \alpha \dot{\theta} \tag{15.22}$$

where H_θ and H_ϕ are the effective-field components. Solving Eqs. (15.22) for $\dot{\theta}$ and $\dot{\phi}$ yields

$$\boxed{\dot{\theta} = -\frac{\alpha\gamma}{1 + \alpha^2} H_\theta - \frac{\gamma}{1 + \alpha^2} H_\phi} \tag{15.23a}$$

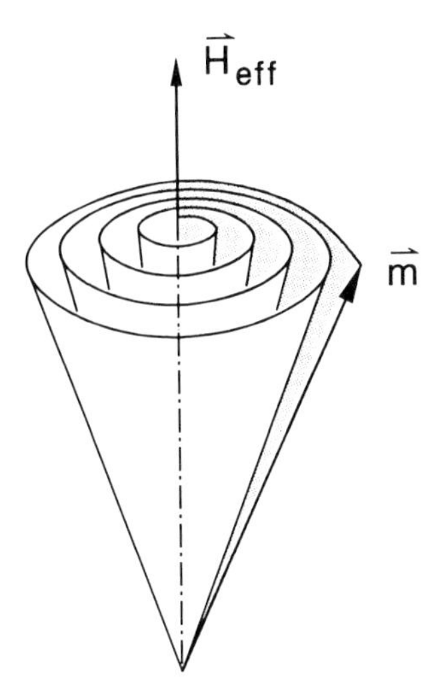

Figure 15.3. Under the influence of an effective magnetic field $\mathbf{H}_{eff}$, the magnetization vector **M** follows a damped gyration path to equilibrium.

$$\boxed{\dot{\phi} \sin \theta = \frac{\gamma}{1 + \alpha^2} H_\theta - \frac{\alpha\gamma}{1 + \alpha^2} H_\phi} \tag{15.23b}$$

These equations form the basis of our computer simulations, to be described in section 15.3. Now, if θ changes by $\Delta\theta$ and ϕ changes by $\Delta\phi$ in a short time interval Δt, the net rotation $\Delta\Psi$ of the dipole moment will be

$$(\Delta\Psi)^2 = (\Delta\theta)^2 + (\Delta\phi)^2 \sin^2\theta$$

Therefore,

$$\dot{\Psi} = \sqrt{\dot{\theta}^2 + \dot{\phi}^2 \sin^2\theta} \tag{15.24}$$

To ensure the convergence of the numerical algorithm used in computer simulations, one must keep $\Delta\Psi$ small. This is achieved by proper selection of the time step Δt.

15.1.3. Energy considerations

The total energy of the system of magnetic dipoles is the sum of contributions from the external field, anisotropy, exchange, and dipole–dipole interactions; that is,

$$\mathscr{E}_{\text{tot}} = \sum_{i,j} \left[\mathscr{E}_{\text{ext}} + \mathscr{E}_{\text{anis}} + \tfrac{1}{2} \left(\sum_{\text{n.n.}} \mathscr{E}_{\text{xhg}} \right) + \mathscr{E}_{\text{dmag}} \right] V \tag{15.25}$$

The outer sum is over all lattice sites (i, j). Individual terms in the above equation are given by Eqs. (15.5), (15.6), (15.17), and (13.24). The factor of one-half multiplying the exchange term signifies that the exchange energy is mutually shared by the interacting moments.

To calculate the time rate of change of energy in this system, notice that $\mathscr{E}_{\text{tot}}$ is a function of θ_{ij} and ϕ_{ij} for all i, j in the lattice. Since θ_{ij} and ϕ_{ij} are the only time-dependent terms, one can write

$$\frac{\mathrm{d}\mathscr{E}_{\text{tot}}}{\mathrm{d}t} = \sum_{i,j} \left\{ \frac{\partial \mathscr{E}_{\text{tot}}}{\partial \theta_{ij}} \dot{\theta}_{ij} + \frac{\partial \mathscr{E}_{\text{tot}}}{\partial \phi_{ij}} \dot{\phi}_{ij} \right\} \tag{15.26a}$$

Using Eqs. (15.2), (15.23), (15.24) in the above equation, one arrives at

$$\boxed{\frac{\mathrm{d}\mathscr{E}_{\text{tot}}}{\mathrm{d}t} = \frac{\alpha V}{\gamma} \sum_{i,j} |\mathbf{M}_{ij}|\, \dot{\Psi}_{ij}^{2}} \tag{15.26b}$$

Since γ in Eq. (15.26b) is negative, the system always loses energy during an adjustment process. The dissipative phenomena whose strength is proportional to the damping coefficient α are responsible for this loss. Observe that when $\alpha \to 0$ the rate of energy loss also approaches zero.

15.2. Domain Wall Structure and Dynamics; Analytic Treatment

It is instructive to analyze the LLG equation in various limits and under different approximations in order to obtain closed-form solutions for comparison with simulated results. In what follows we first derive the static wall structure (including Bloch lines) for both straight and circular walls. We then derive expressions for wall mobility in various limiting regimes.

15.2.1. Static domain wall equations

For a static domain wall, $\dot{\theta} = \dot{\phi} = 0$ at all lattice sites. Consequently, Eqs. (15.23) require that $H_\theta = H_\phi = 0$. Let us ignore the effects of demagnetization and external fields for the time being. Then at each site we must have

$$H_\theta^{(\text{anis})} + H_\theta^{(\text{xhg})} = 0 \qquad\qquad H_\phi^{(\text{anis})} + H_\phi^{(\text{xhg})} = 0$$

Substituting in the above equations from Eq. (15.7) (without random-axis anisotropy) and Eq. (15.16), in the continuum limit ($d \to 0$) we find:

$$\boxed{\begin{aligned}\sin\theta\cos\phi\,\nabla^2(\cos\theta) - \cos\theta\,\nabla^2(\sin\theta\cos\phi)\\ + (K_u/A_x)\sin\theta\cos\theta\cos\phi = 0\end{aligned}} \tag{15.27a}$$

$$\boxed{\sin\phi\,\nabla^2(\sin\theta\cos\phi) = \cos\phi\,\nabla^2(\sin\theta\sin\phi)} \tag{15.27b}$$

Here $\nabla^2 = \partial^2/\partial x^2 + \partial^2/\partial y^2$ is the Laplacian operator in the plane. These equations generalize Eq. (13.5) and form a basic set of equations for the wall structure in two dimensions. Note that spatial variations of M, A_x, and K_u, random-axis anisotropy, demagnetization effects, and the effects of an external field are not included in these equations.

To calculate the wall energy, we use Eqs. (15.6) and (15.17) to find the sum of the anisotropy and exchange energies for an arbitrary lattice site. In the limit when $d \to 0$ we obtain

$$\begin{aligned}\mathcal{E}(x,y) = K_u\sin^2\theta - A_x\big[\cos\theta\,\nabla^2(\cos\theta) + \sin\theta\cos\phi\,\nabla^2(\sin\theta\cos\phi)\\ + \sin\theta\sin\phi\,\nabla^2(\sin\theta\sin\phi)\big]\end{aligned} \tag{15.28}$$

$\mathcal{E}(x,y)$ is the local energy density arising from exchange and anisotropy. Next we solve Eqs. (15.27) for two special cases and derive the wall energy in each case using Eq. (15.28).

15.2.2. Structure and energy density of straight walls

In general, θ and ϕ in Eqs. (15.27) are functions of both x and y. However, for straight walls it seems reasonable to assume that θ is a function of x only, and ϕ is a function of y only. With this assumption Eq. (15.27b) reduces to

$$\frac{d^2\phi(y)}{dy^2} = 0 \tag{15.29}$$

which has the following solution:

$$\boxed{\phi(y) = \phi_0 + (y/\Lambda_0)} \tag{15.30}$$

ϕ_0 and Λ_0 are arbitrary constants. Substituting this in Eq. (15.27a) yields

$$\frac{d^2}{dx^2}\,\theta(x) = \left(\frac{1}{\Lambda_0^{\,2}} + \frac{K_u}{A_x}\right) \sin\theta\cos\theta \tag{15.31}$$

which is similar to the one-dimensional wall equation, Eq. (13.5). Defining the wall-width parameter Δ_w as

$$\boxed{\Delta_w = \left(\Lambda_0^{-2} + (\sqrt{A_x/K_u}\,)^{-2}\right)^{-1/2}} \tag{15.32}$$

Eq. (15.31) can be integrated in the following steps:

$$\frac{d}{dx}\,\theta(x) = \frac{1}{\Delta_w}\sin\theta(x) \tag{15.33}$$

$$\tfrac{1}{2}\left(1 + \tan^2\tfrac{1}{2}\theta\right)\frac{d\theta}{dx} = \frac{1}{\Delta_w}\tan\tfrac{1}{2}\theta \tag{15.34}$$

$$\frac{d}{dx}\left[\ln\,(\tan\tfrac{1}{2}\theta)\right] = \frac{1}{\Delta_w} \tag{15.35}$$

$$\boxed{\tan\tfrac{1}{2}\theta = \exp\left(\frac{x - x_0}{\Delta_w}\right)} \tag{15.36}$$

Equations (15.30) and (15.36) describe the structure of a straight wall in the XY-plane. The arbitrary constant x_0 defines the position of the wall along X. Δ_w is the wall-width parameter given by Eq. (15.32), and Λ_0 is the size parameter for the Bloch line within the wall. If we define Bloch lines as regions through which ϕ rotates by 180°, then the length of each such region (in the Y-direction) will be $\pi\Lambda_0$. Finally, ϕ_0 is an arbitrary constant that defines the position of the Bloch lines along Y. Note that, unlike Δ_w, Λ_0 is independent of A_x and K_u and, in the absence of demagnetizing fields, is solely determined by the initial and boundary conditions. When $\Lambda_0 \to \infty$, the Bloch lines disappear, $\phi(y)$ becomes a constant ϕ_0, and $\Delta_w \to \Delta_0 = \sqrt{(A_x/K_u)}$; this, of course, is the wall without Bloch lines considered in Chapter 13. For finite values of Λ_0, however, Δ_w is always less than Δ_0, i.e., the Bloch lines narrow the wall width.

To calculate the wall's energy density, we substitute for θ and ϕ in Eq. (15.28) from Eqs. (15.30) and (15.36). This yields

$$\mathcal{E}(x,y) = K_u\sin^2\theta(x) + A_x\left\{\left[\frac{d\theta(x)}{dx}\right]^2 + \frac{1}{\Lambda_0^{\,2}}\sin^2\theta(x)\right\} \tag{15.37}$$

Observe that the local energy density is independent of the y-coordinate. Equation (15.37) can be integrated with the aid of Eq. (15.33) to yield the following expression for the wall energy density σ_w:

$$\sigma_w = \int_{-\infty}^{\infty} \mathcal{E}(x, y)\, \mathrm{d}x = 4\sqrt{A_x K_u + (A_x / \Lambda_0)^2} \tag{15.38}$$

In the limit of $\Lambda_0 \to \infty$, the Bloch lines disappear and Eq. (15.38) reduces to the standard expression for wall energy, Eq. (13.10). For finite Λ_0, the wall energy is always greater than that in the absence of Bloch lines.

15.2.3. Structure and energy density of circular walls

The two-dimensional Laplacian operator in cylindrical coordinates is written

$$\nabla^2 = \frac{1}{r}\frac{\partial}{\partial r}\left(r\frac{\partial}{\partial r}\right) + \frac{1}{r^2}\frac{\partial^2}{\partial \beta^2}$$

where r and β are the radial and azimuthal coordinates, respectively. Working in cylindrical coordinates and assuming that θ is a function of r while ϕ is a function of β, Eqs. (15.27) yield

$$\phi(\beta) = \phi_0 + n_0\beta \tag{15.39}$$

$$\left(r^2\frac{\mathrm{d}^2}{\mathrm{d}r^2} + r\frac{\mathrm{d}}{\mathrm{d}r}\right)\theta(r) = \left[n_0^{\,2} + (K_u/A_x)\, r^2\right]\sin\theta(r)\cos\theta(r) \tag{15.40}$$

Here n_0, an integer, is the number of full circle rotations of the in-plane component of wall magnetization. To obtain $\theta(r)$, one must solve Eq. (15.40) subject to the boundary conditions $\theta(0) = 0°$ and $\theta(\infty) = 180°$. This poses two problems. First, the equation is a complicated nonlinear differential equation that does not appear to have a simple closed-form solution. Second, in the absence of coercivity, external field, or demagnetization, the equation will have a unique solution corresponding to the minimum-energy domain for the given n_0. This unique solution (interesting though it may be) is of little value here, since we are interested in circular domains that can have arbitrary radii.

Fortunately, there exists a simple approximation that can solve both problems. Consider a circular domain of radius $r_0 >> \sqrt{(A_x/K_u)}$. Only in the neighborhood of r_0 does the right-hand side of Eq. (15.40) differ significantly from zero, since $\sin\theta \to 0$ when r deviates from r_0 by more than a few wall widths. Thus, replacing r^2 by r_0^2 in the right-hand side of Eq. (15.40) causes only a small error. It is now possible to solve the equation with a technique similar to that used in conjunction with Eq. (15.31). After defining the parameter ν as

$$\boxed{\nu = \left[n_0^2 + \left(r_0/\sqrt{A_x/K_u} \right)^2 \right]^{1/2}} \tag{15.41}$$

one can solve Eq. (15.40) in the following steps:

$$r \frac{\mathrm{d}\theta(r)}{\mathrm{d}r} = \nu \sin\theta(r) \tag{15.42}$$

$$\frac{\mathrm{d}}{\mathrm{d}r}\left[\ln\left(\tan \tfrac{1}{2}\theta\right) \right] = \frac{\nu}{r} \tag{15.43}$$

$$\boxed{\tan \tfrac{1}{2}\theta = (r/r_0)^{\nu}} \tag{15.44}$$

Equations (15.39) and (15.44) describe the structure of a circular wall with radius r_0, provided that $r_0 >> \Delta_0 = \sqrt{(A_x/K_u)}$. The wall width is a function of r_0, n_0, Δ_0 and can be calculated for specific values of these parameters. However, in the limit of small n_0, one can show that Eq. (15.44) is approximately the same as Eq. (15.36) with $\Delta_w = \Delta_0$. Thus, as expected, the wall-width for large circular domains (without too many Bloch lines) is the same as that for straight walls.

To calculate the wall energy we substitute the expressions for $\theta(r)$ and $\phi(\beta)$ in Eq. (15.28). This yields

$$\mathcal{E}(r,\beta) = K_u \sin^2\theta(r) + A_x \left\{ \left[\frac{\mathrm{d}\theta(r)}{\mathrm{d}r} \right]^2 + \frac{n_0^2}{r^2} \sin^2\theta(r) \right\} \tag{15.45}$$

The local energy density is thus independent of β. With the aid of Eq. (15.42) we integrate Eq. (15.45) over the entire plane to obtain the following expression for the wall energy density:

$$\boxed{\sigma_w = \frac{1}{2\pi r_0} \int_0^{\infty} 2\pi r\, \mathcal{E}(r,\beta)\, \mathrm{d}r \simeq \frac{4}{\nu} \left(K_u r_0 + \frac{n_0^2 A_x}{r_0} \right)} \tag{15.46}$$

Again, in the limit of $n_0 = 0$, Eq. (15.46) reduces to the standard formula for the domain-wall energy density, namely, Eq. (13.10).

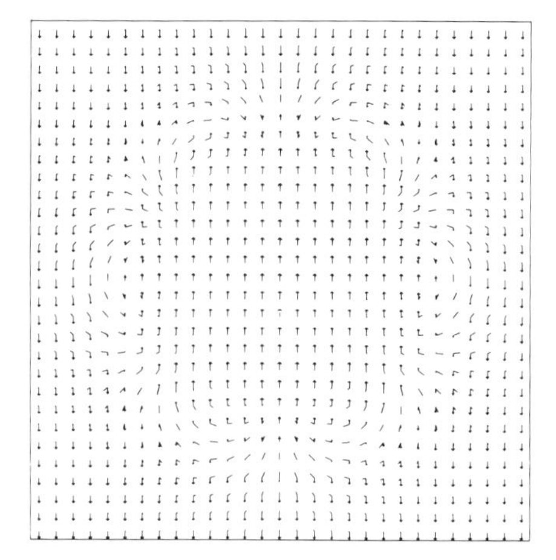

Figure 15.4. Circular domain with internal wall structure. The lattice is 30 × 30 and the domain parameters are $r_0 = 10d$, $\Delta_0 = 2d$, $n_0 = 4$.

15.2.4. Domain walls and the effect of the demagnetizing field

The long-range nature of dipole–dipole interactions makes it difficult to analyze demagnetizing effects without resorting to numerical computations. In this section we present two examples in which numerical calculations of demagnetization energy and field illustrate certain aspects of domain wall behavior in thin-film media.

Example 1. Using the FFT-based techniques of Chapter 13 we compute the demagnetization energy of a circular domain on a 256 × 256 square lattice (see Eq. (13.24)). The distribution of θ, given by Eq. (15.44), is dependent on r_0, Δ_0, and n_0. The distribution of ϕ, given by Eq. (15.39), is dependent on n_0 and ϕ_0. Figure 15.4 shows a typical magnetization distribution for a circular domain, with $r_0 = 10d$, $\Delta_0 = 2d$, and $n_0 = 4$ (d is the lattice constant). The arrows represent the perpendicular component of **M**, while the appendage to each arrow is the corresponding in-plane component. (For instance, ↿_ means that the perpendicular component is along the positive Z-axis and the in-plane component points to the right, while ⌉ means that the perpendicular component is along the negative Z-axis and the in-plane component points to the left.) Now, for symmetry reasons, the demagnetization energy is independent of ϕ_0 except when $n_0 = 1$. For $n_0 = 1$ the energy is a minimum when $\phi_0 = 90°$, i.e., when the wall everywhere around the domain is a Bloch wall; we therefore set $\phi_0 = 90°$ in these calculations.

To enable comparisons with the wall energy density σ_w of Eq. (15.46), we subtract from the computed $\mathcal{E}_{\text{dmag}}$ the value of $2\pi M_s^2$ (which is the demagnetization energy density in the absence of domains), then multiply the difference by the volume of the lattice, and normalize the result by the wall's surface area, $2\pi r_0 \tau$. The final result, denoted by $\sigma_w^{(\text{dmag})}$, is a

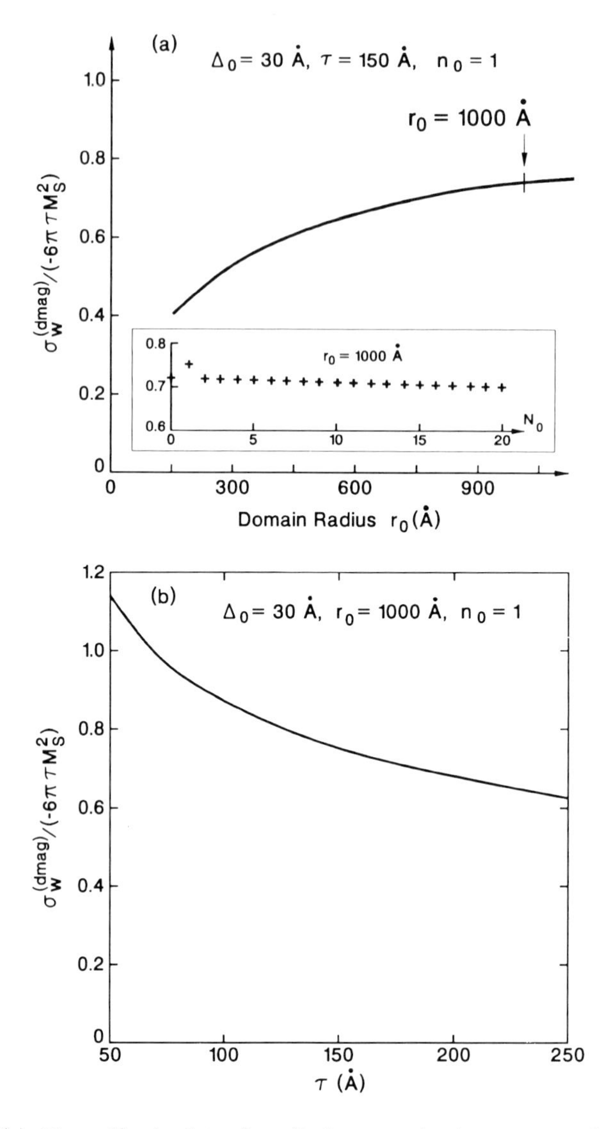

Figure 15.5. (a) Normalized plot of wall demagnetization energy density versus radius for a circular domain without internal structure (i.e., $n_0 = 1$). Fixed parameter values are $\tau = 150$ Å, $\Delta_0 = 30$ Å, and $\phi_0 = 90°$. The inset shows the normalized $\sigma_w^{(\text{dmag})}$ as a function of the number of twists in the wall, for a domain of radius $r_0 = 1000$ Å. (b) Normalized plot of $\sigma_w^{(\text{dmag})}$ versus τ for a circular domain with $r_0 = 1000$ Å. Fixed parameter values are $\Delta_0 = 30$ Å and $n_0 = 1$.

negative wall-surface energy density that represents the savings in the demagnetization energy upon creation of a circular domain. In Chapter 13 it was argued that, under proper conditions, $\sigma_w^{(\text{dmag})}$ will be close to $-6\pi\tau M_s^2$ (see Eq. (13.53)). It is therefore reasonable to compute the dimensionless quantity $\sigma_w^{(\text{dmag})}/(-6\pi\tau M_s^2)$.

Figure 15.5(a) shows a plot of $\sigma_w^{(\text{dmag})}/(-6\pi\tau M_s^2)$ versus domain radius

r_0 for a 150 Å-thick film; here Δ_0 is fixed at 30 Å, and we take $n_0 = 1$, corresponding to a Bloch wall. As expected, the normalized demagnetizing energy density of the wall is in the neighborhood of unity, and is a slowly increasing function of r_0. Next, we chose a domain with $r_0 = 1000$ Å and computed its demagnetization wall energy density as function of n_0; the results are depicted in the inset to Fig. 15.5(a). Clearly the Bloch wall with $n_0 = 1$ produces the largest savings in demagnetization energy. Other values of n_0 shown in the figure, from $n_0 = 0$ to $n_0 = 20$, are not far behind, although there is a small but steady decline in the demagnetization energy savings as the number of VBLs increases. Figure 15.5(b) is a plot of normalized wall demagnetization energy versus film thickness τ; the fixed parameter values are $r_0 = 1000$ Å, $\Delta_0 = 30$ Å, and $n_0 = 1$. The observed deviation of $\sigma_w^{(\mathrm{dmag})}$ from the estimated $-6\pi\tau M_s^2$ is due to the fact that for large τ the assumption $r_0 >> \tau$ is no longer valid, whereas for small τ the requirement of $\Delta_0 << \tau$ breaks down. □

Example 2. Consider the straight wall described by Eqs. (15.30) and (15.36) in the limit of $\Lambda_0 \to \infty$, i.e., in the absence of VBLs. The thickness-averaged demagnetizing field for this simple structure may be computed using the FFT-based method of subsection 13.2.2. It is not difficult to show that the dependence of H_θ and H_ϕ on the in-plane angle of wall magnetization, ϕ_0, may be written as follows:

$$H_\theta^{(\mathrm{dmag})} = 4\pi M_s \left[f_1(x) + f_2(x) \cos^2\phi_0 \right] \tag{15.47a}$$

$$H_\phi^{(\mathrm{dmag})} = 2\pi M_s \sin(2\phi_0)\, f_3(x) \tag{15.47b}$$

For a film with $\tau = 10\Delta_0$ the computed functions $f_1(x)$, $f_2(x)$, and $f_3(x)$ are displayed in Fig. 15.6; also shown here is the relative magnitude of the in-plane magnetization $M_s \sin\theta(x)$. For a Bloch wall ($\phi_0 = \pm 90°$) the coefficients of $f_2(x)$ and $f_3(x)$ in Eqs. (15.47) vanish, leaving the term with $f_1(x)$. For a Néel wall ($\phi_0 = 0°$ or $180°$) we find that $H_\phi^{(\mathrm{dmag})} = 0$, but $H_\theta^{(\mathrm{dmag})}$ has full contributions from both $f_1(x)$ and $f_2(x)$. The Bloch-wall configuration is stable and corresponds to the state of minimum demagnetizing energy. The Néel wall is unstable, because small deviations of ϕ_0 give rise to a field component H_ϕ that tends to move ϕ_0 away from its equilibrium value.

In section 15.2.5 below we shall encounter the case of a Bloch wall slightly perturbed by an applied magnetic field. In preparation for the forthcoming analysis, we note from Eqs. (15.47) and Fig. 15.6 that small deviations of ϕ_0 from $\pm 90°$ cause near-cancellation of the terms containing $f_1(x)$ and $f_2(x)$. Approximating $f_3(x)$ by $\sin\theta(x)$ then leads to

$$H_\theta^{(\mathrm{dmag})} \simeq 0 \tag{15.48a}$$

$$H_\phi^{(\mathrm{dmag})} \simeq 2\pi M_s \sin(2\phi_0) \sin\theta(x) \tag{15.48b}$$

This is a fairly good approximation in the vicinity of the wall center, but a rather poor one near the tails of the wall. □

15.2.5. Wall motion caused by a perpendicular magnetic field

Consider the straight wall described by Eqs. (15.30) and (15.36), and assume that a perpendicular field H_z is applied. We ignore the demagnetizing effects at this stage and postulate that the wall shape remains unchanged. Then, according to Eqs. (15.4), the only fields at work on individual dipoles will be

$$H_\theta = -H_z \sin\theta \qquad H_\phi = 0 \tag{15.49}$$

Substituting in Eqs. (15.23) we obtain

$$\dot{\theta} = \frac{\alpha\gamma}{1+\alpha^2} H_z \sin\theta \qquad \dot{\phi} = -\frac{\gamma}{1+\alpha^2} H_z \tag{15.50}$$

Finally, combining Eqs. (15.30), (15.33), (15.36), and (15.50), we find:

$$\frac{dx_0}{dt} = -\frac{d\theta/dt}{d\theta/dx} = -\frac{\alpha\gamma\Delta_w}{1+\alpha^2} H_z \qquad \frac{d\phi_0}{dt} = -\frac{\gamma}{1+\alpha^2} H_z \tag{15.51}$$

Obviously these uniform motions of x_0 and ϕ_0 do not cause any change in the shape of the wall and, therefore, the results are consistent with earlier assumptions. From Eqs. (15.51) one derives expressions for the wall mobility μ_w and the Bloch line mobility μ_L as follows:

$$\boxed{\mu_w = \frac{\alpha|\gamma|\Delta_w}{1+\alpha^2}} \qquad \boxed{\mu_L = \frac{|\gamma|\Lambda_0}{1+\alpha^2}} \tag{15.52}$$

As for the energy, we may substitute Eqs. (15.50) in Eq. (15.26b) to obtain, for a unit length of the wall,

$$\frac{d\mathcal{E}_{tot}}{dt} = \frac{2\alpha\gamma\Delta_w}{1+\alpha^2} M_s \tau H_z^2 = -2\mathcal{V} M_s \tau H_z \tag{15.53}$$

where $\mathcal{V} = dx_0/dt$ is the wall velocity. The right-hand side of Eq. (15.53) is the expected rate of change of energy for a straight wall moving at velocity $\mathcal{V}$ in a thin-film medium with thickness τ and magnetization M_s. However, we notice that part of this energy is lost to the motion of Bloch lines within the wall, without contributing to the motion of the wall itself. (Compare the terms containing $\dot{\theta}$ and $\dot{\phi}$ in Eq. (15.26b).) We will see later that the demagnetizing field can block the precession of the in-plane component and thereby increase the wall mobility.

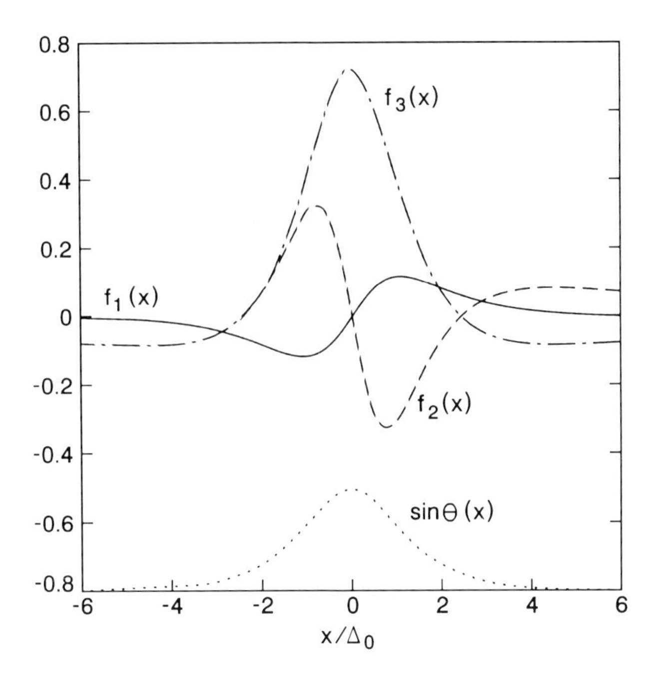

Figure 15.6. The functions $f_1(x)$, $f_2(x)$, and $f_3(x)$, which appear in the expression for the demagnetizing field of a straight wall, Eqs. (15.47).

Similar arguments apply to circular domains. From Eqs. (15.44) and (15.42) one can show that

$$r_0 \frac{d\theta}{dr_0} = -r \frac{d\theta}{dr} = -\nu \sin \theta \tag{15.54}$$

Then, using Eqs. (15.39), (15.50), and (15.54), one obtains

$$\frac{dr_0}{dt} = \frac{d\theta/dt}{d\theta/dr_0} = -\frac{r_0 \alpha \gamma}{\nu(1 + \alpha^2)} H_z \qquad \frac{d\phi_0}{dt} = -\frac{\gamma}{1 + \alpha^2} H_z \tag{15.55}$$

Note that for $n_0 = 0$, Eqs. (15.55) and (15.51) are identical.

Next we look at the effects of demagnetization on wall motion. Only straight walls will be discussed, although the discussion is qualitatively similar for circular walls. The important case turns out to be that of a wall without Bloch lines, for which $\Lambda_0 \rightarrow \infty$. (The corresponding case for circular walls is $n_0 = 1$.) In the absence of the external field, the wall minimizes its demagnetization energy by settling in the Bloch state ($\phi_0 = \pm 90°$). As soon as H_z is applied, the wall begins to move according to Eqs. (15.51). But with increasing ϕ_0 a new field, $H_\phi^{(\mathrm{dmag})}$, develops (see Eq. (15.48b)). Now, if H_z is small enough, a critical value of ϕ_0 is reached for which the following equality holds:

$$H_\theta = -H_z \sin\theta = \alpha H_\phi^{(\mathrm{dmag})} \tag{15.56a}$$

or, equivalently,

$$\boxed{H_z = -2\pi\alpha M_s \sin(2\phi_0)} \tag{15.56b}$$

When Eq. (15.56a) is used in Eqs. (15.23) we find

$$\dot{\theta} = -(\gamma/\alpha) H_\theta \qquad \dot{\phi} = 0 \tag{15.57}$$

As before, Eq. (15.57) can be used to derive the wall velocity. One obtains

$$\boxed{\frac{\mathrm{d}x_0}{\mathrm{d}t} = -\frac{\gamma \Delta_w}{\alpha} H_z \qquad \phi = \phi_0} \tag{15.58}$$

where ϕ_0 is given by Eq. (15.56b).

The wall-mobility formula obtained from Eq. (15.58) is a classical result first derived by Landau and Lifshitz.[9] It represents an upper bound on the mobility, because all the energy provided by the external field is consumed by the motion of θ, while ϕ is constant; see Eq. (15.26b) and the discussion following Eq. (15.53). On the other hand, the mobility formula for μ_w in Eqs. (15.52) is a lower bound, since in deriving it no constraints have been placed on the motion of ϕ.

15.2.6. Döring mass and Walker breakdown

Consider a straight wall without Bloch lines in the presence of demagnetizing fields and a constant external field H_z. For this wall H_θ is given by Eqs. (15.49) and H_ϕ is given by Eq. (15.48b). Substituting in Eqs. (15.23) one obtains

$$\dot{\theta} = \frac{\gamma}{1+\alpha^2}\left[\alpha H_z - 2\pi M_s \sin(2\phi)\right] \sin\theta \tag{15.59a}$$

$$\dot{\phi} = -\frac{\gamma}{1+\alpha^2}\left[H_z + 2\pi\alpha M_s \sin(2\phi)\right] \tag{15.59b}$$

From Eq. (15.59a) one can derive the wall velocity in the same way as in the preceding subsection. The result can then be combined with Eq. (15.59b) to eliminate the term containing $\sin(2\phi)$. The final result is

$$\boxed{\frac{dx_0}{dt} = -\frac{\gamma\Delta_w}{1+\alpha^2}\left[\alpha H_z - 2\pi M_s \sin(2\phi)\right]} \qquad (15.60a)$$

$$\boxed{\dot{\phi} = -\gamma H_z - (\alpha/\Delta_w)\frac{dx_0}{dt}} \qquad (15.60b)$$

Next we differentiate Eq. (15.60a) with respect to time, substitute for $\dot{\phi}$ from Eq. (15.60b), and rearrange the terms to obtain

$$-\left[\frac{1+\alpha^2}{2\pi\gamma^2\Delta_w\cos(2\phi)}\right]\frac{d^2x_0}{dt^2} + \left[\frac{2\alpha M_s}{|\gamma|\Delta_w}\right]\frac{dx_0}{dt} = 2M_s H_z \qquad (15.61)$$

The right-hand side of this equation is the force per unit area of the wall. Therefore, the coefficient of the wall acceleration must be associated with mass. For small H_z, deviation of ϕ from equilibrium (i.e., $\phi_0 = \pm 90°$) is small and, therefore, $\cos(2\phi) \simeq -1$. Also, since we are concerned with walls without Bloch lines, $\Delta_w = \Delta_0$. The mass per unit area of the wall obtained under these conditions is known as the Döring mass and is given by

$$\boxed{\mathcal{M}_D = \frac{1+\alpha^2}{2\pi\gamma^2\Delta_0}} \qquad (15.62)$$

Also, for a step change in H_z, the time constant T_0 for the approach of velocity to equilibrium is obtained from Eq. (15.61):

$$\boxed{T_0 = \frac{1+\alpha^2}{4\pi\alpha|\gamma| M_s}} \qquad (15.63)$$

It is clear from Eq. (15.56b) that there is an upper limit on H_z for Eq. (15.58) to remain valid. The maximum value of H_z is reached when ϕ_0 becomes equal to $\pm 45°$ or $\pm 135°$. This is the critical H_z for the so-called Walker breakdown. Beyond this critical field, the initial precessional motion of the in-plane wall component does not come to a halt, and the wall attains an average velocity somewhere between the previously mentioned upper and lower bounds. When the external field is no longer small enough to justify the approximations that led to Eq. (15.58), one must first solve Eq. (15.59b) to obtain the function $\phi(t)$. One then uses $\phi(t)$ in Eq. (15.60a) and finds the time dependence of the velocity.

15.3. Computer Simulations

There are at least two good reasons for resorting to computer simulations when dealing with complex physical phenomena. The first is that in the absence of reliable analytical methods, one can postulate the underlying mechanisms for the phenomena of interest and, through computer simulations, relate them to the observable macroscopic behavior. By comparing the simulation results with experiment, one can then verify the validity of one's assumptions and/or establish the relative significance of each postulated mechanism when several such mechanisms are simultaneously at work. The second reason is that computer simulations often provide guidance and testing ground for theoretical analysis. Their role in this respect is better appreciated when one considers the fact that theoretical studies are almost always based on simplifying approximations, and that their results in many instances are only indirectly related to experimental observations.

The drawback to computer simulations is that meaningful results usually require an inordinate amount of computation, beyond the reach of even the most powerful machines. The advent of parallel computers in recent years, however, has had a significant impact on the study of physical systems, particularly those with near-neighbor interactions. In the Connection Machine, for example, where 65 536 processors communicate among themselves in a fast and efficient manner,[13,14] one has the ideal environment for micromagnetic simulations.[15,16]

15.3.1. The algorithm

Consider the $N_1 \times N_2$ hexagonal lattice shown in Fig. 15.1. Each cell has diameter d and height τ. To each cell is assigned magnetization $\mathbf{M}$, anisotropy constant K_u, anisotropy axis (θ_0, ϕ_0), and exchange stiffness coefficients A_{x1}, A_{x2}, A_{x3}. The three exchange coefficients represent the interaction between a given cell and the cells immediately to its right, upper right, and lower right. Actually, since the exchange between two cells is a symmetric interaction, the parameters are not associated with the cells as such, but correspond in a one-to-one fashion with the cell borders.†

Any arbitrary distribution of magnetic parameters across the lattice is permissible. For instance, in some of the simulations of subsection 15.3.3 below, θ_0 and ϕ_0 are chosen, randomly and independently for each cell, from the intervals $[0, \Theta]$ and $[0, 2\pi]$, respectively. Θ is the parameter that identifies the range of θ_0. When $\Theta = 0$ the lattice has a perpendicular easy

†The parameters for the left-hand borders of a given cell are assigned to the corresponding neighbors on the left, thus avoiding duplication.

Table 15.1. Parameter sets used in the computer simulations

Sample	M_s (emu/cm³)	K_u (erg/cm³)	A_x (erg/cm)	Θ	γ (Hz/Oe)	α
A	100	10^6	10^{-7}	0°	-10^7	0.5
B	175	0.5×10^6	0.5×10^{-7}	20°	-10^7	0.5
C	100	10^6	10^{-7}	30°	-10^7	0.5

axis with no dispersion among the local axes. In general, Θ represents the half-angle of the cone within which the easy axes are distributed.†

The simulations begin with an initial distribution of dipoles over the lattice. Next, H_θ and H_ϕ are calculated for each cell using Eqs. (15.4), (15.7), and (15.15); the demagnetizing field is obtained with the FFT-based method. These values are inserted into Eqs. (15.23) and the values of $\dot{\Psi}$ everywhere are obtained from Eq. (15.24); these, in turn, yield a value for Δt by constraining $\Delta\Psi$ to be less than or equal to a certain (small) value. Each dipole is then rotated by $\Delta\theta$ and $\Delta\phi$ as prescribed by Eqs. (15.23) for the given Δt. The lattice thus arrives at a new state where each dipole has a slightly modified orientation. The energy of this state is calculated from Eqs. (15.5), (15.6), (15.17), and (13.24), and the process is repeated.

In all the results presented in this chapter we have taken the film thickness τ to be 500 Å and the lattice constant d to be 10 Å. The lattice dimensions are 256×256 except when specified otherwise. As for the parameters of the LLG equation, we chose a gyromagnetic ratio γ of -10^7 Hz/Oe and a damping constant α of 0.5. This value of α is chosen arbitrarily for lack of reliable experimental data; although the specific value of α used in the simulations does not appreciably affect the macroscopic steady states (i.e., equilibrium states) of the lattice in which we are interested, it does leave its imprint on the microstructure of the states. This might be understood in the light of the fact that the system of dipoles as a whole has many states of local minimum energy, and the particular state in which the system comes to rest depends on the exact path it takes upon departing from a previous state. Larger values of α result in rapid approach to equilibrium while small values prolong the simulations. Table 15.1 lists several sets of parameters that have been used in these simulations.

†This particular choice for the distribution of the local axes is merely convenient, since at present there are no definitive measurements to either support or refute it. It is reasonable, however, that in the absence of evidence to the contrary this simplest of models should be fully scrutinized. The simulation algorithm itself is not dependent on the specific distribution functions, nor does correlation among the various parameters alter the algorithm.

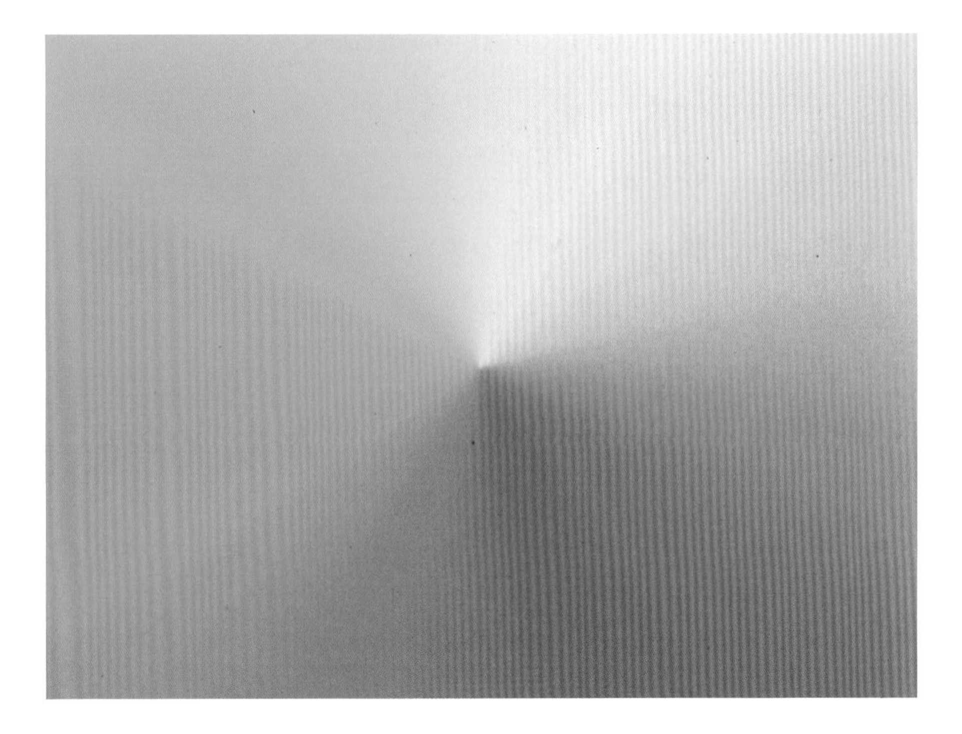

Figure 15.7. The color wheel used to encode the direction of magnetization vector in the plane of the lattice. A red pixel is associated with local magnetization along $+X$, light green corresponds to $+Y$, blue to $-X$, and purple to $-Y$. When a vector is not completely in the plane of the lattice, but has a perpendicular component, its color is obtained by mixing the color of the in-plane component with a certain amount of white (or black), the strength of white (or black) depending on the magnitude of the vertical component along $+Z$ (or $-Z$). A vector fully aligned with the $+Z$ direction is represented by a white pixel, while a vector in the $-Z$ direction is displayed as black.

Color wheel and color sphere. The magnetization pattern displays in this and the following chapters utilize a color coding scheme. Since the magnitude of the dipole moment $\mathbf{m}$ will be fixed throughout the lattice, the color sphere is used to represent its local orientation. The color sphere is white at its north pole, black at its south pole, and covers the visible spectrum on its equator in the manner shown in Fig. 15.7. As one moves from the equator to the north pole on a great circle, the color pales, i.e., it mixes with increasing amounts of white, until it becomes white at the north pole. Moving towards the south pole has the opposite effect as the color mixes with increasing amounts of black. Thus, when the magnetization vector at a given site is perpendicular to the plane of the lattice and along the positive (negative) Z-axis, its corresponding pixel will be white (black). With $\mathbf{m}$ in the plane of the lattice, the pixel is red when it points along $+X$, light green along $+Y$, blue along $-X$, and purple along

Figure 15.8. Color coded plot of magnetization in a lattice containing a stripe domain with two Bloch walls. The in-plane components of the upper and lower walls are along $+X$ and $-X$ respectively. (X is horizontal.)

$-Y$. In the same manner, other orientations of **m** map onto the corresponding color on the color sphere.

15.3.2. Structure and dynamics of simple walls

In section 15.2 we studied the structure and dynamics of straight walls and derived expressions for wall width, energy density, mobility, onset of Walker breakdown, and Döring mass. In the present section we describe computer simulations of these walls, and compare the results with theoretical predictions. This comparison will establish the basic validity and accuracy of the simulations by confirming the predicted wall structure and dynamics.

Example 3. We examine a lattice with the parameter set of sample A in Table 15.1. Initially two walls parallel to the X-axis were created and the initial conditions were chosen such that, after relaxation, both walls were simple straight Bloch walls without any Bloch lines; see Fig. 15.8. The periodic boundary conditions require the creation of pairs of walls instead of isolated walls. For sufficiently large lattices these walls are well separated and their interactions will be negligible. (In these simulations, however, we found that demagnetizing interactions could not always be ignored, as will be pointed out later.) The curves presented below were obtained by proper averaging over the upper half of the lattice and, therefore, correspond to a single wall.

Figure 15.9(a) shows plots of exchange and anisotropy energies during the relaxation process that lead to the magnetization distribution in

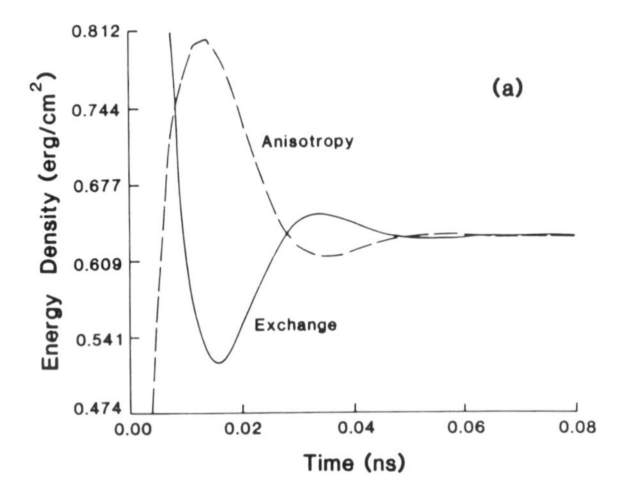

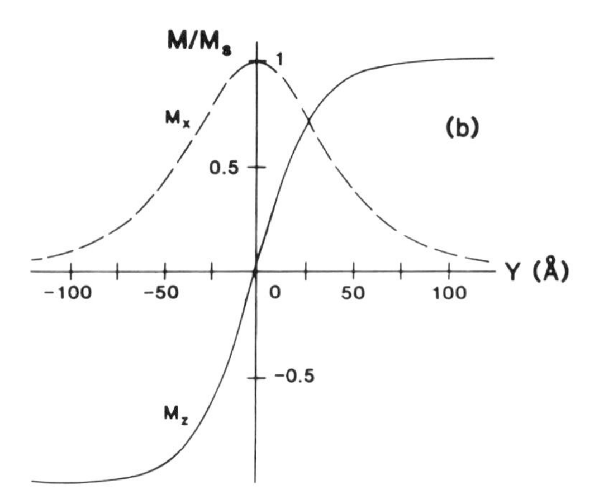

Figure 15.9. Formation of a straight Bloch wall in the upper (or lower) half of a 256×256 hexagonal lattice. The initial state consists of two sharp transitions (zero-width walls) in rows 64 and 192, parallel to the X-axis. (a) Exchange and anisotropy energy densities (per unit wall area) during the relaxation process. (b) The steady state profile of wall magnetization along Y. Both the in-plane and perpendicular components of **M** are shown.

Fig. 15.8. As expected, the two energies are equal at equilibrium and add up to about 1.25 erg/cm², which is the same as $\sigma_w = 4\sqrt{A_x K_u}$. Figure 15.9(b) shows the steady-state profiles of wall magnetization; both the Z-component and X-component of **M** are shown. The wall-width parameter Δ_0 obtained from Fig. 15.9(b) is 32 Å, which is the same as the theoretical value given by $\sqrt{A_x / K_u} = 31.6$ Å.

□

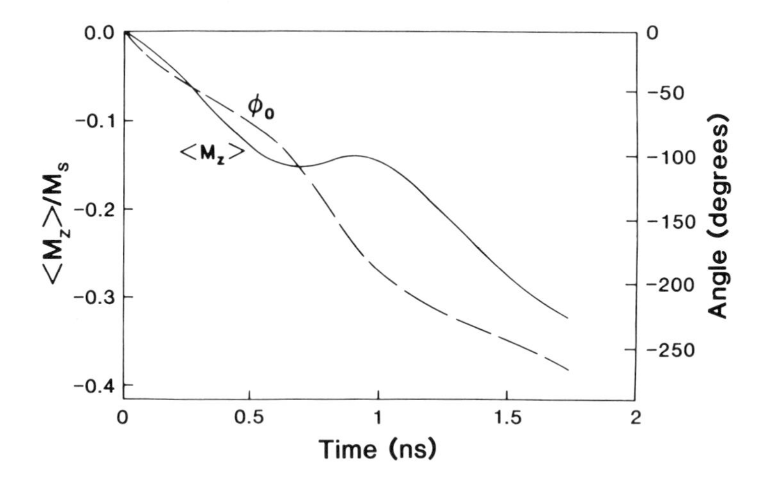

Figure 15.10. Plots of $\langle M_z \rangle$ and angle ϕ_0 during wall motion under H_z = - 500 Oe.

When the lattice in the above example is subjected to an external field along the negative Z-axis, the walls move towards each other and the stripe shrinks. The Walker breakdown field for these walls is $H = 2\pi\alpha M_s$ = 314 Oe. Next we consider several cases with applied fields both greater and less than the breakdown field, and study the dynamics of wall motion.

Example 4. Figure 15.10 corresponds to the case of H_z = - 500 Oe. It shows ϕ_0, the angle of in-plane wall magnetization, as well as $\langle M_z \rangle$, the average lattice moment along Z, both as functions of time. Since H_z is greater than the critical field, the wall does not come to an equilibrium velocity. The maximum velocity obtained from the plot of $\langle M_z \rangle$ versus time is 17 m/s, which, as expected, is less than the steady-state velocity of 31.6 m/s predicted by the Landau–Lifshitz formula, Eq. (15.58). The maximum velocity is reached when ϕ_0 is either 45° or 225°, because at these points the demagnetizing field is opposed to the rotation of the in-plane component. The lowest wall velocity occurs for ϕ_0 = 135° or 315°; at these points the demagnetizing field favors rotation of the in-plane component.

Results of a similar computer experiment with H_z = - 1000 Oe appear in Fig. 15.11; here, instead of ϕ_0, the X-component and the Y-component of **M** are plotted. Qualitatively the situation is similar to that of Fig. 15.10,

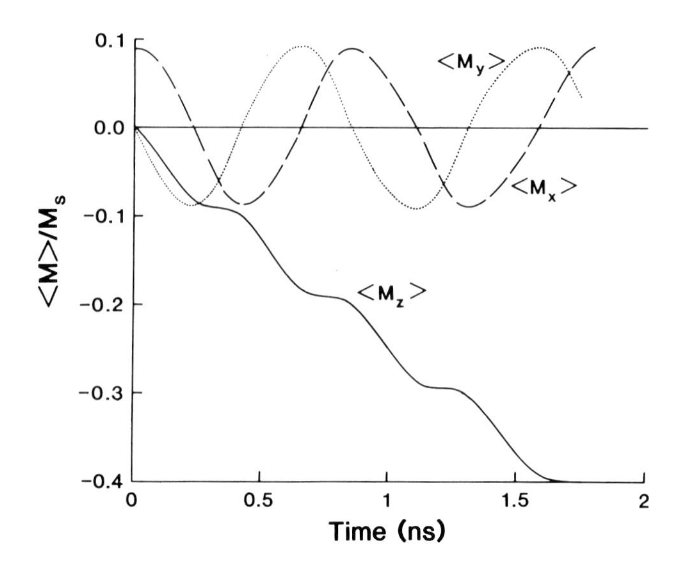

Figure 15.11. Plots of $\langle M_z \rangle$, $\langle M_x \rangle$, $\langle M_y \rangle$ during wall motion under $H_z = -1000$ Oe.

but the numerical values, of course, differ. The maximum wall velocity obtained from the plot of $\langle M_z \rangle$ versus time is now 22 m/s. □

Example 5. Figure 15.12 corresponds to the case of $H_z = -200$ Oe, which is less than the Walker breakdown field. Initially, the angle of the in-plane component moves out and reaches a maximum of 23°. (ϕ_0 as predicted by Eq. (15.56b) is 19.8°.) The wall velocity at this point is about 10.5 m/s, also in good agreement with the value predicted from Eq. (15.58), 12.65 m/s. The settling time constant from Eq. (15.63) is $T_0 = 0.2$ ns. Again this is reasonable, considering that ϕ_0 in Fig. 15.12 reaches its maximum value in about 0.75 ns, i.e., $\simeq 4T_0$. What happens next is not characteristic of isolated walls, and can only be attributed to the interaction of the periodic array of walls under consideration here. Instead of remaining constant at their steady state values, both the wall velocity and the in-plane angle decrease towards zero. Given that the original separation between the walls was about 1100 Å and that the film thickness is 500 Å, this result should not be surprising.

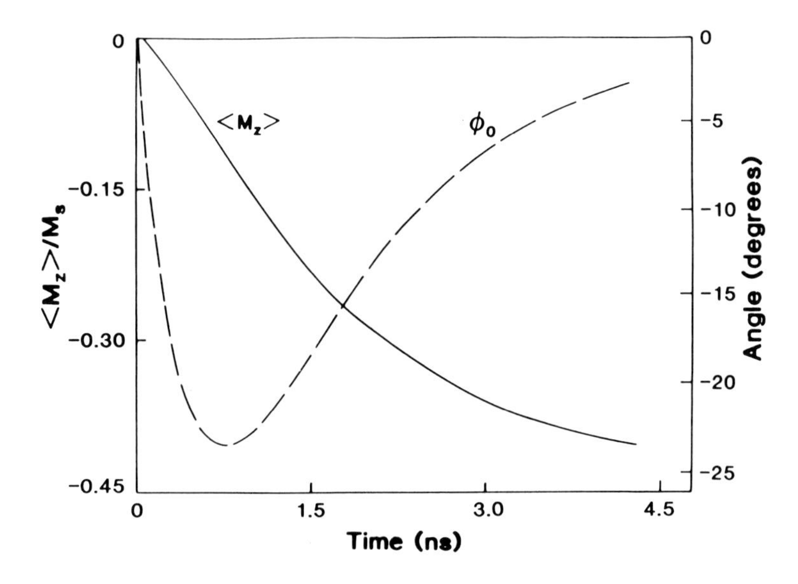

Figure 15.12. Plots of $\langle M_z \rangle$ and angle ϕ_0 during wall motion under $H_z = -200$ Oe.

Similar results for the applied field of $H_z = -100$ Oe are shown in Fig. 15.13(a). The maximum ϕ_0 is now 11°, with a corresponding wall velocity of 5.5 m/s. In order to reduce the effects of wall interaction, this experiment was repeated on a 64 × 512 lattice; the results appear in Fig. 15.13(b). Clearly the in-plane angle and the wall velocity reach the same maximum values as before, but the rate at which they drop towards zero after achieving maximum has been substantially reduced. This is to be expected, since the walls are now placed at an initial distance of 2200 Å. □

Example 6. Another interesting case whose dynamics we shall pursue in the following chapter is that of walls having internal structure. By choosing the initial conditions properly it is possible to create walls with a number of vertical Bloch lines. Figure 15.14 shows the steady-state of two parallel walls in sample A of Table 15.1; each wall has two VBLs with the same chiralities. The presence of VBLs causes the exchange and demagnetizing energies of the system to increase, but, because they have similar chiralities, the two VBLs cannot unwind or annihilate each other. □

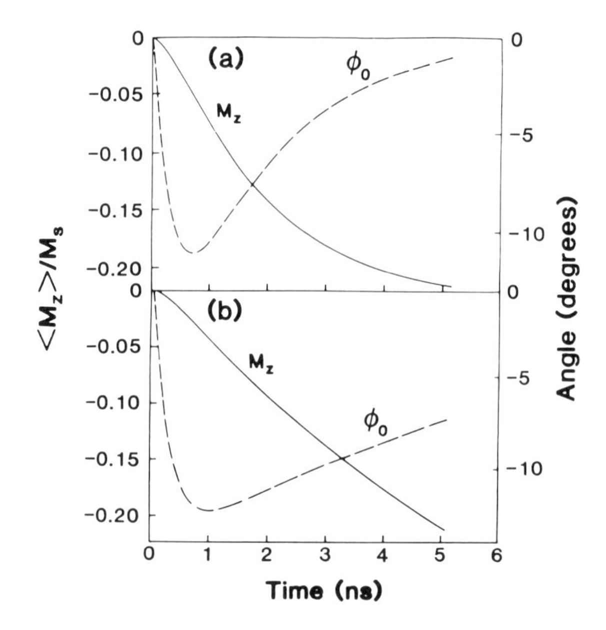

Figure 15.13. $\langle M_z \rangle$ and ϕ_0 during wall motion under $H_z = -100$ Oe. The array of walls created by the periodic boundary conditions does not act as a collection of isolated walls. The initial separation between adjacent walls is 1100 Å in (a) and 2200 Å in (b).

15.3.3. Nucleation coercivity and effects of random anisotropy

In this section we use computer simulations to study the problem of magnetization reversal in the presence of nonuniform parameter distributions.[17-19] The enormous complexity of the magnetic system under consideration makes an analytic treatment of this problem impossible. Computer simulations, on the other hand, are quite versatile, allowing one to vary a number of parameters in order to study their effects on the processes of magnetization reversal. The examples of this section are intended to show the power of large-scale simulations; an in-depth study of coercivity in MO media will be undertaken in the next chapter.

Figure 15.14. Magnetization distribution in a lattice containing a stripe domain with two straight walls. Each wall has two vertical Bloch lines with the same chiralities.

Example 7. The parameter set used in this study is that of sample B in Table 15.1, and the initial state of the lattice is uniformly magnetized along $+Z$. After relaxation in zero field during which the system arrives at equilibrium, the lattice is subjected to a reverse field in order to determine the onset of magnetization reversal. A binary search quickly yields the smallest value of the field at which the reversal begins, i.e., the nucleation coercivity. The various frames in Fig. 15.15 show the process of nucleation under an applied field of 3.16 kOe, which is the coercive field for this sample. Frame (a) shows the initial (remanent) state where all moments are in the vicinity of $+Z$. The landscape is mostly white with some pale colors that indicate slight tilts of magnetization away from the normal. In (b) the time is 1.75 ns and there are indications that nucleation is imminent. Frame (c) corresponds to $t = 1.82$ ns. The first nucleus (with a complex wall structure) has been formed and is growing. Two new domains are also forming in the lower region of the sample. Frame (d) shows that the initial nucleus has become larger, and that the lower two nuclei have merged. Succeeding frames (e)–(l) show the evolution of this process until time $t = 2.65$ ns, at which point the average lattice magnetization has dropped to $0.14 M_s$. (Although not shown in the figure, the domains continue to grow and merge until reversal is complete.) More detailed pictures of the collision at the center of frames (g)–(l) are shown in Fig. 15.16.

Figure 15.17 shows the time dependence of average magnetization $\langle M_z \rangle$ and average energy $\langle \mathscr{E}_{\text{tot}} \rangle$ for three values of the applied field. At $H_z =$ 3.2 kOe, which is somewhat above coercivity, the reversal must take place rapidly, as both $\langle M_z \rangle$ and $\langle \mathscr{E}_{\text{tot}} \rangle$ show a sharp drop at $t \simeq 1$ ns. At $H_z =$ 3.16 kOe, despite the slower rate of the process, complete reversal still occurs; compare the rates of change of $\langle M_z \rangle$ and $\langle \mathscr{E}_{\text{tot}} \rangle$ in (b) with those

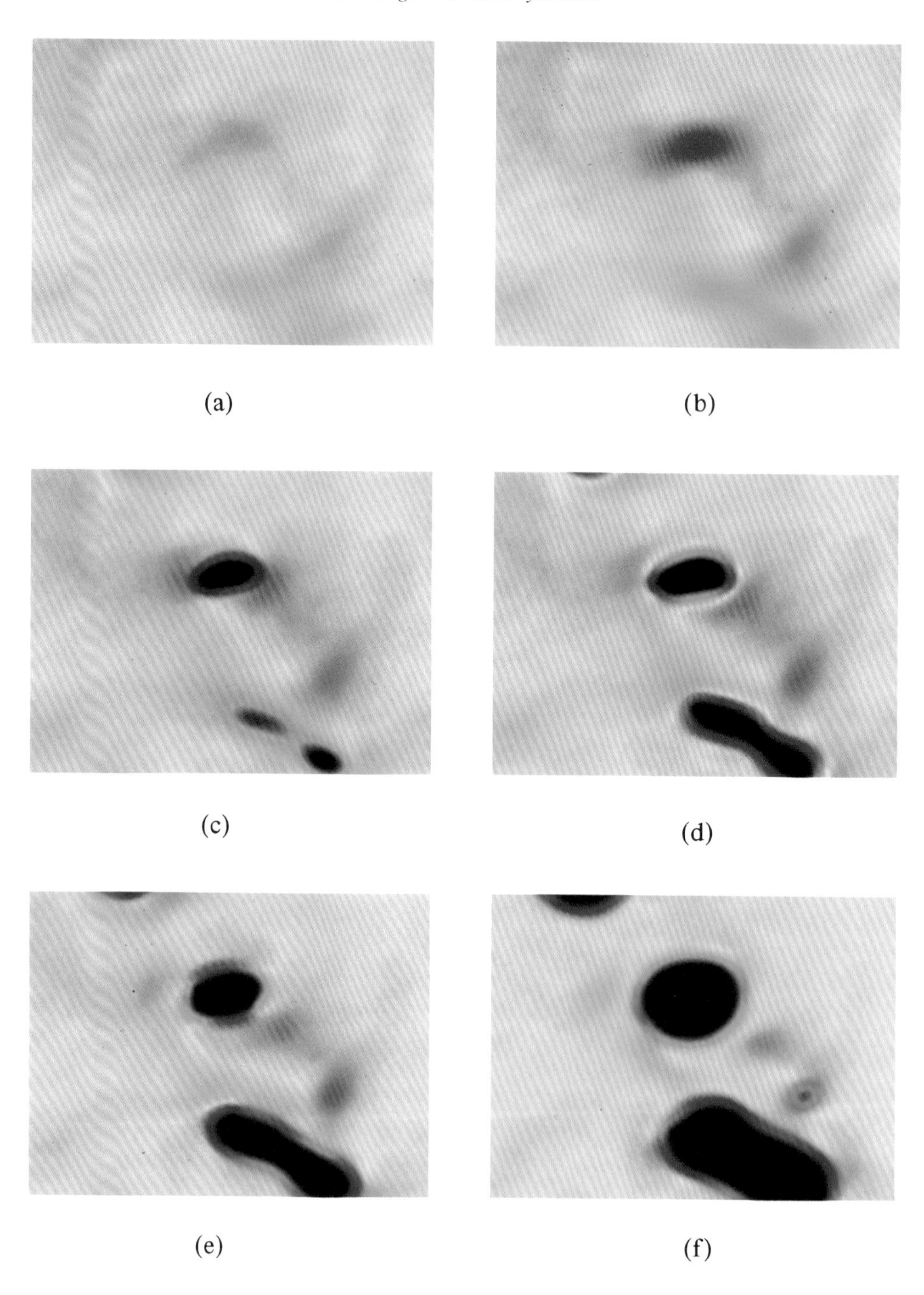

Figure 15.15. Nucleation and growth of reverse domains under an applied magnetic field. The parameters are those of sample B in Table 15.1. (a) Remanent state at $t = 0$; $\langle M_z \rangle = 0.99 M_s$. (b) Beginning of nucleation; $t = 1.75$ ns, $\langle M_z \rangle = 0.95 M_s$. (c) A nucleus has formed near the center; a pair of nuclei are forming at the lower right; $t = 1.82$ ns, $\langle M_z \rangle = 0.89 M_s$. (d) $t = 1.88$ ns, $\langle M_z \rangle = 0.85 M_s$. (e) $t = 1.96$ ns, $\langle M_z \rangle = 0.81 M_s$. (f) $t = 2.15$ ns, $\langle M_z \rangle = 0.66 M_s$.

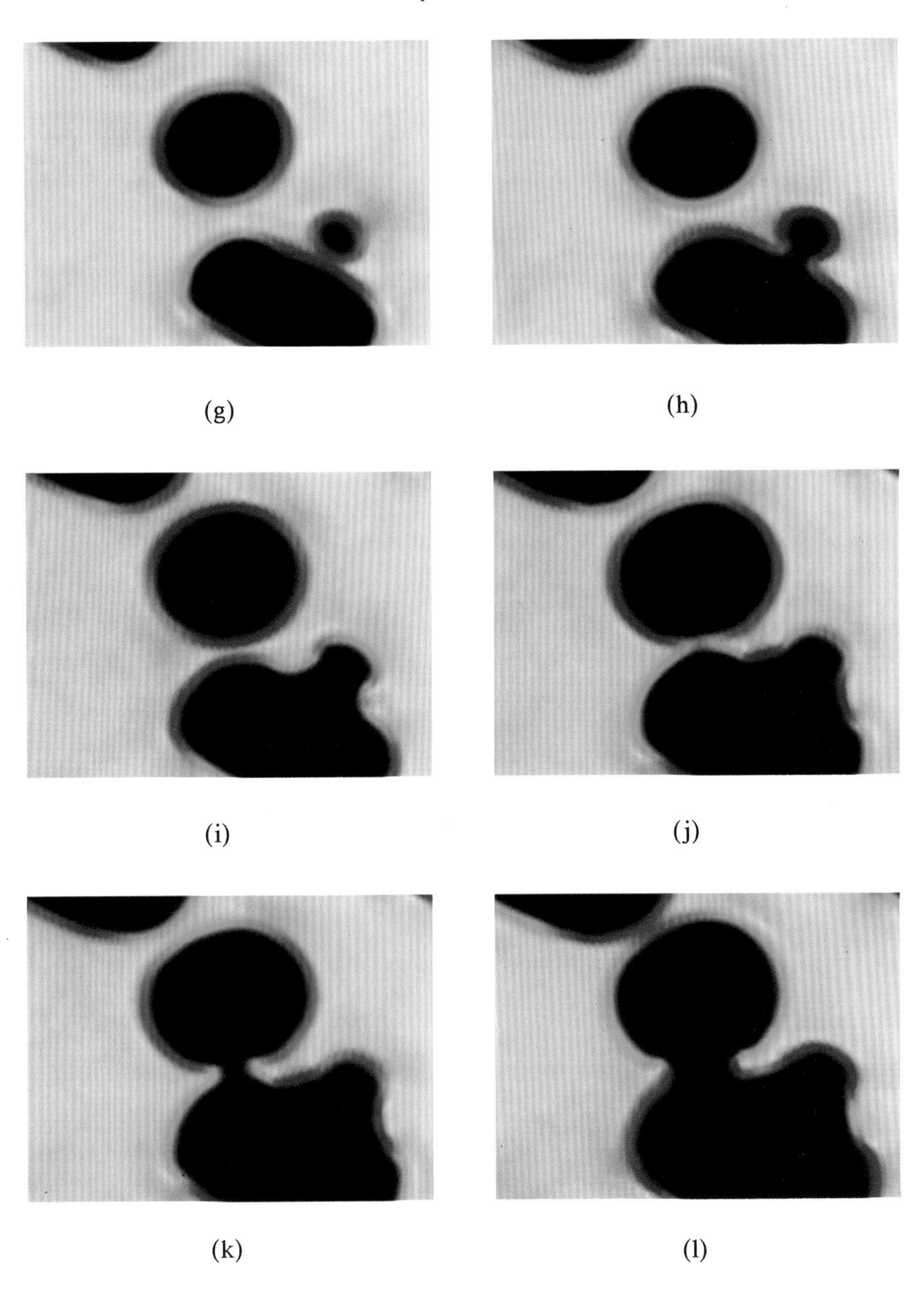

Figure 15.15. (continued)
(g) t = 2.32 ns, $\langle M_z \rangle = 0.49 M_s$. (h) t = 2.38 ns, $\langle M_z \rangle = 0.42 M_s$. (i) t = 2.49 ns, $\langle M_z \rangle = 0.31 M_s$. (j) t = 2.57 ns, $\langle M_z \rangle = 0.21 M_s$. (k) t = 2.60 ns, $\langle M_z \rangle = 0.18 M_s$. (l) t = 2.65 ns, $\langle M_z \rangle = 0.14 M_s$.

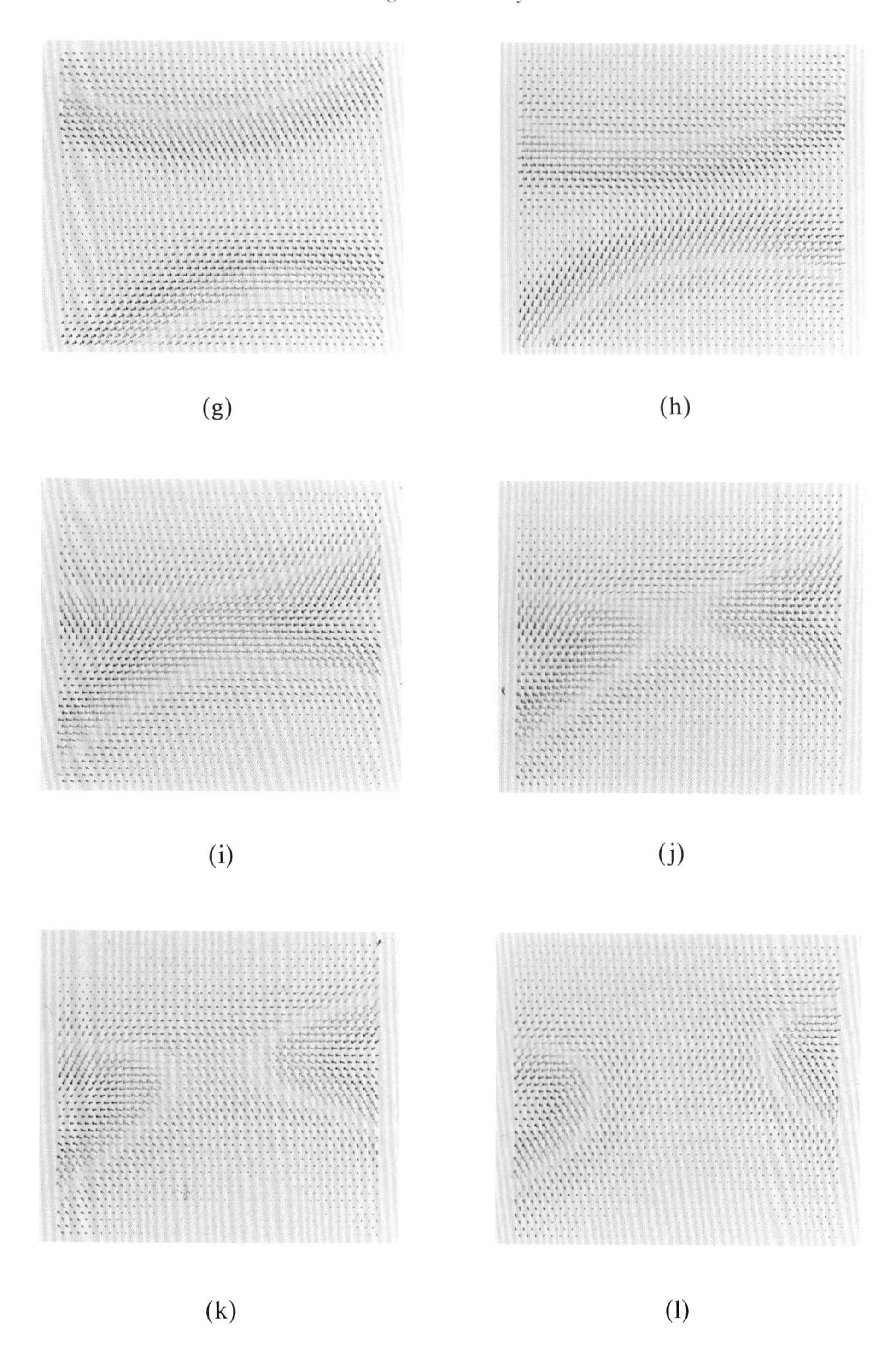

Figure 15.16. Close-up of domains colliding near the center of the frame in the preceding figure. The frames here correspond to Fig. 15.15(g)-(l).

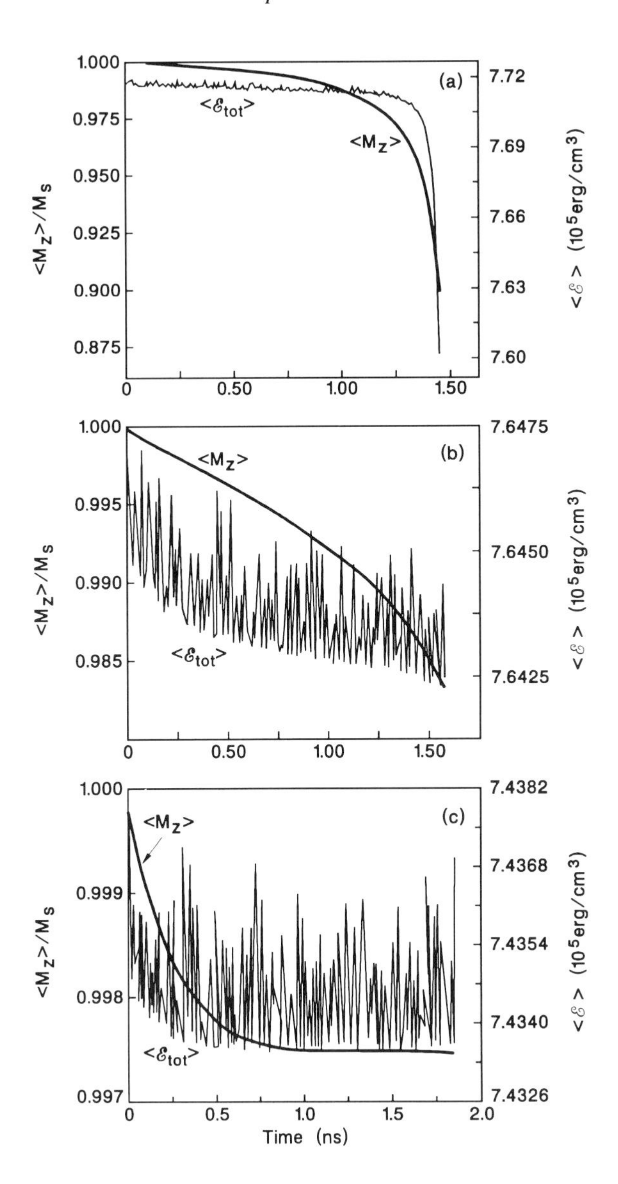

Figure 15.17. Plots of average magnetization $\langle M_z \rangle$ and average energy $\langle \mathcal{E}_{tot} \rangle$ versus time. (a) H_z = 3200 Oe; (b) H_z = 3160 Oe; (c) H_z = 3040 Oe. Note that the vertical scales are different in the three cases.

in (a). The third set of curves, in (c), corresponds to H_z = 3.04 kOe (below coercivity) and shows that the system has no tendency to reverse. □

Example 8. The parameter set used here is that of sample B, which was also the subject of study in the preceding example. The lattice is initially saturated along $+Z$, then allowed to relax to the remanent state under zero

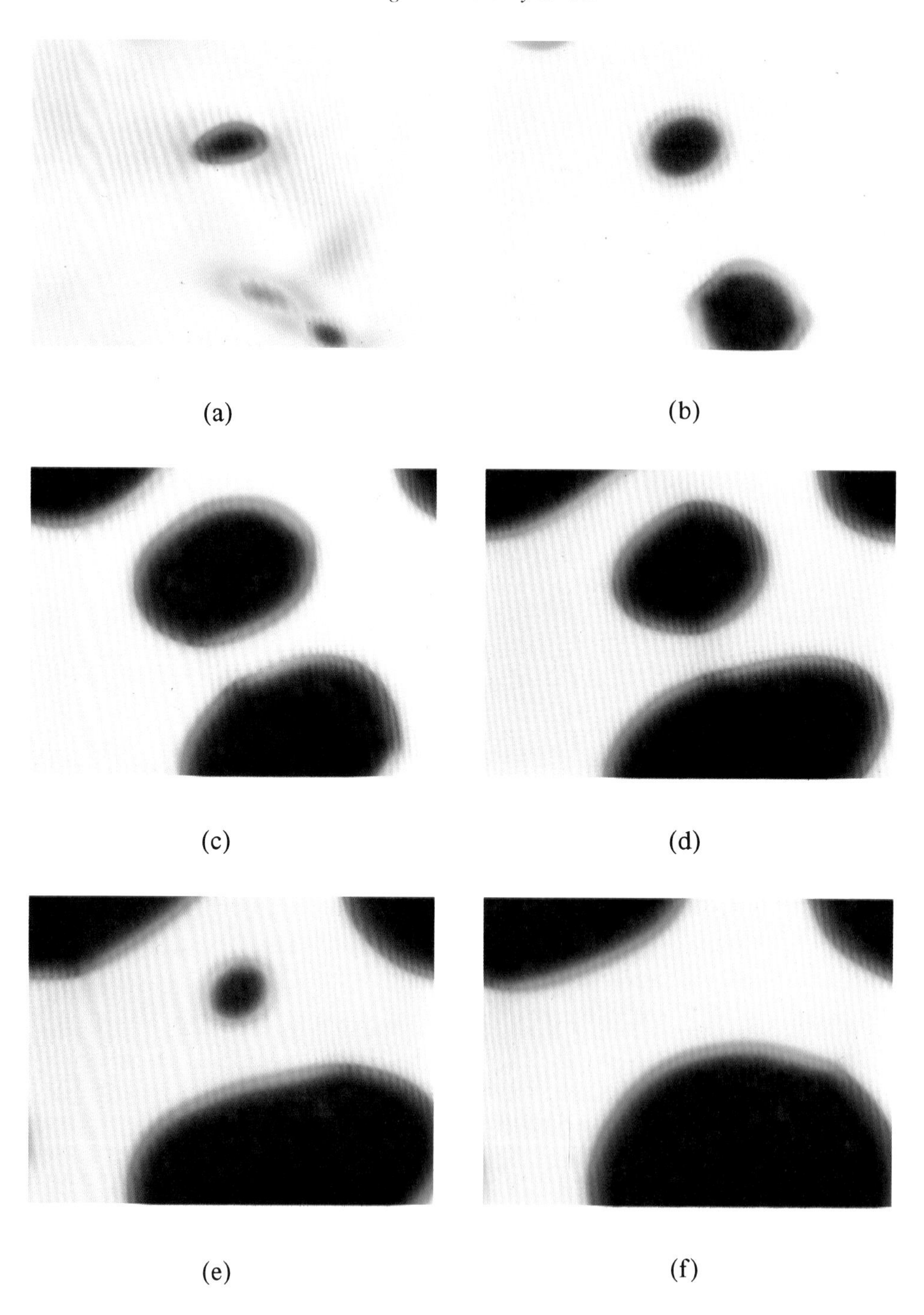

Figure 15.18. Demagnetization in the absence of external fields. The lattice is initially saturated, then briefly exposed to $H_z = -3.16$ kOe in order to create a few small nuclei. The field is then removed and the domains allowed to evolve under internal forces. (a) State of the lattice immediately after removing the external field. (b) $t = 1$ ns; the two nuclei in the lower part of the frame have merged. (c) $t = 5$ ns; the domain in the center has reached its maximum size and, from now on, will shrink. (d) $t = 20$ ns; the small domain in the center is shrinking, while the larger domain continues to grow. (e) $t = 24$ ns; the small central domain is about to burst. (f) The domain is now steady and $\langle M_z \rangle \simeq 0$.

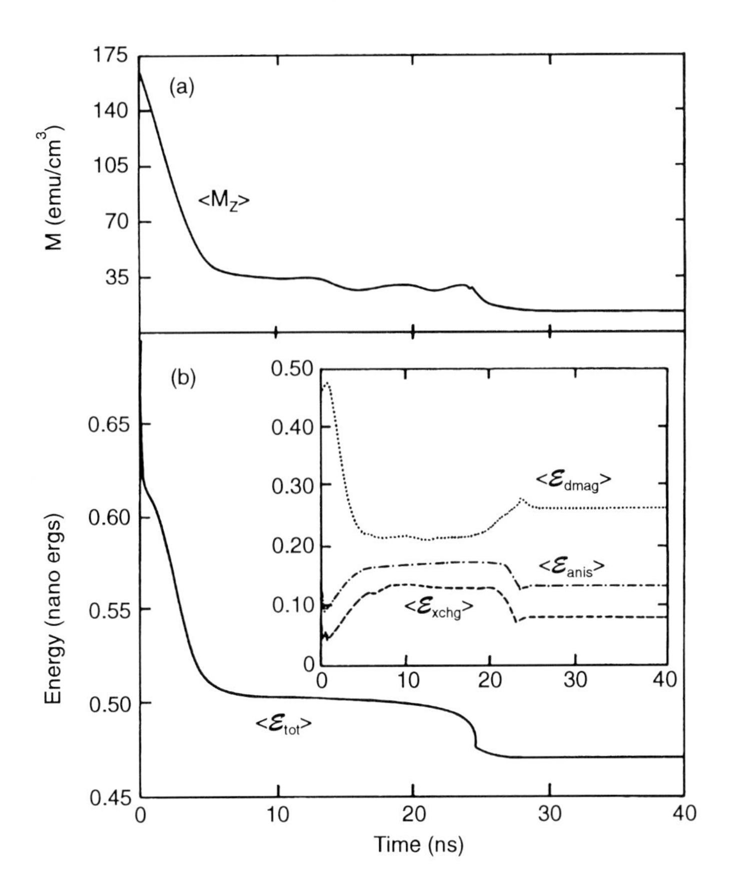

Figure 15.19. Plots of average magnetization and energy during relaxation of the lattice in Fig. 15.18. (a) $\langle M_z \rangle$ versus time. (b) Average energy of the lattice and its various components versus time.

applied field, and then subjected to a reverse field of $H_z = -3.16$ kOe. Once the nuclei have formed, however, the field is reduced to zero and the domains are left to themselves to develop under the pressure of wall energy and the forces of demagnetization. Figure 15.18 shows the various states of this development. In (a) the field has just been turned off, leaving behind three nuclei. Because of the large demagnetizing force, the nuclei tend to expand and eventually demagnetize the sample. First the two nuclei in the lower part of the frame merge; then the remaining domains expand as shown in (b) and (c). Soon, however, the larger domain begins to push the smaller one towards collapse; see (d) and (e). Eventually, the small domain disappears and the lattice reaches equilibrium as shown in (f).

Figure 15.19(a) shows the average lattice magnetization $\langle M_z \rangle$ versus time. The initial sharp drop in $\langle M_z \rangle$ occurs when the early nuclei merge and expand. The plateau corresponds to the time during which one domain expands at the expense of the other. At the end of the plateau, the sudden

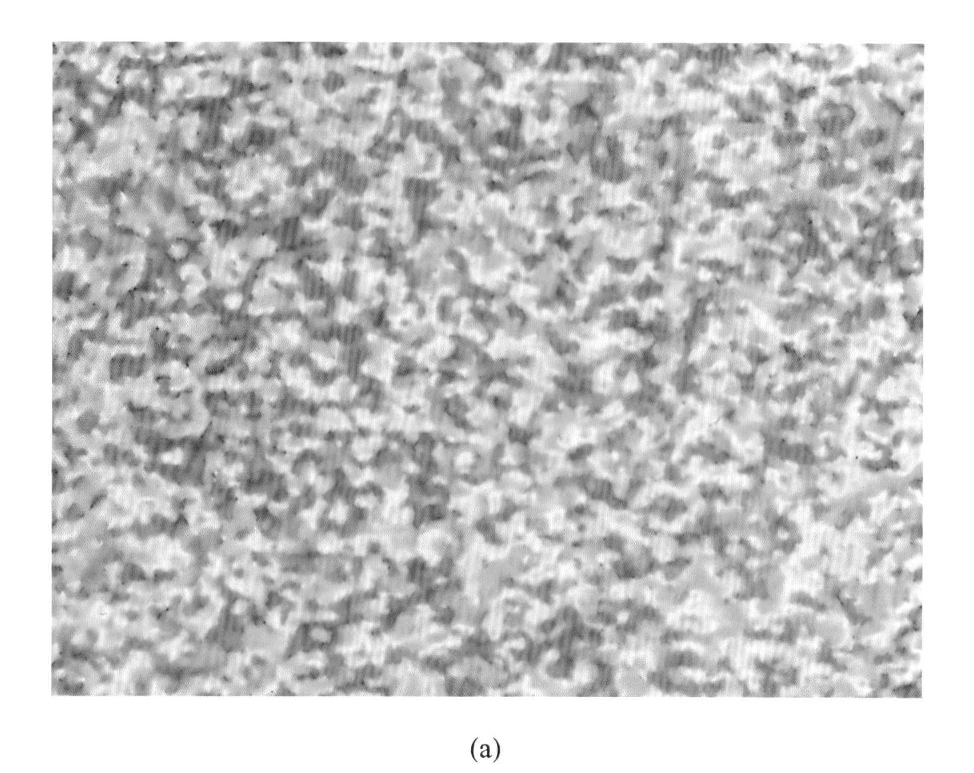

(a)

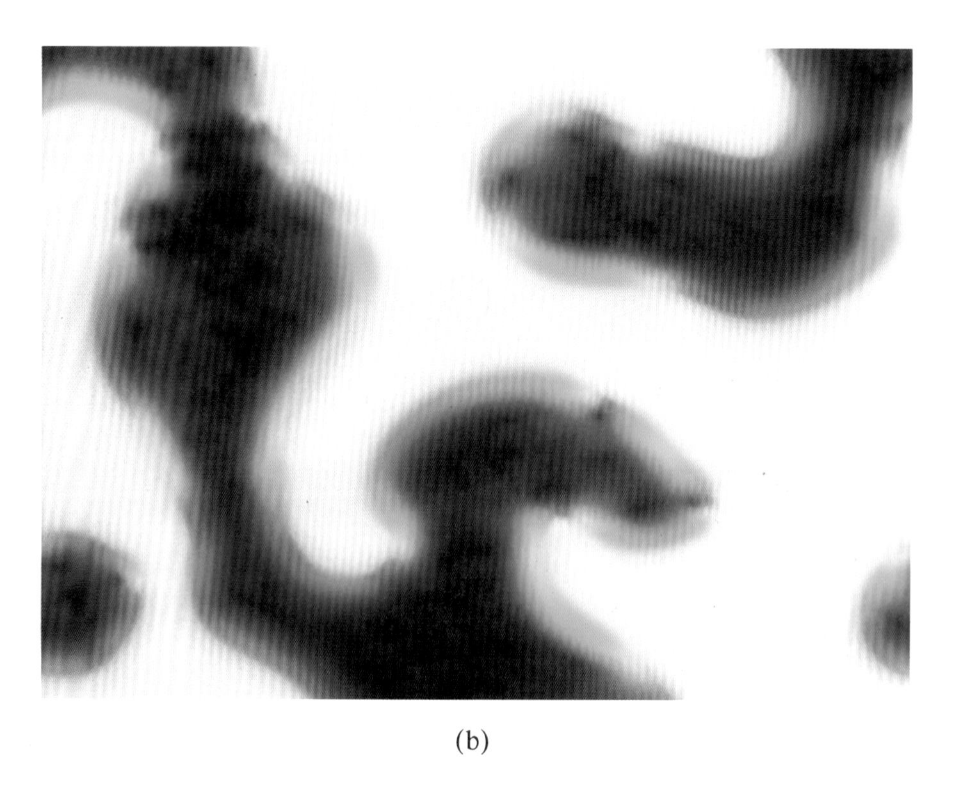

(b)

Figure 15.20. Relaxation from a random initial state in the absence of external fields. (a) Magnetization distribution at t = 5 ps. (b) State of the lattice at t = 1.2 ns.

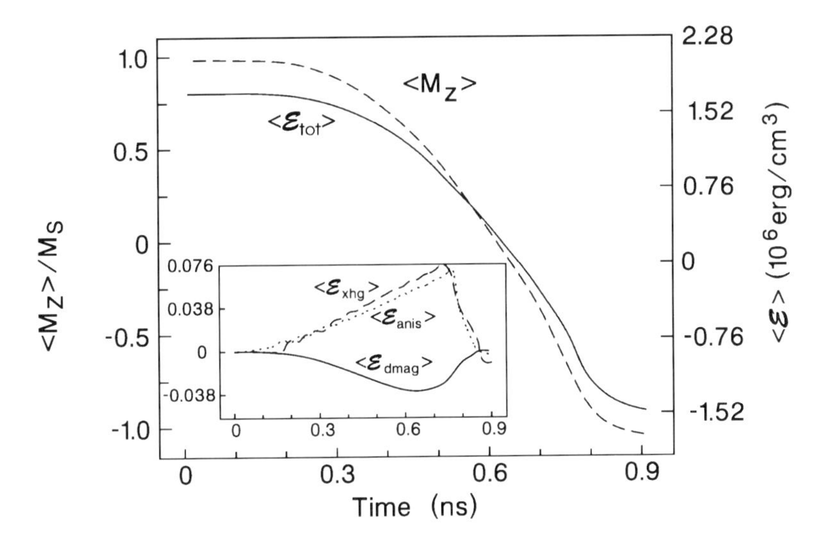

Figure 15.21. Evolution of $\langle M_z \rangle$ and $\langle \mathscr{E}_{\text{tot}} \rangle$ in time when sample C of Table 15.1 is subjected to an external field $H_z = -15.6$ kOe. The inset shows plots of the individual contributions to $\langle \mathscr{E}_{\text{tot}} \rangle$.

collapse of the small domain (similar to a bursting bubble) causes a rapid drop in $\langle M_z \rangle$. Soon afterwards the magnetization reaches an equilibrium value near zero, and the lattice stabilizes. The plot of energy versus time in Fig. 15.19(b) shows similar behavior; the inset shows the contributions of exchange, anisotropy and demagnetization to the total energy. Note how the burst of the small bubble at around $t = 23$ ns causes the demagnetization energy to rise, while at the same time the exchange and anisotropy energies (both associated with the wall energy σ_w) decrease. □

Example 9. The parameter set used in this example is again that of sample B in Table 15.1. The lattice was initialized in a random state, with each dipole equally likely to be either parallel or antiparallel to Z. After about 5 ps of relaxation the system had arrived at the state shown in Fig. 15.20(a). Small domains had clearly formed at this stage, but the system was far from equilibrium. 1.2 ns later, the system had arrived at the equilibrium state shown in Fig. 15.20(b); this final state, which is fully demagnetized, contains a few stripe domains, with several VBLs residing within their walls. The results of this computer experiment are reminiscent of those of actual experiments involving rapid cooling of a magnetic sample through its Curie temperature in zero applied field. □

Example 10. We study a lattice with the parameter set of sample C in Table 15.1. After a binary search, the nucleation coercivity is found to be $H_c \simeq 15.6$ kOe. Figure 15.21 shows plots of average magnetization $\langle M_z \rangle$ and average energy $\langle \mathscr{E}_{\text{tot}} \rangle$ versus time for the applied field of

H_z = 15.6 kOe. The inset shows plots of individual contributions to $\langle \mathscr{E}_{\text{tot}} \rangle$, namely the average exchange, anisotropy, and demagnetization energies. Note that initially the exchange and anisotropy energies increase as new nuclei are formed and the existing nuclei expand. The demagnetization energy, however, decreases during the same period of time. At about t = 0.75 ns the domain walls begin to collide and disappear, at which point the demagnetization energy begins to rise.

The large coercivity of sample C (as compared to sample B) is due to its small M_s and large K_u. We also computed the nucleation coercivity of samples similar to sample C but with different dispersions of the easy axes. For $\Theta = 20°$ the coercivity was found to be about 17 kOe, while for $\Theta = 40°$ it dropped to nearly 14 kOe.

□

Problems

(15.1) A thin magnetic film of thickness τ and saturation magnetization M_s is represented by a square lattice with lattice constant d (cell dimensions = $d \times d \times \tau$). Assuming that the magnetization within each cell is uniform, and that the direction of the moment **m** varies from cell to cell, calculate the magnetostatic interaction energy between a pair of such cells chosen arbitrarily from the lattice.

(15.2) The LLG equation of magnetization dynamics yields to simple geometrical interpretations. For example, when considered as geometrical objects, the vectors $\boldsymbol{\mu} \times \mathbf{H}_{\text{eff}}$ and $\boldsymbol{\mu} \times \dot{\boldsymbol{\mu}}$ appearing on the right-hand side of Eq. (15.21) are both within a plane perpendicular to $\boldsymbol{\mu}$. Consequently $\dot{\boldsymbol{\mu}}$ must be in the same plane; moreover, $\dot{\boldsymbol{\mu}}$ and $\boldsymbol{\mu} \times \dot{\boldsymbol{\mu}}$ must be mutually orthogonal.
(a) Draw a vector diagram that shows the relationship among these vectors.
(b) Determine the trajectory of $\dot{\boldsymbol{\mu}}$ when the damping parameter α goes from 0 to ∞.
(c) For what value of α is the rate of approach of $\boldsymbol{\mu}$ to equilibrium greatest?

(15.3) The original version of the Landau–Lifshitz equation was

$$\dot{\boldsymbol{\mu}} = \gamma_0 \boldsymbol{\mu} \times \mathbf{H}_{\text{eff}} + \lambda \boldsymbol{\mu} \times (\boldsymbol{\mu} \times \mathbf{H}_{\text{eff}})$$

This equation, which is now replaced by the Gilbert version (Eq. (15.21)), has the unphysical property that its predicted rate of approach to equilibrium increases with the increasing of the damping constant λ. Nonetheless, the Landau–Lifshitz form of the equation can be cast in the Gilbert form by cross-multiplication of both sides with $\boldsymbol{\mu}$, and application of the vector identity $\mathbf{u} \times [\mathbf{u} \times (\mathbf{u} \times \mathbf{A})] = -\mathbf{u} \times \mathbf{A}$, where $\mathbf{u}$ is an arbitrary unit vector and $\mathbf{A}$ is arbitrary. What is the relationship between γ_0 and λ of the original Landau–Lifshitz equation and the parameters γ and α of the Gilbert form?

(15.4) Starting from Eq. (15.25) and following the steps outlined in subsection 15.1.3, verify the expression given for the rate of change of energy in Eq. (15.26b).

(15.5) In subsection 15.2.1 we derived a pair of equations, Eqs. (15.27), for the structure of domain walls in two-dimensional space. The starting point for this derivation was the assertion that the net torque on any given dipole in equilibrium must be zero. An alternative approach to the same problem involves energy minimization arguments and the use of variational techniques, similar to those used in section 13.1. Use the latter approach to arrive at Eqs. (15.27).

16
Origins of Coercivity

Introduction

The process of magnetization reversal in thin magnetic films is of considerable importance in erasable optical data storage.[1-12] The success of thermomagnetic recording and erasure depends on the reliable and repeatable reversal of magnetization in micron-sized areas within the storage medium. A major factor usually encountered in descriptions of the thermomagnetic write and erase processes is the coercivity of the magnetic material. Technically, the coercivity H_c is defined for a hysteresis loop† as the value of the applied field at which the net magnetization becomes zero. Coercivity, however, is an ill-defined concept which may be useful in the phenomenology of bulk reversal, but its relevance to the phenomena occuring on the spatial and temporal scales of thermomagnetic recording must be seriously questioned. To begin with, there is the problem of distinguishing the nucleation coercivity from the coercivity of wall motion. Then there is the question of speed and uniformity of motion as the wall expands beyond the site of its origination. Finally one must address issues of stability and erasability, which are intimately related to coercivity, in a framework wide enough to allow the consideration of local instabilities and partial erasure. It is fair to say that the existing theories of coercivity[13-25] are generally incapable of handling the problems associated with thermomagnetic recording and erasure. In our view, the natural vehicle for conducting theoretical investigations in this area is computer simulation based on the fundamental equations of micromagnetics, the basis for which

†In the present context a hysteresis loop is the plot of magnetization versus the applied field that is obtained when the direction of the field is fixed along the easy axis of the sample, while its magnitude is cycled between a large negative and a large positive value. The hysteresis loop is representative of the sample as a whole, rather than of the properties of a particular location on the sample. This assertion, that the loop is a bulk property of the film, is obviously correct when a vibrating sample magnetometer (VSM) is used to monitor the net magnetization, or when the extraordinary Hall effect is employed to trace the loop by monitoring a particular subnetwork within the sample. But even with local measurements, such as when a focused beam of light is used to probe the magneto-optical Kerr effect at a given location, the resulting loop is often "non-local" in the following sense: a domain nucleated anywhere within the sample can expand and sweep under the beam, thus creating the false impression that the area directly under the beam has been reversed by local nucleation. See Chapter 18 for methods of media characterization including vibrating sample magnetometry, extraordinary Hall effect measurements, and magneto-optical loop tracing.

was laid down in the preceding chapter. The purpose of the present chapter is to describe several such studies aimed at clarifying the issues involved in the nucleation stage of the reversal process, as well as those associated with the dynamics of wall motion within microscopically inhomogeneous media.

Details of the computer simulation approach to micromagnetics, with lattices of magnetic dipoles representing thin magnetic films, were described in Chapter 15. Salient features of these simulations may be summarized as follows. A two-dimensional hexagonal lattice of dipoles, mimicking the actual MO medium,[††] obeys the dynamic Landau–Lifshitz–Gilbert equation at every node. The dipoles gyrate under the influence of an effective field that arises from local anisotropy, near-neighbor exchange, classical dipole–dipole interactions, and an externally applied magnetic field. Since the magnetic parameters (such as the magnitude of the dipole moments, the strength of anisotropy, the direction of the local easy axis, the value of the exchange stiffness coefficient, etc.) can be independently assigned to individual lattice sites, it is possible to simulate various defects and inhomogeneities on arbitrarily chosen length-scales.

The focus of this chapter is the mechanism of magnetization reversal in thin films with perpendicular anisotropy, taking into account the very important fact that these films are inhomogeneous on the microscopic scale. The Stoner–Wohlfarth theory of magnetization reversal by coherent rotation provides many clues as to the behavior of real films; this theory is discussed in detail in section 16.1. The rest of the chapter is based on computer simulation results that describe the role of defects and nonuniformities in initiating, sustaining, and controlling the reversal process. The massive parallelism of the Connection Machine on which these simulations have been performed, together with the fast Fourier transform algorithm for demagnetizing field calculations (section 13.2) enabled accurate simulations of a large (256 × 256) hexagonal lattice of dipoles. Since the lattice constant d is chosen to be 10 Å, the total simulated area corresponds to a 0.256×0.222 $(\mu m)^2$ section of the film. Periodic boundary conditions are imposed in the X- and the Y-direction, but the film thickness (along Z) is finite, and the magnetization distribution in the Z-direction is assumed uniform. Unless otherwise specified, the magnetization-pattern displays in this chapter utilize the color-coding scheme described in subsection 15.3.1.

We shall see in section 16.2 that the fields required to initiate the reversal process in a truly uniform material are generally higher than those observed in practice. Various submicron-sized "defects" are then introduced into the magnetic state of the lattice, and the values of nucleation coerci-

[††] Examples of such media are amorphous RE–TM alloys and polycrystalline "superlattices" of Co/Pt consisting of very small crystallites. In either case, the films are known to have uniform composition, and are considered homogeneous on a macroscopic scale. This uniformity and homogeneity, however, does not extend all the way down to the atomic scale of dimensions. On the scale of tens of nanometers, defects and spatial fluctuations of magnetic and/or structural parameters are believed to exist. We shall argue that such small-scale nonuniformities are responsible for the observed coercive behavior of MO media.

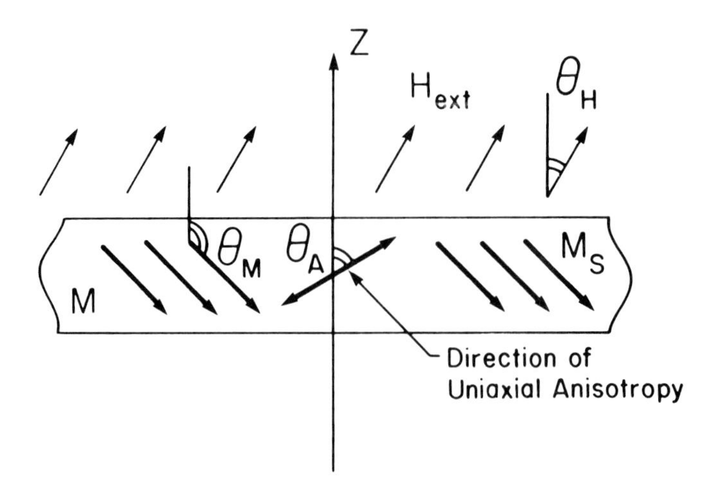

Figure 16.1. Cross-sectional view of a thin magnetic film with uniaxial anisotropy and uniform magnetization under an externally applied field H_{ext}. All angles are measured clockwise from the positive Z-axis. $\mathbf{M}_s$ is the (saturated) magnetization of the film and its angle is denoted by Θ_M. The angle of the applied field is Θ_H, while the angle between Z and the axis of anisotropy is Θ_A. The field, the magnetization, the axis of anisotropy, and the Z-axis are coplanar.

vity corresponding to different types, sizes, and strengths of these defects are computed. Voids, for instance, are found to have an insignificant effect on the value of the nucleation field, but reverse-magnetized seeds, formed and stabilized in areas with large local anisotropy, can substantially reduce the nucleation coercivity. Similarly, the presence of spatial variations in the material parameters, such as random-axis anisotropy,[26,27] will be found to affect the coercivity of nucleation.

Random spatial fluctuations and structural and/or magnetic defects also create barriers to domain wall motion. These barriers are overcome only when sufficiently large magnetic fields (in excess of the so-called wall motion coercivity) are applied. Simulations reveal that the wall coercivity in MO media is generally less than the corresponding nucleation coercivity; this finding is in agreement with the experimentally observed square shape of the hysteresis loops. The strength of the wall coercivity, of course, depends on the type and size of spatial fluctuations and/or defects; the simulation results described in section 16.3 should clarify some of these relationships.

16.1. The Stoner–Wohlfarth Theory of Magnetization Reversal

The theory of magnetization reversal by coherent rotation was developed by Stoner and Wohlfarth in the context of elongated fine particles.[28] Their theory has since been adapted and applied to reversal in thin films.[29–31] In this section we present the Stoner–Wohlfarth theory in a generalized form that also accounts for the demagnetizing effects in thin films.

Consider the uniform film in Fig. 16.1 having magnetization M_s and uniaxial anisotropy constant K_u. The axis of anisotropy has angle Θ_A with the Z-axis. Assuming coherent magnetization processes, we denote by Θ_M the angle between the magnetization vector and Z. The applied field H_{ext} is also uniform and its angle with Z is denoted by Θ_H. All angles are measured clockwise from the positive Z-axis, as indicated. The question we are about to address is as follows. For a fixed set of values of K_u, Θ_A, M_s, and Θ_H, how does the angle Θ_M vary with the magnitude of the applied field, H_{ext}? In particular, what is the equilibrium value of Θ_M when $H_{ext} = 0$, and how does Θ_M change as H_{ext} increases from zero to infinity in the fixed direction given by Θ_H?

To answer this question, we consider the magnetic energy $\mathscr{E}_m$ of the system, which consists of the external field energy, the demagnetization energy, and the anisotropy energy:

$$\mathscr{E}_m = - M_s H_{ext} \cos(\Theta_M - \Theta_H) + 2\pi M_s^2 \cos^2\Theta_M + K_u \sin^2(\Theta_M - \Theta_A) \tag{16.1}$$

The second and third terms in Eq. (16.1) may be combined to yield

$$\begin{aligned}\mathscr{E}_m = & -M_s H_{ext} \cos(\Theta_M - \Theta_H) + \tfrac{1}{2}\left(K_u + 2\pi M_s^2\right) \\ & - \sqrt{\left(\tfrac{1}{2}K_u\right)^2 - \pi M_s^2 K_u \cos(2\Theta_A) + (\pi M_s^2)^2} \\ & \times \cos\left\{2\Theta_M - \tan^{-1}\left[\frac{K_u \sin(2\Theta_A)}{K_u \cos(2\Theta_A) - 2\pi M_s^2}\right]\right\}\end{aligned} \tag{16.2}$$

Next we define an effective internal field H_{eff} and its associated angle Θ_{eff} as follows:

$$H_{eff} = \sqrt{(2K_u/M_s)^2 - 16\pi K_u \cos(2\Theta_A) + (4\pi M_s)^2} \tag{16.3}$$

$$\Theta_{eff} = \tfrac{1}{2} \tan^{-1}\left(\frac{(2K_u/M_s)\sin(2\Theta_A)}{(2K_u/M_s)\cos(2\Theta_A) - 4\pi M_s}\right) \tag{16.4}$$

In evaluating Θ_{eff} from Eq. (16.4) it is imperative that one should take into consideration the signs of both the numerator and the denominator of the arctangent's argument. The value thus obtained for the arctangent should be somewhere in the interval between $[0, 2\pi]$, resulting in a value of Θ_{eff} between 0 and π. Note that both Θ_{eff} and H_{eff} are constants, their values depending only on the internal parameters M_s, K_u, and Θ_A of the film. Later, we will show that in the absence of an external field the equilibrium orientation of magnetization is along this effective field.

Using the above definitions for H_{eff} and Θ_{eff}, the expression for energy in Eq. (16.2) may be rewritten as

$$\mathscr{E}_m = \tfrac{1}{2}(K_u + 2\pi M_s^2) - M_s H_{\text{ext}} \cos(\Theta_M - \Theta_H) - \tfrac{1}{4} M_s H_{\text{eff}} \cos\left[2(\Theta_M - \Theta_{\text{eff}})\right] \tag{16.5}$$

Since H_{eff} and Θ_{eff} are constants, independent of the magnitude and orientation of the applied field, one defines the relative values of Θ_M, Θ_H, and H_{ext} as follows:

$$\hat{\Theta}_M = \Theta_M - \Theta_{\text{eff}} \tag{16.6}$$

$$\hat{\Theta}_H = \Theta_H - \Theta_{\text{eff}} \tag{16.7}$$

$$\hat{H}_{\text{ext}} = H_{\text{ext}}/H_{\text{eff}} \tag{16.8}$$

In terms of these relative parameters, Eq. (16.5) is now written:

$$\mathscr{E}_m = \tfrac{1}{2}(K_u + 2\pi M_s^2) - \tfrac{1}{2} M_s H_{\text{eff}} \left[2\hat{H}_{\text{ext}} \cos(\hat{\Theta}_M - \hat{\Theta}_H) + \tfrac{1}{2}\cos(2\hat{\Theta}_M)\right] \tag{16.9}$$

The problem is now reduced to finding the equilibrium value of $\hat{\Theta}_M$ as a function of $\hat{H}_{\text{ext}}$ for a fixed value of $\hat{\Theta}_H$. To this end, we differentiate $\mathscr{E}_m$ with respect to $\hat{\Theta}_M$, and set the derivative equal to zero in order to find the minima and maxima of the energy function. We find

$$\frac{\partial \mathscr{E}_m}{\partial \hat{\Theta}_M} = \tfrac{1}{2} M_s H_{\text{eff}} \left[2\hat{H}_{\text{ext}} \sin(\hat{\Theta}_M - \hat{\Theta}_H) + \sin(2\hat{\Theta}_M)\right] \tag{16.10}$$

Aside from an irrelevant constant coefficient, the right-hand side of the above equation varies as the difference between the sinusoidal functions

$$F(\hat{\Theta}_M) = 2\hat{H}_{\text{ext}} \sin(\hat{\Theta}_M - \hat{\Theta}_H) \tag{16.11}$$

and

$$G(\hat{\Theta}_M) = -\sin(2\hat{\Theta}_M) \tag{16.12}$$

Figure 16.2 shows plots of these functions with $\hat{\Theta}_H$ arbitrarily set to 45°, with several values of $\hat{H}_{\text{ext}}$ chosen from 0 to 1 in steps of 0.1. For given values of $\hat{H}_{\text{ext}}$ and $\hat{\Theta}_H$, the curves of $F(\hat{\Theta}_M)$ and $G(\hat{\Theta}_M)$ cross in at most four points, at which points the derivative of $\mathscr{E}_m$ is zero. To determine those crossing points that correspond to actual minima of energy, we note in Fig. 16.2 that as one moves from the left to the right of a crossing point

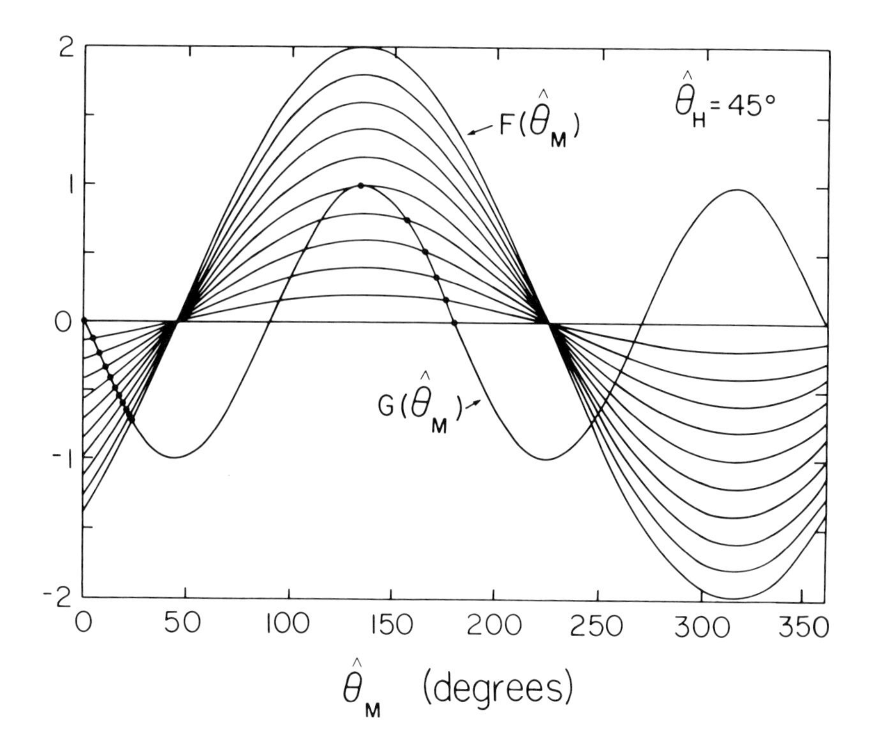

Figure 16.2. Plots of the functions $F(\hat{\Theta}_M)$ and $G(\hat{\Theta}_M)$ defined in Eqs. (16.11) and (16.12). The various $F(\hat{\Theta}_M)$ shown here have $\hat{\Theta}_H = 45°$ and $\hat{H}_{ext} = 0$ to 1 in steps of 0.1. The points $\hat{\Theta}_M$ at which $F(\hat{\Theta}_M)$ crosses $G(\hat{\Theta}_M)$ from below correspond to minima of the magnetic energy $\mathscr{E}_m$. These crossing points are identified on the figure by solid circles.

corresponding to a minimum the slope of $\mathscr{E}_m$, which is proportional to $F(\Theta) - G(\Theta)$, goes from negative to zero, then to positive. In other words, before the crossing point $F(\Theta)$ must be less than $G(\Theta)$, and after the crossing it must be greater. Those crossing points that satisfy this constraint are marked with a small solid circle in Fig. 16.2.

At $\hat{H}_{ext} = 0$ there are always two stable values for $\hat{\Theta}_M$, namely, 0° and 180°, corresponding to $\Theta_M = \Theta_{eff}$ and $\Theta_M = \Theta_{eff} + 180°$. For the situation depicted in Fig. 16.2, $\hat{\Theta}_H = 45°$, that is, $\Theta_H = \Theta_{eff} + 45°$. Now, if the system happens to be in the stable state for which $\Theta_M = \Theta_{eff}$ when the applied field is zero, then, as $\hat{H}_{ext}$ increases, the crossing point moves towards larger values of $\hat{\Theta}_M$ until it reaches $\hat{\Theta}_M = \hat{\Theta}_H = 45°$ for infinitely large $\hat{H}_{ext}$. On the other hand, if originally $\Theta_M = \Theta_{eff} + 180°$, then, as $\hat{H}_{ext}$

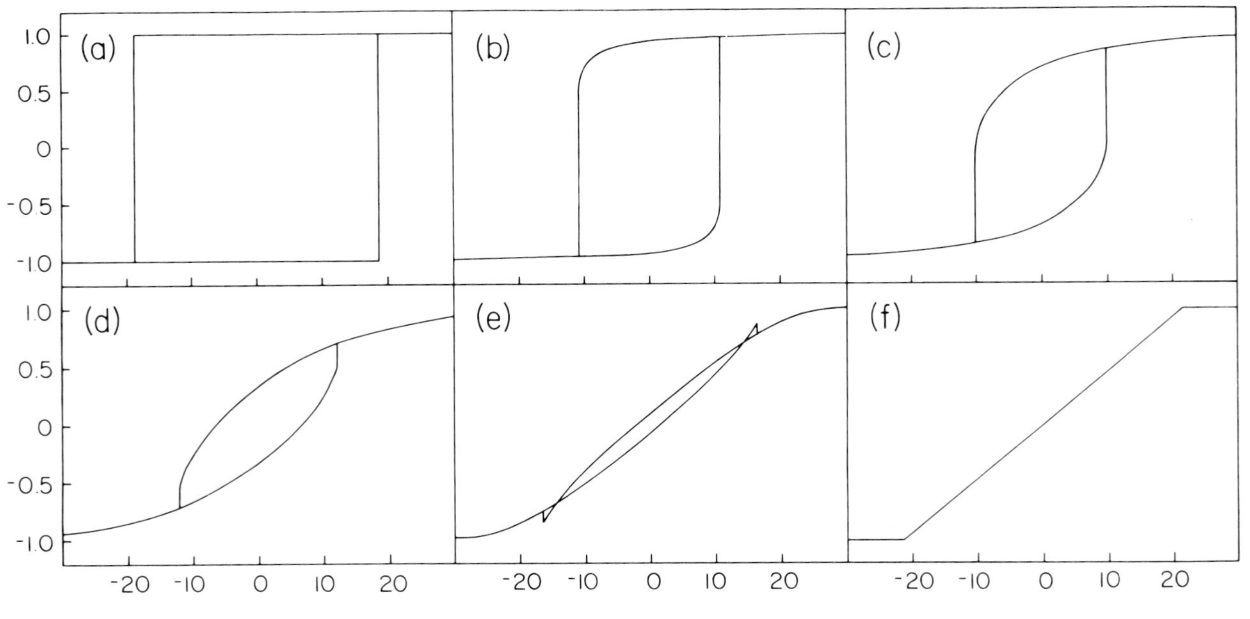

Figure 16.3. Calculated hysteresis loops for a thin-film sample according to the Stoner–Wohlfarth theory. The external field is parallel to Z, the film parameters are $M_s = 100$ emu/cm³ and $K_u = 10^6$ erg/cm³, and the loops in (a)–(f) correspond to $\Theta_A = 0°, 20°, 45°, 70°, 85°$, and $90°$ respectively.

increases, $\hat{\Theta}_M$ decreases until it reaches a critical value of $\hat{\Theta}_C = 135°$ at the critical field value of $\hat{H}_C = 0.5$. At the critical point, the state of minimum energy in which the system has been residing becomes a saddle point. Further increases in $\hat{H}_{ext}$ eliminate this minimum, forcing the system to jump to the only remaining state of minimum energy, which, in the case of Fig. 16.2, is at $\hat{\Theta}_M = 15°$. After the jump, $\hat{\Theta}_M$ increases continuously with increasing $\hat{H}_{ext}$, asymptotically approaching $\hat{\Theta}_H = 45°$.

Qualitatively, the behavior just described for the case of $\hat{\Theta}_H = 45°$ applies to all other values of $\hat{\Theta}_H$ as well, but the values of the critical field $\hat{H}_C$ and the critical angle $\hat{\Theta}_C$ will depend on the value of $\hat{\Theta}_H$. To determine these critical parameters one notes that at the critical point the two curves $F(\Theta)$ and $G(\Theta)$ become tangent to each other, that is,

$$F(\hat{\Theta}_C) = G(\hat{\Theta}_C) \tag{16.13a}$$

$$F'(\hat{\Theta}_C) = G'(\hat{\Theta}_C) \tag{16.13b}$$

Solving the above equations yields

$$\tan \hat{\Theta}_C = -(\tan \hat{\Theta}_H)^{1/3} \tag{16.14}$$

$$\hat{H}_C = -\frac{\cos^3 \hat{\Theta}_C}{\cos \hat{\Theta}_H} \tag{16.15}$$

Now, assuming that the equilibrium state in the absence of the external field occurs at $\hat{\Theta}_M = 0°$, there exist only two possibilities. In the first instance $0° \le \hat{\Theta}_H \le 90°$, in which case $\hat{\Theta}_M$ increases continuously towards $\hat{\Theta}_H$ with increasing H_{ext}; no critical fields will be reached in this case and no discontinuous jumps will occur. In the second instance $90° \le \hat{\Theta}_H \le 180°$, in which case $\hat{\Theta}_M$ initially increases with $\hat{H}_{ext}$ until it reaches $\hat{\Theta}_C$ at $\hat{H}_{ext} = \hat{H}_C$. At the critical field, $\hat{\Theta}_M$ jumps to the other side and suddenly becomes greater than $\hat{\Theta}_H$. The process then resumes its continuous nature, with $\hat{\Theta}_M$ asymptotically approaching $\hat{\Theta}_H$.

Example 1. Let $K_u = 10^6$ erg/cm³, $\Theta_A = 20°$, $M_s = 100$ emu/cm³, and $\Theta_H = 180°$. From Eqs. (16.3) and (16.4) we find $H_{eff} = 19.055$ kOe and $\Theta_{eff} = 21.215°$. Thus $\hat{\Theta}_H = 158.785°$, resulting in $\hat{\Theta}_C = 36.11°$ and $\hat{H}_C = 0.566$. The critical (i.e., switching) field is thus $H_{ext} = H_{eff}\,\hat{H}_C = 10.78$ kOe. □

Example 2. Figure 16.3 shows several hysteresis loops for a thin-film sample having $K_u = 10^6$ erg/cm³ and $M_s = 100$ emu/cm³. The external field is assumed to be along Z, that is $\Theta_H = 0°$ or $180°$. The values of Θ_A corresponding to the different loops in Fig. 16.3(a)–(f) are 0°, 20°, 45°, 70°, 85°, and 90° respectively. When $\Theta_A = 0°$, we find from Eq. (16.4)

Table 16.1. Parameters of the basic lattice used in the simulations

Parameter	Symbol	Numerical value
saturation magnetization	M_s	100 emu/cm^3
anisotropy energy constant	K_u	10^6 erg/cm^3
exchange stiffness coefficient	A_x	10^{-7} erg/cm
damping coefficient	α	0.5
gyromagnetic ratio	γ	-10^7 Hz/Oe
lattice dimensions		256 × 256
lattice constant	d	10 Å
film thickness	τ	500 Å

that $\Theta_{\text{eff}} = 0°$ provided that $K_u > 2\pi M_s^2$, which happens to be the case here. We also find that $H_{\text{eff}} = 2K_u/M_s - 4\pi M_s = 18.744$ kOe from Eq. (16.3). From Eqs. (16.14) and (16.15) one finds $\hat{\Theta}_C = 0°$ and $\hat{H}_C = 1$, leading to a perfectly square loop with a coercivity of 18.744 kOe. The lowest value of coercivity is around 10 kOe, and is reached when $\Theta_A \simeq 45°$. The loop at $\Theta_A = 85°$ has a curious shape: the jump in Θ_M has caused a drop (rather than an increase) in the Z-component of magnetization. Finally, for $\Theta_A = 90°$ we have $\Theta_{\text{eff}} = 90°$ and $H_{\text{eff}} = 2K_u/M_s + 4\pi M_s = 21.256$ kOe. In this case there are no jumps but there is a discontinuity of slope at $H_{\text{ext}} = H_{\text{eff}}$, where the magnetization comes into alignment with the applied field. □

16.2. Nucleation Coercivity

To gain an understanding of the origins of coercivity in thin-film media, we have used computer simulations to investigate the behavior of defects and inhomogeneities.[32-37] The set of parameters in Table 16.1 has been used in the simulations presented in this and the following section. The term "random-axis anisotropy", which we shall use, implies that the anisotropy axes of the lattice are distributed randomly and independently among the lattice cells (or among various groups of these cells). By keeping the deviation angle θ of the anisotropy axis from Z below a certain maximum value Θ, the random assignment of axes preserves the perpendicular nature of the overall anisotropy. θ is selected with uniform probability from the interval $[0, \Theta]$; for brevity, Θ will be referred to as the cone angle. Meanwhile, no constraints are imposed on the distribution of the azimuthal angle ϕ of the local easy axis, its value being selected (also randomly and independently) from the interval $[0, 2\pi]$.

In the presence of random-axis anisotropy, one must be careful to distinguish between the local anisotropy constant K_u and the constant

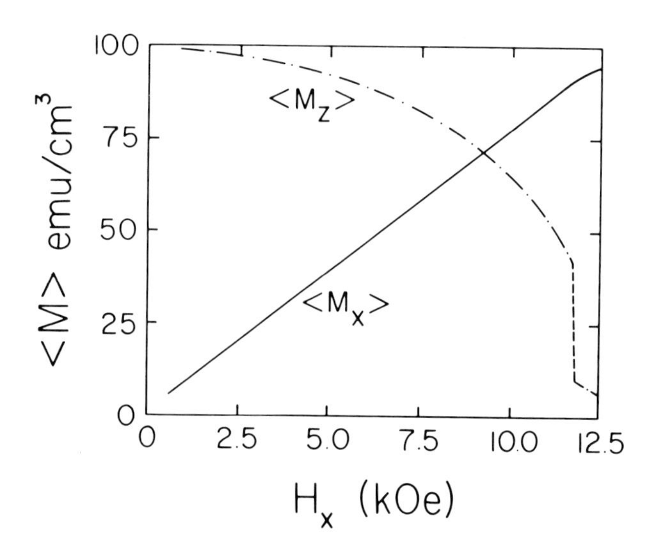

Figure 16.4. Average magnetization components along X and Z versus the magnitude of the applied field H_x. The lattice parameters are given in Table 16.1, and $\Theta = 45°$. The system is relaxed to the steady state for each value of the field.

describing the average (or bulk) anisotropy.† This bulk anisotropy constant should be denoted by $\langle K_u \rangle$, where the angle brackets indicate spatial averaging, in some appropriate sense, over the strength of the local anisotropy associated with each dipole of the lattice. If the cone angle Θ happens to be zero, then $\langle K_u \rangle = K_u$, but when $\Theta \neq 0°$ the bulk constant $\langle K_u \rangle$ drops below K_u.

One possible experiment[38] from which the strength of anisotropy may be determined involves the application of an in-plane field, say H_x along X, to a saturated sample whose magnetization has been aligned with the easy axis Z. By monitoring the normal component $\langle M_z \rangle$ of the magnetization, one can determine the constant of bulk anisotropy $\langle K_u \rangle$ from the curvature of the plot of $\langle M_z \rangle$ versus H_x, in accordance with the Stoner–Wohlfarth model. (Alternatively, one may extract the same information from the slope of the plot of $\langle M_x \rangle$ versus H_x.) Figure 16.4 shows such plots, obtained by computer simulation for the lattice of Table 16.1 with cone angle $\Theta = 45°$. The lattice was initially saturated in the positive Z-direction, then relaxed to the remanent state before the application of H_x. For each value of H_x, starting from zero and increasing

†The bulk anisotropy constant is a measurable quantity that may be determined by a number of techniques, notably the method of torque magnetometry.

in steps of 100 Oe, the lattice was relaxed and then $\langle M_x \rangle$ and $\langle M_z \rangle$ were computed. In the range $0 \leq H_x \leq 12$ kOe the dipoles move more or less coherently and reversibly toward X, in which case the Stoner–Wohlfarth theory becomes applicable and the plots of Fig. 16.4 yield

$$\frac{2\langle K_u \rangle}{\langle M_s \rangle} - 4\pi \langle M_s \rangle \simeq 13 \text{ kOe} \tag{16.16}$$

Since the large value of exchange keeps the dipoles nearly parallel at all times, we find $\langle M_s \rangle \simeq M_s = 100$ emu/cm³. Thus, from Eq. (16.16), $\langle K_u \rangle \simeq 0.7 \times 10^6$ erg/cm³, which, as expected, is less than the local constant K_u.

Digression. It is interesting to know what happens in the above simulation at the critical point where $\langle M_z \rangle$ shows a discontinuity (see Fig. 16.4). Here some dipoles flip over and create regions of reverse magnetization. Figure 16.5(a) shows the steady-state distribution of M_z across the lattice under the applied field of $H_x = 12$ kOe. (Note: The color code used for Fig. 16.5 differs from that used elsewhere; see the caption for a description of this particular coloring scheme. The blue regions in Fig. 16.5(a) have a small positive M_z, while M_z for the yellow regions is small and negative.) Figure 16.5(b) shows the distribution of exchange energy in the steady state; this type of plot emphasizes domain walls and enhances regions with rapid spatial variations of magnetization. □

In the subsections that follow we describe for the lattice of Table 16.1 (with various defects included) the results of computer simulations of the magnetization reversal process. The starting point in all these simulations will be the remanent state along the easy axis Z. A field H_z is applied perpendicular to the lattice and opposite to the magnetization; it is then determined whether or not the magnetization begins to reverse under the influence of H_z. A binary search† for the critical value of H_z yields the nucleation coercivity with any desired accuracy.

16.2.1. Dependence of coercivity on cone angle

Hysteresis loops were traced for several values of the cone angle Θ in the range of 20° to 45°. The loops were always square, indicating that the nucleation coercivity dominates the wall motion coercivity. The coercive

†Here is an example of how a binary search works. Assume that the nucleation coercivity is somewhere between 0 and 20 kOe. The midpoint of this interval is at 10 kOe. Now, if the sample begins to reverse under $H_z = 10$ kOe, we limit the range of search to the interval [0, 10 kOe]. If, on the other hand, the sample does not show any signs of reversal, we conclude that H_c must be between 10 and 20 kOe. In this way, the range of search shrinks by a factor of two in just one step. By going to the midpoint of the newly established range and repeating the above procedure, one can determine the value of H_c with sufficient accuracy in a few steps.

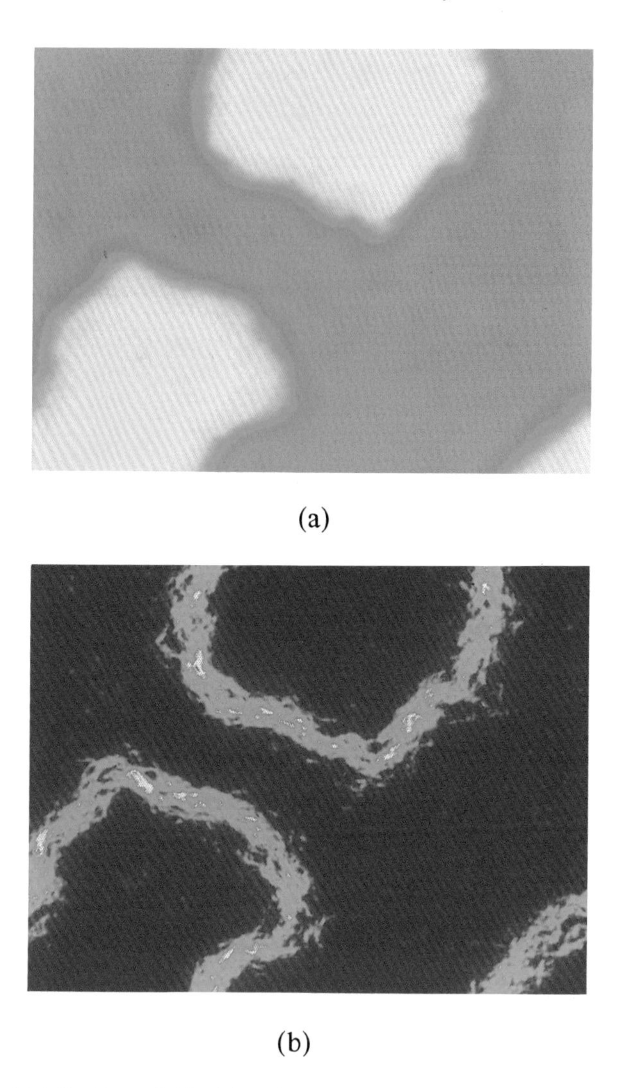

(a)

(b)

Figure 16.5. Distribution of M_z (a) and exchange energy (b) across the lattice in the steady state under H_x = 12 kOe. The color coding here differs from that used elsewhere in the chapter, and is applicable only when a scalar function (such as the local value of M_z across the lattice) is to be displayed. In this scheme the color red is assigned to the minimum value of the function, while purple is used to represent the maximum value. All other values are then mapped onto the color wheel in a linear, one-to-one fashion, starting from red and moving counterclockwise to purple.

field H_c decreased monotonically from 17 kOe at $\Theta = 20°$ to 12.6 kOe at $\Theta = 45°$. Figure 16.6 shows the nucleation phase of the reversal process for a sample with $\Theta = 45°$ under an applied field of 12.6 kOe. The nucleation site is in the lower left corner of the lattice, and its periodic continuation, due to the boundary conditions and the hexagonal symmetry of the lattice, appears in the upper central part.

Figure 16.6. Early stage of nucleation for a sample with cone angle $\Theta = 45°$, under an applied field $H_z = -12.6$ kOe. Because of the hexagonal geometry of the lattice, the periodic continuation of the reverse domain at the lower left corner appears in the upper central region.

The value of H_c for a given cone angle shows a slight dependence on the choice of seed for the random-number generator. For instance, in the case of $\Theta = 45°$, different seeds gave rise to coercivities between 12.5 and 12.8 kOe. This, however, should not be a matter of concern since one expects the computed coercivity to become independent of the random seed once the simulated lattice becomes sufficiently large. By the same token, similar variations in the measured coercivity of real materials can be expected, if microscopically small areas are subjected to the external field.

16.2.2. Dependence of H_c on the strength of exchange

To test the relative significance of anisotropy and exchange in the nucleation process, we varied the exchange parameter A_x, while keeping all other parameters (including the random number seed) fixed. Even when A_x had dropped by as much as 40% of the nominal value given in Table 16.1, the coercivity of the sample remained unchanged within the accuracy of our calculations. We conclude that the nucleation coercivity is principally controlled by the bulk anisotropy field $\langle H_k \rangle$ and that the domain wall formed around the nucleus plays at best a secondary role in the balance of forces. At the time of nucleation, therefore, the average anisotropy field of the nucleating region must be close to the value of the applied field, that is,

$$H_c \simeq \langle H_k \rangle = \frac{2\langle K_u \rangle}{\langle M_s \rangle} - 4\pi \langle M_s \rangle \tag{16.17}$$

Now, in contrast to the above result, the experimentally observed values of coercivity are typically much less than the anisotropy field $\langle H_k \rangle$. Nucleation in real materials, therefore, cannot be attributed to random-axis anisotropy of the type considered thus far. The reversal process must have its origins in what may be termed "defects", be they structural or magnetic in character. To give an example, let us consider a hypothetical sample with the following properties. (i) The bulk of the material has very little dispersion in its local easy axes, i.e., $\Theta \simeq 0°$. (ii) Only a few isolated submicron-sized regions have large values of Θ associated with them. Under these circumstances $\langle K_u \rangle$ for the sample will be nearly equal to its K_u, but, since nucleation takes place in those regions that have large dispersion, the resulting coercivity will be significantly lower than $\langle H_k \rangle$. Such regions with a large dispersion of the local easy axes form but one possible type of defect; several other types of defect are described below.

16.2.3. Voids as defects

Voids are small regions of the sample from which the magnetic material is missing. Figure 16.7 shows a circular void with diameter D = 500 Å in a lattice with random-axis anisotropy (Θ = 45°). We found, within the accuracy of the simulations, that the coercivity of this sample is unaffected by the void, that is, $H_c \simeq 12.6$ kOe.

It should be noted that here the assumed void has not been allowed to influence the magnetic properties at its own boundary; in particular, the saturation magnetization, the anisotropy constant, and the distribution of the easy axes at the periphery of the void have all been left intact. In reality, one expects the presence of the void to alter these parameters, albeit to an extent that is presently unknown. Thus, despite the finding that "ideal" voids cannot affect the value of the nucleation coercivity, the possibility of "real" voids acting as nucleation centers cannot be ruled out.

16.2.4. Resident reverse-magnetized nuclei

This type of defect remains oppositely magnetized to the rest of the film, even in the presence of a saturating field. We conjecture that such resident nuclei must be localized around a high-K_u region, or else they would be

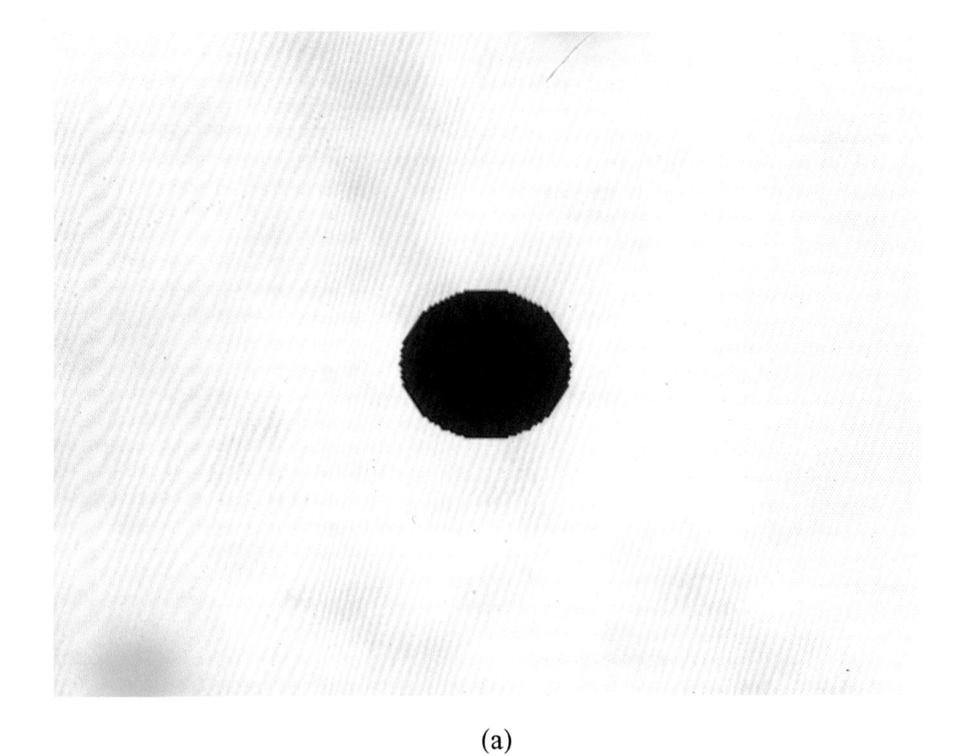

(a)

(b)

Figure 16.7. Nucleation in a lattice containing a void defect at the center. (a) State of magnetization under H_z = - 12.5 kOe, just below coercivity. (b) Nucleation and growth of a reverse domain at H_z = - 12.6 kOe.

unstable. In the simulation results displayed in Fig. 16.8, the value of K_u assigned to the central region of the lattice has been raised tenfold to 10^7 erg/cm^3. Random-axis anisotropy with $\Theta = 45°$ is also assumed throughout the lattice.

A defect with a diameter $D = 100$ Å is found to be incapable of supporting a reverse domain. A stable domain can be formed, however, around a defect with $D = 200$ Å, as shown in the remanent magnetization pattern of Fig. 16.8(a_1). The required field for the expansion of this defect is only 3.45 kOe, which is substantially less than the 12.6 kOe coercivity of the same sample in the absence of any type of defects. Frames (a_2)–(a_4) of Fig. 16.8 show the growth of this resident nucleus under the applied field of 3.46 kOe. Similar results are obtained for another defect with a diameter of 300 Å whose behavior is depicted in frames (b_1) and (b_2); the coercivity in this case is 1.98 kOe. When the defect diameter is raised to 500 Å, the coercivity drops to 1.28 kOe. These results clearly indicate that resident nuclei can control H_c in a major way.

The preceding numerical results are in good agreement with predictions based on a relatively simple theory. Consider a circular domain of radius r_0 in a film of thickness τ, saturation magnetization M_s, and domain wall energy density σ_w. Let an applied perpendicular field H_{ext} favor the direction of magnetization inside the domain. Assuming that $0 << r_0 \lesssim \tau$, the energy of the system (relative to the fully saturated state) is written

$$\mathcal{E} \simeq -2\pi r_0^2 \tau M_s H_{ext} + 2\pi r_0 \tau \sigma_w - \pi(r_0 + 1.5\tau)^2 \tau (2\pi M_s^2) \qquad (16.18)$$

The approximate nature of Eq. (16.18) is due to the last term, which corresponds to demagnetization. The implicit assumption here is that, upon formation of the domain, H_{dmag} in its vicinity (within a radius of $r_0 + 1.5\tau$ from the center) vanishes. Of course, if the domain radius is much less than the film thickness, the above approximation fails because, in that case, H_{dmag} cannot be much affected in locations as far as $r_0 + 1.5\tau$ away from the center. Similarly, if r_0 happens to be much larger than τ the approximation fails once again, because now H_{dmag} is diminished only within the annulus extending from $r_0 - 1.5\tau$ to $r_0 + 1.5\tau$. These are the reasons behind the restriction imposed on Eq. (16.18).

When H_{ext} is sufficiently small, the net pressure on the wall will be inwards and the domain tends to collapse (which, of course, the large K_u within the domain opposes). At the onset of expansion, when H_{ext} is large enough to begin to push the wall outwards, the net pressure must be zero, that is, $d\mathcal{E}/dr_0 = 0$. One can readily determine the critical H_{ext} as

$$H_{ext} \simeq \frac{\sigma_w - 2\pi(r_0 + 1.5\tau)M_s^2}{2 r_0 M_s} \qquad (16.19)$$

The value of the wall energy density is determined from Eq. (13.10) as

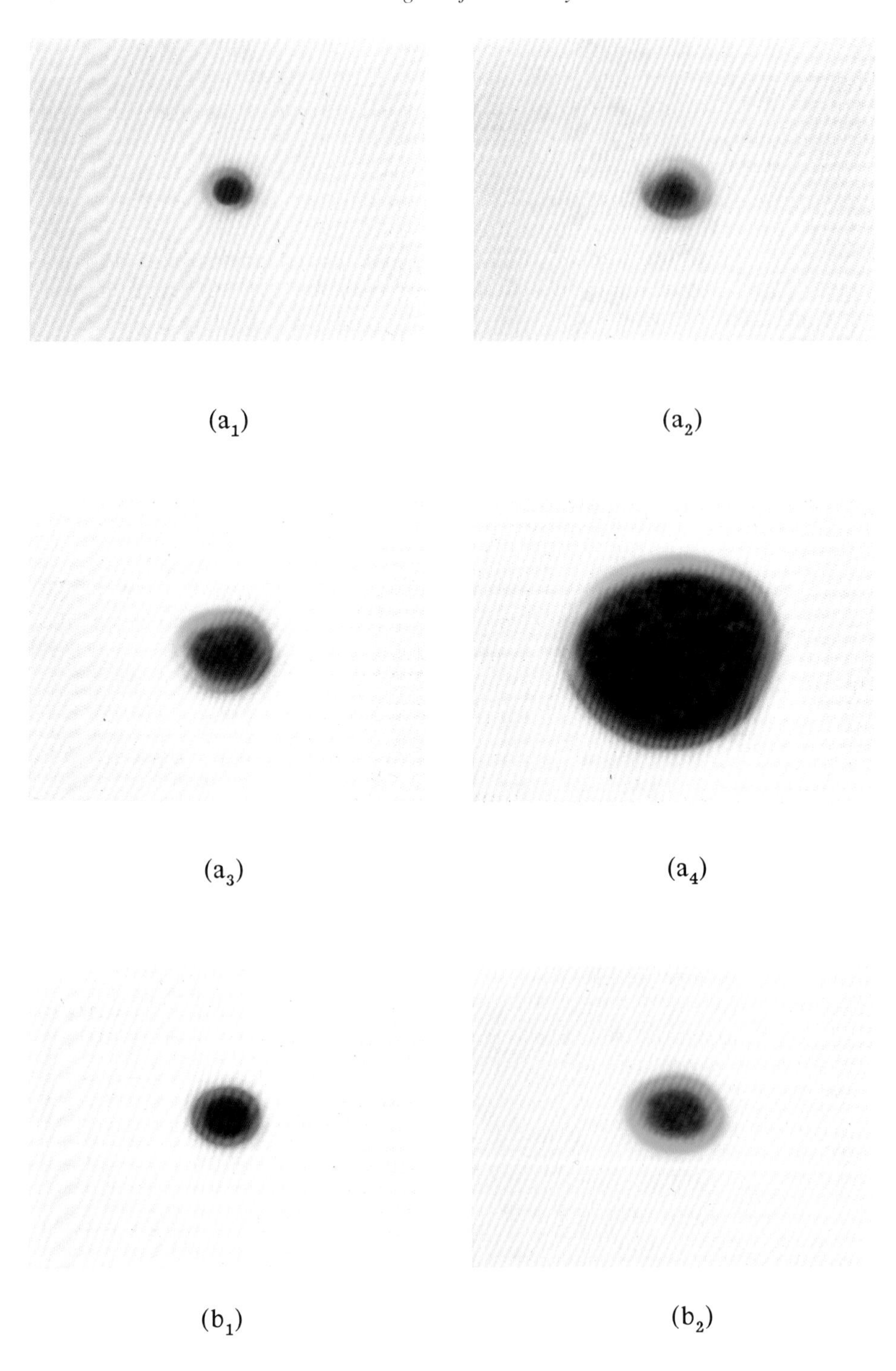

Figure 16.8. Magnetization reversal by growth from a resident nucleus. (a_1) A defect with diameter D = 200 Å in the remanent state. (a_2)–(a_4) Growth of the initial domain under H_z = -3.46 kOe at t = 0.2, 0.5, 0.8 ns after the application of the field. (b_1) A defect of initial diameter D = 300 Å in the remanent state. (b_2) Growth of the initial domain in (b_1) under H_z = -1.98 kOe.

$\sigma_w = 1.265$ erg/cm^2.† Thus for $r_0 = 100$ Å, 150 Å, and 250 Å, corresponding to the above-mentioned simulated defects, the coercivities calculated from Eq. (16.19) are 3.65, 2.33, and 1.27 kOe respectively. These values are in fairly good agreement with earlier simulation results.

Of course, high-K_u regions are not necessarily reverse-magnetized in every situation. Consider, for instance, the case of a completely saturated sample with six small defects shown in Fig. 16.9. The parameters of this lattice are given in Table 16.1, and random-axis anisotropy with cone angle $\Theta = 45°$ is assumed. The six defects are identical and regularly spaced, each having a diameter $D = 400$ Å and anisotropy constant $K_u = 5 \times 10^6$ erg/cm^3. Frame (a) of Fig. 16.9 shows the state of the lattice under the applied field $H_z = -12.5$ kOe, which is slightly less than the coercivity of the lattice. Frames (b) through (f) show the nucleation and growth of a reverse domain under $H_z = -12.6$ kOe. Although the defects act as temporary barriers to the growing nucleus, the walls eventually sweep through the entire sample. At the end, the magnetization is fully saturated in the reverse direction, and resident nuclei (which could have formed around the high-K_u defects) do not materialize.

In contrast to the above situation, Fig. 16.10 shows the case of a sample in which resident nuclei with either polarity can be stable. Here are seven regions of diameter $D = 200$ Å and $K_u = 10^7$ erg/cm^3 embedded in the lattice. The central region, which is initially reverse-magnetized, is capable of supporting a stable domain (i.e., resident nucleus) as shown in frame (a). Under the external field $H_z = -3.5$ kOe (which is slightly greater than the coercivity of this sample) the central nucleus expands and covers the rest of the lattice, with the exception of the high-K_u regions. Frames (b)–(f) follow the growth process in time under the same field. The six unreversed regions in the final frame now act as resident nuclei for future reversals.

One recognizes, of course, that resident nuclei of the type described here are inherently unstable, and could be eliminated by magnetic fields of sufficient strength. The required field for annihilating a particular nucleus depends on its size and on the strength of its anisotropy. In reality, if the coercivity were controlled by this type of defect, then one would expect to find a dependence of H_c on the magnetization history of the sample and, in particular, on the largest field applied to saturate the sample prior to the beginning of the reversal process. Such dependences have indeed been observed in practice.[38]

16.2.5. Weakly anisotropic defects

Let us assume that the anisotropy constant K_u within the central region of the lattice has only half the value of K_u elsewhere, all other parameters

†Due to random anisotropy and the presence of magnetization twists (e.g., vertical Bloch lines), the effective value of σ_w in these simulations must be slightly different from that of 1.265 erg/cm^2 derived from the nominal A_x and K_u values. To keep the analysis simple these differences have been ignored.

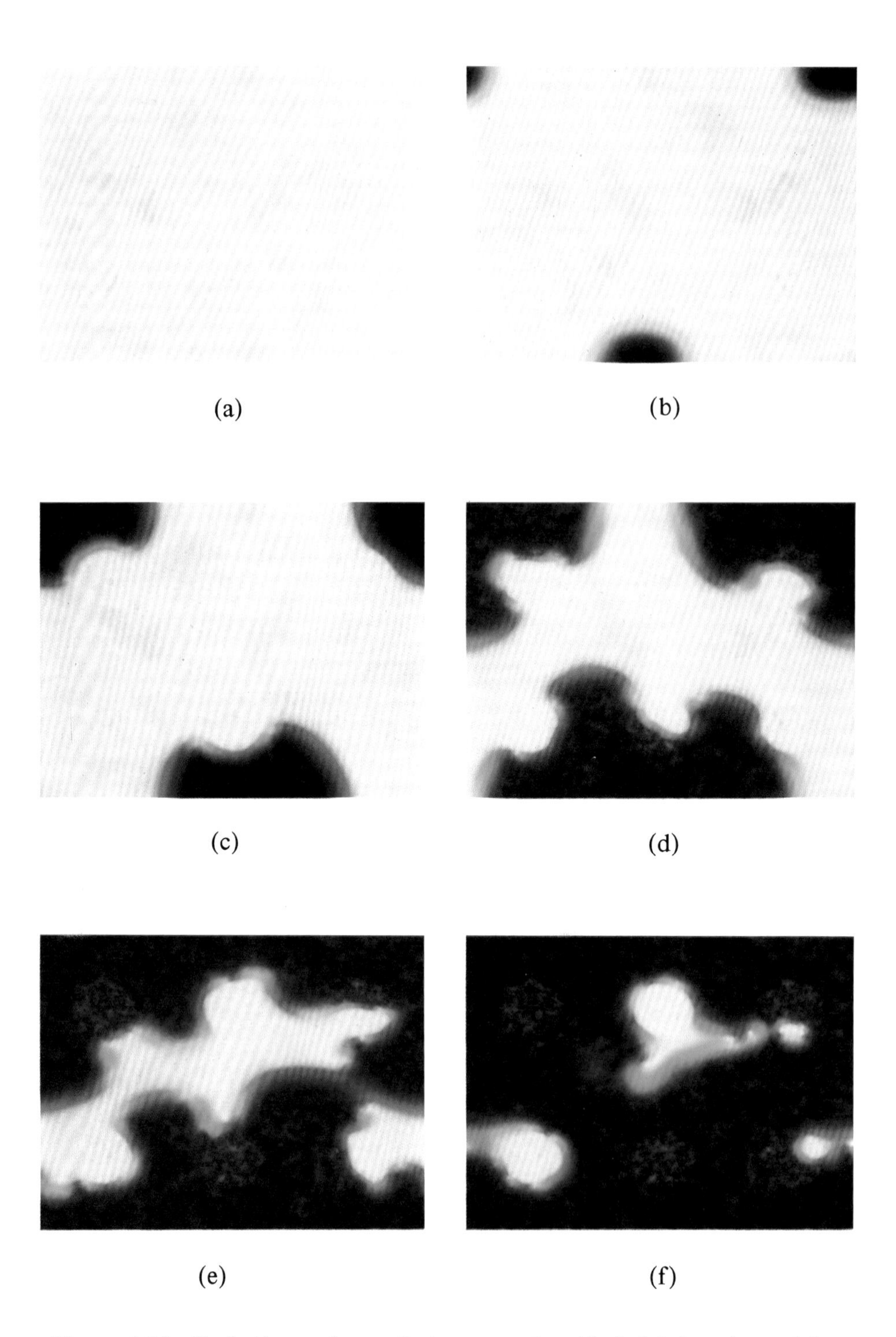

Figure 16.9. Nucleation and growth in a sample with isolated regions of large anisotropy. Six disk-shaped areas of diameter 400 Å within the lattice were assigned a K_u value five times greater than that for the rest of the lattice. The magnetization of the entire lattice was then saturated along $+Z$ and relaxed to the remanent state. Frame (a) shows the state of the lattice under H_z = - 12.5 kOe, just below coercivity. In frame (b) the applied field is - 12.6 kOe and there is nucleation. Frame (c) shows how the growth of the initial nucleus is hampered by three of the defects. The remaining frames follow the growth process in time, and show the way in which the magnetization manages to reverse the high-K_u regions.

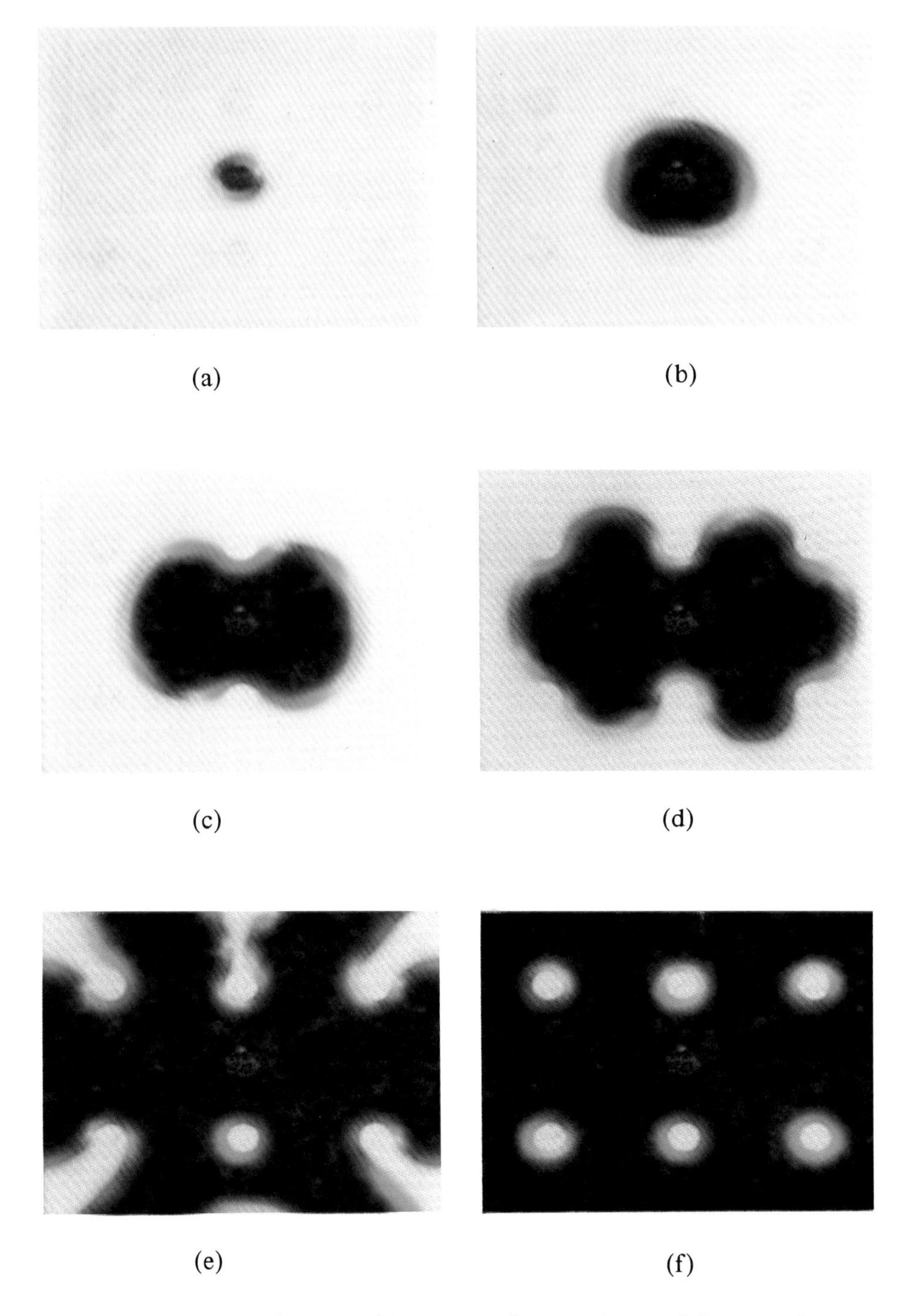

Figure 16.10. Growth from a resident nucleus in a sample containing seven isolated regions of large anisotropy. Each such region has diameter D = 200 Å and K_u = 10^7 erg/cm^3. The central "defect" is the home of the resident nucleus; the rest of the lattice is saturated along $+Z$ and relaxed to the remanent state, shown in frame (a). Frame (b) shows the state of the lattice under a reverse field H_z = -3.5 kOe, which is slightly greater than the coercivity for this sample. The remaining frames (c) to (f) follow the growth process in time and show how the magnetization fails to reverse in high-K_u regions. The unreversed regions now become nuclei for future reversals.

being the same as in previous cases. The entire sample (including the defect) is initially magnetized along $+Z$, and the system allowed to relax and settle down into the remanent state. The various frames in Fig. 16.11 correspond to defects of different sizes, each showing an early state of magnetization during the reversal process under a field that is only slightly greater than the corresponding sample's computed coercivity. Frames (a)–(d) correspond to defect diameters D = 200, 600, 800, and 1000 Å respectively. The corresponding coercive fields for these samples are computed as H_c = 12.6, 11.8, 9.4, and 8.7 kOe. The 200 Å defect does not seem to have much influence on the coercivity, but the other defects have an appreciable effect. Note that in frames (b), (c), and (d) the defect itself is the nucleation site.

16.2.6. Defects with tilted easy axis

For this type of defect we assume that the axes of anisotropy (within the defective region) are uniformly tilted away from the normal. For several defects of this type the various frames of Fig. 16.12 show the states of the lattice both before and after nucleation. Except for the directions of the local easy axes within the defects, all lattice parameters in these simulations were the same as before. Frames (a_1) and (a_2) correspond to a defect of diameter D = 1000 Å and a uniform tilt angle of 10° from normal within the defect. In (a_1) the applied field is 12.32 kOe, which is just below coercivity, whereas in (a_2) the field is 12.34 kOe. Compared to the same sample with no defects, the coercivity has dropped only slightly, but the nucleation site is now on the boundary of the defect. Frames (b_1) and (b_2) correspond to a similar defect with a tilt angle of 20° ; the coercive field in this case has dropped to 10.45 kOe. For a smaller defect of diameter 400 Å and 20° tilt angle, shown in frames (c_1) and (c_2), the coercivity is about 11.33 kOe. Apparently, in order to affect the coercivity in a significant way, this type of defect must be fairly large and greatly tilted. The Stoner–Wohlfarth theory of section 16.1 is readily applicable to this type of defect, provided that the defect is not too small. The above results concerning the 1000 Å defect with 20° tilt are in good agreement with those of Example 1, section 16.1.

16.3. Coercivity of Domain Wall Motion

Having studied the process of nucleation in some detail, we now turn to the subject of domain wall structure and its associated coercivity. In a truly homogeneous magnetic material the domain walls can be readily moved with the application of a small magnetic field. For example, in certain defect-free crystals of magnetic garnets a field of less than one oersted is strong enough to move the domain walls around. On the other hand, in the amorphous films of TbFeCo used in MO recording, the strength of the applied field must typically reach several kilo-oersteds before any significant domain wall movement occurs. The reasons for the strong resistance to wall motion exhibited by these (seemingly homogeneous) amorphous materials are manifold, but they can ultimately be traced to the presence of submicron-sized defects and nanostructure. The present section explores possible sources of domain wall coercivity by examining the results of computer simulations of lattices that contain various types of defects and random spatial fluctuations.

16.3.1. Walls and random-axis anisotropy

Figure 16.13 shows the structure of domain walls in a medium having the parameters of Table 16.1 and cone angle $\Theta = 45°$. Initially the central band of the lattice was magnetized along $+Z$ while the remaining part was magnetized along $-Z$, as shown in frame (a). When the lattice was allowed to relax for 0.8 ns, the pattern in frame (b) was obtained. Notice that there are now three vertical Bloch lines (2π VBLs) in each wall and that the walls are no longer straight. By allowing the lattice to relax for another 0.9 ns we obtain the pattern of frame (c), which shows significant VBL movements along the walls. Finally, frame (d) shows the steady-state situation at $t = 4.56$ ns. Both walls are now considerably straightened, but the number of VBLs in each wall has remained intact; no amount of relaxation could unwind a 2π Bloch line.

The curves in Fig. 16.14 show the average magnetization $\langle M_z \rangle$ and total energy $\langle \mathscr{E}_{\text{tot}} \rangle$ of the system during the relaxation process depicted in Fig. 16.13. The inset in Fig. 16.14(b) shows the various components of energy. Obviously, the demagnetization energy does not change much during the process of wall formation. This result should be expected since, in this particular example, the film thickness τ is several times greater than the wall width. On the other hand, the anisotropy energy drops sharply in the early phase as the moments throughout the lattice move closer to the local easy axes. In fact, this reduction is large enough to overwhelm the modest increase in anisotropy energy at the walls. For the same reasons the exchange energy of the entire system rises, albeit very slightly, despite a sharp reduction in exchange energy at the walls.

16.3.2. Motion of domain walls

A perpendicular field $H_z = -200$ Oe moves the two walls in Fig. 16.13(d) somewhat closer together, but fails to eliminate the stripe of reverse magnetization. The steady state of the lattice under this applied field is shown in Fig. 16.15. The corresponding curves of $\langle M_z \rangle$ and $\langle \mathscr{E}_{\text{tot}} \rangle$ in Fig. 16.16 indicate that the time needed to arrive at the steady state is about 2 ns. In this experiment, the force of demagnetization opposes the external field in collapsing the reverse-magnetized stripe.

The stripe domain shown in Fig. 16.13(d) will collapse, however, under

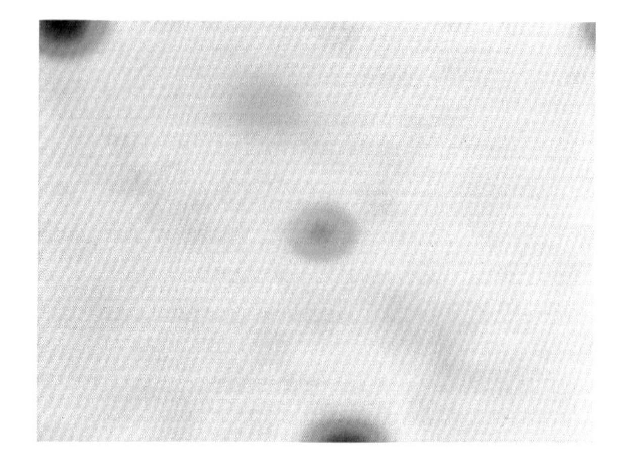

(a)

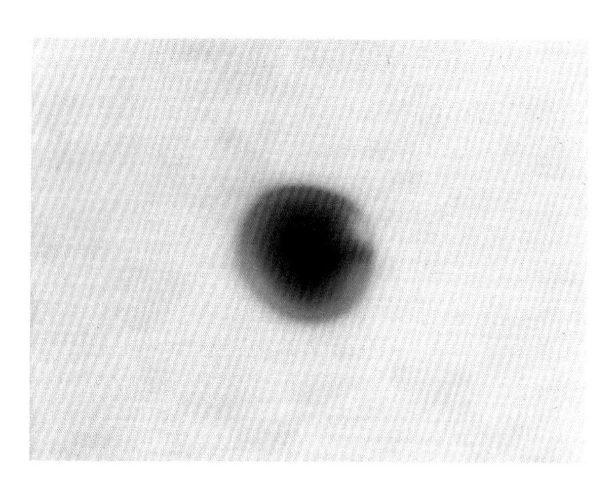

(b)

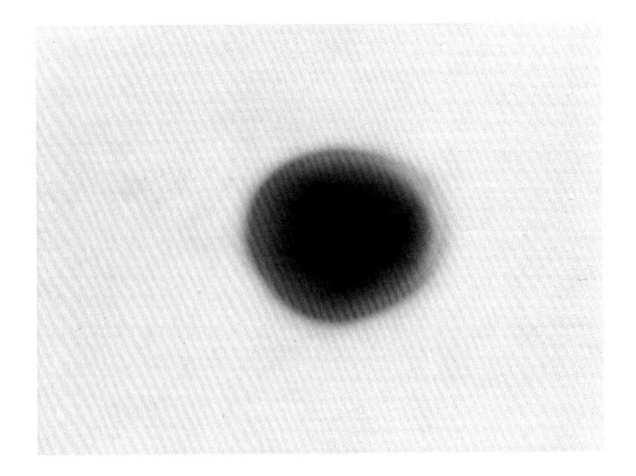

(c)

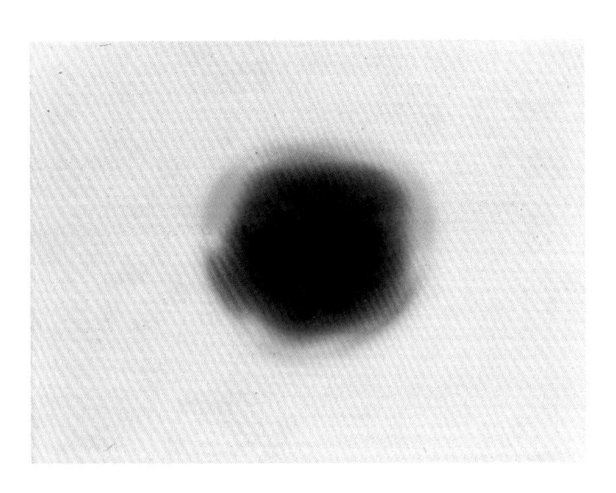

(d)

Figure 16.11. Nucleation in a sample with a weakly anisotropic defect at the center; K_u within the defect has only half its value elsewhere. (a) Defect of diameter 200 Å under $H_z = -12.64$ kOe. (b) Defect of diameter 600 Å subjected to $H_z = -11.75$ kOe. (c) The defect diameter is 800 Å and the external field is $H_z = -9.4$ kOe. (d) The defect diameter is 1000 Å and the applied field is $H_z = -8.7$ kOe.

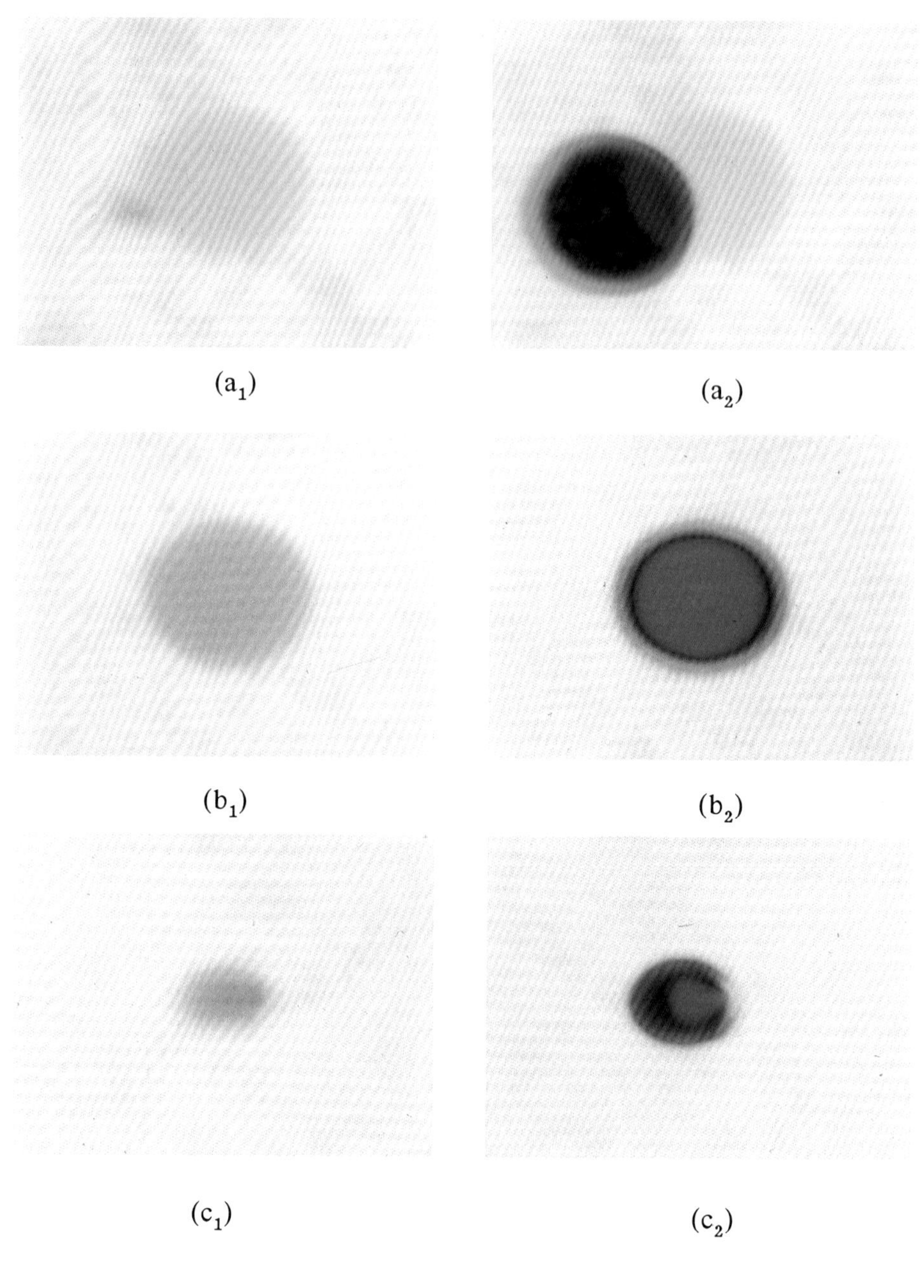

Figure 16.12. Nucleation in a sample with a tilted-axis defect. (a_1) Defect with diameter of 1000 Å and tilt angle of 10°, subject to H_z = - 12.32 kOe. The defect is visible as the orange-colored region in the center of the lattice. Although a red spot near the left boundary has formed at this stage, the applied field is not yet strong enough to reverse the magnetization. (a_2) Same as (a_1) but with H_z = - 12.34 kOe. This is a snapshot of the reversal process; the nucleated domain grows until the entire sample is reversed. (b_1) Defect with diameter of 1000 Å and anisotropy axis tilt of 20°, subject to H_z = - 10.44 kOe. (b_2) Same as (b_1) but with H_z = - 10.46 kOe; this is a snapshot of the reversal process. (c_1) Defect with diameter of 400 Å and anisotropy axis tilt of 20°, subject to H_z = - 11.32 kOe. (c_2) Same as (c_1) but with H_z = - 11.34 kOe. Again, this is a snapshot of the reversal process.

an applied field of H_z = - 1000 Oe, as shown in Fig. 16.17. Frames (a) and (b) in this figure correspond, respectively, to t = 0.96 ns and t = 3.58 ns, where t is the time that has elapsed since the application of the field. Although not shown in the figure, the walls proceed to collide and annihilate each other. The curves of $\langle M_z \rangle$ and $\langle \mathcal{E}_{tot} \rangle$ in Fig. 16.18 show the rate of reduction of the average magnetization and energy during this collapse process.

In the above example, the randomness in the lattice parameters is clearly too weak to cause significant resistance to wall motion. On the other hand, experimental evidence (such as that obtained by observations of domains under polarized light) clearly indicates that MO media have a wall motion coercivity in the range of several hundred to several thousand oersteds. For the simulated lattice to exhibit large coercivities, therefore, one must enhance the role of spatial fluctuations in hampering the movements of the walls. This we do by increasing the correlation length of these fluctuations through the incorporation of "patches" into the lattice. The patches, whose dimensions are large compared to the width of the domain wall, will then present nonuniformities on a scale that is too great to be averaged out over the width of the wall. This feature is the key to the effectiveness of patches as barriers to wall motion. Figure 16.19 shows a typical lattice sectioned into 346 patches of random shape and size. These patches are created by selecting at random a number of lattice sites as seeds, and growing outward from them (in a random fashion) until every site in the lattice belongs to one patch or another. By assigning different attributes to different patches one can thus create spatial variations in the structure and/or magnetic properties of the lattice over length scales comparable to the average patch size.

16.3.3. Wall coercivity and patch-to-patch random anisotropy

Figure 16.20 shows a strip of reverse magnetization in the patchy lattice of Fig. 16.19. For the sake of clarity, the patch boundaries are highlighted in black. Each patch is assigned an axis of anisotropy, randomly and independently of all the other patches, with a cone angle of $\Theta = 45°$. The walls in Fig. 16.20 are more jagged than those in Fig. 16.13, where the patches were basically the size of an individual lattice cell. Under an applied field H_z of +1.5 kOe the walls in Fig. 16.20(a) moved slightly and came to equilibrium, as shown in frame (b). A plot of $\langle M_z \rangle$ versus time is shown in Fig. 16.20(c). This plot indicates that the initial $\langle M_z \rangle$, which is slightly above zero, has increased to about 20% of its saturation value in the first 0.5 ns after the application of the field, but has stopped growing at that point. Compare this situation with that depicted in Fig. 16.17, where a field of 1 kOe was sufficient to annihilate the walls. Clearly it is the presence of the patches (and not the demagnetizing force) that is now responsible for blocking the wall motion. Thus the wall motion coercivity has increased as a result of increased correlation among the local easy axes.

When the field was further raised to H_z = 2 kOe, it became possible to

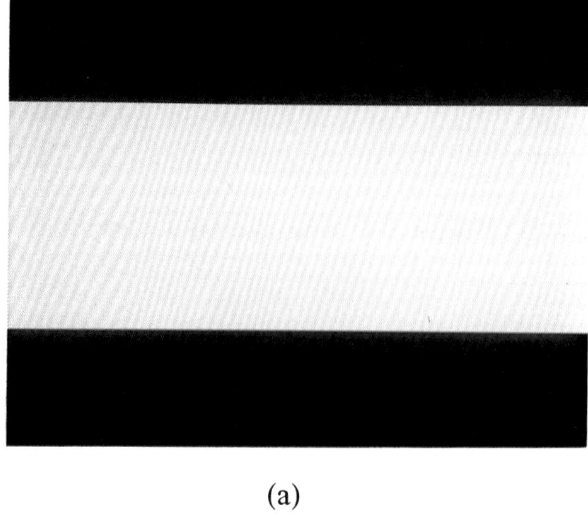

(a)

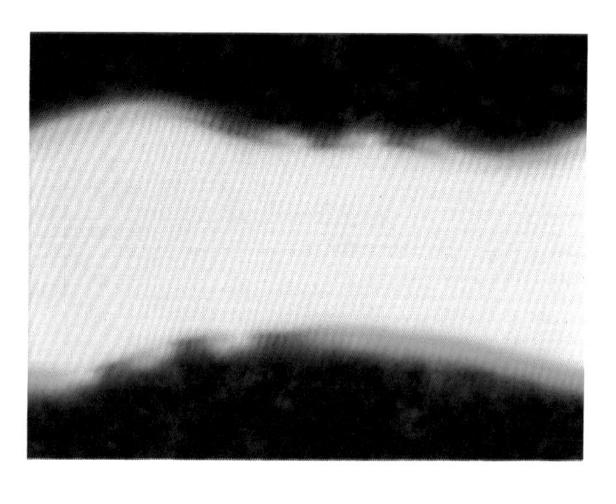

(b)

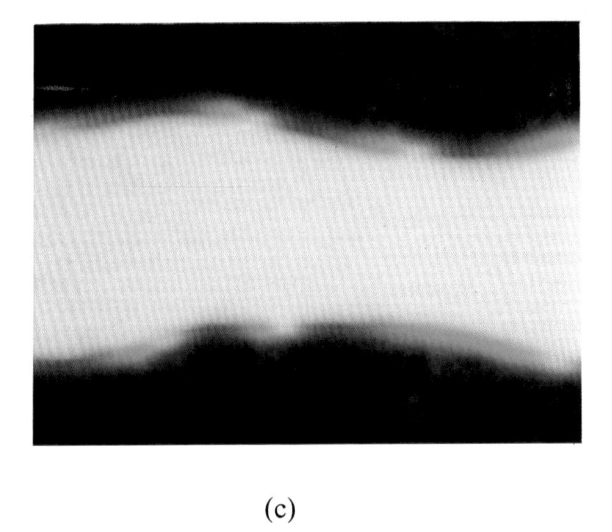

(c)

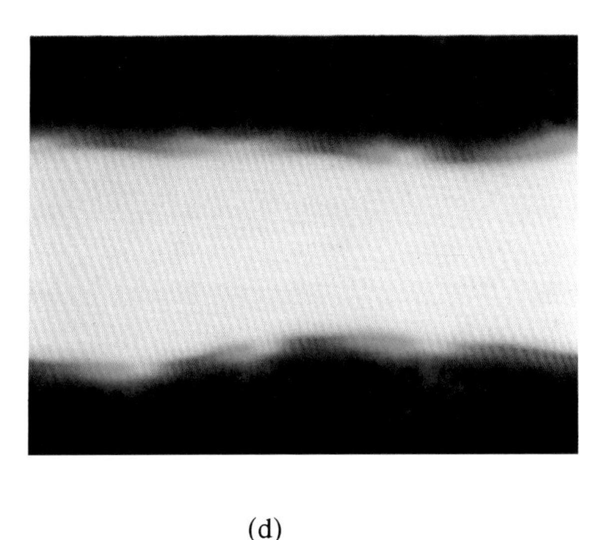

(d)

Figure 16.13. Formation of domain walls in a sample with a cone angle of 45°, in the absence of an applied field. (a) Dipoles in the white region are initialized along $+Z$, while those in the dark region are initialized along $-Z$. (b) The state of the lattice at t = 0.8 ns. Each wall contains three vertical Bloch lines at this stage. (c) The state of the lattice at t = 1.7 ns; the number of VBLs has not changed since the previous frame, but they have moved along the walls. (d) The steady state of the lattice at t = 4.56 ns. The number of VBLs in each wall is still three.

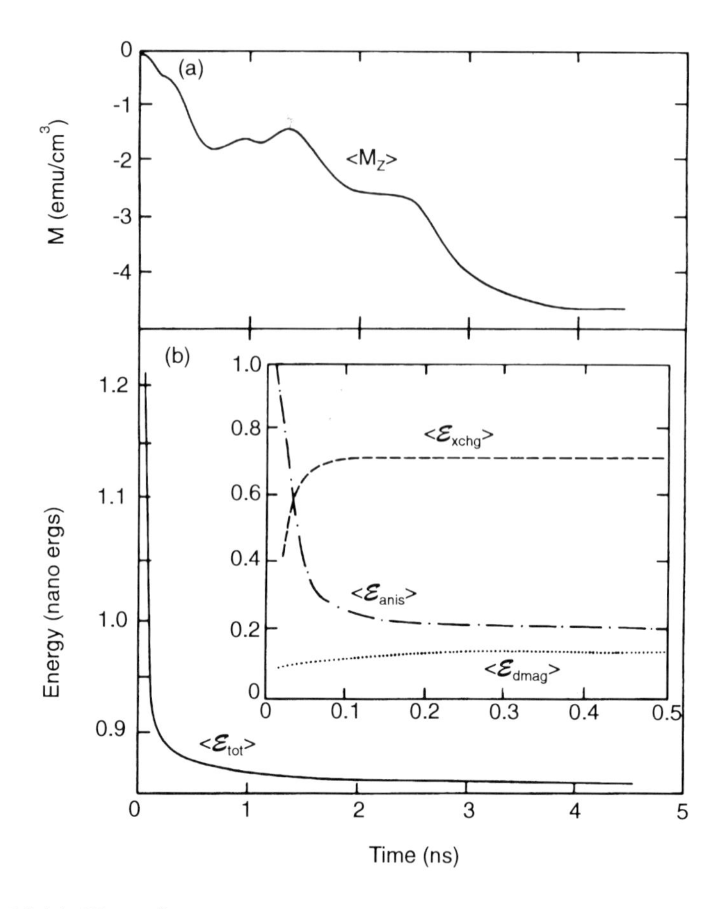

Figure 16.14. Plots of average magnetization and energy in the process of domain wall formation corresponding to Fig. 16.13. (a) $\langle M_z \rangle$ versus time. (b) Total energy of the lattice versus time. The inset shows the evolution of exchange, anisotropy, and demagnetization energies during the initial phase of the process.

Figure 16.15. Steady state of the lattice (shown at t = 3.66 ns) when the stripe domain of Fig. 16.13(d) is subjected to H_z = -200 Oe.

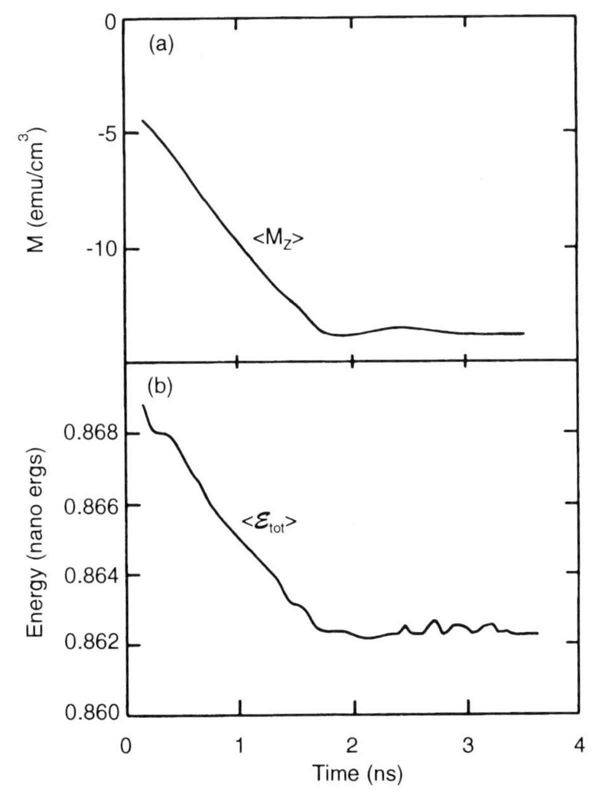

Figure 16.16. Plots of average magnetization and energy when the stripe domain of Fig. 16.13(d) shrinks under an external field H_z of -200 Oe.

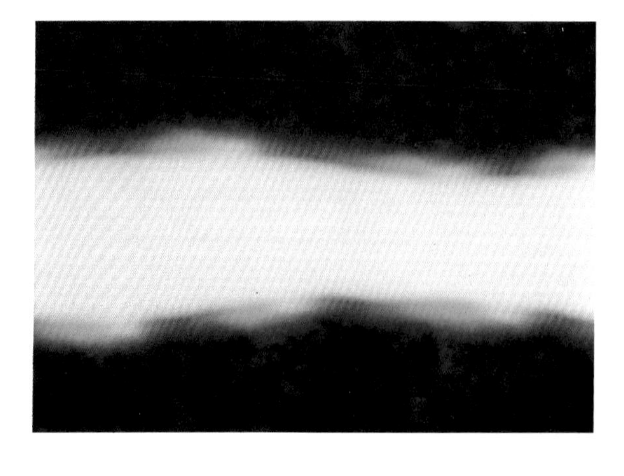

(a)

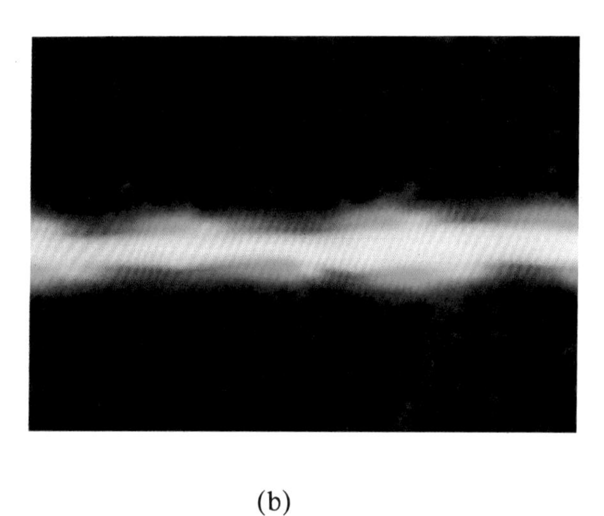

(b)

Figure 16.17. Collapse of the stripe domain of Fig. 16.13(d) under $H_z = -1000$ Oe. Shown are the states of the lattice at $t = 0.96$ ns (a) and at $t = 3.58$ ns (b).

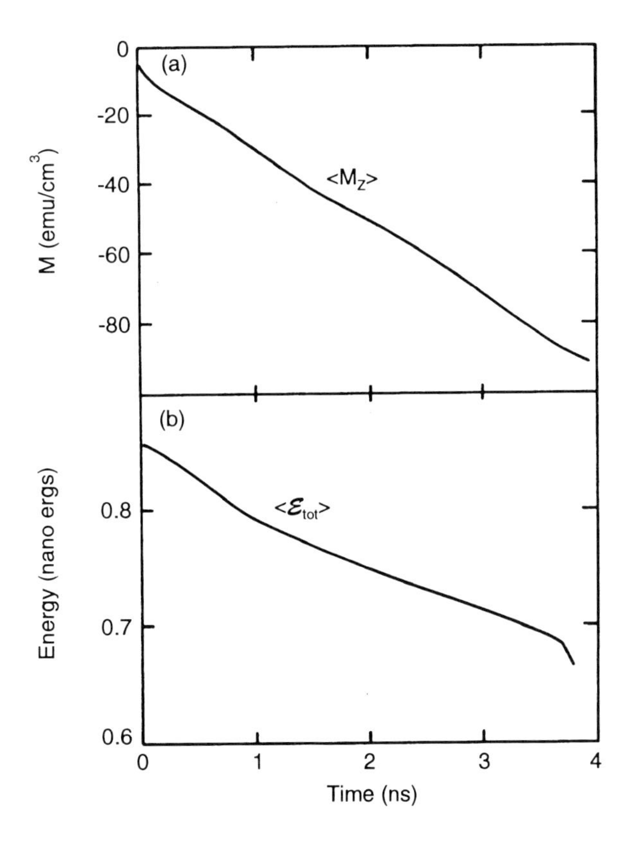

Figure 16.18. Plots of average magnetization and energy when the stripe domain of Fig. 16.13(d) collapses under $H_z = -1000$ Oe.

push the walls beyond the barriers and force them to collide and annihilate. (Because of the periodic boundary condition, the walls collide after wrapping around the lattice.) Figure 16.20(d) shows an advanced state of domain expansion under the 2 kOe field ($t = 1.154$ ns). Note that the lower wall has remained almost intact, while the upper wall has moved substantially. The plot of $\langle M_z \rangle$ versus time in Fig. 16.20(e) (starting with the application of the 2 kOe field) reveals that the movement is slow in the beginning, as the field struggles to overcome the pinnings. Once released, the wall moves rapidly for a period of about 1 ns until either another pinning occurs or the demagnetizing force begins to push the two walls apart (remember the periodic boundary condition). In any event, the motion slows down at $t = 1.2$ ns and the growth rate of $\langle M_z \rangle$ drops by a factor of nearly two. The walls continue to move out, however, wrap around at the boundary, and eventually annihilate.

In order to understand the effect of patch size on the coercivity, we repeated the preceding simulation for another lattice that had the same set of parameters as the lattice in Fig. 16.20, but whose total number of

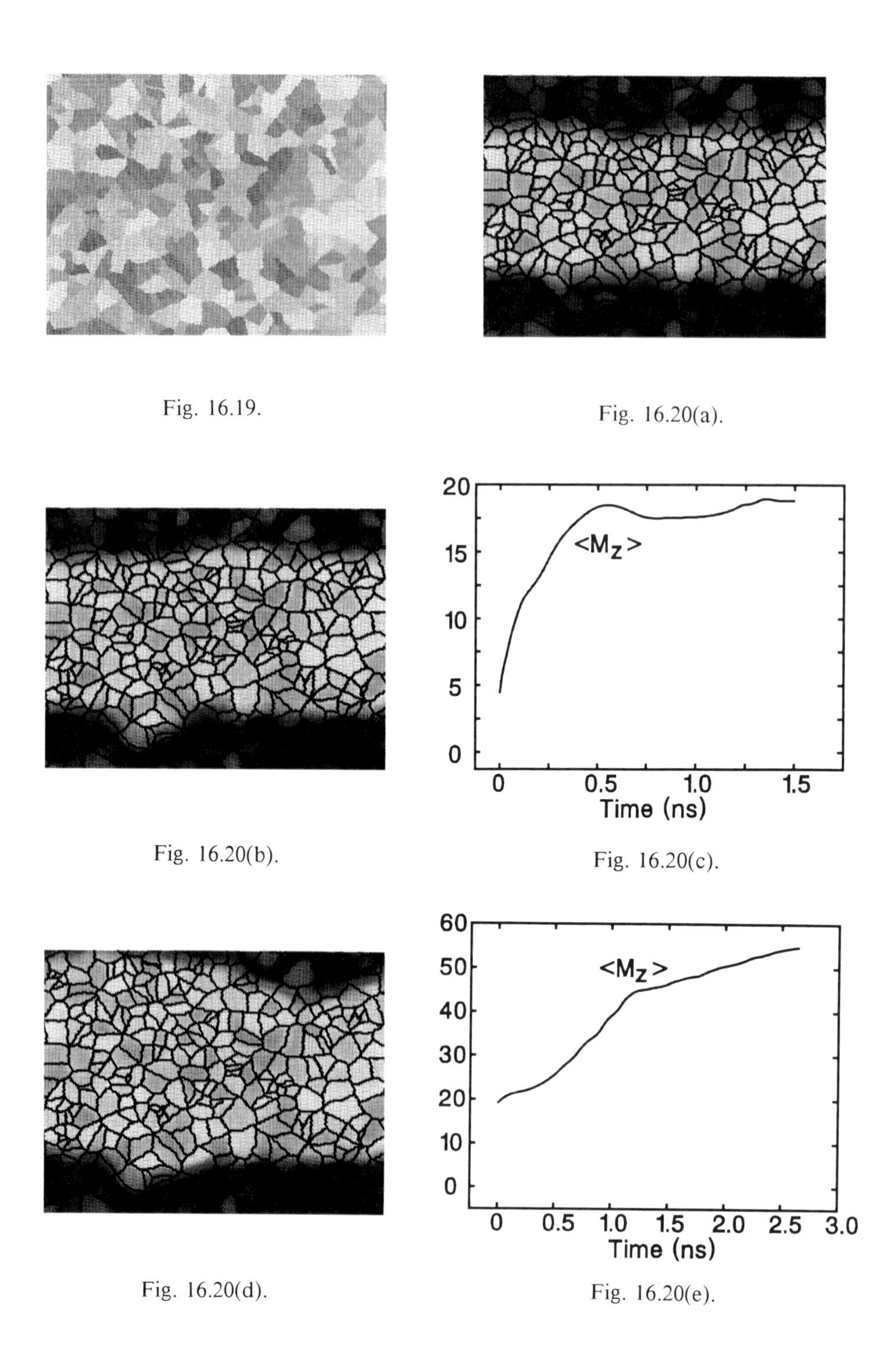

Fig. 16.19.

Fig. 16.20(a).

Fig. 16.20(b).

Fig. 16.20(c).

Fig. 16.20(d).

Fig. 16.20(e).

Figure 16.19. Patchy lattice with 346 patches. The colors are used to identify different patches; they have no other significance.

Figure 16.20. Stripe of reverse magnetization in the patchy lattice of Fig. 16.19. (a) The state of the lattice at $H_z = 0$. (b) At $t = 1.5$ ns after the application of $H_z = 1.5$ kOe. (c) Average magnetization $\langle M_z \rangle$ versus time under $H_z = 1.5$ kOe. (d) The state of the lattice 1.154 ns after the application of $H_z = 2$ kOe. (e) Average magnetization $\langle M_z \rangle$ versus time under $H_z = 2$ kOe.

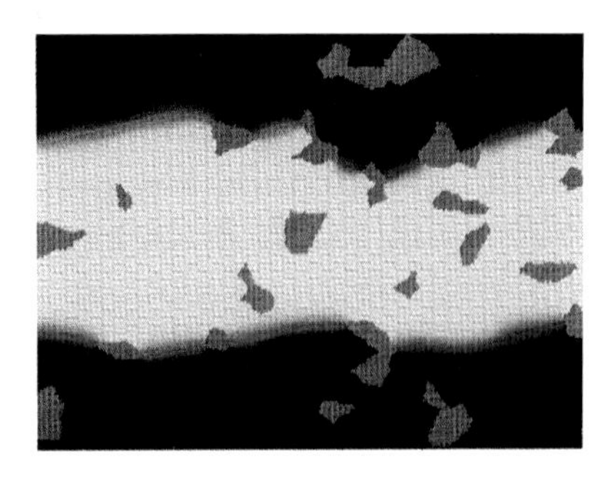

(a)

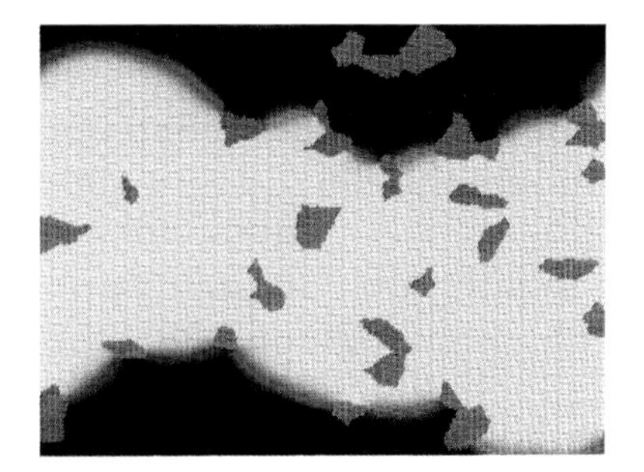

(b)

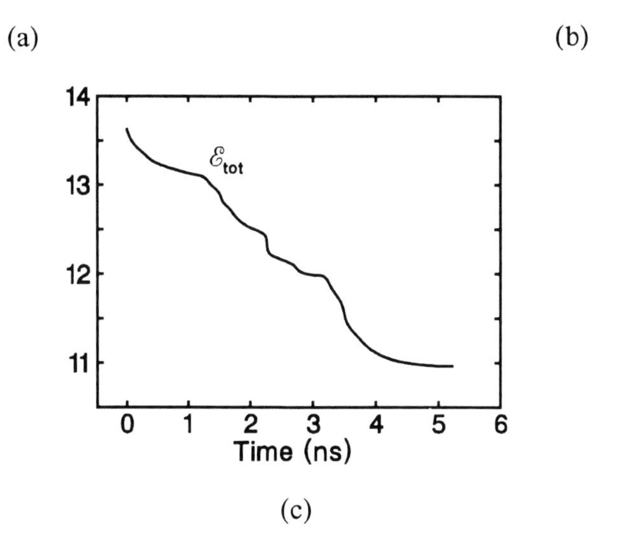

(c)

Figure 16.21. Stripe of reverse magnetization in a patchy lattice with voids. (a) At $H_z = 0$. (b) After $t = 5$ ns under $H_z = 1.5$ kOe. (c) Total magnetic energy versus time under $H_z = 1.5$ kOe.

patches had increased to 1300. Again we found that $H_z = 1.5$ kOe could not move the walls significantly, whereas $H_z = 2$ kOe could. It is probably safe, therefore, to assert that the average size of the patch does not affect the wall coercivity in a substantial way, so long as it is larger than the characteristic width of the domain wall.

16.3.4. Pinning of domain wall by voids

Figure 16.21 shows another strip of reverse magnetization in the patchy lattice of Fig. 16.19. This time, however, a few of the patches have been made void by assigning the value of zero to their magnetic parameters. These void patches are shown as solid gray regions in the figure. (The boundaries of the remaining patches are not shown here.) The remaining patches are all identical in their magnetic properties except for the value of the anisotropy constant K_u, which fluctuates randomly and independently from patch to patch. (The standard deviation of these fluctuations is 20% of the nominal K_u.) No other spatial variation of parameters has been assumed and, in particular, all axes of anisotropy are perpendicular (i.e., $\Theta = 0°$). The wall in Fig. 16.21(a) has automatically adjusted itself and minimized its length by attaching to the voids in its neighborhood.

Figure 16.21(b) shows the state of the lattice under an applied field H_z of +1.5 kOe at $t \simeq 5$ ns. Apparently, the walls have continued to seek voids to which to attach, while expanding in response to the external field. Figure 16.21(c) is a plot of the total magnetic energy of the lattice during this growth period. It is marked by the slow declines characteristic of continuous wall motion and rapid drops corresponding to detachments from or attachments onto the voids. It is thus observed that void-like defects can create jagged domain boundaries and cause discontinuities in the propagation process.

16.3.5. Lattice with in-plane defects

Consider the case of a patchy lattice having a different kind of defect, shown in Fig. 16.22. The 662 patches in this case are divided into two groups. The first group, consisting of 638 patches, were assigned axes of anisotropy nearly perpendicular to the plane of the lattice, as has been our practice so far. Each one of the patches in this group is assigned an axis, randomly and independently of the others, with a cone angle of $\Theta = 10°$. The second group of 24 patches is assigned axes of anisotropy that were nearly parallel to the plane of the lattice. The randomly and independently chosen values of θ and ϕ for these patches belonged to the intervals [80°, 90°] and [0°, 360°], respectively. (In the color-coded distribution of the anisotropy axes shown in Fig. 16.22, the patches with prominent colors are those with nearly in-plane axes.) The remaining parameters of the lattice are the same as before.

The lattice was initially saturated along $+Z$ and then allowed to relax in zero field until the remanent state, shown in Fig. 16.23(a), was reached. The strong exchange interaction has clearly forced the magnetization of the in-plane patches towards $+Z$, but the tendency towards the plane is still visible. We subjected the remanent state to a reverse field along $-Z$, and searched for the critical magnitude of this field that would initiate the reversal process; the critical nucleation field was found to be 8910 Oe. It is important to note that in the absence of the in-plane patches the same sample had a nucleation coercivity close to 18 kOe. The in-plane regions, therefore, facilitate the nucleation process. In Fig. 16.23(b) we show the pattern of magnetization under the applied field $H_z = -8900$ Oe, which is only slightly weaker than the critical field. The defects have been pushed toward the plane, yet the field is not quite strong enough to reverse them. When the field was raised to $H_z = -8920$ Oe, the reversal began, as the

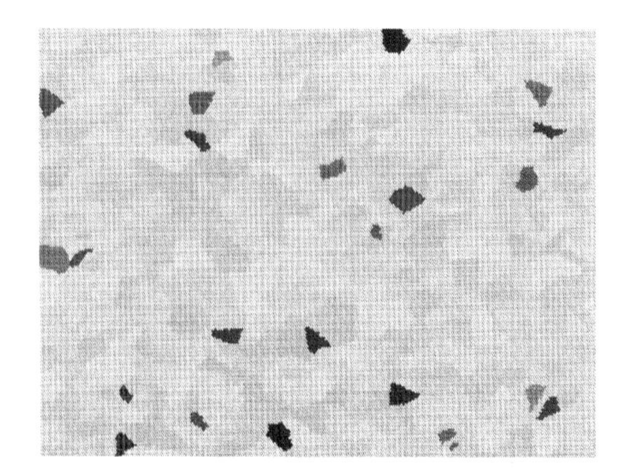

Figure 16.22. Lattice with 662 patches, of which 24 are "defective". A normal patch is assigned anisotropy axis within a cone of $\Theta = 10°$ from the perpendicular direction, whereas the easy axis of a defective patch is nearly in the plane of the lattice. The color depicts the direction of anisotropy for individual patches.

time evolution series in Figs. 16.23(c)–(h) shows. First the defect at the lower left corner of the lattice (with its periodic extension at the upper right) nucleated. Then the nucleus grew until the entire sample was reversed. Note how the defects seem to attract the domain wall as it approaches them, and then try to pin the wall to prevent its further progress. The applied field, however, is strong enough to overcome the pinning and bring the reversal to completion. A hysteresis loop with a high degree of squareness is the hallmark of this type of reversal.

In order to investigate the pinning process and the phenomenon of wall coercivity, we took the state of Fig. 16.23(e) as the initial state for another simulation and set the external field to zero. Note that because of periodic boundary conditions imposed on the hexagonal lattice, the various black regions in Fig. 16.23(e) are in fact different pieces of one and the same reverse-magnetized domain. This (roughly circular) domain must begin to collapse immediately after the removal of the external field.† Figure 16.24 shows the time evolution of the shrinking process, with frames (a)–(d) corresponding to $t = 0.44$, 1.12, 1.16, and 1.93 ns after the external field has been reset to zero. In the absence of defects the collapse would have been complete, but in the present situation, after an initial period of shrinking, the defects trap the domain wall and prevent its collapse. Note in Fig. 16.24(c), for instance, how the defect in the lower central part

†The reason for the collapse of the domain is that the wall energy density, $\sigma_w \simeq 1.265$ erg/cm^2, is greater than the corresponding demagnetization energy density, $\sigma_w^{(\mathrm{dmag})} \simeq 6\pi\tau M_s^2 \simeq 0.943$ erg/cm^2, causing energy minimization to favor shrinking of the domain.

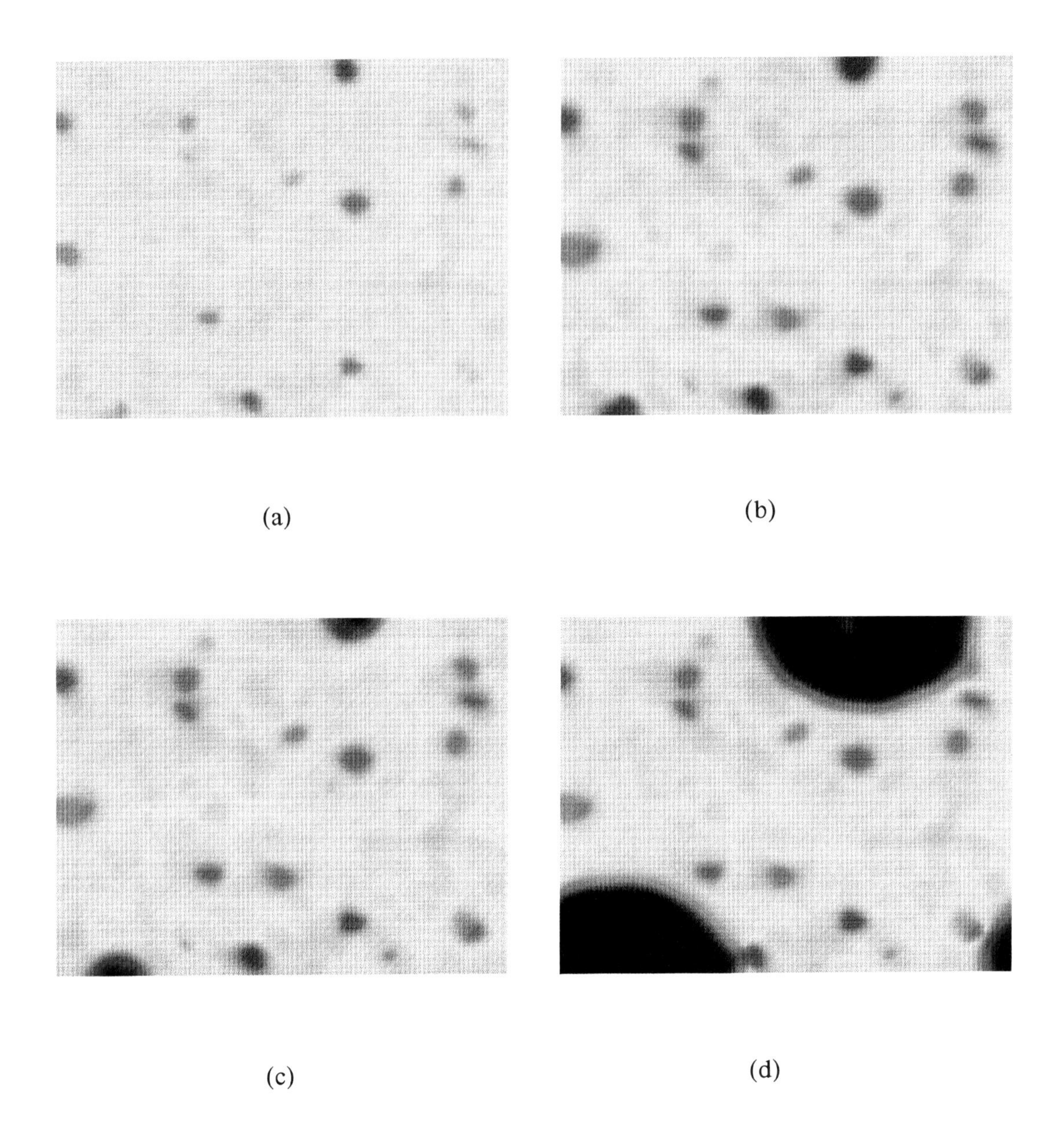
(a)
(b)
(c)
(d)

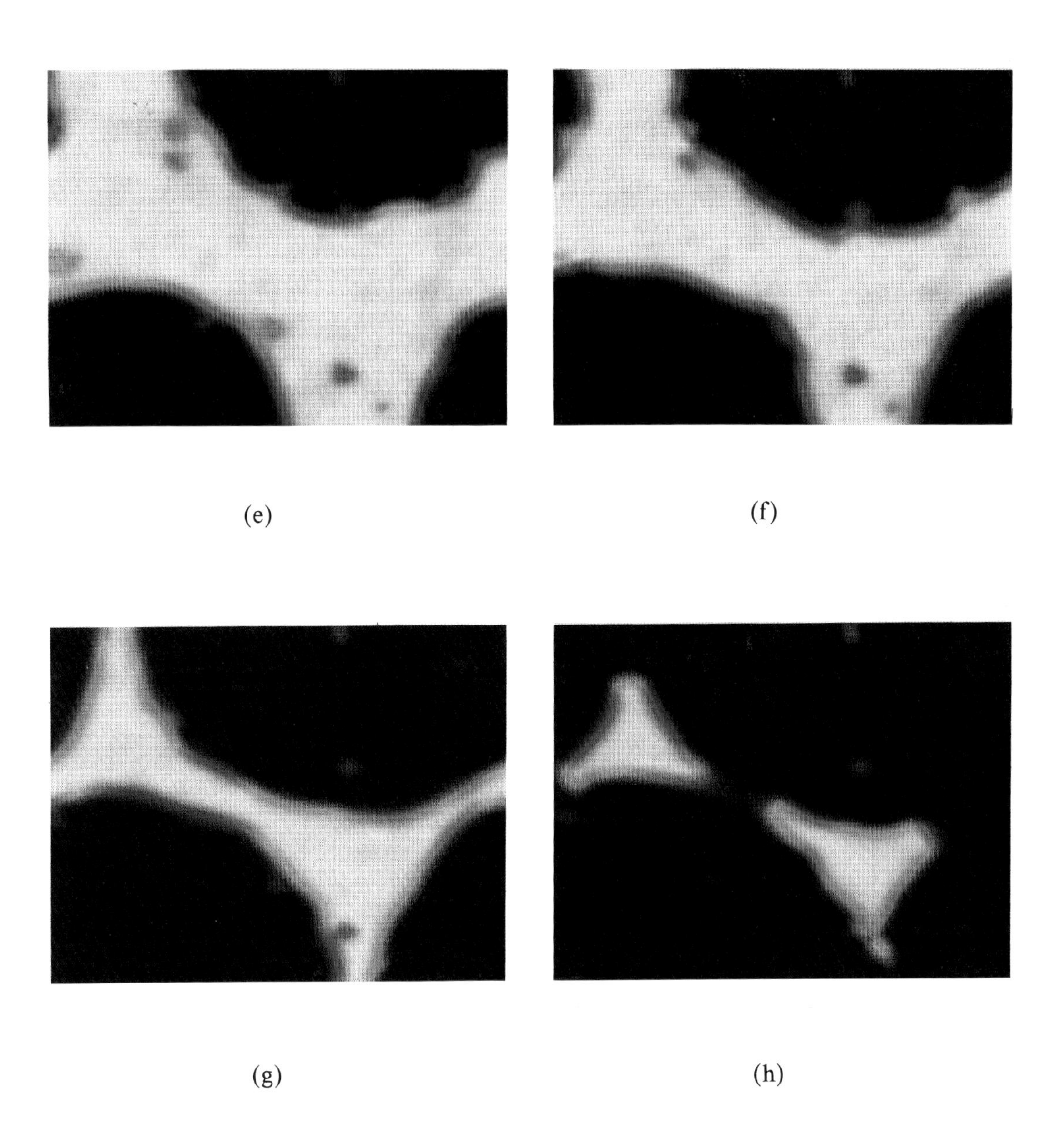

Figure 16.23. Magnetization patterns in the patchy lattice of Fig. 16.22. (a) The remanent state at $H_z = 0$. (b) The steady state under $H_z = -8.9$ kOe. The remaining frames show the evolution in time under $H_z = -8.92$ kOe. (c) At $t = 0.24$ ns reversal is under way with a nucleus forming in the lower left corner, its periodic extension appearing in the upper central region. (d)–(h) State of the lattice at $t = 0.72$, 1.00, 1.08, 1.25, 1.32 ns respectively.

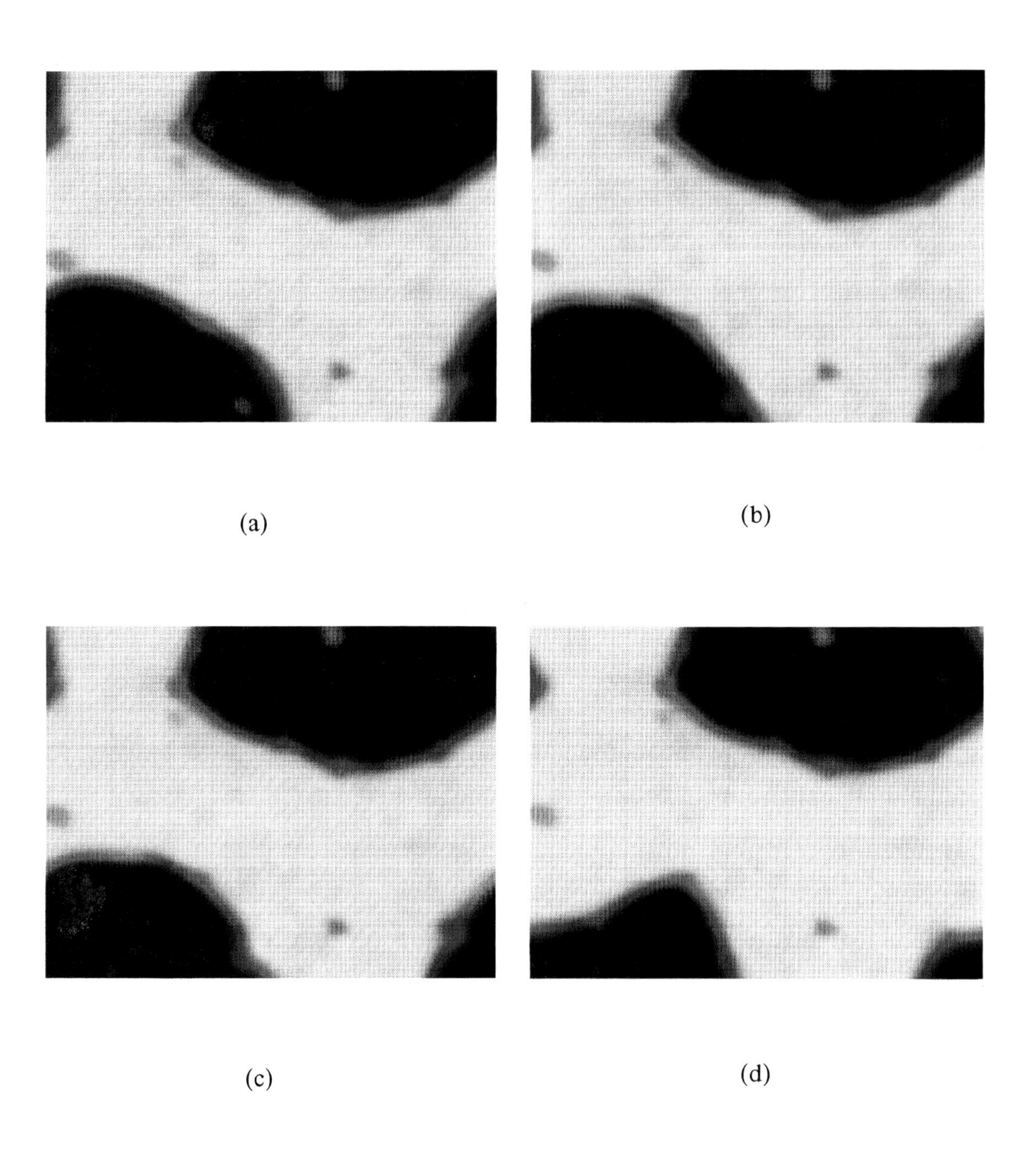

Figure 16.24. Evolution of the state of Fig. 16.23(e) after the removal of the external field. (a)–(d) State of the lattice at t = 0.44, 1.12, 1.16, 1.93 ns respectively.

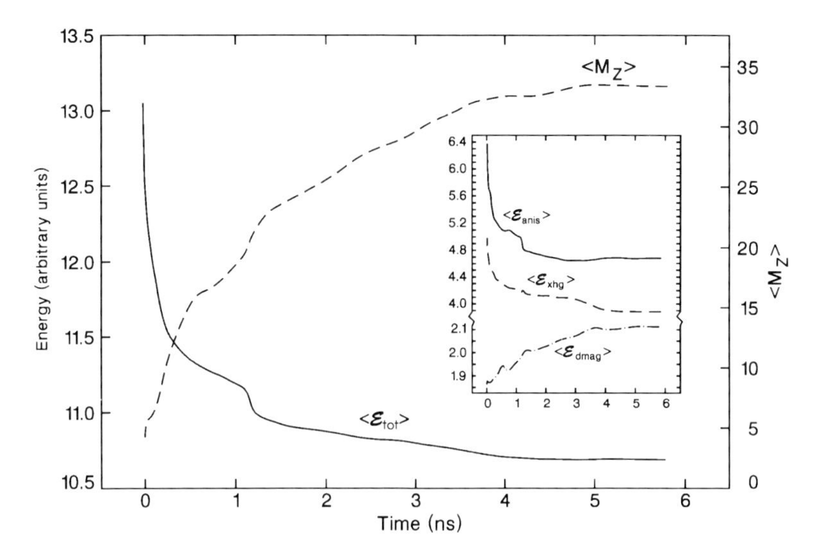

Figure 16.25. Total magnetic energy $\mathscr{E}_{\text{tot}}$ and average magnetization $\langle M_z \rangle$ of the lattice during the relaxation process of Fig. 16.24. $\mathscr{E}_{\text{tot}}$ is the sum of the exchange, anisotropy, and demagnetization energies, which are shown separately in the inset.

attracts the wall and keeps it pinned there afterwards. Therefore, like the other types of defect described in previous examples, defects with in-plane anisotropy may also be responsible for the observed wall coercivity of MO media. A comparison with Fig. 16.22 indicates that the domain in Fig. 16.24(d) is fully anchored on a number of defects. The fact that the domain in Fig. 16.24(d) is stable has practical significance for data storage, since it provides one possible mechanism for the stability of very small thermomagnetically recorded marks.

Figure 16.25 shows plots of energy $\mathscr{E}$ and average magnetization $\langle M_z \rangle$ during the relaxation process depicted in Fig. 16.24. As expected, the total energy $\mathscr{E}_{\text{tot}}$ of the system decreases, while the average magnetization along Z increases. A jump in the energy curve is an indicator of the capture of the wall by a defect. Plots of the various components of energy in the inset show that the demagnetization energy increases while the exchange and anisotropy energies decrease. This is to be expected, since during relaxation the total length of the domain wall decreases.

Finally, starting from the state of Fig. 16.23(f), we show in Fig. 16.26 the result of relaxation under zero applied field. The magnetization pattern in Fig. 16.26(a) is the steady state of the lattice obtained after 4 ns of relaxation. The reverse domain here is somewhat larger than the one in Fig. 16.24(d), simply because it started from a larger initial domain. The jaggedness of the wall in Fig. 16.26(a) is particularly striking. Figure 16.26(b) shows the difference between the initial state in Fig. 16.23(f) and the final relaxed state in Fig. 16.26(a), using a different coloring scheme

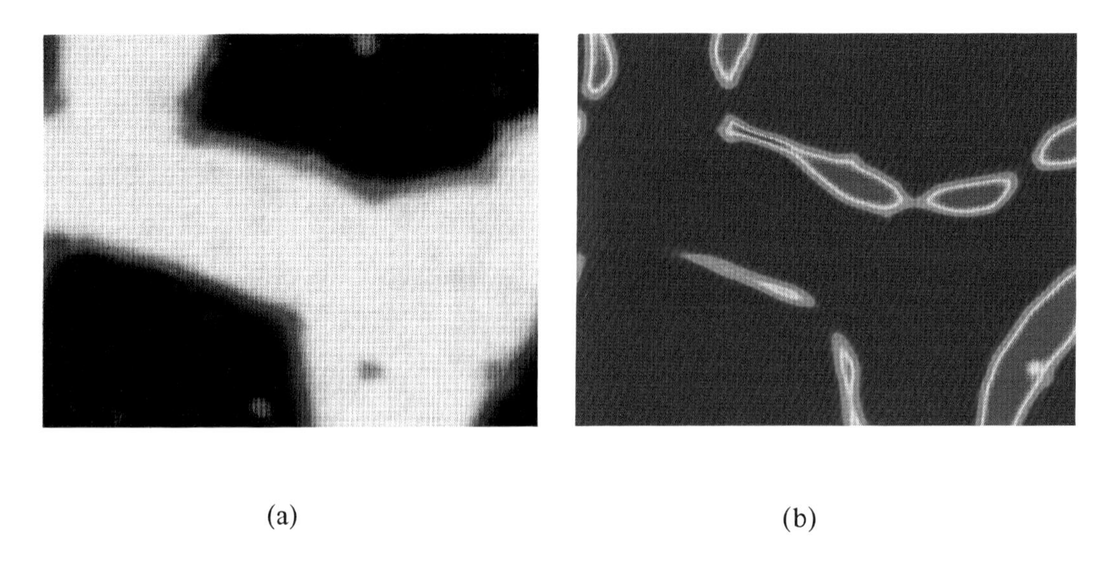

(a) (b)

Figure 16.26. (a) Magnetization pattern corresponding to the zero-field relaxation of the state in Fig. 16.23(f). (b) Dot-product of the initial state in Fig. 16.23(f) and the final (relaxed) state shown in frame (a) of the present figure. Each pixel represents a scalar value in the interval [- 1, +1], corresponding to the dot product of the two unit vectors. In this coloring scheme +1 corresponds to purple and - 1 to red, and the values in between are mapped continuously onto the color wheel.

(see figure caption). What is depicted here is the dot product between the magnetization directions of the initial and final states. This figure clearly shows the nonuniform motion of the wall during zero-field relaxation: while some regions of the wall have not moved at all, others have travelled by as much as a few hundred angstroms. In general, the wall seems to have remained pinned on several defects, while regions of the wall in between those defects have relaxed toward a state of minimum energy. Once again, the observed behavior confirms that small patches with in-plane anisotropy are effective in capturing and stabilizing small domains.

16.3.6. Isolated or weakly-coupled patches

Patch formation is a commonplace occurrence in thin-film growth processes: different patches originate from different seeds and form patch-borders wherever they meet. Though little is known about the properties of these borders, the exchange interaction there is likely to be weaker than that within the patch interiors, either due to the presence of small gaps or because of impurity concentration at the borders. We describe how the coercivity is influenced by these patch-borders.

The simulation results for a lattice containing 37 patches of random shape and size are shown in Figs. 16.27 and 16.28 (patch-borders are highlighted in grey).† Each patch has its own axis of anisotropy, which is oriented randomly with a cone angle of $\Theta = 45°$ (relative to the normal). The average diameter of each patch is about 400 Å, which is large compared to the width of domain walls ($\ell \simeq 100$ Å). Five resident nuclei are set within different patches; to stabilize these nuclei, it has been necessary to boost the anisotropy constant at the central disk of each such patch by a factor of ten over the nominal value of K_u.

Different values are assigned to the exchange stiffness coefficient at the patch-borders: A_x has 50% of its nominal value at all the borders except at the borders of four of the patches that contain nuclei; these four have A_x values that are 10%, 20%, 30%, and 40% of the nominal (see the caption to Fig. 16.27). This we have done to increase the number of cases studied in one simulation.

Figure 16.27(a) shows the initial state of magnetization with five resident nuclei built into five different patches. By following the changes of color around the domain walls, it is readily seen that the accumulated winding angle for each domain is 360°. Now, when this artificially set initial state is allowed to relax for 0.5 ns, the remanent state of Fig. 16.27(b) is obtained. In the process four out of five domains have remained more or less intact, but the one on the lower left side has minimized its exchange energy by expanding and sticking to the upper border of the patch, where the exchange coefficient is low (10%). Note also that the accumulated winding angle for this domain has dropped to zero.

Next, in order to study the expansion of these nuclei, we applied a field in the negative Z-direction. Figure 16.28(a) shows the steady state of the lattice under $H_z = -1$ kOe. The domain on the lower left side has now expanded and filled the entire patch, partly due to its proximity to the border, and partly because of the low exchange coefficient there. Figure 16.28(b) shows the steady state under $H_z = -1.5$ kOe. Here the patches with values of 10% and 20% exchange at their borders confine their resident nuclei to within the patch: the coupling to the outside is too weak to allow these domains to expand any further. The upper-left patch with 30% exchange at its border has more links to the outside, but there is still some degree of resistance, and the domain remains pinned to one side of the patch. The borders of the patch in the upper central part (with 40% exchange) are too far from the corresponding resident nucleus to give it an incentive to move. Similarly, the nucleus on the right-hand side, surrounded by strongly-coupled patch borders (50% exchange), is yet to be compelled to move.

Finally, when H_z is raised to -2 kOe, the lattice responds by fully reversing its magnetization; Figs. 16.28(c)–(e) are snapshots of this reversal process at $t = 0.95$, 1.25, and 1.55 ns after the field has been stepped up. Observe that the domain with the weakest exchange coupling to the outside

†Due to the periodic boundary conditions, some of the patches appearing at the edges of the lattice are actually different parts of the same patch.

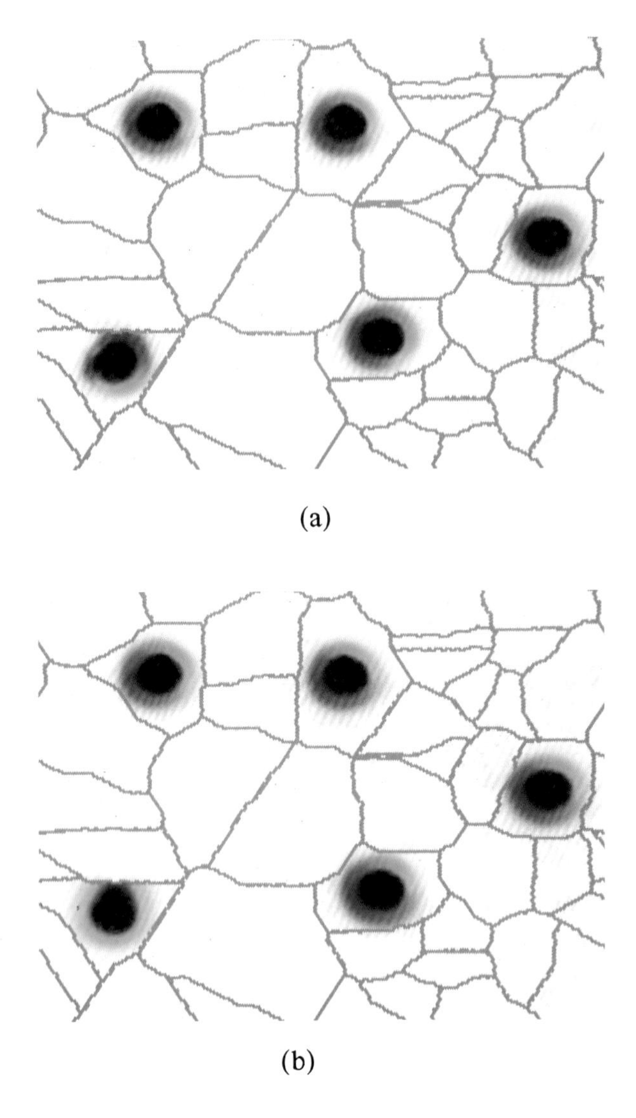

Figure 16.27. Resident nuclei within weakly exchange-coupled patches. The patches have reduced exchange coefficients at their borders. The initial state of the lattice, shown in (a), has five reverse-magnetized domains. The borders surrounding the lower left domain have a stiffness coefficient which is only 10% of the nominal value. Other borders have their A_x value reduced to 20% (lower central), 30% (upper left), 40% (upper central), and 50% (all others). (b) Remanent state of the lattice after the initial state has been fully relaxed.

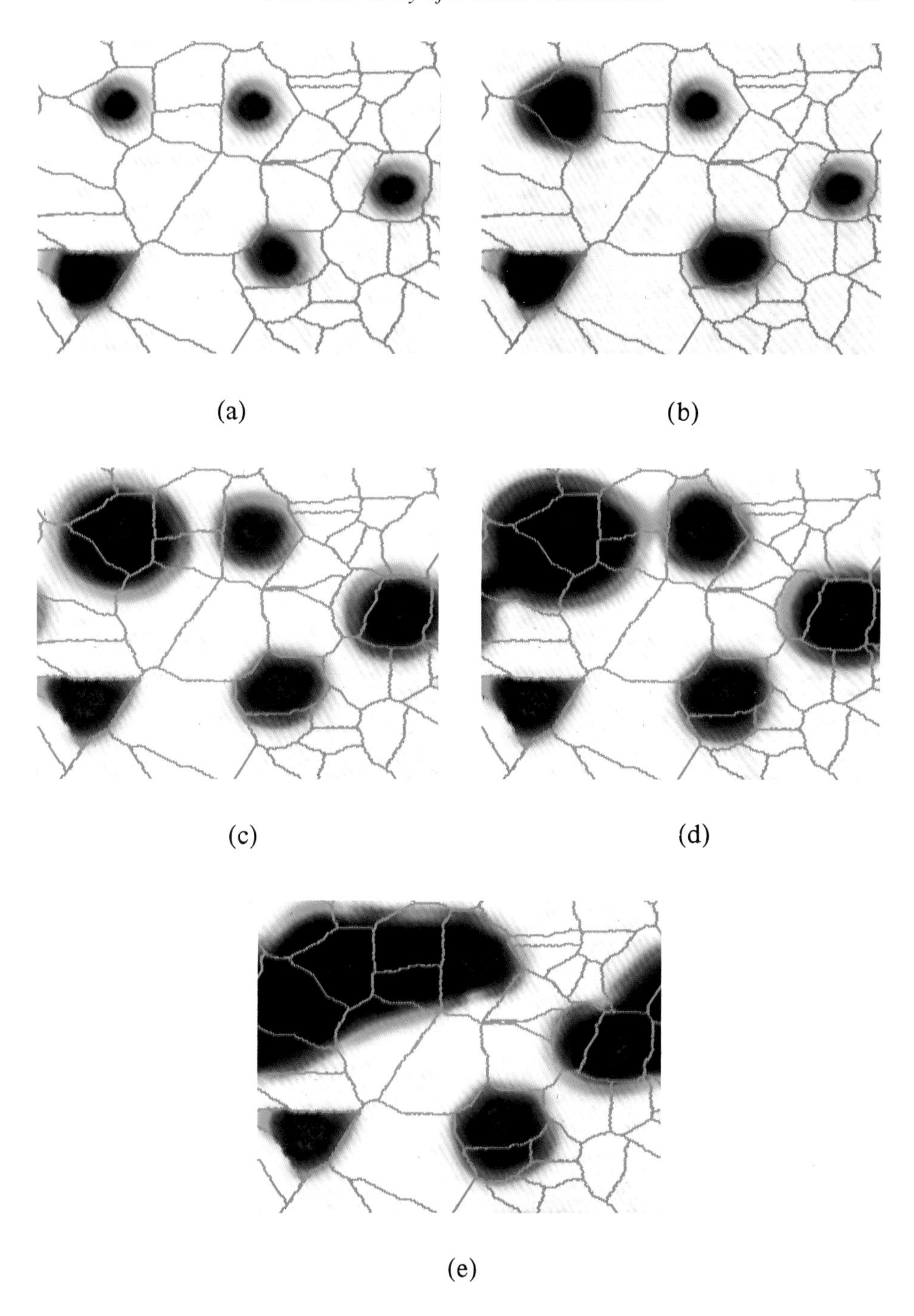

Figure 16.28. Evolution of the magnetization state of Fig. 16.27(b) under an applied magnetic field. (a) The steady state under $H_z = -1$ kOe. (b) The steady state under $H_z = -1.5$ kOe. (c), (d), (e) Snapshots of the magnetization pattern under $H_z = -2$ kOe, at $t = 0.95$, 1.25, and 1.55 ns respectively.

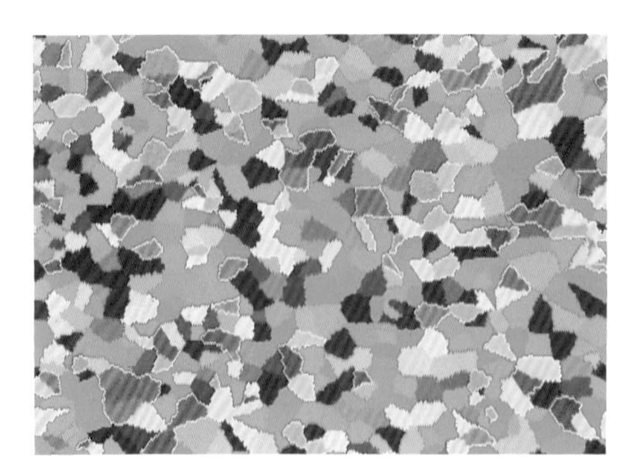

Figure 16.29. Distribution of the anisotropy constant K_u on a lattice containing 719 patches. K_u is selected randomly from the interval [5 × 10^4, 2 × 10^6] erg/cm^3 and assigned to the various patches. Patches with the lowest K_u are red, while those with the highest are purple. Intermediate values of K_u are then mapped continuously onto the color wheel (see Fig. 15.7).

has remained within the confines of its patch, the domain with 20% exchange at its border is being partially released, and the remaining three domains are crossing the borders and merging freely. The growth process in this case continued until the entire lattice was reversed.

16.3.7. *Patches with different anisotropy constants*

Figure 16.29 shows the distribution of K_u in a lattice with 719 patches. Values of K_u are selected from the interval [5 × 10^4, 2 × 10^6] erg/cm^3 and assigned randomly to the various patches. The color code uses red for the patch with the lowest anisotropy and purple for the highest, with a continuous mapping of the intermediate values onto the color wheel. The exchange stiffness coefficient has the nominal value of 10^{-7} erg/cm within each patch, but the patch-borders are assigned reduced values of A_x, somewhere in the range from 0% to 50% of the nominal, in random fashion. The distribution of the axes of anisotropy is random from cell to cell† with a cone angle $\Theta = 45°$. The remaining parameters are the same as those in Table 16.1.

The lattice was initialized in the saturated state and allowed to relax (in the absence of external fields) to the remanent state; the remanent magnetization was greater than $0.99M_s$. When an external field was applied to initiate reversal, the onset of nucleation was at H_z = 4.16 kOe. Figures 16.30(a)–(f) show snapshots of the reversal process under this critical field, starting with the nucleation of the weakest link in the lattice and

†Randomness from cell to cell is distinct from randomness from patch to patch. The former implies that each dipole of the lattice has its own local easy axis, whereas in the latter case all dipoles within a given patch have the same easy axis.

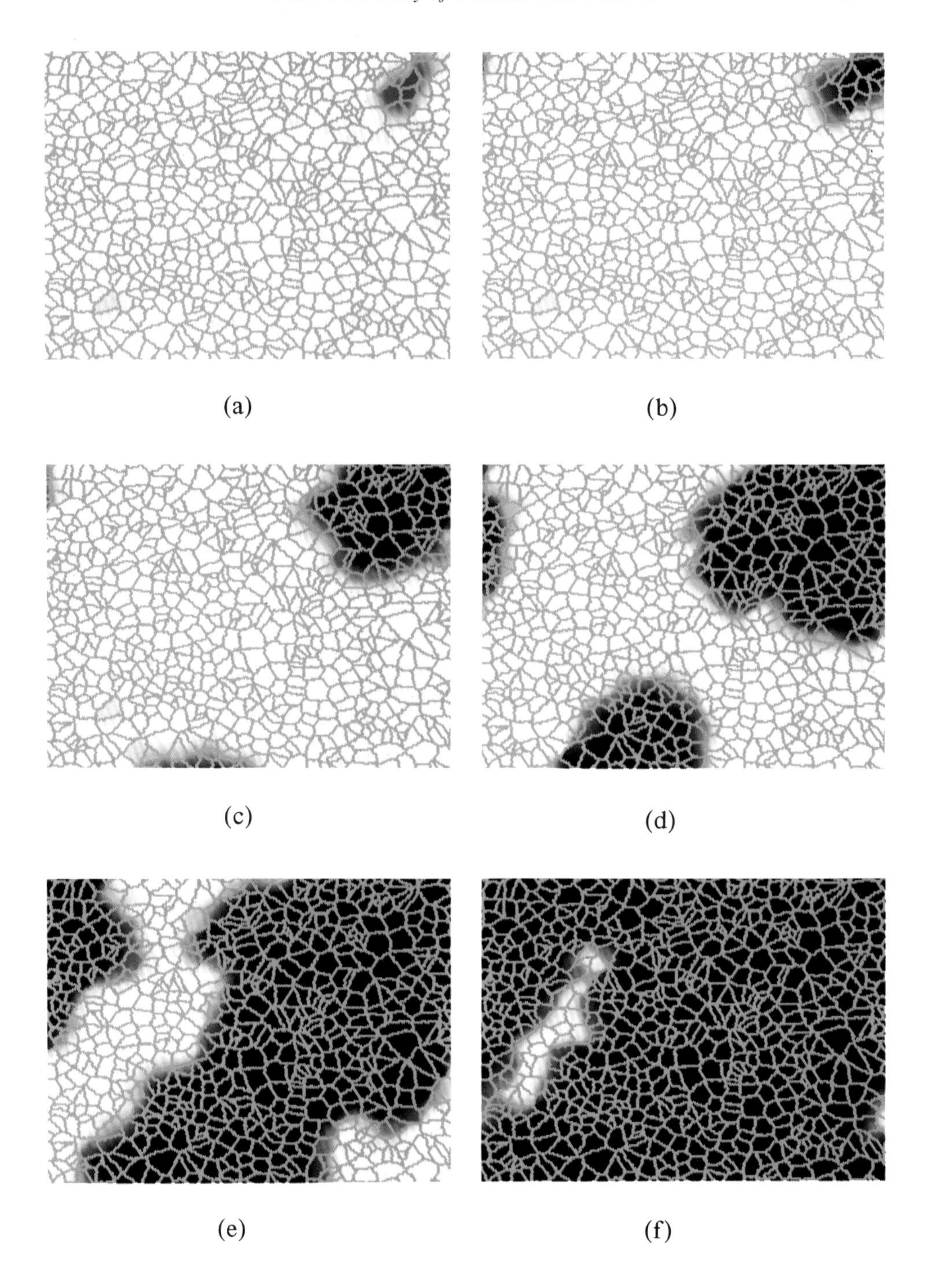

Figure 16.30. Nucleation of a reverse domain in a saturated sample. The applied field of 4.16 kOe is perpendicular to the plane of the lattice, and frames (a)–(f) show the evolution of the reversal process in time. Although not shown here, the final state of the sample is fully reversed.

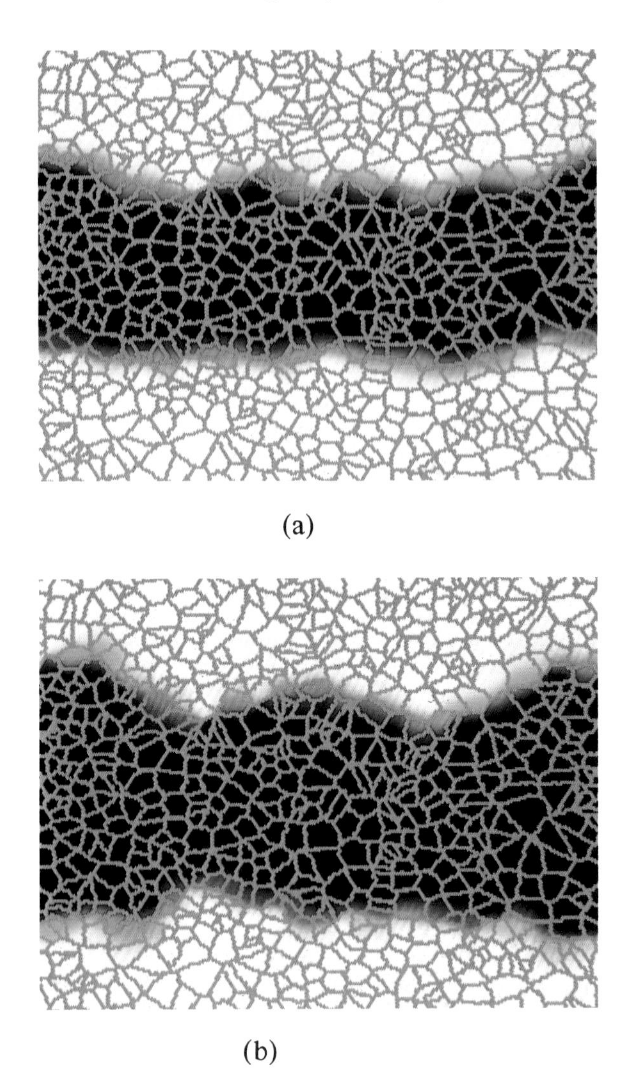
(a)
(b)

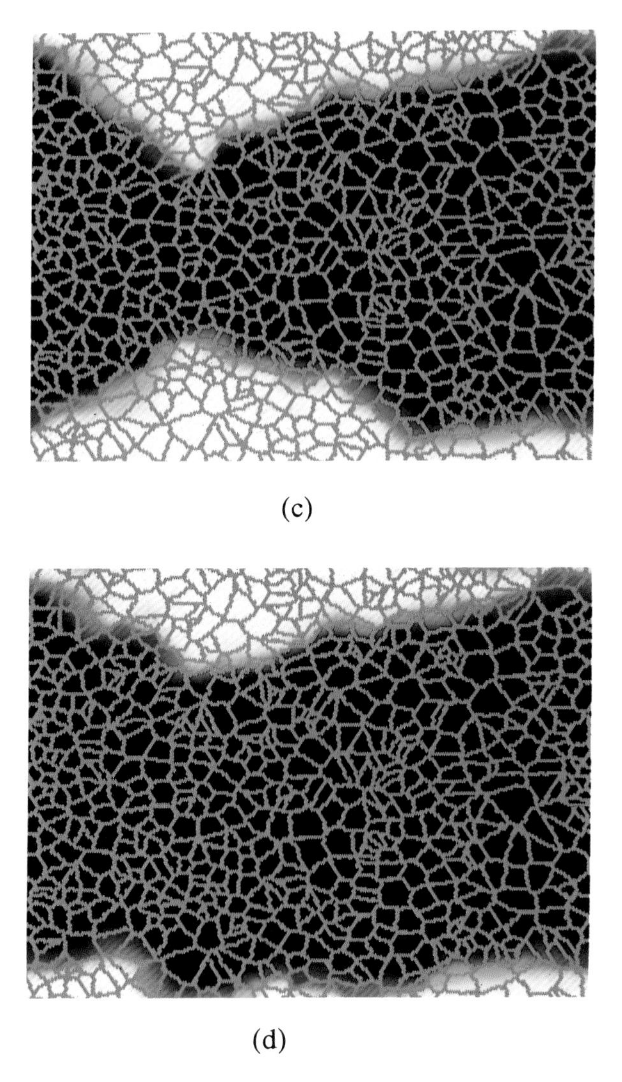

Figure 16.31. Wall motion under an applied field. The stripe in (a) is obtained by initializing the lattice with a band of reverse dipoles and allowing it to relax in zero field. (b), (c), (d) Steady state under $H_z = -1, -1.4, -1.8$ kOe respectively.

continuing with the growth of that nucleus. Comparison with the anisotropy map of Fig. 16.29 reveals the weakest link as a fairly large area with small anisotropy constant. In Fig. 16.30 the growing domain has a jagged boundary, since patches with large K_u and patch-borders with small A_x initially resist the propagation of the wall. As the wall finds its way around the barriers, however, it gathers strength and pulls itself through until all the barriers are overcome and the entire lattice is reversed.

To investigate the coercivity of wall motion in this lattice we initialized it with a reverse stripe, and allowed it to relax until the stable pattern of Fig. 16.31(a) was obtained. A field of $H_z = -1$ kOe expanded the stripe to

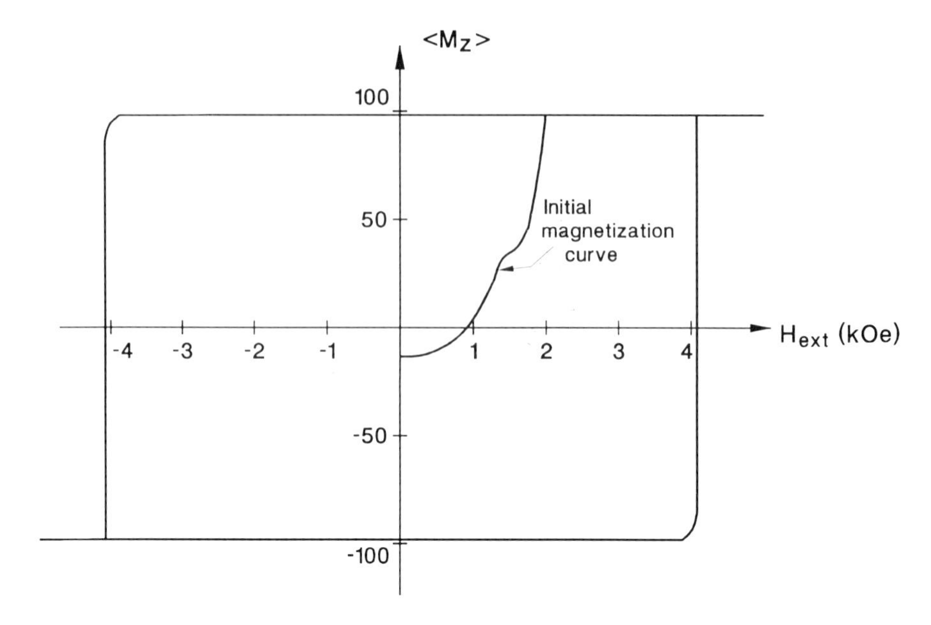

Figure 16.32. Computed hysteresis loop for the lattice of Fig. 16.29; the initial curve is obtained by the process outlined in Fig. 16.31.

the position shown in frame (b), but the expansion stopped at that point. The field was then raised to - 1.4 kOe, which succeeded in pushing the walls to the new position shown in frame (c). Even a field of - 1.8 kOe could not force the walls to annihilate each other (see frame (d)); full saturation of the present lattice required at least - 1.9 kOe. The complete hysteresis loop including the initial magnetization curve for this lattice is shown in Fig. 16.32. It is remarkable that the hysteresis loops for many amorphous films of RE–TM alloys resemble this computed loop not only qualitatively in terms of general features but also quantitatively.

16.4. Concluding Remarks

In this chapter we have examined several hypothetical mechanisms of coercivity in thin-film MO media. Using computer simulations we found that regions as small as a few hundred angstroms in diameter with unusually large or small magnetic parameters could act as nucleation centers and initiate the reversal process. These small regions, when certain of their structural and/or magnetic attributes deviate from the rest of the film, could trap domain walls and cause significant changes in the coercivity of wall motion. The coercive field values obtained by simulation are comparable with those observed in practice. Based on indirect observations, therefore, it is quite reasonable to surmise that these sources, individually or in combination, exist in real materials. For a definitive answer, however, one must await further progress in experimental nano-magnetics to enable the direct, high-resolution observation of magnetic media.[39-41] Among the existing tools for observation of the magnetic state in thin films, Lorentz electron microscopy and magnetic force microscopy have the potential to clarify the situation in the near future. We shall describe the principles of operation of Lorentz and magnetic force microscopy in Chapter 18.

Problems

(16.1) In torque magnetometry a magnetic sample is placed in a uniform magnetic field that rotates around a perpendicular axis, as shown in Fig. 16.33. (The rotation of the magnet is generally slow enough to allow one to ignore dynamical gyromagnetic effects.) Assuming the sample is a thin film of some MO material with perpendicular anisotropy constant K_u and saturation magnetization M_s, and also assuming coherent rotation of the dipole moments, determine the dependence of the torque experienced by the sample on the angle θ between the normal to the film and the direction of the field.

(16.2) One method of measuring the constant of magnetic anisotropy for thin-film samples consists of the application of a uniform field $\mathbf{H}_{\text{ext}}$ at $\theta = 45°$ to the film normal, and monitoring the torque exerted on the sample as a function of the strength of the applied field (see Fig. 16.34). Assuming the sample is a thin film of some MO material with perpendicular anisotropy constant K_u and saturation magnetization M_s, and also assuming coherent rotation of the dipole moments, determine the dependence of the torque on the magnitude of the field, H_{ext}.

(16.3) Figure 16.35 shows a thin film of some MO material with perpendicular anisotropy constant K_u and saturation magnetization M_s, subjected to an in-plane field $\mathbf{H}_{\text{ext}}$. A normally incident laser beam monitors the MO Kerr angle, which is proportional to the component of magnetization perpendicular to the sample. Assuming coherent rotation of the dipole moments under the influence of the applied field, determine the dependence of the Kerr signal on H_{ext}.

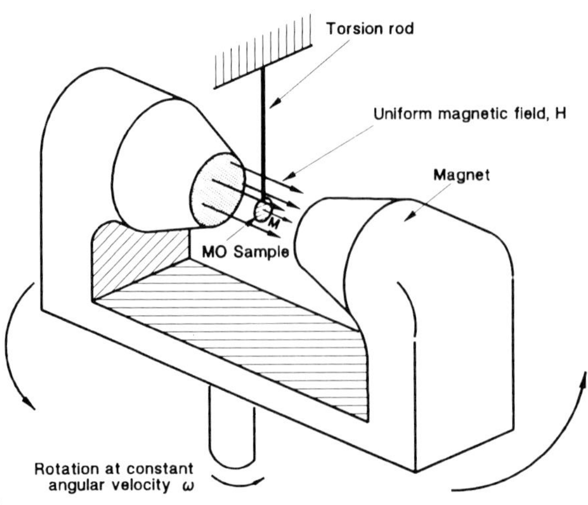

Figure 16.33.

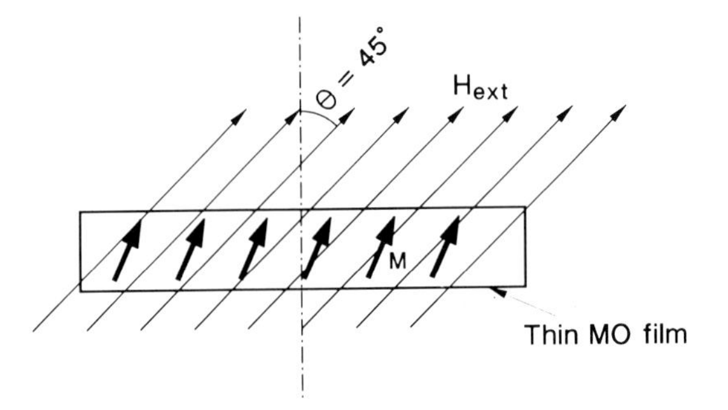

Figure 16.34.

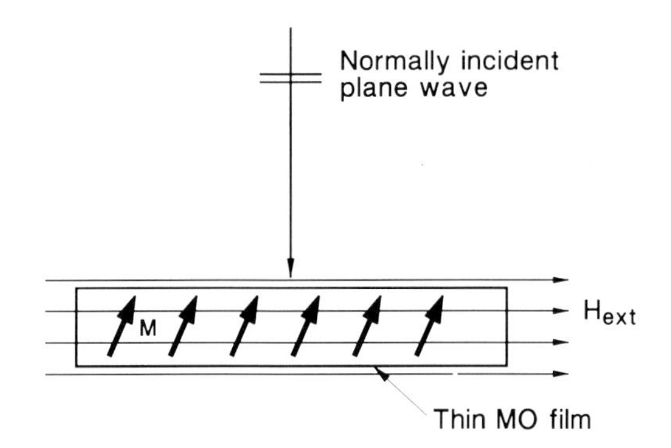

Figure 16.35.

17

The Process of Thermomagnetic Recording

Introduction

In thermomagnetic recording a focused laser beam creates a hot spot and allows an external magnetic field to reverse the direction of local magnetization.[1-13] Erasure is similar to recording, except for a reversing of the external field, which enables the magnetization within the heated region to return to its original state. The technology of high-density magneto-optical data storage owes a large measure of its success to the accuracy, reliability, and repeatability of the thermomagnetic process. The recorded domains are highly regular and uniform, fully reversed, and free from instabilities. What is more, a given area of the storage medium can be erased and rewritten several million times without any degradation.

The present chapter is devoted to the analysis of the thermomagnetic process based on the physical principles of micromagnetics and domain dynamics. In section 17.1 we review some of the facts and experimental observations concerning the storage media and the recording process. This will help familiarize the reader with the variety of phenomena and the order of magnitude of the parameters involved, and will set the stage for in-depth analyses of thermomagnetic recording in subsequent sections. The discussion in section 17.2 revolves around the energetics of domain formation and the forces acting on the domain wall in its formative stages. This treatment of the problem, being based on the arguments described in section 13.3, is essentially a quasi-static treatment that ignores the dynamics of wall motion; moreover, it fails to properly account for the role of coercivity in the recording process. These shortcomings aside, the quasi-static model is an invaluable tool for obtaining rough estimates of the recorded domain size, exploring the range of parameters of the write/erase processes, evaluating the sensitivity of media to the recording conditions, etc. In one form or another, the quasi-static model has been the cornerstone of many theoretical arguments involving the nature of thermomagnetic recording for over two decades.[14-17]

Recent research in thermomagnetic recording has focused on the dynamic processes involved in the nucleation and growth (or contraction) of submicron domains in thin film MO media.[18,19] Some of the more promising investigations in this area are based on dynamic simulations of large arrays of interacting dipoles, similar to those described in Chapter 15 (and used extensively in the discussion of coercivity mechanisms in Chapter 16). In section 17.3 we describe certain modifications of the algorithm of Chapter 15 that are necessary for the study of the recording process in rare earth - transition metal alloy films. The parameters α and γ are obtained in subsection 17.3.1 from a generalization of the LLG equation to ferrimagnetic systems with strongly coupled subnetworks. The presence of a nonuniform temperature distribution, of course, adds to the complexity of the simulations, particularly since the magnetic properties of MO materials are strongly temperature dependent. Some parameters (such as M_s and K_u) can be measured directly, while others (such as A_x) are "hidden" and must be obtained from model calculations. The mean-field theory will be employed in subsection 17.3.3 to match the model parameters with the available experimental data, and to extract from them the hidden parameters and their temperature-dependences. In the remainder of section 17.3 we discuss the results of dynamic computer simulations.

Exchange-coupled magnetic multilayers have emerged in recent years as media with new and improved characteristics.[20-24] The use of an exchange-coupled magnetic capping layer, for instance, has produced media with enhanced sensitiviy to the external magnetic field, while exchange-coupled trilayer designs have ushered in such novelties as directly overwritable media (DOW) and magnetically induced super-resolution (MSR). Exchange-coupled magnetic multilayer media and their properties will be discussed in section 17.4.

17.1. Facts and Observed Phenomena

Thermomagnetic recording is the process in which a focused laser beam assisted by an applied magnetic field creates a reverse-magnetized domain in a thin MO film. This is a complex dynamical phenomenon involving the nucleation of an initial domain, followed by its expansion and/or contraction, and final stabilization as the heating and then cooling progresses. Not every combination of material characteristics and recording conditions is suitable for recording. If the material's coercivity is low, for instance, the domains will be unstable and tend to collapse before the end of the recording cycle. If the applied field is too weak, or the laser pulse exceedingly short, or the MO film thick and the disk velocity too high, the recorded domains may be only partially reversed. A strong laser pulse could burn an amorphous material into a crystalline state, while a weak pulse might produce irregular and noisy domains. Figure 17.1 is a collection of micrographs that show domains with a variety of internal structure in a number of different MO media; the circumstances under which these domains have been formed are described in the caption. In view of the fact that high-density MO recording requires repeatable, smooth, fully reversed, and precisely positioned domains, it must be clear that optimization of the recording process is an important aspect of media and system design.

For a typical MO disk system plots of the measured carrier-to-noise ratio (CNR) in readout versus the laser power used during writing are

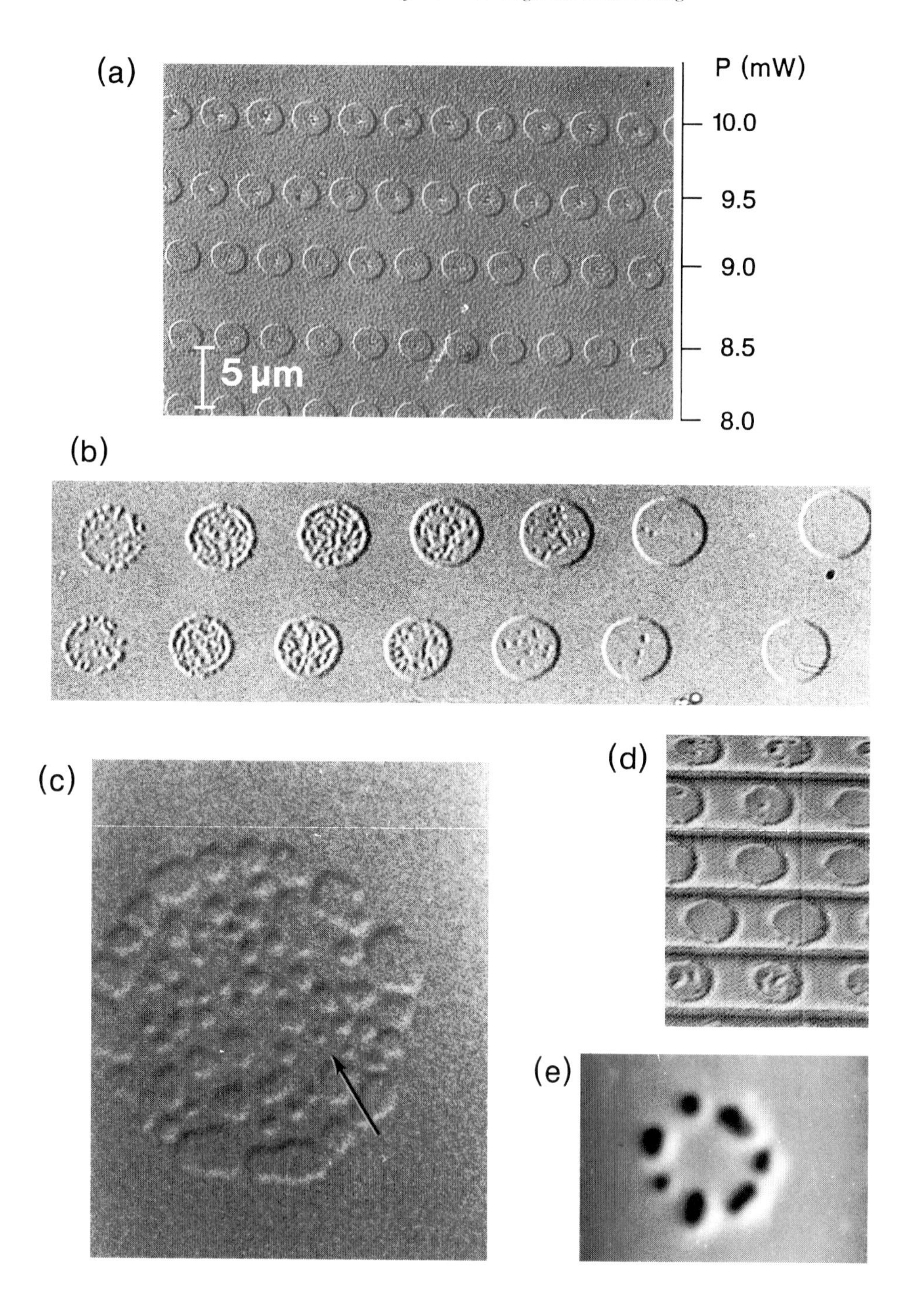
(a)
P (mW)
10.0
9.5
9.0
8.5
8.0
5 µm
(b)
(c)
(d)
(e)

Figure 17.1. Images of thermomagnetically recorded domains in different media and under various conditions. (a) Lorentz electron micrograph of domains recorded at different laser powers in a Co/Pt thin film. The applied magnetic field is 315 Oe, the laser pulse duration is 200 ns, and the pulse power is as indicated. Source: F.J.A.M. Greidanus, W.B. Zeper, B.A.J. Jacobs, J.H.M. Spruit, and P.F. Carcia, Magneto-optical recording in Co/Pt multilayers, *Japn. J. Appl. Phys.* **28**, supplement 28-3, 37–44 (1989). Courtesy of Bas Zeper, Philips Research Laboratories.

(b) Lorentz micrographs of domains written in $Tb_{0.19}Fe_{0.81}$ film using 0.5 μs laser pulses. The bias field for the domains on the far right is +90 Oe, decreasing in steps of 30 Oe and reaching the value of -90 Oe at the far left. Source: J.C. Suits, D. Rugar, and C.J. Lin, Thermomagnetic writing in TbFe: modeling and comparison with experiment, *J. Appl. Phys.* **64**, 252 (1988). Courtesy of D. Rugar, IBM Almaden Research Center.

(c) Lorentz micrograph of an unsaturated mark on a film 500 Å thick of $Tb_{32}Fe_{68}$. The writing is done with a 6 mW 0.5 μs laser pulse under a 135 Oe field (erasing direction). The smallest domain indicated has a diameter of 70 nm. Source: C.J. Lin, Materials for magneto-optic data storage, *Mat. Res. Soc. Symp. Proc.* **150**, 15–31 (1989).

(d) Magnetic force micrographs of domains recorded on a Kr-sputtered Co/Pt disk. The cobalt layers are 4.1 Å thick, separated by layers 14.7 Å thick of platinum, in a multilayer film with a total thickness of 188 Å. The recording laser power is 5.6 mW and, starting from the bottom track, the bias field values are 0, 452, 226, and 113 Oe respectively. Source: H.W. van Kesteren, A.J. den Boef, W.B. Zeper, J.H.M. Spruit, B.A.J. Jacobs and P.F. Carcia, Scanning magnetic force microscopy on Co/Pt magneto-optical disks, *J. Mag. Soc. Japan* **15**, Supplement S1, 247–250 (1991). Courtesy of H.W. van Kesteren, Philips Research Laboratories.

(e) Magnetic force micrograph of a broken-up domain, written on a TbFeCo sample 800 Å thick with a small bias field in the erasing direction. The mark diameter is approximately 2 μm. Source: P. Grütter, H.J. Mamin and D. Rugar, Magnetic force microscopy, in *Scanning Tunneling Microscopy II*, R. Wiesendanger and H.J. Güntherodt, eds., Surface Sciences Series, Vol. 28, Springer. Courtesy of D. Rugar, IBM Almaden Research Center.

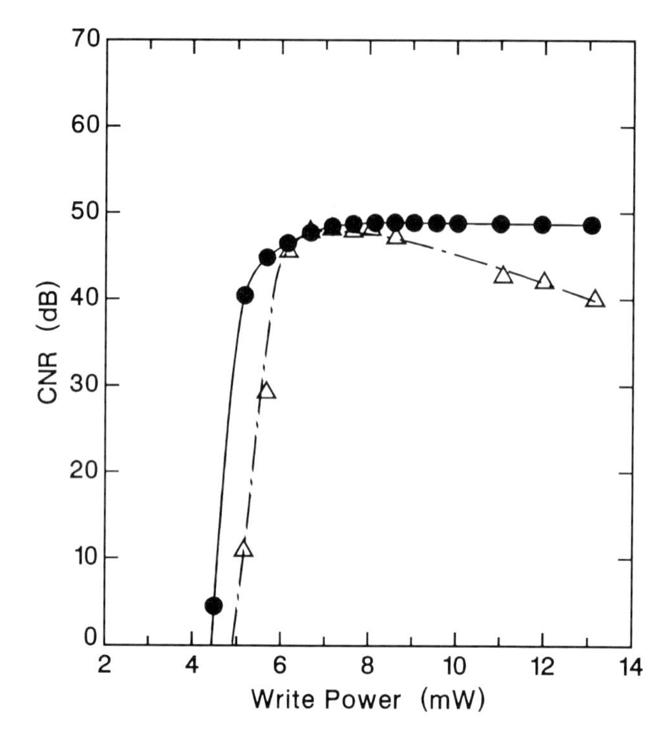

Figure 17.2. Carrier to noise ratio (CNR) in readout as a function of the write laser power using the techniques of laser power modulation (triangles) and magnetic field modulation (solid circles). The disk, consisting of a $Tb_{27}Fe_{63}Co_{10}$ layer, sandwiched between two SiN_x dielectric layers and overcoated with a reflecting heat-sinking aluminum layer, is supported by a glass substrate. The applied magnetic field was ±300 Oe, and the track velocity and carrier frequency during recording were $V = 11.3$ m/s and $f = 7.4$ MHz, resulting in domains 0.75 μm long.[25]

given in Fig. 17.2. The two curves in this figure correspond to recording with laser power modulation at fixed magnetic field and recording with magnetic field modulation at fixed laser power. The pulse train was adjusted to yield domains 0.75 μm long at a disk velocity V of 11.3 m/s; the applied field H_{ext} was ±300 Oe. There is a clear threshold power (approximately 5 mW in both cases) below which recording fails. Beyond this threshold the CNR for the field modulation technique reaches a plateau where the recorded domain pattern seems independent of the write laser power. With the light power modulation scheme, on the other hand, the CNR reaches a maximum after threshold but declines subsequently. As a practical matter, the insensitivity of the field modulation scheme to power level fluctuations is quite attractive.

Another example of write-process characterization is given in Fig. 17.3. Here both carrier and noise levels of the read signal are shown versus the strength of the applied field during writing. The disk velocity V is 9 m/s, and the 1.77 MHz recording pulse train that modulates the laser beam

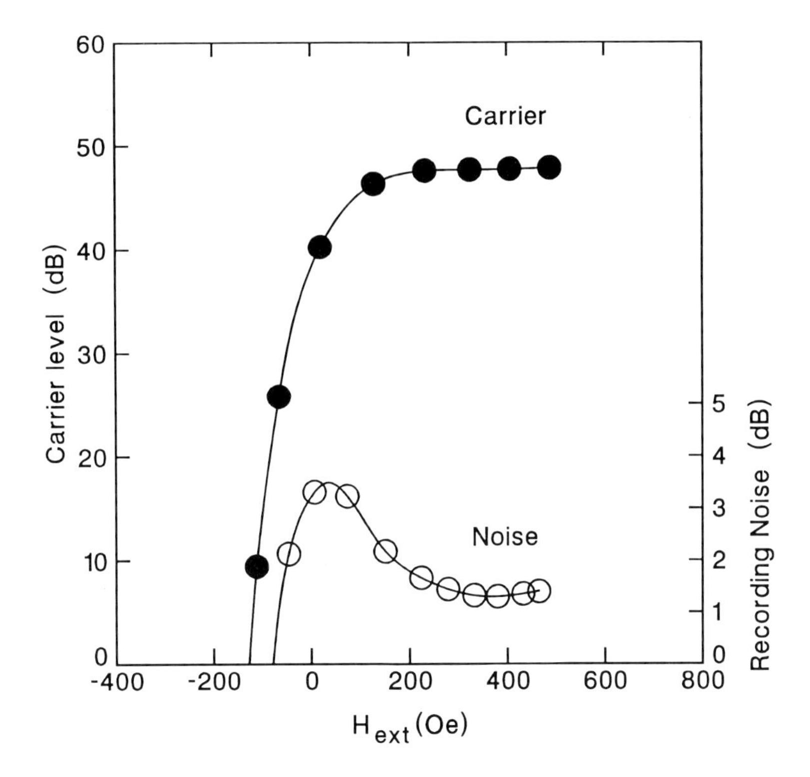

Figure 17.3. Carrier and noise levels in readout as functions of the applied magnetic field during writing. The disk consists of a $Tb_{27}Fe_{61}Co_{12}$ magneto-optical layer, sandwiched between two SiN_x dielectric layers and supported on a glass substrate. The MO layer's compensation point and Curie point temperatures are 90 °C and 200 °C respectively. The 1.77 MHz carrier was recorded with 8 mW 100 ns pulses at a linear track velocity of 9 m/s. The measured noise is the difference between the noise levels of the erased and recorded tracks.[26]

consists of 8 mW 100 ns pulses. Note the relative insensitivity of the domain pattern to the applied field in the range 200 Oe $\leq H_{ext} \leq$ 400 Oe. The peaking of the recording noise in the vicinity of the threshold is an indication that domains in this regime are partially reversed and highly nonuniform.

The process of thermomagnetic erasure is similar to the process of writing, except of course for the direction of the applied field.† One possible mechanism for erasure is that as the temperature rises and the

†In the case of recording by laser power modulation, writing is done with short laser pulses, whereas erasure is achieved with a continuous beam. A block consisting of several thousand domains within a sector is erased by a laser beam that is switched on at the beginning of the block and turned off at the end. Thus the temperature of the MO layer where a domain is to be written is somewhat different from that in the vicinity of a domain about to be erased. This is another difference between writing and erasure.

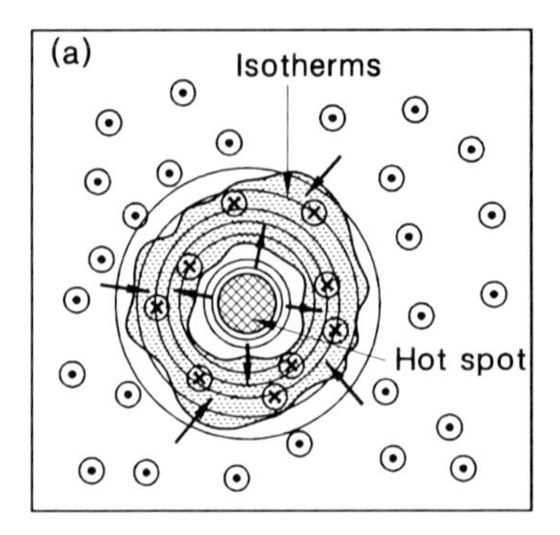

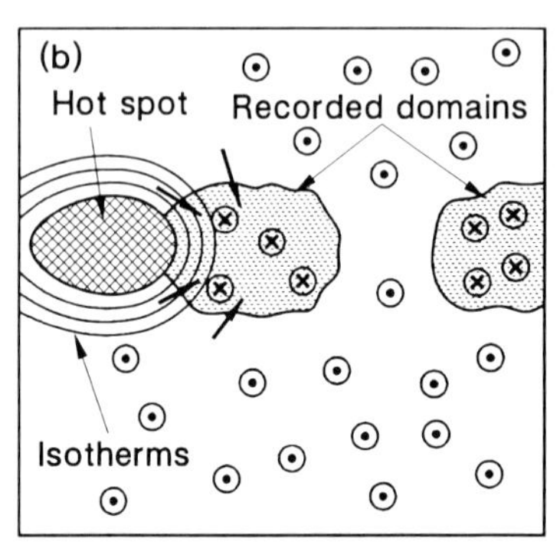

Figure 17.4. (a) In a stationary medium erasure of a written domain begins by nucleation from within and/or wall collapse from without. (b) When the medium moves with fixed velocity under a focused CW beam, the nucleating and/or destabilizing hot spot moves along the track and erases the recorded domains.

coercivity drops, the forces of the wall energy (σ_w) and the external field overcome the forces of demagnetization and coercivity, thus forcing the inward collapse of the domain. Another possibility is that a new domain nucleates within the old one and expands until their walls meet and mutually annihilate. Both mechanisms can also be at work simultaneously; see Fig. 17.4. Stroboscopic observations of the thermomagnetic process have verified these mechanisms,[12] but quantitative details are still lacking.

In the case of recording by light power modulation, the requisite magnitude of the erasure field, $|\mathbf{H}_{ext}|$, may be the same as, greater than, or less than the field used for writing, the difference depending on the MO material, disk structure, laser power level, disk velocity, etc. In some types of material it has been possible to do both writing and erasure without an external field. In these media a high-energy laser pulse creates a reverse domain and a low-energy pulse erases that domain. This is the basis for one direct overwrite (DOW) scheme proposed by Shieh and Kryder.[27,28] Unfortunately, the restrictions imposed on the media by this approach are incompatible with the requirements of high-density MO recording; to date, this DOW scheme has not been employed in commercial products.

Figure 17.5. Polarized-light microphotograph of four expanded domains in a $Tb_{20.3}(FeCo)_{79.7}$ sample. The MO film, which is sputter-deposited on a glass substrate and coated with a dielectric layer for protection, has a square hysteresis loop and a coercivity close to 4 kOe. Initially, four small domains were recorded on this medium with a short laser pulse and a small magnetic field. These micron-sized domains were subsequently expanded to the diameter shown in the figure of about 50 μm under an applied field of 3.6 kOe. The field was returned to zero to stop the domains growing, and the final, stable domains were photographed. Note the various degrees of jaggedness in these identically written and expanded domains.

Thermomagnetically recorded domains in a truly uniform and homogeneous medium are unstable and will either expand or collapse at the end of the write cycle.[29-31] In contrast, real MO materials are microscopically nonuniform; the large coercivities exhibited by them are but one manifestation of this nonuniformity (see Chapter 16). Figure 17.5 shows images of fairly large domains obtained by polarized-light microscopy on a typical MO medium. These domains, which were initially recorded with a focused beam and a small magnetic field, have subsequently been expanded in a large field. The degree of domain-boundary jaggedness visible in this photograph varies among the domains, and the fact that such jaggedness exists gives credence to the commonly held belief that the underlying magnetic structure in these media is highly nonuniform. Later, when we describe the results of computer simulations in section 17.3, we shall return to this topic and study the effects of media inhomogeneities on the recording process.

Certain applications of exchange-coupled magnetic multilayers will be described in section 17.4. In preparation for this discussion let us briefly review the physics of exchange coupling between adjacent magnetic layers. Figure 17.6(a) shows a simple bilayer structure in which the top layer has perpendicular magnetic anisotropy, while the bottom layer is magnetized in the plane of the film. Because of exchange coupling between the two layers

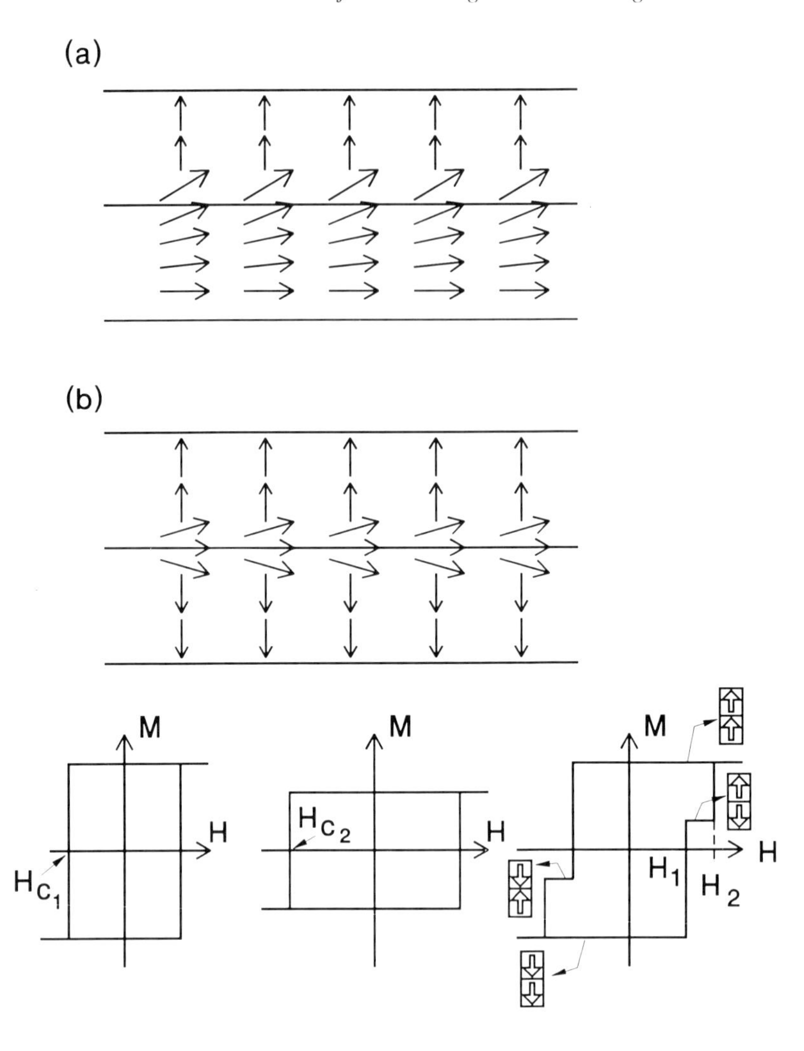

Figure 17.6. Exchange-coupled magnetic multilayers offer additional degrees of freedom for the design of MO media. The bilayer in (a) consists of a perpendicular film in contact with a capping layer that is magnetized in the plane of the medium. Exchange coupling between the two layers creates a 90° magnetic wall at the interface. The structure in (b) consists of two perpendicularly magnetized layers. Each layer has its own (intrinsic) coercivity in the absence of the other layer, but coupling can shift the individual layer coercivities. Typical hysteresis loops of the individual layers and the loop of the bilayer in this case are shown at the bottom of the figure.

the magnetization cannot undergo a sharp transition and, therefore, a 90° wall exists at the interface. As might be expected, the width and the energy density σ_w of this wall may be determined from a consideration of the balance between the various sources of energy including exchange, anisotropy, demagnetization, and the external field.

Another example of exchange coupling in bilayers is shown in Fig. 17.6(b). Here both layers have perpendicular anisotropy and, when their magnetizations are oppositely oriented, there exists a 180° wall at their interface. Again, it is not difficult to determine the width and the energy density σ_w of this interfacial wall using energy minimization arguments. In general, each layer in the absence of the other one will have its own hysteresis loop, but the loop for the bilayer is somewhat different from the sum of the individual loops, as shown at the bottom of Fig. 17.6(b). For example, consider the case of two RE–TM films, coupled as in Fig. 17.6(b), both on the same side of the compensation point (i.e., both RE-rich or both TM-rich). Now, when the net magnetizations of the two layers are parallel, their corresponding subnetwork magnetizations will also be parallel, and there will be no interfacial wall. The wall appears when one of the layers is switched, in which case both RE and TM subnetworks of the two layers will develop interfacial walls. Trapped in the bilayer configuration, the film with the lower intrinsic coercivity H_{c1} will reverse at $H_1 > H_{c1}$ because creating the interfacial wall requires additional energy. On the other hand, the film with the larger coercivity H_{c2} switches below H_{c2} because its reversal is accompanied by the annihilation of the interfacial wall.

In the above example, had the two films been on opposite sides of the compensation point (i.e. one RE-rich and the other TM-rich) then the interfacial wall would have existed when the net magnetizations of the two layers became parallel, in which case we would have found $H_1 < H_{c1}$ and $H_2 > H_{c2}$. Clearly, exchange-coupled multilayer media can exhibit a variety of types of behavior, and the study of these can be fascinating as well as rewarding.

17.2. Magnetostatic Model of the Recording Process

In this section we apply the methods of section 13.3 to determine the variations of the magnetic energy in a ferrimagnetic film during a typical recording cycle. We consider a 15 nm-thick film incorporated in a quadrilayer structure and subjected to a 5 mW 50 ns pulse from an 840 nm laser. The beam is focused to a Gaussian spot of diameter 1 μm at the 1/e point, and, for simplicity, we assume that the disk is stationary. The temperature distribution in the MO film is calculated numerically (using the methods of Chapter 11) and is shown in Fig. 17.7(a). Also calculated (with the aid of the mean-field theory of Chapter 14) are the functions $M_s(T)$, $A_x(T)$, and $K_u(T)$ for an RE–TM alloy with T_c = 400 K and T_{comp} = 293 K; plots of these functions are shown in Fig. 17.7(b). The ambient temperature is 295 K, and an external field H_{ext} = 400 Oe is applied in a direction opposite to the initial magnetization.†

The procedure used in the following calculations is as follows. At any

† In the notation of Chapter 13 the applied field is B_0.

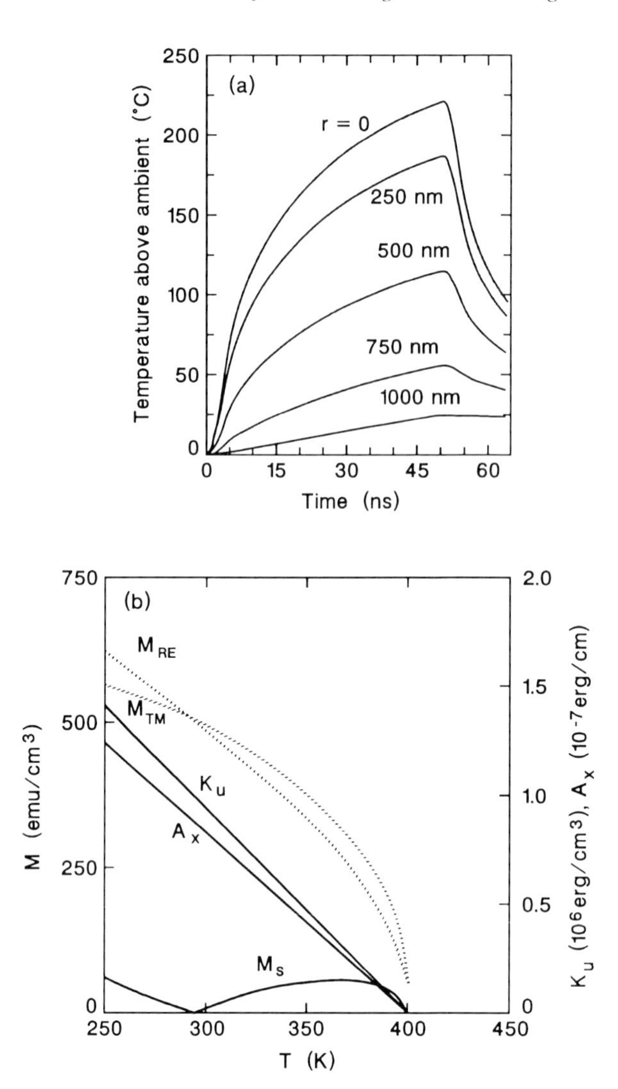

Figure 17.7. (a) Temperature versus time within the magnetic layer of a quadrilayer MO device, calculated at various distances r from the beam center. The laser power is delivered in a 5 mW 50 ns pulse, and the focused Gaussian spot diameter is 1 μm at the 1/e point. (b) Saturation moment $M_s(T)$, exchange stiffness coefficient $A_x(T)$, and anisotropy constant $K_u(T)$ for a typical ferrimagnetic RE–TM alloy with compensation point around the ambient temperature of 300 K. The mean-field theory of Chapter 14 has been used to compute the temperature-dependence of these parameters. Also shown are the individual subnetwork magnetizations.

given instant of time t, knowing the temperature profiles, we determine the distributions of the magnetization $M_s(r)$, the exchange coefficient $A_x(r)$, and the anisotropy constant $K_u(r)$ in the MO layer; (r is the radial distance from the beam center). We then assume the existence of a domain of radius r_0 and wall-width parameter Λ, and compute the magnetic energy of the domain $\Delta\mathcal{E}_{tot}(r_0; \Lambda)$, as described in Chapter 13. This energy can be minimized by varying Λ until the optimium wall width and the equilibrium energy $\Delta\mathcal{E}_{tot}(r_0)$ are determined.

Figure 17.8 shows the change in total energy $\Delta\mathcal{E}_{tot}$ produced upon creation of a domain of radius r_0 at different stages of the recording process. At $t = 0$, before the laser is turned on, $\Delta\mathcal{E}_{tot}$ is an increasing function of r_0, and therefore no reverse domain of any radius can be formed. At $t = 8$ ns, however, there is a minimum of energy around $r_0 = 500$ nm. Whether or not the nucleation begins at this stage depends on the local value of the coercivity, but from energy considerations it is a possibility. At $t = 50$ ns, the end of the heating cycle, the region $0 \leq r_0 \leq 520$ nm is above T_c, and the minimum of energy has moved to $r_0 = 1200$ nm. In this case, whether or not the domain boundary reaches this minimum depends on the wall coercivity and its temperature-dependence. The derivative of $\Delta\mathcal{E}_{tot}$ with respect to r is a measure of the strength of the forces acting on the domain wall, and if the wall is to proceed with its expansion beyond a certain point, these forces must overcome the wall coercivity at that point.

Plots of $\Delta\mathcal{E}_{tot}$ for the cooling phase of the recording cycle are shown in Fig. 17.8(b). It is seen that, depending on the position of the wall at a given time, the forces acting on the wall can be expansive or contractive. It is interesting, however, to note that just after the laser is turned off, when coercivities are low and the wall is somewhere between the beam center and the radius of minimum energy, the forces tend to be expansive and thus to prevent the written spot from collapsing. This behavior can be seen in the curves for $t = 60$ ns and $t = 85$ ns, and to a lesser extent even at $t = 100$ ns. By the time the temperature returns to ambient, the coercivities have increased dramatically and, despite the presence of contractive forces, there will no longer be any danger of collapse.[††]

The relative significance of the various contributions to energy may be deduced from Fig. 17.9, which is obtained at $t = 50$ ns. It is seen that $\Delta\mathcal{E}_{dmag}$ is much smaller than the other terms. This, in fact, is the case at all stages of the process, and the relation between $\Delta\mathcal{E}_{ext}$ and $\Delta\mathcal{E}_{wall}$ essentially determines the behavior of $\Delta\mathcal{E}_{tot}$ in this example. For materials with different characteristics, of course, this balance could be different. For thick magnetic films, for instance, one can no longer neglect demagnetizing effects (the material considered here would have to be thicker than 100 nm if demagnetization were to play a role). The

[††]Throughout these calculations the wall-width parameter Λ remained close to $\sqrt{A_x/K_u}$ at all radii and was not sensitive to temperature, being about 30 ± 2 Å in the entire range.

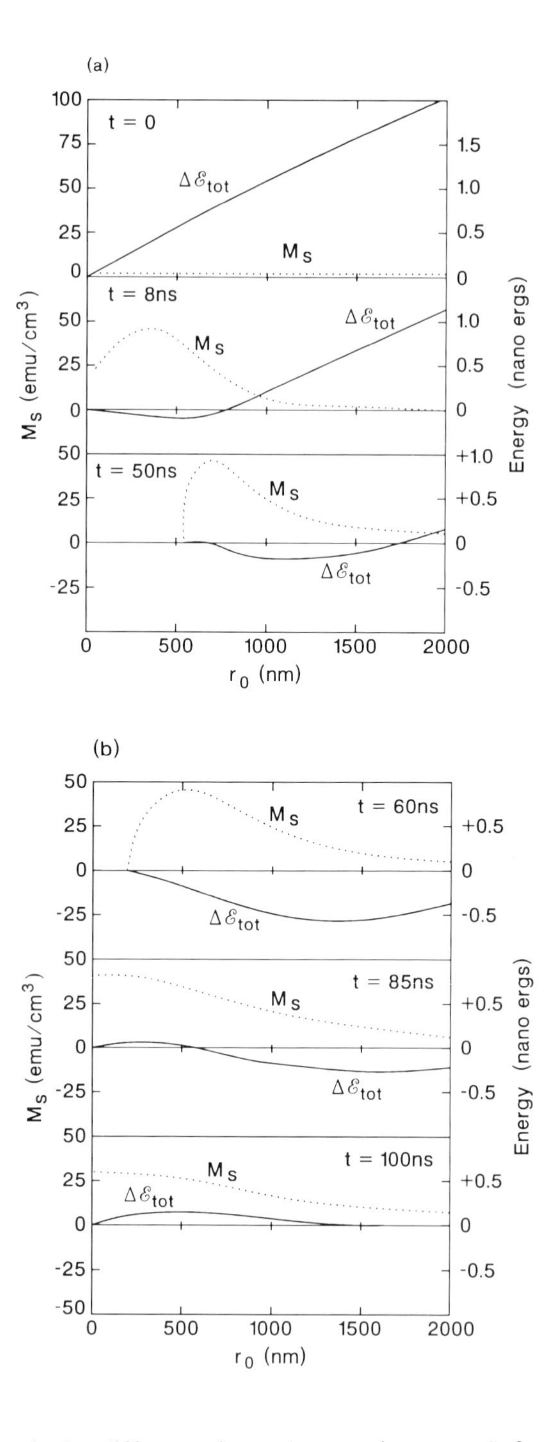

Figure 17.8. $\Delta\mathscr{E}_{\text{tot}}$ is the difference in total magnetic energy before and after the formation of a reverse domain of radius r_0 in the MO film whose temperature profile and magnetic parameters are described in Fig. 17.7. r_0 is the distance from the center of the focused spot, around which circular symmetry has been assumed. The dotted curves show the magnetization M_s versus r_0 at each stage. The three frames in (a) correspond to t = 0, 8, and 50 ns and represent the heating cycle. The frames in (b) correspond to t = 60, 85, and 100 ns and represent the cooling cycle.

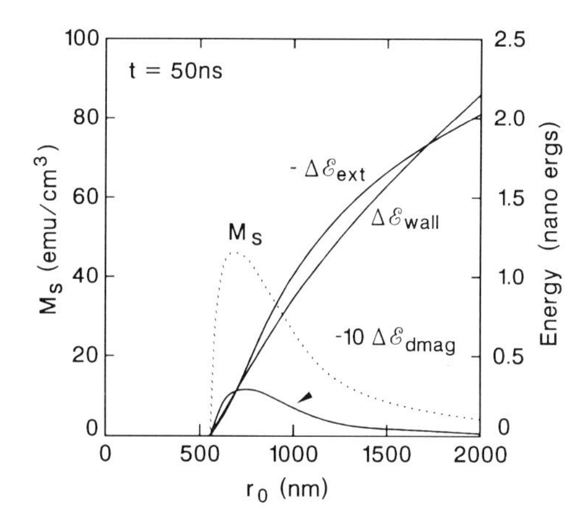

Figure 17.9. Individual contributions to the total magnetic energy $\Delta\mathscr{E}_{tot}$ as functions of domain radius r_0 at t = 50 ns. $\Delta\mathscr{E}_{wall} = \Delta\mathscr{E}_{anis} + \Delta\mathscr{E}_{xhg}$ is the wall energy, $\Delta\mathscr{E}_{ext}$ is the external field energy (H_{ext} = 400 Oe), and $\Delta\mathscr{E}_{dmag}$ is the demagnetization energy. The magnetization M_s is also shown.

demagnetizing force is generally in favor of domain formation and, therefore, large values of $\Delta\mathscr{E}_{dmag}$ translate into deeper and broader minima for the curves of Fig. 17.8. One advantage of having a large demagnetization contribution is that recording can be achieved with smaller external fields. Another situation where the demagnetization energy becomes important is when T_{comp} happens to be well below T_c. The saturation moment thus attains large values in the temperature range between the two, giving more weight to the demagnetization energy and allowing it to play a more active role in the domain formation process.

Similar considerations apply to the process of erasure. If the demagnetization energy is negligible, it may be possible to erase the recorded spots without an external field simply by raising the local temperature until the coercivity drops below the contractive force arising from the wall energy. If the demagnetization effect is large, however, we will have to use an appreciable magnetic field in conjunction with local heating to achieve erasure. The erasure process in this case is likely to be a combination of two processes: nucleation of a new domain within the recorded spot, and destabilization and contraction of the existing domain walls (see Fig. 17.4). If the field is not large enough, the process could lead to partial erasure and the creation of multiple ring-like domains.[3]

17.3. Dynamic Simulation of the Recording Process

In Chapter 15 we described the principles of the computer simulation of magnetization dynamics, and showed that two-dimensional lattices of magnetic dipoles could faithfully reproduce the bahvior of thin MO films. Later, in Chapter 16, we used computer simulations to clarify the notion of coercivity and describe its origins. These simulations can also be used to describe reversal dynamics during thermomagnetic recording and erasure. Since the temperature of the MO film varies (both in time and space) during the thermomagnetic process, it is necessary to incorporate the temperature-dependence of the various parameters in the computer model. The mean-field theory of Chapter 14 can be used for this purpose, but first it must be augmented to include the dynamic parameters α and γ, in order to allow the calculation of their temperature-dependences. In subsection 17.3.1 below we extend the Landau–Lifshitz–Gilbert (LLG) equation to the case of ferrimagnets containing two magnetic subnetworks, and derive expressions for the effective values of γ and α in terms of the corresponding parameters of the individual subnetworks. The subsections that follow discuss the simulation algorithm, describe the material parameters and recording conditions for a typical MO film, analyze the conditions under which reversal may or may not take place, and finally present a few simulation results.

17.3.1. The LLG equation for the strongly coupled ferrimagnet

Consider a thin magnetic film consisting of two antiferromagnetically coupled subnetworks. Let $\mathbf{M}_1$ denote the magnetization of the first subnetwork. The magnitude of this vector will be denoted by M_1 (where $M_1 \geq 0$) and the unit vector parallel to $\mathbf{M}_1$ will be called $\boldsymbol{\mu}_1$. The gyromagnetic ratio for the first subnetwork is γ_1 and the corresponding damping parameter is α_1. Similarly $\mathbf{M}_2$, $\boldsymbol{\mu}_2$, M_2, γ_2, and α_2 represent the second subnetwork ($M_2 \geq 0$). Let $h\mathbf{M}_2$ (where $h \leq 0$) be the effective local exchange field of the second subnetwork on the first. Similarly, $h\mathbf{M}_1$ is the effective local exchange field of the first subnetwork on the second. The remaining effective fields on the moments of the two subnetworks will be denoted by $\mathbf{H}_1$ and $\mathbf{H}_2$ respectively. Under these circumstances the LLG equations for the local dipole moments of the two subnetworks are

$$\dot{\mathbf{M}}_1 = -\gamma_1 \mathbf{M}_1 \times (\mathbf{H}_1 + h\mathbf{M}_2) + \alpha_1 \mathbf{M}_1 \times \dot{\boldsymbol{\mu}}_1 \quad (17.1a)$$

$$\dot{\mathbf{M}}_2 = -\gamma_2 \mathbf{M}_2 \times (\mathbf{H}_2 + h\mathbf{M}_1) + \alpha_2 \mathbf{M}_2 \times \dot{\boldsymbol{\mu}}_2 \quad (17.1b)$$

The superscript dot here and henceforth indicates differentiation with respect to time. If h is sufficiently large, the two moments $\mathbf{M}_1$ and $\mathbf{M}_2$ will remain strongly coupled and antiparallel at all times. In this case we define a net moment for the ferrimagnetic material along the unit vector $\boldsymbol{\mu}$, of magnitude $M_s = M_1 - M_2$. Obviously $\boldsymbol{\mu} = \boldsymbol{\mu}_1 = -\boldsymbol{\mu}_2$. Equations 17.1 are now combined to yield

$$\left(\frac{M_1}{\gamma_1} - \frac{M_2}{\gamma_2}\right)\dot{\boldsymbol{\mu}} = -\boldsymbol{\mu} \times (M_1\mathbf{H}_1 - M_2\mathbf{H}_2) + \left(\frac{\alpha_1 M_1}{\gamma_1} + \frac{\alpha_2 M_2}{\gamma_2}\right)\boldsymbol{\mu} \times \dot{\boldsymbol{\mu}} \qquad (17.2)$$

It is now possible to define net effective values for γ, α, and $\mathbf{H}$ as follows:

$$\gamma_{\text{eff}} = \frac{M_1 - M_2}{M_1/\gamma_1 - M_2/\gamma_2} \qquad (17.3)$$

$$\alpha_{\text{eff}} = \frac{\alpha_1 M_1/\gamma_1 + \alpha_2 M_2/\gamma_2}{M_1/\gamma_1 - M_2/\gamma_2} \qquad (17.4)$$

$$\mathbf{H}_{\text{eff}} = \frac{M_1\mathbf{H}_1 - M_2\mathbf{H}_2}{M_1 - M_2} \qquad (17.5)$$

Substituting in Eq. (17.2) the effective values just defined, one obtains

$$\dot{\boldsymbol{\mu}} = -\gamma_{\text{eff}}\,\boldsymbol{\mu} \times \mathbf{H}_{\text{eff}} + \alpha_{\text{eff}}\,\boldsymbol{\mu} \times \dot{\boldsymbol{\mu}} \qquad (17.6)$$

At first glance there appears to be two singularities associated with the above equation. The first singularity occurs at the angular momentum compensation, where $M_1/\gamma_1 = M_2/\gamma_2$. Since both γ_{eff} and α_{eff} are infinite at this point, $\boldsymbol{\mu}$ aligns itself with the local effective field instantaneously and without gyration. The second singularity arises at the magnetization compensation, where $M_1 = M_2$. Here $\gamma_{\text{eff}} = 0$ and H_{eff} is infinite. However, as will be shown below, the product of γ_{eff} and H_{eff} turns out to be finite. The LLG equation for the tightly coupled ferrimagnet in Eq. (17.6) is thus free from physical singularities and describes unambiguously the dynamic behavior of the net magnetic moment.

We proceed to express the effective field in Eq. (17.5) in terms of its four components, namely, the externally applied field $\mathbf{H}^{(\text{ext})}$, the demagnetizing field $\mathbf{H}^{(\text{dmag})}$, the anisotropy field $\mathbf{H}^{(\text{anis})}$, and the exchange field $\mathbf{H}^{(\text{xhg})}$. This effective field is written

$$\mathbf{H}_{\text{eff}} = \frac{M_1}{M_1 - M_2}\left\{\mathbf{H}_1^{(\text{ext})} + \mathbf{H}_1^{(\text{dmag})} + \mathbf{H}_1^{(\text{anis})} + \sum_{\substack{\text{nearest}\\ \text{neighbors}}} \mathbf{H}_1^{(\text{xhg})}\right\}$$

$$- \frac{M_2}{M_1 - M_2}\left\{\mathbf{H}_2^{(\text{ext})} + \mathbf{H}_2^{(\text{dmag})} + \mathbf{H}_2^{(\text{anis})} + \sum_{\substack{\text{nearest}\\ \text{neighbors}}} \mathbf{H}_2^{(\text{xhg})}\right\} \tag{17.7}$$

The external field is the same for both subnetworks, that is, $\mathbf{H}_1^{(\text{ext})} = \mathbf{H}_2^{(\text{ext})}$. Similarly, $\mathbf{H}_1^{(\text{dmag})} = \mathbf{H}_2^{(\text{dmag})}$. As for the anisotropy field, we assume that both subnetworks have uniaxial magnetic anisotropy along the same unit vector $\mathbf{n}$, but different anisotropy constants, K_{u1} and K_{u2}. Thus,

$$\mathbf{H}_1^{(\text{anis})} = \frac{2K_{u1}}{M_1}(\boldsymbol{\mu}_1 \cdot \mathbf{n})\,\mathbf{n} \qquad \mathbf{H}_2^{(\text{anis})} = \frac{2K_{u2}}{M_2}(\boldsymbol{\mu}_2 \cdot \mathbf{n})\,\mathbf{n} \tag{17.8}$$

Next, let us assume that the magnetization of the film is represented by a two-dimensional hexagonal lattice of dipoles. It was shown in Eq. (15.15) that the effective exchange field exerted on a given dipole by a near-neighbor dipole may be written as

$$\mathbf{H}_1^{(\text{xhg})} = +\frac{4A_{x1}}{3M_1 d^2}\boldsymbol{\mu}_{\text{n. n.}} \qquad \mathbf{H}_2^{(\text{xhg})} = -\frac{4A_{x2}}{3M_2 d^2}\boldsymbol{\mu}_{\text{n. n.}} \tag{17.9}$$

where $\boldsymbol{\mu}_{\text{n. n.}}$ is the direction of the magnetic moment of the near-neighbor dipole, and d is the constant of the hexagonal lattice. We designate $K_{u1} + K_{u2}$ as the net anisotropy energy constant K_u, and $A_{x1} + A_{x2}$ as the net exchange stiffness coefficient A_x of the ferrimagnetic material. Substituting the above results and definitions in Eq. (17.7) yields

$$\boxed{\mathbf{H}_{\text{eff}} = \mathbf{H}^{(\text{ext})} + \mathbf{H}^{(\text{dmag})} + \frac{2K_u}{M_1 - M_2}(\boldsymbol{\mu} \cdot \mathbf{n})\,\mathbf{n} + \frac{4A_x}{3(M_1 - M_2)d^2}\sum_{\substack{\text{nearest}\\ \text{neighbors}}} \boldsymbol{\mu}_{\text{n. n.}}} \tag{17.10}$$

Equation (17.10) is the complete expression for the effective field, which, together with Eqs. (17.3) and (17.4), provides a complete set of parameters for the LLG equation.

17.3.2. The simulation algorithm

The essential features of the micromagnetic simulation algorithm have already been described in section 15.3. The Landau–Lifshitz–Gilbert (LLG) equation of magnetization dynamics is the basis of these simulations where each cell of a 256×256 hexagonal lattice is assigned a magnetic dipole moment. The constant of the hexagonal lattice d is 10 Å, the assumed film thickness τ (corresponding to the height of each cell in the lattice) is 50 nm and, in the XY-plane of the lattice, periodic boundary conditions apply. (The total volume of the simulated lattice is therefore $256 \times 222 \times 50$ nm^3.) The main difference between earlier simulations and those described in the present chapter is that we have now imposed a temperature profile on the lattice to mimic the writing conditions. The magnetic parameters of the LLG equation, being temperature dependent, thus vary in time. However, once these dependences have been incorporated into the model there will be no technical differences between these and the previous simulations.

There are two issues with regard to the present simulations that should be brought up at once. The first issue concerns the accuracy of the mean-field model used to derive the temperature-dependence of the various LLG parameters. Mean-field theory is notoriously poor in the vicinity of the Curie point, and the simulated behavior in this region must be regarded with a measure of skepticism. Second, we have used the same color coding scheme as in preceding chapters to display the magnetization distribution patterns (see the color wheel in Fig. 15.7). The color pixels represent the direction of magnetization only, even though in reality the magnitude of **M** varies across the lattice as well. While studying subsection 17.3.5, the reader should keep this fact in mind, although, knowing the thermal profile and the functional dependence of $|\mathbf{M}|$ on T, it is not difficult to visualize the distribution of $|\mathbf{M}|$ across the lattice.

17.3.3. Temperature profile and material parameters

The temperature distribution imposed on the lattice has a Gaussian spatial profile, with exponential rise and fall times (see Fig. 17.10). The following equation succinctly describes this temperature profile:

$$T(r,t) = T_{amb} + (T_{max} - T_{amb}) \exp\left[-(r/r_0)^2\right] f(t) \qquad (17.11)$$

In this equation T_{amb} is the ambient temperature (300 K), T_{max} is the maximum temperature reached during the heating cycle ($T_{max} = 500$ K), and r_0 is the $1/e$ radius of the Gaussian hot spot ($r_0 = 180$ nm), which is also related to the full width at half-maximum (FWHM) of the spatial distribution (FWHM $= 1.665 r_0 = 0.3$ μm). $f(t)$ is the time-dependence of the temperature profile, given by

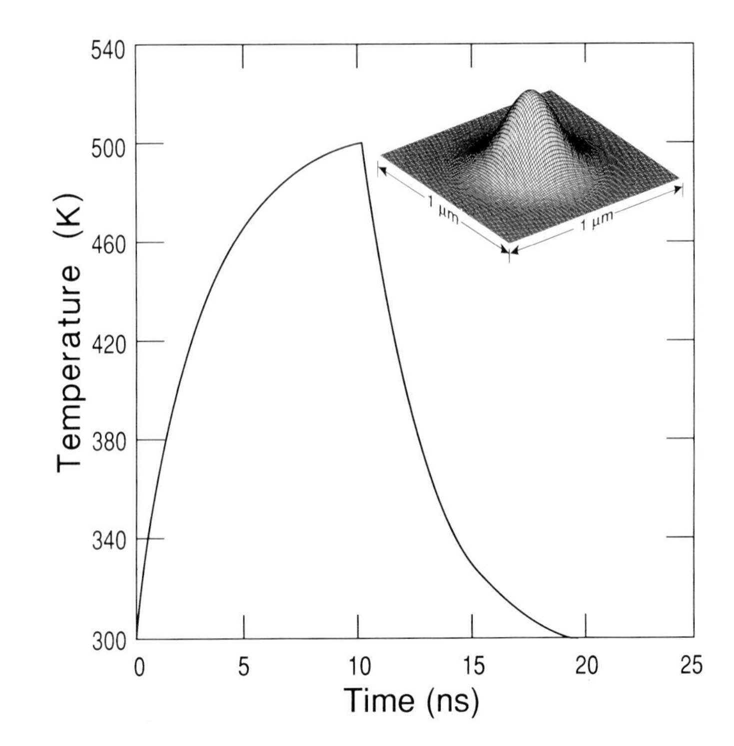

Figure 17.10. Temperature distribution imposed on the lattice for dynamic simulations of the thermomagnetic process. The spatial temperature profile in the inset is Gaussian with FWHM = 0.3 μm, and the temporal profile $f(t)$ consists of exponentially rising and falling sections.

$$f(t) = \begin{cases} \dfrac{1 - \exp(-t/\tau_1)}{1 - \exp(-t_{\text{peak}}/\tau_1)} & 0 \le t \le t_{\text{peak}} \\[2ex] 1 - \dfrac{1 - \exp\left[(t_{\text{peak}} - t)/\tau_2\right]}{1 - \exp\left[(t_{\text{peak}} - t_{\text{end}})/\tau_2\right]} & t_{\text{peak}} \le t \le t_{\text{end}} \end{cases}$$

The parameter values are t_{peak} = 10 ns, t_{end} = 20 ns, and $\tau_1 = \tau_2$ = 3 ns. This time-dependence is such that at $t = 0$ the lattice is at the uniform temperature of T_{amb}, by $t = t_{\text{peak}}$ the temperature has peaked everywhere with a maximum of T_{max} at the center of the lattice, and by $t = t_{end}$ the entire lattice has returned to the ambient temperature. For the sake of simplicity we have ignored the effects of heat diffusion and assumed that the spatial temperature profile remains intact. Note also that the assumed profile is stationary with respect to the lattice (i.e., no relative motion of the beam and the medium is assumed). These characteristics, of course, are somewhat artificial, considering the nature of laser-induced heating in actual media. What is more, the assumed pulse duration is relatively short compared to those currently used in thermomagnetic recording. These details, however, are of little consequence when the basic processes of magnetization reversal are concerned.

The imposed temperature distribution with FWHM = 0.3 μm is wider than the simulated lattice but, as it turns out, magnetization reversal is confined near the center of the hot-spot, and the phenomena of interest occur within the boundary of the lattice. Since the Curie temperature of the material under consideration is T_c = 441 K, at the peak of the pulse the radius r_c of the region above the Curie temperature (i.e., the Curie disk) is 106 nm. The Curie disk first appears at t = 3.4 ns and expands until t = 10 ns, at which point it begins to shrink rapidly and disappears at t = 11 ns. (In the simulation results that follow the Curie disk will be displayed in gray pixels.)

The temperature dependences of the various material parameters used in these simulations are shown in Fig. 17.11. For concreteness, we have confined our attention to a $Tb_{22}(FeCo)_{73}Ar_5$ alloy material. The saturation magnetization M_s and the individual subnetwork moments M_{Tb} and M_{FeCo} are shown in Fig. 17.11(a). This material has compensation point T_{comp} = 250 K and Curie temperature T_c = 441 K. Figure 17.11(b) shows the material's anisotropy constant K_u and its exchange stiffness coefficient A_x, both versus temperature. The inset is a plot of the effective anisotropy field $H_k = 2K_u/M_s$ in the vicinity of the Curie point; we shall return to this curve later and point out its significance in conjunction with the nucleation process.

From knowledge of $K_u(T)$ and $A_x(T)$ one can readily derive the plots of Fig. 17.11(c), which show the wall energy density σ_w and the wall width $4\Delta_w$ as functions of T. Although σ_w and Δ_w are not needed for the simulations, they are important characteristics of the material and we shall have occasion to use them in the analysis of the results. Note in particular that the wall width is rather weakly dependent on T, going from about 200 Å at room temperature to about 250 Å just below the Curie point. Finally, Fig. 17.11(d) shows plots of the effective gyromagnetic ratio γ_{eff} and damping coefficient α_{eff} versus temperature.

17.3.4. Observations concerning the nature of nucleation

Computer simulations have revealed that reversal begins by the nucleation of a domain wall at the rim of the Curie disk. A simple analysis shows that such nucleation is possible only when the applied field H_{ext} is greater than the effective anisotropy field H_k within a certain annulus surrounding the Curie disk. Let T_c^- denote a temperature slightly below the Curie temperature, and let $M_s(T_c^-)$ be the saturation magnetization at that point. Similarly denote the corresponding wall energy density by $\sigma_w(T_c^-)$ and that of the wall width by $\Delta_w(T_c^-)$. Ignoring demagnetizing effects and considering only the balance between wall energy and external field energy, the formation of a wall at the Curie rim requires that the following condition be satisfied:

$$\sigma_w(T_c^-) \leq H_{ext} M_s(T_c^-) \Delta_w(T_c^-) \tag{17.12}$$

Substituting for σ_w and Δ_w in terms of K_u and A_x, we find the necessary condition for nucleation as follows:

$$\boxed{H_{ext} \geq \frac{2K_u(T_c^-)}{M_s(T_c^-)} = H_k(T_c^-)} \tag{17.13}$$

Typically H_k is a decreasing function of T that rapidly goes to zero at the Curie point; see the inset in Fig. 17.11(b). Thus, as one moves away from the Curie rim, one finds a sharp increase in H_k. If, however, H_k remains below H_{ext} in the immediate neighborhood of the Curie rim (say, within an annulus of width Δ_w), then the preceding analysis indicates that wall formation is energetically favorable and the nucleation process may therefore commence. Conversely, when H_k at the radius of $r = r_c + \Delta_w$ is greater than H_{ext}, a wall becomes costly and there can be no nucleation.

In light of the above discussion it is clear that the temperature gradient at the Curie rim plays an important role in the nucleation stage of the reversal process. A small gradient at the rim provides for a slow rise in H_k, thus facilitating nucleation. A large gradient, on the other hand, makes the immediate neighborhood of the Curie disk less hospitable to domain walls. In the case of a Gaussian temperature profile with $1/e$ radius of r_0, the maximum gradient occurs at $r = (1/\sqrt{2})r_0$. This corresponds in our simulations to $r = 127$ nm, which is somewhat greater than the largest radius of the Curie disk attained. Now, a domain formed during the heating cycle (when $t \leq t_{peak}$) must expand rapidly or else it will be consumed by the advancing Curie disk. Nucleation followed by rapid growth during the heating cycle is not impossible, but it is unlikely, especially when the applied field is relatively weak. The more likely scenario is the formation of a wall (at the Curie rim) during the cooling period, $t \geq t_{peak}$. In this event the newly created wall can remain stationary, or even shrink slowly, and yet survive. In addition, the cooling period has the desirable feature that with passage of time the temperature gradient at the Curie rim declines, thus rendering nucleation more likely.

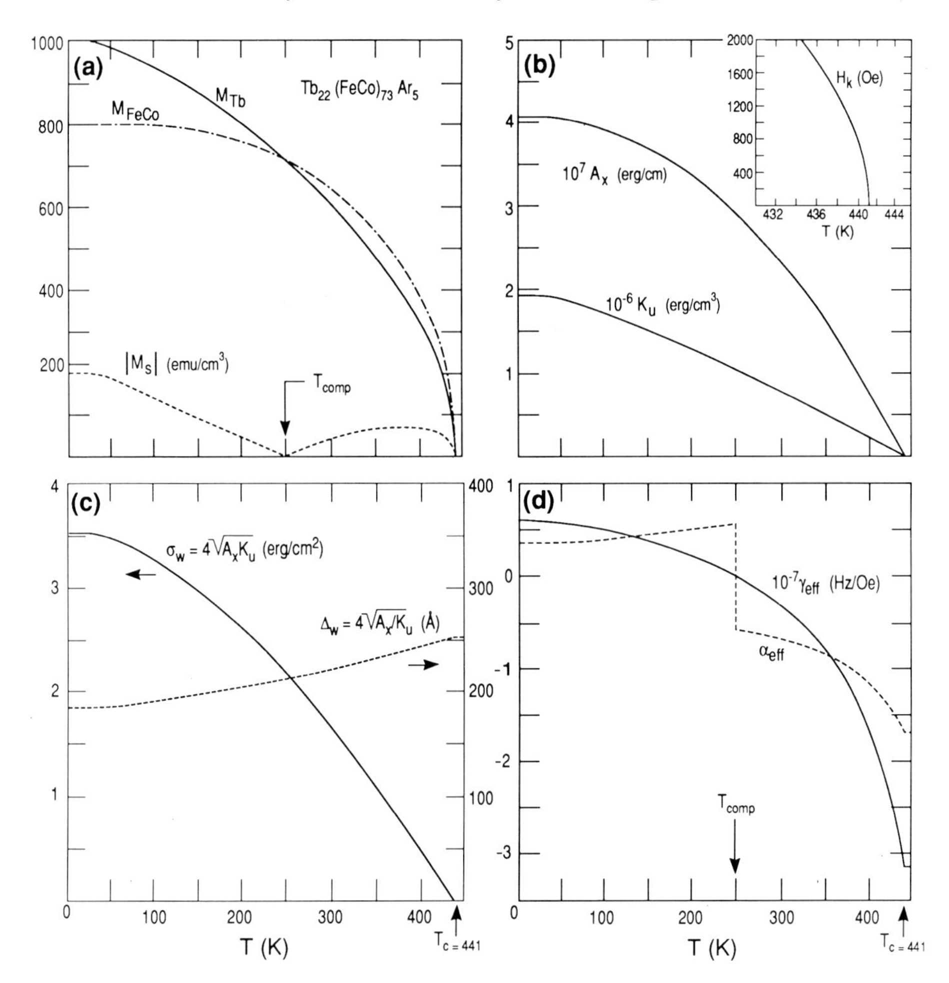

Figure 17.11. Temperature-dependence of the magnetic parameters of amorphous $Tb_{22}(FeCo)_{73}Ar_5$, used as the recording medium in the dynamic simulations. (a) Saturation and subnetwork magnetizations. (b) Anisotropy and exchange coefficients. The inset shows the effective anisotropy field H_k in the vicinity of the Curie point. (c) Domain-wall energy density σ_w and wall width $4\Delta_w$. (d) Effective gyromagnetic ratio γ_{eff} and Gilbert damping parameter α_{eff}. These were calculated from Eqs. (17.3) and (17.4) assuming $\gamma_1 = g_1\mu_B/\hbar$, $\gamma_2 = g_2\mu_B/\hbar$ and $\alpha_1 = \alpha_2 = 0.1$. Note that the subnetwork magnetizations $M_1(T)$ and $M_2(T)$ required by Eqs. (17.3) and (17.4) are readily available from mean-field calculations.

In the following subsection we discuss simulation results pertaining to the formative stage of domain nucleation and the early phase of adjustment (via expansion or contraction) in the thermomagnetic process.

17.3.5. Simulation results and discussion

In this section we present several examples of the recording/erasure process in amorphous RE–TM films, obtained by dynamic simulations of the LLG equation on a Connection Machine. The various combinations of lattice parameters used in these examples are not chosen arbitrarily; their selection has been based on actual observations in thermomagnetic recording experiments. Roughly speaking, the first example represents soft or low-coercivity materials; the second, third, and fourth examples are typical of the so-called "wall motion dominated" media, and the fifth example exhibits behavior characteristic of "nucleation-dominated" media.[32]

Example 1. The lattice used in this example is fairly uniform, having random, cell-to-cell variations of the easy axis within a 45° cone around the normal. These fluctuations, however, having a short coherence length compared to the wall width, cause minimal pinning and offer no resistance to wall motion. Consequently, the lattice is incapable of supporting domains and, as we shall see, domains formed at the end of the recording cycle will collapse. Figure 17.12 shows the state of magnetization of the lattice at several instants of time during the cooling period, with H_{ext} = 500 Oe in the negative Z-direction. Frame (a) shows the lattice at t = 10.68 ns. The gray region at the center is above the Curie temperature (i.e., it corresponds to the Curie disk). The colored pixels around the rim are the dipole moments that have rotated into the plane of the lattice, in anticipation of wall formation. The remaining white pixels are the dipoles that are still in the remanent state, being perpendicular to the lattice and pointing in the positive Z-direction. The next frame, (b), shows the lattice at t = 10.92 ns when, in addition to the colored pixels representing a continuous domain wall, there is a thin black ring surrounding the Curie disk. The black pixels correspond to fully reversed dipoles, that is, dipoles perpendicular to the lattice and pointing along $-Z$. In frame (c) we show the situation at t = 11 ns, when the Curie disk is about to disappear and is leaving a reverse domain behind. The final frame in this sequence, (d), shows a shrinking domain at t = 11.58 ns; apparently the wall energy density σ_w is too large for the external and demagnetizing fields to overcome, thus forcing the collapse of the domain.

When the same simulation is repeated with an applied field of 1000 Oe the patterns in Fig. 17.13 are obtained. Frames (a)–(d) in this case correspond to t = 10.20, 10.60, 10.96, and 11.58 ns respectively. Comparing the results with the previous case, we note that the larger field causes nucleation to begin earlier. Also, the domain formed in the latter case does not seem intent on shrinking; the larger H_{ext} must have successfully counteracted the inclination of the wall to collapse. Upon removal of the external field, however, this domain will also collapse.†

□

†Any domain in a soft (i.e., low-coercivity) material whose wall energy exceeds the savings in the demagnetization energy tends to collapse. In practice, defects and inhomogeneities of the medium can prevent this collapse by pinning the wall and thus stabilizing the domain, even in the absence of the external field.

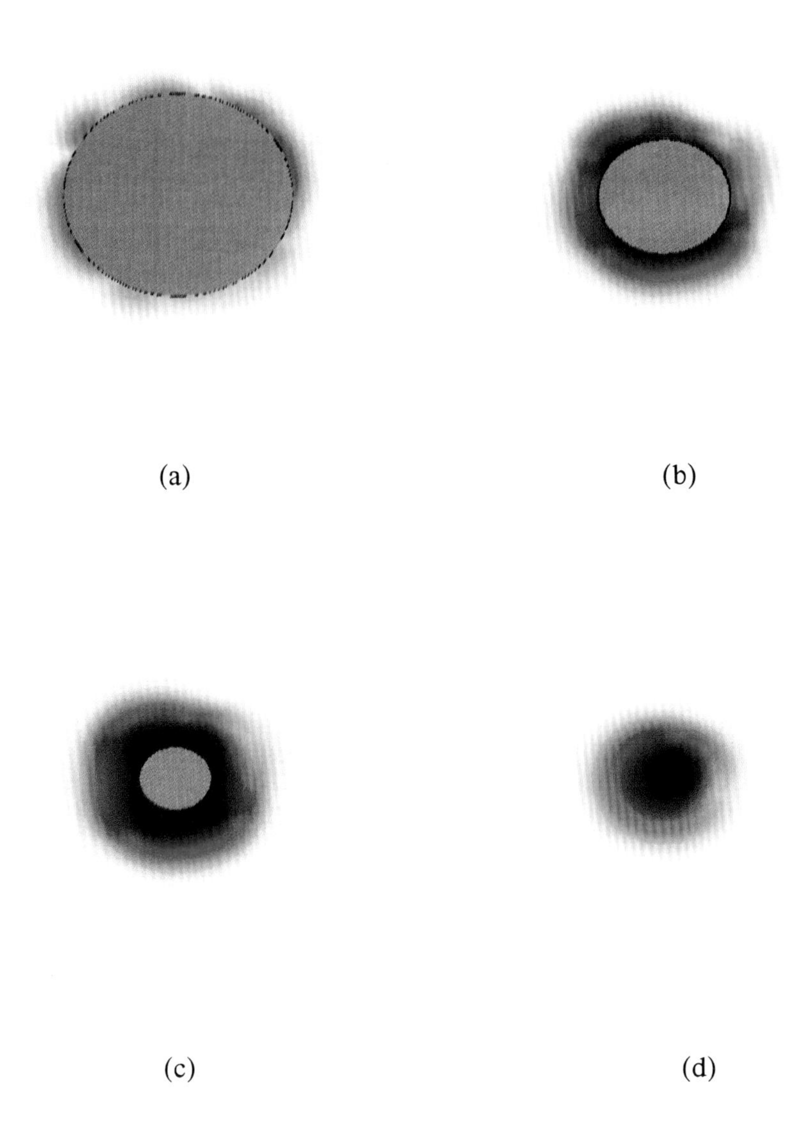

Figure 17.12. Snap-shots from the state of magnetization of the lattice described in Example 1, during the cooling period. The magnitude of the external field H_{ext} is 500 Oe. The white (black) pixels are magnetized perpendicular to the plane of the lattice along the positive (negative) Z-axis, the central gray region is above T_c, and the colored pixels have magnetization in the plane (or inclined towards the plane) of the lattice. (a)–(d) t = 10.68, 10.92, 11, and 11.58 ns respectively.

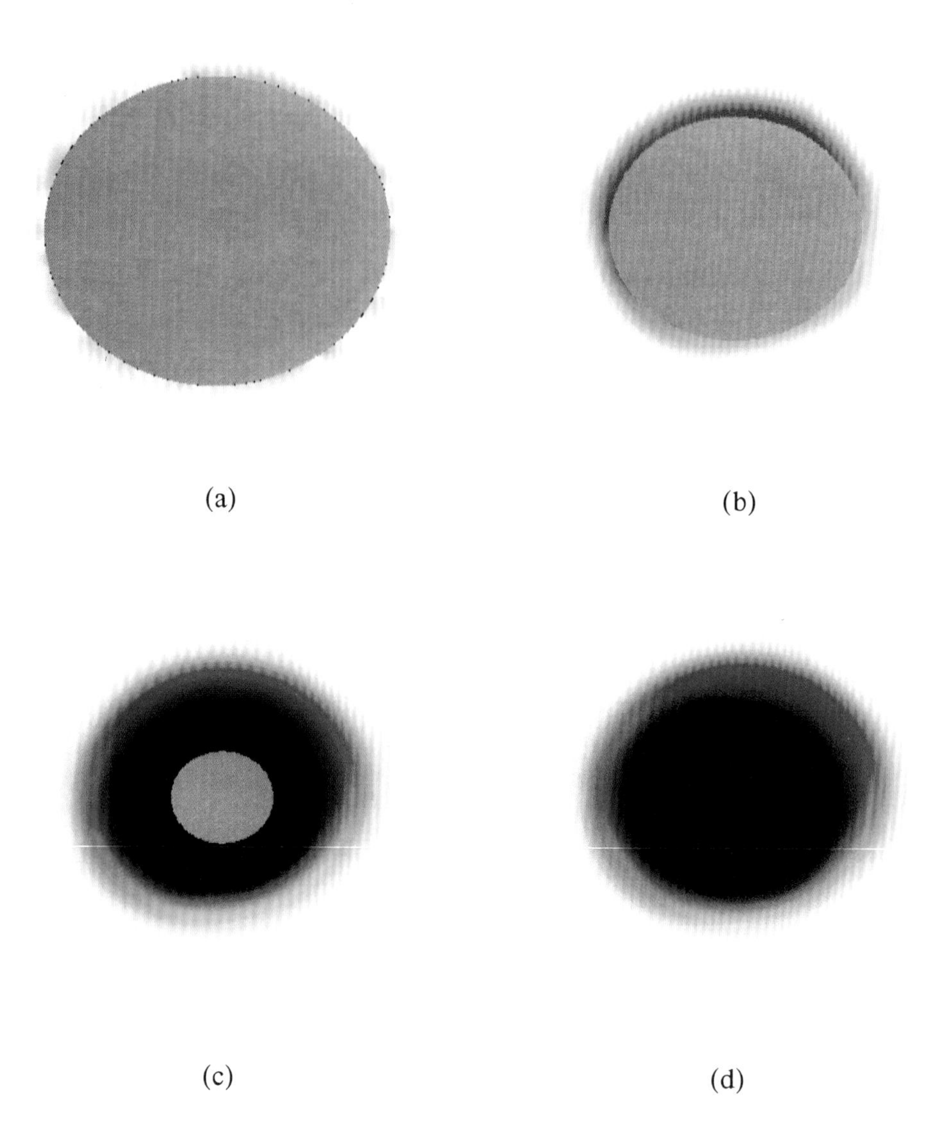

Figure 17.13. Same as Fig. 17.12 but with H_{ext} = 1000 Oe. (a)–(d) t = 10.20, 10.60, 10.96, 11.58 ns respectively.

Example 2. The lattice in this case consists of 632 patches, with an average patch diameter of 95 Å. Each patch has its own easy axis, and the axes corresponding to the various patches are distributed randomly within a cone of 45° from the normal. Also, the exchange stiffness coefficient at the patch borders is reduced to 20% of its nominal value. These nonuniformities in the magnetic properties of the lattice have been introduced in order to create local potential wells and trap the domain walls. The applied field H_{ext} during recording is 500 Oe, but at t = 21 ns it is set to zero to determine whether the recorded domain would be stable in the absence of an external field.

Figure 17.14 shows the process of domain nucleation and contraction during the cooling cycle. Frames (a)–(f) correspond to t = 10.7, 10.9, 10.98, 11.5, 16.7, and 28 ns respectively. The first and the last frame display the entire lattice, but the intermediate frames are close-ups of the boxed region in (a). In the early phase of cooling the domain shrinks somewhat, as the transition from (d) to (e) indicates, but the domain in (e) remains stable afterwards. When the field is removed at t = 21 ns there is a slight change of shape of the domain (compare the lower left side of the domain in (e) with that in (f)) but, by and large, the recorded domain survives.

This example brings to the fore some of the interesting issues with regard to the recording process on amorphous RE–TM media. We conclude from it that stable domains as small as 80 nm in diameter can be written under realistic conditions and that the jaggedness of the domain boundary may be attributed to the inhomogeneities inherent in the material. The existence of these inhomogeneities is essential if stable domains are to be recorded; on the other hand, very small domains can become excessively jagged because of the inhomogeneities, in which case noise in readout may become a performance-limiting factor. The scale of the jaggedness is related to the patch-size, which may be a controllable parameter during media fabrication. (Presumably, the underlayer roughness, the sputtering gas pressure, the bias voltage on the growing film during sputtering, the rate of deposition, etc., control the magnitude and the correlation distance of inhomogeneities.) However, if the patches become smaller than the wall width, then their effectiveness as barriers or traps to wall motion will be reduced and, consequently, domain instability may occur. □

Example 3. The domain created at the end of the recording process in the preceding example is now subjected to the same laser pulse as before, but in the absence of an external field. Frames (a)–(d) in Fig. 17.15 show close-ups of the lattice at t = 2.4, 2.8, 3.4, and 3.5 ns respectively. The domain begins to collapse even before the temperature reaches the Curie point: by the time the center of the lattice is at T_c the domain has shrunk to about one-fifth of its original size; see frame (c). By t = 3.6 ns the domain will have all but disappeared.

A similar sequence of events is observed during the laser-induced heating of a somewhat larger domain, shown in Fig. 17.16. The initial domain diameter in this case is 180 nm, the applied field is zero, and the induced temperature profile is the same as before. Frames (a)–(d) correspond to t = 0, 3.4, 3.6, and 3.8 ns respectively. Again the collapse has started before the arrival of T_c; by t = 4 ns the advancing Curie disk meets the shrinking domain and thus completes the erasure process.

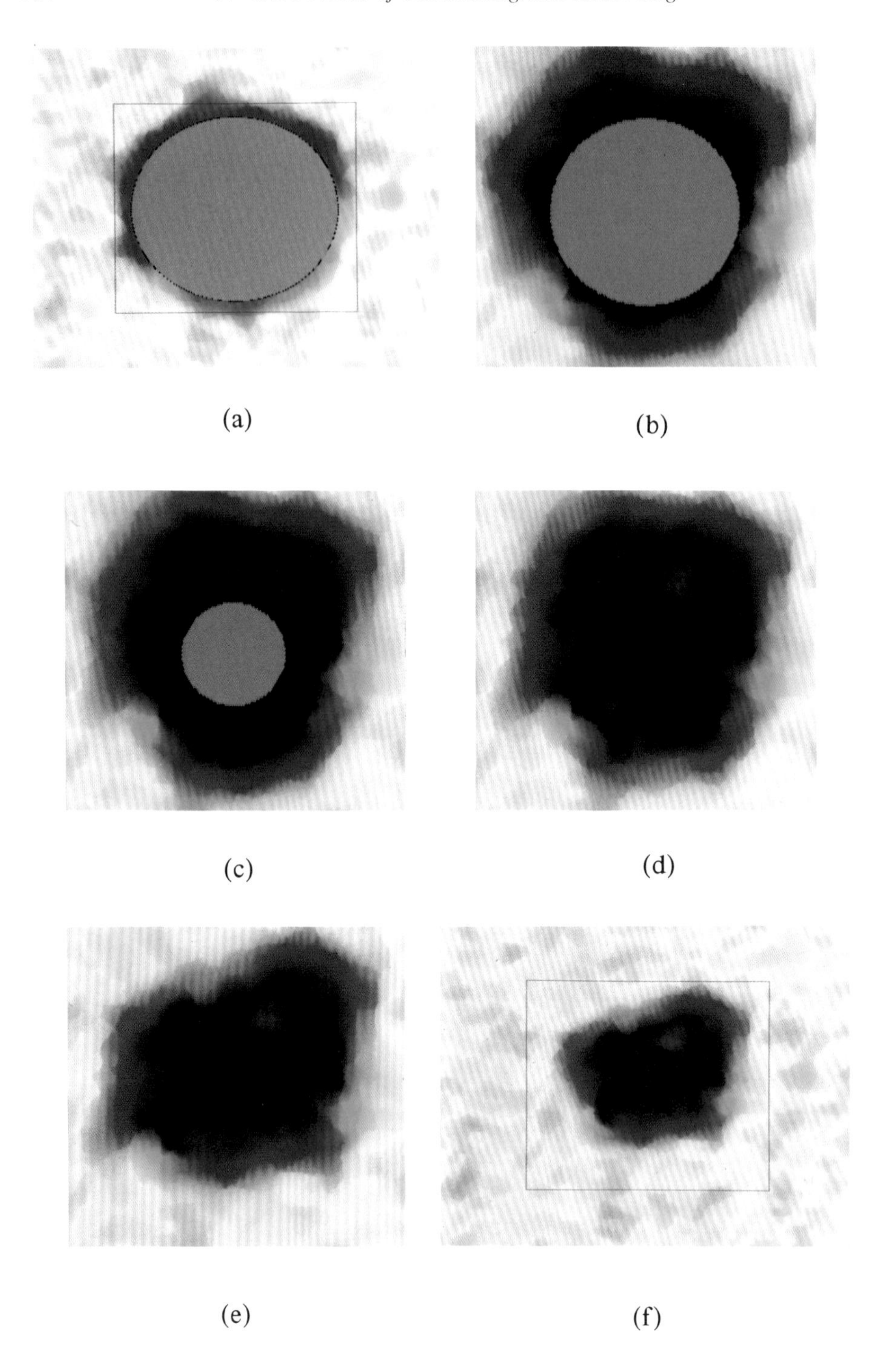

Figure 17.14. Thermomagnetic writing in a patchy lattice with average patch diameter of 95 Å. Each patch has its own easy axis, but the axes of the various patches are distributed within a 45° cone around the normal. The exchange stiffness coefficient at the patch borders is reduced to 20% of its nominal value within the patch, and the externally applied field is 500 Oe. Frames (a)–(f) correspond to t = 10.7, 10.9, 10.98, 11.5, 16.7, and 28 ns respectively. The first and the last frame display the entire lattice; the other frames are close-ups of the boxed region.

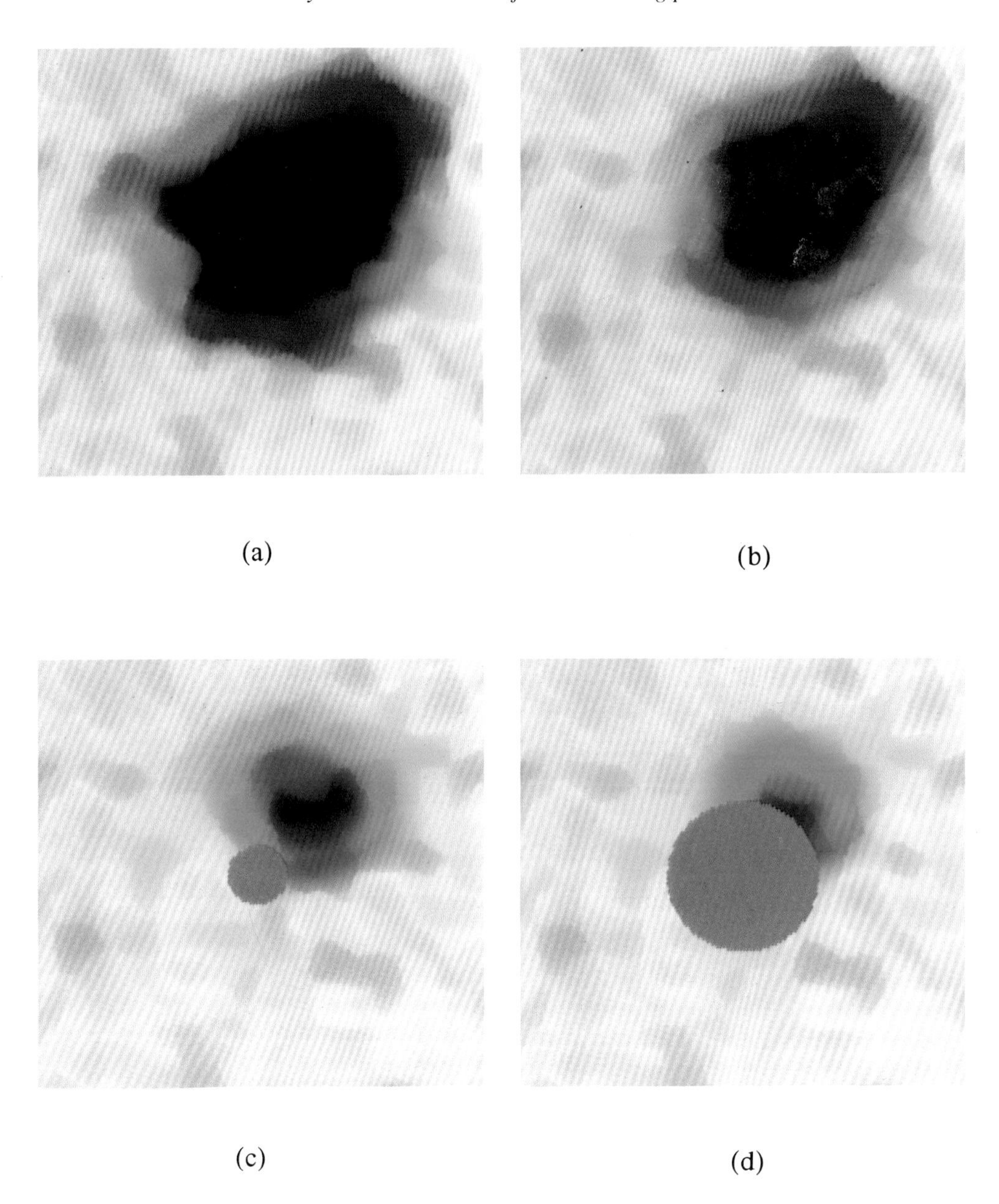

Figure 17.15. Thermomagnetic erasure of the domain created at the end of the recording process, shown in Fig. 17.14(f). The imposed temperature profile is the same as before, i.e., that described by Eq. (17.11) and shown in Fig. 17.10, but no external magnetic field is applied. Frames (a)–(d) correspond to t = 2.4, 2.8, 3.4, and 3.5 ns respectively.

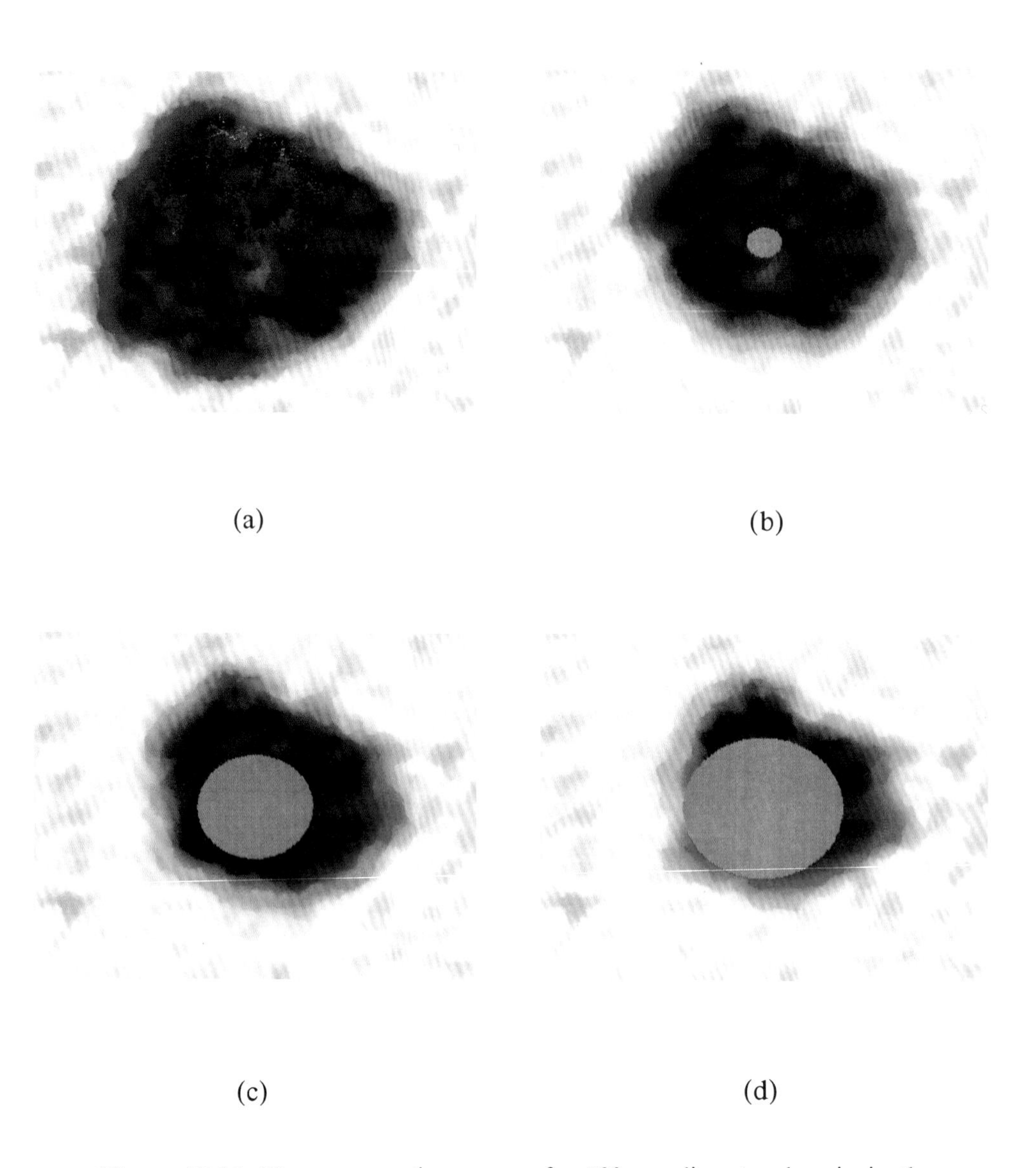

Figure 17.16. Thermomagnetic erasure of a 180 nm-diameter domain in the same lattice as in Fig. 17.15, with the same temperature profile and no external magnetic field. Frames (a)–(d) correspond to t = 0, 3.4, 3.6, and 3.8 ns respectively.

These examples show the potential for erasing very small domains without an external field, since the inward pressure caused by the wall energy is sufficiently high. On the other hand, such small domains are only marginally stable and may be perturbed during the read process; also, changes in the environment may easily bring about their collapse. □

Example 4. Figure 17.17 is another example of recording on a patchy lattice similar to that of Example 2. As before, the reversal occurs during the cooling cycle, but now there are several independent nuclei, which eventually combine to form a fully reversed domain. □

Example 5. We consider a patchy lattice with 632 patches, with easy axes randomly assigned to each patch within a 45° cone, as in Example 2. This time, however, we set the exchange at patch borders equal to zero, thus totally decoupling the patches from each other. Frames (a)–(c) in Fig. 17.18 show the state of magnetization during cooling under H_{ext} = 1000 Oe at t = 10.4, 10.8, and 22 ns. Note that the final domain has a few unreversed patches, perhaps due to the action of the demagnetizing field during cooling. Almost all patches that reached the Curie point are reversed in this example, since, as the hot spot began to shrink, there was no exchange coupling to the outside that would force them to remain up-magnetized. The domain's shape and position are therefore precisely determined by the Curie radius, and not, as in previous examples, by the balance between magnetic energy and local coercivity.

When the same computer experiment was carried out with H_{ext} = 500 Oe, the imperfectly reversed domain in Fig. 17.19 was obtained. Here the demagnetizing field overcomes the external field over many patches during the cooling cycle, thereby preventing them from reversing. □

17.4. Exchange-coupled Magnetic Multilayers

Up to this point we have considered "simple" media structures only; the word simple in this context implies that although the medium might contain a stack of dielectric and metallic layers, its magnetic constituent is a single, uniform film with a fixed chemical composition. Such media designs are highly practical and, in fact, single-magnetic-layer constructs are the only type of media that are commercially available at the time of writing. For the purpose of studying the physics of thermomagnetic recording, we have relied on single-magnetic-layer media to the exclusion of all others, since these simplest of structures exhibit the most essential features of the recording and erasure processes. There exist, however, more complex media designs with two or more magnetic layers in contact with each other. Such designs provide additional degrees of freedom and create possibilities for enhancing performance or introducing new features to the media. In this section we shall describe three types of exchange-coupled magnetic multilayers, and discuss the specific improvements and new modes of operation that these designs engender.

17.4.1. Magnetic capping layer for lowering the write/erase field

A double-layer magnetic structure has been proposed for reducing the required external field for writing and erasure.[20] This double-layer design,

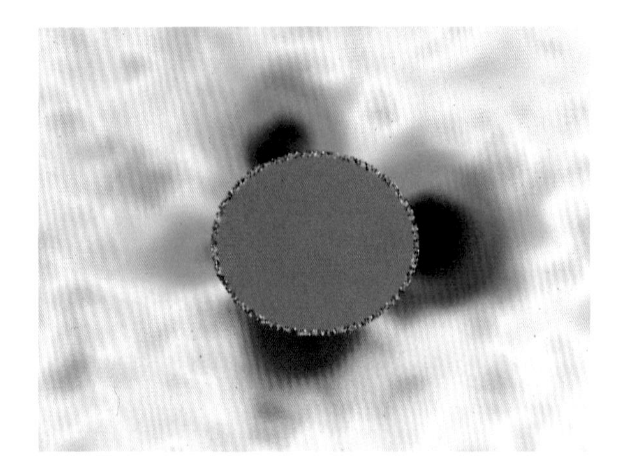

(a)

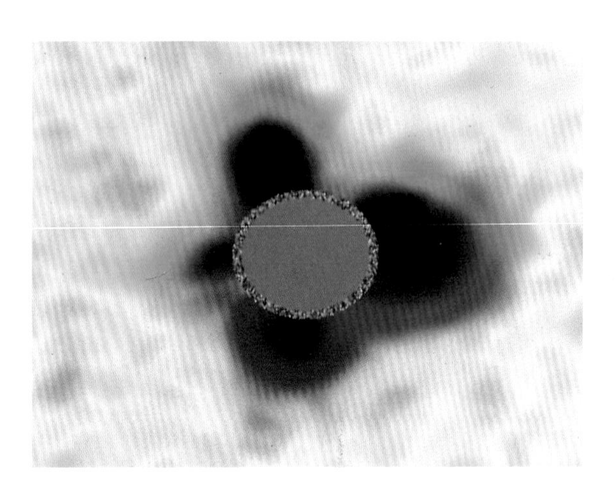

(b)

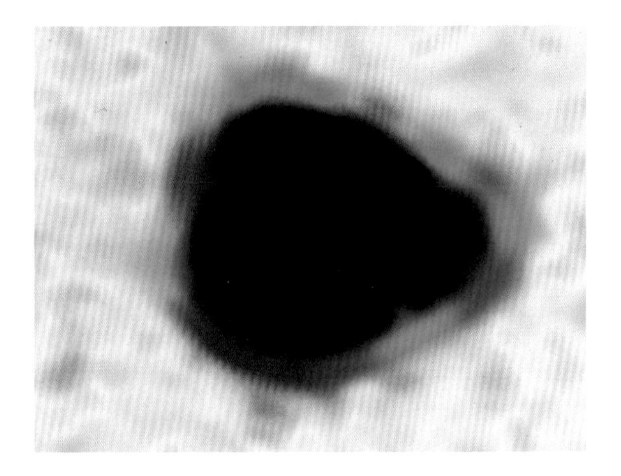

(c)

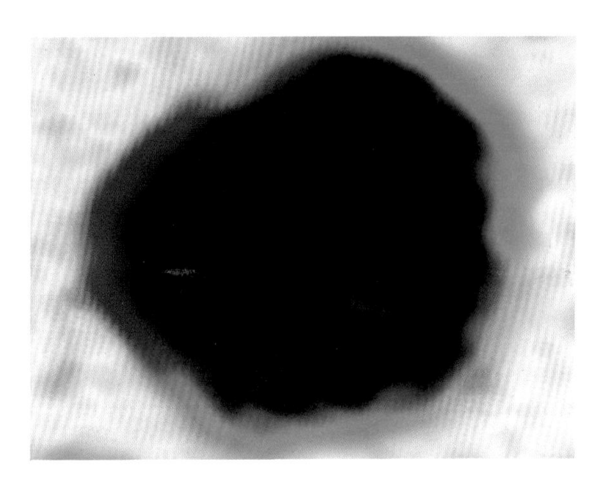

(d)

Figure 17.17. Thermomagnetic writing in a patchy lattice. There are several nucleation sites, but eventually they combine and form a single domain. Once formed, the domain expands under the influence of the applied field.

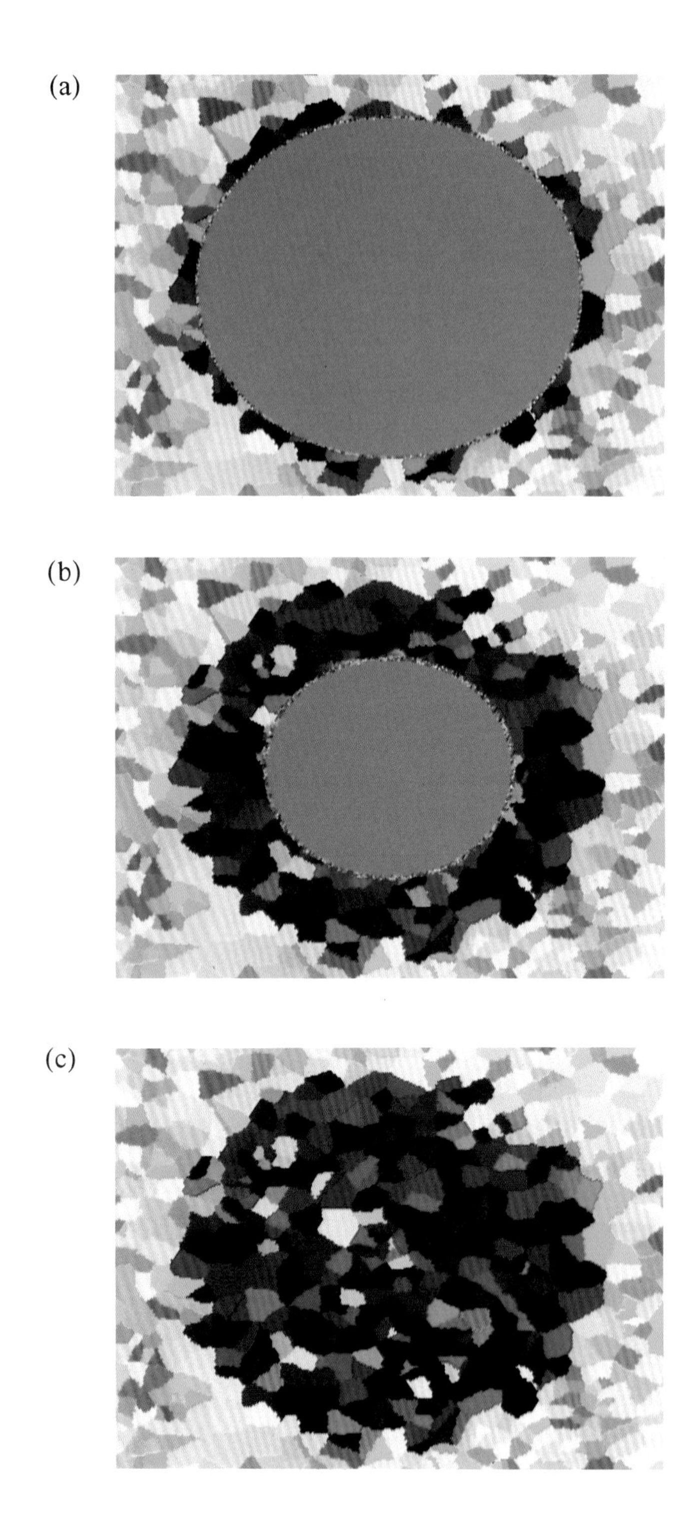

Figure 17.18. Thermomagnetic writing in a patchy lattice with average patch diameter of 95 Å. Each patch has its own easy axis, but the axes of the various patches are distributed within a 45° cone around the normal. The exchange stiffness coefficient at the patch borders is set to zero and the externally applied field is 1000 Oe. Frames (a)–(c) correspond to t = 10.4, 10.8, and 22 ns respectively.

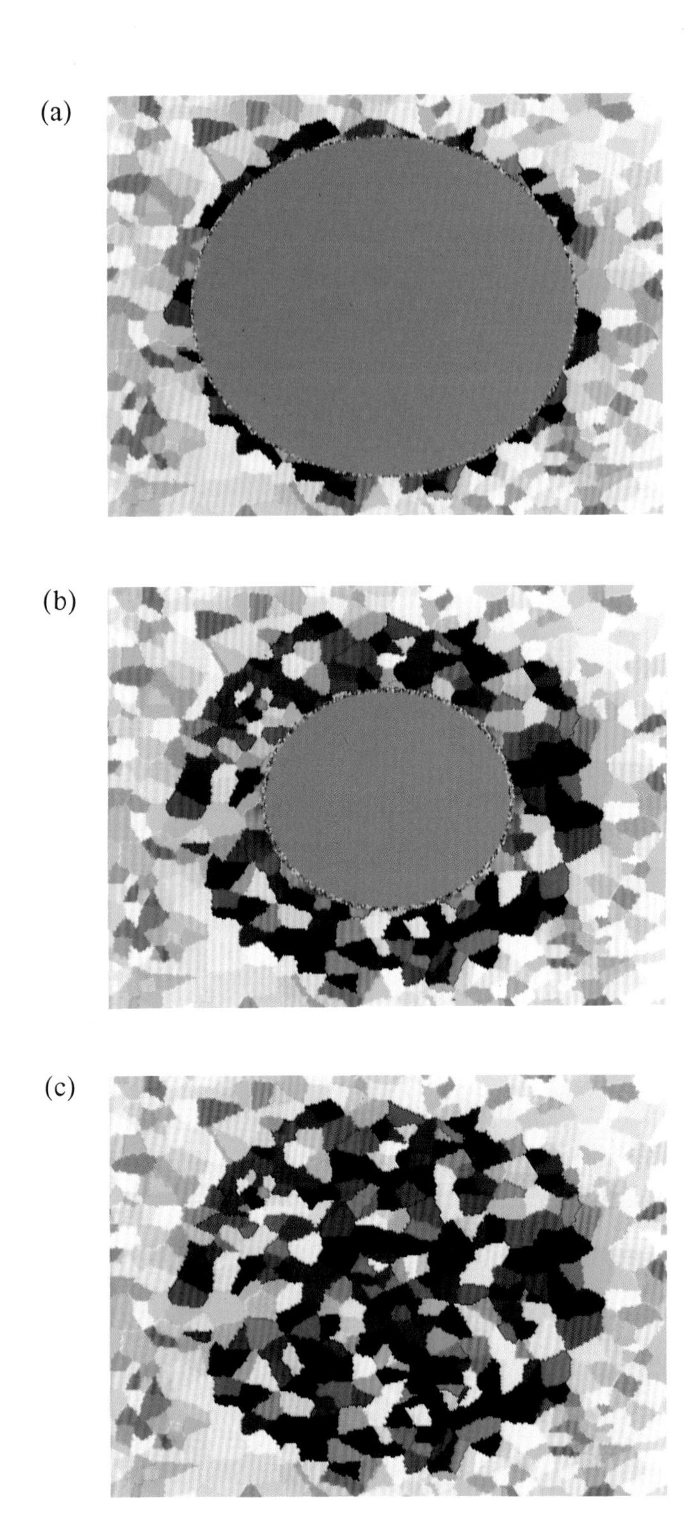

Figure 17.19. Same as Fig. 17.18 but with H_{ext} = 500 Oe.

shown in Fig. 17.20(a), consists of the usual TbFeCo layer with perpendicular magnetization, and a second layer consisting of either CoPt alloy or amorphous RE–TM alloy with in-plane magnetization. The two layers are coupled by mutual exchange interaction at the interface. During writing and erasure, the external field lifts the magnetization of the in-plane layer, which in turn will pull or push the magnetization of the perpendicular layer in the desired direction (Fig. 17.20(b)). This assistance from the forces of exchange helps reduce the external field required for recording and erasure. Accordingly, it has been reported that optimum recording by magnetic field modulation can in this case be achieved with as little as ± 50 Oe of external field. Figure 17.20(c) shows measured data of carrier and noise in readout versus the applied magnetic field for writing for two disks referred to as A and B. The disks were identical except for a 20 Å capping layer deposited on the MO layer of disk B. The transition region between fully recorded and fully erased states is much narrower in disk B, a characteristic that enables magnetic-field-modulation recording on this disk to be achieved with a small, low-power magnetic head and at a high recording frequency.

17.4.2. Direct overwrite in exchange-coupled multilayer

Direct overwrite (DOW) has been the subject of extensive research over the past few years. Aside from recording by magnetic field modulation (subsection 1.6.2), the most promising solutions have originated from the concept of exchange-coupled magnetic multilayers.[21–23] Figure 17.21 shows the essential features of one such scheme based on a triple-magnetic-layer MO disk structure. In this configuration, one magnetic layer is the storage layer, another is the "assist" layer, and a third is simply there to facilitate magnetic transitions between the other two layers. The recording scheme requires two permanent magnets, one for erasing (or initializing) the assist layer, and the other for writing. In every path under the initializing magnet the assist layer is fully erased, but the storage layer, thanks to its high coercivity, retains its pattern of recorded domains.

For readout, the laser power is set at a low level, causing negligible heating of the magnetic layers. For writing, the laser is pulsed, either with moderate or high peak power, depending on whether the intended domain's magnetization is up (↑) or down (↓). The high-power pulses heat all three magnetic layers and create ↓ domains in both the storage layer and the assist layer; (the ↓ direction being set by the write magnet). Subsequent to formation of these domains, the track passes under the initializing magnet, which erases the assist layer but leaves the ↓ domain in the storage layer intact. This creates a 180° magnetic wall between the two layers, an energetically unfavorable configuration (see the inset to Fig. 17.21). Now, when a moderate pulse raises the temperature of the storage layer (but not the assist layer), this interlayer magnetic wall provides a destabilizing force and pushes the ↓ domain to annihilation. In the only other possible situation where both layers are ↑, the moderate pulse is too weak to cause

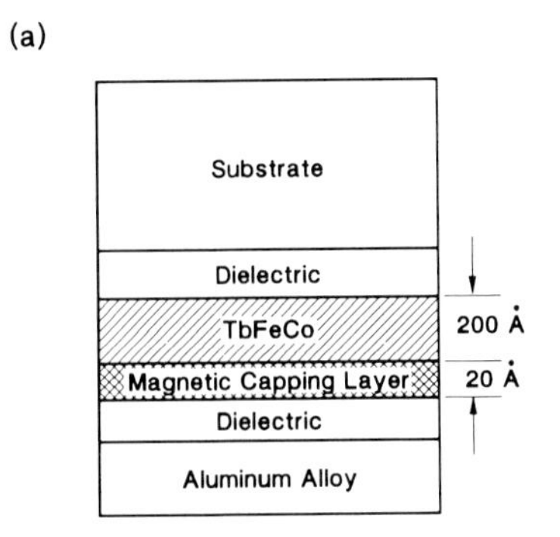

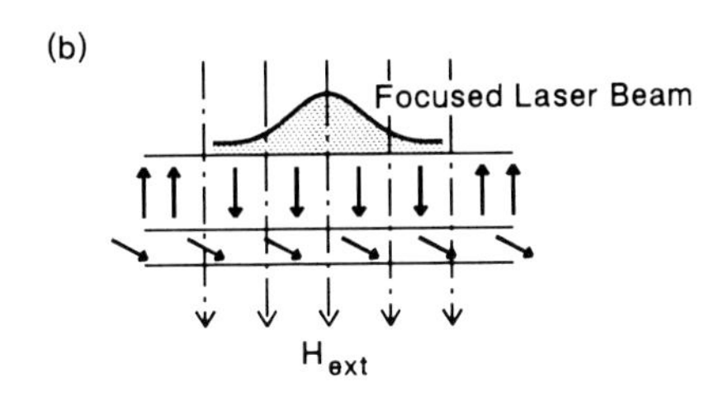

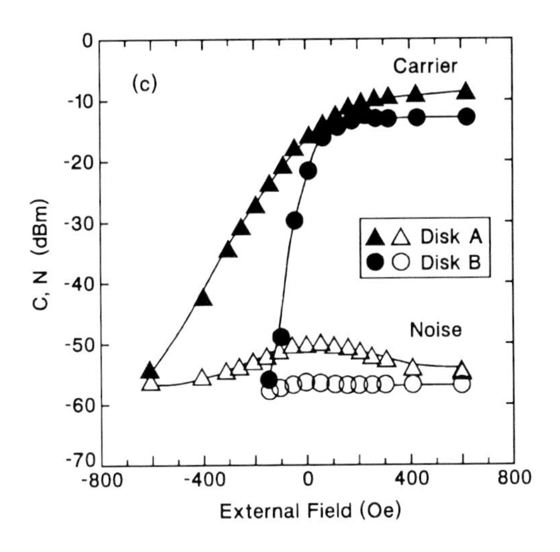

Figure 17.20. (a) Schematic diagram of a multilayer magneto-optical disk incorporating a magnetic capping layer in conjunction with the regular MO layer. The capping layer has in-plane magnetization. (b) During thermomagnetic recording the applied field forces the magnetization of the capping layer out of the plane, which pulls the magnetization of the MO layer along, thus facilitating the reversal process. (c) Measured carrier and noise levels as functions of the write field for two disks, one without the capping layer (disk A) and one with the capping layer (disk B). The disk with the capping layer has a sharper threshold and can operate with smaller magnetic fields.[20]

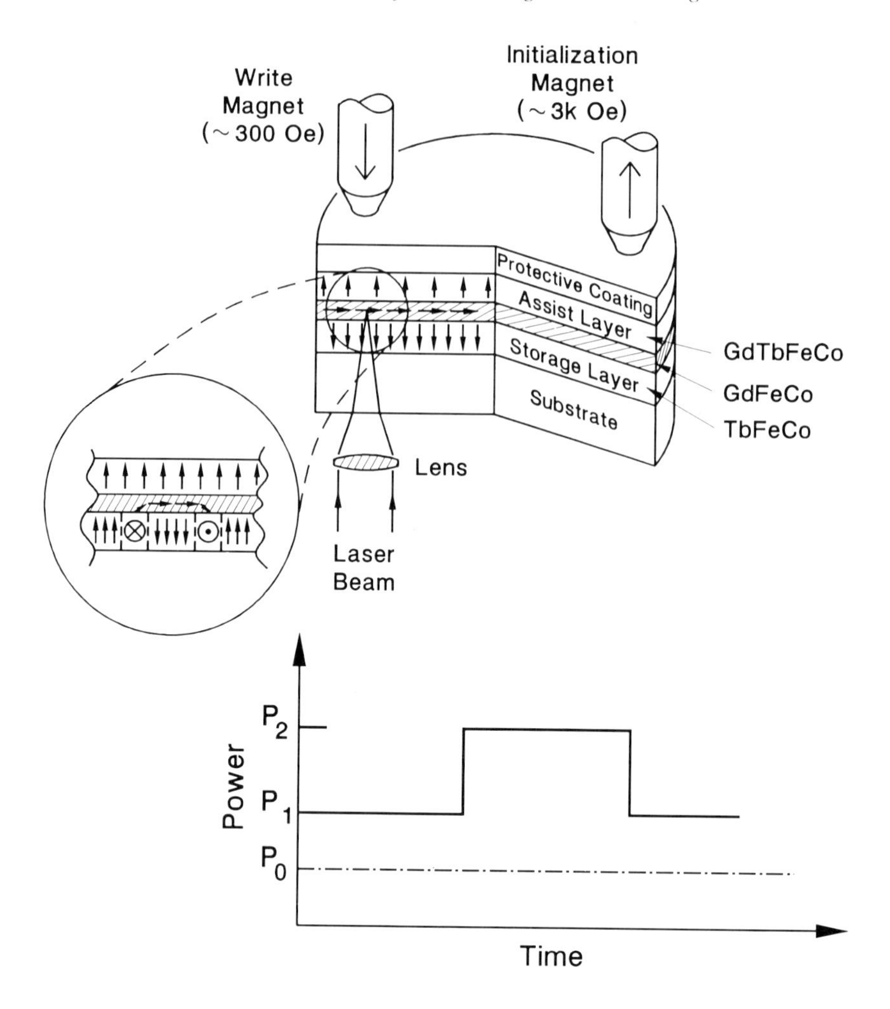

Figure 17.21. Direct overwrite in exchange-coupled triple-layer MO media. The initialization magnet can erase only the assist layer, since its field is weaker than the coercivity of the storage layer. The write magnet works in conjunction with high-power laser pulses, magnetizing both the assist layer and the storage layer in the "down" direction. The inset shows a recorded "down" domain in the storage layer, after the assist layer has been erased. In this case, a 180° wall between the two layers creates a quasi-stable state, which responds to a medium-power laser pulse by annihilating the "down" domain.

any magnetization reversals. Thus the moderate-power and high-power pulses create in the storage layer the desired ↑ and ↓ domains, regardless of any previously recorded data. This is the essence of direct overwrite.

The above scheme requires a complex media-fabrication process, and perhaps the sacrifice of a few dB of CNR in exchange for DOW. There are other, more complex exchange-coupled designs for direct overwrite with certain additional advantages (such as elimination of the initializing magnet); these designs will not be reviewed here but the reader is encouraged to consult the literature.[23]

17.4.3. Magnetically induced super resolution (MSR)

We close this chapter with a brief description of the MSR scheme, first proposed by researchers at Sony Corporation.[24] Although MSR is not directly related to the subject of this chapter, it arose in the course of investigations involving exchange-coupled structures for DOW, and shares certain operational features with the DOW scheme described in the preceding subsection. There are two slightly different ways of achieving MSR, known as front-aperture detection, FAD, and rear-aperture detection, RAD. Both FAD and RAD require trilayer exchange-coupled magnetic media similar to those used for direct overwrite. As before, the role of the intermediate layer is to make the magnetic transition between the other two layers smooth. The storage layer is written onto at the time of writing, and maintains a faithful copy of the recorded data at all times. The read layer, on the other hand, receives a copy of "selected" domains from the storage layer and presents this modified version to the readout beam. It is this selective presentation of recorded domains to the read beam that achieves super-resolution, since it removes the adjacent domains at the time of reading and essentially allows the read beam to "see" one domain at a time. Selective copying is activated by the rise in media temperature induced by the read beam itself. Details of operation of the two MSR methods will now be given separately.

MSR by front-aperture detection. In this method, shown schematically in Fig. 17.22(a), both the storage layer and the read layer normally contain identical copies of the recorded domains. Within a small area in the rear side of the focused spot, however, the rise of temperature weakens the coupling between the layers. At this point the magnetization in the rear aperture aligns itself with the applied field $\mathbf{H}_r$, which is always in the same direction during readout; the beam thus sees the magnetization pattern in the front aperture only. Super-resolution is achieved by virtue of the fact that intersymbol interference from the domains in the rear aperture has been eliminated. Once the disk moves away and the temperatures return to normal, interlayer exchange regains its strength and the magnetization of the read layer reverts to its original orientation.

MSR by rear-aperture detection. Figure 17.22(b) shows the schematic diagram of MSR by RAD. Initially the data is recorded on both the storage layer and the read layer, but the latter is erased prior to readout by the initialization field $\mathbf{H}_i$. During readout, thermal effects of the read beam reduce the coercivity of the read layer within the rear aperture. As a result the underlying domains in the storage layer copy themselves onto the hot area of the read layer by the force of interlayer exchange (and, if necessary, by assistance from an applied field $\mathbf{H}_r$). Thus super-resolution is

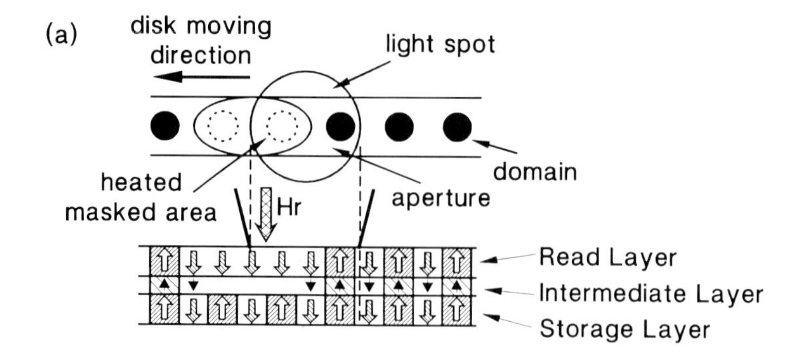

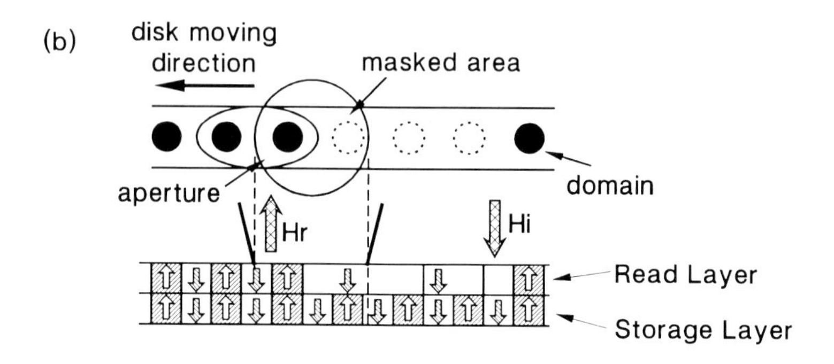

Figure 17.22. The principle of magnetic super-resolution (MSR) in exchange-coupled triple-layer structures. (a) MSR by front-aperture detection. (b) MSR by rear-aperture detection. Courtesy of M. Kaneko, Sony Research Laboratories.

achieved because the front aperture is erased and the only domain that is being read is within the rear aperture. When the disk moves away and the temperatures return to normal, the transferred domains persist in the read layer until such time as they are erased again by the initializing field $\mathbf{H}_i$.

Needless to say, for the above readout mechanisms to succeed, the coercivities of the read and storage layers, the read field $\mathbf{H}_r$, the initialization field $\mathbf{H}_i$, the various wall energies σ_w, and the temperature-dependence of several magnetic parameters involved must all be carefully selected. With MSR it has been possible to increase the recording densities beyond the cutoff frequency normally allowed by an objective lens. For instance, recorded mark lengths of 0.4 μm have been read with good CNR at λ = 780 nm with an objective lens having NA = 0.53.[33] The drawbacks to MSR, aside from the complexity of media manufacture, are the inevitable sacrifices one will have to make in desirable media characteristics such as minimum jitter, maximum noise immunity, environmental stability, and so forth. The systems demonstrated in the laboratories to date, however, seem quite promising.

Problems

(17. 1) The exchange-coupled magnetic bilayer in Fig. 17.6(a) has exchange stiffness coefficients A_{x1} and A_{x2}, and anisotropy constants K_{u1} and K_{u2}, for the top and bottom layers respectively. The perpendicular axis is the hard axis for the second layer and, therefore, $K_{u2} < 0$. Investigate the structure and energy density of the wall in the transition region between the two layers.

(17. 2) Figure 17.6(b) shows the individual and combined hysteresis loops of an exchange-coupled magnetic bilayer. The top layer has saturation magnetization M_{s1}, coercivity H_{c1}, and thickness τ_1; the corresponding parameters for the bottom layer are M_{s2}, H_{c2}, and τ_2. Determine the relationship between the various coercivities and the energy density σ_w of the interfacial wall.

(17. 3) Consider the balance of energy at the rim of the Curie disk in the early stage of nucleation, and verify the inequality given in Eq. (17.12). Substitute for σ_w and Δ_w in terms of A_x and K_u and verify Eq. (17.13).

(17. 4) One method of direct overwrite in exchange-coupled magnetic multilayer media was described in subsection 17.4.2. In terms of the coercivities, magnetizations, thicknesses, and wall energies of the various layers (including the interfacial wall) write down the necessary conditions for the successful operation of this DOW scheme.

18

Media Characterization

Introduction

The conventional media of magneto-optical data storage, the rare earth-transition metal alloys, are generally produced by radio frequency (rf) sputtering from an alloy target onto a plastic or glass substrate. For protection against the environment as well as for optical and thermal enhancement, the RE–TM alloy films are sandwiched between two dielectric layers (such as SiN_x or AlN_x) and covered with a metallic reflecting and heat-sinking layer, before finally being coated with several microns of protective lacquer. The properties of the MO layer are determined not only by the composition of the alloy, but also by the sputtering environment and by the condition of the substrate's surface. The temperature at which the film grows, the sputtering gas pressure, the substrate bias voltage (self or applied), the rate of deposition, the surface roughness of the substrate, the quality of underlayer and overlayer can all affect the properties of the final product. It is therefore necessary to have accurate characterization tools with which to measure the various properties of the media and to establish their suitability for application as the media of erasable optical data storage.

The most widely used method of characterization for magnetic materials is vibrating sample magnetometry (VSM). With VSM it is possible to measure the component of net magnetization M_s along the direction of the applied field. For MO media, one obtains the hysteresis loop when the field is perpendicular to the plane of the sample. When the sample holder is rotated by 90° and the field is applied in the plane of the film, one measures the deviation of M_s from the easy axis, which allows determination of the constant of magnetic anisotropy K_u. These measurements can be performed at various temperatures, thus yielding plots of $H_c(T)$, $M_s(T)$, and $K_u(T)$ for the sample. From these results it is then possible to obtain the values of the Curie and compensation-point temperatures.

Another device used routinely for media characterization is the magneto-optical loop-tracer.[1,2] In the loop-tracer a laser beam is used either in reflection (MO polar Kerr effect) or in transmission (Faraday effect) to monitor the magnetization of the sample in the presence of a magnetic field. The state of polarization of the beam after interaction with the sample determines the state of magnetization of the MO layer. In the visible and near-infrared range of wavelengths typically used in loop-tracers, the beam interacts mainly with the transition metal element(s);

therefore, the measured Kerr angle θ_k and ellipticity ϵ_k (or their Faraday counterparts) are proportional to M_{TM}, the magnetization of the TM subnetwork. These measurements can be performed in a range of temperatures, yielding plots of $\theta_k(T)$, $\epsilon_k(T)$, and $H_c(T)$. If the field is applied in the plane of the sample (or in fact in any direction other than the perpendicular one) then the deviation of M_{TM} from the easy axis as a function of the field may be used to determine K_u.

Galvanomagnetic phenomena have been known for many years and have found practical applications in the area of advanced storage technology.[3–13] These phenomena offer yet another approach to measuring certain electrical, thermal, and magnetic properties of the media. The extra-ordinary Hall effect and the electrical resistivity are used routinely for galvanomagnetic characterizations. The physical origins of these phenomena are described at the start of section 18.1. The remainder of section 18.1 is devoted to a discussion of measurement results that, in addition to the Hall effect and magnetoresistance, include results of vibrating sample magnetometry and Kerr loop-tracing on amorphous RE–TM films and compositionally modulated TM/TM films.

Observations of domains and magnetic structure in MO films constitute a fruitful approach to media characterization. Polarized-light microscopy has been used extensively for this purpose, and many important properties of the media have been deduced from these observations.[14,15] Section 18.2 describes the basic elements of polarization microscopy and presents results of observations on several samples. Since the resolution of optical microscopy is limited by the wavelengths of visible light, the detailed observation of domains using this technique has become increasingly difficult, with advances in optical recording and the accompanying refinement of the media. In those instances where high-resolution imaging of magnetic structure has been required, Lorentz electron microscopy and magnetic force microscopy have provided suitable solutions.

Lorentz microscopy is performed in a transmission electron microscope (TEM) where the electron beam, passing through the magnetic specimen, interacts with the **B**-field in and around the sample.[16–21] The various rays of the beam are bent in accordance with the Lorentz law of force and in proportion to the strength of the field at their various locations. Lorentz microscopy is capable of imaging magnetic structure in thin-film samples with a resolution of a few hundred angstroms. In section 18.3 we describe the physical principles of Lorentz microscopy and present examples of domain micrographs obtained with this technique.

Section 18.4 is devoted to a discussion of magnetic force microscopy (MFM)†, which is capable of resolving magnetic features as small as a few hundred angstroms.[23–32] In MFM a sharp magnetic needle interacts with the field pattern established by the sample near its surface. A cantilever then converts the force on the needle to a displacement, which is measured interferometrically or otherwise. A model for magnetic force microscopy

† MFM is an offshoot of scanning tunneling microscopy (STM).[22]

will be described in section 18.4, along with examples of force computation and a micrograph of magnetic domains in MO media.

18.1. Magnetic, Magneto-optical, and Galvanomagnetic Measurements

The measurements of magnetization described in this section have been performed with a commercial vibrating sample magnetometer.† In this instrument a small sample is placed between the pole pieces of an electromagnet whose maximum field capability is 14 kOe. The sample is brought mechanically to vibration so that an oscillating magnetic field is established on the two pick-up coils, symmetrically located with respect to the sample on the pole pieces. The voltage induced in the coils is proportional to the oscillating field, which, in turn, is proportional to the magnetic dipole moment of the sample. The VSM is calibrated with a standard nickel sample, and the constant of proportionality between the magnetic moment and the pick-up signal is established. Since the magnetization M_s is obtained by normalizing the measured moment with the volume of the magnetic material, inaccurate measurement of the area and/or thickness of the sample as well as errors of calibration will introduce corresponding errors in the measured value of M_s. When the moment is not aligned with the applied field, the component of $\mathbf{M}_s$ parallel to the field will be the only one measured; the signals induced in the two coils by the perpendicular component are identical and will, therefore, cancel out.

The measurements of the MO Kerr effect, of magnetoresistance, and of the Hall effect, described below, have been performed with the apparatus shown in Fig. 18.1(a). The electromagnet has a maximum field capability of 20 kOe, and its rotating base allows the field to be applied either perpendicular to or in the plane of the sample. The Kerr effect signal is measured using normally incident beam from a HeNe laser (λ = 633 nm). The differential detector module monitors the polarization state of the reflected beam and produces a signal proportional to the Kerr angle θ_k (when the quarter-wave plate is removed) or to the ellipticity ϵ_k (when the plate is inserted).

In galvanomagnetic measurements with the so-called "four-point probe", illustrated in Fig. 18.1(b), three different geometries are possible. In the perpendicular geometry the field is normal to the sample and the current flows between point contacts 1 and 2; the Hall effect is measured between terminals 3 and 4. In the longitudinal geometry the field is in the plane of the sample and parallel to the direction of current, which flows between terminals 1 and 2. The transverse geometry is similar to the longitudinal one, with the exception that the current terminals are 3 and 4. When the exact values of electrical resistivity and Hall resistivity are

†*Digital Measurement Systems* VSM Model 1660.

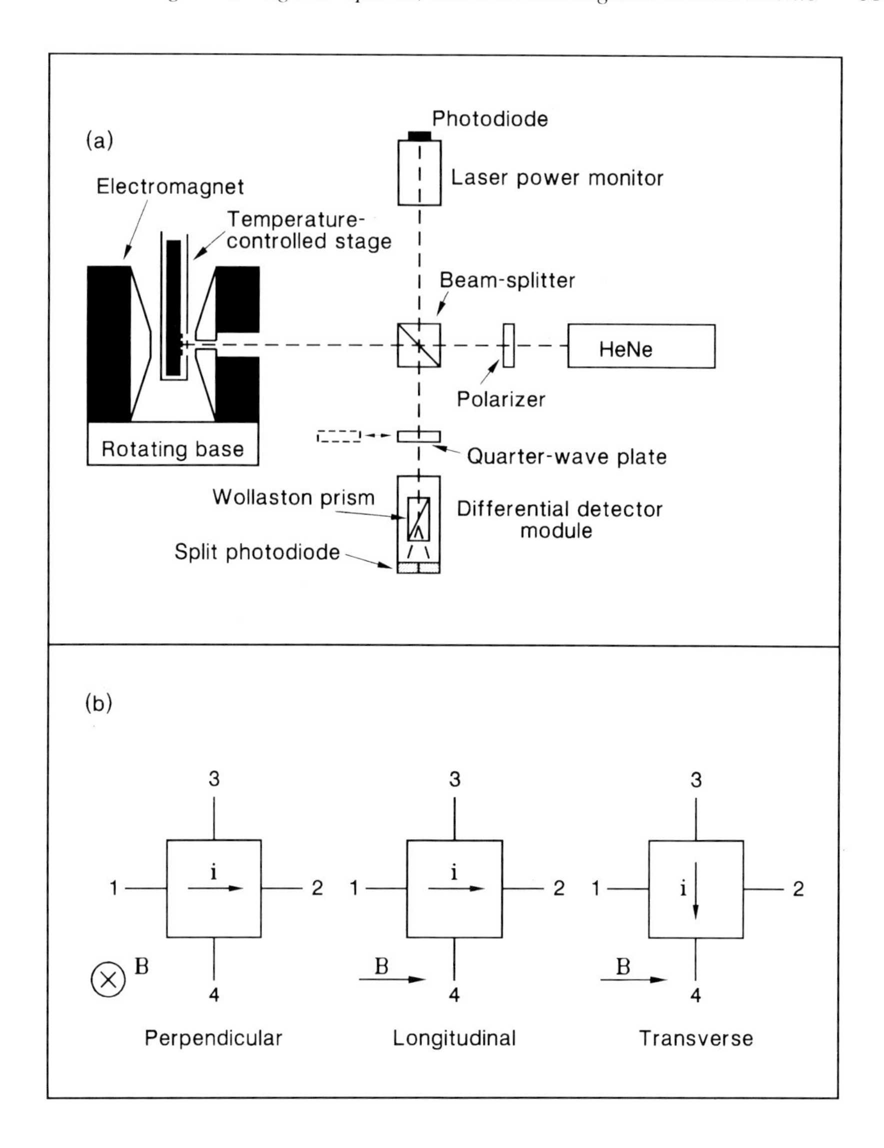

Figure 18.1. (a) Schematic diagram of a magneto-optical loop-tracer. Light from the HeNe laser is linearly polarized and is reflected from the MO sample, which is placed inside the dewar. The Wollaston prism and the split detector constitute the differential detection module, and the $\lambda/4$ plate is inserted in the optical path whenever the measurement of ellipticity is desired. The laser power monitor is needed for the measurement of sample's reflectivity. The magnet sits on a rotating base that allows the application of both in-plane and perpendicular fields. Inside the dewar, a four-point probe attached to the sample enables the measurement of magnetoresistance and the Hall voltage. (b) Three geometries for galvanomagnetic measurements.

desired, one must apply the formulas of van der Pauw[33] to the four-point probe data. The physics of magnetoresistance and the Hall effect is discussed briefly in subsection 18.1.1, followed by a detailed analysis of the measurement results on several MO samples.

18.1.1. Magnetoresistance and the Hall effect

The galvanomagnetic effects may be separated into those that depend on the magnetic induction **B** and those that depend on the magnetization **M**. Let us denote by ρ_0 the ordinary DC resistivity of an otherwise isotropic material in the absence of **B** and **M**. Thus ρ_0 is the resistivity with the exclusion of magnetic contributions and, under normal circumstances, it is a function of temperature only. The contributions of the **B**-field to the transport properties are usually classified under the "ordinary" effects, whereas those stemming from the magnetization are referred to as "anomalous" or "extraordinary".

The most common galvanomagnetic effects are the ordinary magneto-resistance (MR) and the ordinary Hall effect.[6] These effects are caused by the Lorentz force acting on conduction electrons. A simple description of ordinary magnetoresistance, for example, maintains that the **B**-field gives the conduction electrons a curved path, thereby increasing their frequency of collisions with phonons and impurities and resulting in an increased electrical resistance.[12] Ordinary MR is best observed in non-magnetic metals and semiconductors; the effect is appreciable if the mean free path of the conduction electron is large compared with its radius of curvature in the magnetic field. In amorphous materials and in polycrystalline specimen composed of very small crystallites, the mean free path is short and the ordinary MR is consequently small. Ordinary MR in thin films is further reduced due to the surface scattering of the conduction electrons, which tends to shorten the mean free path. In ferromagnetic metals, the exchange interaction between conduction electrons and the magnetic ions reduces the mean free path, resulting in a lower ordinary MR. This may not be true at low temperatures, however, because of the highly ordered state of the magnetic lattice.

At fields that are not too large and temperatures that are not too low, the ordinary MR is usually a parabolic function of the applied field, its magnitude increasing with the square of the effective **B**-field. There are exceptions to this rule, however. For example, in certain circumstances when the field and the current are both in the plane of a thin-film sample and parallel to each other, the spiraling electrons could avoid the film surfaces and result in a decreasing resistivity with an increasing field. Ordinary magnetoresistance is longitudinal when the **B**-field is parallel to the direction of the current, and transverse when the two are orthogonal. In ordinary MR the transverse effect is generally observed to be larger than the longitudinal effect.

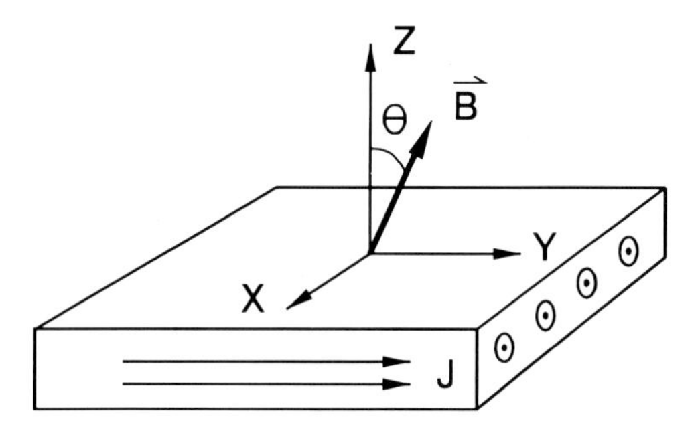

Figure 18.2. External magnetic field **B** applied to a current-carrying sample. The field is in the YZ-plane at angle Θ to the Z-axis, and the current **J** is parallel to the Y-axis.

The following resistivity tensor describes in compact form the ordinary magnetoresistance and the ordinary Hall effect:

$$\Delta\rho = \begin{bmatrix} \rho_{xx} & \rho_{xy} & 0 \\ -\rho_{xy} & \rho_{xx} & 0 \\ 0 & 0 & \rho_{zz} \end{bmatrix} \tag{18.1}$$

$\Delta\rho$ is the incremental resistivity, contributed by a **B**-field applied along the Z-axis, to the isotropic base resistivity ρ_0 of the material. ρ_{xx} is the transverse magnetoresistivity, ρ_{zz} is the longitudinal magnetoresistivity, and ρ_{xy} is the Hall resistivity. These elements of the resistivity tensor depend on the magnitude of **B**. At $B = 0$ they are all equal to zero; in low and moderate fields ρ_{xx} and ρ_{zz} are usually quadratic functions of B, whereas ρ_{xy} is linear in B. For arbitrary orientations of **B**, the MR tensor in Eq. (18.1) must be transformed with the aid of appropriate rotation matrices. For instance, if **B** happens to be in the YZ-plane at an angle Θ to Z, as in Fig. 18.2, the rotation matrix R will be

$$R = \begin{bmatrix} 1 & 0 & 0 \\ 0 & \cos\Theta & -\sin\Theta \\ 0 & \sin\Theta & \cos\Theta \end{bmatrix} \tag{18.2}$$

Consequently, the resistivity tensor becomes

$$\Delta\rho' = R^{\mathrm{T}}(\Delta\rho)\,R$$

$$= \begin{bmatrix} \rho_{xx} & \rho_{xy}\cos\Theta & -\rho_{xy}\sin\Theta \\ -\rho_{xy}\cos\Theta & \rho_{xx} + (\rho_{zz} - \rho_{xx})\sin^2\Theta & \frac{1}{2}(\rho_{zz} - \rho_{xx})\sin 2\Theta \\ \rho_{xy}\sin\Theta & \frac{1}{2}(\rho_{zz} - \rho_{xx})\sin 2\Theta & \rho_{xx} + (\rho_{zz} - \rho_{xx})\cos^2\Theta \end{bmatrix} \tag{18.3}$$

Example 1. Figure 18.2 shows a current-carrying film under an applied magnetic field **B**. The field makes angle Θ with respect to the normal to the plane of the film, and the current density **J** is along the Y-axis, that is, $\mathbf{J} = (0, 1, 0)J$. The incremental electric field $\Delta\mathbf{E}$ produced by this current and by the resistivity tensor in Eq. (18.3) is thus given by

$$\begin{bmatrix} \Delta E_x \\ \Delta E_y \\ \Delta E_z \end{bmatrix} = \Delta\rho' \begin{bmatrix} 0 \\ 1 \\ 0 \end{bmatrix} J = \begin{bmatrix} \rho_{xy}\cos\Theta \\ \rho_{xx} + (\rho_{zz} - \rho_{xx})\sin^2\Theta \\ \frac{1}{2}(\rho_{zz} - \rho_{xx})\sin 2\Theta \end{bmatrix} J$$

Accordingly, the resistivity along Y is $\rho_0 + \rho_{xx} + (\rho_{zz} - \rho_{xx})\sin^2\Theta$, and the Hall resistivity along X is $\rho_{xy}\cos\Theta$. □

Negative slope and s–d scattering. Another galvanomagnetic effect, usually observed in transition metals and their alloys, is the reduction of resistivity with increasing magnetic field. This effect was explained by Mott as related to the scattering of the conduction s electrons into the d band.[4] The effect has been mainly studied in ferromagnetic materials, but, d band structure permitting, it can appear in non-ferromagnetic metals as well.[6] Mott assumes that electrical conduction in transition metals is mainly due to s electrons, since d electrons are more strongly bound with the ions and, consequently, have a large effective mass. He also assumes that the scattering of the conduction electrons is a spin-preserving process. Now, in a ferromagnetic transition metal (or alloy) the d band is exchange-split between the spin up (d↑) and spin down (d↓) electrons. With reference to Fig. 18.3, let the Fermi level be close to the upper edge of the d↑ band. At low temperatures this band (d↑) is almost full and has no vacancies for additional spin up electrons. Thus, if an s electron with spin up is to be scattered during a collision, it can only go to another (available) level in the s band as dictated by the Pauli exclusion principle. At elevated temperatures, however, some of the electrons near the Fermi surface in d↑

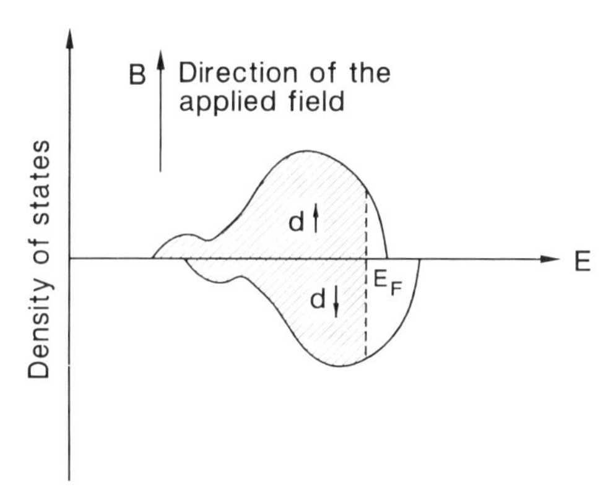

Figure 18.3. Schematic diagram of the split d band in transition metals and their alloys. Application of an external magnetic field in the up direction enhances the splitting, thus reducing the number of empty states in the d↑ band. With fewer states into which the conduction s↑ electrons may scatter, the probability of scattering reduces and, therefore, the electrical resistivity of the material declines.

move to the d↓ band. This creates vacancies in d↑, thus raising the probability of scattering from the conduction band into the d↑ band, and resulting in a higher resistivity.

If we now allow the temperature to be fixed at a reasonably high value and apply a field in the direction of magnetization, we shall observe the following phenomenon: as the field increases some of the electrons move back from d↓ to d↑ and the resistivity subsequently declines.† For a given electron with spin magnetic moment μ_B, the difference in energy between spin up and spin down states in a magnetic field **B** is $2B\mu_B$. At $B \simeq 20$ kgauss this energy difference is only 4×10^{-16} ergs, which is two orders of magnitude less than $k_B T$ at room temperature. Thus at ordinary temperatures and fields we expect the reduction in resistivity to be small (though by no means inconsequential). Note also that according to this argument the filling of the d↑ band (at $B \simeq 20$ kG) will not result in a substantial increase in the saturation magnetization of the material.

Kasuya[5] observed that negative MR also occurs in rare earth metals, although transitions between s and f bands in these media are unlikely. He argued that the f electron moments, while localized at their lattice sites, interact with the conduction electrons via exchange and spin–orbit mechanisms. At the absolute zero of temperature, all the magnetic dipole

†Obviously, for this mechanism to work the material need not be magnetically ordered. It can happen in a non-magnetic metal or in a ferromagnet above the Curie temperature, provided that the Fermi level is sufficiently close to the upper edge of the d band.

moments of the f electrons are ordered on the periodic lattice of the crystal, and the magnetic scattering must therefore vanish. At temperatures above zero, this order is disturbed and conduction-electron scattering from the magnetic moments leads to an increase in resistivity. Under these circumstances, the application of a magnetic field tips the balance in favor of magnetic alignment, causing the resistivity once again to decline.

Galvanomagnetic effects in magnetically ordered media. A phenomenon specific to magnetically ordered media is the magneto-resistance and Hall effect induced by scattering from the resident magnetic moments. This effect, whose magnitude depends on the relative orientation of the magnetization and the current, has been analyzed by Smit[8] and by Kondo[9] who attributed it to the interaction between the spin system and the lattice via spin–orbit coupling. These authors showed that the conduction electrons moving parallel (or antiparallel) to the direction of magnetization are more easily scattered than those traveling in a transverse direction. When the spin of the magnetic electron is aligned with the current, its orbital moment is likewise aligned due to the spin–orbit coupling. Since the plane of the orbit is perpendicular to the angular momentum, in this position the magnetic electron exhibits the largest cross-section towards the conduction electrons, thus creating the largest resistivity (see Fig. 18.4).

The extraordinary galvanomagnetic effects may be analyzed phenomenologically, using a resistivity tensor similar to that described earlier in conjunction with the ordinary effects. The extraordinary tensor has the same form as Eq. (18.1), but, for single-domain specimens, the Z-axis must be designated as the direction of magnetization. ρ_{xy} is now the coefficient of the extraordinary Hall effect. In contrast to ordinary MR, the longitudinal extraordinary resistivity ρ_{zz} is generally larger than the corresponding transverse resistivity ρ_{xx}.

18.1.2. Measurements on Co/Pt sample

We describe various measurements performed on a Co/Pt superlattice sample, fabricated by electron beam evaporation on a glass substrate. The sample consists of twenty periods of alternating layers of Co (5 Å) and Pt (10 Å), with a total thickness of 300 Å. The VSM measurement results at room temperature are shown in Fig. 18.5(a). The easy axis is perpendicular, and the sample has saturation moment $M_s = 410$ emu/cm^3 and coercive field $H_c = 1.3$ kOe. The reversal process takes place by a rapid nucleation of domains at $H = H_c$, resulting in the steep part of the hysteresis loop, followed by a rather slow growth of these domains, an additional field of approximately 1 kOe being required for the completion of reversal.

The VSM curve obtained with an applied in-plane field (dashed curve in Fig. 18.5(a)) shows a steep rise in the vicinity of $H = 0$. Since initially the sample is in a demagnetized state, this steep rise is probably due to domain wall motion and/or alignment of the in-plane components of

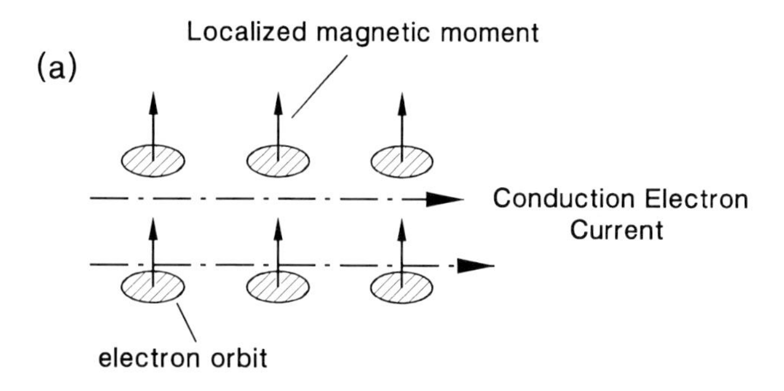

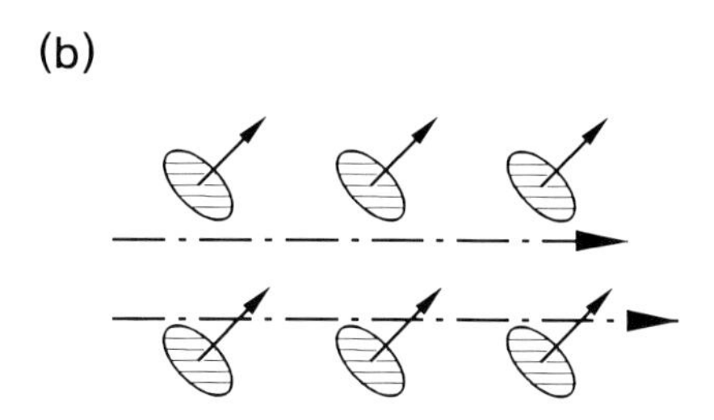

Figure 18.4. (a) Within a magnetically ordered medium, conduction electrons travelling perpendicular to the direction of magnetization see a small cross-section from the magnetic ions. The spin–orbit coupling of the "localized" magnetic electrons is responsible for orienting their orbital planes perpendicular to their spin magnetic moments. (b) When the magnetization is tipped in the direction of the current, the cross-section for collision between conduction electrons and the magnetic ions increases; as a result, the resistivity of the material begins to rise.

magnetization. At larger fields the slope of the curve is caused by the rotation of magnetization away from the (perpendicular) easy axis and towards the direction of the applied field. Extrapolating from the slope of this curve at intermediate values of the applied field, one obtains the anisotropy field, $H_k = 2K_u/M_s - 4\pi M_s \simeq 12$ kOe. The uniaxial anisotropy energy constant for this sample is thus $K_u \simeq 3.5 \times 10^6$ erg/cm^3.

Starting from a demagnetized state and applying a perpendicular field to the sample, we obtained the Hall loop of Fig. 18.5(b). The Hall resistivity in the saturated state is $\rho_{xy} = 1.2$ $\mu\Omega$ cm. The coercive field value and other features of this loop are similar to the VSM loop. The Hall effect signal, when the sample is initially saturated in the perpendicular direction and the applied field is in-plane, is shown in Fig. 18.5(c). The initial slow decrease of the Hall voltage (which was found to be reversible) is due to the rotation of **M** towards the field. Following this initial phase, the rapid, irreversible drop in the Hall voltage indicates breakup of the

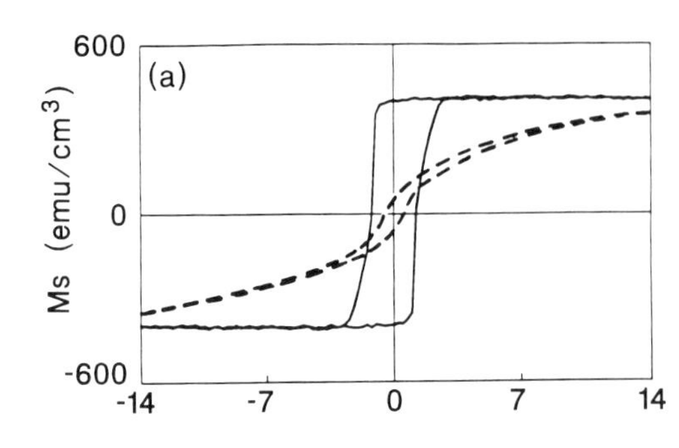
600
(a)
Ms (emu/cm^3)
0
-600
-14
-7
0
7
14

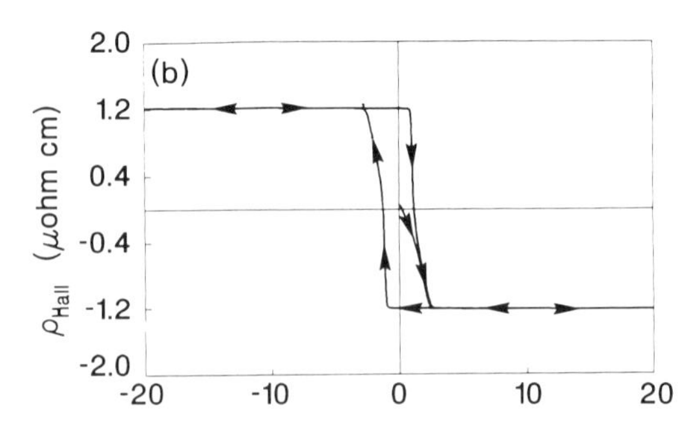
2.0
(b)
1.2
0.4
-0.4
-1.2
-2.0
ρ_{Hall} (μohm cm)
-20
-10
0
10
20

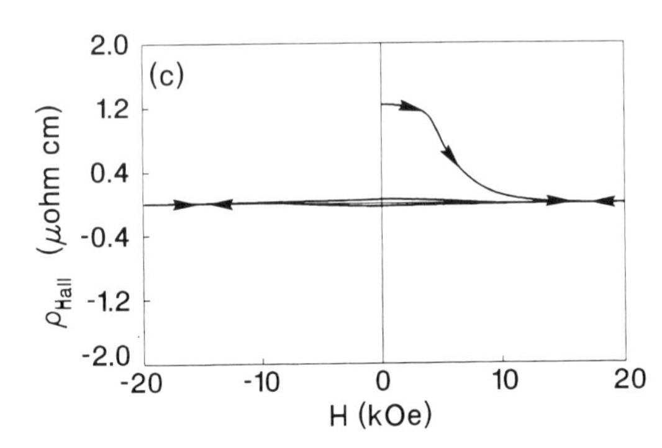
2.0
(c)
1.2
0.4
-0.4
-1.2
-2.0
ρ_{Hall} (μohm cm)
-20
-10
0
10
20
H (kOe)

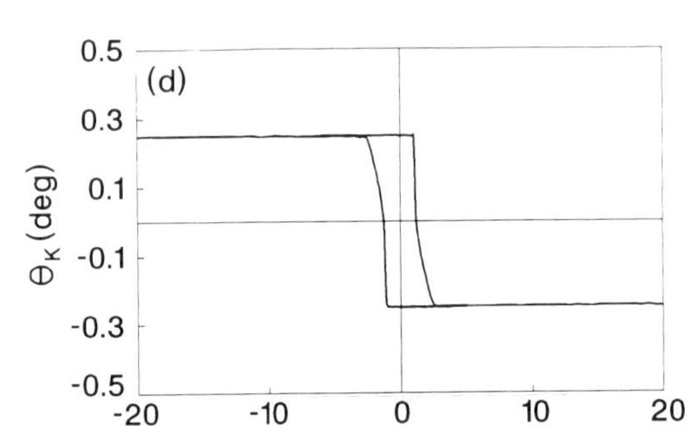
0.5
(d)
0.3
0.1
-0.1
-0.3
-0.5
Θ_K (deg)
-20
-10
0
10
20

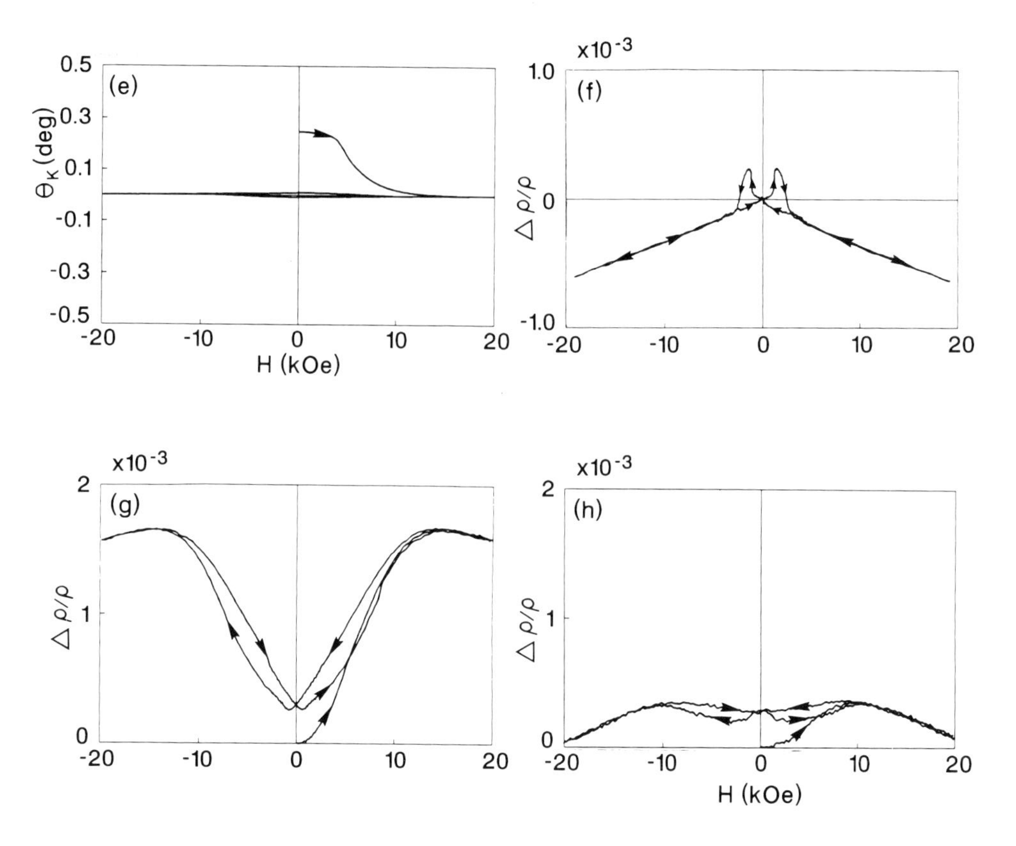

Figure 18.5. Various measurements on a 300 Å-thick sample of Co/Pt (5 Å-thick cobalt layers alternating with 10 Å-thick platinum layers). (a) VSM measurement with the magnetic field perpendicular (solid curve) and in the plane of the sample (broken curve). (b) Hall resistivity versus the magnitude of an applied perpendicular field (initial state = demagnetized). (c) Hall resistivity versus the magnitude of an applied in-plane field (initial state = saturated along the easy axis). (d) Polar Kerr signal versus the magnitude of an applied perpendicular field (initial state = saturated along the easy axis). (e) Polar Kerr signal versus the magnitude of an applied in-plane field (initial state = saturated along the easy axis). (f) Resistivity variation as a function of the magnitude of an applied perpendicular field. (g) Resistivity variation with the magnitude of the applied field; longitudinal geometry (initial state = saturated along the easy axis). (h) Resistivity variation with the magnitude of the applied field; transverse geometry.

sample into oppositely magnetized domains.† Finally, at sufficiently large fields, the magnetization comes into alignment with the field, at which point the Hall voltage becomes negligibly small. Reducing the magnetic field now back to zero does not yield the original Hall voltage, simply because the sample is no longer saturated. The small hysteresis in Fig. 18.5(c) is repeatable and probably due to the motion of domain walls in the demagnetized state. Alternatively, this hysteretic behavior might be related to variations of the local easy axes across the sample.

The polar MO Kerr effect was also measured for this sample. The hysteresis loop (obtained in a perpendicular field) is shown in Fig. 18.5(d). The maximum Kerr angle is $\theta_k = 0.25°$ and the loop is identical with the Hall loop. When the saturated film (along the easy axis) was subjected to an in-plane field, the polar Kerr signal behaved as in Fig. 18.5(e), once again in agreement with the corresponding Hall effect measurement. The value of H_k estimated from the curvature of the initial part of this curve (near the top) is about 11 kOe, in fair agreement with the VSM estimate. The similarity between the Kerr effect and the Hall effect signals is remarkable, considering the fact that the laser beam illuminates only a small fraction of the total area of the sample.

Figure 18.5(f) shows a plot of $\Delta\rho/\rho$ versus H, where H is the magnitude of the applied perpendicular field and ρ is the initial resistivity at $H = 0$. Using the method of van der Pauw[33], the value of ρ in the remanent state is found to be 58 $\mu\Omega$ cm. The peaks centered around the coercive field are caused by the scattering of conduction electrons from the domain walls that appear during the formation and growth of reverse domains. The width of the peak corresponds to the transition region in the vicinity of H_c in the hysteresis loops. The linear part of the curve in Fig. 18.5(f), with a negative slope of 3.2×10^{-8} per oersted, originates from the s-d scattering phenomenon described earlier.

Figure 18.5(g) shows the measured values of $\Delta\rho/\rho$ versus H, where H is the magnitude of the longitudinal field applied in the plane of the sample and parallel to the direction of the current. At first, the sample is saturated along the easy axis. The initial increase of the resistance with H is due mainly to the alignment of the magnetization vector **M** with the current. As expected, the maximum is reached around $H = H_k \simeq 12$ kOe. Thus, ignoring the ordinary MR effects, one can associate the maximum rise of 1.65×10^{-3} in $\Delta\rho/\rho$ with $(\rho_{zz} - \rho_{xx})/\rho$ corresponding to the extraordinary effect. Once the magnetization and the field have been aligned, further increases in H cause a linear decrease of $\Delta\rho/\rho$ that is, as before, caused by s-d scattering. It must be emphasized that the s-d mechanism becomes effective only *after* the complete alignment of **M** with **H**; a justification for this statement will be given shortly.

†This breakup into domains is not predicted by the coherent rotation theory of Stoner and Wohlfarth (see section 16.1), but occurs in practice probably due to inhomogeneities in the sample that might arise, for example, from regions with differing directions of the easy axis. An alternative explanation for the breakup into domains might involve the changing balance between domain wall energy and demagnetization energy.

Let us denote by the angle Θ the deviation of magnetization from the easy axis. Assuming uniaxial anisotropy energy density given by

$$\mathcal{E}_{\text{anis}} = K_u \sin^2\Theta \tag{18.4}$$

the Stoner–Wohlfarth theory predicts a linear dependence of sin Θ on the magnitude of the applied in-plane field, that is,

$$\sin\Theta = H/H_k \tag{18.5}$$

(see section 16.1.) Thus, in accordance with the analysis presented in subsection 18.1.1., we must have

$$\Delta\rho = (\rho_{zz} - \rho_{xx})\,(H/H_k)^2\,; \qquad 0 \leq H \leq H_k \tag{18.6}$$

The shape of this function, however, does not quite agree with the experimental data: the relevant part of the curve in Fig. 18.5(g) lies somewhat above the quadratic function of Eq. (18.6). We speculate that the disagreement between the coherent rotation theory and experiment in this instance is caused by the breakdown of magnetization into domains, which occurs at $H \simeq 4$ kOe.

Earlier in the section we asserted that the s-d mechanism becomes effective only *after* **M** has aligned itself with the applied in-plane field. The explanation relies on the fact that the effective field acting on **M** is

$$\mathbf{H}_{\text{eff}} = (H_k \cos\Theta)\,\hat{\mathbf{z}} + H\,\hat{\mathbf{x}} \tag{18.7}$$

where the first term is the effective anisotropy field acting along the easy axis Z, and the second term is the applied field (assumed to be along X). Replacing H in Eq. (18.7) by $H_k \sin\Theta$ from Eq. (18.5) yields a net effective field of magnitude H_k along the direction of magnetization. Thus the total effective magnetic field acting on **M** during its rotation from the easy axis towards the plane is a constant, independent of Θ. Consequently, the s-d scattering probabilities should be unaffected by the magnitude H of the applied field until such time as the magnetization reaches the plane. This occurs at $H = H_k$, which, as corroborated by the data in Fig. 18.5(g), is the onset of the contribution of the s-d effect to magnetoresistance.

It is seen in Fig. 18.5(g) that when H is brought back to zero the resistance is somewhat larger than its initial value. This is to be expected, simply because in the end the sample is no longer saturated. The difference in $\Delta\rho/\rho$ between the initial (saturated) state and the demagnetized state is about 3.0×10^{-4}, which is slightly greater than the height of the peaks in Fig. 18.5(f). The two demagnetized states, one created by a perpendicular field and the other by an in-plane field, are therefore expected to have somewhat different structures.

In Fig. 18.5(h) we show the measured values of $\Delta\rho/\rho$ versus H, where H is the magnitude of the transverse field applied in the plane of the sample and perpendicular to the direction of current. Starting from the magnetization state saturated along the easy axis, the resistance is seen to increase at first, peaking at $H \simeq 10$ kOe before it begins to decline. As in the previous case, the linear decrease, with a slope of 3.1×10^{-8} per oersted for $H > H_k$, is due to s-d scattering. The initial increase and then decrease of $\Delta\rho/\rho$ is partly due to the breakup of the saturated state into domains. There is also the fact that point-probe geometry permits a fraction of the current to flow parallel to the applied field, thus enabling the current to sense the magnetization **M** as it moves into the plane.

18.1.3. Measurements on $Tb_{28}Fe_{72}$ sample

The measurement results for this 1350 Å-thick sample, which was sputter-deposited onto glass and has a perpendicular easy axis, are shown in Fig. 18.6. The net magnetization of the film is $M_s = 115$ emu/cm^3; its anisotropy field extracted from the VSM data is $H_k \simeq 16$ kOe. When saturated along the easy axis, the sample exhibits a zero-field resistivity ρ of 580 $\mu\Omega$ cm, a Hall resistivity ρ_{xy} of 9.2 $\mu\Omega$ cm, and a Kerr rotation angle θ_k of 0.32°. The curvature of the Hall voltage versus the applied in-plane field in Fig. 18.6(c) yields $H_k \simeq 20$ kOe, which is somewhat larger than the VSM estimate. (Not shown in Fig. 18.6 are the Kerr effect data, which were similar to the Hall effect data.)

The main difference between the present sample and the Co/Pt sample discussed in subsection 18.1.2 is the disappearance of the negative slope in the perpendicular MR data. The slope due to s-d scattering in iron should be, if anything, positive here, simply because this is a Tb-rich alloy and the iron moment is opposite to the saturation moment. The lack of any measurable s-d effect is due to the particular band structure of iron, which is significantly different from that of cobalt. (The Fermi level in iron is well below the edge of the d band.) Also, there is always the possibility that the s-d effect is masked by something else, such as the alignment of terbium moments with the applied field, which, according to Kasuya's model,[5] could result in reduced resistivity.

18.1.4. Measurements on $Tb_{24}Fe_{76}$ sample

The measurement results for this 1040 Å-thick sample, which is sputtered onto a glass substrate, are shown in Fig. 18.7. The net magnetization of the film M_s is 54 emu/cm^3, and its anisotropy field H_k as extracted from the VSM data is $\simeq 90$ kOe. When saturated along the easy axis, the sample exhibits a zero-field resistivity ρ of 370 $\mu\Omega$ cm, a Hall resistivity ρ_{xy} of 9.4 $\mu\Omega$ cm, and a Kerr rotation angle θ_k of 0.27°. The value of the anisotropy field H_k derived from the in-plane Hall signal in Fig. 18.7(c) is approximately 105 kOe.

The perpendicular MR curve has the usual peaks in the vicinity of H_c, but does not have a significant slope. This might indicate that the dispersion of Tb moments, which, in this Tb-rich sample, could give rise to a negative slope, is fairly small. Alternatively, the ordinary MR effects may be present, masking the minute effects of Tb moment alignment.

18.2. Polarized-light Microscopy

The physical phenomenon that makes possible the readout of domains from magneto-optical disks also enables the observation of magnetic patterns and domain structure under a polarized-light microscope. Optical microscopy has thus become an invaluable tool in MO media research.[14,15] The schematic diagram of one such microscope is shown in Fig. 18.8. Here a white light source is used to illuminate the sample through the objective lens, and the reflected light, after passing through the same objective, carries its polarization information to the viewer. The viewer may be looking through the eye-piece or at a monitor hooked up to the TV camera at the end of the microscope tube. Since the light source is an extended incoherent source, and because the objective has a reasonably wide field of view, the illuminated area at the sample will be fairly large, and a high-quality image of this area will be projected into view. Depending on the objective used, the field of view can be tens or hundreds of microns in diameter. The polar MO Kerr effect being the basis of operation of this device, the incident polarization must be linear and the reflected beam must pass through a crossed analyzer in order to produce high-contrast images of magnetic domains.

In addition to the standard elements of a commercial microscope, the system depicted in Fig. 18.8 contains a few elements that have been designed specifically for MO media research. These include a laser diode, an electromagnet, and a hot plate. The hot plate allows control of the ambient temperature, so that domain patterns at elevated temperatures may be observed. The electromagnet produces a perpendicular field at the sample,† thus enabling the writing and erasure of domains as well as the expanding and contracting of them. The laser beam, reflecting off a dichroic mirror and brought to focus through the objective, is used for thermomagnetic writing on the sample. The system has been automated by placing the hot plate, the electromagnet, the diode laser, and the TV camera under computer control.

18.2.1. Observations and discussion

Typical images of demagnetized MO films as seen through a polarized-light microscope are shown in Fig. 18.9; the black and white regions correspond to "up" and "down" magnetized domains. The applied field in

† A simple iron core wrapped with a few layers of wire can easily produce a field around 5 kOe with 10–20 amperes of current.

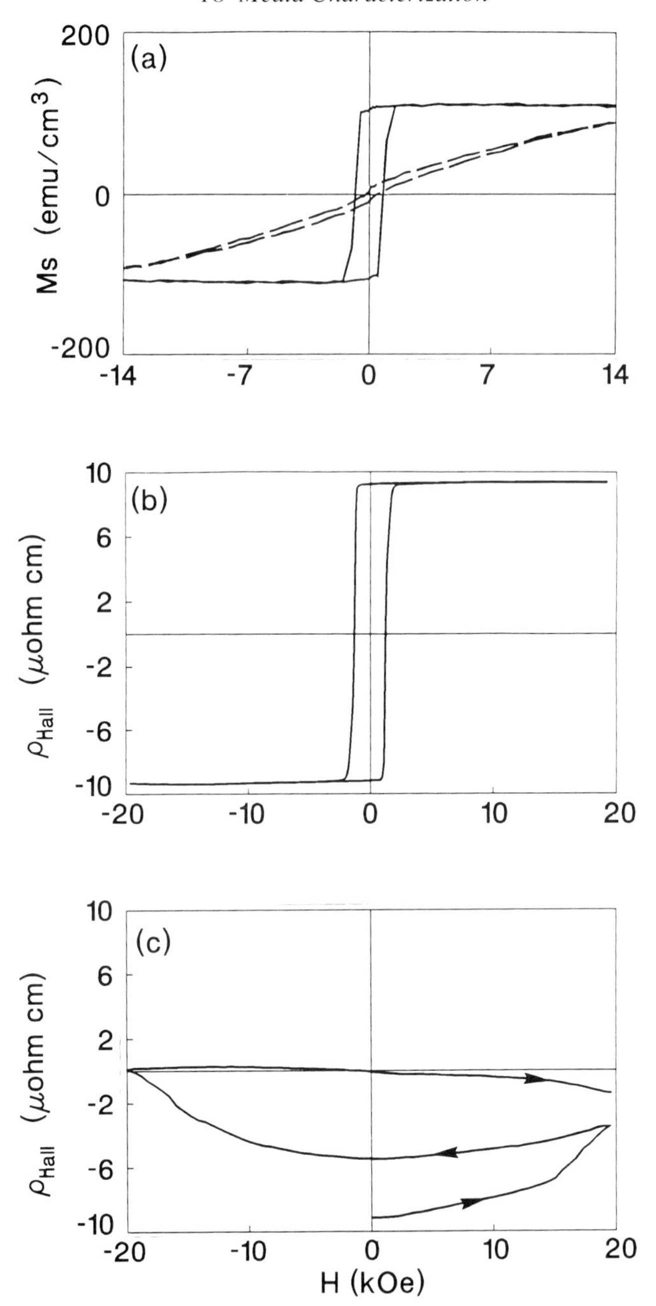
(a)
Ms (emu/cm^3)
200
0
-200
-14
-7
0
7
14
(b)
ρ_{Hall} (μohm cm)
10
6
2
-2
-6
-10
-20
-10
0
10
20
(c)
ρ_{Hall} (μohm cm)
10
6
2
-2
-6
-10
-20
-10
0
10
20
H (kOe)

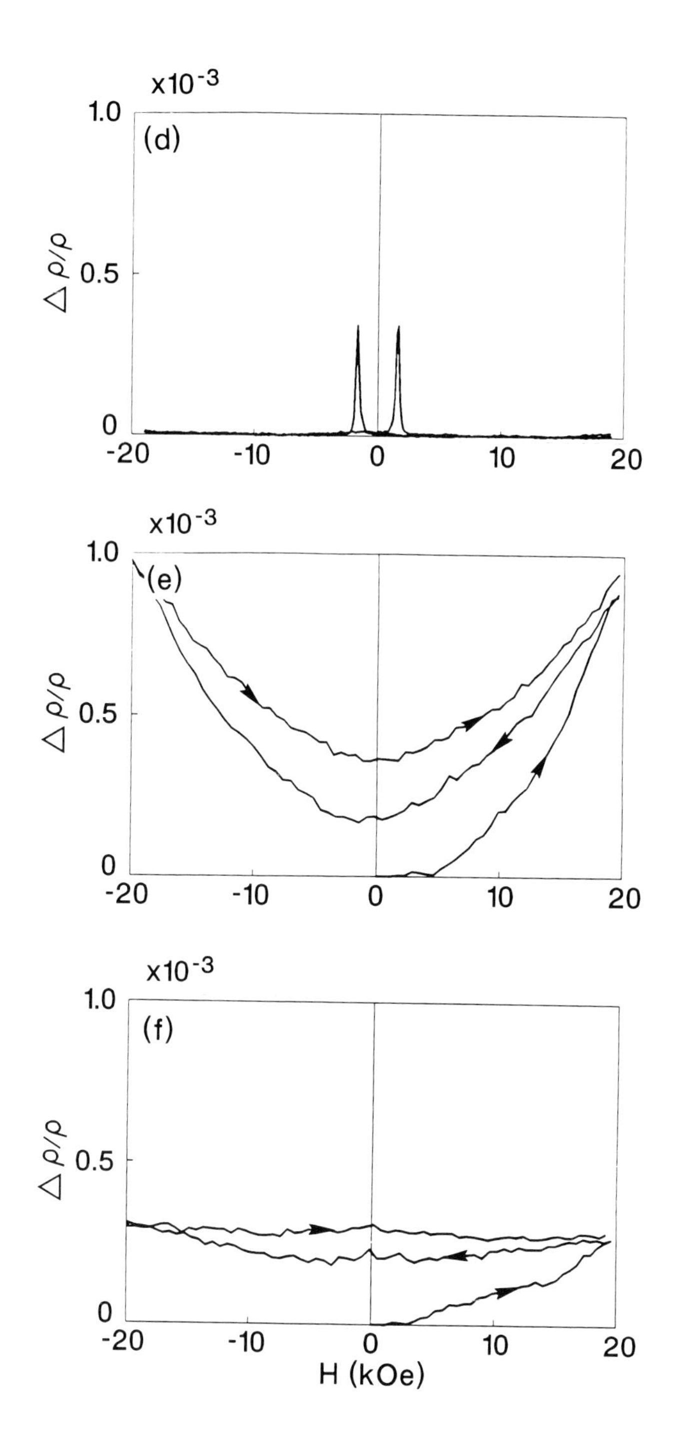

Figure 18.6. Various measurements on a 1350 Å-thick $Tb_{28}Fe_{72}$ sample. (a) VSM measurement with the magnetic field perpendicular (solid curve) and in the plane of the sample (broken curve). (b) Hall resistivity versus the magnitude of an applied perpendicular field. (initial state = saturated along the easy axis). (c) Hall resistivity versus the magnitude of an applied in-plane field (initial state = saturated along the easy axis). (d) Resistivity variation as a function of the magnitude of an applied perpendicular field. (e) Resistivity variation with the magnitude of the applied field; longitudinal geometry (initial state = saturated along the easy axis). (f) Resistivity variation with the magnitude of the applied field; transverse geometry.

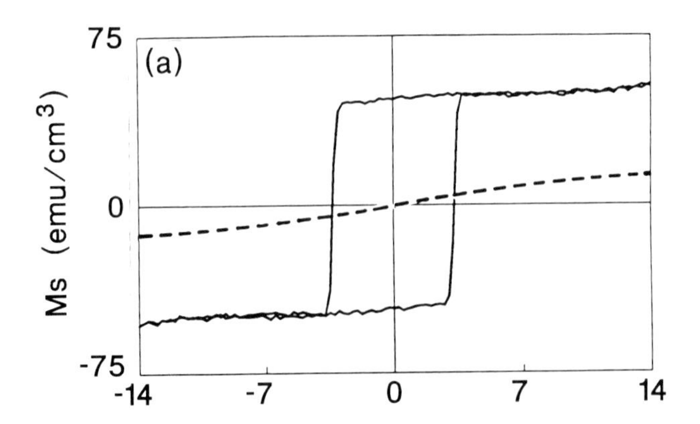
75
(a)
Ms (emu/cm³)
0
-75
-14
-7
0
7
14

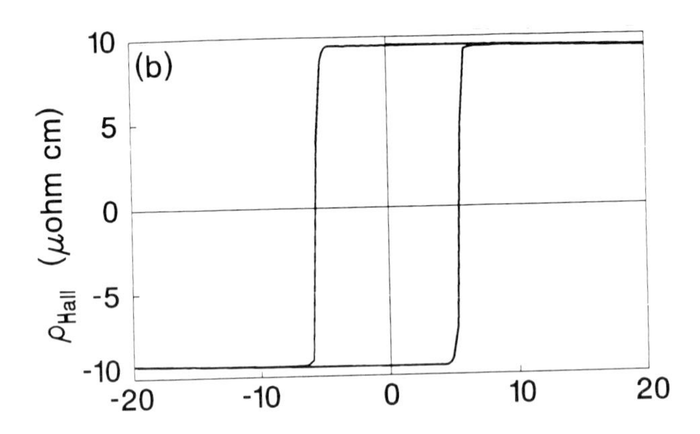
10
(b)
5
0
-5
-10
ρHall (μohm cm)
-20
-10
0
10
20

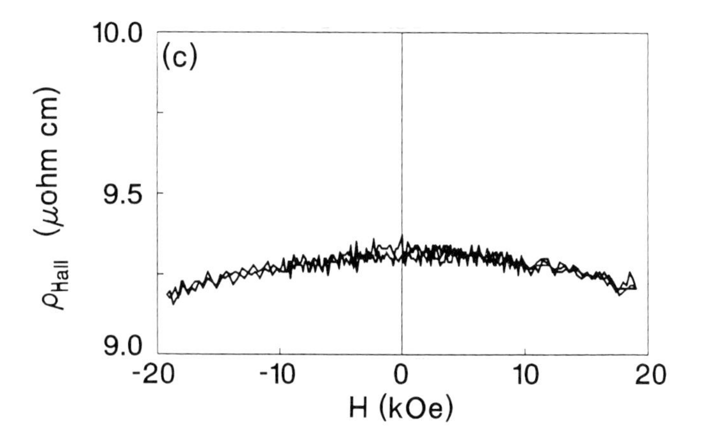

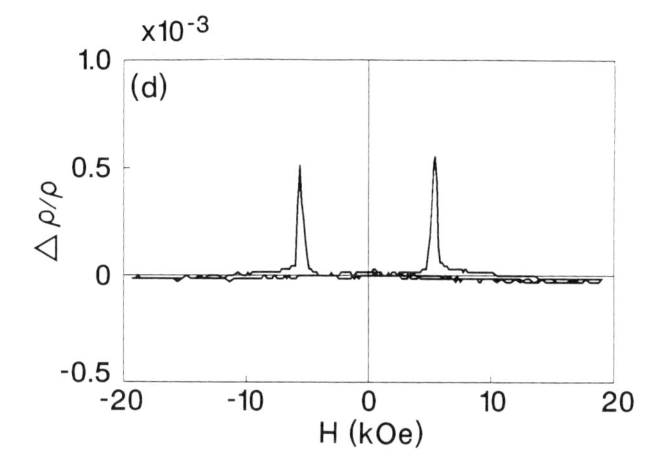

Figure 18.7. Various measurements on a 1040 Å-thick $Tb_{24}Fe_{76}$ sample. (a) VSM measurement with the magnetic field perpendicular (solid curve) and in the plane of the sample (broken curve). (b) Hall resistivity versus the magnitude of an applied perpendicular field. (c) Hall resistivity versus the magnitude of an applied in-plane field (initial state = saturated along the easy axis). (d) Resistivity variation as a function of the magnitude of an applied perpendicular field.

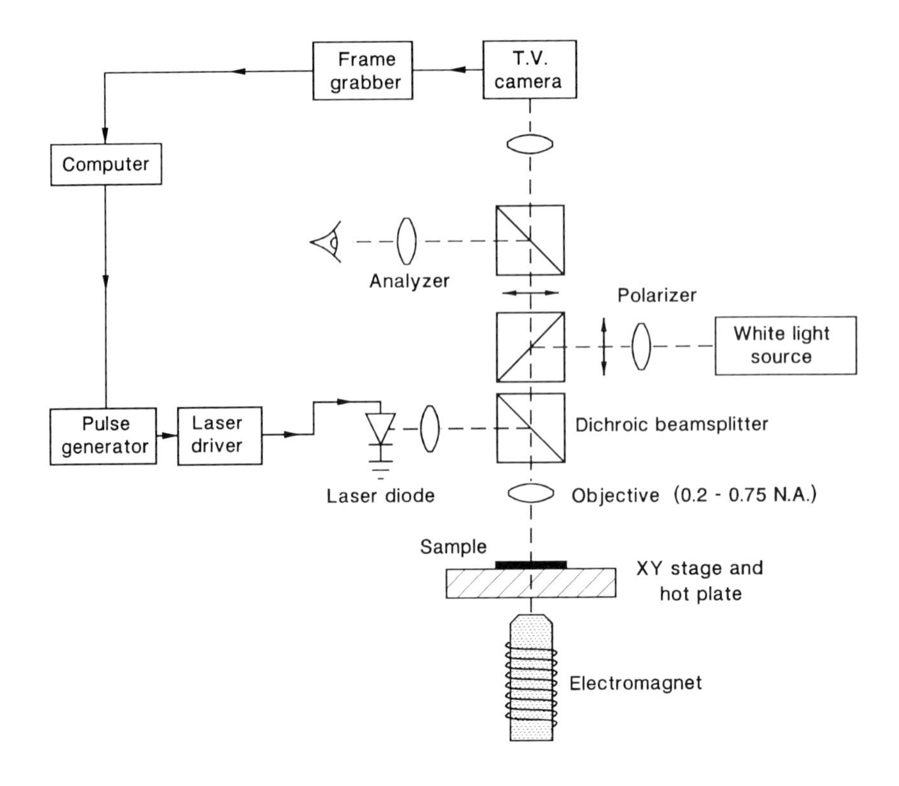

Figure 18.8. Schematic diagram of a modified polarization microscope used for observation of domain growth and domain structure.

each case was close to the coercivity of the sample, the NA of the microscope objective was 0.85, and the analyzer was adjusted for maximum contrast. In general, optical microscopy is incapable of resolving features of domains and domain boundaries smaller than 0.5 μm, but is quite appropriate for larger features. Other examples of polarization microscopy on MO media are given in Fig. 18.10. Here small domains, initially recorded with the laser beam, have been expanded under an applied field from the electromagnet. The field is slowly increased until the domain wall expands to a diameter of several tens of microns; it is then reduced to zero and the stable domain remaining behind is photographed. The four pictures in Fig. 18.10 have been obtained in a similar fashion and from the same area of a $Tb_{20}(FeCo)_{80}$ sample, but at the ambient temperatures of 25, 50, 75, and 100 °C respectively. The differences between these domains can be attributed to the varying competition between demagnetization energy and domain wall energy.

An interesting property of MO media is that samples that are similar in terms of composition, thickness, hysteresis loop, coercivity, Kerr angle, etc., sometimes exhibit markedly different degrees of domain jaggedness

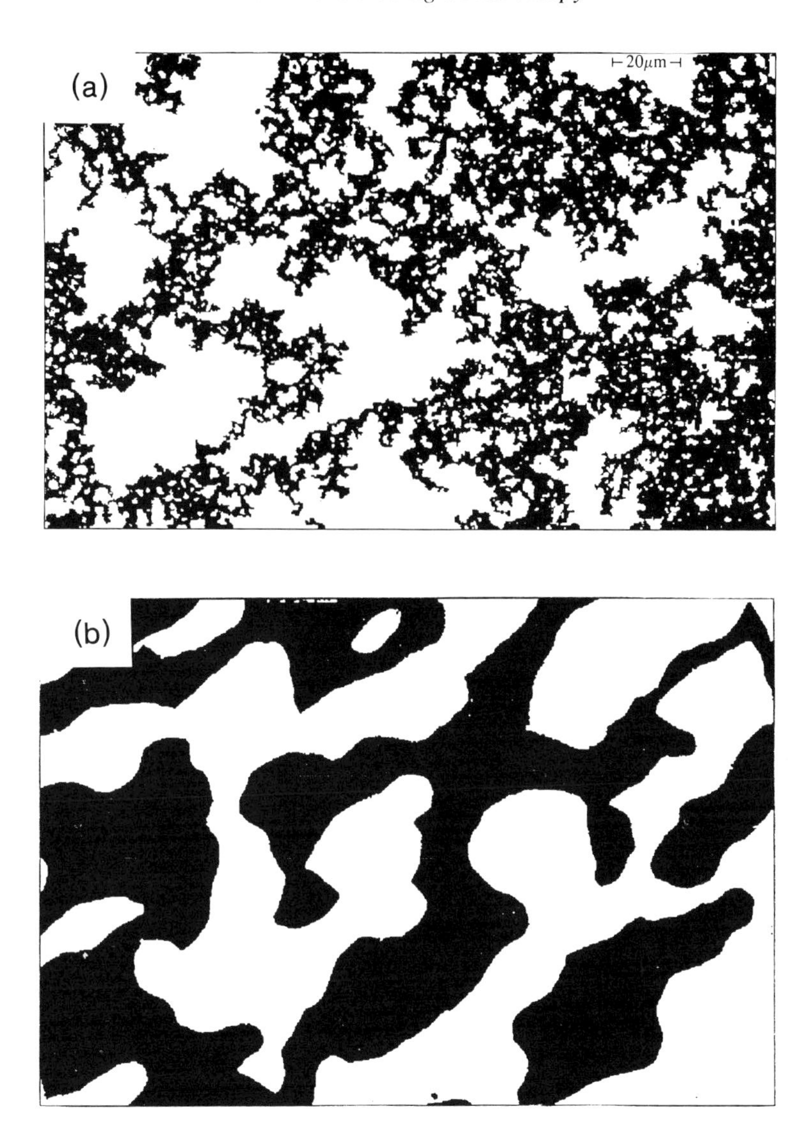

Figure 18.9. Demagnetized states of (a) Co(3 Å)/Pt(8 Å) and (b) $Tb_{26}(FeCo)_{74}$ thin-film samples.

under a microscope. This feature is related to their magnetic microstructure and arises from differences in the deposition environment, the substrate condition, and the post-deposition treatment of the samples. For instance, in Fig. 18.11, which shows two expanded domains side by side, the domain in the $Tb_{17}Fe_{60}Co_7Ar_{16}$ film is fairly smooth whereas that in the $Tb_{23}Fe_{58}Co_9Ar_{10}$ sample is not. Although the balance between the

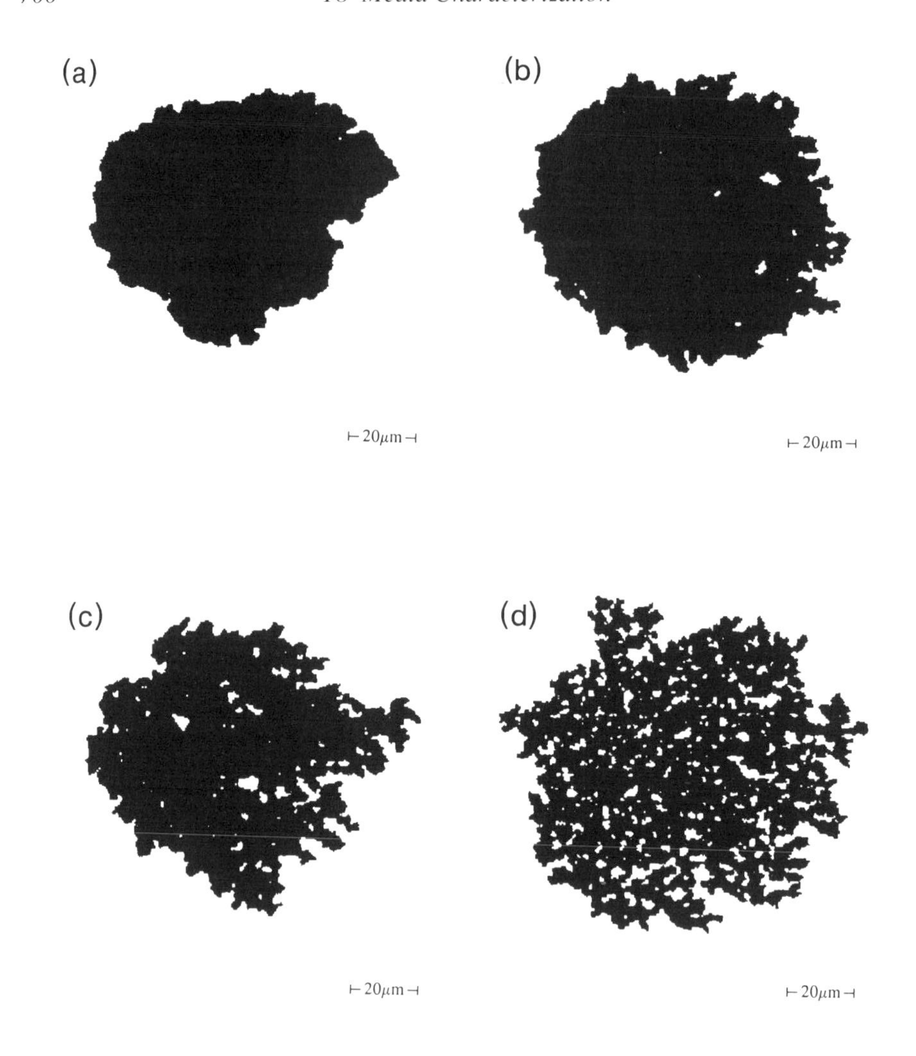

Figure 18.10. Expanded domains in a $Tb_{20.3}(FeCo)_{79.7}$ sample. (a) T_{amb} = 25 °C; minimum field required for domain growth = 3.87 kOe. (b) T_{amb} = 50 °C; minimum field = 2.61 kOe. (c) T_{amb} = 75 °C; minimum field = 1.67 kOe. (d) T_{amb} = 100 °C; minimum field = 0.99 kOe.

demagnetization and wall energies is partly responsible for the observed behavior in this case, differing fabrication conditions (i.e., sputtering gas pressure, substrate bias, underlayer roughness, etc.) cannot be ruled out as the source of observed differences, since these factors contribute to the formation of local barriers to domain wall propagation.

A possible way to quantify the degree of domain jaggedness is suggested by the concepts of fractal geometry.[34] Consider measuring the length of a domain wall with a ruler of fixed length ℓ. If the ruler is short,

Figure 18.11. Expanded domains in two different samples of TbFeCo. In (a) the material composition is $Tb_{17}Fe_{60}Co_7Ar_{16}$, while in (b) it is $Tb_{23}Fe_{58}Co_9Ar_{10}$.

it can follow the turns and twists of the wall and yields a large value for the perimeter L. A long ruler, on the other hand, will miss the zigzags and yields a smaller value for L. Within a certain range of ℓ, the plot of $L(\ell)$ on a log–log scale is usually linear with a negative slope of $-\alpha$, as shown in Fig. 18.12. In other words,

$$L \simeq L_0(\ell/\ell_0)^{-\alpha} \tag{18.8}$$

A fairly smooth domain will have $\alpha \simeq 0$, while a highly jagged domain will have $\alpha \simeq 1$. The fractal dimension is defined as $D = 1 + \alpha$. The more jagged the domain, the higher its fractal dimension will be. For example, the domain on the left-hand side in Fig. 18.11 has dimension $D = 1.05$, whereas that on the right has $D = 1.2$.

18.3. Lorentz Electron Microscopy

Lorentz microscopy is a powerful tool for high-resolution studies of magnetic structure in thin films.[16–21] Figure 18.13 shows Lorentz images of domain structure observed in MO media. (See also Figs. 1.17(b) and 17.1(a), (b), (c).) The underlying physical mechanism of the various modes of Lorentz microscopy is the interaction between the propagating electron wave and the magnetic vector potential field. For a given electron trajectory, the interaction, known as the Aharonov–Bohm effect, results in a phase delay directly proportional to the path integral of the vector potential.[35] Lorentz microscopy is therefore a branch of phase-contrast

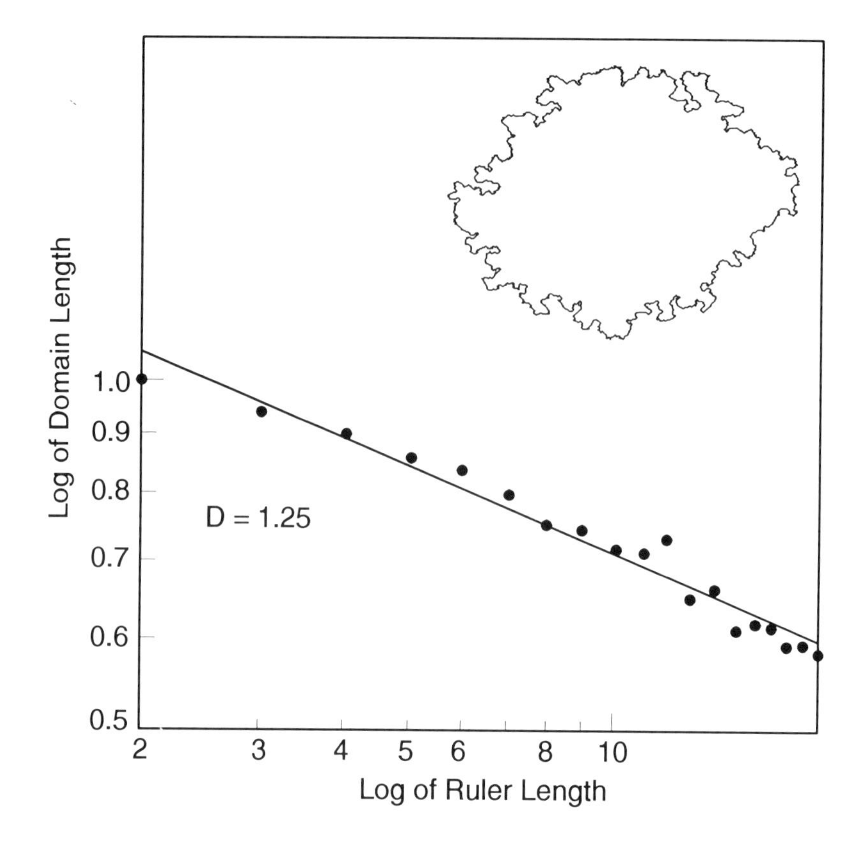

Figure 18.12. The log–log plot of perimeter L versus ruler length ℓ for the domain boundary shown in the inset. The solid line is the best fit to the data points and the extracted fractal dimension is $D = 1.25$.

microscopy whose various modes (e.g., Fresnel, Foucault, differential phase contrast, small angle diffraction, electron interference and holography) simply represent different designs for capturing the information contained in the phase of the beam after passage through a magnetic specimen. In this section we describe a technique for computing the phase imparted to the electron beam by an arbitrary, two-dimensional pattern of magnetization, and present numerical results of this analysis.

18.3.1. Mathematical analysis

Consider a magnetic film parallel to the XY-plane with magnetization distribution $\mathbf{M}(x, y)$ and thickness τ, as shown in Fig. 18.14. The vector potential $\mathbf{A}(x, y, z)$ corresponding to this magnetic pattern was calculated in Chapter 13; see Eq. (13.21). The phase modulation imparted to the electron beam after passage through the film is obtained by integrating $\mathbf{A}(x, y, z)$

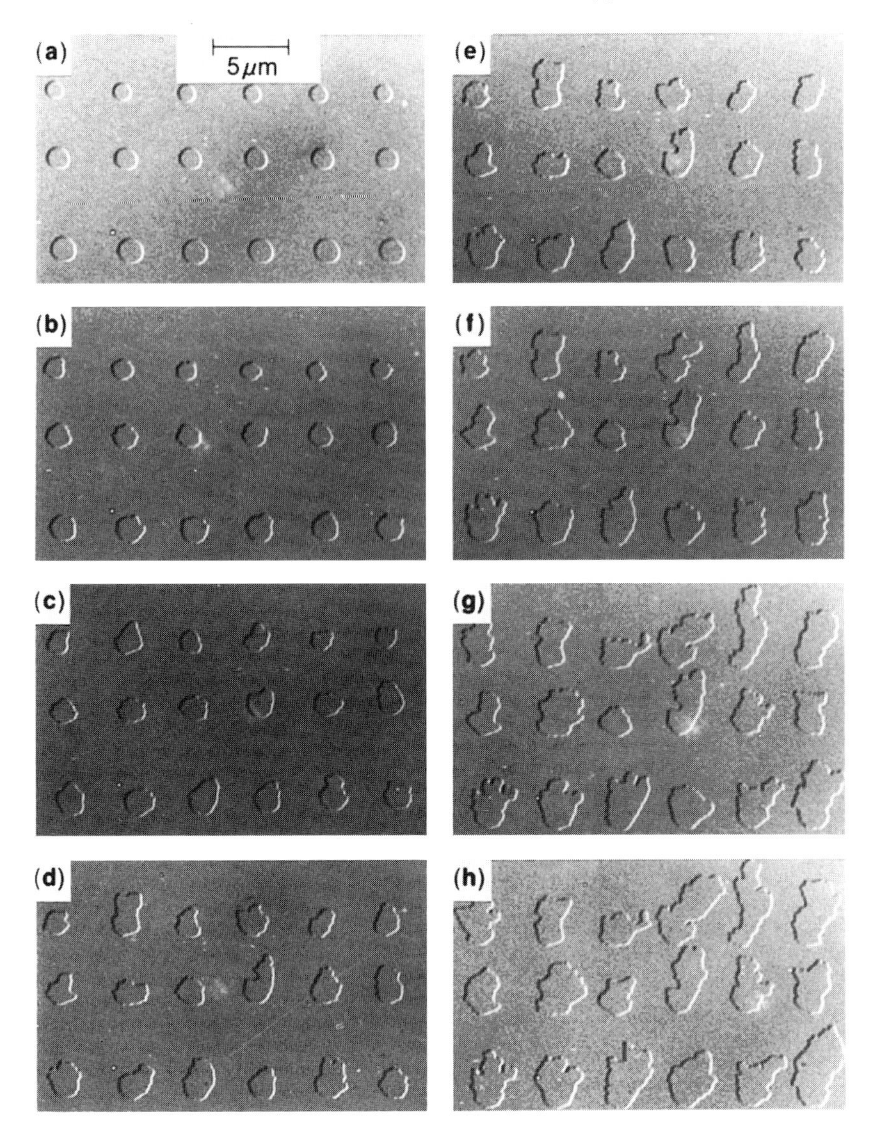

Figure 18.13. A sequence of Lorentz micrographs showing expansion of an array of written domains in $Tb_{19}Fe_{81}$. Source: C.J. Lin and D. Rugar, Observation of domain expansion and contraction in TbFe films by Lorentz microscopy, *IEEE Trans. Mag.* **24**, 2311 (1988). Courtesy of C.J. Lin, IBM Almaden Research Center.

along the electron trajectories. Denoting the direction of propagation by the unit vector **p**, the imparted phase $\Phi(x, y)$ will be

$$\Phi(\mathbf{r}) = \frac{e}{\hbar c} \int_{-\infty}^{\infty} \mathbf{p} \cdot \mathbf{A}(\mathbf{r} + \ell\mathbf{p})\, d\ell \tag{18.9}$$

Here $\mathbf{r} = (x\hat{\mathbf{x}} + y\hat{\mathbf{y}})$ is the intersection of an arbitrary electron trajectory with the XY-plane at $z = 0$, e is the electronic charge, $\hbar$ is Planck's constant divided by 2π, and c is the speed of light. Substituting for $\mathbf{A}$ from Eq. (13.21), and evaluating the integral in Eq. (18.9) yields

$$\Phi(x,y) = \frac{2e}{\hbar c} \sum_{m=-\infty}^{\infty} \sum_{\substack{n=-\infty \\ m,n \neq 0}}^{\infty} i \frac{\tau}{s} G_{\mathbf{p}}(\tau s)\, (\boldsymbol{\sigma} \times \hat{\mathbf{z}}) \cdot \left[\mathbf{p} \times (\mathbf{p} \times \mathcal{M}_{mn})\right] \times \exp\left[i2\pi\left(\frac{mx}{L_x} + \frac{ny}{L_y}\right)\right] \tag{18.10a}$$

where

$$G_{\mathbf{p}}(\tau s) = \frac{1}{(\mathbf{p}\cdot\boldsymbol{\sigma})^2 + (\mathbf{p}\cdot\hat{\mathbf{z}})^2} \; \frac{\sin\left(\dfrac{\pi\tau s\,\mathbf{p}\cdot\boldsymbol{\sigma}}{\mathbf{p}\cdot\hat{\mathbf{z}}}\right)}{\dfrac{\pi\tau s\,\mathbf{p}\cdot\boldsymbol{\sigma}}{\mathbf{p}\cdot\hat{\mathbf{z}}}} \tag{18.10b}$$

This is the general expression for the phase of the beam after passage through the sample. Some interesting features of this function are summarized below.

(i) The factor $1/s$ in Eq. (18.10a) appears to discriminate against high spatial frequencies. This appearance, however, is deceiving. The local deflection of the electron beam is proportional to the gradient of the phase function. Since for a sinusoidal function the gradient is proportional to the frequency, the $1/s$ factor maintains the balance amongst the various Fourier components in their contributions to the deflection of the electrons.

(ii) In the expression for $\Phi(x,y)$ all Fourier components $\mathcal{M}_{mn}$ of the magnetization distribution appear in $\mathbf{p} \times (\mathbf{p} \times \mathcal{M}_{mn})$. Since $\mathbf{p}$ is a constant unit vector, one might as well begin the analysis by Fourier transforming $\mathbf{p} \times (\mathbf{p} \times \mathbf{M}(x,y))$ instead of $\mathbf{M}(x,y)$. In this way it immediately becomes clear that the projection of $\mathbf{M}(x,y)$ along the propagation direction $\mathbf{p}$ makes no contribution to $\Phi(x,y)$. What is more, since in Eq. (18.10a) the vector $\boldsymbol{\sigma} \times \hat{\mathbf{z}}$ is dot-multiplied by the Fourier coefficients of $\mathbf{p} \times (\mathbf{p} \times \mathbf{M}(x,y))$, the latter vector's component along Z plays no role in the outcome and may also be discarded. One thus retains only the components of $\mathbf{p} \times (\mathbf{p} \times \mathbf{M}(x,y))$ along X and Y for further processing.

(iii) The part of the magnetization distribution $\mathbf{M}(x,y)$ that survives the initial processing steps described above may still fail to contribute to $\Phi(x,y)$. For instance, if $\mathbf{p} \times (\mathbf{p} \times \mathcal{M}_{mn})$ happens to be parallel to its corresponding frequency vector $\boldsymbol{\sigma}$, the dot product in Eq. (18.10a) vanishes.

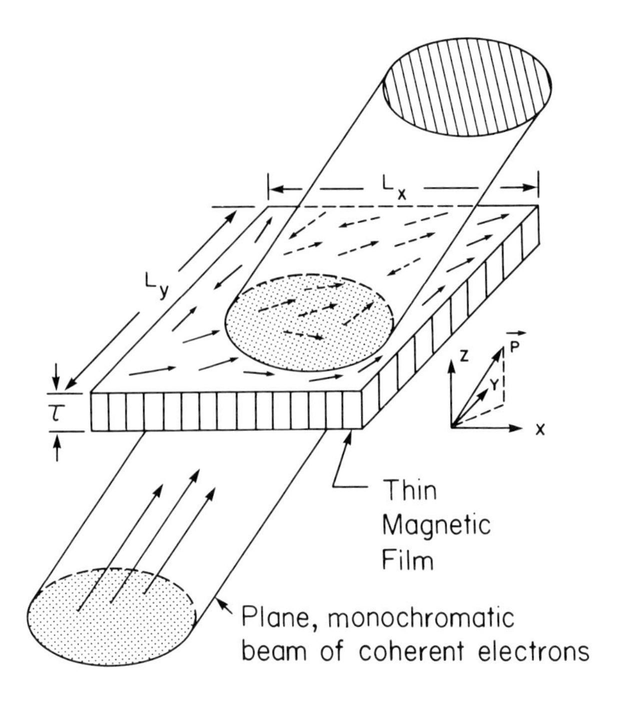

Figure 18.14. Schematic depiction of the magnetic film and the electron beam in Lorentz microscopy. The plane of the film is XY, and the unit vector $\mathbf{p}$ signifies the beam's propagation direction.

Thus the various spatial frequencies contained in the magnetization pattern receive unequal representation in the phase function.

(iv) When the electron beam is normally incident on the sample (i.e., when $\mathbf{p} = \hat{\mathbf{z}}$) the function $G_{\mathbf{p}}(\tau\mathbf{s})$ is equal to unity for all spatial frequencies $\mathbf{s}$. At all other angles of incidence, $G_{\mathbf{p}}(\tau\mathbf{s})$ attenuates certain spatial frequencies relative to others. Notice that the film thickness τ appears only in $G_{\mathbf{p}}(\tau\mathbf{s})$ as a scaling factor for $\mathbf{s}$. Thus for a given magnetization distribution and a given (oblique) direction of incidence, thinner films convey the information content of the high-frequency terms better than thick films.

(v) The zero-frequency term $\mathscr{M}_{00}$ does not appear in Eq. (18.10), indicating that a uniformly magnetized film will cause no deflection of the electron beam. This is contrary to our expectations based on the Lorentz law of force, which predicts a net deflection angle proportional to the in-plane component of magnetization. Inspection of the zero-frequency term $\mathcal{A}_{00}$ in Eq. (13.21f), however, reveals that this term is an odd function of z. Thus the part of the path within the region $z < 0$ cancels the contribution to the phase made by the part that lies in $z > 0$. Of particular interest here is the case of normal incidence (i.e., $\mathbf{p} = \hat{\mathbf{z}}$) where the zero-frequency vector potential everywhere is orthogonal to the path, making no contribution whatsoever to the phase function. The absence of the zero-frequency term from the phase function, however, has no practical significance, since in practice the sample dimensions are always finite and the zero-frequency term is inevitably replaced by low-frequency terms, which continue to obey the Lorentz law of force.

18.3.2. Numerical results and discussion

To gain a better understanding of Lorentz microscopy, let us first consider the simple case of a magnetic film whose in-plane magnetization (oriented along Y) has the following distribution:

$$\mathbf{M}(x, y) = M_s \cos(2\pi x / L_x)\, \hat{\mathbf{y}} \tag{18.11a}$$

Provided that L_x is sufficiently large, this magnetization will appear uniform in the vicinity of the origin. For the above distribution, $\Phi(x, y)$ is readily computed from Eqs. (18.10). For a normally incident beam we have

$$\Phi(x, y) = -\frac{2e\tau M_s}{\hbar c} \frac{\sin(2\pi x / L_x)}{1/L_x} \tag{18.11b}$$

If the electron beam is confined to the neighborhood of the origin, one may replace the sine function in Eq. (18.11b) by its argument to obtain

$$\Phi(x, y) \simeq -\frac{4\pi e\tau M_s}{\hbar c}\, x \tag{18.11c}$$

In the Gaussian system of units $e = 4.80325 \times 10^{-10}$ esu, $c = 2.99793 \times 10^{10}$ cm/s, and $\hbar = 1.05459 \times 10^{-27}$ erg s. For a film of thickness $\tau = 60$ nm and saturation moment $M_s = 1000$ emu/cm^3, the phase of the electron beam upon transmission through the sample is therefore given by

$$\Phi(x, y) = -1.146 \times 10^6 x \tag{18.11d}$$

(Φ is in radians and x in centimeters.) Now, let the beam's kinetic energy be $E_k = 100$ keV. The electron wavelength λ is computed from the formula

$$\lambda = \frac{h}{\sqrt{2mE_k + (E_k/c)^2}} \tag{18.12}$$

to be 0.037 Å.† The phase function in Eq. (18.11d) may now be written:

$$\Phi(x,y) = \frac{2\pi}{\lambda}(67.5 \times 10^{-6})\,x \qquad (18.13)$$

Accordingly, the deflection angle for a uniform, normally incident beam on this sample is 67.5 microradians.

In the following example a case of practical interest is explored. Here $L_x = L_y = 1.28\ \mu\text{m}$, $\tau = 60$ nm, $M_s = 1000$ emu/cm³, and $\lambda = 0.037$ Å. The magnetization distribution is defined on a 256×256 square lattice with a lattice constant of 5 nm; periodic boundary conditions apply. The diffraction calculations are based on the scalar diffraction theory of Chapter 3, and require two Fourier transforms for the Fresnel mode of Lorentz electron microscopy.

Example 2. Figure 18.15(a) shows the magnetization distribution for a circular domain in a perpendicular medium. Within the domain the magnetization **M** is along $-Z$; outside it is along $+Z$. In the state of minimum magnetostatic energy shown here the wall moments everywhere are parallel to the wall. The computed $\Phi(x,y)$ for a domain of this type with a diameter of 0.64 μm and wall-width parameter Δ_w of 100 Å is shown in Fig. 18.15(b), and the corresponding Fresnel pattern with a defocus of 0.37 mm is shown in Fig. 18.15(c). The Fresnel pattern in this case is a bright ring whose radius, depending on the sense of magnetization within the wall, is either slightly larger or smaller than the radius of the magnetic domain itself. When the electron beam is incident at an oblique angle with $\mathbf{p} = (0.5, 0, 0.866)$, Figs. 18.15(d) and (e) are obtained. These figures show, respectively, the phase function $\Phi(x,y)$ and the Fresnel pattern with 0.37 mm of defocus. Note that the Fresnel pattern is no longer circularly symmetric. The brightness of the ring has increased on one side and decreased on the other, in agreement with experimental observations.[18]

A more complex wall structure for circular domains is depicted in Fig. 18.16(a). Here the wall is not in a state of minimum magnetostatic energy and shows several twists (i.e., vertical Bloch lines). The corresponding normal-incidence phase function and the Fresnel pattern with 0.37 mm of defocus are shown in Figs. 18.16(b) and (c) respectively. Frame (d) shows the Fresnel pattern at oblique incidence with $\mathbf{p} = (0.5, 0, 0.866)$. Compared with Fig. 18.15(e), the bright and dark halves of the ring are somewhat more pronounced in the present case. □

† In Eq. (18.12) h is Planck's constant and m is the electron mass. The denominator is the momentum of the electron, including the relativistic correction to the classical momentum–energy relation.

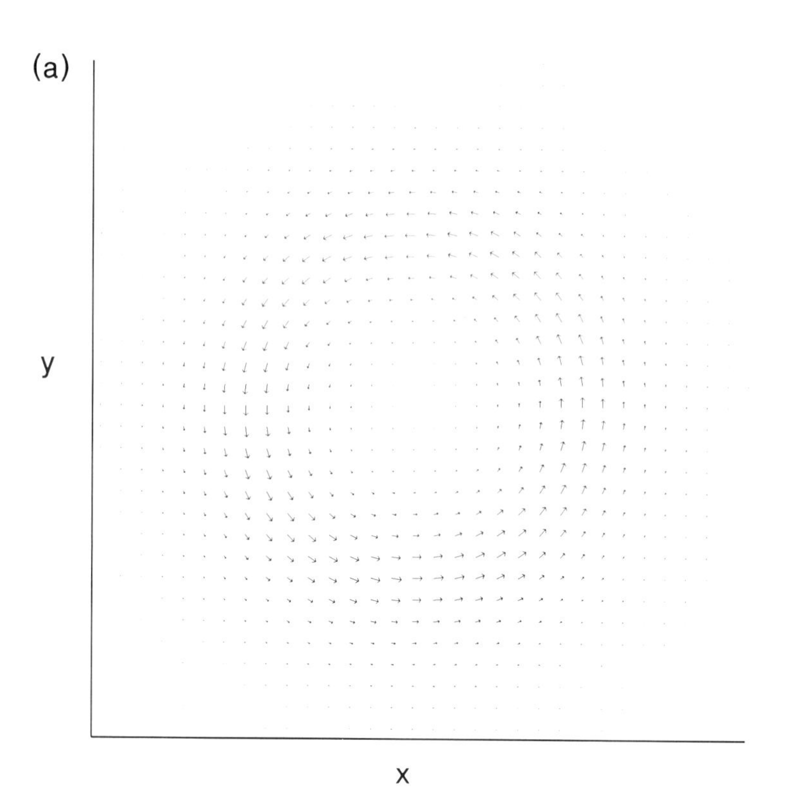
(a)
y
x

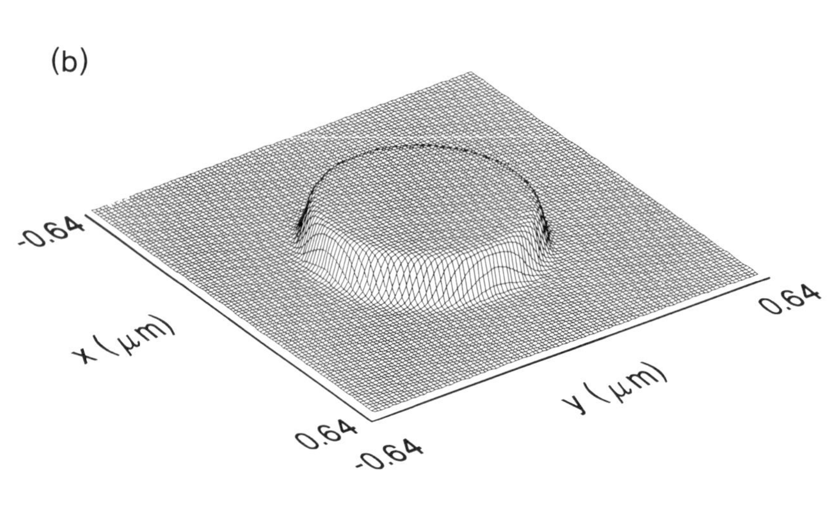
(b)
-0.64
0.64
x (μm)
y (μm)
0.64
-0.64

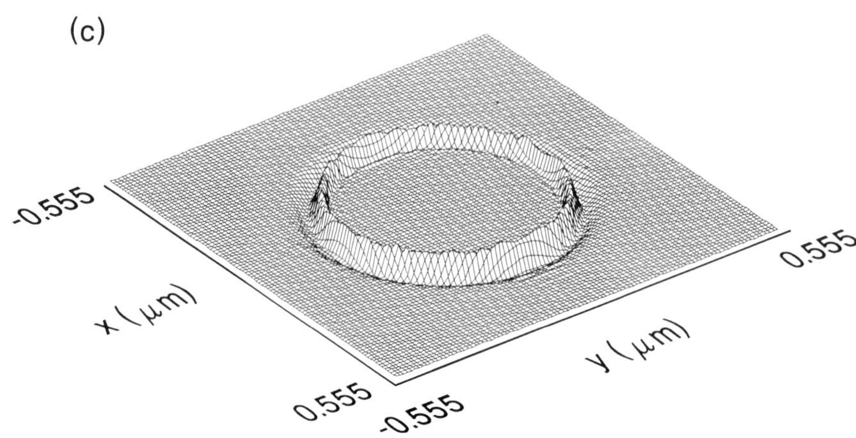
(c)
-0.555
0.555
x (μm)
y (μm)
0.555
-0.555

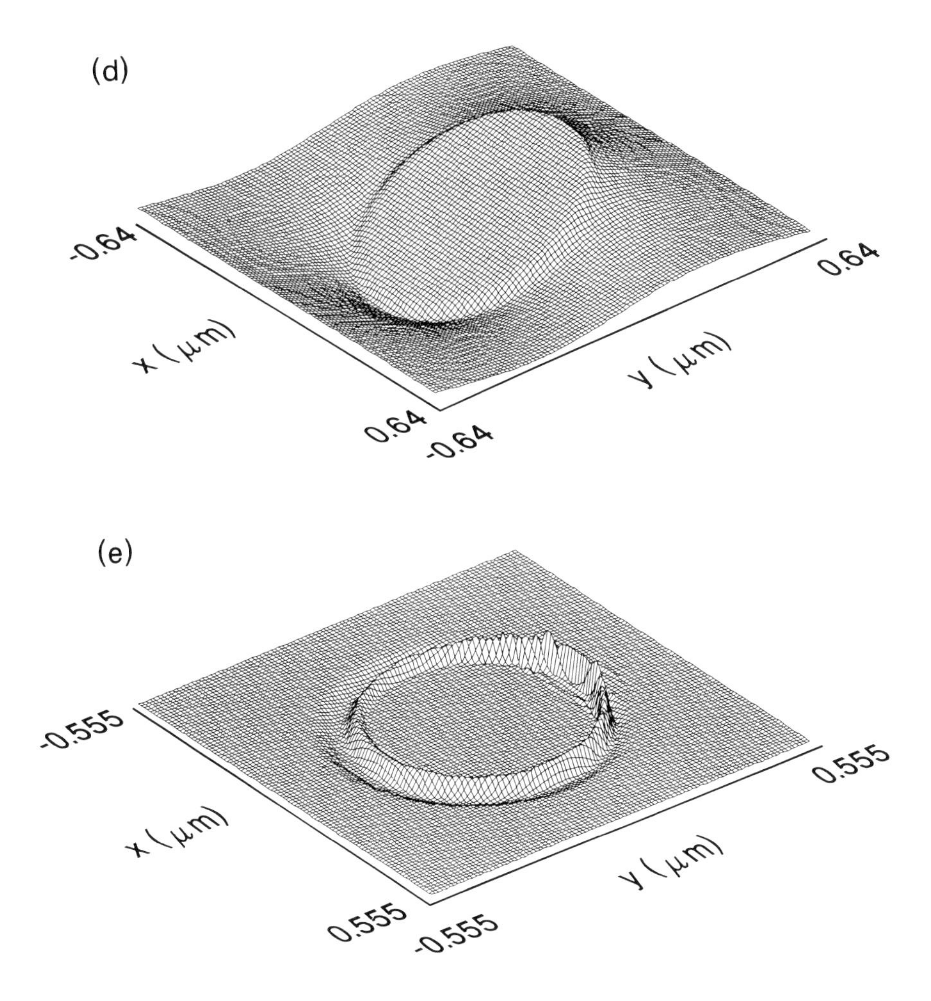

Figure 18.15. (a) Circular domain in a film with perpendicular magnetization. The wall magnetization everywhere is parallel to the wall. (b) Computed $\Phi(x,y)$ at normal incidence for a domain diameter of 0.64 μm and a wall-width parameter Δ_w of 100 Å. The minimum and maximum values of the function are -0.7 and +2.9 radians respectively. (c) Intensity distribution for the Fresnel pattern at 0.37 mm of defocus. (d) Computed $\Phi(x,y)$ at oblique incidence with $\mathbf{p}$ = (0.5, 0, 0.866). (e) Fresnel image at oblique incidence. The defocus distance is 0.37 mm.

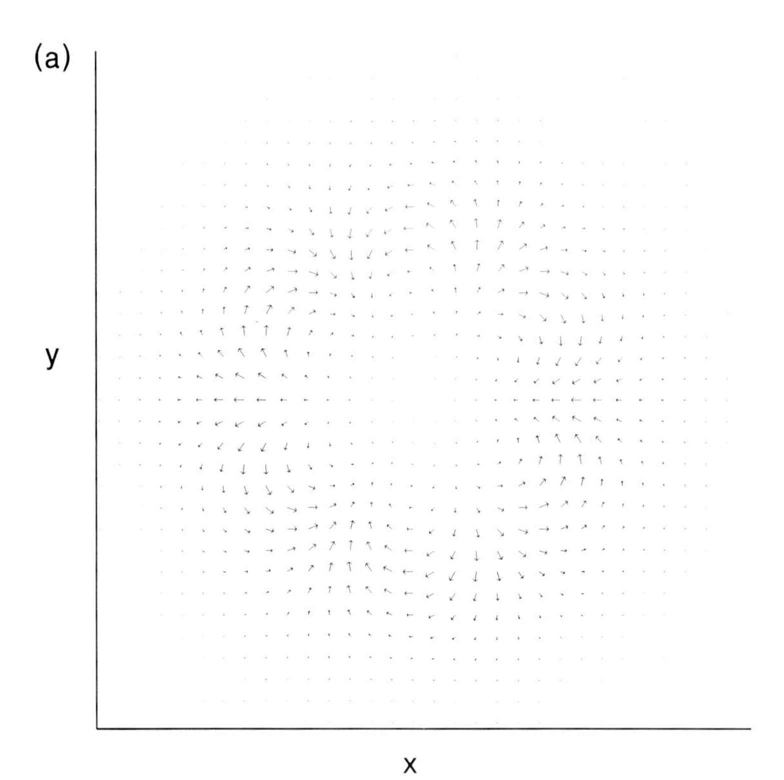
(a)
y
x

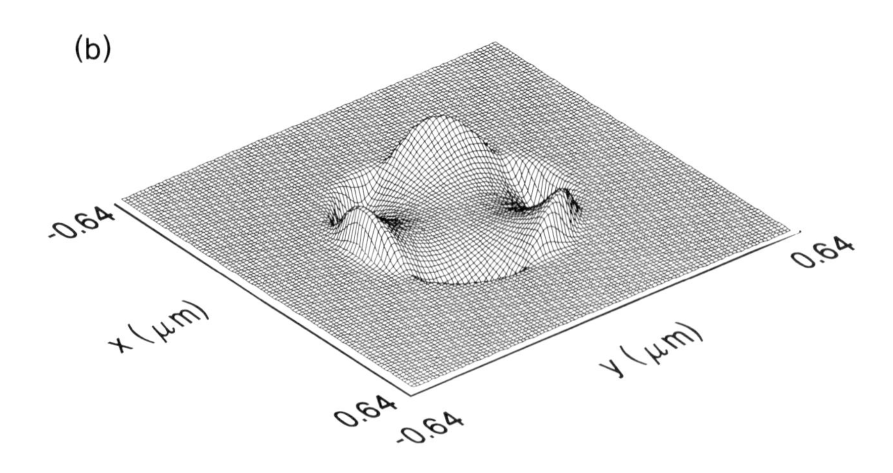
(b)
-0.64
x (μm)
0.64
-0.64
y (μm)
0.64

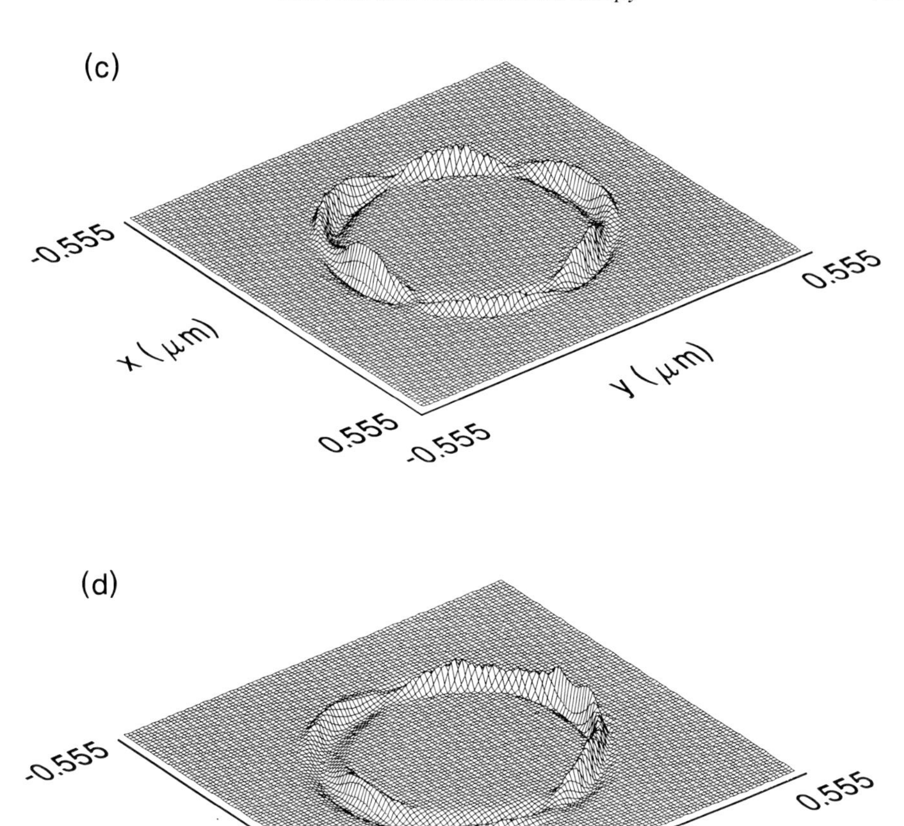

Figure 18.16. (a) Circular domain in a perpendicular film; the wall magnetization contains several vertical Bloch lines. (b) Computed $\Phi(x, y)$ at normal incidence for a domain diameter of 0.64 μm and a wall-width parameter $\Delta_w = 100$ Å. The minimum and maximum values of the function are ± 2.67 radians. (c) Fresnel pattern at normal incidence with 0.37 mm of defocus. (d) Fresnel image at oblique incidence; $\mathbf{p} = (0.5, 0, 0.866)$, defocus = 0.37 mm.

18.4. Magnetic Force Microscopy (MFM)

The trend in MO data storage is towards higher recording densities and faster transfer rates. In order to achieve high bit densities several factors must be taken into consideration. First, reverse-magnetized domains must be recorded whose dimensions are small and whose positions are precise. Second, the domains must be stable in the presence of thermal gradients and external magnetic fields. Third, the magnetization distribution within these domains must be uniform and their boundaries must be as smooth as possible if readout noise and jitter are to be minimized. Understanding thin-film magnetization processes with nanometer spatial resolution and nanosecond temporal resolution has thus become the focus of research activities in recent years.

The magnetic force microscope (MFM) is one of the few instruments capable of providing resolution below the wavelengths of visible light. In this instrument a sharp needle of some magnetic material (typically cobalt or nickel) is used to scan the surface of a magnetic film at a distance of several nanometers. Through the deflections of the needle, which are proportional to the magnetic force exerted on the tip, the MFM provides a picture of the sample's stray-field pattern. The relation between the measured deflections and the magnetization pattern is usually complicated, and useful quantitative information can be extracted only after extensive analysis of the data. In this section we present a model for magnetic force microscopy that can be used to interpret the measurement results.

18.4.1. Experimental observations

Figure 18.17 is a magnetic force micrograph, showing MO marks recorded on a Co/Pt disk with laser power modulation at various frequencies; the 1.6 μm track-pitch provides the scale of spatial dimensions. The image is obtained with a MFM whose needle is a cobalt-coated tungsten wire sharpened at one end by a chemical etching process. In operation, the needle is vibrated and its vibration phase, having been affected by the local force gradient, is used in a feedback loop to control the tip-to-sample separation with a piezo actuator. Thus a constant force gradient is maintained at the tip as the needle scans the surface. To ensure an attractive force during imaging, a small voltage is applied between the tip and the sample. As a consequence of the resulting electrostatic force gradient the image shows some surface topography (e.g., grooves) as well. The fine structure of the domains is clearly visible in this picture, a feat that could not be achieved with optical microscopy. (Other MFM images of MO media may be seen in Figs. 17.1(d), (e).)

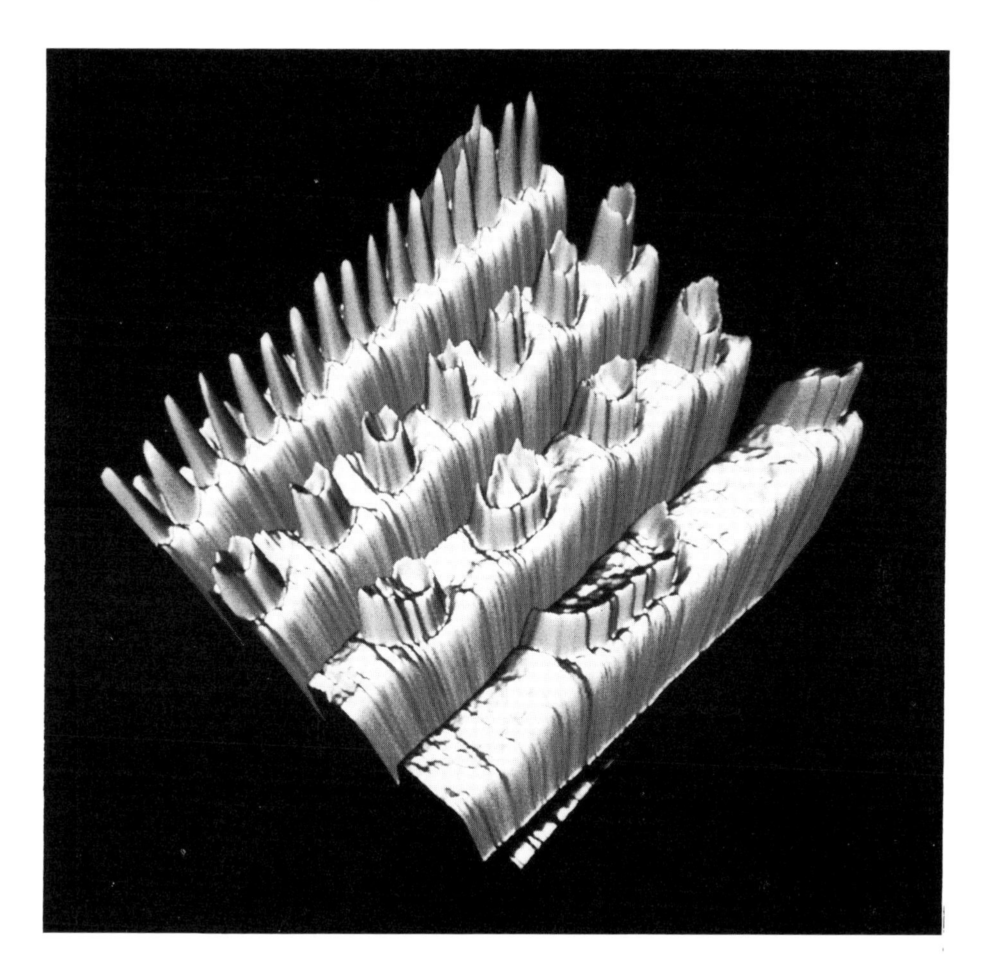

Figure 18.17. Magnetic force microscope image of a section of a Co/Pt disk, showing recorded domains with laser power modulation at various frequencies. From top to bottom, the domain periods are 0.5 μm, 1.25 μm, 2 μm, and 5 μm. Source: H.W. van Kesteren, G.J.P. van Engelen and B.A.J. Jacobs, Characterization of magneto-optical media by magnetic force microscopy, *J. Mag. Soc. Japan* **17**, supplement S1, 23–28 (1993). Courtesy of H.W. van Kesteren, Philips Research Laboratories, Eindhoven, The Netherlands.

18.4.2. A model for the needle in MFM and the method of force calculation

The model of the needle is shown in Fig. 18.18. The tip and the stem are treated separately as indicated. The cubes comprising the needle's tip are arranged in layers parallel to the XY-plane. Thus the needle's axis is along Z and the centers of the cubes in the kth layer are at z_k where

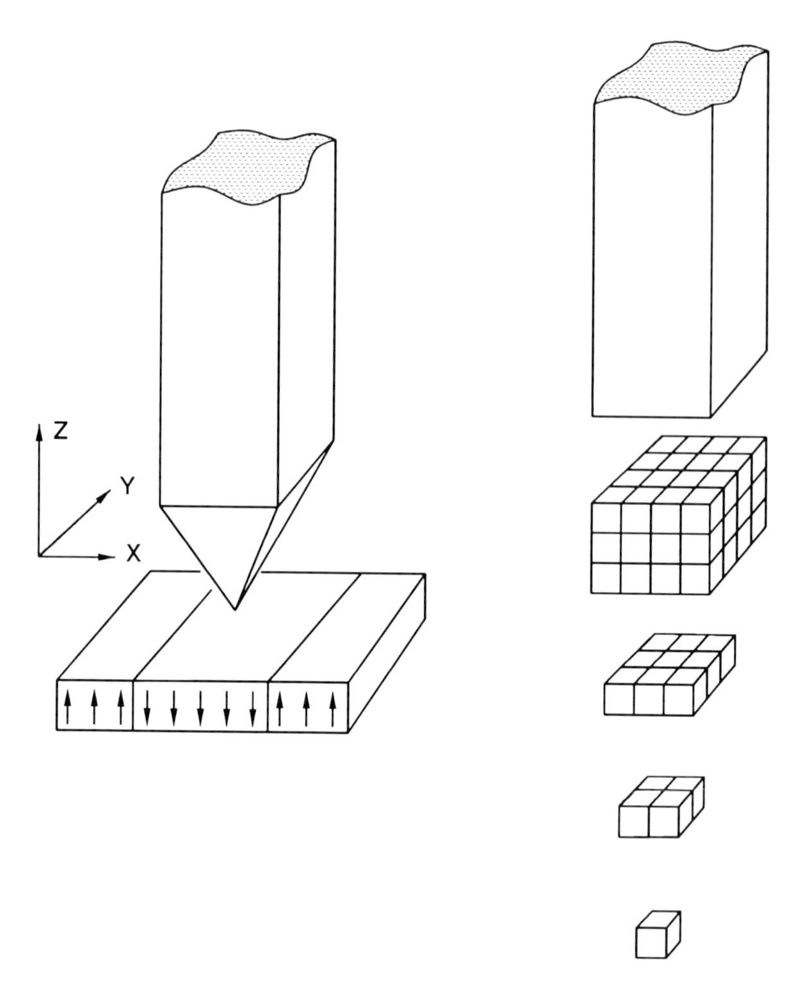

Figure 18.18. Magnetic needle scanning the surface of a magnetic film. The micromagnetic model treats the tip as a collection of cubical grains, assigning a dipole moment to each cube and adjusting the orientation of each dipole under the influence of an effective field. The stem of the needle is either non-magnetic, in which case it has no influence on the magnetization of the tip, or it is magnetized along the long axis of the needle, in which case it helps align the tip dipoles through the forces of exchange and the classical dipole–dipole interactions.

$$z_k = z_1 + (k - 1)\Delta \tag{18.14}$$

Here z_1 is the vertical coordinate of the center of the cube at the sharp end of the needle, and Δ is the linear dimension of the cubes. Within each layer the cubes are arranged in a square, centered on the needle's axis. The number of cubes in layer k will be N_k^2. Although no restrictions are

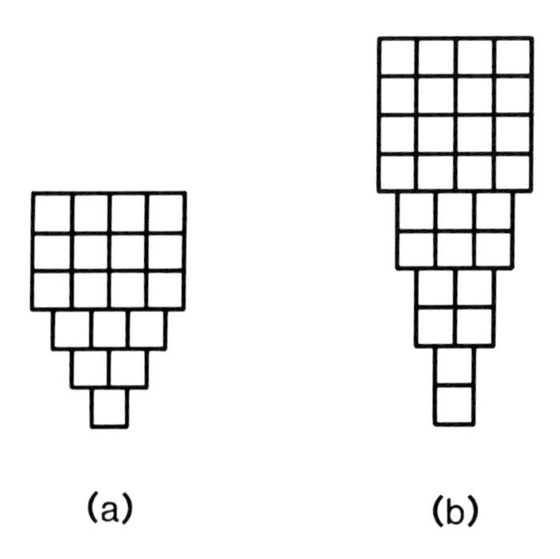

Figure 18.19. Two possible cross-sections of the needle tip.

imposed on the values of Δ and N_k, in practice Δ should correspond to the average grain (crystallite) size of the material, and N_ks must be chosen in accordance with the geometry of the tip. Figure 18.19 shows cross-sections of the tip for two sets of values of N_k. In (a) $N_1 = 1$, $N_2 = 2$, $N_3 = 3$, $N_4 = N_5 = N_6 = 4$, for a total of 62 cubes and a cone angle of $\Theta \simeq 53°$. In (b) $N_1 = N_2 = 1$, $N_3 = N_4 = 2$, $N_5 = N_6 = 3$, $N_7 = N_8 = N_9 = N_{10} = 4$, for a total of 92 cubes and a cone angle of $\Theta \simeq 28°$. Each cube corresponds to a crystallite in the actual needle; therefore, a single dipole of strength $\mathbf{m}$ is assigned to each cube. Assuming that the needle has saturation magnetization M_s, the assigned dipole moment's strength will be

$$|\mathbf{m}| = M_s \Delta^3 \tag{18.15}$$

The orientation of these moments in space is not arbitrary and depends on the following factors.

Local crystalline anisotropy. Each grain has one or more easy axes depending on its crystalline structure and shape. For the sake of simplicity, we shall assume that each dipole has uniaxial anisotropy of strength K_u. Both K_u and the direction of the easy axis for individual lattice sites are parameters that can be arbitrarily assigned.

Exchange at grain boundaries. Another factor that plays a role in the overall magnetization of the needle is the exchange interaction between neighboring grains at their common boundaries. This interaction is represented by an equivalent exchange field H_{xhg} that tends to align a given dipole with its nearest neighbor. When only a fraction of the adjacent surfaces of the two cubes make contact, the exchange field is proportionately reduced. In the absence of direct methods for measuring exchange, H_{xhg} will be treated as an adjustable parameter.

Classical dipole–dipole interactions. These interactions make a significant contribution to the forces that determine the state of magnetization of the needle. The dipolar field is long range and must be computed between each pair of dipoles. The fact that cubes of uniform magnetization are replaced by point dipoles is a source of some inaccuracy in the calculation of dipole–dipole interactions.[36] Nonetheless, given the nature of the model and the degree of approximation brought about by our other assumptions, the errors involved in these demagnetization calculations should be acceptable.

Magnetic field of the stem. The tip's magnetization is influenced by the magnetic charges within the stem. One may assume that the stem is magnetically hard and saturated along Z, in which case the density of the uniform charge distribution on the stem's bottom surface will be M_s (saturation moment) and the field $\mathbf{H}$ at the lattice site (x_i, y_i, z_i) will be

$$\mathbf{H}(x_i, y_i, z_i) = \iint_{\substack{\text{bottom}\\ \text{surface}}} M_s \frac{(x_i - x)\hat{\mathbf{x}} + (y_i - y)\hat{\mathbf{y}} + (z_i - z)\hat{\mathbf{z}}}{\left[(x - x_i)^2 + (y - y_i)^2 + (z - z_i)^2\right]^{3/2}} \, dx dy \tag{18.16}$$

The other contribution of the stem is through its exchange interaction with the uppermost layer of cubes in the tip. With the stem's magnetization saturated along Z, the exchange field acting on the lattice sites just below the stem will be $H_{xhg}\hat{\mathbf{z}}$.

Stray field from the sample. Let the sample probed by the needle be a film of thickness τ and magnetization distribution $\mathbf{M}(x, y)$, uniform through the thickness. The stray field $\mathbf{H}(x, y, z)$ and the field gradients produced by the film can be computed using the method of section 13.2. The average field over the volume of a cube is then obtained with only a small additional effort. Consider a cube of dimension Δ centered at (x_0, y_0, z_0) with sides parallel to X, Y, and Z. The average field is given by

$$\mathbf{H}_{\text{ave}}(x_0, y_0, z_0) = \frac{1}{\Delta^3} \int_{z_0 - \frac{1}{2}\Delta}^{z_0 + \frac{1}{2}\Delta} \int_{y_0 - \frac{1}{2}\Delta}^{y_0 + \frac{1}{2}\Delta} \int_{x_0 - \frac{1}{2}\Delta}^{x_0 + \frac{1}{2}\Delta} \mathbf{H}(x, y, z) \, dx dy dz \tag{18.17}$$

This field can be determined from Eqs. (13.22), provided that the right-hand side of Eq. (13.22b) is multiplied by the coefficient $\mathcal{C}_{mn}$ given below:

$$\mathcal{C}_{mn} = \left(\frac{\sin(\pi s_x \Delta)}{\pi s_x \Delta}\right)\left(\frac{\sin(\pi s_y \Delta)}{\pi s_y \Delta}\right)\left(\frac{\sinh(\pi s \Delta)}{\pi s \Delta}\right) \tag{18.18}$$

Now, a stable distribution of moments within the needle's tip may be obtained by first assigning an initial direction to each dipole. (For instance, the simulation might begin with all dipoles parallel to Z.) Then the effective field $\mathbf{H}_{\text{eff}}$ at each site is computed and the system of dipoles relaxed in accordance with the LLG equation, Eq. (15.21). The procedure must continue until a stable configuration (i.e., a local minimum of energy) is obtained. Once the state of magnetization is determined, the net force on the needle is computed by adding up the forces exerted on individual grains by the stray field of the sample. The force $\mathbf{F}$ on a point dipole $\mathbf{m}$ located in the field $\mathbf{H}(x, y, z)$ is then given by

$$\mathbf{F} = \nabla\left[\mathbf{m}\cdot\mathbf{H}(x,y,z)\right] \tag{18.19a}$$

which, in matrix notation, is written

$$\begin{pmatrix} F_x \\ F_y \\ F_z \end{pmatrix} = \begin{pmatrix} \partial H_x/\partial x & \partial H_y/\partial x & \partial H_z/\partial x \\ \partial H_x/\partial y & \partial H_y/\partial y & \partial H_z/\partial y \\ \partial H_x/\partial z & \partial H_y/\partial z & \partial H_z/\partial z \end{pmatrix}\begin{pmatrix} m_x \\ m_y \\ m_z \end{pmatrix} \tag{18.19b}$$

Since according to Maxwell's equations $\nabla \times \mathbf{H} = 0$, the 3×3 matrix of gradients in Eq. (18.19b) is symmetric (i.e., $\partial H_y/\partial x = \partial H_x/\partial y$ and so on); therefore only six out of the nine elements of this matrix need be evaluated. Using Eq. (13.22) one can derive the Fourier-series representation for each row of the gradient matrix as follows:

$$\frac{\partial \mathbf{H}}{\partial x} = \sum_{m=-\infty}^{\infty}\sum_{n=-\infty}^{\infty} i2\pi s_x \mathcal{H}_{mn}(z)\exp\left[i2\pi\left(\frac{mx}{L_x}+\frac{ny}{L_y}\right)\right] \tag{18.20a}$$

$$\frac{\partial \mathbf{H}}{\partial y} = \sum_{m=-\infty}^{\infty}\sum_{n=-\infty}^{\infty} i2\pi s_y \mathcal{H}_{mn}(z)\exp\left[i2\pi\left(\frac{mx}{L_x}+\frac{ny}{L_y}\right)\right] \tag{18.20b}$$

$$\frac{\partial \mathbf{H}}{\partial z} = \sum_{m=-\infty}^{\infty} \sum_{n=-\infty}^{\infty} -2\pi s \mathcal{H}_{mn}(z) \exp\left[i2\pi\left(\frac{mx}{L_x} + \frac{ny}{L_y}\right)\right] \tag{18.20c}$$

Since in the model each dipole represents a cube of uniform magnetization, the force components in Eq. (18.19b) must be averaged over the volume of the cube. **m** being constant throughout the cube, the averaging is done over the field gradients. The procedure is similar to that described earlier in conjunction with the averaging of **H**, and so is the result: the average field gradients are still obtained from Eqs. (18.20) provided that the $\mathcal{H}_{mn}(z)$ of Eq. (13.22b) are first multiplied by the $\mathcal{C}_{mn}$ of Eq. (18.18).

Finally, the force exerted on the stem by the external field must be added to the force experienced by the tip. Since the stem is uniformly magnetized along Z, the magnetic charges accumulated at its bottom surface will have a uniform density of $\pm M_s$ ($\pm$ sign for magnetization parallel/antiparallel to Z). The force $\mathbf{F}_s$ on the stem is given by

$$\mathbf{F}_s = \pm M_s \iint_{\substack{\text{bottom} \\ \text{surface}}} \mathbf{H}(x, y, z_s)\, dx dy \tag{18.21}$$

where z_s is the position of the stem's bottom surface. In calculating the average field using Eq. (13.22) we must multiply the Fourier components $\mathcal{H}_{mn}(z_s)$ by the following coefficient:

$$\mathcal{C}'_{mn} = \left(\frac{\sin(\pi s_x D)}{\pi s_x D}\right)\left(\frac{\sin(\pi s_y D)}{\pi s_y D}\right) \tag{18.22}$$

where D is the linear dimension of the stem's bottom surface.

18.4.3. Results of numerical simulations

The model described in this section has been used to simulate the operation of a magnetic force microscope. The following examples provide insight into the nature of magnetization distribution within the tip, and also give quantitative information about the forces on the needle as it scans a domain wall.

Example 3. Consider a needle with the tip cross-section of Fig. 18.19(a). Let M_s = 1500 emu/cm³ (both tip and stem), $K_u = 10^7$ erg/cm³ and H_{xhg} = 1 kOe. Also allow the anisotropy axes to have random directions in space. The other parameters of the model are Δ = 25 nm, α = 0.5, and $\gamma = -10^7$ Hz/Oe. The simulation begins with all dipoles aligned in the positive Z-direction, and continues by relaxing them according to the LLG

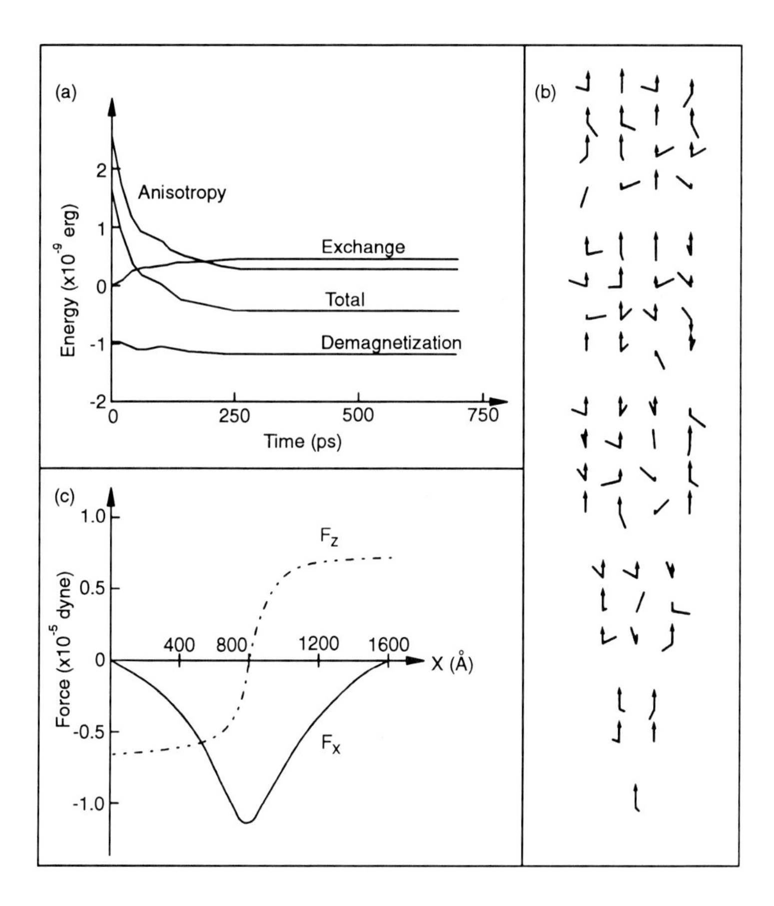

Figure 18.20. (a) The various magnetic energies of the needle during the relaxation process. At $t = 0$ all dipoles are along $+Z$, but by $t = 300$ ps they have settled into a locally stable configuration. (b) Steady-state magnetization of the tip. Each dipole is shown as an arrow with an appendage: the arrow is the component of **M** along Z and the appendage is its component in the XY-plane. The different blocks of arrows correspond to the different layers of cubes in the tip model of Fig. 18.18. (c) Horizontal and vertical force components F_x, F_z versus the tip's position along X. The scanned magnetic film has a straight Bloch wall at $x = 800$ Å.

equation. The energy of the system is computed during relaxation and shown in Fig. 18.20(a). It is observed that the contributions of both anisotropy and demagnetization to the total energy decrease, while the contribution of exchange increases. As expected, the total energy decreases with time.

After nearly 500 iterations (corresponding to 200 ps in Fig. 18.20(a)) the tip reaches the stable configuration of Fig. 18.20(b). The state of each dipole is shown by an arrow with an appendage. The arrow is the projection of the moment along Z, while the appendage is its component within the XY-plane. (For instance, ↿_ means that the perpendicular component is along the positive Z-axis and the in-plane component points to the right, while ⌉ means that the perpendicular component is along the negative Z-axis and the in-plane component points to the left.) The average magnetization along Z has dropped from $M_s = 1500$ emu/cm³ to $\langle M_z \rangle = 886$ emu/cm³.

Next we give the field and field gradients calculated for a magnetic film with thickness $\tau = 30$ nm and saturation moment $M_{s0} = 100$ emu/cm³. The film is perpendicularly magnetized and contains two straight domain walls, both parallel to Y. The first wall is at $x = 80$ nm while the second one is at $x = 240$ nm. The basic pattern of magnetization is defined between $x = 0$ and $x = L_x = 320$ nm. (Since the pattern is one-dimensional, L_y is irrelevant and therefore chosen arbitrarily.) The wall-width parameter Δ_w is 7.5 nm and its structure is that of a simple Bloch wall. The field and field-gradients are computed numerically, using a sampling interval of 2.5 nm.

When the needle is brought to the vicinity of the film, the tip's dipoles are allowed to relax further in order to conform to the stray field of the sample. For the particular set of parameters used in these calculations, we found that the effect on the magnetization of the tip was small: depending on the position of the needle along X, the average Z-component, $\langle M_z \rangle$, varied only within $\pm 1\%$ of its zero-field value. Once the needle reached a stable state, we calculated the force and then repeated the procedure for the next point along X. Figure 18.20(c) shows F_x and F_z as functions of x in the interval $0 \leq x \leq 160$ nm. (The wall is at the midpoint of this interval.) The separation between the end of the tip and the sample was 10 nm. Note that the peak of F_x and the zero-crossing of F_z occur at the wall center. The slight asymmetry in the curves is due to the asymmetry of magnetization distribution in the needle, which is attributable to its random axis anisotropy.

□

Example 4. When the tip is a continuation of the magnetic stem of the needle the stem plays an important role in orienting the dipole moments of individual grains within the tip. On the other hand, in the absence of a magnetic stem (such as when the tip is electroplated onto the end of a tungsten needle) the tip magnetization tends to form flux closure paths in order to reduce its demagnetization energy. Two examples of the computed magnetic pattern of the needle are shown in Fig. 18.21; dipole configurations in the first six layers of the tip are displayed in both cases. There are a total of 91 grains of cobalt ($M_s = 1450$ emu/cm³) in each needle-tip, and each grain is assumed to be $25 \times 25 \times 25$ nm³ in size. In Fig. 18.21(a) the stem is magnetic and the average tip magnetization is found to be 527 emu/cm³. In Fig. 18.21(b), where the stem is non-magnetic, the net computed magnetization is 79 emu/cm³.

Next we assumed that a thin magnetic film is placed under the needle; the saturation moment M_{s0} of the film is 100 emu/cm³ and its thickness τ is 50 nm. A straight domain wall of the Bloch type with a wall width of approximately 10 nm was assumed in this sample. The calculated values of

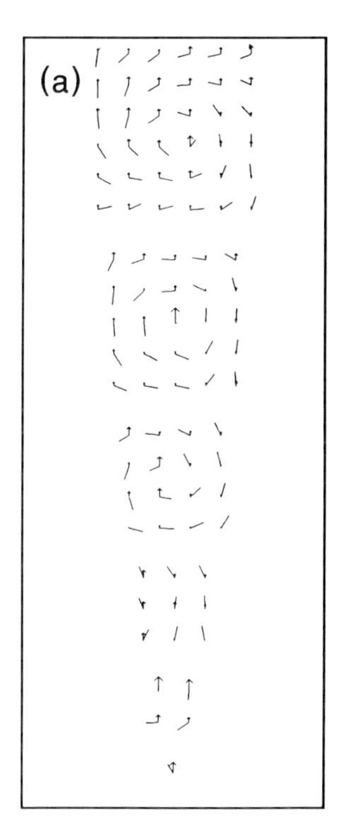

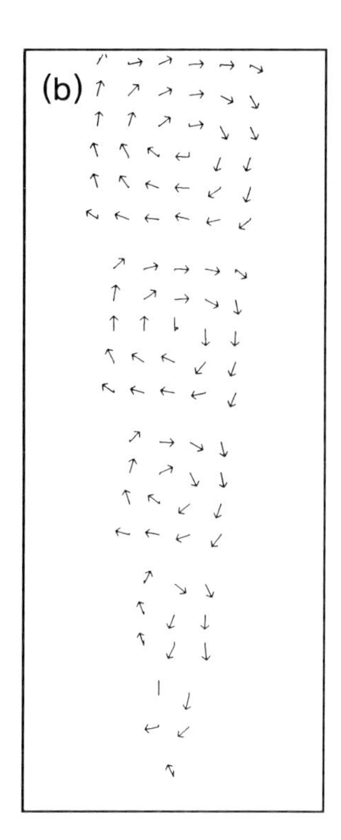

Figure 18.21. Distribution of magnetization in the various layers of the tip. The needle material is cobalt with $M_s = 1450$ emu/cm³. Each grain has a randomly oriented c-axis with anisotropy constant $K_u = 4 \times 10^6$ erg/cm³; the effective exchange field between neighboring grains is 1 kOe. (a) The stem is magnetic and saturated along the easy axis. Each dipole in the steady state is represented by an arrow with an appendage; as in the preceding figure, the arrow is the component of **M** along Z and the appendage is its component in the XY-plane. In the remanent state of the needle shown here, the average tip magnetization $M_z = 527$ emu/cm³. (b) The stem is non-magnetic. In the magnetization distribution shown here each dipole is represented by an arrow with an appendage; this time, however, the arrow shows the component of **M** in the XY-plane, while the appendage is its component along Z. The average tip magnetization $M_z = 79$ emu/cm³.

the force components F_x and F_z along X and Z are shown in Fig. 18.22; to obtain these plots we allowed the magnetization of the needle to relax in the stray field of the sample, while the needle scanned the surface along X. The abscissa in Fig. 18.22 is the position of the needle's axis relative to the wall center. In Fig. 18.22(a) the tip has the magnetization pattern shown in Fig. 18.21(a), that is, both the tip and the stem are assumed magnetic; the plots of Fig. 18.22(b) correspond to the tip of Fig. 18.21(b), which is associated with a non-magnetic stem. Z_0 is the tip-to-sample spacing. Note that F_z switches sign as the wall is crossed, whereas F_x reaches a maximum at that point. Clearly, the needle with the magnetic stem produces a larger

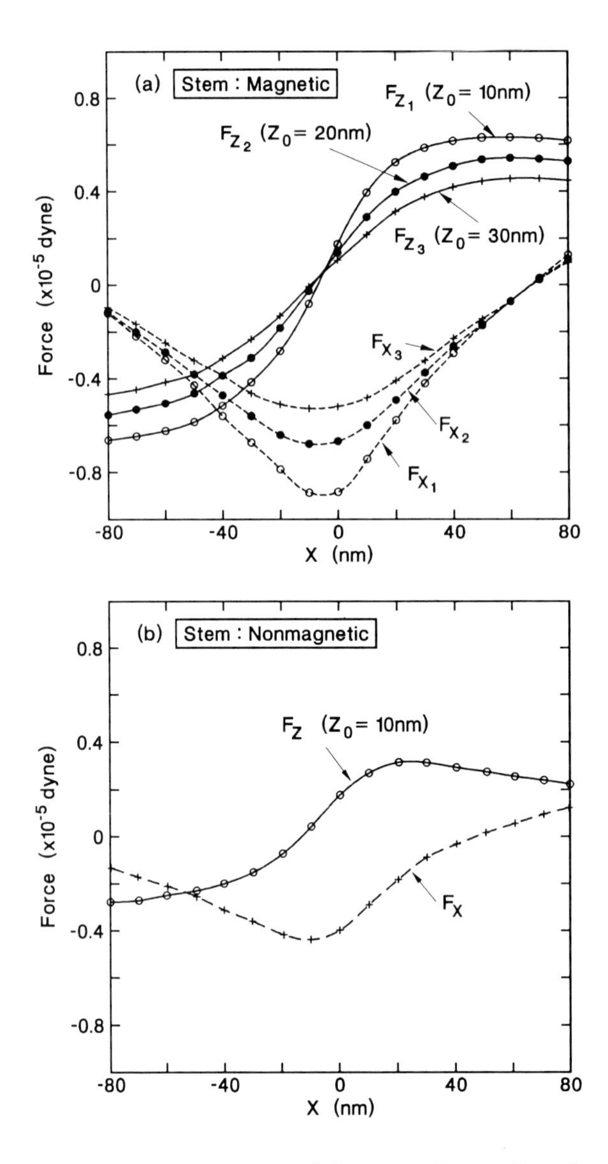

Figure 18.22. Components F_x and F_z of force on the needles whose magnetization patterns are shown in Figs. 18.21(a) and (b) respectively. The sample scanned is a 50 nm-thick film with M_s = 100 emu/cm³, containing a straight Bloch wall parallel to Y (wall width $\simeq$ 10 nm). The scan direction is perpendicular to the wall in both cases and Z_0 is the tip-to-sample spacing.

signal, even though the measured width of the transition region (i.e., the resolution of the microscope) is about the same in the two cases. We conclude that the needle with the magnetic stem is a more sensitive probe of the stray field of the sample, mainly due to the fact that its dipole moments do not form flux-closure paths.

When the tip-to-sample separation increased from 10 nm to 20 nm (see F_{z1} and F_{z2} in Fig. 18.22(a)) the strength of the signal F_z dropped by as much as 15% (F_x dropped by more than 20%). This result reveals the importance of maintaining a constant tip-to-sample spacing during a scan.

□

Problems

(18.1) In the loop-tracer system of Fig. 18.1 the Kerr angle θ_k can be obtained from the split-detector signals S_1 and S_2.

(a) What combination of S_1 and S_2 yields the value of θ_k?

(b) Is the value of θ_k thus obtained sensitive to laser power fluctuations?

(c) The quarter-wave plate (QWP) is inserted in the optical path in order to measure the ellipticity ϵ_k of the reflected polarization. What is the correct orientation of the QWP relative to the incident polarization?

(d) Devise a strategy for measuring the reflectivity R of the sample using the split-detector signals S_1 and S_2, as well as the signal S_0 from the laser monitor.

(18.2) With the field applied in the plane of the sample, the loop-tracer of Fig. 18.1 can yield the constant of magnetic anisotropy K_u either from the Kerr effect or from the Hall effect measurements. (Typical data obtained in such measurements are shown in Figs. 18.5(c), 18.5(e), 18.6(c), and 18.7(c).) What is the procedure for extracting K_u from the measured data?

(18.3) The measured plots of magnetoresistance $\Delta\rho/\rho$ versus the magnitude of the perpendicularly applied field H show strong peaks in the neighborhood of the coercivity H_c (see Figs. 18.5(f), 18.6(d), and 18.7(d)). In this region the magnetization breaks up into opposite domains, and the Hall effect creates circulating currents within the sample. These currents in turn produce a Hall voltage that is at least partially responsible for the observed peaks. Estimate the order of magnitude of these peaks in terms of the Hall resistivity ρ_{xy} and the ordinary resistivity ρ.

(18.4) Starting from Eq. (18.9) and using the Fourier series expansion of the vector potential **A** given in Eqs. (13.21), verify the expression for $\Phi(x, y)$ in Eq. (18.10).

(18.5) The average stray field $\mathbf{H}_{\text{ave}}$ over the volume of a cube is defined in Eq. (18.17). Using the Fourier series expansion of **H** in Eq. (13.22), derive the coefficients $\mathcal{C}_{mn}$ in Eq. (18.18). Similarly, for the average field on the bottom surface of the stem verify the coefficients $\mathcal{C}'_{mn}$ in Eq. (18.22).

References

References for Chapter 1

A. Overview

1. A.B. Marchant, *Optical Recording*, Addison-Wesley, Massachusetts (1990).
2. P. Hansen and H. Heitmann, Media for erasable magneto-optic recording, *IEEE Trans. Mag.* **25**, 4390-4404 (1989).
3. M.H. Kryder, Data-storage technologies for advanced computing, *Scientific American* **257**, 116-125 (1987).
4. G.A.N. Connell, Magneto-optics and amorphous metals: an optical storage revolution, *J. Mag. Magnet. Mater.*, **54-57**, 1561-1566 (1986).
5. F. Guterl, Compact Disc, *IEEE Spectrum*, Special 25th Anniversary Issue, **25** (11), 102-108 (1988).
6. R.M. White, Magnetic disks: storage densities on the rise, *IEEE Spectrum*, 32-38 (August 1983).
7. A.E. Bell, Critical issues in high density magnetic and optical data storage, in *SPIE Proc.* Vol. 382, 2-15 (1983).
8. W.H. Meiklejkohn, Magneto-optics: a thermomagnetic recording technology, *Proc. IEEE* **74**, 1570-1581 (1986).
9. L. Fujitani, Laser optical disk: the coming revolution in on-line storage, *Comm. ACM* **27**, 546-554 (1984).
10. C.M. Goldstein, Optical disk technology and information, *Science* **215**, 862-868 (1982).
11. A.E. Bell, Optical data storage technology: status and prospects, *Computer Design*, 133-146 (January 1983).
12. D.S. Bloomberg and G.A.N. Connell, Prospects for magneto-optic recording, in *Proc. IEEE 1985 COMPCON spring*, IEEE Computer Society, 32-38 (1985).
13. S. Miyaoka, Digital audio is compact and rugged, *IEEE Spectrum*, 35-39 (March 1984).
14. M. Mansuripur, Disk storage: magneto-optics leads the way, *Photonics Spectra*, 59-62 (October 1984).
15. M. Mansuripur, M.F. Ruane, and M.N. Horenstein, Erasable optical disks for data storage: principles and applications, *Ind. Eng. Chem. Prod. Res. Dev.* **24**, 80-84 (1985).

16. P. Chen, The compact disk ROM: how it works, *IEEE Spectrum*, 44-49 (April 1986).
17. M.H. Kryder, Data storage in 2000 - trends in data storage technologies, *IEEE Trans. Mag.* **25**, 4358-4363 (1989).

B. Diffraction, Gaussian Beams, Push–Pull Methods

1. G. Bouwhuis and J.J.M. Braat, Recording and reading of information on optical disks, Chapter 3 in *Applied Optics and Optical Engineering*, Volume IX, R.R. Shannon and J.C. Wyant, eds., Academic Press, New York (1983).
2. H.H. Hopkins, Canonical and real-space coordinates used in the theory of image formation, Chapter 8 in *Applied Optics and Optical Engineering*, Volume IX, R.R. Shannon and J.C. Wyant, eds., Academic Press, New York (1983).
3. H. Kogelnik, Propagation of laser beams, Chapter 6 in *Applied Optics and Optical Engineering*, Volume VII, R.R. Shannon and J.C. Wyant, eds., Academic Press, New York (1979).
4. G. Bouwhuis, J. Braat, A. Huijser, J. Pasman, G. Van Rosmalen, and K.S. Immink, Chapters 2 and 3 in *Principles of Optical Disk Systems*, Adam Hilger, Bristol (1985).
5. Special issue of *Applied Optics* on the subject of video disks, July 1, 1978.
6. M.V. Klein, *Optics*, 1st edition, Wiley, New York (1970).
7. M. Born and E. Wolf, *Principles of Optics*, 6th edition, Pergamon Press, Oxford (1980).
8. M. Mansuripur, Certain computational aspects of vector diffraction problems, *J. Opt. Soc. Am. A* **6**, 786-805 (1989).
9. M. Mansuripur, Detecting transition regions in magneto-optical disk systems, *Appl. Phys. Lett.* **55**, 716-717 (1989).
10. M. Mansuripur, Analysis of astigmatic focusing and push–pull tracking error signals in magneto-optical disk systems, *Appl. Opt.* **26**, 3981-3985 (1987).
11. E. Wolf, Electromagnetic diffraction in optical systems, I. An integral representation of the image field, *Proc. Roy. Soc. A* **253**, 349-357 (1959).
12. A. Hardy and D. Treves, Structure of the electromagnetic field near the focus of a stigmatic lens, *J. Opt. Soc. Am.* **53**, 85-90 (1973).
13. P. Sheng, Theoretical considerations of optical diffraction from RCA video disk signals, *RCA Review* **39**, 513-555 (1978).
14. H.H. Hopkins, Diffraction theory of laser read-out systems for optical video discs, *J. Opt. Soc. Am.* **69**, 4-24 (1979).
15. J.R. Benford, Microscope objectives, Chapter 4 in *Applied Optics and Optical Engineering*, Volume III, R. Kingslake, ed., Academic Press, New York (1965).

C. Interaction of Light with Magnetic Matter, Enhancement by Multilayering, Magneto-optic Kerr and Faraday Effects

1. D.O. Smith, Magneto-optical scattering from multilayer magnetic and dielectric films, *Opt. Acta* **12**, 13 (1965).
2. R.P. Hunt, Magneto-optic scattering from thin solid films, *J. Appl. Phys.* **38** 1652-1671 (1967).
3. P.S. Pershan, Magneto-optic effects, *J. Appl. Phys.* **38**, 1482-1490 (1967).
4. M. Mansuripur, Figure of merit for magneto-optical media based on the dielectric tensor, *Appl. Phys. Lett.* **49**, 19-21 (1986).
5. R.W. Wood, *Physical Optics*, 3rd edition, Chapter 21, Optical Society of America, Washington DC (1988).
6. F.A. Jenkins and H.E. White, *Fundamentals of Optics*, 4th edition, Chapter 32, McGraw Hill, New York (1976).
7. G.J. Sprokel, Reflectivity, rotation, and ellipticity of magnetooptic film structures, *Appl. Opt.* **23**, 3983-3989 (1984).
8. R.S. Weis and T.K. Gaylord, Electromagnetic transmission and reflection characteristics of anisotropic multilayered structures, *J. Opt. Soc. Am. A* **4**, 1720-1740 (1987).
9. K. Balasubramanian, A.S. Marathay, and H.A. Macleod, Modeling magneto-optical thin film media for optical data storage, *Thin Solid Films* **164**, 341-403 (1988).
10. Y. Tomita and T. Yoshini, Optimum design of multilayer-medium structures in a magneto-optical readout system, *J. Opt. Soc. Am. A* **1**, 809-817 (1984).
11. K. Egashira and T. Yamada, Kerr effect enhancement and improvement of readout characteristics in MnBi film memory, *J. Appl. Phys.* **45**, 3643-3648 (1974).
12. G.A.N. Connell, Interference enhanced Kerr spectroscopy for very thin absorbing films, *Appl. Phys. Lett.* **40**, 212 (1982).
13. K.Y. Ahn and G.J. Fan, Kerr effect enhancement in ferromagnetic films, *IEEE Trans. Mag.* **2**, 678 (1966).

D. Light Absorption, Heat Diffusion, and Thermal Analysis

1. H.S. Carslaw and J.C. Jaeger, *Conduction of Heat in Solids*, Oxford University Press, London (1954).
2. S.C. Shin, Thermal analysis of magneto-optical thin films under laser irradiation, *J. Mag. Magnet. Mater.* **61**, 301-306 (1986).
3. H. Wieder and R.A. Burn, Direct comparison of thermal and magnetic profiles in Curie point writing on MnGaGe films, *J. Appl. Phys.* **44**, 1774 (1973).
4. P. Kivits, R. deBont, and P. Zalm, Superheating of thin films for optical recording, *Appl. Phys.* **24**, 273-278 (1981).

5. H. Wieder, Novel method for measuring transient surface temperatures with high spatial and temporal resolution, *J. Appl. Phys.* **34**, 3213 (1972).
6. M.K. Bhattacharyya and Z. Cendes, Finite-element modeling of laser beam heating of magnetic films, *J. Appl. Phys.* **57**, 3894 (1985).
7. M. Mansuripur, G.A.N. Connell, and J.W. Goodman, Laser-induced local heating of multilayers, *Appl. Opt.* **21**, 1106 (1982).
8. M. Mansuripur and G.A.N. Connell, Laser-induced local heating of moving multilayer media, *Appl. Opt.* **22**, 666 (1983).
9. M. Mansuripur and G.A.N. Connell, Thermal aspects of magneto-optical recording, *J. Appl. Phys.* **54**, 4794 (1983).
10. W.A. Michael and D. Treves, The heat problem in magneto-optic readout, *J. Appl. Phys.* **40**, 303 (1969).

E. Noise in Optical Recording Systems and Media

1. J.P.J. Heemskerk, Noise in a video disk system: experiments with an (AlGa)As laser, *Appl. Opt.* **17**, 2007 (1978).
2. G.A. Acket, D. Lenstra, A.J. DenBoef, and B.H. Verbeek, Influence of feedback intensity on longitudinal mode properties and optical noise in index-guided semiconductor lasers, *IEEE J. Quant. Electron.* **QE-20**, 1163 (1984).
3. F. Inoue, A. Maeda, A. Itoh, and K. Kawanishi, The medium noise reduction by intensity dividing readout in the magnetic-optical memories, *IEEE Trans. Mag.* **21**, 1629 (1985).
4. A. Arimoto, M. Ojima, N. Chinone, A. Oishi, T. Gotoh, and N. Ohnuki, Optimum conditions for the high frequency noise reduction method in optical video disk players, *Appl. Opt.* **25**, 1398 (1986).
5. M. Ojima, A. Arimoto, N. Chinone, T. Gotoh, and K. Aiki, Diode laser noise at video frequencies in optical video disk players, *Appl. Opt.* **25**, 1404 (1986).
6. J.W. Beck, Noise considerations of optical beam recording, *Appl. Opt.* **9**, 2559 (1970).
7. D. Treves and D. S. Bloomberg, Signal, noise, and codes in optical memories, *Opt. Eng.* **25**, 881 (1986).
8. D. Treves, Magneto-optic detection of high-density recordings, *J. Appl. Phys.* **38**, 1192 (1967).
9. B.R. Brown, Readout performance analysis of a cryogenic magneto-optical data storage system, *IBM J. Res. Dev.*, 19-26 (January 1972).
10. R.L. Aagard, Signal to noise ratio for magneto-optical readout from MnBi films, *IEEE Trans. Mag.* **9**, 705 (1973).
11. M. Mansuripur, G.A.N. Connell, and J.W. Goodman, Signal and noise in magneto-optical readout, *J. Appl. Phys.* **53**, 4485 (1982).
12. G.A.N. Connell, D. Treves, R. Allen, M. Mansuripur, Signal-to-noise ratio for magneto-optic readout from quadrilayer structures, *Appl. Phys. Lett.* **42**, 742 (1983).

13. D. S. Bloomberg and G. A. N. Connell, Magnetooptical recording, Chapter 6 in *Magnetic Recording*, Volume II, C.D. Mee and E.D. Daniel, eds., McGraw-Hill, New York (1988).

F. Micromagnetics; the Thermomagnetic Recording Process

1. A.H. Morrish, *The Physical Principles of Magnetism*, Wiley, New York (1965).
2. S. Chikazumi and S.H. Charap, *Physics of Magnetism*, Wiley, New York (1964).
3. A.H. Eschenfelder, *Magnetic Bubble Technology*, Springer, New York (1980).
4. B.G. Huth, Calculation of stable domain radii produced by thermomagnetic writing, *IBM J. Res. Dev.* **18**, 100-109 (1974).
5. A.A. Thiele, Theory of the static stability of cylindrical domains in uniaxial platelets, *J. Appl. Phys.* **41**, 1139 (1970).
6. A.P. Malozemoff and J.C. Slonczewski, *Magnetic Domain Walls in Bubble Materials*, Academic Press, New York (1979).
7. E. Bernal G., Mechanism of Curie point writing in thin films of manganese bismuth, *J. Appl. Phys.* **42**, 3877 (1971).
8. W.F. Brown, *Magnetostatic Principles in Ferromagnetism*, Interscience, New York (1962).
9. M. Prutton, *Thin Ferromagnetic Films*, Butterworths, London (1964).
10. D.J. Craik and R.S. Tebble, *Ferromagnetism and Ferromagnetic Domains*, North Holland, Amsterdam (1965).
11. R.F. Soohoo, *Magnetic Thin Films*, Harper and Row, London (1965).
12. E.C. Stoner and E.P. Wohlfarth, *Phil. Trans. Roy. Soc. A* **240**, 599 (1948).
13. B.K. Middleton, Magnetic thin films and devices, Chapter 11 in *Active and Passive Thin Film Devices*, T.J. Coutts, ed., Academic Press, New York (1978).

G. Error Correction and Modulation Codes

1. A.M. Patel, Signal and error-control coding, Chapter 5 in *Magnetic Recording*, Volume II, C.D. Mee and E.D. Daniel, eds., McGraw-Hill, New York (1988).
2. R.E. Blahut, *Theory and Practice of Error Control Codes*, Addison-Wesley, Reading MA (1983).
3. R.L. Adler, D. Coppersmith, and M. Hassner, Algorithms for sliding block codes, *IEEE Trans. Information Theory* **29**, 5-22 (1983).
4. J.K. Wolf and G. Ungerboeck, Trellis coding for partial response channels, *IEEE Trans. Commun.* **34**, 765 (1986).
5. D.G. Howe, Signal-to-noise ratio for reliable data recording, in *Proc. SPIE* Vol. 695, 255-261 (1986).

6. D.G. Howe, The nature of intrinsic error rates in high-density recording, in *Proc. SPIE* Vol. 421, 31-42 (1983).
7. K.A.S. Immink, Coding methods for high-density optical recording, *Philips J. Res.* **41**, 410-430 (1986).

H. Film Deposition and Characterization

1. L.I. Maissel and R. Glang, eds., *Handbook of Thin Film Technology*, McGraw-Hill, New York (1970).
2. G.L. Weissler and R.W. Carlson, eds., *Vacuum Physics and Technology*, Volume 14 of *Methods of Experimental Physics*, Academic, New York (1979).
3. R.V. Stuart, *Vacuum Technology, Thin Films, and Sputtering*, Academic Press, New York (1983).
4. R. Hasegawa, ed., *Glassy Metals: Magnetic, Chemical and Structural Properties*, CRC Press, Boca Raton, FL (1983).

References for Chapter 2

1. H. Kogelnik, On the propagation of Gaussian beams of light through lenslike media including those with a loss and gain variation, *Appl. Opt.* **4**, 1562 (1965).
2. H. Kogelnik and T. Li, Laser beams and resonators, *Proc. IEEE* **54**, 1312 (1966).
3. S.A. Self, Focusing of spherical Gaussian beams, *Appl. Opt.* **22**, 658 (1983).
4. A. Yariv, *Optical Electronics*, 4th edition, Holt, Rinehart and Winston, Philadelphia (1991).

References for Chapter 3

1. M. Born and E. Wolf, *Principles of Optics*, 6th edition, Pergamon Press, Oxford (1980).
2. M.V. Klein, *Optics*, 1st edition, Wiley, New York (1970).
3. E. Wolf, Electromagnetic diffraction in optical systems, I. An integral representation of the image field, *Proc. Roy. Soc. A* **253**, 349-357 (1959).
4. A. Hardy and D. Treves, Structure of the electromagnetic field near the focus of a stigmatic lens, *J. Opt. Soc. Am.* **53**, 85-90 (1973).
5. M. Mansuripur, Certain computational aspects of vector diffraction problems, *J. Opt. Soc. Am. A* **6** 786-805 (1989).

References for Chapter 4

1. G. Bouwhuis, J. Braat, A. Huijser, J. Pasman, G. Van Rosmalen, and K. S. Immink, *Principles of Optical Disk Systems*, Chapters 2 and 3, Adam Hilger, Bristol and Boston (1985).
2. G. Bouwhuis and J.J.M. Braat, Recording and reading of information on optical disks, Chapter 3 in *Applied Optics and Optical Engineering*, Volume IX, R.R. Shannon and J.C. Wyant, eds., Academic Press, New York (1983).
3. P. Sheng, Theoretical considerations of optical diffraction from RCA video disk signals, *RCA Review* **39**, 513-555 (1978).
4. H.H. Hopkins, Diffraction theory of laser read-out systems for optical video discs, *J. Opt. Soc. Am.* **69**, 4-24 (1979).
5. M. Mansuripur, Detecting transition regions in magneto-optical disk systems, *Appl. Phys. Lett.* **55**, 716-717 (1989).

References for Chapter 5

1. P. Yeh, *Optical Waves in Layered Media*, Wiley, New York (1988).
2. H.A. Macleod, *Thin Film Optical Filters*, Macmillan, New York (1986).
3. D.O. Smith, Magneto-optical scattering from multilayer magnetic and dielectric films, *Opt. Acta* **12**, 13-45 (1965).
4. R.P. Hunt, Magneto-optic scattering from thin solid films, *J. Appl. Phys.* **38**, 1652-1671 (1967).
5. P. Yeh, Electromagnetic propagation in birefringent layered media, *J. Opt. Soc. Am.* **69**, 742-756 (1979).
6. G.J. Sprokel, Photoelastic modulated ellipsometry on magneto-optic multilayer films, *App. Opt.* **25**, 4017-4022 (1986).
7. K. Balasubramanian, A. Marathay, and H.A. Macleod, Modeling magneto-optical thin film media for optical data storage, *Thin Solid Films* **164**, 391-403 (1988).
8. R.S. Weis and T.K. Gaylord, Electromagnetic transmission and reflection characteristics of anisotropic multilayered structures, *J. Opt. Soc. Am. A* **4**, 1720-1740 (1987).
9. P. Wolniansky, S. Chase, R. Rosenvold, M. Ruane and M. Mansuripur, Magneto-optical measurements of hysteresis loop and anisotropy energy constants on amorphous $Tb_x Fe_{1-x}$ alloys, *J. Appl. Phys.* **60**, 346-351 (1986).

References for Chapter 6

1. M. Mansuripur, F.L. Zhou, and J.K. Erwin, Measuring the wavelength dependence of magneto-optical Kerr (or Faraday) rotation and ellipticity: a technique, *Appl. Opt.* **29**, 1308-1311 (1990).
2. R.W. Wood, *Physical Optics*, 3rd edition, Optical Society of America (1988).
3. W.L. Wolf, Properties of Optical Materials, in *Handbook of Optics*, W.G. Driscoll and W. Vaughan, eds., McGraw-Hill, New York (1978).
4. M. Mansuripur, Detecting transition regions in magneto-optical disk systems, *Appl. Phys. Lett.* **55**, 716-717 (1989).
5. M.D. Levenson and R. Lynch, Edge detection for magneto-optical data storage, *Appl. Opt.* **30**, 232-252 (1991).
6. M. Mansuripur, Figure of merit for magneto-optical media based on the dielectric tensor, *Appl. Phys. Lett.* **49**, 19-21 (1986).

References for Chapter 7

1. E. Wolf, Electromagnetic diffraction in optical systems: an integral representation of the image field, *Proc. Roy. Soc. A* **253**, 349-357 (1959).
2. B. Richards and E. Wolf, Electromagnetic diffraction in optical systems: structure of the image field in an aplanatic system, *Proc. Roy. Soc. A* **253**, 358-379 (1959).
3. A. Boivin, J. Dow, and E. Wolf, *J. Opt. Soc. Am.* **57**, 1171 (1967).
4. A. Hardy and D. Treves, Structure of the electromagnetic field near the focus of a stigmatic lens, *J. Opt. Soc. Am.* **63**, 85-90 (1973).
5. H. Kubota and S. Inoue, Diffraction images in the polarizing microscope, *J. Opt. Soc. Am.* **49**, 191-198 (1959).
6. M. Mansuripur, Certain computational aspects of vector diffraction problems, *J. Opt. Soc. Am. A* **6**, 786-805 (1989).
7. J. W. Goodman, *Introduction to Fourier Optics*, McGraw-Hill, New York (1968).
8. M. Mansuripur, Analysis of multilayer thin film structures containing magneto-optic and anisotropic media at oblique incidence using 2×2 matrices, *J. Appl. Phys.* **67**, 6466 (1990).
9. A.B. Marchant, *Optical Recording*, Addison-Wesley, Massachusetts (1990).
10. A. Takahashi, M. Mieda, Y. Murakami, K. Ohta, and H. Yamaoka, Influence of birefringence on the signal quality of magneto-optic disks using polycarbonate substrates, *Appl. Opt.* **27**, 2863 (1988).
11. I. Prikryl, Effects of disk birefringence on a differential magneto-optic readout, *Appl. Opt.* **31**, 1853 (1992).

References for Chapter 8

1. G. Bouwhuis, J. Braat, A. Huijser, J. Pasman, G. van Rosmalen, and K.S. Immink, *Principles of Optical Disk Systems*, Adam Hilger, Bristol (1985).
2. G. Bouwhuis and J.J.M. Braat, Recording and reading of information on optical disks, in *Applied Optics and Optical Engineering*, Volume IX, R. Shannon and J.C. Wyant, eds., Academic Press, New York (1983).
3. H.H. Hopkins, Diffraction theory of laser readout systems for optical video disks, *J. Opt. Soc. Am.* **69**, 4-24 (1979).
4. M. Mansuripur, Distribution of light at and near the focus of high numerical aperture objectives, *J. Opt. Soc. Am. A* **3**, 2086 (1986).
5. M. Mansuripur, Certain computational aspects of vector diffraction problems, *J. Opt. Soc. Am. A* **6**, 786-805 (1989).
6. J.S. Hartman, M. Lind, and M. Mansuripur, Read channel optical modeling for a bump-forming dye-polymer optical data storage medium, SPIE **1078**, *Proceedings of the Optical Data Storage Meeting in Los Angeles CA*, 308-323 (1989).
7. A.B. Marchant, *Optical Recording*, Addison-Wesley, Reading MA (1990).
8. M. Mansuripur, Detecting transition regions in magneto-optic disk systems, *Appl. Phys. Lett.* **55**, 716-717 (1989).
9. J.C. Lehureau and J.Y. Beguin, Analysis of magneto-optic readout with circular polarized light, *IEEE Trans. J. Magnet. Japan* **4**, 258-262 (1989).
10. R.T. Lynch and M.D. Levenson, Edge detection for magneto-optical data storage, in *Proc. SPIE* Vol. 1316, 168-173 (1990).
11. N. Fukushima, K. Miura, and I. Sawaki, Detection of magneto-optic signals using an external cavity laser diode, in *Proc. SPIE* Vol. 1078, *Proceedings of the Optical Data Storage Meeting in Los Angeles CA*, 90-93 (1989).
12. T. Suhara and H. Nishihara, Possibility of super-resolution in integrated-optic disk pickup, Paper 27D-16, presented at the International Symposium on Optical Memory (ISOM), Kobe, Japan (September 1989).
13. J.J.M. Braat and G. Bouwhuis, Position sensing in video disk readout, *Appl. Opt.* **17**, 2013 (1978).
14. D. Cohen, W. Gee, M. Ludeke, and J. Lewkowicz, Automatic focus control: the astigmatic lens approach, *Appl. Opt.* **23**, 565 (1984).
15. M.. Yamammoto, A. Watabe, and H. Ukita, Optical pregroove dimensions: design considerations, *Appl. Opt.* **25**, 4031 (1986).
16. M. Mansuripur, Analysis of astigmatic focusing and push–pull tracking error signals in magneto-optical disk systems, *Appl. Opt.* **26**, 3981-3986 (1987).

17. S.L. DeVore, Radial error signal simulation for optical disk drives, *Appl. Opt.* **25**, 4001 (1986).
18. S. Nakamura, T. Maeda, and Y. Tsunoda, Autofocusing effect due to wavelength change of diode lasers in an optical pickup, *Appl. Opt.* **26**, 2549-2553 (1987).
19. K. Matsumoto and T. Maeda, Acousto-optic accessing in optical disks, Paper 28D-4, presented at the International Symposium on Optical Memory (ISOM), Kobe, Japan (September 1989).
20. A. Ohba, Y. Kimura, S. Sugama, Y. Urino, and Y. Ono, Holographic optical element with analyzer function for magneto-optical disk head, Paper 27D-24, presented at the International Symposium on Optical Memory (ISOM), Kobe, Japan (September 1989).
21. R. Linnebach, K. Gillessen, J. Frohlich, and S. Greenard, Application of holographic optical elements in optical memory devices, Paper 27D-20, presented at the International Symposium on Optical Memory (ISOM), Kobe, Japan (September 1989).
22. R. Katayama, K. Yoshihara, Y. Yamanaka, M. Tsunekane, K. Yoshida, and K. Kubota, Multi-beam magneto-optical disk drive for parallel read/write operation. in *Proc. SPIE* Vol. 1078, *Proceedings of the 1989 Optical Data Storage Conference*, 98-104.
23. T. Murakami, K. Taira, and M. Mori, Magneto-optic erasable disk memory with two optical heads, *Appl. Opt.* **25**, 3986 (1986).
24. P. Sheng, Theoretical considerations of optical diffraction for RCA video disk signals, *RCA Review* **39**, 512-555 (1978).
25. J.G. Dil and B.A.J. Jacobs, Apparent size of reflecting polygonal obstacles of the order of one wavelength, *J. Opt. Soc. Am.* **69**, 950-960 (1979).
26. D. Kuntz, Specifying laser diode optics, *Laser Focus/Electro-Optics*, 44-54 (March 1984).

References for Chapter 9

1. J. Smith, *Modern Communication Circuits*, McGraw-Hill, New York (1986).
2. A. Papoulis, *Probability, Random Variables, and Stochastic Processes*, 2nd edition, McGraw-Hill, New York (1984).
3. A. Arimoto, M. Ojima, N. Chinone, A. Oishi, T. Gotoh, and N. Ohnuki, Optimum conditions for the high frequency noise reduction method in optical videodisc players, *Appl. Opt.* **25**, 1398-1403 (1986).
4. M. Ojima, A. Arimoto, N. Chinone, T. Gotoh, and K. Aiki, Diode laser noise at video frequencies in optical videodisc players, *Appl. Opt.* **25**, 1404-1410 (1986).
5. G.A. Acket, D. Lenstra, A.J. Den Boef, and B.H. Verbeek, Influence of feedback intensity on longitudinal mode properties and optical

noise in index-guided semiconductor lasers, *J. Quant. Electron.* **QE-20**, 1163-1169 (1984).
6. A.G. Dewey, Optimizing signal to noise ratio from a magneto-optic head, paper EO8.2, presented at the meeting of the Optical Society of America, Boston MA (1990).
7. A.G. Dewey, Optimizing the noise performance of a magneto-optic read channel, in *Proc. SPIE* Vol. 1078, 279-286 (1989).
8. J.W. Beck, Noise considerations in optical beam recording, *Appl. Opt.* **9**, 2559-2564 (1970).
9. J.P.J. Heemskerk, Noise in a video disk system: experiments with an (AlGa)As laser, *Appl. Opt.* **17**, 2007-2012 (1978).
10. F. Inoue, A. Maeda, A. Itoh, and K. Kawanishi, The medium noise reduction by intensity dividing readout in the magneto-optical memories, *IEEE Trans. Mag.* **21**, 1629-1631 (1985).
11. D. Treves and D.S. Bloomberg, Signal, noise, and codes in optical memories, *Optical Engineering* **25**, 881-891 (1986).
12. A.G. Dewey, Measurement and modeling of optical disk noise, in *Proc. SPIE* Vol. 695, 72-78 (1986).

References for Chapter 10

1. A.M. Patel, Signal and error control coding, in *Magnetic Recording, Volume II*, C.D. Mee and E.D. Daniel, eds., McGraw-Hill, New York (1988).
2. P.A. Franaszek, Run-length limited variable length coding with error propagation limitation, US Patent 3 689 899 (1972).
3. D.G. Howe, The nature of intrinsic error rates in high-density digital optical recording, in *Proc. SPIE*, Vol. 421, *Optical Disk Systems and Applications*, 31-42 (1983).
4. D.G. Howe, Signal to noise ratio (SNR) for reliable data recording, in *Proc. SPIE*, Vol. 695, *Optical Mass Data Storage II*, 255-261 (1986).
5. K.S. Immink, Coding methods for high-density optical recording, *Philips J. Res.* **41**, 410-430 (1986).
6. R.L. Adler, D. Coppersmith and M. Hassner, Algorithms for sliding block codes, *IEEE Trans. Inform. Theory* **29**, 5-22 (1983).
7. A.M. Patel, Zero-modulation encoding in magnetic recording, *IBM J. Res. Dev.* **19**, 366-378 (1975).
8. D.T. Tang and L.R. Bahl, Block codes for a class of constrained noiseless channels, *Information and Control* **17**, 436-461 (1970).
9. C.E. Shannon, *A* mathematical theory of communication, *Bell Syst. Tech. J.* **27**, 379 (1948).
10. B. Fitingof and Z. Waksman, Fused trees and some new approaches to source coding, *IEEE Trans. Inform. Theory*, **IT-34**, 417-424 (May 1988).

11. T. Cover, Enumerative source coding, *IEEE Trans. Inform. Theory* **IT-19**, 73-76 (January 1973).
12. A.J. Viterbi and J.K. Omura, *Principles of Digital Communication and Coding*, McGraw-Hill, New York (1979).
13. G.D. Forney, Convolutional codes II: maximum likelihood decoding, and convolutional codes III: sequential decoding, *Information and Control*, **25**, 222-297 (1974).
14. M. Mansuripur, *Introduction to Information Theory*, Prentice-Hall, New Jersey (1987).
15. I.S. Reed and G. Solomon, Polynomial codes over certain finite fields, *J. SIAM* **8**, 300-304 (1960).
16. R.E. Blahut, *Theory and Practice of Error Control Codes*, Addison-Wesley, Reading MA (1983).
17. J.K. Wolf and G. Ungerboeck, Trellis coding for partial-response channels, *IEEE Trans. Comm.* **COM-34**, 765-773 (1986).

References for Chapter 11

1. P. Kivits, R. deBont, and P. Zalm, Superheating of thin films for optical recording, *Appl. Phys.* **24**, 273-278 (1981).
2. A.E. Bell and F. W. Spong, *IEEE J. Quant. Electron.* **QE-14**, 487 (1978); A.E. Bell, R.A. Bartolini, and F.W. Spong, *RCA Rev.* **40**, 345 (1979).
3. M.R. Madison and T.W. McDaniel, Temperature distribution produced in an N-layer film structure by static or scanning laser or electron beam with application to magneto-optical media, *J. Appl. Phys.* **66**, 5737 (1989).
4. M. Mansuripur, G.A.N. Connell, and J.W. Goodman, *Appl. Opt.* **21**, 1106 (1982); also *Appl. Opt.* **22**, 666-670 (1983); also *J. Appl. Phys.* **54**, 4794-4798 (1983).
5. S.C. Shin, Thermal analysis of magneto-optical thin films under laser irradiation, *J. Mag. Magnet. Mater.* **61**, 301-306 (1986).
6. M. Bhattacharyya and Z. Cendes, Finite element modeling of laser beam heating of magnetic films, *J. Appl. Phys.* **57**, 3894-3896 (1985).
7. M. Chen and K. Rubin, Progress of erasable phase-change materials, in *Proc. SPIE* Vol. 1078, Optical Data Storage Topical Meeting, 150-156 (1989).
8. J. Halter and N. Iwamoto, Thermal-mechanical modeling of reversible dye-polymer media, in *Proc. SPIE* Vol. 899, Optical Storage Technology and Applications (1988).
9. H.S. Carslaw and J.C. Jaeger, *Conduction of Heat in Solids*, Oxford University Press, London (1954).
10. G. Birkhoff, R.S. Varga, and D. Young, Alternation direction implicit methods, in *Advances in Computers*, F.L. Alt and M. Rubinoff, eds., Academic Press, New York (1962).

11. W. F. Ames, *Numerical Methods for Partial Differential Equations*, Barnes & Noble, New York (1969).
12. M. Mansuripur, G.A.N. Connell, and J.W. Goodman, Signal and noise in magneto-optical readout, *J. Appl. Phys.* **53**, 4485 (1982).

References for Chapter 12

1. J.D. Jackson, *Classical Electrodynamics*, 2nd edition, Wiley, New York (1975).
2. N.W. Ashcroft and N.D. Mermin, *Solid State Physics*, Holt, Rinehart and Winston, Philadelphia (1976).
3. C. Kittel, *Introduction to Solid State Physics*, 5th edition, Wiley, New York (1976).
4. F. Hund, *Z. Physik* **33**, 855 (1925).
5. S.V. Vonsovskii, *Magnetism*, Keter Publishing House, Jerusalem (1974).
6. S. Chikazumi and S.H. Charap, *Physics of Magnetism*, R.E. Krieger, Florida (1964).
7. S. Gasiorowics, *Quantum Physics*, Wiley, New York (1974).
8. P. Weiss, *J. Phys.* **6**, 661 (1907).
9. C. Herring, Direct exchange between well-separated atoms, in *Magnetism*, Vol. IIB, G.T. Rado and H. Suhl, eds., Academic Press, New York (1965).
10. H.A. Kramers, *Physica* **1**, 182 (1934).
11. P.W. Anderson, *Phys. Rev.* **79**, 350 (1950).
12. J.H. Van Vleck, *Theory of Electric and Magnetic Susceptibilities*, Clarendon Press, Oxford (1932).
13. F.H. Spedding and A.D. Daane, *The Rare Earths*, Wiley, New York (1961).
14. F.H. Spedding, S. Legvold, A.D. Daane, and L.D. Jennings, *Progr. Low Temp. Phys.* **2**, 368, North-Holland, Amsterdam (1957).
15. R.M. Bozorth and J.H. Van Vleck, *Phys. Rev.* **118**, 1493 (1960).
16. C. Zener, *Phys. Rev.* **81**, 446 (1951); **82**, 403 (1951); **83**, 299 (1951); **85**, 324 (1951).
17. L. Pauling, *Proc. Nat. Acad. Sci. USA* **39**, 551 (1953).
18. T. Kasuya, *Progr. Theor. Phys.* **16**, 45 (1956).
19. W.C. Koehler, *J. Appl. Phys.* **32**, 20S (1961).
20. W.M. Hubbard and E. Adams, *J. Phys. Soc. Japan* **17**, Suppl. B-I, 143 (1962).
21. R.M. Bozorth, B.T. Matthias, H. Suhl, E. Corenzwit, and D. D. Davis, *Phys. Rev.* **115**, 1595 (1959).
22. J. Kanamori, Anisotropy and magnetostriction of ferromagnetic and antiferromagnetic materials, in *Magnetism*, Vol. I, G.T. Rado and H. Suhl, eds., Academic Press, New York (1963).

References for Chapter 13

1. A.P. Malozemoff and J.C. Slonczewski, *Magnetic Domain Walls in Bubble Materials*, Academic Press, New York (1979).
2. A.H. Eschenfelder, *Magnetic Bubble Technology*, Springer, New York (1980).
3. A.A. Thiele, *Bell System Technical Journal* **48**, 3287 (1969).
4. N. Heiman, K. Lee, and R.I. Potter, in *AIP Conf. Proc.* Vol. 29, 130 (1975).
5. A. Gangulee and R. Kobliska, *J. Appl. Phys.* **49**, 4896 (1978).
6. T. Mizoguchi and G. Cargill, *J. Appl. Phys.* **50**, 3570 (1979).
7. H-P.D. Shieh and M. Kryder, *IEEE Trans. Mag.* **24**, 2464 (1988).
8. B.G. Huth, *IBM J. Res. Dev.* **18**, 100 (1974).
9. M. Mansuripur and G.A.N. Connell, *J. Appl. Phys.* **55**, 3049 (1984).
10. A.H. Morrish, *The Physical Principles of Magnetism*, Wiley, New York (1965).
11. M. Mansuripur and R. Giles, *IEEE Trans. Mag.* **24**, 2326 (1988).
12. M. Mansuripur, *J. Appl. Phys.* **66**, 3731 (1989).
13. I. Gradshteyn and I. Ryzhik, *Table of Integrals, Series and Products*, Academic Press, New York (1965).
14. R.N. Bracewell, *The Fourier Transform and its Applications*, McGraw-Hill, New York (1978).
15. M. Mansuripur, *J. Appl. Phys.* **66**, 6175 (1989).
16. R.A. Hajjar and H-P.D. Shieh, *J. Appl. Phys.* **68**, 4199 (1990).

References for Chapter 14

1. P. Chaudhari, J.J. Cuomo, and R.J. Gambino, Amorphous metallic films for magneto-optic applications, *Appl. Phys. Lett.* **22**, 337-339 (1973).
2. Y. Mimura, N. Imamura, T. Kobayashi, A. Okada, and Y. Kushiro, Magnetic properties of amorphous alloy films of Fe with Gd, Tb, Dy, Ho, and Er, *J. Appl. Phys.* **49**, 1208-1215 (1978).
3. T. Chen, D. Cheng, and G.B. Charlan, An investigation of amorphous Tb-Fe thin films for magneto-optic memory applications, *IEEE Trans. Mag.* **MAG-16**, 1194-1196 (1980).
4. Y. Togami, Magneto-optic disk storage, *IEEE Trans Mag.* **MAG-18**, 1233-1237 (1982).
5. G.A.N. Connell, R. Allen, and M. Mansuripur, Magneto-optical properties of amorphous terbium-iron alloys, *J. Appl. Phys.* **53**, 7759-7761 (1982).
6. M. Urner-Wille, P. Hansen, and K. Witter, Magnetic, magneto-optic, and switching properties of amorphous GdFeSn-alloys, *IEEE Trans. Mag.* **MAG-16**, 1188-1193 (1980).

7. D. Chen, G.N. Otto, and F.M. Schmit, MnBi films for magneto-optic recording, *IEEE Trans. Mag.* **MAG-9**, 66-83 (1973).
8. E. Bernal G., Mechanism of Curie-point writing in thin films of MnBi, *J. Appl. Phys.* **42**, 3877-3887 (1971).
9. B.G. Huth, Calculations of stable domain radii produced by thermomagnetic writing, *IBM J. Res. Dev.* **18**, 100-109 (1974).
10. R. Hasegawa, Temperature and compositional dependence of magnetic bubble properties of amorphous GdCoMo films, *J. Appl. Phys.* **46**, 5263-5267 (1975).
11. N. Heiman, K. Lee, R.I. Potter, and S. Kirkpatrick, Modified mean-field model for rare earth - iron amorphous alloys, *J. Appl. Phys.* **47**, 2634-2638 (1976).
12. A. Gangulee and R.J. Kobliska, Mean-field analysis of the magnetic properties of amorphous transition metal - rare earth alloys, *J. Appl. Phys.* **49**, 4896-4901 (1978).
13. R.C. Taylor and A. Gangulee, Magnetic properties of amorphous GdFeB and GdCoB alloys, *J. Appl. Phys.* **53**, 2341-2342 (1982).
14. R.C. Taylor and A. Gangulee, Magnetic properties of the 3d transition metals in the amorphous ternary alloys: $Gd_{0.2}(Fe_x Co_{1-x})_{0.8}$, $Gd_{0.2}(Co_x Ni_{1-x})_{0.8}$, and $Gd_{0.2}(Fe_x Ni_{1-x})_{0.8}$, *Phys. Rev. B* 22, 1320-1326 (1980).
15. A. Gangulee and R.C. Taylor, Mean-field analysis of the magnetic properties of vapor-deposited amorphous FeGd thin films, *J. Appl. Phys.* **49**, 1762-1764 (1978).
16. C. Kittel, *Introduction to Solid State Physics*, 5th edition, Wiley, New York (1976).
17. M. Cyrot, ed., *Magnetism of Metals and Alloys*, North Holland, Amsterdam (1982).
18. E. Callen, Y.J. Liu, and J.R. Cullen, Initial magnetization, remanence, and coercivity of the random anisotropy amorphous ferromagnet, *Phys. Rev. B* **16**, 263-270 (1977).
19. T. Mizoguchi and G.S. Cargill, Magnetic anisotropy from dipolar interactions in amorphous ferromagnetic alloys, *J. Appl. Phys.* **50**, 3570-3582 (1979).
20. M. Mansuripur and G.A.N. Connell, Energetics of domain formation in thermomagnetic recording, *J. Appl. Phys.* **55**, 3049-3055 (1984).
21. P. Chaudhari and D. Cronemeyer, The temperature dependence of the uniaxial anisotropy of $Gd_{1-x-y} Co_x Mo_y$ amorphous alloy films on glass substrates, in *AIP Conf. Proc.* Vol. 29, 113-114 (1976).
22. Y. Mimura, N. Imamura, and T. Kobayashi, Magnetic properties and Curie point writing in amorphous metallic films, *IEEE Trans. Mag.* **MAG-12**, 779-781 (1976).
23. J.J. Rhyne, J.H. Schelleng, and N.C. Koon, Anomalous magnetization of amorphous $TbFe_2$, $GdFe_2$, and YFe_2, *Phys. Rev. B* **10**, 4672-4679 (1974).
24. H. Takagi, S. Tsunashima, S. Uchiyama, and T. Fujii, Stress-induced anisotropy in amorphous GdFe and TbFe sputtered films, *J. Appl. Phys.* **50**, 1642-1644 (1979).

25. A. Itoh, H. Uekusa, Y. Tarusawa, F. Inoue, and K. Kawanishi, Magnetostriction and internal stress in GdFe amorphous films with perpendicular anisotropy prepared by rf diode sputtering, *J. Mag. Magnet. Mater.* **35**, 241-242 (1983).
26. Hong Fu, M. Mansuripur and P. Meystre, Generic source of perpendicular anisotropy in amorphous rare earth - transition metal films, *Phys. Rev. Lett.* **66**, 1086 (1991).

References for Chapter 15

1. W. Reim and D. Weller, Thermomagnetically written domains in TbFeCo thin films, *IEEE Trans. Mag.* **24**, 2308 (1988).
2. P. Hansen, New type of compensation wall in ferrimagnetic double layers, *Appl. Phys. Lett.* **55**, 200 (1989).
3. C.J. Lin and D. Rugar, Observation of domain expansion and contraction in TbFe films by Lorentz microscopy, *IEEE Trans. Mag.* **24**, 2311 (1988).
4. T. Sato, K. Nagato, A. Kawamoto, Y. Yoneyama, and T. Yorozu, Domain observation and its compositional dependence in RE-TM magneto-optical media, *IEEE Trans. Mag.* **24**, 2305 (1988).
5. K. Ohashi, H. Takagi, S. Tsunashima, S. Uchiyama, and T. Fujii, *J. Appl. Phys.* **50**, 1611 (1979).
6. K. Ohashi, H. Tsuji, S. Tsunashima and S. Uchiyama, *Jpn. J. Appl. Phys.* **53**, 7759 (1982).
7. L. Landau and E. Lifshitz, *Physik A*, *Soviet Union* **8**, 153 (1935).
8. T.L. Gilbert, *Phys. Rev.* **100**, 1243 (1955).
9. A.P. Malozemoff and J.C. Slonczewski, *Magnetic Domain Walls in Bubble Materials*, Academic Press, New York (1979).
10. C.C. Shir. Computations of the micromagnetic dynamics in domain walls, *J. Appl. Phys.* **49**, 3413 (1978).
11. M. Mansuripur, Magnetization reversal dynamics in the media of magneto-optical recording, *J. Appl. Phys.* **63**, 5809 (1988).
12. M.E. Schabes and A. Aharoni, Magnetostatic interaction fields for a three-dimensional array of ferromagnetic cubes, *IEEE Trans. Mag.* **23**, 3882 (1987).
13. W.D. Hillis and G. Steele, Data-parallel algorithms, *Comm. ACM* **29**, 1170 (1986).
14. L.W. Tucker and G.G. Robertson, Architecture and applications of the Connection Machine, *IEEE Computer Magazine* **10**, 26 (1988).
15. M. Mansuripur and R. Giles, Simulation of the magnetization reversal dynamics on the Connection Machine, *Computers in Physics* **4**, 291 (1990).
16. R. Giles, P. Alexopoulos, and M. Mansuripur, Micromagnetics of thin film media for magnetic recording, *Computers in Physics* **6**, 53 (1992).
17. R. Harris, M. Plischke, and M.J. Zuckermann, New model for amorphous magnetism, *Phys. Rev. Lett.* **31**, 16 (1973).

18. J.M.D. Coey, Amorphous magnetic order, *J. Appl. Phys.* **49**, 1646 (1978).
19. K. Moorjani, Magnetic order in disordered media, in *Mott Festschrift*, Plenum Press, New York (1985).

References for Chapter 16

1. P. Hansen and H. Heitmann, *IEEE Trans. Mag.* **25**, 4390 (1989).
2. P. Chaudhari, J.J. Cuomo, and R.J. Gambino, *Appl. Phys. Lett.* 22, 337 (1973).
3. R.J. Gambino, P. Chaudhari, and J.J. Cuomo, in *AIP Conf. Proc.* Vol. 18, Part 1, 578-592 (1973).
4. T. Chen, D. Cheng, and G.B. Charlan, *IEEE Trans. Mag.* **16**, 1194 (1980).
5. Y. Mimura, N. Imamura, and T. Kobayashi, *IEEE Trans. Mag.* **12**, 779 (1976).
6. Y. Mimura, N. Imamura, T. Kobayashi, A. Okada, and Y. Kushiro, *J. Appl. Phys.* **49**, 1208 (1978).
7. F.E. Luborsky, *J. Appl. Phys.* **57**, 3592 (1985).
8. H. Tsujimoto, M. Shouji, A. Saito, S. Matsushita, and Y. Sakurai, *J. Mag. Magnet. Mater.* **35**, 199 (1983).
9. G.A.N. Connell, R. Allen, and M. Mansuripur, *J. Appl. Phys.* **53**, 7759 (1982).
10. M. Urner-Wille, P. Hansen, and K. Witter, *IEEE Trans. Mag.* **16**, 1188 (1980).
11. T.C. Anthony, J. Burg, S. Naberhuis, and H. Birecki, *J. Appl. Phys.* **59**, 213 (1986).
12. Y. Sakurai and K. Onishi, *J. Mag. Magnet. Mater.* **35**, 183 (1983).
13. S. Chikazumi and S.H. Charap, *Physics of Magnetism*, R.E. Krieger, Florida (1964).
14. S.V. Vonsovskii, *Magnetism*, Wiley, New York (1974).
15. R. Friedberg and D. I. Paul, *Phys. Rev. Lett.* **34**, 1234 (1975).
16. D.I. Paul, *Phys. Lett.* **64A**, 485 (1978).
17. D.I. Paul, *J. Appl. Phys.* **53**, 2362 (1982).
18. B.K. Middelton, Magnetic thin films and devices, Chapter 11 in *Active and Passive Thin Film Devices*, T.J. Coutts, ed., Academic Press, New York (1978).
19. A. Sukiennicki and E. Della Torre, *J. Appl. Phys.* **55**, 3739 (1984).
20. K. Ohashi, H. Tsuji, S. Tsunashima, and S. Uchiyama, *Jpn. J. Appl. Phys.* **19**, 1333 (1980).
21. K. Ohashi, H. Takagi, S. Tsunashima, S. Uchiyama, and T. Fujii, *J. Appl. Phys.* **50**, 1611 (1979).
22. M.C. Chi and R. Alben, *J. Appl. Phys.* **48**, 2987 (1977).
23. J.M.D. Coey, *J. Appl. Phys.* **49**, 1646 (1978).
24. J.M.D. Coey and D.H. Ryan, *IEEE Trans. Mag.* **20**, 1278 (1984).
25. E. Callen, Y.J. Liu, and J.R. Cullen, *Phys. Rev. B* **16**, 263 (1977).

26. R. Harris, M. Plischke, and M.J. Zuckermann, *Phys. Rev. Lett.* **31**, 160 (1973).
27. R. Harris, S.H. Sung, and M.J. Zuckermann, *IEEE Trans. Mag.* **14**, 725 (1978).
28. E.C. Stoner and E.P. Wohlfarth, *Phil. Trans. Roy. Soc. A* **240**, 599 (1948).
29. D.O. Smith, *J. Appl. Phys.* **29**, 264 (1958).
30. E.M. Bradley and M. Prutton, *J. Electronics and Control* **6**, 81 (1959).
31. S. Middelhoek, Ph.D. Thesis, University of Amsterdam (1961).
32. M. Mansuripur, *J. Appl. Phys.* **63**, 5809 (1988).
33. M. Mansuripur and R. Giles, *IEEE Trans. Mag.* **24**, 2326 (1988).
34. M. Mansuripur, *J. Appl. Phys.* **66**, 3731 (1989).
35. M. Mansuripur and R. Giles, *Computers in Physics* **4**, 291 (1990).
36. R. Giles and M. Mansuripur, *Computers in Physics* **5**, 204-219 (1991).
37. R. Giles and M. Mansuripur, *J. Mag. Soc. Japan* **15**, 17-30 (1991).
38. P. Wolniansky, S. Chase, R. Rosenvold, M. Ruane, and M. Mansuripur, *J. Appl. Phys.* **60**, 346 (1986).
39. C. J. Lin and D. Rugar, *IEEE Trans. Mag.* **24**, 2311 (1988).
40. D. Rugar, H.J. Mamin, and P. Guthner, *Appl. Phys. Lett.* **55**, 2588 (1989).
41. R.C. O'Handley, *J. Appl. Phys.* **62**, R15 (1987).

References for Chapter 17

1. G.W. Lewicki, Curie point switching in MnBi films, *IEEE Trans. Mag.* **5**, 298 (1969).
2. H. Wieder, S. Lavenberg, G. Fan, and R. Burn, *A* study of thermomagnetic remanence writing on EuO, *J. Appl. Phys.* **42**, 3458 (1971).
3. H. Wieder and R. Burn, Self-demagnetizing effects in thermomagnetic writing on MnAlGe films, *Appl. Phys. Lett.* **22**, 188 (1973).
4. P. Chaudhari, J. Cuomo and R. Gambino, Amorphous metallic films for magneto-optic applications, *Appl. Phys. Lett.* **22**, 337 (1973).
5. Y. Mimura, N. Imamura, and T. Kobayashi, Curie point writing in amorphous magnetic films, *Jpn. J. Appl. Phys.* **15**, 933 (1976).
6. S. Matsushita, K. Sunago, and Y. Sakurai, Thermomagnetic writing in GdCo sputtered films, *IEEE Trans. Mag.* **11**, 1109 (1975).
7. T. Chen, D. Cheng, and G.B. Charlan, An investigation of amorphous TbFe thin films for magneto-optic memory application, *IEEE Trans. Mag.* **16**, 1194 (1980).
8. F. Inoue, A. Itoh, and K. Kawanishi, Thermomagnetic writing in magnetic garnet films, *Jpn. J. Appl. Phys.* **19**, 2105 (1980).
9. S. Honda, K. Ueda, and T. Kusuda, Dynamic behavior of small bits written by laser irradiation on GdFe films, *J. Appl. Phys.* **52**, 2295 (1981).

10. F. Tanaka, Y. Nagao, and N. Imamura, Dynamic read/write characteristics of magneto-optical TbFeCo and DyFeCo disk, *IEEE Trans. Mag.* **20**, 1033 (1984).
11. P. Hansen, Thermomagnetic switching in amorphous rare earth - transition metal alloys with high compensation temperature, *J. Appl. Phys.* **63**, 2364 (1988).
12. H-P.D. Shieh and M. Kryder, Dynamics and factors controlling regularity of thermomagnetically written domains, *J. Appl. Phys.* **61**, 1108 (1987).
13. T.W. McDaniel, Simulation of bit jitter in magneto-optic recording, *J. Appl. Phys.* **63**, 3859 (1988).
14. E. Bernal G., Mechanism of Curie point writing in thin films of manganese bismuth, *J. Appl. Phys.* **42**, 3877 (1971).
15. B.G. Huth, Calculations of stable domain radii produced by thermomagnetic writing, *IBM J. Res. Dev.* **18**, 100 (1974).
16. M. Mansuripur and G.A.N. Connell, Energetics of domain formation in thermomagnetic recording, *J. Appl. Phys.* **55**, 3049 (1984).
17. J.C. Suits, D. Rugar, and C.J. Lin, Thermomagnetic writing in TbFe: modeling and comparison with experiment, *J. Appl. Phys.* **64**, 252 (1988).
18. R. Giles and M. Mansuripur, Dynamics of magnetization reversal in amorphous films of rare earth - transition metal alloys, *J. Mag. Soc. Japan* **15**, Suppl. S1, 299 (1991).
19. M. Hasegawa, K. Moroga, M. Okada, O. Okada, and Y. Hidaka, Computer simulation of direct overwrite scheme in the exchange-coupled bilayer for magneto-optical memory, *J. Mag. Soc. Japan* **15**, Suppl. S1, 307 (1991).
20. Y. Yamada, M. Yoshihiro, N. Ohta, H. Sukeda, T. Niihara, and H. Fujiwara, Highly power sensitive and field sensitive MO disk for 8 MB/s data transfer, *J. Mag. Soc. Japan* **15**, Suppl. S1, 417 (1991).
21. J. Saito, M. Sato, H. Matsumoto, and H. Akasaka, Direct overwrite by light power modulation on magneto-optical multilayered media, in *Proc. Internat. Symp. on Optical Memory*, 1987, published as Supplement 26-4 in *Jpn. J. Appl. Phys.* **26**, 155 (1987).
22. K. Aratani, M. Kaneko, Y. Mutoh, K. Watanabe, and H. Makino, Overwrite on a magneto-optical disk with magnetic triple layers by means of the light intensity modulation method, in *Proc. SPIE* Vol. 1078, 258 (1989).
23. K. Tsutsumi, Y. Nakaki, T. Fukami, and T. Tokunaga, Directly overwritable magneto-optical disk with light power modulation method using no initializing magnet, in *Proc. SPIE* Vol. 1499, 55 (1991).
24. K. Aratani, A. Fukumoto, M. Ohta, M. Kaneko, and K. Watanabe, Magnetically induced super resolution in novel magneto-optical disk, in *Proc. SPIE*, Vol. 1499, 209 (1991).
25. H. Miyamoto, T. Niihara, H. Sukeda, M. Takahashi, T. Nakao, M.

Ojima and N. Ohta, Domain and read-write characteristics for magnetic field modulated magneto-optical disk with high data transfer rate, *J. Appl. Phys.* **66**, 6138 (1989).

26. M. Takahashi, T. Niihara and N. Ohta, Study on recorded domain characteristics of magneto-optical TbFeCo disks, *J. Appl. Phys.* **64**, 262 (1988).
27. H-P.D. Shieh and M.H. Kryder, Magneto-optical recording materials with direct overwrite capability, *Appl. Phys. Lett.* **49**, 473 (1986).
28. P. Hansen, Direct overwrite in amorphous rare earth - transition metal alloys, *Appl. Phys. Lett.* **50**, 356 (1987).
29. A.A. Thiele, *Bell Sys. Tech. J.* **48**, 3287 (1969).
30. A.P. Malozemoff and J.C. Slonczewski, *Magnetic Domain Walls in Bubble Materials*, Academic Press, New York (1979).
31. A.H. Eschenfelder, *Magnetic Bubble Technology*, Springer, New York (1980).
32. C.J. Lin, Thermomagnetically induced nucleation and growth of domains in GdTbFeCo, *Jpn. J. Appl. Phys.* **S28-3**, 23 (1989); C.J. Lin and H. Notarys, The importance of domain nucleation and growth during thermomagnetic recording in GdTbFeCo disks, *IEEE Trans. Mag.* **25**, 3533 (1989).
33. M. Ohta, A. Fukumoto, K. Aratani, M. Kaneko, and K. Watanabe, Readout mechanism of magnetically induced super resolution, *J. Mag. Soc. Japan* **15**, Supplement S1, 319 (1991).

References for Chapter 18

1. P. Wolniansky, S. Chase, R. Rosenvold, M. Ruane, and M. Mansuripur, Magneto-optical measurements of hysteresis loop and anisotropy energy constants on amorphous $Tb_x Fe_{1-x}$ alloys, *J. Appl. Phys.* **60**, 346 (1986).
2. R.A. Hajjar, F.L. Zhou, and M. Mansuripur, Magneto-optical measurement of anisotropy energy constants on amorphous rare earth - transition metal alloys, *J. Appl. Phys.* **67**, 5328 (1990).
3. W. Thomson, *Proc. Roy. Soc.* **8**, 546 (1857).
4. N.F. Mott, *Proc. Roy. Soc. A (London)* **153**, 699 (1936), and **156**, 368 (1936).
5. T. Kasuya, *Progr. Theor. Phys. (Kyoto)* **16**, 58 (1956).
6. J.P. Jan, *Solid State Physics* **5**, Academic Press, New York (1957).
7. S.V. Vonsovskii, *Magnetism*, Keter Publishing, Jerusalem (1974).
8. J. Smit, *Physica* **17**, 612 (1951).
9. J. Kondo, *Progr. Theor. Phys. (Kyoto)* **27**, 772 (1962).
10. T.R. McGuire and R.I. Potter, *IEEE Trans. Mag.* **11**, 1018 (1975).
11. S. Yumoto, K. Toki, O. Okada, and H. Gokan, *IEEE Trans. Mag.* **24**, 2793 (1988).
12. A.B. Pippard, *Magnetoresistance in Metals*, Cambridge University Press, Cambridge UK (1989).

13. D.A. Thompson, L.T. Romankiw, and A.F.Mayadas, *IEEE Trans. Mag.* **11**, 1039 (1975).
14. Y. Takeno and Y. Iwama, Structure and magnetic properties of MnBi thin films, *Jpn. J. Appl. Phys.* **18**, 269 (1979).
15. B.E. Bernacki and M. Mansuripur, Characterization of magneto-optical recording media in terms of domain boundary jaggedness, *J. Appl. Phys.* **69**, 4960 (1991).
16. J.N. Chapman, *J. Phys. D: Appl. Phys.* **17**, 623 (1984).
17. J.N. Chapman and G.R. Morrison, *J. Mag. Magnet. Mater.* **35**, 254 (1983).
18. J.C. Suits, R.H. Geiss, C.J. Lin, D. Rugar, and A.E. Bell, *J. Appl. Phys.* **61**, 3509 (1987).
19. T. Nguyen, P. Alexopoulos, C. Hwang, S. Lambert, and I. Sanders, *IEEE Trans. Mag.* **24**, 2733 (1988).
20. S. Tsukahara, in *JARECT* **15**, *Recent Magnetics for Electronics*, Y. Sakurai, ed. (1984).
21. I.R. McFadyen, in *Proc. 47th Annual Meeting of the Electron Microscopy Society of America*, G.W. Bailey, ed., San Francisco Press (1989).
22. G. Binnig, H. Rohrer, C. Gerber, and E. Weibel, Tunneling through a controllable vacuum gap, *Appl. Phys. Lett.* **40**, 179 (1982).
23. H.W. van Kesteren, A.J. den Boef, W.B. Zeper, J.H.M. Spruit, B.A.J. Jacobs, and P.F. Carcia, Scanning magnetic force microscopy on Co/Pt magneto-optical disks, *J. Mag. Soc. Japan* **15**, supplement S1, 247-250 (1991).
24. Y. Martin, C. Williams, and H.K. Wickramasinghe, *J. Appl. Phys.* **61**, 4723 (1987).
25. J. Saenz and N. Garcia, Observation of magnetic forces by the atomic force microscope, *J. Appl. Phys.* **62**, 4293-4295 (1987).
26. Y. Martin, D. Rugar, and H. Wickramasinghe, High-resolution magnetic imaging of domains in TbFe by force microscopy, *Appl. Phys. Lett.* **52**, 244 (1988).
27. D. Rugar, H.J. Mamin, R. Erlandsson and B. Terris, Force microscope using fiber-optic displacement sensor, *Rev. Sci. Instr.* **59**, 2337 (1988).
28. P. Hobbs, Y. Martin, C. Williams, and H.K. Wickramasinghe, Atomic force microscopy: implementations, in *Proc. SPIE* Vol. 897, 26-30 (1988).
29. P. Grütter, D. Rugar, H.J. Mamin, G. Castillo, C.J. Lin, I. McFadyen, R. Valletta, O. Wolter, T. Bayer and J. Greschner, Magnetic force microscopy with batch-fabricated force sensors, *J. Appl. Phys.* **69**, 5883-5885 (1991).
30. D. Rugar, H.J. Mamin, and P. Guethner, Improved fiber-optic interferometer for atomic force microscopy, *Appl. Phys. Lett.* **55**, 2588 (1989).
31. M. Mansuripur, Computation of fields and forces in magnetic force microscopy, *IEEE Trans. Mag.* **25**, 3467 (1989).
32. D. Rugar and P. Hansma, Atomic force microscopy, *Physics Today*, 23-30 (October 1990).

33. L.J. van der Pauw, *Philips Technical Review* **20**, 220 (1958).
34. B.B. Mandelbrot, *The Fractal Geometry of Nature*, W.H. Freeman, New York (1983).
35. Y. Aharonov and D. Bohm, *Phys. Rev.* **115**, 485 (1959).
36. M. Schabes and A. Aharoni, Magnetostatic interaction fields for three-dimensional array of ferromagnetic cubes, *IEEE Trans. Mag.* **23**, 3882-3888 (1987).

Index